1994 Hybrid Electric Vehicle Challenge

SP-1103

All SAE papers, standards, and selected books are abstracted and indexed in the Global Mobility Database.

Published by:
Society of Automotive Engineers, Inc.
400 Commonwealth Drive
Warrendale, PA 15096-0001
USA
Phone: (412) 776-4841
Fax: (412) 776-5760
February 1995

TL
220
.A14
1995

Permission to photocopy for internal or personal use, or the internal or personal use of specific clients, is granted by SAE for libraries and other users registered with the Copyright Clearance Center (CCC), provided that the base fee of $6.00 per article is paid directly to CCC, 222 Rosewood Drive, Danvers, MA 01923. Special requests should be addressed to the SAE Publications Group. 1-56091-653-2/95$6.00.

No part of this publication may be reproduced in any form, in an electronic retrieval system or otherwise, without the prior written permission of the publisher.

ISBN 1-56091-653-2
SAE/SP-95/1103
Library of Congress Catalog Card Number: 94-74755
Copyright 1995 Society of Automotive Engineers, Inc.

Positions and opinions advanced in this paper are those of the author(s) and not necessarily those of SAE. The author is solely responsible for the content of the paper. A process is available by which discussions will be printed with the paper if it is published in SAE Transactions. For permission to publish this paper in full or in part, contact the SAE Publications Group.

Persons wishing to submit papers to be considered for presentation or publication through SAE should send the manuscript or a 300 word abstract of a proposed manuscript to: Secretary, Engineering Meetings Board, SAE.

Printed in USA

PREFACE

The papers in this SAE special publication, <u>1994 HEV Challenge</u> (SP-1103), represent some of the most advanced thinking on hybrid electric vehicles (HEVs) anywhere. In this volume are the winning entries in a U.S. Department of Energy (DOE)-sponsored HEV Paper Design Contest and design reports from the 1994 Hybrid Electric Vehicle Challenge sponsored by the Saturn Corporation and DOE.

The HEV Paper Design competition was conducted in 1992. Aimed at college- and university-level engineering classes, this competition was an attempt to generate innovative ideas for HEVs from the creative minds of students unburdened with preconceived notions of how future vehicles should look. It was intended that this paper design competition would involve those schools that did not have the resources to build a HEV to compete in DOE-sponsored vehicle competitions. The results of the competition were a number of very impressive designs that give a glimpse into future HEV configurations. Included here are the winning papers from some of the top engineering schools in the United States.

In addition to these state-of-the-art design papers, vehicle design reports from engineering schools participating in the 1994 Hybrid Electric Vehicle Challenge are presented. As one of the judged events in the Challenge, teams must submit a design report that explains the design choices applied to and details the content of their vehicles. Papers from three distinct vehicle classes are represented: the Ground-Up class, comprised of scratch-built HEVs; the Escort class, composed of HEVs based on converted Escort station wagons, and the Saturn class, HEV conversions based on Saturn SL sedans. These papers reflect the range of choices faced by engineers as they grapple with trade-offs for range, performance, efficiency, and cost. The Ground-Up and Escort entries in the 1994 HEV Challenge had to demonstrate at least a 40.2 km (25 mi) electric-only (ZEV) range; the Saturn class, designed as Power-Assist hybrids, had to demonstrate a 8 km (5 mi) ZEV range. The Saturn class had the additional requirement of not recharging their batteries from an external source of electricity, meaning they needed to be charge-sustaining. The rules for all classes emphasized safety considerations but were intentionally left relatively open to encourage creativity and innovation. As a result, many different approaches to constructing HEVs were explored and tested during the competition itself. Tables A, B, and C summarize the results of the HEV Challenge by vehicle class. Congratulations to all the schools that participated in this competition. For an analysis of the 1994 HEV Challenge results, see SAE papers 950176 and 950177, available separately from this special publication.

Listed in Table D are all the sponsors who made this competition possible. Our sincere thanks for their support. Finally, thanks go to all the students and faculty advisors whose efforts are so well documented here. Their energy, creativity, and enthusiasm transformed into working vehicles is the reason that these competitions are so successful. These are truly events where all the participants, students and sponsors alike, win.

Robert P. Larsen, Director
1994 HEV Challenge
Center for Transportation Research
Argonne National Laboratory

1994 Hybrid Electric Vehicle Challenge

1994 HEV Challenge
Overall Results

Table A. Overall Results for the Ground-Up Class.

Place	Car No.	Team	Vehicle Class	Emissions Event	Design Event	Acceleration Event	Range Event	Road Rally Event	Commuter Challenge Event	Energy Economy Event	Bonus	Penalties	Total Score
1	40	University of California - Davis	Grd-up	100.00	230.58	80.00	92.42	70.00	50.00	234.00	114.00	53.00	918.00
2	37	Cal State Poly Univ - Pomona	Grd-up	0.00	189.24	63.81	100.00	69.34	14.68	164.07	116.00	13.00	704.15
3	5	Lawrence Technological University	Grd-up	0.00	208.69	56.22	93.01	40.86	12.60	128.37	87.00	30.00	596.75
4	16	Cornell University	Grd-up	0.00	247.03	73.55	97.43	0.00	10.94	121.23	106.00	91.00	565.18
5	39	University of California - Santa Barbara	Grd-up	0.00	171.35	52.61	37.55	31.69	11.80	143.19	109.00	3.00	554.19
6	19	Michigan State University	Grd-up	0.00	227.90	59.10	89.68	24.14	0.00	91.58	115.00	78.00	529.40
7	27	University of Idaho / Wash State Univ	Grd-up	0.00	218.27	0.00	87.21	24.14	15.30	78.01	94.00	88.00	428.93
8	15	New York Institute of Technology	Grd-up	0.00	188.96	50.40	32.14	27.92	0.00	78.37	80.00	49.00	408.78
9	21	University of Tennessee	Grd-up	0.00	210.81	33.17	10.00	0.00	10.51	12.50	102.00	51.00	328.00
10	20	University of Tulsa	Grd-up	0.00	175.05	29.26	10.00	24.14	8.69	25.00	105.00	75.00	302.13
11	25	Cal State Poly Univ - San Luis Obispo	Grd-up	0.00	169.65	18.03	10.00	0.00	10.09	12.50	79.00		299.26
12	32	University of Texas, Arlington	Grd-up	0.00	132.98	39.44	0.00	0.00	17.48	0.00	34.00	50.00	173.90

1994 HEV Challenge
Overall Results

1994 Hybrid Electric Vehicle Challenge

Table B. Overall Results for the Escort Class.

Place	Car No.	Team	Vehicle Class	Emissions Event	Design Event	Acceleration Event	Range Event	Road Rally Event	Commuter Challenge Event	Energy Economy Event	Bonus	Penalties	Total Score
1	6	Weber State University	Escort	90.00	183.13	53.30	100.00	16.06	40.25	250.00	105.00	2.00	835.74
2	8	University of Illinois	Escort	0.00	238.87	58.34	85.61	70.00	45.31	146.19	112.00	58.00	698.32
3	33	University of Alberta	Escort	70.00	173.84	40.61	87.96	19.28	46.92	157.74	116.00	50.00	662.35
3	34	University of Wisconsin	Escort	0.00	219.16	55.97	88.59	14.61	50.00	129.34	104.00		661.66
5	12	Jordan College	Escort	0.00	154.66	21.64	88.45	11.97	29.91	149.69	101.00	3.00	554.32
6	11	Concordia University	Escort	0.00	211.24	21.88	56.49	12.40	20.01	147.91	119.00	72.00	516.93
7	28	West Virginia University	Escort	0.00	212.31	25.22	10.00	13.59	34.11	85.26	118.00	25.00	473.49
8	23	Washington University	Escort	0.00	184.62	28.72	89.64	11.17	0.00	93.49	97.00	100.00	404.64
9	29	Texas Tech University	Escort	0.00	153.87	29.20	26.09	10.67	45.49	49.31	114.00	55.00	373.62
10	10	Colorado School of Mines	Escort	0.00	189.84	21.32	10.00	0.00	44.90	12.50	90.00	50.00	318.56
11	14	Stanford University	Escort	0.00	222.53	17.61	10.00	10.67	0.00	25.00	96.00	77.00	304.81
12	31	United States Naval Academy	Escort	0.00	197.19	27.02	10.00	0.00	0.00	12.50	105.00	50.00	301.71
13	1	Pennsylvania State University	Escort	0.00	180.47	31.94	10.00	0.00	33.46	12.50	64.00	52.00	280.37
14	7	University of California - Irvine	Escort	0.00	158.65	0.00	0.00	0.00	0.00	0.00	45.00		203.65
15	30	California State University - Northridge	Escort	0.00	0.00	0.00	0.00	0.00	0.00	0.00	18.00		18.00
16		Seattle University	Escort	0.00	0.00	0.00	0.00	0.00	0.00	0.00	73.00		73.00

1994 Hybrid Electric Vehicle Challenge

1994 HEV Challenge
Overall Results

Table C. Overall Results for the Saturn Class.

Place	Car No.	Team	Vehicle Class	Emissions Event	Design Event	Acceleration Event	Range Event	Road Rally Event	Commuter Challenge Event	Energy Economy Event	Bonus	Penalties	Total Score
1	24	University of Maryland	Saturn	60.00	250.00	63.92	100.00	70.00	39.35	244.42	78.00	150.00	755.68
2	22	University of Western Ontario	Saturn	0.00	139.34	80.00	75.06	37.14	43.07	65.55	55.00	205.00	290.16
3	17	GMI Engineering & Mgmt Inst	Saturn	0.00	163.35	24.86	8.20	9.37	22.09	64.78	72.00	75.00	289.66
4	2	University of Texas at Austin	Saturn	0.00	150.35	54.62	89.82	11.43	42.82	71.67	60.00	255.00	225.71
5	35	Western Michigan University	Saturn	0.00	157.81	29.10	45.26	0.00	28.61	63.98	64.00	250.00	138.76
5	9	Cedarville College	Saturn	0.00	205.21	21.16	27.62	27.14	47.00	66.49	81.00	338.00	137.62
7	18	California State University - Chico	Saturn	0.00	139.06	23.59	41.12	11.43	0.00	43.49	59.00	260.00	57.69
8	26	Wentworth Institute of Technology	Saturn	0.00	136.55	0.00	0.00	0.00	0.00	0.00	48.00	175.00	9.55
9	3	Ecole de Technologie Superieure	Saturn	0.00	124.69	0.00	0.00	0.00	47.69	40.84	58.00	205.00	66.22
10	36	Alfred University	Saturn	0.00	150.57	0.00	0.00	0.00	0.00	0.00	37.50		188.07
11	38	California State University - Fresno	Saturn	0.00	0.00	0.00	0.00	0.00	0.00	0.00			0.00
11	41	Illinois Institute of Technology	Saturn	0.00	0.00	0.00	0.00	0.00	0.00	0.00			0.00

Table D. 1994 HEV Challenge Sponsor List

Primary Sponsors
U.S. Department of Energy
Saturn Corporation
Natural Resources - Canada
SAE International
Primary Organizers
Argonne National Laboratory
National Renewable Energy Laboratory
Gold Level Sponsors
British Petroleum Oil
Chrysler Corporation
Detroit Edison
EDS
Ford Motor Company
General Motors
General Motors of Canada
Lawrence Technological University
Unique Mobility
Silver Level Sponsors
Firestone/Bridgestone
Goodyear Tires
Hitachi
Johnson Controls
Bronze Level Sponsors
Allied Signal Automotive
IIT Automotive
Lear Seating Corporation
McLaren Engines
NSK Corporation
Siemens Automotive
Volvo Cars of North America

TABLE OF CONTENTS

HEV Challenge

University of Alberta 1

Alfred University 15

California State Polytechnic University, Pomona 39

California Polytechnic State University, San Luis Obispo 47

California State University, Fresno 55

University of California-Davis 61

University of California-Irvine 73

University of California-Santa Barbara 83

Cedarville College 89

Colorado School of Mines 101

Concordia University 113

Cornell University 129

GMI Engineering & Management Institute 145

University of Illinois at Urbana-Champaign 153

Jordan College Energy Institute 165

Lawrence Technological University 173

University of Maryland 181

Michigan State University 195

New York Institute of Technology 207

The Pennsylvania State University 217

Seattle University 233

Stanford University 249

University of Tennessee 265

Texas Tech University 279

University of Texas at Arlington 285

University of Tulsa 293

US Naval Academy 299

Washington State University and University of Idaho 315

Washington University in St. Louis 329

Weber State University 335

West Virginia University 347

University of Wisconsin-Madison 357

HEV Paper Design

California State University, Chico 371

Concordia University 383

Lehigh University 405

Texas A & M 429

Western Washington University 451

Development of the University of Alberta Entry for the 1994 HEV Challenge

M.D. Checkel, V.E. Duckworth
J.B. Lybbert, V. Yung
Faculty of Engineering, University of Alberta

ABSTRACT

Because of the limitations of their storage batteries, electric cars have always suffered from short range, high weight, and high cost. New battery technologies will provide a significant improvement but all-electric vehicles will still tend to be heavy, costly, and severely limited in range compared with their combustion engine counterparts. Despite these inherent disadvantages, there is a huge impetus for electric car development because of the pollution disadvantages of the combustion engine. Given the weight/cost/range problems of purely electric cars, it is desirable to develop hybrid cars which have the capability of operating as zero-emission electric cars in urban areas and which use a small internal combustion engine to extend the operating range. The internal combustion engine and its fuel are far lighter, cheaper, and more effective at extending range than carrying enough battery capacity to give an all-electric vehicle a suitable range.

The U.S. Department of Energy, Natural Resources Canada, Saturn Corporation, and the Society of Automotive Engineers have organized a second student design competition to highlight the possibilities of hybrid electric cars. The University of Alberta, along with 41 other North American university teams, are developing and building safe, practical, road-licensed cars with hybrid electric drive systems. The car developed by the University of Alberta team demonstrates the near-term feasibility of the hybrid electric concept and is an evolution of their winning design from the 1993 HEV Challenge Competition. This paper describes the major design choices and the development process used to produce the 1994 University of Alberta vehicle.

INTRODUCTION

The challenges of developing a fully functional, practical electric automobile have been around since the first electric vehicles were challenging the early gasoline and steam vehicles for supremacy. At that time, internal combustion engine capabilities developed rapidly while there was a lack of progress in developing suitable storage batteries for electric cars. The result has been a century of domination by combustion-driven vehicles. Now, however we are finding limitations on using internal combustion engines to power virtually the entire transportation system in large cities. The problems became obvious in sensitive areas such as the California coastline more than four decades ago, and led to controls on vehicle emissions. Rigorous pollution controls now limit new vehicles to less than 10% of pre-control emission levels. However, with growing population, increasing size of urban areas, increasing use of personal transportation, and increasing traffic congestion, the problem simply cannot be eliminated. For example, despite the continuing renewal of the California vehicle fleet with ultra-clean vehicles, transportation continues to dominate the pollutant emissions in California. In fact, despite the tighter emission standards, the total emissions in California, which have been declining steadily since the 1970's are expected to bottom out and start increasing again due to continuing population and urban traffic increases into the next century [1]. This problem is not isolated to the California coastline; urban areas outside California have now grown to the point where the pollution can no longer be effectively dispersed and are seeing increased smog, haze and ozone.

These problems have brought renewed interest in electric cars to provide clean transportation in the congested cities of the twenty first century. Electric vehicles can, in theory, use relatively clean nuclear or hydroelectric power for transportation and thus virtually eliminate the combustion-generated pollutants. Even if the electric power is generated by oil or coal combustion, the pollutants can be generated outside the city centre, minimizing the direct population exposure and reducing contributions to the urban smog problem. Also, pollutants generated at a point location like a power plant can have sophisticated pollution controls applied with a reasonable level of success.

Are electric vehicles ready? Enormous development progress has been made in electric motors, motor controllers, and storage batteries over the past century. However, the energy storage density of an electric battery is still poor compared with the energy storage of a similar mass or volume of hydrocarbon fuel. Table 1 shows some practical energy storage densities for vehicle applications. The fuels have a huge advantage over the batteries in total energy storage. This is not surprising considering that the fuel energy is released by reacting that fuel with several times its mass of air, utilizing all the chemical bond energy available in the combined mass, and then dumping the fuel and its products into the atmosphere. By contrast, battery energy release

Table 1.
Typical On-Board Energy Density of Fuels and Batteries
(Incl. Tankage, Mounting Hardware, etc.)

FUEL OR BATTERY	ENERGY DENSITY	
	kJ/L	kJ/kg
GASOLINE	33 000	40 000
METHANOL	16 000	18 000
NATURAL GAS	8 000	6 000
Na-S	72	355
Ni-Cd	36	111
Pb-Acid	18	72

Batteries: Na-S = sodium-sulphur, Ni-Cd = nickel cadmium, Pb-Acid = lead-acid

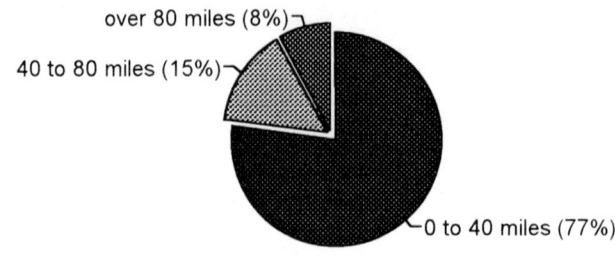

Figure 1. Typical Daily Driving Distance of Private Vehicles

involves chemical reaction of only a small fraction of the battery mass, with no outside reactants, and the chemical reactions are ones which can be reversed simply by applying a reverse voltage across the reactants when they are returned to a charging station. It can be argued that the electrical energy stored in a battery is converted to propulsive power much more efficiently than the chemical energy in a fuel. However, this still only reduces the effective propulsive energy of a fuel by four or five times in comparison with a battery which contains two or three orders of magnitude less energy.

To illustrate the problem, consider that a typical 1100 kg (2425 lb) car requires about 35 L (9.25 USgal) of gasoline for a 500 km (311 mi) driving range. The fuel and tank adds a negligible mass and volume to the existing vehicle. Only about 20% of the energy in the fuel is used and the rest is wasted through the cooling system and the exhaust as heat. By contrast, a highly efficient vehicle which converts the battery energy to propulsive energy with 90% efficiency would need about 725 kg (1599 lb) of sodium-sulphur (Na-S) battery or 2300 kg (5071 lb) of nickel-cadmium (Ni-Cd) battery to contain enough energy for the same driving range. And, since the battery is large and heavy compared with the car, the whole vehicle would have to be made larger, requiring more energy and hence more battery power to travel the same distance. The result is that electric vehicles are normally highly specialized to accommodate their power source. They use light weight materials to allow for battery weight and improve energy efficiency; they have limited interior space due to battery volume; and they still have a very short range of continuous operation compared with a fueled vehicle.

Another significant difference between battery energy and fuel energy for a vehicle is the rate at which it can be replenished. Typical battery recharging schedules call for six to ten hours of charging at maximum energy input rates of a few kW. Filling up with gasoline replenishes a vehicle's energy supply at a rate measured in tens of megawatts! For a commuter vehicle with only limited daily use, this difference can be ignored since there is adequate time available for overnight recharging. However, most commuter vehicles are periodically required to do more than just routine commuting. Examples would include running a series of errands after work, driving up to the mountains for a weekend, or driving across the country for a holiday. A vehicle which requires several charging hours per hour of operation is fine for commuting but it simply cannot fulfill these occasional requirements. That makes it much less useful to a typical commuter and severely limits the possible market penetration of purely electric cars. Some future developments to resolve this problem are foreshadowed by current research. Solutions may include quick recharge batteries and high power charging stations or removable battery packs and battery exchange stations. However, the commercialization of those systems and the enormous infrastructure development which would be required to make them practical on a large scale is still far in the future.

A more feasible solution for near term introduction of electric automobiles is to develop hybrid electric vehicles. Such vehicles could operate as purely electric vehicles over a range adequate to cover most normal daily operation. Then, rather than carry enough extra batteries to provide normal vehicle range, the vehicles would carry a small internal combustion engine adequate to run the vehicle. This allows normal operation without the worry of stranding the vehicle when the battery runs down. It also allows virtually limitless extended range operation similar to normal automobiles because the vehicle can fuel up in the normal way during days of continuous operation or on cross-country trips. What is a reasonable range for electric power? Studies of commuter driving patterns show the average private vehicle will drive less than 65 km (40 mi) in one day approximately 77% of the time. A pie chart with a typical daily driving range breakdown is shown in Figure 1.

The U.S Department of Energy, Natural Resources Canada, Saturn Corporation, and the Society of Automotive Engineers are continuing to stimulate interest and research into hybrid vehicles by sponsoring the 1994 Hybrid Electric Vehicle (HEV) Challenge. This is a student design competition to develop automobiles that are capable of running a "reasonable" distance on battery power and extended distances using conventional liquid fuels, (gasoline, M85 or E100). This report details the technical design of the University of Alberta's hybrid electric vehicle, "DARDANUS".

The competition allows Universities to either build a complete vehicle from the ground up, convert a Ford Escort station wagon to hybrid drive, or convert an existing Saturn passenger vehicle to an electric assist hybrid. The U of A team chose to convert the Escort station wagon. Throughout the design process, the emphasis of the project team was to develop a car which would demonstrate the near-term practicality of hybrid electric vehicles. That meant the car, while meeting the rules and intents of the competition, should also be attractive to every segment of the automotive industry. It should be easily and economically manufactured, should be easy to sell (to a significant target market who are currently driving gasoline-powered vehicles), and should also be easy to service. The paradigm was that this "car of the future" should look and feel like it belonged in a local driveway in 1994. This was one reason for choosing to convert the Escort wagon rather than building a futuristic automobile. Another consequence of this philosophy is that most of the components used in the hybrid conversion are readily available production items and the car has retained all of the interior trim and other appointments consistent with a marketable vehicle. Another illustration of the philosophy is that the U of A car uses a 240 V, 20 A supply for battery charging and regular unleaded gasoline for its combustion engine. The option of using a higher electrical voltage for greater efficiency was rejected since most homes and businesses are only supplied with 240 V power. The option of using alcohol fuels for emission reduction was likewise rejected on the basis that the fuels are not commonly available. The objective was always to try and match the capability, flexibility, and ease of use of the base car while adding a Zero Emission Vehicle (ZEV) operating capability.

The first section of this paper covers the powertrain design strategy, component selection, and integration. This is followed by a section on operating modes, drive train control strategy, emissions control, and fuel and electrical consumption of the hybrid power plant. The next section of the report covers the design of powertrain mounts, battery box, suspension modifications, and other necessary changes to the car. The last section of the report summarizes the features of the car and the projects achievements.

POWERTRAIN CONFIGURATION
There are two primary coupling methods available when configuring the powertrain of a Hybrid Electric Vehicle: series and parallel.

SERIES CONFIGURATION - This configuration is common to many locomotives, mine trucks, and commuter buses. The premise is simple. An internal combustion engine drives an electric generator which supplies the required electrical power to an electric drive motor which actually drives the vehicle. A battery system may be used to store excess electrical output. The main advantage of this system is its simplicity. The power transfer between the combustion engine and the electric drive motor is independent of each respective unit. The internal combustion engine can be sized and tuned to operate at its optimum speed and efficiency while supplying the average required vehicle power. The battery system can supply excess power during vehicle acceleration and absorb the available excess power when vehicle loads are low. This allows a substantial reduction in combustion engine size compared with a normal vehicle. However, because the vehicle is driven only by the electric motors, the electrical system must be oversized to give appropriate peak performance. Also, using an internal combustion engine to drive a generator, the generator to charge batteries, and the batteries to run motors is less efficient than using the internal combustion engine to supply power directly to the drive train.

PARALLEL CONFIGURATION - This configuration allows for either the electric motor or the internal combustion engine to drive the vehicle independently. In addition, the capability to drive with both combustion engine and electric motors simultaneously allows the designer to size each power system for high efficiency in normal operation while still providing high performance in combined mode driving. Driving the vehicle directly with the internal combustion engine increases efficiency in hybrid mode since there are no electrical losses in series with the engine's mechanical output. However, the parallel configuration is not as simple to implement as series since some variable mechanical linkage is required between the combustion engine, the electric motor(s), and the vehicle's final drive system.

CONFIGURATION DECISION - The series configuration appeared easier to implement. Its advantages included simplicity of component selection, simplicity of construction and installation, and simplicity of control. Its main disadvantages were the requirement for very high power electric motors for competitive performance and efficiency loss during hybrid operation. The parallel configuration required more specialized component manufacture and controls to integrate the two power systems. However, it gave higher efficiency in normal operation and higher peak performance. Two additional factors were instrumental in making the basic powertrain decision. These were the predicted reliability of the powertrain and the acceptability of the powertrain operating characteristics as perceived by the driver. It was felt that reliability of the series configuration would be poorer since, by definition, every component must be functional to keep the system operating and, if any component is degraded, the vehicle can only operate at the level determined by its weakest link. In contrast, either the electrical system or the internal combustion engine could fail in a parallel hybrid system and the vehicle would remain functional on the other system. The only critical components in the parallel system are the coupler linking the engine and electric motor to the transmission and the transmission/final drive itself.

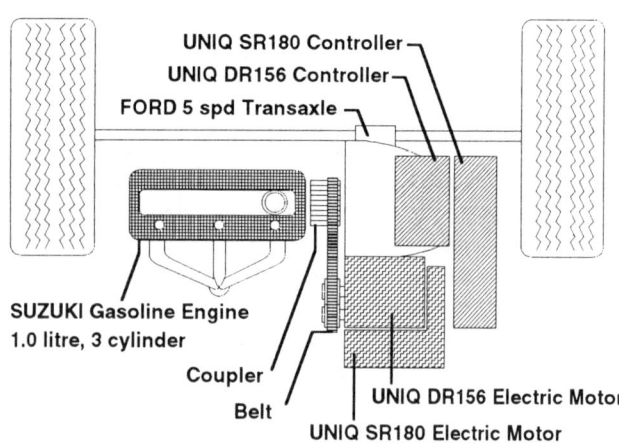

Figure 2. Schematic of Hybrid Powertrain Layout for the 1994 U of A HEV.

The other question was the operational characteristics of the vehicle with each drive train configuration. A car driven by a series electrical system would drive like an electric car and would have essentially the same operating characteristics whether it was in electric-only or hybrid mode. A car with a parallel system could potentially have different driving characteristics and different operating controls depending on which of three possible modes it was in. It was felt that this would be very undesirable to a typical customer but could be overcome if an effort was made to keep the driving controls and vehicle operation as consistent as possible between operating modes.

Based on the above considerations, the University of Alberta team selected a parallel hybrid electric drive for their vehicle because of higher efficiency, higher performance, and better reliability. The actual implementation of this drive, shown schematically in Figure 2, was to combine the power from two electric motors and the internal combustion engine with a coupler at the clutch input. This meant that, regardless of the operating mode, the driving power would pass through the clutch and transmission in the same fashion. This allowed the driver to perform all control operations the same as in a normal gasoline-powered car regardless of whether the vehicle was in electric-only, gasoline-only, or hybrid modes. It also allowed for the greatest possible operating flexibility.

The next question was how much power would be required. A vehicle simulation model was used to answer this question, as well as related questions about battery capacity, peak current, component duty cycle, et cetera. The model was based on a simulation program which ran vehicles through various transient cycles and calculated energy consumption, peak power demand, and other parameters of interest [2]. Basically, the model calculates the tractive power requirement due to rolling resistance, aerodynamic resistance, grade, and acceleration for each point in a second-by-second description of the operating cycle. A similar model was described by Sovran and Bohn [3]. The FTP-78 Urban and Highway test schedules [4] were used to define the minimum required vehicle performance. A cycle simulation of the FTP-78 Urban Driving Cycle determined a minimum tractive power requirement of 36 kW. It should be noted that, on the FTP cycles, acceleration and deceleration are limited to 1.5 m/s^2. Acceleration rates during normal urban driving vary considerably but typical values rise to more than 2.5 m/s^2 at low speeds [5]. The FTP cycles thus represent a minimum acceptable performance but typical driving demands more power. More aggressive driving cycle simulations including lane change accelerations, and hill climb simulations were also performed to define the desirable peak power and the required high power duty cycles. To determine the required transmission gear ratios and closely define the engine operating conditions, further simulation subroutines which included engine torque limits and gear shifting were used. It became apparent that the vehicle could perform adequately with about 45 kW (60 hp) of tractive power but a multi-ratio transmission was highly desirable to keep the engine/motors near that peak power point over a range of vehicle speeds. It also became apparent very early that the battery requirements for a 72 km (40 mi) range would raise the Escort's operating weight to very close to the registered gross vehicle weight (GVW). For this reason, most simulations were performed assuming the vehicle was at GVW.

ELECTRIC MOTOR SELECTION - Many different types of electric motors were investigated prior to the final selection. These included DC Shunt Wound, DC Series Wound, and AC Synchronous motors. The critical factors considered were output power, efficiency, durability, availability, weight, and cost. Some consideration was given to finding a motor or motors which could directly drive the vehicle axles. This would have given greater electric drive efficiency at some cost in operating complexity since control operations would have been different for the engine and the electric motors. In fact, no acceptable motors were identified which would give adequate torque and speed range to drive a car of this size without a transmission.

The electric motors chosen were one SR180 and one DR156 DC brushless, permanent magnet motor produced by Uniq Mobility. These motors were relatively compact and light weight having a combined mass of only 31.8 kg (70 lb). The SR180 was rated at 32 kW (43 hp) continuous duty with a peak power output of 62.7 kW (84 hp), and the DR156 was rated at 15.8 kW (21.2 hp) continuous duty with a peak power output of 25.1 kW (33.6 hp). The peak power output was rated on a 33% duty cycle. Since continuous power is not required, this meant the vehicle would have 87.7 kW (118 hp) electric power available. Motors of two different sizes were chosen in a effort to increase the efficiency through staging. The SR180 would be employed in conjunction with the DR156 during acceleration but would be taken off-line once near constant speed was achieved. Also, the motors have a peak efficiency of 94% which is critical to obtaining the optimum range with a given size battery pack. An overview of the overall system efficiency is given later in the report.

Each of the motors has its own controller. The motors operate on three phases of modulated DC current and have a maximum draw of 220 A. For efficiency, the power is regulated by pulse width modulation, essentially switching the nominal battery voltage of 180 V on for longer or shorter times in each of three motor inputs. The motors/controllers have an integral regenerative braking capability. Because the motors are permanently linked to the clutch input, the regenerative braking can be used in all three operating modes. Lightly operating the brake pedal switches the controllers into regenerative mode and the kinetic energy of the vehicle is converted to electrical energy by running the electric motors as DC generators. The generated power is used to recharge the battery pack. Heavier application of the brake pedal maximizes the regeneration rate and activates the normal hydraulic brakes as well.

These motors and their controllers were expensive due to both the low production volume (<200 units/year) and to the high price of developing compact, high flux density, permanent magnets.

INTERNAL COMBUSTION ENGINE - The design philosophy for a hybrid electric vehicle (HEV) is to operate on electric power in urban areas and to use the internal combustion engine mostly to extend the range of the vehicle for highway driving. Since the engine in a HEV will operate mainly under steady load at highway speed, it can be sized smaller than an engine in a conventional vehicle. One benefit of a smaller engine is that it can run more efficiently, with additional drive power supplied by the electric motors when needed. Also, a smaller engine provides much needed space in the engine compartment for the electric motors and controllers.

The U of A HEV Project chose to retain the Suzuki 1.0 L engine in the powertrain for its combination of compact size, light weight, efficiency, and power. The three cylinder, four stroke gasoline engine used in the 1993 HEV provided reasonable power for the vehicle in highway operation and was adequate as a backup to the electric drive in urban driving. The engine, constructed with an aluminum block and cylinder head, has a mass of 68 kg (150 lb) less than the stock iron block Ford 1.9 L engine. The 0.65 kW/kg power to weight ratio of the Suzuki is significantly higher than the 0.49 kW/kg ratio of the stock engine.

The Suzuki engine was designed for a much lighter car than the U of A HEV, so significant changes are needed for the engine to operate optimally. Time limitations meant that emissions analysis did not go beyond the testing stage for the 1993 HEV. The objective for the 1994 project is to improve the emissions performance of engine as much as possible with the goal of reaching levels similar to those mandated by the Ultra Low Emission Vehicle (ULEV) standards.

COUPLING MECHANISM & TRANSMISSION - The primary difficulty with the parallel configuration was the design of a coupling mechanism which would allow

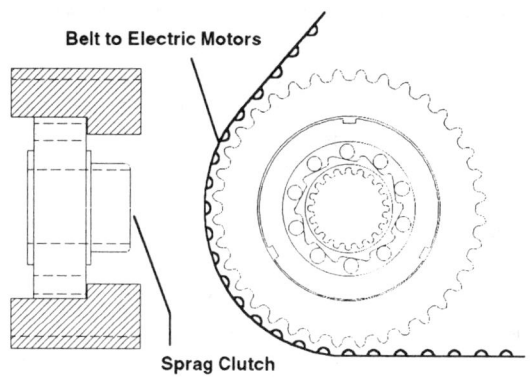

Figure 3. Simplified Layout of 1994 Coupling Mechanism

switching between the three driving modes; electric, gasoline, and hybrid without introducing complicated operating procedures for the driver. A coupling mechanism that met these requirements was invented, engineered and developed by the U of A student team. This coupler fits on the end of the engine crankshaft between the flywheel and the clutch. It incorporates a toothed belt drive which permanently connects the electric motors to the clutch input and a one-way clutch mechanism that connects the internal combustion engine while it is driving but slips to allow the engine to stop for electric-only mode. Because the electric motors are permanently connected, regenerative braking is possible in all driving modes.

The U of A coupler design uses a sprag type over-running clutch mechanism normally found in the hub of a torque converter. The clutch was created to replace traditional ramp and roller clutches and is therefore of similar dimensions (approximately 90 mm [3.54 in] diameter.). The clutch consists of an inner ring and an outer ring with cylindrical elements between. When the inner ring rotates in one direction, torque is transmitted by the ramps on the inner ring wedging the cylindrical elements out against the outer ring. When the outer ring rotates in the other direction the ramps release the cylindrical elements and thus no torque is transmitted.

The coupler employed in the 1994 HEV is an evolution of the 1993 model. In the 1993 HEV, the length of the coupler forced the engine and transaxle to be removed as one piece when the engine, transaxle, or coupler is to be serviced or inspected. The previous design was 88 mm long, had a diameter of 151 mm (5.96 in) and a 100 mm (3.94 in) engine/transaxle separation. The new design, shown in Figure 3 is 66 mm (2.60 in) long, has a diameter of 113 mm (4.45 in) and an engine/transaxle separation of 66 mm (2.60 in). This reduction in coupler length allows sufficient clearance for the engine to be removed independent of the transaxle which reduces service time from six hours in 1993 to one hour in 1994.

The other aspects of the coupler that were improved were the mass of the coupler and the problem with lubricant leakage. The new coupler has a mass of 4 kg (8.8 lb), down from the 7.5 kg (16.5 lb) of the old coupler. The lubricant leakage problem was remedied through the use of a smaller seal diameter which increases the critical seal speed from 3679 rpm to 9023 rpm. This higher critical seal speed allows much higher oil pressure within the coupler at the expected operating speeds.

To optimize the electric motors for city performance and the gasoline engine for highway operation, the gear ratio incorporated in the toothed belt drive runs the electric motors at 1.4 times engine speed. This had the effect of derating the engine since its peak speed was limited to 5000 rpm rather than 6500 rpm. However, driving operations were not affected since drivers do not typically use that part of the power band and rpm range above the torque peak.

The transmission chosen for this design was the stock Escort five speed manual transmission. The five ratios were suitable for both the individual and combined powerplants and the high efficiency and low weight of the transmission (less than 37 kg [80 lb]) were very attractive. Compared with a direct electric drive, computer simulations of the vehicle's zero emissions mode showed an improvement of 22.7% in the 0-100 m (0-328 ft) acceleration time and also an 82.8 km/hr (51.5 mph) higher top speed when using the five speed transmission. The Ford transmission met all the needs of the design and was easily integrated using existing mountings, shift linkages, and axles. The transmission is located on the driver side of the engine bay (its original location) directly below the electric motor controllers.

BATTERY PACK - In 1993, an indepth search was conducted for available batteries to power the HEV. This search showed that a nickel-cadmium battery pack was the best choice. Rapid advancement in battery technology necessitated a new investigation. One system that was considered was an advanced lead-acid battery which is slightly lower in energy density than our Ni-Cd pack, but at approximately one quarter the cost. As no performance gains would made from the purchase of a new battery pack, the decision was made to use the 1993 Ni-Cd pack in the new vehicle.

The battery system for the 1994 U of A HEV thus consists of STM 1.60 nickel-cadmium cells manufactured by SAFT NIFE Corporation in France and are specifically designed for automotive uses.. The cell dimensions are 85 mm (3.45 in) wide, 45 mm (1.77 in) deep, and 278 mm (10.94 in) tall It utilizes a light plastic casing and features a very high depth of discharge. This is important in an automobile where a constant power availability is desirable throughout the discharge cycle. The entire battery pack is composed of 142 cells yielding a nominal pack voltage of 180 V. The total weight of the battery pack is 284 kg (626 lb). These cells are approximately 1.5 times as energy dense as conventional lead-acid cells providing 50% greater vehicle range with similarly sized battery packs. The battery

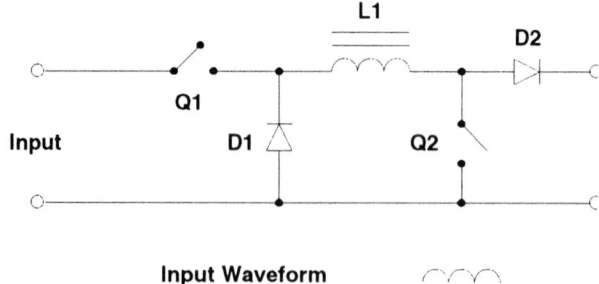

Figure 4. Simplified Battery Charger Schematic

pack is rated at 61 A-hr which yields a range of approximately 72 km (45 mi) of city driving on a single charge. The battery pack is situated in the space originally occupied by the rear seats. A new rear seat was constructed with the seat cushion integrated into the battery box cover. A detailed description of the battery box is given in the section on vehicle structural modifications.

The cells for the battery pack were purchased at a cost over Cdn$25,000.00. As with the electric motors, this high cost is due mainly to the low production volume (<100 packs/year) and the high price of cadmium. Some additional features of these nickel cadmium cells which were designed for automotive use include: self-watering ability, high energy density, and recyclable internals. The self-watering system attaches to the top of each cell with reservoirs located along the sides of the battery box. Additionally, these cells perform better than normal lead-acid cells at low temperatures (a concern in the Canadian climate). The nickel and cadmium found inside the cells are recyclable and they can be reconstituted and reused in another cell. Only minimal gases are emitted during a normal charging cycle and Ni-Cd cell are much more robust/durable than lead-acid, with less damage to the cells if they are overcharged. These cells have a life of between 2000 and 3000 charge-discharge cycles which translates into five to eight years of daily charging.

BATTERY CHARGER - A new battery charger has been constructed for the 1994 HEV Challenge. The charger power section, shown schematically in Figure 4, consists of a basic buck converter (parts labeled Q1, D1, and L1). By adding another diode, D2, and high speed switch, Q2, the same system can be used in the boost mode. This mode is used for current shaping while the rectified AC input voltage is below the battery voltage. By using ultra-fast insulated gate bipolar transistors (IGBT) and a ferrite main inductor, switching losses in the converter have been reduced by several percent over the charger used in the 1993 competition. In addition to the basic circuit shown above, power factor and current shaping is controlled with an adaptive algorithm in a micro-controller. The charger has numerous safety features, reflex or burp mode charging, multiple-stage charging, and improved efficiency.

CONTROL STRATEGY

The U of A HEV design allows the vehicle to be propelled in three distinct modes, selected using a multi-position rotary switch mounted on the dashboard. The switch has four positions (in clockwise order); OFF, electric only mode (EM), hybrid mode (HEV), and combustion only mode (ICE). The OFF position is obvious. The others are described below.

ELECTRIC MODE - The ZEV operating mode is the primary mode of operation for the vehicle. To engage, the driver must turn the ignition key to the ON position and the mode switch to EM position. The Suzuki engine computer, fuel injector, and starter circuit are locked out. The electric motor output and hence the vehicle speed is controlled by the accelerator pedal. The accelerator pedal is linked to two potentiometers (pots) sending signals to each of the two motor controllers. The pots are staged so that the first portion of pedal travel sends a signal to only the small motor controller. Depressing the pedal further then signals the larger motor controller as well. The signal indicates to the controllers the desired speed. The controllers will supply maximum current to the motors to accelerate the vehicle until the required speed is achieved. This control scheme has the advantage of running the small motor at close to its peak efficiency during normal driving while the added power of the large motor is then available for acceleration. When changing gears, the clutch pedal and accelerator pedal are operated as in any other manual transmission vehicle.

It should be noted that the electric motors do not idle - no sound is heard from the engine bay when the vehicle is stopped. A low electrical whine (quieter than the normal combustion rumble) is noticeable when the vehicle is in motion. There is some belt noise from the toothed belt which is reduced by an insulated belt cover.

HYBRID MODE - The hybrid mode of operation is useful for situations like overtaking or driving up steep grades. As well as the electric motor control pots, the accelerator pedal is connected to the combustion engine throttle and thus determines the output of that engine when it is running.

Hybrid mode can be engaged with the vehicle moving or stopped and from either electric mode or gasoline mode. In switching from electric mode, the operator will turn the selector switch from the EM position to the HEV position. The Suzuki engine computer, fuel injector, and starting circuit are now enabled and all of the electric motor systems remain active and operational. The internal combustion engine is then started as in a conventional vehicle. (It was felt that an auto-start feature would add complexity and failure modes while not significantly enhancing the vehicle. Drivers already know how to start their engine!) If the car is already in motion, the combustion engine will not couple with the transmission until its output shaft speed matches the rotational speed of the clutch driven by the electric motors.

Once the combustion engine has coupled with the transmission input shaft, the torque from both power plants is delivered to the transmission.

In switching from gasoline (ICE) mode to hybrid mode, the operator turns the selector switch from the ICE position to the HEV position. The gasoline engine is already running and the electric motors are permanently coupled so the additive torque from the electric motors is instantaneous. The performance with both power plants running is dramatically improved.

GASOLINE MODE - This mode of operation is intended for situations like long distance highway driving. The operator has the option of either starting the vehicle in this mode or switching to this mode while the vehicle is in motion. To start the vehicle in this mode, the operator sets the Selector Switch to ICE position. The gasoline engine systems are enabled and the accelerator potentiometer inputs to the electric motor controllers are disabled in his position. The operator starts the engine using the ignition key and drives off, just as with a conventional automobile.

To engage this mode while the vehicle is in motion (assuming that the car is currently operating in electric only mode), a short transition through the hybrid mode is recommended. This allows the internal combustion engine to be started without the loss of power while driving. Once the gasoline engine is engaged, a transition from hybrid to ICE mode can safely be made.

BRAKING STRATEGY AND ON-BOARD CHARGING - Regenerative braking is available in all three operating modes with the U of A HEV and uses the energy normally wasted in braking to recharge the battery. The electric motors are automatically switched to generators when the driver presses the brake pedal, with the first portion of the pedal travel being electrical braking and the last portion applying the hydraulic brakes. This allows for the majority of braking energy to be reclaimed into the battery, while still maintaining the full braking capacity of the hydraulic brakes. No regeneration occurs while the vehicle is in neutral or when the clutch pedal is depressed. This is to prevent possible damage to the motors as a result of them coming to a halt too rapidly.

It should be noted that there is no engine braking from the internal combustion engine. With the U of A coupler in place, closing the throttle will allow the engine to drop to idle speed while the sprag clutch overruns. This is a change from normal vehicle operation but not a large one. It has the advantage of saving all the vehicle kinetic energy for the regenerative brake system.

In ICE mode, a separate regeneration control allows the internal combustion engine to be used to recharge the battery pack. The driver depress a red push-button on the vehicle dashboard and adjusts a control knob to select the level of regeneration. With the prototype system, this is useful for recharging the batteries in areas where there are no suitable charging facilities. On a production vehicle, the

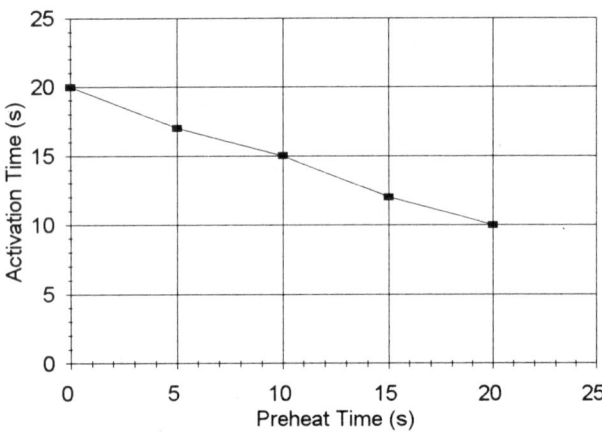

Figure 5. Effects of Electrical Heating on Oxygen Sensor Activation Time.

capability could be especially useful in a situation where certain traffic zones are designated zero emission zones. When driving between these zones, this feature could be used to charge the battery pack and extend the electric range available. While this capability can be very useful in adding flexibility to the car, it must be emphasized that it is generally much more cost-efficient to use grid electrical power for battery charging than on-board power produced from gasoline.

EMISSIONS CONTROL

Three important changes have been made to the internal combustion engine system. A heated exhaust gas oxygen (HEGO) sensor has been installed, Emitec GmbH and Engelhard Corporation have supplied the HEV Project with an electrically-heated catalytic converter (EHC) system, and a programmable engine control module (ECM) has been obtained from Suzuki Motor Corporation.

HEATED EXHAUST GAS OXYGEN SENSOR - Tests reveal that the exhaust gas from the Suzuki engine after start-up would not consistently reach the temperature required to activate the oxygen sensor if the engine was simply allowed to idle. The emissions testing at the 1994 HEV Challenge will follow the United States Federal Test Program (FTP-78) Urban Driving Cycle. This tests allows an initial 20 s of idling before the first vehicle acceleration.

The HEGO sensor from the Ford Escort 1.9 L engine was installed in the Suzuki engine. By electrically heating the oxygen sensor, closed loop operation can begin much sooner after engine start-up than with an unheated sensor. Tests were performed to determine the effect of the HEGO sensor, and the results are shown in Figure 5. With heating beginning at engine start-up, activation of the HEGO sensor occurred in 20 s. As would be expected, preheating the sensor before start-up reduced the post-start activation time.

ELECTRICALLY HEATED CATALYTIC CONVERTER - The catalytic converter is the key to reducing the emission of pollutants to levels below those attainable by modern electronic engine control systems. If the emission levels mandated by the upcoming ULEV standards are to be met, selection of the catalytic converter system must be done with a special emphasis on minimizing emissions during cold starts, when a large percentage of vehicle emissions occur.

By electrically heating the catalyst in the EHC conversion of unburned hydrocarbons (HC) and nitrous oxides (NO_x) can much more rapidly after startup. Post start heating compensates for the heat energy lost to the cool exhaust gases until the exhaust temperature rises high enough to sustain converter operation. The net effect of electrical heating is to reduce the time required for the converter to reach its full effectiveness as compared to an unheated system.

ENGINE CONTROL MODULE RE-CALIBRATION - The Suzuki engine control module was originally calibrated for operation in a 775 kg (1709 lb) Suzuki Swift. Significant changes to the ECM calibration is required to attain low emission levels in the HEV which has over double the mass of the Swift. The programmable Suzuki ECM allows the engine control parameters to be optimized for the new service conditions.

Re-calibration of the ECM focused on the fuel enrichment and spark advance tables. Fuel enrichment was significantly reduced in the range between 2400 rpm and 4000 rpm where simulations showed that most vehicle driving is expected to occur. The result of decreasing the fuel enrichment is to lower emissions of HC and improve the performance of the catalytic converter which requires a stoichiometric mixture for proper operation. At higher engine loads, combustion temperature rises causing an increased level of NO_x emission. The effect of the higher load is minimized by re-mapping the ECM with less spark advance.

These changes to the ECM calibration, while having the effect of lowering emission levels of controlled pollutants, lower the torque output of the engine. However, the engine should still provide adequate performance for operation at normal highway speed. The electric motors are available when more power is required. (Note: Peak power is not affected as we will leave the calibration at high engine speed (<4000 rpm) unchanged.)

ENERGY CONSUMPTION

The Ford Escort station wagon has a published urban fuel economy rating of 6.73 L/100 km (35 mpg) with a curb weight of 1180 kg (2600 lb). The U of A hybrid conversion has increased the weight about 40% and this raises the tractive energy requirement due to increased inertial and rolling resistance. A prediction based on tractive energy analysis of an urban emissions test cycle shows that the energy requirement should rise by 29% due to the greater vehicle mass. The HEV would thus be expected to use about 29% more fuel when driving the same cycle as a (lighter) stock Escort wagon. However, the 40% increase in vehicle weight is partially offset by running a smaller engine at a higher, more efficient power level.

In the 1993 competition, using the smaller Suzuki engine, the U of A HEV had a fuel consumption of 6.78 litre/100 km (35 mpg) over an urban/highway split. It might be assumed that the fuel efficiency would improve by using the smaller engine at a higher, more efficient output level. However, the combination of high power enrichment and higher operating speeds balanced this effect giving virtually standard fuel consumption. Through engine control modifications, an improved fuel economy has been achieved in the 1994 vehicle.

The electrical drive train efficiency of the U of A vehicle was 53.8% or 9.96 km/kW-hr as measured at the 1993 HEV Challenge. This was on a driving cycle that was roughly comparable to a highway cycle (some 'stop & go' driving with a majority of 90 km/hr running). The electrical energy consumption during a highway cycle would be relatively unaffected by the energy recovery associated with regenerative braking. On a city cycle simulation, this should allow up to 15% reduction in electrical energy consumption. This feature is also available in hybrid and combustion-only modes so that a similar savings in total energy consumption (but no savings in fuel consumption) can also be expected in these modes.

The major losses of electrical power was in the charging system. A new charger has been built incorporating power factor correction (PFC) circuitry and a more efficient isolation transformer has been obtained. Also, PFC circuitry is being added to the 1993 charger which will be used as a back-up unit. The Uniq motors have a slightly lower peak efficiency rating than the Solectria motors of the 1993 vehicle. However, this should be offset by the staging of the Uniq motors, as described in the "Electric Motors Selection" section, under typical driving conditions.

From a marketing standpoint, the consumer wants a comparison between the operating economy of the gasoline engine and the electric motors. The cost of gasoline in Edmonton, Canada is Cdn$0.45/L (Cdn$1.70/USgal) and the retail cost of electricity is approximately Cdn$0.07/kW-hr. The range of the U of A HEV, on battery power alone, is approximately 72 km (45 mi) and the battery pack energy capacity is approximately 10 kW-hr. The fuel cost to travel 72 km in the HEV on gasoline power, at 7 L/100 km would be about Cdn$2.25. This same distance traveled on electric power alone would have a "fuel" cost about Cdn$0.70 (35% of gasoline cost). However, because of the considerable cost and limited life of the battery, combined with limited daily mileage available in electric mode, the actual economics would have to include some charge for battery amortization. Since practical battery life has not been determined and our own battery costs are unrealistic for future production vehicles, a cost comparison including battery amortization will not be presented.

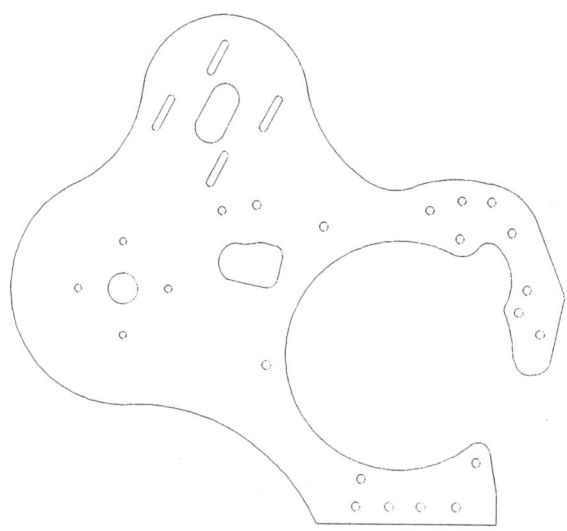

Figure 6. Electric Motor Mounting Plate (Viewed from Driver Side of Vehicle)

VEHICLE STRUCTURE MODIFICATIONS
The U of A HEV is one of eighteen Ford Escort conversion vehicles in the upcoming competition. During the conversion process, some necessary structural modifications were made to the Escort chassis in the engine bay and the rear passenger seat area.

DRIVETRAIN MOUNTING SYSTEM - The drivetrain mounting system is a fundamental component of any automobile. It provides the foundation used to support the entire powertrain assembly. It must support both static and dynamic loads, and control vibration transfer from the drivetrain to the chassis.

To design the mounting system, full scale models were used to help visualize pertinent tolerances and space limitations. This trial and error approach allowed a number of different arrangements to be evaluated. After extensive testing, only one feasible configuration was found. Space had proven to be a critical constraint.

The electric motor mounting plate which bolts directly onto the face of the transmission bell housing was the focus of extensive design because it is the sole support of the electric motors. It is also required to maintain the alignment of the electric motors with the input shaft of the transmission. The shape of the plate, given in Figure 6, was dictated by the profile of the electric motors thus the top and front edges of the plate were contoured to match the curvature of the small and large electric motors, respectively. Further, a 26 mm (1 in) adjustment of the small electric motor (DR156) was required to tension the belt.

Optimization of the plate design was conducted using Cosmos/M, a finite element analysis (FEA) package. This analysis provided a method to monitor both stress and deflection in the transmission plate. Using the FEA, it was

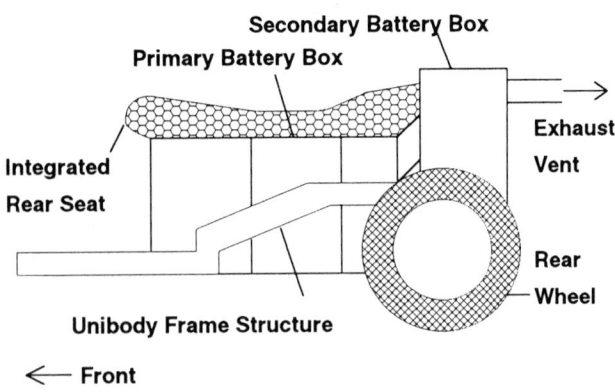

Figure 7. Side View of Battery Box Location in Vehicle

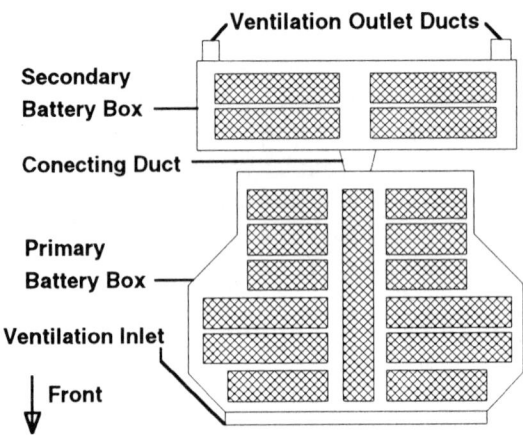

Figure 8. Top View of Battery Box Showing Cell Layout

determined that the minimum plate thickness needed to meet system constraints under maximum loading was 4.76 mm (3/16 in) using 1020 As-Rolled steel plate. The maximum loading was defined as a 4g downward acceleration combined with peak torque of the electric motors (126.6 N-m [93.4 lb-ft]) and nominal belt tension. These conditions represent abnormally severe loading and would rarely be encountered during typical operation of any vehicle.

The maximum stress 120 MPa (17.4 ksi), occurred at the bolt locations of the large electric motor, resulting in a factor of safety of 2.75. The maximum deflection, 1.8 mm (0.071 in), occurred at the forward edge of the plate. Plate deflections larger than 5 mm (0.197 in) would cause unacceptable belt wear.

The second component of the mounting system was the drivetrain-to-chassis mounts. A four point mounting system using flexible rubber bushings was developed. By reducing the number of mounting locations from seven (1993 HEV), to four (1994 HEV), the drivetrain is supported much less rigidly, enabling a reduction in the amount of vibration transmission that can occur. These four mounts were required to function in two ways. First, structural integrity demands that the system be capable of supporting loads resulting from the drivetrain's weight, from ICE and EM torque, from vehicle handling and acceleration, and from shock loads due to uneven road conditions. Three mounts were used to support the transmission while one mount was used to support the IC engine. The two lateral transmission mounts were used predominantly to carry the torque load, while the rear transmission and IC mounts primarily supported the drivetrain's weight. Second, for a smooth ride under normal operating conditions, the mounts must minimize vibrations transmitted from the drivetrain to the vehicle chassis. In doing so, adequate rigidity must be maintained to prevent a collision between the oscillating drivetrain and stationary equipment within the engine bay.

In order to optimize the amount of vibration isolation of the four point system, a dynamic analysis was performed. From this analysis, modes of vibration and system resonance characteristics were investigated, allowing the bushing stiffnesses to be optimized. No such analysis was done on the 1993 vehicle, but from vibration characteristics it was obvious there was a resonance problem in the 2500 rpm range. As a result, stock bushings have been incorporated in the final 1994 design. Using stock Ford bushings for the transmission mounts provide excellent torque response due to the fact that the HEV torque magnitude, 151.8 N-m (112 lb-ft), is very similar to the torque produced by the original 1.9 L engine, 146.4 N-m (108 lb-ft). Vibration transfer from the inherently rough running 3 cylinder Suzuki engine is minimized using the more flexible Suzuki bushing.

PASSENGER COMPARTMENT - The major modifications in this area were a box to accommodate the battery and a roll cage to meet competition safety requirements. The battery and its enclosure weigh approximately 330 kg (730 lb) and are located beneath and behind the rear passenger seat. This central location was chosen to optimize vehicle dynamic performance while retaining a back seat usable by two people.

The floor pan beneath the rear seat was removed to make room for the battery container. Since the side rails and rear cross member were unaltered, the vehicle floor could be removed without compromising the structural integrity of the vehicle.

The box is built in two sections connected by a duct for venting and electrical connections. The front box is below the rear seat with its bottom approximately 5 cm (2 in) below the floor of the car. The back box sits behind the back seat on the floor above the rear suspension as can be seen in Figure 7. The front box contains 106 cells and the back box contains 36. Figure 8 shows the layout of the Ni-Cd cells.

The exterior of the box is manufactured from 20 gauge steel closely matching the grade and gauge of the original Escort floor pan. The ideal material for the containment box is stainless steel which ensures corrosion resistance, however the cost was prohibitive. To provide increased corrosion resistance, the interior of the box is coated with a thick layer of water resistant epoxy and the exposed underside of the box is painted with a protective layer of undercoating. The battery box is attached to the vehicle with continuous seam welds running the full length of all sides. These welds are predicted to withstand a 15g deceleration of the vehicle in the event of a collision.

The lids of both boxes are fibreglass. A flange from the lids extends down below the level of the battery terminals and ensures that a loose cell cannot inadvertently contact a conducting surface in the box.

The battery box was designed to house 142 Nickel Cadmium cells in 17 rows as shown in Figure 8. Each row is supported on a tray and slotted into the side walls of the box. These trays are welded to two longitudinal rails along each side of the box. The battery is vented by drawing outside air from below the vehicle, through both boxes, and then exhausting through vents located in the top of the rear box lid. Two fans draw air from the outlet of the vents and exhaust it to the outside of the vehicle in the rear bumper area. The box is completely sealed to the vehicle interior to prevent odour or other problems and the seals are located in areas with sufficient structural integrity to withstand collision deformation.

Important structures within the battery enclosure itself are the battery trays. Stainless steel was used for the construction of these trays. The need for these trays to remain undamaged by potential battery fluid spills offset the higher cost associated with using stainless steel.

ADDITIONAL MODIFICATIONS - Additional changes to the passenger compartment were made for increased safety. Since the battery box is permanently mounted to the vehicle and may crush the vehicle in the event of a rollover, a rollbar was installed to increase occupant safety. Two styles of roll bar bracing were investigated; cross bracing and parallel bracing. Cross bracing is generally the most effective. However, with the U of A HEV design, finite element analysis showed that parallel bracing was adequate. As with standard roll bar arrangements, the main hoop would support the roof of the vehicle in the event of a rollover. The main hoop is braced against the rear suspension mounts to provide stability and increased structural stiffness for the whole vehicle. As an additional safety measure for the occupants, a reinforcing structure was added to the front of the battery box. This structure protects the occupants by ensuring that in the event of a severe frontal impact, the batteries will break out of the bottom of the box under the car instead of entering the passenger compartment.

SUSPENSION AND BRAKE MODIFICATIONS

The modified Escort has a curb weight of approximately 1406 kg (3100 lb) which is 272 kg (600 lb) more than the stock Escort. The weight distribution has changed from a 59%/41% split to an approximate 55%/45% split on the front and rear axles respectively. The vertical centre of gravity of the vehicle has been maintained close to that of the unconverted Escort. The increased weight of U of A HEV prompted some modifications to maintain the original ground clearance and to increase spring stiffness to account for greater mass. The front spring coefficient was increased from 26.6 kN/m to 29.3 kN/m (153 lb/in to 167 lb/in) while the rear spring coefficient was increased from 15.6 kN/m to 18.8 kN/m (89 lb/in to 107 lb/in).

Since the U of A HEV can operate as a fully electric vehicle, it will not always have engine vacuum to operate the standard vacuum-boosted power brakes. Rather than add another failure mode (i.e. a vacuum pump), a direct hydraulic system was used. The stock vacuum boost assembly was removed and the master cylinder was bolted to the fire wall. To account for the loss of vacuum assist, the mechanical advantage of the pedal was increased by 30%.

With the increase in rear axle weight, consideration was given to replacing the rear drum brakes with disk brakes and/or changing the front/rear proportioning valve. Stock components were identified for such conversions but it was felt that the additional cost was not justified by any deficiencies in the braking performance. Regenerative braking of the motors provide an equivalent amount of braking assist as standard power brakes, and the modifications to the hydraulic system allow adequate braking performance if regenerative braking is not available.

BODY STYLING & ERGONOMICS

Body styling and ergonomics was a major emphasis in the design of the 1994 University of Alberta Hybrid Electric Vehicle. Modifications to the vehicle's appearance were limited to those needed for a hybrid operation, and were also made to enhance the visibility of the vehicle for public presentations. The project aimed to keep the interior and exterior of the vehicle as close to stock as possible to show that a vehicle may be converted to an HEV without having to change its overall appearance.

VEHICLE EXTERIOR - No exterior vents are visible on the 1994 HEV. The battery box ventilation ducts exhaust the unwanted gases to a vent on the rear quarter panel which is hidden by the rear bumper. Fresh air is drawn into the battery box through a serpentine vent on the underside of the vehicle. This vent was designed to prevent water from entering the battery box.

To allow the connection of the charger to the vehicle, a charge port was added to the rear passenger side quarter panel. An Escort fuel door and its associated hardware were provided by a local auto-recycler and fitted to the panel opposite the stock fuel door. The charge door is opened by pulling a lever situated next to the driver seat. This offers increased safety as access to the high voltage port is restricted to the operator.

Replacement wheels for the HEV were obtained from an Acura Integra. These 14" alloy wheels have a higher load rating than the stock 13" steel wheels. Given the increased mass of the HEV, the higher load rating provides a additional factor of safety. As well, the aesthetics of the wheel design was desirable.. The U of A HEV logo was placed on the wheel centre caps to give the vehicle added visibility.

REAR SEAT/BATTERY BOX COVER - Excluding the rollover protection mandated by the rules, the interior is difficult to distinguish from a stock escort. The original front seats were kept while the original upholstery was used in the construction of the integrated rear seat. The primary battery box is built below the rear seat while the secondary box is located directly behind the backrest of the rear seat. The rear box occupies an eight inch depth at the front of the cargo area and, along with the ventilation ducts, is upholstered to match the vehicle interior. The rear seat squab is integral with the cover of the primary battery box. Built of rigid urethane foam and ergonomically contoured for comfort, this seat/cover can be easily removed by one person. A woven fiberglass reinforced coating makes this cover extremely rigid and electrically insulates the batteries from the remainder of the box. Incorporating a secondary box allows the vehicle to retain the original rear legroom while increasing the size of the battery pack to provide the desired vehicle range.

HYBRID CONTROLS - Three additional controls necessitated for the hybrid conversion. They are located in at the top of the centre console and are within easy reach. The mode switch is closest to the driver, putting all four modes of operation on a single rotary switch. Manually controlled regeneration is turned on or off by a push-button switch and adjusted by a rotary control to its right.

HEADS UP DISPLAY - Additional instrumentation has been incorporated into a heads up display (HUD) system integrated into the upper dash surface. The driver views the display through a reflection in the lower area of the windshield directly ahead. A computer monitors vehicle conditions and activities. A menu operated by a single four position joystick switch allows the driver to view a wide selection of detailed information on the HUD. The menu system allows easy access to information on any one of the numerous systems monitored by the computer. In addition to the alphanumeric data, a bar of light emitting diodes graphically displays levels and voltages.

SUMMARY AND CONCLUSION

Imagination, design, creation, fabrication, and testing by University of Alberta students have resulted in an improved design from their 1993 HEV Challenge vehicle. The University of Alberta HEV runs on electricity, gasoline, or a combination of the two and would satisfy a market that wants an environmentally friendly form of transportation but will not sacrifice the comfort and practicality offered by today's automobiles.

Electric power is supplied by a nickel cadmium battery pack which can be charged overnight from a 220 V outlet. The combustion engine uses commonly available regular unleaded gasoline. The vehicle has a range of approximately 72 km (45 mi) on electric power which would satisfy the needs of 77% of daily North American automobile travel. If additional range is needed, the use of gasoline should give a total range of over 375 km (233 mi) without stopping. Performance is similar and adequate to keep up with traffic in either electric or gasoline mode. When added performance is required for passing or hill climbing, hybrid mode gives adequate power for spirited driving. The fuel cost to the consumer is predicted to be significantly less on electricity than gasoline so it should encourage users to travel in electric mode as much as possible.

The U of A vehicle, DARDANUS, drives very much like the standard Escort. All of the common controls, instrumentation, and features of conventional automobiles are found in its design and the additional operating capabilities have been incorporated without significant complication for the driver. Additionally, this vehicle has maintained all of the standard occupant safety features and is structurally secure in the event of a collision. By reprogramming the Suzuki engine management system, emissions have been significantly improved.

The U of A HEV was designed to be put into production as a practical and marketable automobile. Wherever possible, the modifications made to this Escort have incorporated ideas and methods that will lead to simple manufacture at a production level. The majority of the components used in the conversion are readily available and currently being produced. The more exotic and expensive components including the electric motors and batteries could become commonplace with increased demand. The market appeal of this vehicle is that it looks and drives much like a conventional automobile with all of the usual comforts and functional aspects retained. It has stock production headlights, horn, steering wheel, accelerator pedal, brake pedal, and shifter, all in their stock locations, allowing the operator to feel comfortable when driving. The additional controls and instruments have been kept to a minimum and have been added in a way which does not increase the driver's workload.

The electric technology of the future has been successfully integrated into an automobile for today. The U.S. Department of Energy, Natural Resources Canada, SAE, and Saturn Corporation have tapped into the resources and potential available within the North American University and College system to find a solution to a pressing environmental problem: air pollution caused by automobiles. The hybrid concept may not be the final solution to this problem but it is a necessary step in the right direction and will help ease the public into an emissions free automotive future.

REFERENCES

1. "California Air Quality, A Status Report", California Air Resources Board, Sacramento, 1991

2. M.D. Checkel, "NGV Conversion Effects on Vehicle Emissions and Fuel Economy, Appendix A 'Development of Multi-mode Test Schedule' ", Report for Alberta Transportation and Utilities, R & D Division, January, 1992

3. "Emission Test Driving Schedules", SAE Information Report, SAE J1807, SAE Handbook

4. G. Sovran and M.S. Bohn, "Formulae for the Tractive Energy Requirements of Vehicles Driving the EPA Schedules", SAE paper 810184, Annual Congress, Detroit, 1981

5. P. Wasielewski, L. Evans and M.F. Chang, "Automobile Braking Energy, Acceleration and Speed in City Traffic", SAE paper 800795, Passenger Car Meeting, Dearborn, 1981

APPENDIX

MEMBERS OF THE 1994 UNIVERSITY OF ALBERTA HYBRID ELECTRIC VEHICLE PROJECT

Alan Eyre	Darren Lloyd	Laura Walker	Rob Simpson
Alan Wharmby	Darwin Li	Leo Specht	Robert Walker
Alan Montpellier	Dave Sutherland	Leon Chow	Romeo Zoldan
Alice Cox	David Amero	Lik Wu	Ron Alton
Alvin Wong	David Kwong	Lindsay Sheppard	Russell McKinnon
Andrew Christenson	Don McClatchie	Lorie Barton	Sam Jayashankar
Andrew Fulcher	Duane Kruger	Louis Garner	Scott Martin
Andrew Tack	Edward Cheng	Mark Hamlyn	Sean Britton
Andy Cej	Elaine Laflamme	Micheal Zhou	Sean Langford
Aparjit Singh	Ellen Kim	Mike Melnick	Sereke Tesfaye
Apoorva Sharma	Eric Ottenhof	Myrl Tanton	Serge Lemay
Barb Naber	Glenn Lutz	Neall Booth	Sheldon Kofoed
Ben Chan	Greg MacIntyre	Norm Li	Stephan Pare
Ben Vandenberg	Greg Sargent	Orrin Lind	Steven Sutankayo
Bill Bizuk	Hiten Mistry	Patrick Marcotte	Tina Hoffman
Bing Shen	Janice Nicholson	Patrick Scorer	Tom Prokop
Brad Johnston	Jason George	Phil Hanoski	Tony Fang
Brian Stachniak	Jason Joly	Puneet Arora	Tony Wong
Cameron Clark	Jeremy Sewall	Rae Parnmuich	Trevor Hilderman
Carol Whitney	Jim MacDonald	Ralph Tessman	Trevor Sawatzky
Chad Haustein	Joe Seitz	Ravi Meadows	Victor Yung
Cory Heringa	John Roberts	Ray Lehtiniemi	Vince Duckworth
Cory Houston	Joyce Theng	Remko Brouerius Van Nidek	Vincent Chou
Cory Sutela	Justin Lybbert		Xinyang Qiu
Curt Collie	Kent Parnwell	Riaz Tejani	Zulifikar Shivji
Curtis Machida	Kevin Gans	Richard Stirling	
Dale Ulan	Kurt Littlewood	Richard Yeomans	

SPONSORS OF THE 1994 UNIVERSITY OF ALBERTA HYBRID ELECTRIC VEHICLE

A G T Limited
Acklands Limited
Alarmex Ltd.
Alberta Motor Association
Alberta Power Limited
Argyll Suzuki
Association of Professional Engineers, Geologists, and
    Geophysicists of Alberta
BP Oil Company
Black Gold Import Auto Parts Ltd.
British Steel Alloys
Burndy Inc.
Comcept Microsystems
Corning
Edmonton Motor Dealers Association
Edmonton Power
Emitec GmbH
CanMET
Engelhard Corporation
High-Q Design Ltd.
Highland/CoRod Inc.
Highline Minuteman
Honeywell Microswitch
Kentwood Ford

MAC Tools, Division of Stanley Canada
Micor Industries
Midas
Noramco Wire & Cable
Northern Alberta Institue of Technology
Northwest Airlines
Personal Publishing Ltd.
SAFT NIFE Corp.
Stanley Technology Group Inc.
Structural Research & Analysis Corporation
Suzuki Canada Incorporated
Taylor Industrial Software
TransAlta Utilities Corporation
TriLine
TriMotive
Wirtanen Electric Ltd.

Alfred University
1994 Hybrid Electric Vehicle Challenge

Abstract

Working with a 1991 Saturn SL2, a four wheel drive, parrallel, hybird electric vehicle has been developed. An electric drivetrain consisiting of 100 Ni-Cd batteries, and four 7.5 hp 3-pahse AC electric motors has been incorporated into the rear of the car. In the front of the car, a conventional drive system was implimented. The chosen engine was a four cylinder, four cycle, double overhead cam engine. The two drive systems are syncranized electriconically, not mechanically, by torque matching. Substantial vehicle modifications and reinforcements were required to impliment the two systems. Also, a new independant rear suspension was added to the car. The vehicle has gained approximately 1000 lbs. in total weight from the conversion. The intended purpose for the vehicle is to experiment and prove the validity of Hybrid Electric Vehicles. A team from Alfred University intened to raise the overall efficiency of the vehicle from 15% to 45% with two systems.

I Overall Concept

The overall concept of Alfred University's Hybrid Electric Vehicle (HEV), is centered around flexibility in design. It is intended that the vehicle be suited for a variety of driving styles and situations; ranging from econcomical to performance driving styles to urban and rural driving conditions. After researching previous HEV's, and establishing our vehicle's flexability requirement, a parrellel drive system was the chosen solution. The electric motors for the vehicle were to have the role of providing the driving force for light duty situations; such as maintaining highway speeds, and driving in urban and suburban areas. MOOG 300A motors were chosen from the driving requirements and the following criteria:

- Regenerative capability
- Small size
- High starting torque
- Low rpm range
- Backed by a company with substanital customer support and services

It was desired for the engine to have sufficient horsepower for fast acceleration, hill climbing, and other high-load situations. Also, it was desired to select an engine that was small in size, lightweight, and fuel economical. Availability of both the engine and spare parts would play a crucial role in selecting the engine. Satisfying all the criteria for the engine was a Suzuki Kantana, 750cc, 4-cylinder, 4-cycle, double overhead cam engine. An added benefit from the engine was the ease of conversion for running the required methonal fuel. Aftermarket carburators and a special conversion kit were purchased for the alchahol conversion. The engine weighs approximately 150 lbs. and produces 125 horsepower. The engine was fitted with a dual exhaust system, containing two catalytic converters and two mufflers, which gives an acceptable balance between performance and emmissions.

The Saturn chassis did not allow for both the electric and conventional drive trains to be implimented in the engine compartment of the car. The vehicle's drive systems would have to be split between the front and rear sections of the car. Keeping the batteries with the electic motors minimized electical cable losses from the batteries to the motors. Because of the geometery and volume required, the electical system was fitted in the rear of the car.

The four wheel drive strategy for the car solved many problems. There would be no neccessary mechanical couplings of the two drive systems while supplying excellent traction in adverse conditions, such as rain and snow. Simply put, the rear of the car would be electric and front of the car would be conventional. The motor controllers from MOOG simplified things even more, due thier torque matching capabilities. Torque matching allows for the two drive systems to be operated simutaneously with out one ever overpowering the other.

Figures 1 and 2 are provided on the following pages for a graphical representation of the control strategy and operational concept for the vehicle.

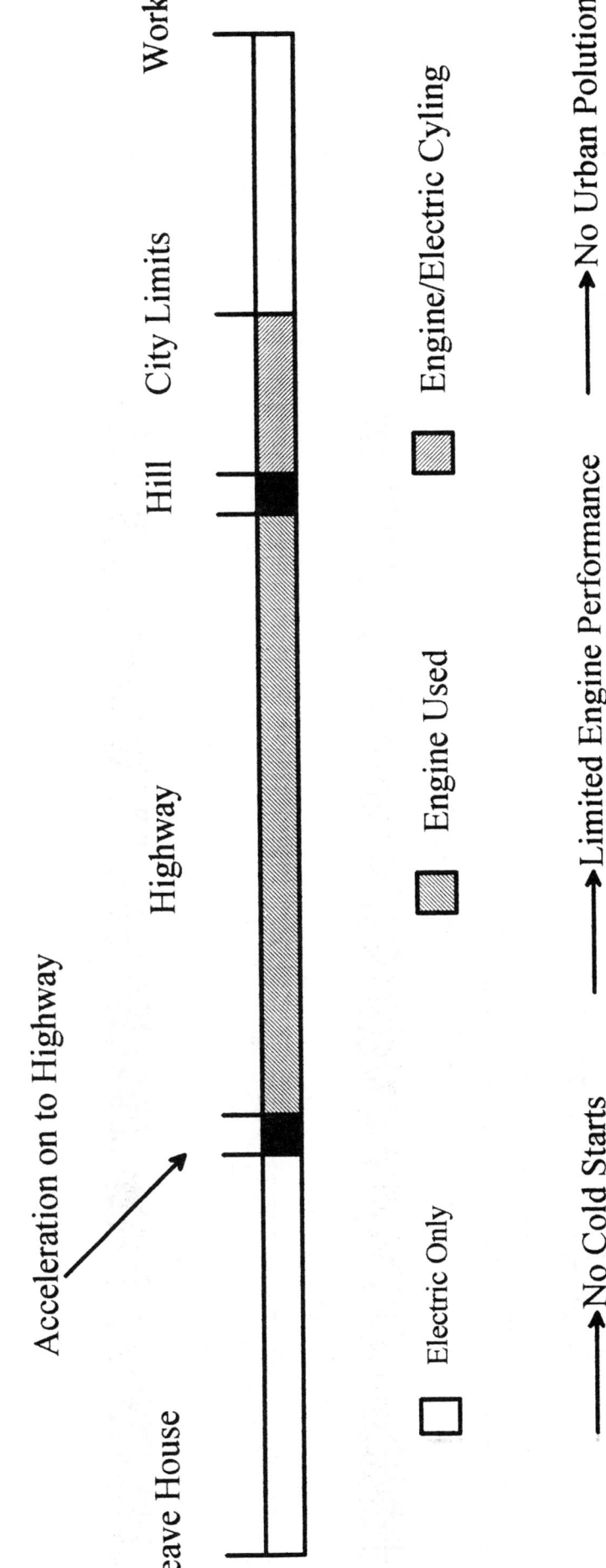

figure 1. Hybrid Control Concept

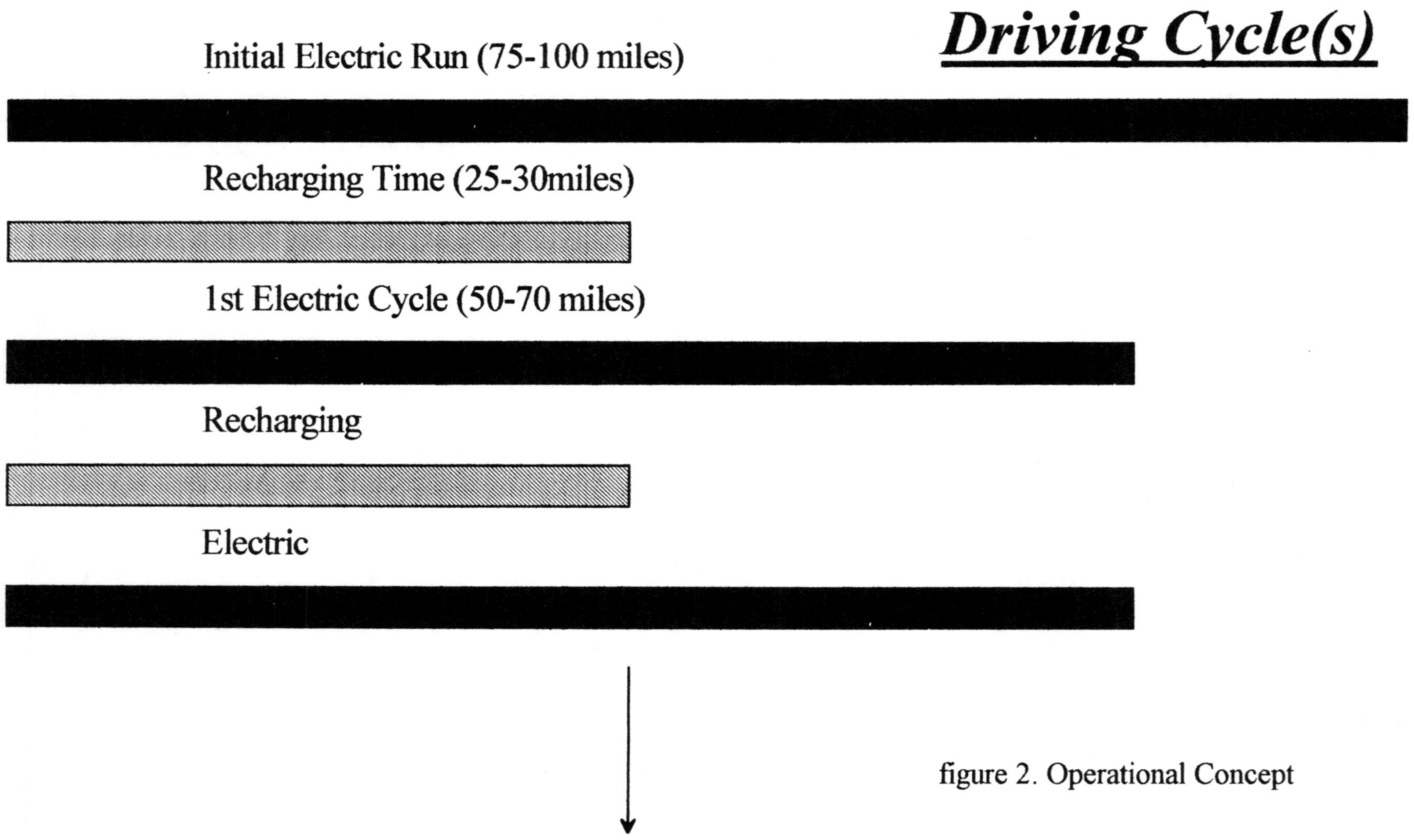

Driving Cycle(s)

Initial Electric Run (75-100 miles)

Recharging Time (25-30miles)

1st Electric Cycle (50-70 miles)

Recharging

Electric

figure 2. Operational Concept

II Rear Section of the Car

In the rear of the car we decided to use an idependant suspension rear axle. Specifically an axle from a 1985 Subuaru was implimented. The differential was needed to bring the driving force of the motors to the wheels. The first step was to design a bracket to hold the differential and the hubs in place. This bracket mounts to the car through one of the main 2" x 2" frame sections which is shown in the figure below.

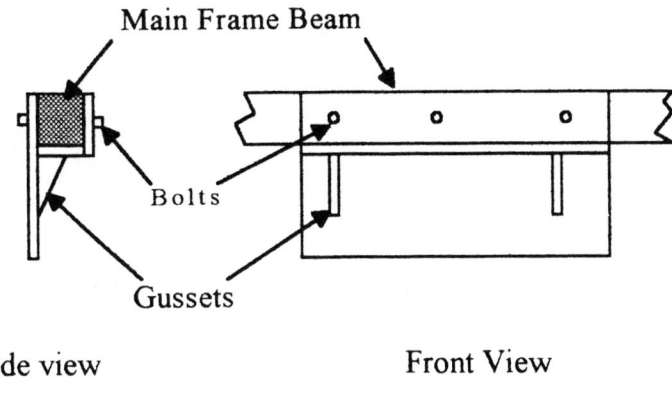

Figure
Figure 3. A schematic of the rear differential bracket

This bracket is made out of 3/16" plate steel and 3/8" grade 8 fine thread bolts were used to fasten the braket to Saturn's structure. As you can see we designed the bracket to fit around the frame beam to prevent damage to the existing frame. With this design, substantial torques could be applied to the bolts, anchoring the braket in place. Gussets were made and welded in the places shown for added support.

To properly secure the front of the differential, original mounting ears for the Subaru differential were welded directly to the bracket. On the other end of the differential, special parts were fabricated. Figure 4 shows the pulley shaft assembly that was fabricated.

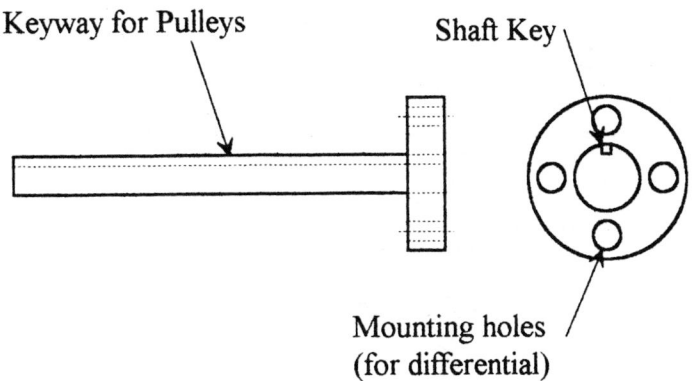

Figure 4. Rear Pulley Shaft

19

The shaft mounts directly to the differential with the four mounting holes. On the other end of the shaft, a pillow block bearing was mounted from the rear bumper supports. The bearing gives both lateral and logitudinal support. The shaft assembly was fabricated at a local machine shop with 4140 steel as the material. Refer to appendix "A", "Detailed Drawings," for a complete drawing of the shaft assembly. Also a detailed drawing is provided of the bearing support.

The next step was to some how adapt the hubs of the Suburau to the rear suspension of the Saturn. It was desired to uses the same points of attachment for the control arms and struts. To do this the hubs of the Saturn were cut and ground down and a special weldment was made with 1/2" steel plating. The weldment provided a unique adaptation of the Subaru rear axle to the Saturn chassis. The weldment allowed for simular suspension mounting points as the original Saturn. A six link suspension was used, simular to that of the stock SL2. The suspension arms and attachments were all made to handle high loads, reinforced, and grade 8 fasteners were used.

Changing the rear spring rates was a must, because of the heavier load from the batteries. These new springs' wire diameter was double that of the original springs. The new springs gave the capability of supporting the anticapated weight.

Mounting the electric motors was the next required step. The motors have mounting holes on the front face of the casing, where the output shaft is located. Face plates were made out of 1/2" aluminum and were fabricated at a local machine shop. Slots were milled in the mounting plated for proper adjustment of the motors. A detailed figure is provided in appendix "A".

Using 2" x 3" x 3/16" angle iron, the motors were mounted to the frame of the Satrun. The angle iron was welded across the rear of the car so that the top of the angle was level with the bottom of the trunk. The two inside motors were bolted to the opposite sides of this beam to give a required offset for the pulleys. The rest of the offset was made up on where the pulley was mounted on the motors output shaft. Again, grade 8 fine thread bolts and lock tight nuts were used for mounting.

The next step was the installation of the battery rack. Figure 5 shows an overhead view of the battery support rack. We used the angle iron supports for the electric motors as one of the main support beams. Then three more of these angle iron supports were welded to the frame of the car. A 3/4" plywood floor was fitted to the trunk area for the batteries to sit on.

Figure 5. Overhead view of the Trunk Compartment

After the battery floor was installed, the next step dealt with fastening the sides of the battery box to the floor. We used 1/4" plywood for all the sides. The sides were 13" high and were fastened around the floor edges with wood glue and staples. Once the walls were installed, fiberglass was used to reinforce the battery box. Then, 1/4" plywood was used to cover the box. These pieces were also glued and stapled with fiberglass reinforcing. An exhaust fan was mounted in the rear of the battery compartment for venting purposes.

With the batteries and motors installed, room needed to be made for the motor controllers. Two controllers were installed in the back window decklid. Sixteen gage sheet metal was used to mount the controlers. The other two controllers were mounted in the under side of the trunk with 10 gage sheet metal. After the implimentation of the controlers, wiring and electical connections completed the implimentation of the electric drive system.

III Front Section of the Car

To mount the engine in the existing engine compartment of the Saturn, only minor modifcations were neccessary. A cross member on the Saturn cadle was cut and removed to make room for the the Katana 750 motorcycle engine. Another cross member (1.5" x 1.5" angle iron) was added to maintain structural integrity. The engine itself is was also used as solid reinforcement in the lateral direction.

For simplicity and time, both the steering and brake systems were converted to manual units. Proper brake lines were fitted to the car in order to adapt the rear Subaru brakes to the Saturn's master cylinder.

Continuing with the idea of changing as little as possible, we decided to use the existing half-shafts and differential from the Saturn. This solved the problem of producing mating parts with matching splines. A cover for the differential was fabicated and a special adapter plate was desgined for a motorcyle sproket. The adaptor for the sprocket was designed first, see appendix "A", and was constructed in such a way that it mounted the sprocket directly to the existing differential. Figure 6 below is a schematic of the fabircated parts for the differential.

To cover the differential's gears, a piece was constructed from sheet metla. This cover, was designed to be mounted to the sprocket adaptor through a series of allen head bolts. The whole thing was then sealed with a liquid gasket, a semi-permanent adhesive. Axle grease was added to the gears through an installed grease fitting.

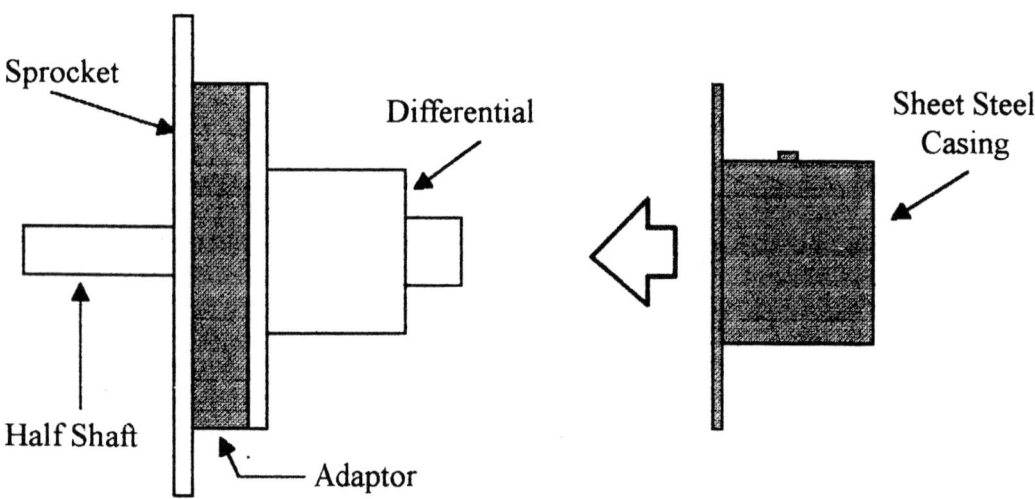

The engine itself was mounted on a series of sliding brackets mounted to the frame and fastened with bolts and welds. All of these brackets were designed and constructed by the team members in the laboratory. The engine mounts were welded directly to the existing frame. All four of the main mounting brackets were constructed so that the engine could be positioned for proper chain tightening.

Converting the engine from gasoline to alchahol was a simple process. Aftermarket carbutators were purchased and calibrated for running the chosen M85 fuel. Along with the carburators a special conversion kit and fuel pump were also purchased.

The next major trick in getting the front end operational was to connect the clutch, accelerator, and shifting mechanism from the existing Saturn to the motorcycle engine. A combination of brakets and cables was implimented to connect the Saturn controls to the motorcycle engine. Because they operated in simular fassions, the accelerator cable was easily connected to the motorcycle engine. The clutch and shifting cables presented a more difficult challenge. It was decided to make use of the existing hydralic clutch and shifting cables. Their ability to push and pull provided more flexibility in designing the connections.

The clutch connection was the first problem we looked at. The Katana clutch was engaged through the tensioning of a cable by the opertors left hand. On the Saturn, when the driver presses the clutch pedal, the hydralic cable extends not contracts. Essentially we were looking at opposing motions. To solve this problem we designed a lever type device. This was a device that took the "pushing" force of the Saturn clutch and converted it to the "pulling" force required by the motorcycle engine. A diagram of this is shown in figure 7 below.

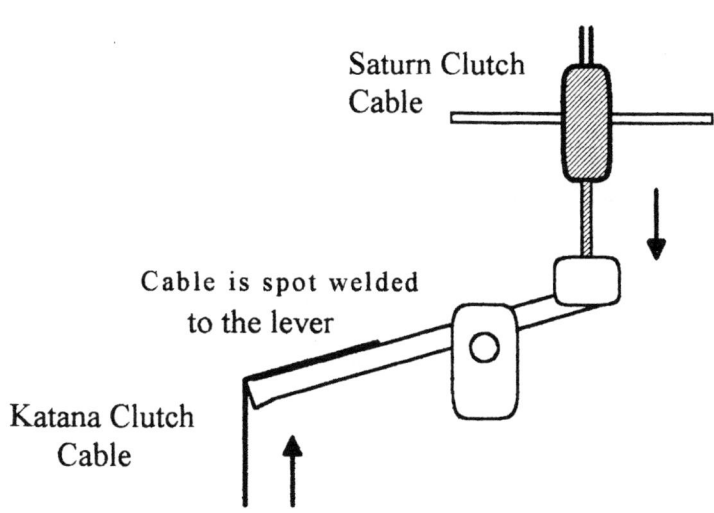

Figure 7. Clutch schematic

The connection of the shifting mechanism was more of a direct connection. The shifting of gears in a motorcycle is accomplished through the manipulation of a lever operated by the rider's left foot. In the Saturn this was accomplished by the operation of the gear shifter by the driver's right hand. The Saturn had two cables to accomplish this task. One was directed by the lateral motion of the "stick" the second was directed by the forward or backward motion of the "stick". Together these two cables controlled the operation of the car's transmission. To convert the Saturn's gear shifter to control the motorcycle transmission, the lateral cable was removed. The gear shifter's range of the Saturn was limited to a straight forward and backward motion. This inline motion provided neccessary movement for shifting the motorcylce's transmission. At this point, a direct connection from the Saturn cable to the shifting lever on the Katana was made. The pushing motion of the hydraulic cable would "up" shift the transmission in the motorcycle and the pulling motion would "down" shift the transmission. With the completion of the shifting mechanism, the engine was now fully operational from within the Saturn engine compartment, with no other required operations except those of driving a standard shift vehicle.

The fuel storage was implimented under the hood of the Saturn next to the engine. For this application a five gallon race car fuel cell, specially made to hold achahol derived fuels was used. These fuel cells are commonly used in racing applications, and are capable of maintaining structural integrity at collisions of up to 150 mph. Mounting the fuel cell was a relatively simple task. As it was designed for the fuel cell was mounted to the frame of the Saturn with straps. A HALON fire extinguishing system was also installed under the hood of the Saturn. Figure 8 is provided to display the arrangement of the components in the engine compartment.

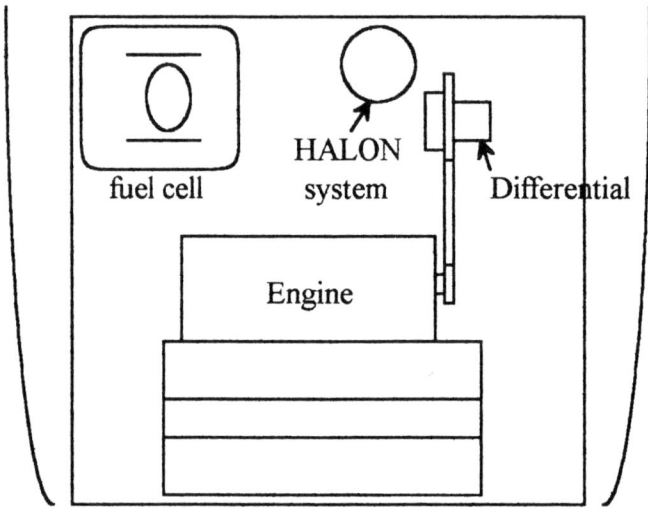

Figure 8. Engine Compartment Layout

Appendix A

Detailed Drawings

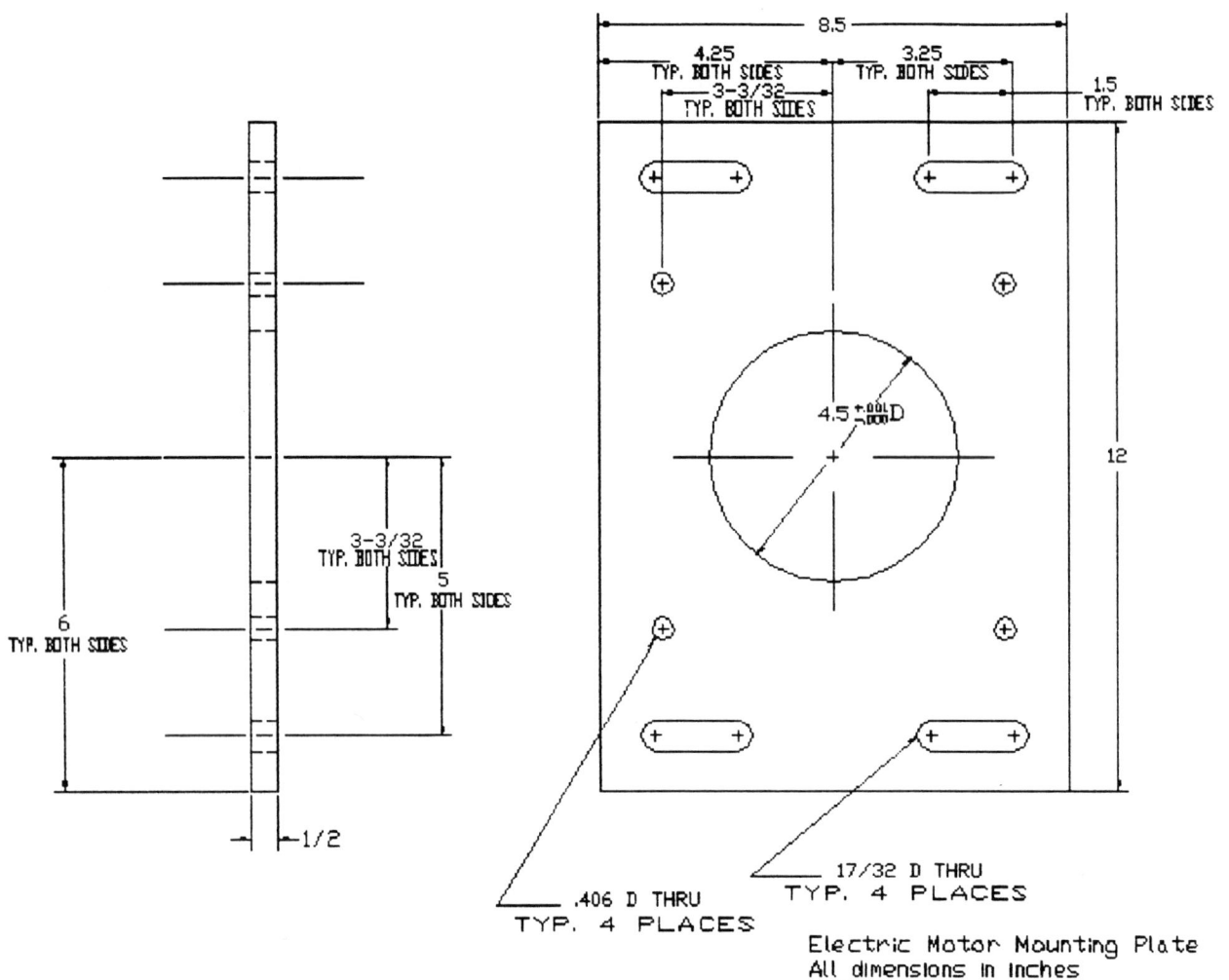

Figure

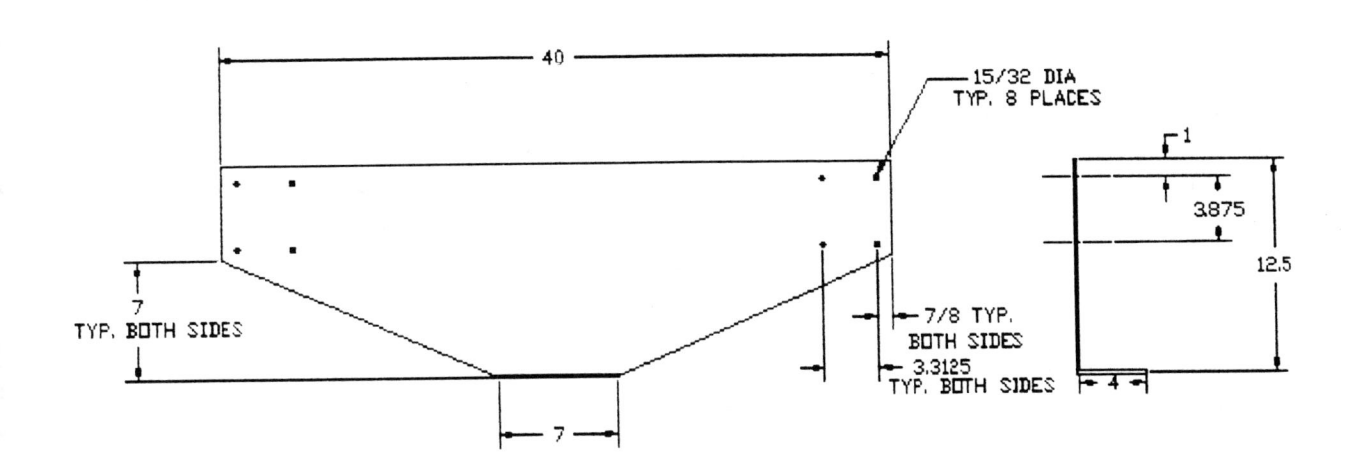

BUMPER PLATE DESIGN
ALL DIMENSIONS IN INCHES
1/4" 36A PLATING

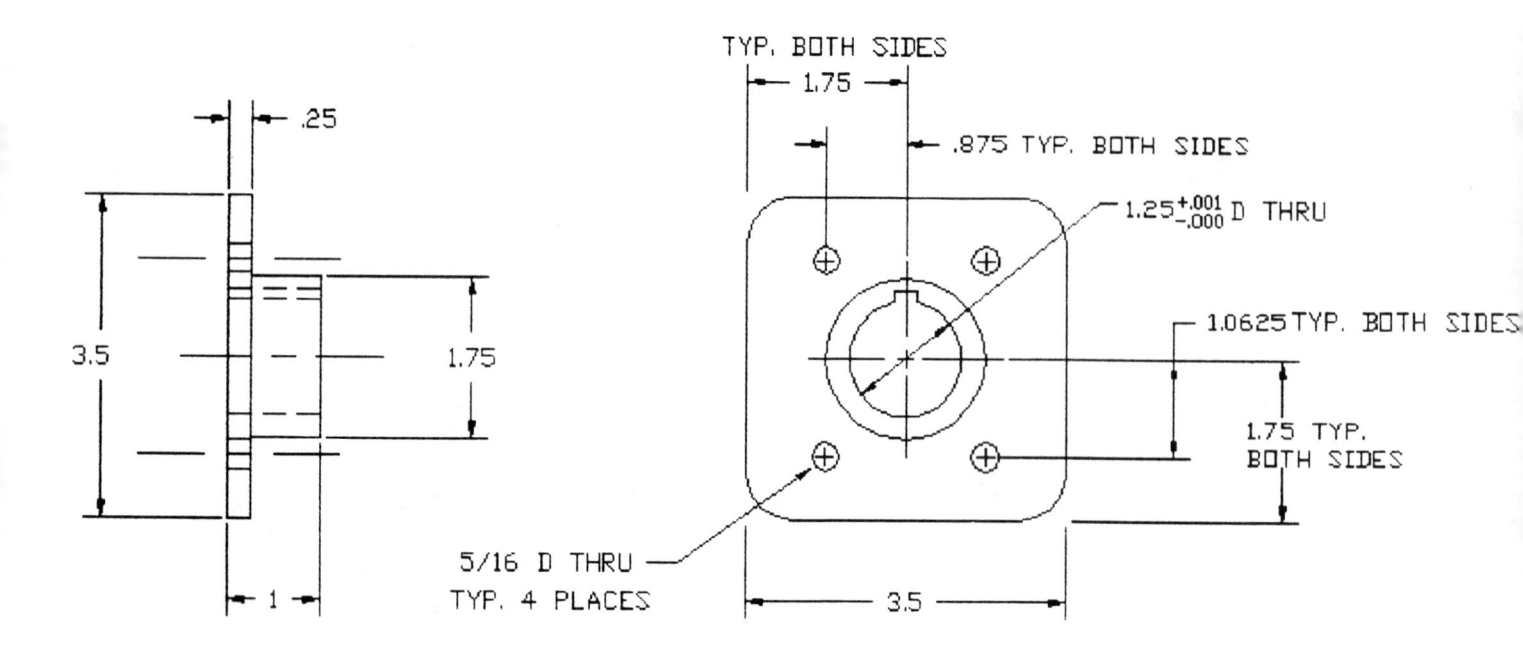

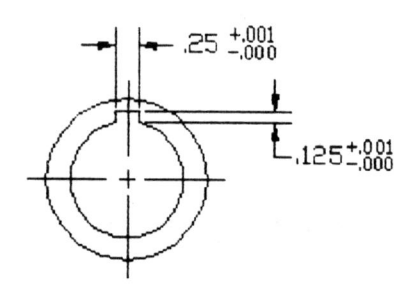

Collar Design
All dimensions in inches

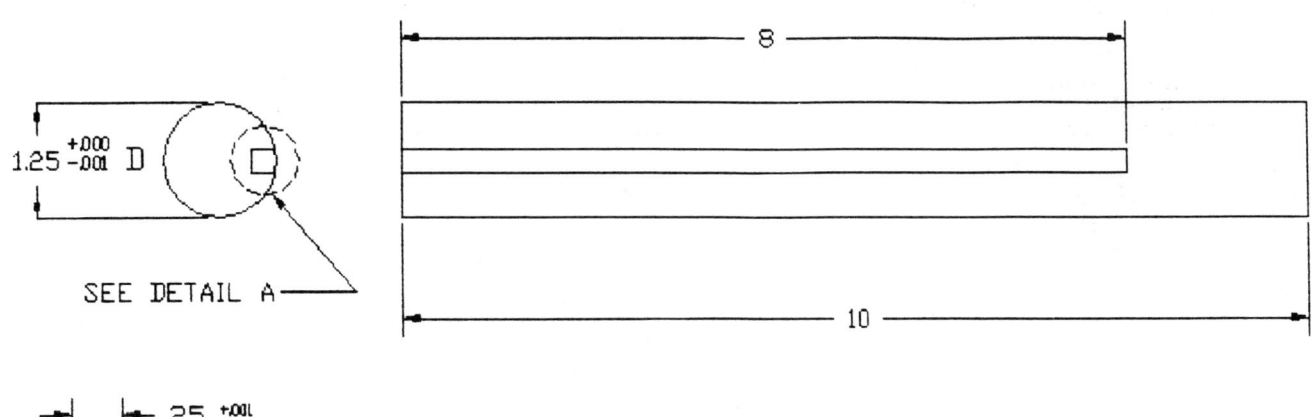

1.25 +.000/-.001 D

SEE DETAIL A

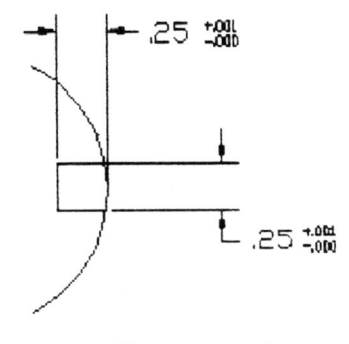

8

10

.25 +.001/-.000

.25 +.001/-.000

DETAIL A

Rear Drive Shaft Design
All dimensions in inches
(4140 Steel)

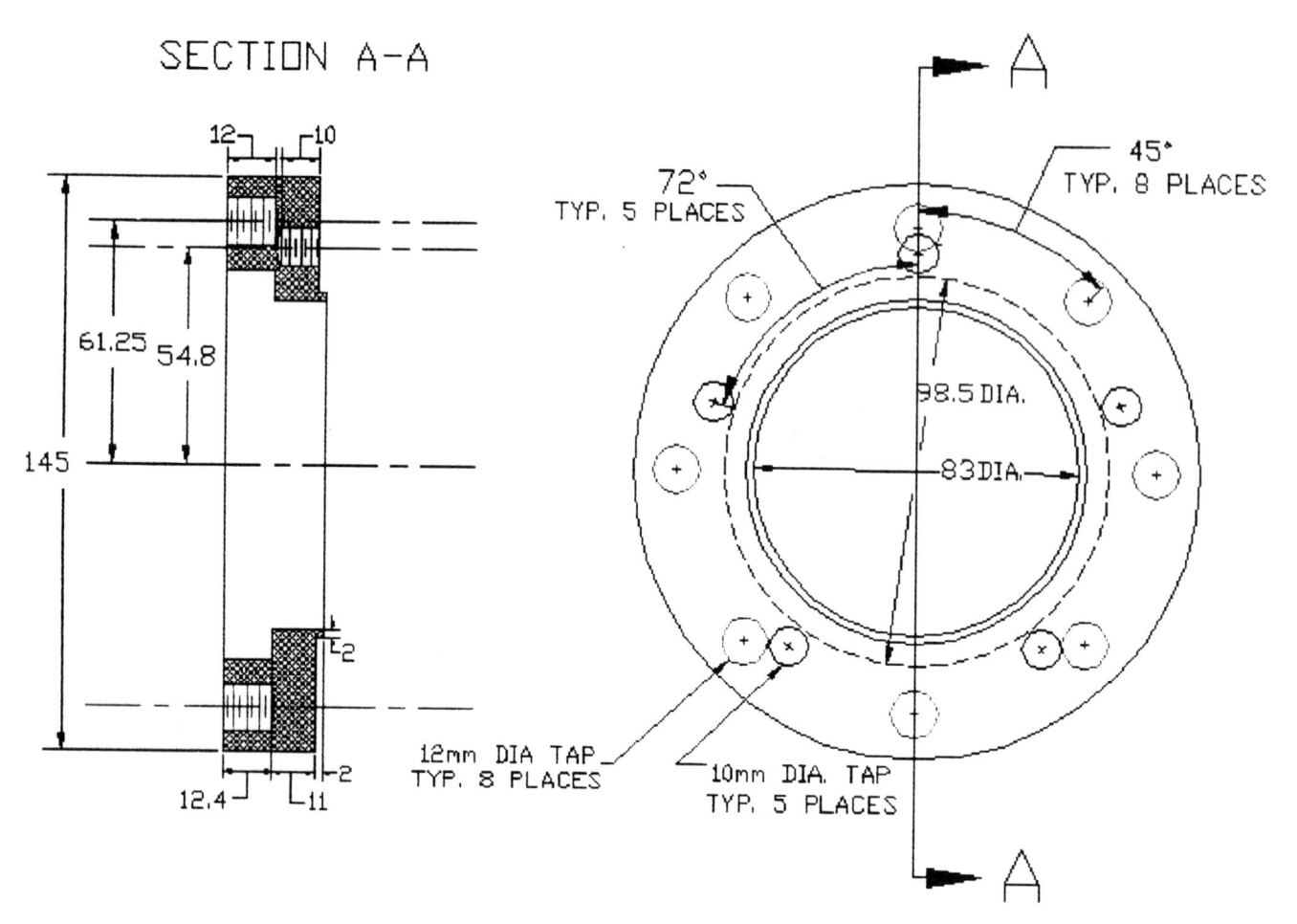

Figure

SPROCKET ADAPTER DESIGN
ALL DIMENSIONS IN MILLIMETERS

Appendix: Project Management

The project organization was divided into two sections, the organization of engineering and that of project support. Naturally, certain issues pertain to both of them.

One of the first steps to organizing the engineering team was to set up a team hierarchy, with a Project Leader (see Figure 4 for organizational charts), to be responsible for all teams, and several team leaders. This worked well in many ways because it assigned responsibility for each vehicle sub-system to one or two people, making it easier to work on several systems concurrently. It was understood that the teams would work together if some had some spare time and others were overloaded. The teams for 1994 essentially consisted only of the team leaders; however, this type of structure lends itself well to the addition of more team members in future teams.

Project scheduling was attempted, but was not successful and was dropped by the second semester. Although it was originally thought that, given the complexity and tight time constraints of this project, tight scheduling would be essential, it turned out not to be. The work done on the vehicle is highly specialized and depended on the expertise of one or two people. Day-to-day scheduling turned out to be simply too cumbersome and hard to stick to, especially given the many design changes. Also, there were many unforseen part delays which caused frequent project delays and the need to re-schedule. A tight, formal schedule would have had to be re-generated daily in order to be at all useful. As it was, the schedule served as a constant reminder of delays, was not followed, and was bad for team morale.

A type of scheduling which did work was a looser "things to do" list, made approximately bi-weekly. Everything which needed to get done was listed, prioritized, ordered, and divided up among team members. The only "deadlines" were those needed to meet major project milestones. Although this is in many ways the same exact thing as the previous schedules, presenting the information in this way seemed to work better. Probably the most important thing to remember about time/activity management is that there are many ways to go about it, and the way that seems simplest and fastest for a given situation is usually the best way. Also, this ensured that the schedule would be approved and understood by all involved. If, at a later time, the project teams grow to more than one or two people apiece, then traditional Gannt chart scheduling, to be used by each team, might be necessary. However, in order for the "to do" list method to work well, the team needs to make a "whole project to do list" in the beginning of the project. The list does not need to be detailed, but major activities, milestones, and parts needed to achieve them, need to be on it. The team can divide up each major activity into its details in weekly or bi-weekly planning sessions. This is essentially what was done this year; however, frequent design changes and rework rendered several "to do" lists useless. A suggested schedule, intended to provide a starting point for next year, is presented in Figure 5.

Tight scheduling is necessary for parts procurement; again, the to do list can be used here also, but a Gannt chart might be useful here, just to graphically present the lead time for parts, when they need to be ordered to meet deadlines, and to coordinate parts with activities. Each team leader was responsible for his own part procurement, which worked well since each team leader knew best which parts he needed. The only problems with that were unavoidable delays on the part of the vendors, or delays in getting the necessary funding for parts purchase.

Of course, in order for this type of team structure to work, good communication is vital. Each team leader wrote weekly progress reports, which was only occasionally cumbersome. However, since many activities took longer than one week, it might be possible to change to bi-weekly written progress reports. The general project meetings and engineering team meetings provided a good forum for team communication.

There is still plenty of room for communication improvement, and one way to do that is through more central reporting and updates. The team leader should be apprised of all activities, and a weekly work schedule should be planned when possible. The engineering meetings should try each week to plan what times they intend to work, and on what, so that the teams can coordinate with one another. This doesn't have to be rigid; it just has to be more specific than "we'll be working on it every night". This weekly work schedule can be subject to change if necessary, but it is important that all team leaders tell one another of any changes, so that the teams can re-coordinate their actions as required. If a team is working alone on the vehicle, they could post a brief note before they leave, telling what they did and what needs to be done next. This provides a written record of work done, as well as ensuring that the message will get to the next team (since it is often difficult to get messages to people by phone in time).

The engineering team's communication within itself is not the only area which needs improvement; the engineering support's communication with the engineers also needs work. Marketing especially had many problems with mis-communication.

One way to help solve this problem is to have all people in a given team report to one person only, and that is the team leader. Each team leader can then report to the Project Co-leader in charge of engineering support. The Project Co-leader on the engineering support side has the same duties as the Co-leader on the technical side--to report to the other Co-leader, and to oversee all actions of sub-teams. The Co-leaders also help on any team which needs them, and work with faculty and administration. Although this system was attempted this year, the Strategic Action Committee (SAC) especially had many problems this year with members taking it upon themselves to take unapproved actions or pass along inaccurate information.

Because business students and engineers are not used to working together, the engineering support meetings should be run more formally than the engineering meetings, with agendas and minutes. At the end of the meeting, the engineering support teams should repeat and summarize any plans or decisions so that any misunderstandings can be cleared up immediately. As a last double check, the engineering support teams should also write up brief minutes, to be submitted to the SAC leader by the next day and compared to his or her minutes.

The information systems team worked out very well, mainly because there was a team leader at every single class (it was done as a class project). This is not always possible, however. If an engineering support team is using the HEV Challenge as a class project, then one person should take notes at every class and send those notes to the Co-leader. The leader will then have up to the minute information on every important class decision, and can clear up misunderstandings before the team takes large steps in the wrong direction. Also, in order to facilitate working together, the engineering support teams should also attend general meetings. Even if all the information discussed is not directly relevant to their actions, it helps them get a better idea of what exactly is going on with the project.

Another problem with engineering support was giving the Business School a sense of ownership in the project. Incorporating their work into class assignments worked well, while the first plan of a volunteer marketing team worked badly. Engineering Support also worked best with a combination of engineers and business students on the team, as was shown by the information systems team. Therefore, the SAC, which handles publicity and fundraising and consists of engineers (mostly underclass) and business students, was formed. The current hierarchy consists of the SAC, information systems, and accounting, (see Figure 5 for organization chart).

BOne of the goals of the HEV Challenge is to make it into a University wide project, with as many students participating as possible. However, it is probably best to make sure that one group is working well before bringing in another, and also best not to rush things. One of the main problems this year was that the project had such a short time period--publicity, fundraising, and design all had to happen concurrently in the first few months of the project. This created a certain sense of urgency and stress among team members, prompting some to take unconsidered actions, or to rush ahead with plans which conflicted with the main plan laid out by the team leaders. Also, since there was no fundraising/publicity team in the beginning of the year, the engineers were forced to become involved in fundraising, an activity which took them away from their real function. An advisable course of action for the next two years, now that the SAC is established, is to leave all the fundraising and publicity to the engineering support teams. Naturally, there is some overlap when it comes to gifts in kind and parts procurement, since the engineers are responsible for getting their own parts, but that overlap can be made less conflicting by good communication among all teams.

Communication is probably the single most important part of the Project structure, and it is not enough to simply go through the motions of meetings and reports. Each team member has to make a concentrated effort to communicate vital information in a timely and organized fashion. If the lines of communication are kept open, and if each person does only his or her job unless asked to help someone else out, the project should proceed with relative smoothness.

Overall, the most important part of Project management is solid organization combined with flexibility. It's best to start by establishing a definite Project structure--but with built in "room to move". Tailor the organization to meet the individual needs of all those in it. If some part of the structure seems to make more work than is justified by the results, then change it right away to something less cumbersome.

Given the fact that this is the first time AU has attempted a student project of this scope and magnitude, the project progressed really quite well. The project progress and structure have been documented, so that records exist to make the way for the next two years smoother. Most importantly, the team met its goals-- the necessary monies and equipment were acquired, and the vehicle was successfully completed on time.

California State Polytechnic University, Pomona

Robert Augenstein (Editor)
Mechanical Engineering

Tricia Chee
Electrical Engineering

Scott Kysar
Mechanical Engineering

Dave Mitchell and Hans Wenzell
Engineering Technology - Mechanical

Kevin Winter and Bill Wong
Mechanical Engineering

ABSTRACT

Team APEX from California State Polytechnic University, Pomona, returns to the 1994 HEV Challenge with its second-generation ground-up vehicle, "APEX 2". This car combines a DC electric motor drive system and a constant-speed auxiliary power unit (APU) in a series configuration. The vehicle contains many practical features such as automatic onboard charging, alternator protection, a high strength to weight ratio frame, and an environmentally friendly Lexan body. These design attributes are brought together to make APEX 2 safe, efficient, and easily mass produced.

INTRODUCTION

Cal Poly is one of twelve schools given the opportunity to participate in the ground-up class of the HEV Challenge. Design considerations and choices made by Team APEX are presented within this paper. Results of these decisions are included whenever possible.
A hybrid electric vehicle (HEV) is a realistic step in the journey towards practical electric vehicles. Team APEX strove to build a vehicle that meets consumers' expectations for performance and cost. The Team emphasized three areas in the design of APEX 2: weight reduction, economy, and environmental impact reduction. At the same time, we were unwilling to sacrifice practicality, producability, or performance.
During the initial design stages, our goals appeared to contain features which were at odds with each other. These obstacles were overcome by careful research into various drive configurations, electric motors and controllers, APU's, energy storage devices, as well as a host of other areas. Transforming this information into a functional vehicle proved both challenging and rewarding.

POWER TRAIN CONFIGURATION

Team APEX selected a series drive train configuration for its simplicity and reliability. There are four major drivetrain components, plus battery charger, whose relationship is shown in Figure 1. Component location dictated a front-wheel drive vehicle. The series configuration requires only one transmission which provides several advantages. First, this configuration allows for design flexibility as it can accommodate

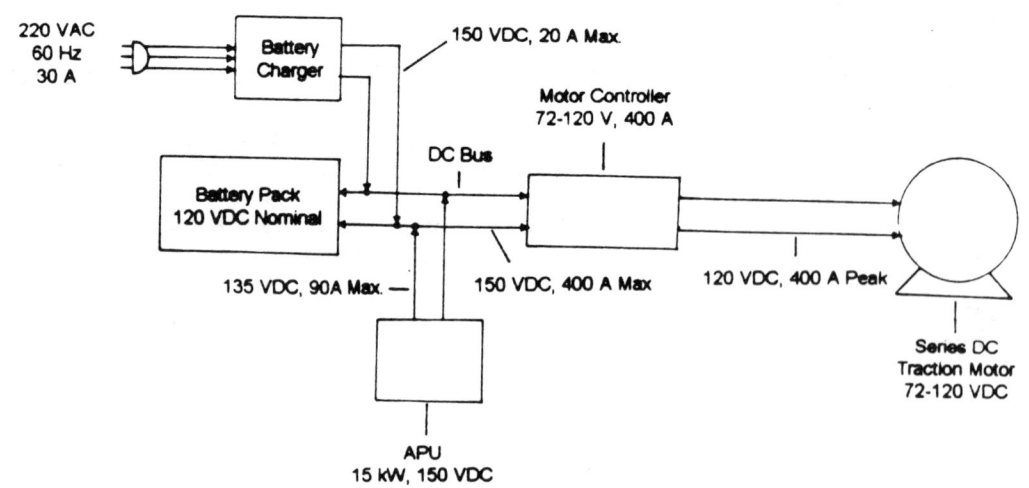

Figure 1. Series Hybrid Powertrain. This figure displays the relationship between major powertrain components.

various auxiliary power units without modification to the drive train. Second, vehicle efficiency is increased due to weight savings and a decrease in inertial losses. Third, costs are reduced by eliminating the second transmission required in parallel configurations. In addition having only one transmission simplifies production of the vehicle. Finally, a series configuration reduces the possibility of technical complications.

ELECTRIC MOTOR - The weight reduction program allows APEX 2 to be powered by a smaller motor. APEX 2 uses a 19 hp Advanced DC, series-wound, brush motor. This motor's operating voltage range is from 0 to 120 volts, with a specified rating of 72 to 120 volts. The continuous rating of the Advanced motor is 19 horsepower with a peak of 60 horsepower.

CONTROLLER - Team APEX chose a pulse-bandwidth Curtis motor controller. The decision to use the Curtis controller was based on a history of compatibility and reliability with the Advanced motor.

BATTERY - Lead-acid batteries are used based on a trade-off study of availability, cost, and quality. The battery pack consists of ten, 12 volt, lead-acid, deep cycle batteries. The pack is connected in series for a nominal voltage of 120 volts. As the primary source of motive power, the battery pack can supply 400 amperes during peak power demand from the motor. The batteries have an 11 Kw-hr energy capacity, with a maximum 30 kW-hr power draw. This is enough energy to power the vehicle at 35 MPH for 55 miles.

APU - The APU is a Briggs & Stratton Vanguard engine close coupled to a Fisher Electric alternator. The engine is a 480cc, air-cooled, overhead valve, V-twin design producing 16 horsepower. APU selection was made according to vehicle power requirements at a constant speed of 72 km/h. The alternator provides a maximum of 12kW of power at 3600 RPM.

CONTROL STRATEGY FOR ALL MODES OF OPERATION

APEX 2 was envisioned from the start as a series hybrid design with an emphasis on ZEV operation. As stated earlier, our team felt this was the best configuration in terms of simplicity of design and component reliability. We also visualized an almost completely passive electrical control system to address some of the peculiarities of hybrid electric vehicles.

In ZEV operation, under typical commuting conditions, the vehicle's range is 58 kilometers. Under steady cruising (56 km/h), the range increases to approximately 89 kilometers. To extend the range, control strategy switches modes from

ZEV to HEV. In the HEV mode, the APU output equals the energy being used to drive the car under normal conditions.

To protect the alternator, we designed a circuit to interrupt charging when the power draw exceeds 10 kW. When the power draw returns to below 10 kW, the circuit reengages the alternator. The APU may be turned on either manually or by a passive control circuit when in the HEV mode. The passive control circuit senses when battery voltage drops below 110 volts. When this occurs the circuit will activate the APU. The charging is self regulating, depending on battery voltage, and stops when the battery voltage is above 150 volts.

EMISSIONS CONTROL STRATEGY

The Briggs engine provides a reliable power plant. Unfortunately, it produces high quantities of emissions in its stock condition. To reduce emissions Team APEX modified the engine by installing a catalyst, a mixture controller, adjusting the ignition timing and rejetting the carburetor. These modifications will reduced tailpipe emissions by at least fifty percent from stock emissions.

CATALYST - The catalyst is designed specifically for the Briggs, to be the largest contributor to reducing emissions. The catalyst reduces hydrocarbons, nitrous oxides, and carbon monoxides emissions. The team designed and crafted the catalyst housing and stainless steel exhaust system.

MIXTURE CONTROL - Prior to the installation of the catalyst, the primary modification was to change the carburetor in favor of a more efficient design. Carburetors designed for Briggs & Stratton engines tend to run rich air-fuel mixtures. To compensate, the Team chose to install a "lean-limit" controller on the engine. This device allows the engine to operate at the leanest tolerable mixtures, reducing emissions considerably. It operates by injecting air beneath the carburetor, which effectively reduces the air-fuel ratio. The "lean-limit" controller compensates for changes in temperature, pressure, and loading conditions. For example, the controller is bypassed when wide open throttle conditions are required, allowing full engine power. As an additional improvement, the team manufactured an interface plate for the lean-limit controller that creates a "swirl-effect" to increase volumetric efficiency. Re-jetting the carburetor was also an effective modification.

TIMING - Leaner combustion conditions require the ignition timing of the engine to be changed. This adjustment was complicated by the fixed-timing magneto-ignition system Briggs uses on this engine. To achieve the desired degrees of offset, the Team fabricated Woodruff keys that allowed timing advancement in discrete amounts.

FUEL AND ELECTRICAL POWER EFFICIENCY CONSIDERATIONS

Regardless of power mode the easiest and most practical way to increase efficiency is to reduce the weight of the vehicle. To this end, Team APEX concentrated on weight reduction in all design aspects. Several weight saving measures are addressed in this paper. These measures were taken specifically to increase efficiency. Additionally, the Team scrutinized and modified vehicle components to maximize efficiency.

FUEL - The Team used the techniques discussed in the Emission Control section of this paper to address fuel efficiency. Re-jetting the carburetor and adjusting the ignition timing not only reduced emissions, but also increased fuel efficiency. Testing developed load curves, allowing the engine to be tuned to run at its most fuel-efficient speed.

Unleaded fuel was selected because it did not require engine modifications and is readily available. Ethanol and M85 require extensive engine modifications and pose additional fuel storage problems.

ELECTRICAL SYSTEM - The Team addressed electrical power efficiency in several ways. Electrical components were selected for their operating efficiency as well as compatibility with other components. Wire size was determined based on peak demand and wire length was minimized to reduce resistive losses. The alternator has a 90% DC efficiency, and a full wave bridge type rectifier is used so that as much power as possible can be converted to DC. Based on dynamometer tests, the vehicle is 85% efficient from batteries to motor.

VEHICLE STRUCTURE

The vehicle structure is a ladder frame constructed of chromoly steel. The ladder configuration provides stiffness and light weight with adaptability, as evidenced by our use of production components for both the front and rear suspensions. This frame allows easy modification to all systems and can be mass produced.

FRAME - The frame is fabricated with normalized 4130 steel. Design calculations indicated the frame be constructed of rectangular tubing with a sidewall thickness of .15 cm and a top and bottom wall thickness of .30 cm. Rectangular tubing with these dimensions is unavailable commercially. Therefore, rectangular beams were constructed from 0.15 cm thick normalized 4130 steel sheets. The sheets were bent into C-sections. Two C-sections were slipped over one another and welded together to form rectangular beams, measuring 7.62 cm X 3.81 cm as shown in Figure 2. The beams were then cut to appropriate lengths and assembled to create the frame. The roll cage adds rigidity to the frame while providing for occupant safety. The entire roll cage utilizes normalized 4130 seamless tubing, 3.51 cm in diameter with a wall thickness of .24 cm. Side impact members protect the passenger and battery compartments in the event of a collision. The entire structure was designed using finite element and traditional analysis methods. This design provides adequate strength and rigidity at a minimum weight. In fact, the entire chassis weighs only 50 kg!

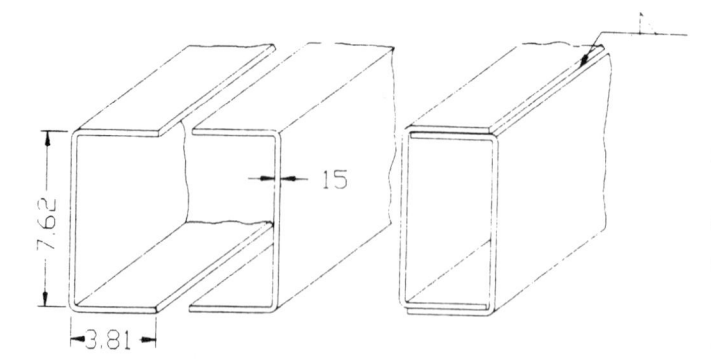

Figure 2. C-Section Assembly. Rectangular beams were made from two c-sections.

SUSPENSION - The suspension was adapted from a Honda CRX. This configuration is comprised of torsion bars, shock absorbers, and an anti-roll bar at the front. The rear suspension consists of a dead beam axle with coil springs, shock absorbers and Panhard rod. The simplicity of our ladder frame design allowed us to incorporate the front and rear suspension assemblies easily.

Power is transmitted to the wheels by a five-speed manual transaxle. Steering is standard rack and pinion. Braking is handled by a dual circuit hydraulic system with front discs and rear drums. The emergency brake is a manually operated cable type.

HANDLING

The weight distribution of APEX 2 is close to that of the CRX, and the suspension performs well. Handling is predictable, instilling confidence in the driver. The vehicle experiences mild understeer in tight corners and is near neutral in sweeping turns. No torque steer is experienced during hard accelerations from a stop or leaving

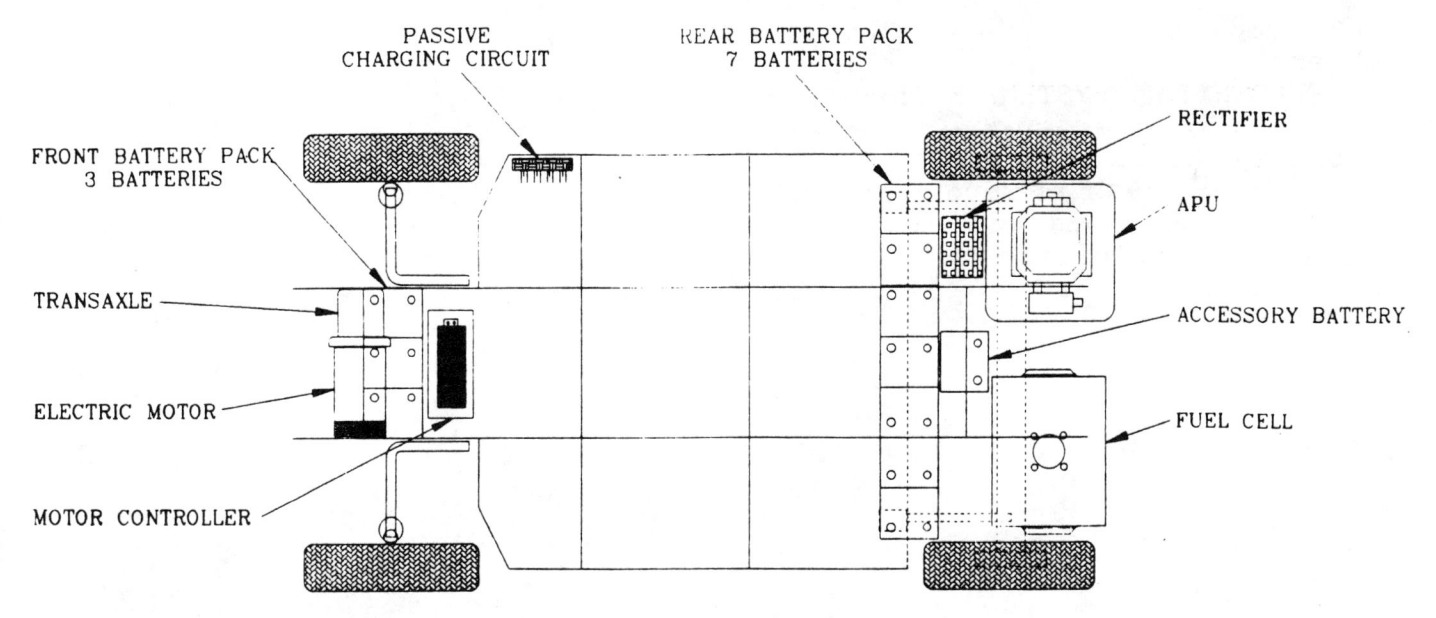

Figure 3. Component Placement. This schematic shows the layout of all major vehicle components.

a corner. Body roll is minimized by mounting the rear batteries in a low frame position. This location shortens the distance between the centroid of the sprung mass and the rear roll center. The front anti-roll bar also helps to minimize lateral weight transfer, controlling body roll.

ERGONOMICS

Standard seats have been used to provide for driver and passenger comfort. Automotive controls have been sized and located in such a manner as to be similar to those of a conventional vehicle. Body styling provides excellent visibility. Instrumentation has been selected to make the driver comfortable and convey the necessary information to safely operate the vehicle. Interior volume is exceptional for this vehicle which is otherwise dimensioned similar to other compact cars.

VEHICLE SAFETY

There are two levels of designed-in safety. Level One handles all the major production concerns about safety. Front and rear crumple zones are incorporated into the frame's design. Safety also influenced component placement. Major components, such as batteries, APU, and fuel tank, are isolated from the passenger compartment and each other to reduce the chance of injury to the occupants in the event of a collision, see Figure 3.

Level Two safety designs provide the additional protection necessary in vehicle competitions. This includes a five-point racing harness restraint system. In the event of a rollover, the vehicle occupants are protected by the roll cage. Finally, a halon fire extinguishing system has been installed to contain any vehicle fires within the APU or battery compartments.

BODY STYLING

Team APEX envisioned a 2 passenger, 3-door hatchback body style, with an interior and exterior fabricated of polycarbonate. With emphasis on weight reduction, economy, and minimal environmental impact, the use of fiberglass or carbon fiber would not have been appropriate. Figure 5 is pictures of our vehicle as it appeared in

Figure 4. APEX 2. The car, as it appeared in January 1994.

January. The styling features of our body include: contemporarily styled, aluminum-alloy wheels, and formed polycarbonate windshield.

The exterior body panels as well as floorboards, fender wells, battery boxes, and dashboard, are hand fabricated from .76 cm thick corrugated polycarbonate sheet. This material is very light weight at .375 lb/sq.ft, and it has good impact strength (400 psi in compression). Furthermore this material resists acid, solvents, and corrosion. It is self-extinguishing should it be exposed to flame, and is non-conductive, both electrically and magnetically.

CHOICE OF MATERIALS

Team APEX chose materials based on design criterion, cost, and availability. Material choices have been presented in the appropriate sections.

VEHICLE MANUFACTURABILITY

Team APEX stressed simplicity to aid manufacturability since this project's inception. The ladder frame is perhaps the most straightforward frame layout. Several body styles and many suspension configurations may be placed on the same platform. All major components of our vehicle are off-the-shelf designs proving their production viability. Cost is also a major subject when discussing manufacturability. We made every attempt to control costs while still producing a quality vehicle. The DC motor system is significantly less expensive than an AC system at this time; and the cost of lead acid batteries is more reasonable than other exotic types. We approximate that our vehicle would cost approximately twenty thousand dollars to purchase if it were in production.

VEHICLE EVALUATION

ELECTRIC MOTOR - A 13 hp DC motor was tested hoping to improve efficiency. However, acceleration and driveability suffered. By contrast, the 19 hp motor is able to accelerate the vehicle to 72 km/h in 15 seconds.

APU - Team APEX tests the APU using an artificial load bank to simulate battery load. The load bank consists of two 4.8 kW variable carbon resistors, along with sensors and measurement devices for data acquisition. This test rig allows the Team to quantify the changes made to the APU in terms of emissions, specific fuel consumption, temperature, power, and efficiency. Improvements and changes that have been tested have resulted in an 80% reduction in CO emissions, 55% less HC, and 10% reduced NO_x over the stock engine configuration. Fuel economy also improved by 35%.

ROAD TESTING - APEX 2 has been evaluated in many different situations and has proven to be a capable performer. Most recently, Team APEX participated in two events, the APS Electric 500, in Arizona and the Clean Air Road Rally to Disneyland. Each event had a different focus. The Electric 500 was primarily a race performance evaluation. Speed and acceleration events were stressed, while range was minimized. However, at the Clean Air Road Rally, the events focused on more practical aspects of electric vehicles. The event was over public roads and tested the car's "real world" capability in terms of range and handling. Our car performed well in each of these situations.

SUMMARY

APEX 2 is a DC series hybrid vehicle using recyclable polycarbonate and existing components extensively. We modified these components to increase efficiency and reduce emissions. Our frame is simple yet light and strong. The vehicle brings together all of these design attributes to make APEX 2 safe, efficient, and easily mass produced.

Technical Report
HEV Team at California Polytechnic State University, San Luis Obispo

Eric Boettcher

ABSTRACT

The Hybrid Electic Vehicle Team at Califonia Polytechnic State University, San Luis Obispo has worked hard to optimize and improve a ground-up vehicle for the 1994 HEV Challenge. An impressive student effort has resulted in the optimization of many custom vehicle features. Original designs include a parallel powertrain, high voltage vehicle & charging systems, computer control, tubular frame, front & rear suspensions, and an aluminum body. This innovation is a direct result of student volunteers working toward a common goal.

POWER CONFIGURATION

The coupling of two powerplants for the HEV Challenge was heavily investigated, examining the advantages between the prominent configurations of series and parallel. Parallel, although more complicated than a given series layout, was chosen for the powertrain. The comparative efficiencies between the two made the decision obvious. Additionally, the entire concept of designing and building a ground-up vehicle was used to encourage everyone to build the best and most efficient HEV possible, allowing the acceptance of added complexity.

A comparison between a series and a parallel powertrain layout began with a general conception of how each system appears. Research into past efforts helped to illustrate the simplicity and desirability of a series system. However, these perceptual benefits were taken at the price of efficiency, when viewing the entire power scheme. The parallel configuration allowed for each powerplant to directly move the vehicle where efficiency is saved because of fewer energy conversions. In a series powertrain an engine drives a generator, which then feeds batteries, which than eventually feed the motor, driving a powertrain. Assuming optimistic efficiencies still places the series layout at nearly a 10% deficit to the parallel, in full HEV operation.

The components for the powertrain were chosen for efficiency and estimated power requirements that the vehicle would need. Assumptions about average velocity and potential energy capacity led to the selection of the Solectria motor and controller units. The motor/controller combination enables the motor to develop 18.75 kW of power. This purchase, along with rules at that point in the competition, made the APU selection favor the Geo Metro 1 liter engine. This 36.75 kW engine is more than able to move the car during average driving or in the acceleration event.

ELECTRIC MOTOR/CONTROLLER - The 18.75 kW AC induction motor (AC GTx20) and matching AC300 controller from Solectria rounded out the system. What was needed to get from the motor and controller was roughly 20 kW for acceleration, operation from a relatively low voltage - so that the battery weight was kept down, and a motor that could operate continually at high cruising speeds without needing excessive cooling. This system was the closest match found from any company and it was far cheaper (over $10,000) from the most desirable solution from Unique. Unique manufactures a line of much more powerful motors, but these also require much higher operating voltages which mean more weight. With the AC300 controller a 98% efficiency in the conversion of the DC voltage to AC was achieved while getting a myriad of features that add safety and convenience as well. These include high temperature current limiting to prevent motor damage while allowing some limited amount of power to drive, several emergency interrupt capabilities, and the ability to change on the fly from a highly efficient delta winding to a wye winding which provides double the torque at low speeds. This last feature allows the vehicle to cruise with the highest efficiency while using the high torque mode to accelerate up a hill or onto a freeway. This switching is performed in a very similar way to that used for entering the four wheel drive mode on a truck. A lever in the cockpit is pulled which disables power from getting to the motor while a three pole switch is turned that re-taps the motor connections. By the time the handle has reached its final position, the motor controller has been switched into the proper operating mode and then the motor interrupt is released. If higher efficiency operation is desired, the handle should be left in the delta position. There is no absolute need to use the wye mode and the use of that mode is opaque to the operation of the rest of the vehicle.

AUXILLARY POWER UNIT - In determining what alternative power plant would be optimal, a minimum of 12kW was established. Group discussion attempted to come up with alternatives. The engine had to be:

- lightweight
- fuel efficient
- have low emissions
- be either easily available or manufacturable.

Design discussions noted that the available power plants that fit the design best were those for use in automobiles. These already had extensive design in the areas of emissions and fuel economy. Potential negatives were mostly due to size; a car engine that makes only 12 kW is hard to come by. Three different possibilities presented themselves in the discussion:

- 12 kW industrial-use engine
- 600cc automobile engine imported from Japan
- 1000cc engine from a Geo Metro.

In comparison, the industrial-use engine in theory emmited close to if not more than the 600cc engine because of designs inherent to an auto engine and not to an industrial-use engine, such as fuel injection, electronic ignition and catalytic converters. If the necessary time, equipment and experienced personnel were available, an existing industrial-use engine could have been modified to fit the design. The final choice became the Geo Metro 1000cc engine. This engine, unlike the 600cc engine, was available locally and as an entire system (all wiring, electronics and supporting structure included). Once this decision was made it became a goal to optimize its efficiency.

Possible engine modification involved many areas. The areas looked at were intake flow, exhaust backpressure, lubrication and a few modifications found on race engines. One of the most exciting new technologies in auto engines is the use of ceramic thermal coatings. This type of coating was used to coat the tops of the pistons, the combustion chambers and the valve faces. This reduces the amount of heat lost to the cooling system during the power stroke thereby increasing the amount of pressure on the piston. The coating also acts to raise the knock limit and cushion the piston if knocking occurs, allowing higher compression ratios. Initial estimates were in the order of a 10% power gain. Attempts were made to reduce to a minimum all intake and exhaust restrictions. In the area of emissions, insulation was placed on the exhaust line before the catalytic converter in order to cut down the cold start emissions by causing the catalytic convertor to fire off much sooner than it presently does. Finally, synthetic lubricants were used in order to cut down on internal friction.

BATTERIES

When the process of selecting the batteries and the electric motor began, there were an overwhelming number of considerations to be weighed against each other. The first step consisted of measuring the relative impact of each of these considerations against how many points were being awarded in the specific events. The main factors for the design specifically included:

- finding a battery technology that would provide the voltage and current needed without being excessively bulky or heavy
- finding a motor and controller capable of the acceleration requirements
- matching these two systems together such that these would make a coherent system.

The design team set baseline performance requirements including:

- the maximum weight
- the minimum ZEV range desired
- the expected cruising power requirements.

From these parameters attention was focused on components capable of meeting these goals. Determining batteries that could provide a minimum of 120 volts and over 22 kW in energy when at a low state of charge received priority. Battery life had to meet the 300 life cycle limit specified while meeting a high enough level of safety and recyclability to be in tune with the broader goals of the competition.

Ultimately, the decision fell upon a sealed, recombinant lead acid battery manufactured in the U.S. by the Concorde Battery Corporation. The batteries have a small footprint (approx. 10" x 7.6") which saves on space and a 24 volt nominal potential. The design places two groups of five series connected batteries in parallel on opposing sides of the car. This combination provides about 6.7 KWHr of energy at the three hour rate. This was originally intended to be 10 KWHr through modification. This modification would have greatly increased the ZEV range as well as the weight of the batteries, but at this time calculations indicate a 40 km ZEV range with better acceleration due to the lower weight with 275 kg of batteries.

Given the designing of an electric vehicle, there were several reasons the parallel battery pack approach seemed desirable. This configuration of the batteries provides an even weight distribution on the chassis, it lowers the currents flowing through the wires and batteries reducing electrical power losses, it provides twice the available power to accelerate the vehicle at low states of charge, it extends the life of the batteries by cutting the charging rates and discharge rates to much lower levels, and it means high rate discharges will have a smaller effect on battery voltage variations. It is true that some of these troublesome effects could have been avoided by using a different type of battery; however, the only other battery technology that seems to be fully developed today is nickel-cadmium which costs twice as much, requires many more interconnections, is not recyclable, and uses cadmium which is highly toxic, non-recyclable and harmful to the environment. Thus, it was concluded that nickel-cadmium batteries are not very desirable for high volume production. Using sealed batteries means that the batteries produce no gasses during operation, are 100 percent maintenance free for life, will not corrode the environment or surroundings and the batteries will not spill if tipped (the batteries can operate perfectly upside-down)! Once the batteries are considered to be at the end of the expected life, the pack can be brought back to the manufacturer and completely recycled. From a mass production standpoint, a consumer should be able to return batteries for a refurbished set making an initial investment only once.

Two companies stand out as being superior choices for supplying electric vehicle motors and controllers, Unique Mobility (Englewood, CO) and Solectria (Arlington, MA). Both offer motors with very high efficiencies and power to weight ratios as well as solid state controllers with interfacing already designed with electric vehicles in mind. Both companies oblige by providing very expensive systems that are truly state of the art, but these units are custom made to best suit the needs of the vehicle in hand. If the batteries are not matched to the controller or the controller isn't well matched to the motor, not one of the three components will perform properly. The biggest challenge in choosing these three components is compiling a number of "complete systems" that are weighed against each other as independent solutions.

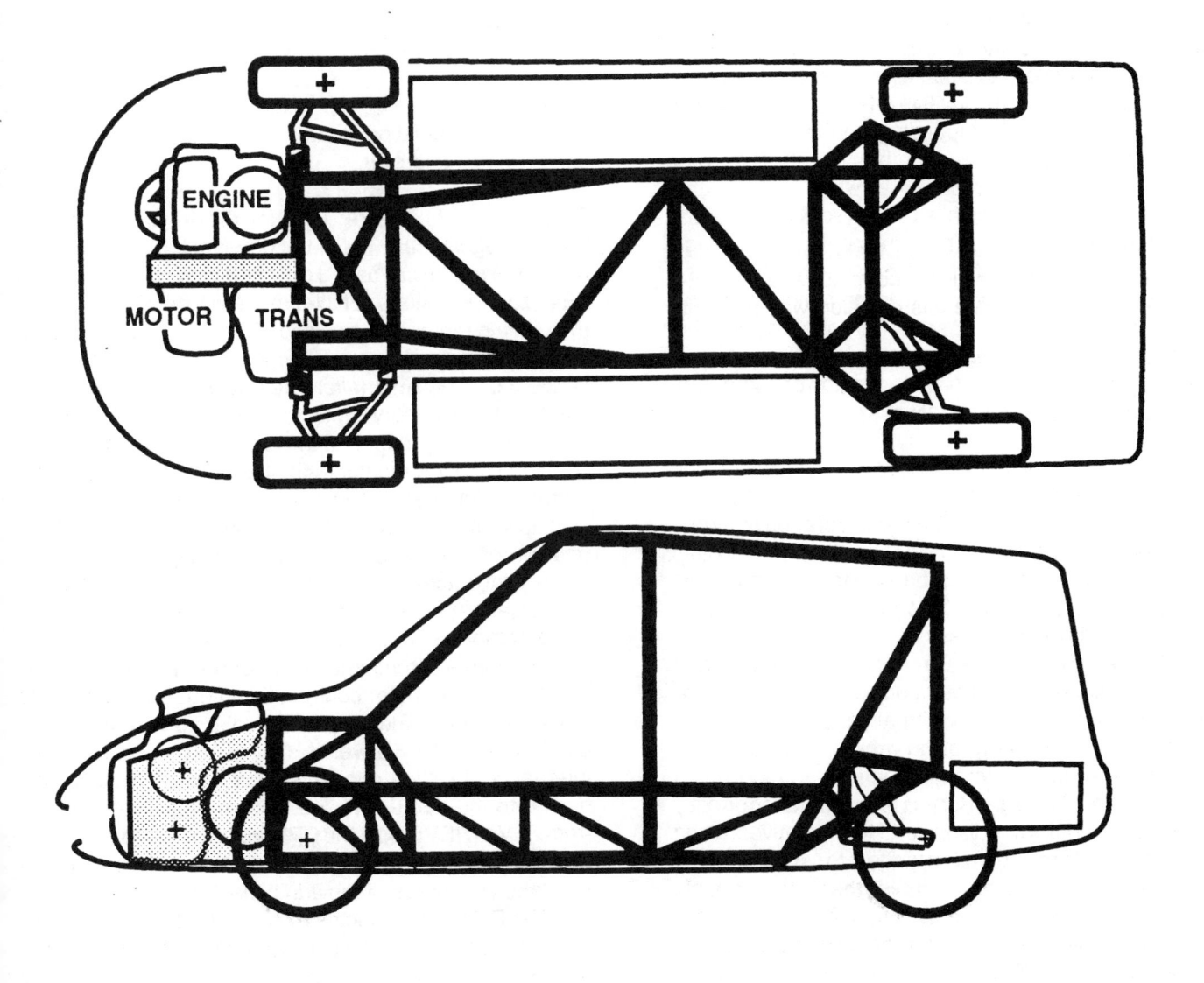

Figure 1. HEV LAYOUT

CONTROL STATAGY

CONTROL STRATAGY MODES - The operation of the HEV will be controlled through the mode switch of the vehicle. The mode switch allows the operator to choose between ZEV, HEV, and APU modes. Further detail has been developed for the HEV mode, consisting of efficiency, range, and power sub-modes. These modes work to allow the driver versatility in HEV operation.

Zero Emission Vehicle Mode - The ZEV mode simulates the operation of an electric vehicle. Here, the battery packs are the sole energy source for the locomotion, made possible by an electric motor directly coupled with the powertrain. The input from the driver, through the throttle, is directly relayed to the motor controller, which then commands the motor in its function. Added features to the electrical operation include regenerative braking and low speed torque capabilities. These two options are available anytime the electric motor is activated, with the regenerative braking also functional during APU operation. These features will be discussed in more detail later.

Hybrid Electric Vehicle Mode - HEV mode enables both powerplants to directly drive the vehicle, supplying mechanical power to a common powertrain. As mentioned previously, this mode has been divided into three parts. Each sub-mode allows the driver to effectively favor one or both of the powerplants, depending on the driving environment. The efficiency sub-mode has the electric motor operating, similar to ZEV. However, when the operator demands more power, which is conveyed through pedal position, the APU activates. The same applies to the range sub-mode, which begins with the gas engine. For these sub-modes, when the high power demand ends, the powerplant that is considered secondary shuts down. This system response of placing both units into the system is effectively jumping the control up to the power sub-mode, consisting of both powerplants always on line and ready to perform.

Auxillary Power Unit Mode - The alternative power unit has been designed to power the vehicle through the APU mode. This operation is very characteristic of a conventional vehicle, combusting fossil fuel for mechanical power. Points of operation for this mode would be after the battery supply had been depleted or when the performance of a single, more powerful drive unit is desired.

The regenerative breaking and low speed torque are product features of the motor/controller. The regenerative braking, which can be used virtually any time, allows for some of the dynamic energy of the vehicle to be converted into electrical energy rather than been lost to friction in a braking situation. The signal for this feature is taken directly from the brake pedal, when displaced. Regenerative braking and conventional braking work in tandem to slow or stop the vehicle. The low speed torque enables the motor to double its torque output through a wiring change. However, this torque is only available at low motor speeds and is very inefficient to develop. This feature is activated by a second shifter, with only two positions.

CONTROL STATEGY IMPLEMENTATION - The HEV is controlled by a 386SX-16MHz industrial computer system with a math coprocessor. The computer system has:

- 32 analog input ports
- 4 analog output ports
- 50 bi-directional digital lines
- a dual axis servo controller card
- a floppy disk
- a monitor/keyboard interface card
- a LCD display mounted in the dash.

A computer was chosen due to its versatility. Computer control allows for adjustment of any parameter during testing. The computer also provides the ability to control and monitor other systems on the car. The computer system could also record the response of the vehicle to different driving conditions. This information could then be taken via a floppy disk to a desktop computer and analyzed.

All of the software will be read from this floppy disk. The control strategy can be easily changed by swapping floppy disks. This provides an attractive option: a customer could chose different modes of operation depending on their needs.

The driver of the car has 4 inputs to the CPU:

- power
- accelerator pedal position
- brake pedal position
- mode selection.

The computer is powered up when the driver turns the key like in a regular car. Then, the driver selects the mode of operation for the car based on the driving conditions. The ZEV (zero emissions vehicle) mode will be selected when the electric motor is to be used. The HEV (hybrid electric vehicle) mode will be selected when a greater power output is needed. The APU (alternate power unit) mode will be used when the battery capacity is very low. The APU mode is used primarily to extend the range of the vehicle.

The CPU will distribute the load between the electric motor and internal combustion engine based on the mode selected, the position of the accelerator pedal, and the position of the brake pedal. The computer will turn the power on to each power plant based on the mode selection. The accelerator pedal and brake pedal will send a voltage to the computer. If ZEV or HEV is selected, the computer will then turn a potentiometer, via a servo motor, to control the motor. The motor controller's output is based on the resistance of the 10 kohm pot; 5k = idle, < 5k = regenerative braking, and > 5k = drive. If HEV or APU is selected, the internal combustion engine is controlled by a servo connected to the throttle linkage. When the driver switches from ZEV to HEV mode, the CPU will be required to start the engine and bring it to the drive train RPM before the driver can take advantage of the additional torque.

The RPM of the electric motor, engine, and drive train will be monitored in order to engage the engine, determine the velocity of the car, and determine the gear. In order to determine the RPM of the engine and motor, magnetic pickups are placed close to the teeth on the flywheel; one flywheel is on the engine, the other is on the input to the transmission. These pickups will generate a 4 Vp-p sine wave which will be converted to a voltage by LM2917 frequency to voltage converters. The drive train RPM is determined by converting the pulse from the speedometer cable output that is used for the speedometer on a GEO Metro. The voltages from these sensors are input through the analog input card and interpreted by the computer. The three RPM's and the velocity of the car will be displayed on the dash. These RPM's will also be used to optimize the load distribution.

For safety reasons the CPU also monitors the temperature of all ten batteries as well as the motor, engine, motor controller and computer. The temperature is sensed by LM 335 temperature sensors, possessing a 10mV/°K output. This voltage is amplified 2.5 times in order to reduce the effects of noise. These temperatures are displayed on the dash at all times. If any temperature approaches the maximum temperature for that component, the computer will display a warning sign on the display with instructions on how to proceed.

In order to operate efficiently, the CPU must know the energy storage on the HEV. The RS232 output of the Watt-Hour meter supplied by the Challenge is connected to the RS232 port of the computer. This meter gives current and voltage readings which can be translated into power. The fuel capacity is monitored by using the standard fuel sender on GEO Metro; this sender varies its resistance based on fuel level.

This controller system consumes 100 w. The power for this system comes from a 13.2Vdc battery. This battery is charged by one of two means: 120Vdc to 13.4Vdc DC to DC converter or alternator on the engine.

In order to combat the problem of "drive by wire", the control for the motor and engine is readily switchable. The power and ignition for the engine can be changed back to the key switch as in conventional cars by the flip of a switch. The mechanical linkage for the fuel injection can then be attached to the accelerator pedal. The electric motor will be switched by switching a plug. This will convert the potentiometer at the accelerator pedal from a voltage source going to the computer to a resistance going to the motor controller. The power for the motor will come from a switch located on the dash. In the event of an emergency, the two power plants can be turned off by an emergency shutoff switch located on the dash.

CHARGER

The development of a practical on-board charger required meeting the guideline of having an isolation transformer in the charging loop. The best solution would be to use a pulse width modulated, high frequency battery charger that takes advantage of high switching speeds to reduce the size of the transformer to something suitably sized for the application. This solution is also very expensive. To reduce cost and complexity, the design of a simple battery charger based on a full wave bridge rectification scheme was chosen for efficiency and extreme reliable. In the development of the charger, scavenging of power transformers and rectifiers of a prototype Chloride Spegel charger began. The Chloride charger was originally intended for a 250 Volt 50 amp input with a maximum output of over 40 amps at 230 volts, intended for the G-van electric vehicles currently being used by utility companies.

With the specifications provided by Chloride, the design of new charging electronics around existing hardware eliminated the need to acquire expensive components. In the first stages of testing, the taps of the primary isolation / step down transformer were manually changed during the charge cycle to match the requirements of the batteries. Use of the control electronics provided on the existing charger could not be used since the sealed batteries are not designed for the over voltage charging method found with vented wet cells. Instead, a method known as constant potential charging was used; however, development of custom control electronics to perform the transformer tapping was implemented, automating the process initially done by hand to prevent the charger from drawing excessively high currents at the beginning of the charge. The addition of a second transformer was warrented to step 220 volts down to 110 volts. This was required because even with a near unity power factor we could not get a full 6.7 KWHr's worth of energy out of a 15 amp 120 volt outlet in the six hours given to charge. Though this came very close, it was necessary to use the 30 amp 220 volt option for charging but at only a slightly higher rate than before. There is no hurry to charge the batteries since the battery life and charging efficiency will be higher at slower rates. The objective is to charge in about five and a half hours as opposed to three or four hours which is now an option. The addition of the second transformer modestly drops the power factor of the charger but not to the extent that it will cause any problems at slow charge rates. This whole exercise brings up the point that a commercial charging station for the broad range of EV's that will eventually come to the market must be designed to accommodate a variety of charging methods, voltage ranges, electrical adaptors and probably incorporate a power factor correction.

Another primary cause for concern was in obtaining suitable components to complete the high voltage system. The requirement stating a need for a switch rated to disrupt full load battery current proved to be quite costly. In fact, care must be taken to work with the manufactures on every part of the high voltage system to compensate for the differences between high voltage DC systems from the AC ratings these devices are specified with. Since a DC system has no phase lag between the voltage and current, disrupting a high current DC signal under load without arc welding the contacts in place becomes quite a challenge. Electroswitch Inc. was the only company that could provide a switch rated to disrupt 200 amps at 200 volts under load in a package suitably sized for vehicular use. JTD fuses from LittelFuse are used for the main fuses of the battery pack. The need for a number of isolated DC-DC converters, which are high frequency electronics that provide the power for the 12 volt electronics without electrically connecting the two systems together, were necessary for safety reasons.

CHASSIS

Among the considerations used in the layout of the chassis were the Rules & Regulation set by 1993 Challenge and the basic layout that was initially conceptualized. The major criteria that affected the chassis design were:

- a tandem seating arrangement for better aerodynamics
- twin battery packs to be mounted on each side of the passenger compartment
- a 240 cm wheelbase to achieve a 9.15 m turning radius
- a modular powertrain system for the engine, motor, transmission and related accessories
- a single gull wing door on the left side of the car.

To determine weight distribution and total weight of the car, the locations and weights of individual components were entered into a spreadsheet with moment equations to find the center of gravity and total weight of the vehicle. The front to rear weight bias is 61/39 and the side-to-side bias is only 0.4% off center.

The forces used to analyze the frame are:

Roll-Over	1.5g load on top of A-pillar
Front Impact	29g load on front of frame
Side Impact	23g load on side of frame
Bump	3.2g load through suspension
Braking	0.9g load through suspension
Cornering	0.9g load through suspension
Twisting	205 N*m torque on front

The roll-over case came from NHTSA for passenger vehicle testing. To pass the test, a maximum of 12.7 cm of deflection is allowed. The front impact case was derived from assuming an impact with a solid object at 48 kmph and stopping in the length of the crush zone (30 cm). The side impact test was derived from a solid object hitting the side of the car at 48 kmph and stopping in 46 cm (the width of the battery boxes).

Harsh driving conditions were also considered. The worst conditions were determined to be a 3.2g bump which is hitting a 10 cm bump at 96 kmph, a 0.9g braking acceleration, and a 0.9g cornering acceleration. The accelerations that the chassis will absorb from the road will be substantially less than those listed above due to the shocks absorbing a certain percentage of the forces involved. The twist case was done with the rear of the frame restrained solidly.

Computer simulation of the frame was done on CAEDS finite element analysis software. The initial simulation of the frame was done with no cross bracing in the chassis. The cross bracing seen in the final design was added as needed to reduce deflections and forces on individual members and to reduce twisting.

MATERIALS - There were two steps to making the decision on frame materials. The first was choosing what kind of frame to build: a space frame of tubing or monocoque system using aluminum honeycomb. Since cost was a major component in the competition, the cheaper and easier to obtain tubing needed for the space frame was the deciding factor. Then, the decision between carbon steel, alloy steel, or aluminum was made based on three criteria:

- cost
- manufacturability
- weight.

Aluminum was removed because it is substantially more difficult to weld than steel, and early rules would not

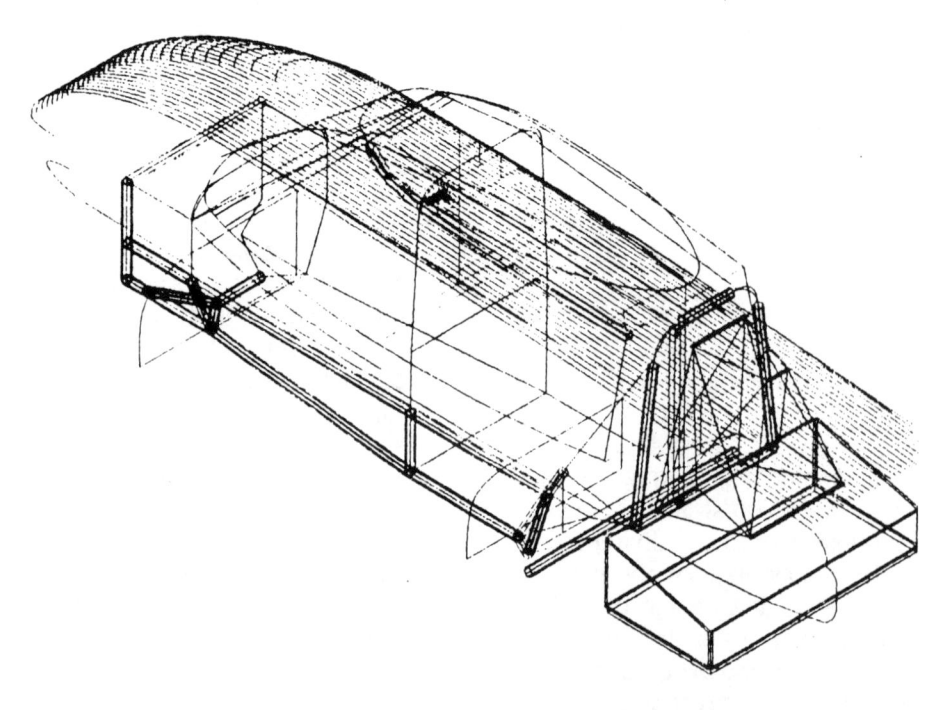

Figure 2. COMPUTER STRUCTURE DESIGN

allow the use of aluminum roll hoops. Carbon steel was chosen over alloy steel for two reasons. The first is cost. Since carbon steel is much cheaper than alloy steel, it would be much easier to get the material donated. The other reason is that to get the most out of alloy steel, it must be heat treated and no facilities nearby were large enough to accommodate the frame.

MANUFACTURABILITY - Once the decision to use carbon steel was made, manufacturability became a big plus. The frame can easily be built on a production line with MIG welders and simple jigs. The jig only needs to consist of pipe clamps and a flat surface to mount them. The only possible problem area would be in the rear portion of the frame. The fish mouthing is relatively complex, but the design could be changed slightly and jigs added to facilitate the cutting of the tubes.

BATTERY BOXES - The battery boxes for the HEV were to be a deviation of a past project that employed aluminum honeycomb structures as the main support. This would also appear on this vehicle, due to reduced frame requirements, lightweight, and availability. This material was also used in the rear structure, taking advantage of the materials energy absorption characteristics in a rear collision situation.

SUSPENSION DESIGN & HANDLING

FRONT SUSPENSION - The front suspension of the vehicle is composed of Geo Metro components modified to function as a double A-arm suspension system. Key modifications to the Metro components include replacing the steering knuckle's upper shock mount to accommodate a ball joint for the upper A-arm. The shock absorber and spring were then attached to a custom lower A-arm. The stock Geo Metro steering system was altered, but the stock steering ratio was maintained.

The geometry of the suspension linkages was designed with considerations to caster, camber, roll center, ackerman steer, bump steer, and how these vary with vehicle ride height and roll angle. Stock Geo Metro caster angle and steering angles were maintained in the design, and the roll center was designed to be slightly above ground level.

The material chosen for the A-arms is chromoly steel tube, due to its high strength, toughness, weldability, and ease of fabrication. For the suspension arm bushings, graphite-impregnated urethane bushings are used to maintain suspension geometry while still providing for high frequency vibration isolation.

REAR SUSPENSION - From the very beginning, packaging was a major concern in the design of the vehicle. The shape of the vehicle called for the luggage container to be located below and behind the rear seat. As a result, a trailing arm design, instead of an A-arm design, was chosen for the rear suspension. The suspension also had to have geometry to enable the proper attachment of a coil-over shock to the chassis. The suspension had to withstand a 5-g vertical load and two 1.5-g loads in the horizontal plane, all at the center of the contact patch of the tire. 4130 steel was chosen from the beginning due its high strength to cost ratio.

The first step taken was to choose a geometry that would meet the packaging criteria and allow for adjustment in camber, toe-in, and toe-out. Another issue that was very important was manufacturability. Due to time constraints, the suspension design was easy to build and the amount of welding and machining was kept to a minimum. Financially, there was not much of a concern because the difference in the cost of materials between the trailing arm designs was minimal. The geometry was rather simple. The needed points of the chassis were drawn in 3-D on a CAD system together with other known points such as the tire center. After the rear portion of the chassis was recreated, different geometries were drawn. The present HEV rear suspension configuration was chosen due to its simple yet effective design. Other design considerations were much heavier, more difficult to design, and much more time consuming to build.

Adjustments to the system also had to be rather simple and quick. This was accomplished by analyzing each direction of adjustment on their own. Toe-in and toe-out simply required plates or "shims" of proper thickness between the interface of the suspension brackets and the chassis to achieve the desired angle. The camber adjustment however needed more attention. High loads at the chassis/suspension bracket interface was a concern for obvious safety reasons. It was decided, after looking at many designs, to pivot the trailing arms about a vertical axis traveling through the center of the outside brackets. The actual axis would be a steel sleeve placed through and welded to the chassis with a 1.27 cm grade 8 bolt through the sleeve attaching the bracket to the chassis. The inside brackets would be attached to the chassis with two 9.525 mm grade 8 bolts. Careful scrutiny of the statics of the system using high loads and safety factors showed that the bolts were well within safety standards.

BODY

The HEV would often operate as a typical electric vehicle. The power available for such operation is very limited and not to be wasted. For this reason, many steps were taking to reduce the power requirements of the vehicle. The first and most obvious is the passenger layout. Tandem seating has been used to reduce the frontal area of the vehicle. Secondly, the body has been aerodynamically refined to reduce drag. The basic vehicle layout is a combination of two aerodynamics shapes, the lower body and the occupant canopy. However, a certain statement could be made with the vehicle, in an attempt to break away from the stereotypical electric vehicle appearance. This resulted in the addition of a hood scope placed over the engine. This added feature does increase the drag on one side of the vehicle, but it also helps to bring in a familiar shape to the entire vehicle.

The occupants will gain access to the vehicle through a single gull-wing door located on the left side of the vehicle. This side selection allowed the gear shifter not to interfere with the ingress and egress of the driver. Once inside the passengers would be seated just like in any other vehicle. The driver would have an instrument cluster to observe, standard controls to manipulate, and a mirror on each side for rear vision.

MATERIALS - The body for the HEV was a major design undertaking, that would ultimately consist of complex curves and flat planes. The easiest material choice for a quick, 'one off' body would be fiber glass. However, this material has several disadvantages to it beyond fabrication properties. The first shortcoming is the lack of recyclability. The challenge rules imply a theme of recyclability, and this material would not fall into such a category. Secondly, the amount of time necessary to produce a quality product potentially involved many hours of labor, through possible mold making and a lot of sanding. The final item that struck down a fiberglass body was the involvement of resin in the construction process. Resins require that many precautions be taken when used. These reasons helped to accept the challenge of building an aluminum body panel system. Projections were that an aluminum body would require an equal amount of time as would a fiberglass body, would weigh the same, and fell into the category of being recyclable.

The windows for the vehicle were allowed to be safety glass or 1/4" plastic. Plastic was chosen as the desired material based on past experiences in fabrication and availability. Optical integrity was insured by carefully creating precise and extremely smooth molds.

MANUFACTURABILITY - The body, being made from aluminum, would require the knowledge of forming and joining techniques. Fortunately, a sufficient background in both these areas existed to construct a fine product. Other areas such as the electronics and computer system provided little in the way of choices.

CONCLUSION

Modification and improvement of the vehicle for the 1994 Challenge was only possible due to past accomplishments. Weight reduction and refined assembly precedures have made the current vehicle 13.9% more efficient in operation and function. Continued testing and Challenge performance will illustrate the high degree of dedication and commitment that was needed to develop this HEV into the car of the future.

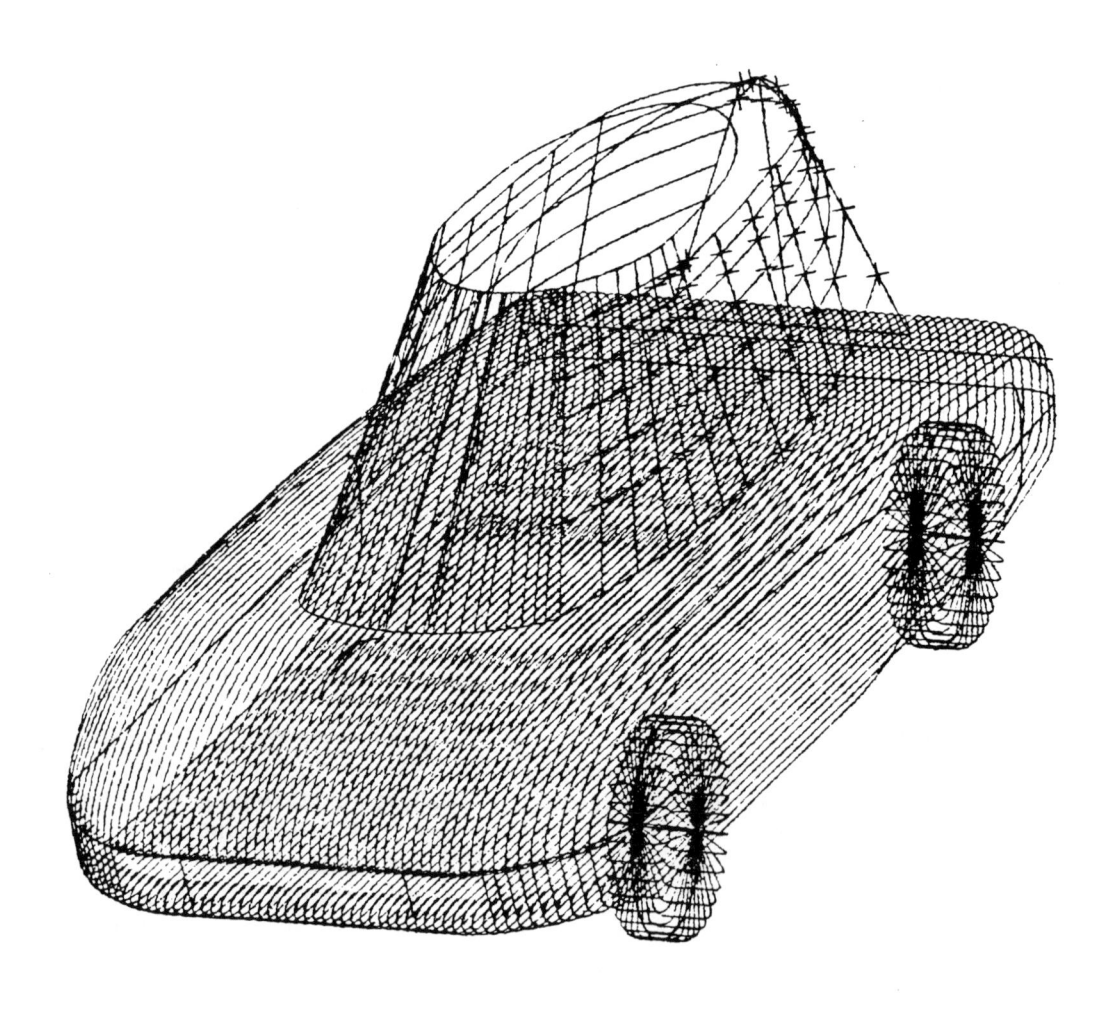

Figure 3. COMPUTER BODY DESIGN

HYBRID ELECTRICAL VEHICLE DESIGN SPECIFICATIONS FOR 1993-94 SATURN HYBRID ELECTRIC VEHICLE CHALLENGE

CALIFORNIA STATE UNIVERSITY FRESNO

DESIGN TEAM:
Scott Bemis	Mechanical Engineering
Lou Chacon	Electrical Engineering
Ron Kibby	Industrial Technology
Roger Moore	Electrical Engineering
Greg Panico, Coordinator	Industrial Technology
George Stevens	Mechanical Engineering
Dong Vang	Mechanical Engineering
Charles Wright	Mechanical Engineering

FACULTY ADVISORS:
Leslie Aldrich	Industrial Technology
Robert Hecht	Electrical Engineering
Walter Loscutoff	Mechanical Engineering

ABSTRACT

A power train of a 1991 Saturn SL2 was modified to a "hybrid electric" configuration as required for the 1994 Saturn Hybrid Electric Vehicle Challenge. Advantages of the hybrid electric configuration include reduced emissions by operating in the "electric only" mode during critical conditions such as stop and go freeway traffic, city traffic, and IC engine warmup. The IC engine serves to compensate for acceleration and range performance losses normally associated with electric cars. Another advantage of the hybrid configuration is the utilization of low cost clean (West coast) hydroelectric energy as opposed to fossil fuel energy.

The CSUF design consisted of a 100 hp internal combustion engine and 40 hp electric motor connected in a parallel configuration. The power source for the electric motor was a 168 volt battery pack consisting of fourteen twelve volt batteries connected in series. The electric motor was linked to the input shaft of a manual 5-speed transmission via chain drive. To enhance emission performance, the IC engine was modified to operate on ethanol fuel.

The powertrain was designed to operate in either electric only, IC only, or combined modes. The configuration allows for energy regeneration during periods of braking or light engine loads. A microcomputer was used to monitor battery level and engine loading to determine the appropriate level of electric boost or regeneration.

POWER TRAIN CONFIGURATION

The CSUF configuration consisted of a Suzuki inline 4 cylinder internal combustion engine connected in parallel with a Unique Mobility DC pulsed brushless electric motor. The system function diagram is shown in figure. 1.

The system was designed for a horsepower of 140 hp with the IC engine contributing 100 hp and the electric motor 40 hp. The parallel configuration allows for the operation of electric motor and ICE independently or combined. The power source for the electric motor was a 168 volt battery pack consisting of fourteen 12- volt lead acid batteries connected in series. The electric motor was linked to the input shaft of a manual 5-speed transmission via chain drive. To enhance emission performance, the IC engine was modified to operate on ethanol fuel.

PARALLEL POWER CONFIGURATION-Both series and parallel HEV configurations were considered for this competition. Determined individuals with convincing arguments represented both sides of the issue. The parallel HEV was agreed to have the following characteristics:

1. Smaller electric motor
2. Fewer batteries required
3. More efficient I.C. drive
4. Shallow discharge cycles
5. Variable IC engine speed required
6. More use of existing conventional equipment

The series HEV was agreed to have the following characteristics:

1. Smaller IC engine
2. Constant speed IC engine
3. Completely dependent on electric motor/controller
4. Possible use of single speed transaxle

There was early agreement that the series HEV would eventually, as technology advances, become superior. Ultimately the importance of maximum fuel economy in the IC engine drive, was the driving factor in selecting a parallel HEV for this year. It was determined that a small conventional engine transaxle supplemented by some electric power and regenerative braking would have higher energy

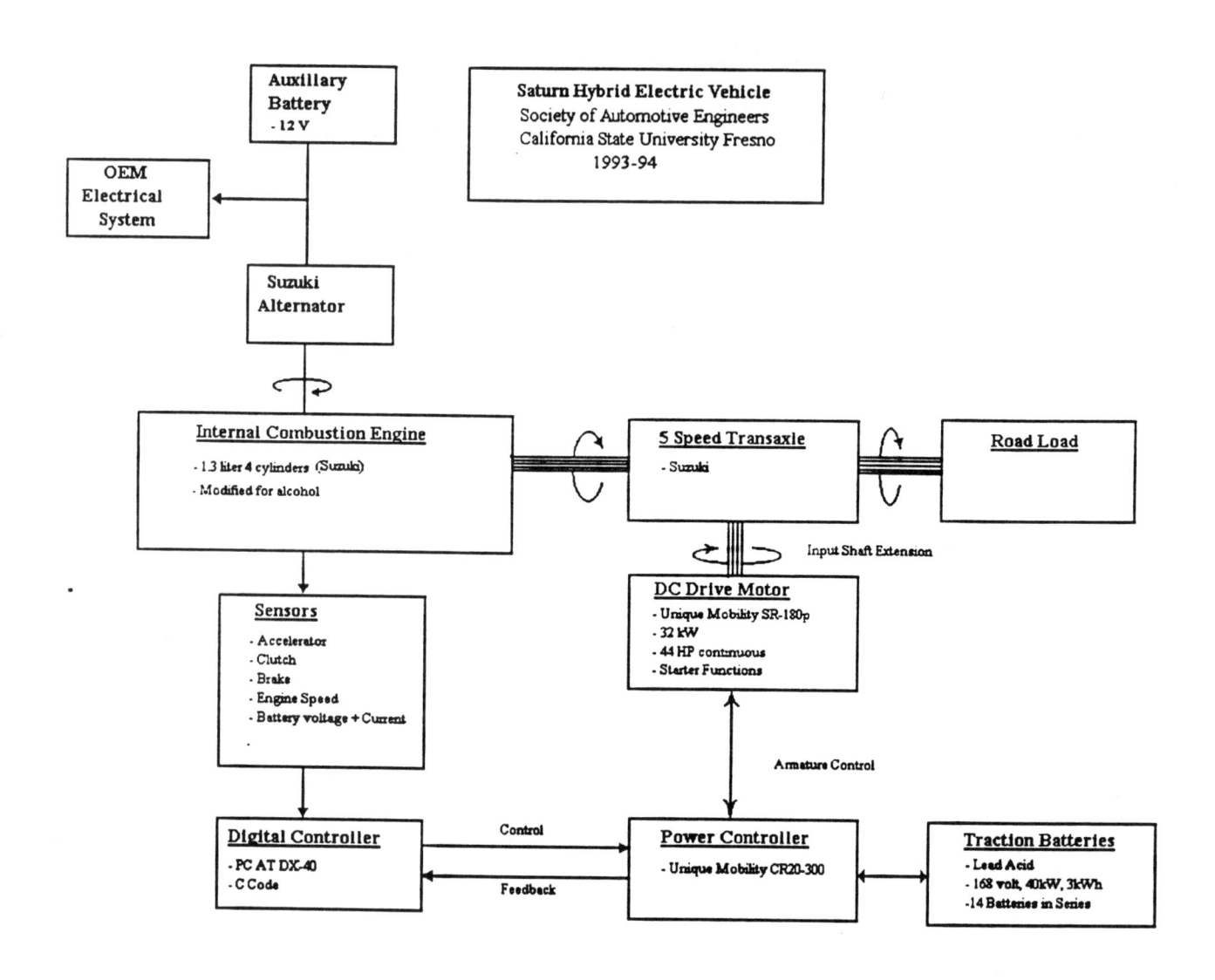

Figure 1. HEV Parallel Functional Configuration

efficiency than the series HEV's combination of the IC engine/generator, controller, and electric motor.
    This may not be true of the most efficient electric motor/controllers when operating at their peak efficiency however a fuel mileage driving cycle will require substantial operation outside of it's most efficient range. The use of a CVT could reduce this effect but brings with it high frictional losses within the CVT.

  BATTERIES- The electrical storage components for this vehicle are the Exide DC-9 deep cycle, wet cell, lead acid batteries. Fourteen of these batteries are connected in series to provide a nominal pack voltage of 168 volts. The relevant specifications of individual batteries are shown below:

Size:
    Length   22.2 cm
    Width   13.2 cm
    Height   18.6 cm
Weight: 9.5 kg
Nominal Voltage: 12V
Cranking Amps (0° deg. F): 225A
Capacity (C/1 rate): 18 Ah
Electrolyte Quantity: 2.2 l
Cycle Life (80% DOD): 200 cycles

Battery Selection- The batteries were selected with four major design considerations. The first consideration was to provide sufficient energy capacity to finish the required 5 miles at 30 miles per hour. The calculated power requirement was 6.2 kW to maintain 30 mph in the Saturn. It follows that 5 miles at this speed at a system efficiency of 82% would require about 1.26 kWh. When a maximum of 80% DOD is considered, the energy requirements increase to 1.58 kWh. The battery pack energy capacity is 3.02 kWh at the C/1 rate. The capacity will decrease at higher discharge rates but it should be sufficient to meet the requirements.

Minimization of battery weight was the second major design consideration. Weight was a major concern for two reasons. First, the gross vehicle weight could not be exceeded. Secondly, increased weight adversely affects acceleration and fuel economy.

The third major design consideration was the maximum power rating. Since the electric motor had been chosen, the batteries were selected with the criteria of supplying the motors rated power requirement of about 200 A. With a rating of 225A at zero Fahrenheit, the Exide batteries should be able to handle the electric motors power demands.

The last major design consideration was cost. Unfortunately, limited budget requirements restricted the selection to lead acid batteries. The most weight and cost efficient batteries available were obtained.

ELECTRIC MOTOR- The electric motor selected was the Unique Mobility SR-180P. It was decided to spend a large portion of the budget on this motor because of its' high efficiency, high power, and relatively light weight. These characteristics were essential to the overall vehicle efficiency and driveability. Because the system (motor and controller) has a high maximum efficiency of 92%, the demands on battery capacity and fuel can be minimized. The rated power of 32kW was also important to maintain reasonable driveability in the ZEV mode and to provide sufficient assist power in the HEV mode.

ELECTRIC MOTOR CONTROLLER- The controller function in the vehicle is actually performed by two separate units. The first unit is the power controller which is a Unique Mobility CR20-300 control unit. This is the standard controller used with the SR-180P motor in electric vehicles. Since the CSUF hybrid vehicle uses a parallel configuration, a second logic controller is required to drive the power controller. A 386 DX-40 IBM PC AT compatible computer was used as the logic controller. Initially, Motorola 68705 microcontrollers with integral RAM and EPROM were considered, but the time required to program and reprogram these devices was excessive. The 386 DX-40 interfaces to sensors and the power controller through CTI analog and digital I/O cards. The system provides a maximum of 48 lines of digital I/O, 16 A to D inputs, and 1 D to A output.

CONTROL STRATEGY- The control strategies are different in all three modes of operation. In the Power On mode, the electrical system is disabled and the control of the engine is left to the OEM Suzuki engine computer. In the ZEV mode the engine ignition is disabled and the throttle position voltage is sensed by the logic controller and a signal proportional to the throttle position voltage is sent to the power controller.

A power assist strategy was chosen for the HEV mode. The combination of power assist control strategy and parallel hybrid configuration were chosen to minimize conversion losses found in the generators, motors, and batteries. In this mode, values for the throttle position voltage, brake pedal position voltage, battery voltage, battery current, and the engine speed are sampled and then values for engine load and battery state are determined. From these values, the amount of power assist or regeneration is determined.

ELECTRIC MOTOR AND DRIVE TRAIN

The power of the electric motor is transmitted through a chain drive to the transmission input shaft as shown in Fig. 2. The system is simple, lightweight and easy to fabricate. The most difficult part of fabrication was the modification of the transaxle shaft and end housing.

The stiff member and pillow block bearing assembly was used to prevent deflection of the power shafts. This will : 1) keep a constant center distance, 2) prevent misalignment of the sprockets, and 3) prevent over stressing of the bearings. The stiff member was machined from 6061 T6 aluminum which is light in weight and has the strength and rigidity to handle the stresses found in this application.

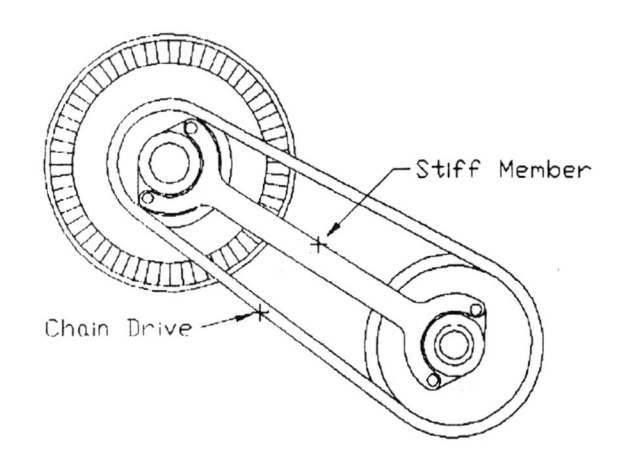

Fig. Drive train assembly for electric motor power transmission

The peak power the motor is capable of producing is 50 kW and the maximum intermittent torque is rated at 90 Nm. Therefore, a size 530 motorcycle chain was selected because it exceeds these requirements and provides and adequate safety margin.

A gear ratio of 1.5:1 was selected to connect the electric motor to the IC engine. This provides torque multiplication for the electric motor and allows the electric motor to run at higher speeds where efficiency can be maximized.

EMISSIONS CONTROL STRATEGY

The emissions goal of CSUF HEV project are based on maximum emissions as oulined in the 1994 California Emmisions requirements. The CSUF-SAE HEV minimum performance and emissions goals for 1993-94 are as follows:

FUEL EFFICIENCY CONSIDERATIONS-It was decided that E95 was the most appropriate fuel for the hybrid. E95 is environment friendly, it has less NOx than gasoline, and it's less toxic than gasoline or M85. It generates better fuel economy than M85, (10%) but higher fuel economy compared to gasoline (30%). The most attractive attribute is that it is a renewable resource that can be produced from surplus grain production.

VEHICLE STRUCTURAL MODIFICATIONS

COMPONENT LAYOUT AND ROUTING- The IC engine was mounted to the far right and rear of the engine compartment to allow space for the electric motor. The drive axles were modified to fit.

The main power cables were routed through the left fender wells and through left structural frame member to the controller mounted behind the seat. 1 gauge power cables were used to ensure insulation from the frame member. Control and signal cables were routed similarly down the right side of the car. The layout is shown in Figure 3.

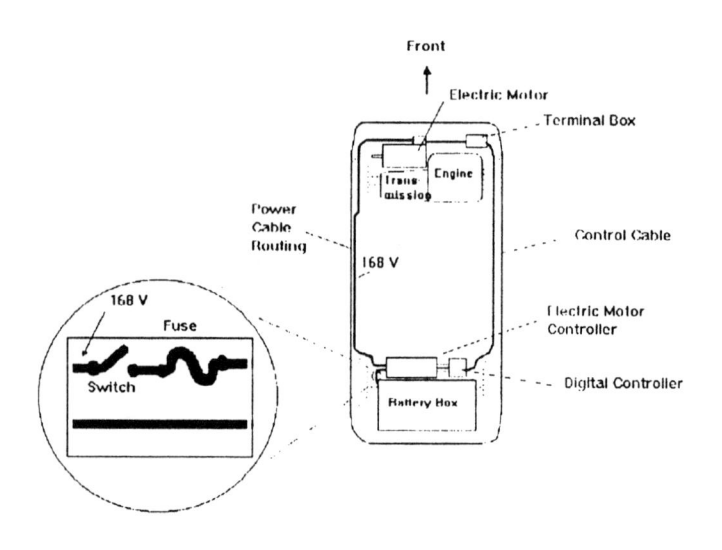

Figure 3. Component Layout and Routing

BATTERY BOX-The battery box was made from 5052 0.063" sheet aluminum. Aluminum was chosen because it is easily formed and light weight. The box design maximizes all available space in the lower trunk area (spare tire area) and allows some unused trunk space for luggage. The battery box seams were welded closed and the box was coated with an epoxy sealant primer and then coated with rubberized underbody sealant to prevent any possible acid leakage.

Two sections were incorporated in the battery box to allow for easy removal of the batteries and because there wasn't enough room in the spare tire compartment for all 14 batteries. Both sections were designed to be removable from the car.

The box design has two parallel air flow passages for battery cooling. A fresh air intake comes through the floor at the back of the lower battery box. Two intakes are routed to the two sides of the controller via two sealed enclosures and air ducts. A main fan draws air through the electric motor controller heat sink and then blows it through the top and bottom section of the battery box. The air flow exits at the back of the lower battery box section which faces out the back of the car. The box is shown in Figure 3.

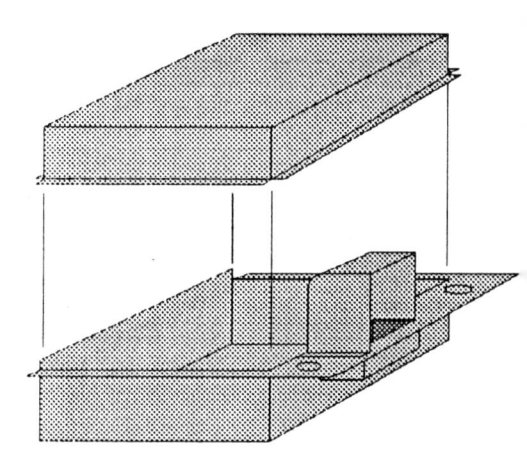

Figure 3. Battery Box Design

OTHER MODIIFICATIONS - Modifications were performed on the following subsystems. Most involve simple bending of metal to fit component parts.

Exhaust Systems - The stock exhaust system was used with the exception of a smaller muffler. The stock header and tailpipe were heat treated and bent into a proper fit.

Clutch linkage - The stock brake pedal was removed and a clutch pedal and a smaller brake pedal were installed to facilitate the 5-speed transmission.

Manual transmission - A manual 5-speed Suzuki transmission replaced the stock Saturn automatic 4-speed. The manual 5-speed was chosen because it allowed more shift control to accommodate the input from the electric motor and to allow more control of the engine speed for the best possible economy.

Radiator modification - The passenger side surge tank had a fitting placed/welded to the upper right hand corner of the radiator to accommodate the Suzuki engines cooling flow path.

Motor mounts - New motor mounts for the Suzuki engine and the electric motor were fabricated to properly fit the motors into their spaces.

Hybrid Electrical Vehicle Design Specifications For 1993-94 Saturn Hybrid Electric Vehicle Challenge

CALIFORNIA STATE UNIVERSITY FRESNO

AUTHORS:
Scott Bemis	Mechanical Engineering
Roger Moore	Electrical Engineering
George Stevens	Mechanical Engineering
Dong Vang	Mechanical Engineering
Charles Wright	Mechanical Engineering

VEHICLE SPECIFICATIONS

Vehicle Description
- Class: Saturn Conversion 1991 SL-2
- Original Weight: 1076 kg Curb Weight: 1076 kg
- Expected Conversion Weight: 1530 kg
- Wheelbase: 260.14 cm Overall Length: 447.8 cm
- Width: 171.8 cm
- Height: 133.4 cm
- Coast down-time, from 50-40 mph: unknown
- Frontal Area: 2.292 m^2
- Drag Coeff.: .35 Drag area product 0.802 m^2
- Power to Weight Ratio: 0.0682 kW/kg
- Power train configuration: Parallel

Electric Motor Specification
- SR180P Unique Mobility
- DC pulsed brushless
- Peak Power: @6700 rpm, 32.0 kW, 68 hp
- Max. Torque: @0 rpm, 57.6 N-m, 42.5 ft-lb
- Rated System Efficiency: 92% @6700 rpm
- Max. RPM: 6700 rpm
- Weight: 52 lbf 23.57 kg

Electric Motor Controller
- Unique Mobility CR20-300
- MOSFET
- Input voltage: 168 volts   Max. current: 300 amps
- Rated System Efficiency: 92%
- Dimensions: 60.96 x 29.46 x 14.2 lwh. cm

Tire Specifications
- Manufacturer: Firestone Model: Radial
- Size & speed: P185/60R14
- Tire pressure: Front:35 psi Back:35psi

Traction Batteries
- Exide DC-9 Lead Acid Pack Wt: 133 kg
- 14 series connected batteries 12 volts/battery
- Pack voltage: 168 volts
- Capacity per battery:
    18 amp-hrs (C/1 hr rate) .216 kWh
- Cycle life at a 80% discharge: 200 cycles

Digital Controller
- Logic - IBM AT compatible 386 DX-40
- I/O CTI analog and digital interface cards

IC Engine/Drivetrain
- Engine type: Suzuki in-line 4 cyl. DOHC, 16 valve
- Displacement: 1298cc (79.2 cu. in.)
- Horsepower: 100 hp @ 6500 rpm
- Torque: 83 lb-ft @ 5000
- Induction sys.: Electronic multipoint fuel injection
- Fuel: E95 Fuel
- Compression: 10:1
- Clutch: Dry, single disc
- Transmission: 5 speed OEM with removed 5th
- Gear Ratios (:1) 3.146,1.894,1.28,.914,.757, rev 3.272

Other Specifications
- 600V 1-gauge power cables
- 530 RK Chain and gearing
- Chain Drive ratio: 1.5:1

Hybrid-Electric Vehicle Development at the University of California, Davis: The Design of AfterShock

Marten Byl, Paul Cassanego, Gregory Eng, Troy Herndon, Keith Kruetzfeld, Andrew McKee, Gregory Reimers, and Faculty Advisor Dr. Andrew Frank

University of California, Davis

ABSTRACT

This paper describes the design and fabrication of AfterShock. AfterShock is a high performance, hybrid-electric vehicle built by students at the University of California, Davis. This two passenger, full-size car has a maximum 110 km Zero Emission Vehicle range while attaining a measured fuel economy of only 1.3 liters per 100 km (300 mpg). These impressive figures are the result of an innovative parallel powertrain, light-weight vehicle design, and applied aerodynamics. Furthermore, this environmentally friendly vehicle satisfies current California emissions requirements and utilizes recyclable materials.

INTRODUCTION

The approaching enactment of the 1990 Clean Air Act, mandating Zero Emission Vehicle (ZEV) sales in California, is the driving force behind the search for an environmentally friendly vehicle. Among these alternatively powered vehicles is the hybrid-electric vehicle (HEV) concept. This vehicle is accepted by state legislators and the general public for its ZEV capabilities, the attainable Ultra-Low Emission Vehicle (ULEV) rating, and performance characteristics similar to conventional automobiles.

With such an intriguing vehicle concept, the United States Department of Energy (DOE) is organizing the second annual Hybrid Electric Vehicle Challenge (Challenge) sponsored by the US DOE, Natural Resources Canada, Saturn Corporation, and the Society of Automotive Engineers. The first Challenge (June 1993) set benchmarks for HEVs throughout the world. Hence, the US DOE has utilized the Challenge to collect information about different HEV configurations from the forty participating universities located in the US and Canada. The information gathered will assist in HEV concept evaluation and dictate its future.

The Challenge is comprised of three vehicle categories in which the HEVs compete. Two are conversion categories where teams convert an existing Ford Escort or a Saturn SL2. The University of California at Davis, however, is competing in the Ground-Up Class. Those in the Ground-Up Class design and build the entire vehicle. Ground FX was the first HEV developed by UC Davis.[1] This paper, however, outlines the design of AfterShock, the second generation HEV designed and fabricated by the UC Davis HEV Program.

HYBRID ELECTRIC VEHICLE DEFINITION

A HEV is defined, according to the 1994 HEV Challenge Regulations, as a vehicle powered by two independent power sources. The primary power source must be an electric motor (EM). The secondary power source or auxiliary power unit (APU) must be an internal combustion engine (ICE) running on unleaded gasoline, methanol (M85), or ethanol (E95). These two power sources can be coupled in any fashion (series or parallel). The vehicle must also employ components common to conventional automobiles.

UC DAVIS DESIGN GOALS

The Challenge has several vehicle restrictions and performance requirements. The car must accommodate a driver and passenger with designated cargo space while employing current automotive standards (four wheels, conventional occupant seating positions, headlights, etc.). A 40 km range at 48 kph as a ZEV and a 480 km range as a HEV were among the performance requirements.

The UC Davis HEV Team supplemented the basic requirements with additional performance goals. This allows complete compliance with the Challenge Regulations while reflecting the anticipated driver needs in California. Thus, the following goals, shown in Table 1, were created.

Table 1: Design Goals

1) 160 km range at 100 kph as a ZEV.
2) 2.9 liters per 100 km (80 mpg) or higher fuel economy.
3) 0 - 100 kph acceleration in under 15 seconds as both a ZEV and HEV.
4) Unlimited range as a HEV.
5) Transitional Low Emissions Vehicle (TLEV) rating as a HEV.
6) Safety, driveability, and reliability comparable to a conventional automobile.

Although these design goals appear demanding, they are attainable. The UC Davis HEV Team has proven such a vehicle is possible with no significant compromises. To attain these goals, the vehicle's design concentrates on an innovative, powertrain, a sleek aerodynamic body, and a strong aluminum monocoque/space-frame integrating a high performance suspension for excellent handling. Furthermore, weight, manufacturability, and recyclable materials were constantly considered during vehicle design and fabrication phases.

UC DAVIS HEV DESIGN CONCEPT

To achieve the goals outlined in Table 1, AfterShock's design must first emphasize efficiency. The force equation below identifies the variables associated with vehicle drag. Aerodynamic drag and rolling resistance produce the most significant drag components.

Equation 1: Force Balance

$$Force_{drag} = F_a + F_{rr}$$
$$Force_{drag} = 1/2\, \rho\, A_f\, C_d\, V^2 + C_{rr}\, M\, g$$

F_a : aerodynamic drag force
F_{rr} : rolling drag force
ρ : air density
A_f : maximum frontal area
C_d : drag coefficient
V : velocity
C_{rr} : coefficient of rolling resistance
M : vehicle mass
g : gravity

Therefore, efficient vehicle design must minimize the coefficients of aerodynamic drag and rolling resistance, frontal area, and vehicle mass. The UC Davis HEV Team has attempted to reduced these values as much as possible. Table 2 summarizes AfterShock's parameters which dictate the vehicle's performance.

Table 2: HEV Design Parameters

A_f	1.8 m²
C_d	0.24
M	1050 kg
C_{rr}	0.0095
Road Load @ 80 kph	6.1 kW
EM Efficiency	85%
Drivetrain Efficiency	95%
BSFC	0.304 kg/kWh

AFTERSHOCK'S SPECIFICS

The design phase of the program began July 1, 1993 and concluded Sept. 1, 1993. Shortly thereafter, fabrication commenced. The vehicle was completed just prior to the 1994 HEV Challenge located in Detroit, Michigan (June 14 - 20, 1994). Approximately, seventy-five undergraduates and six graduate students have contributed to AfterShock's development.

AfterShock is a two passenger vehicle with designated cargo space. The car is best described as a commuter sports-car due to its excellent handling, low weight, and high fuel economy.

The chassis is an aluminum monocoque/space-frame with four-wheel independent suspension incorporating a mid-engine, rear drive powertrain. Battery enclosures are conveniently located on the vehicle's sides for easy access and safety.

Encapsulating the chassis is a sleek, aerodynamic body. The body is constructed from pre-impregnated fiberglass. A 0.24 coefficient of drag is achieved due to the fully faired wheels, complete under-belly, and airfoil shape.

Propelling AfterShock is a distinctive, parallel powertrain. A Unique Mobility EM is coupled to a Briggs & Stratton ICE through a common input shaft to a 5-speed, Chevrolet Sprint transaxle. Powering the EM is a 156 V (nominal), 100 Ah (C/5), 15.6 kWh, SAFT Nickel-Cadmium energy storage system (ESS).

Vehicle operation is similar to a conventional automobile with an additional mode switch. The driver manually chooses the appropriate mode for the particular driving situation. AfterShock is currently equipped with a ZEV, HEV1, HEV2, and APU mode. The dual HEV modes allow different powertrain control strategies and greater flexibility for testing. Future testing will determine the best HEV mode (optimum fuel economy and exhaust emissions).

The design parameters in Table 2 dictate excellent performance. The measured fuel economy is 2.9 liters per 100 km for HEV1 mode and 1.3 liters per 100 km for HEV2 mode. ZEV range is 200 km during city driving and 130 km at 100 kph. AfterShock satisfies current California emissions standards and nearly meets TLEV requirements.

POWERTRAIN DEVELOPMENT

PARALLEL VS SERIES POWERTRAIN - The powertrain configuration, series or parallel, is the stimulus of many debates and the first major design decision associated with the HEV concept. Generally speaking, a series hybrid uses the APU to generate electricity for the EM, while a parallel hybrid uses the APU to directly transfer torque to the drive wheels. The UC Davis HEV Team studied this topic extensively, investigating the advantages and disadvantages of each configuration.

The series configuration is the most widely accepted HEV powertrain due to its low emissions capability, good fuel economy, low weight, simple powertrain integration, and APU management. To obtain low emissions and good fuel economy, the ICE can be load leveled and therefore point tuned for maximum efficiency. The non-cycled APU produces a lower average specific fuel consumption and reduces emissions. Furthermore, a series hybrid does not require extensive hardware to couple the APU and EM. Consequently, a series hybrid may minimize weight and complexity. This straightforward powertrain allows simplified APU management.

The greatest disadvantage associated with a series configuration is its inherent powertrain losses. Although maximized fuel economy is possible, the concept suffers from the numerous energy conversion cycles necessary to transform the combustible fuel into motive energy. Thus, fuel economy suffers.

Although a series hybrid is appealing, a properly designed and controlled parallel configuration has advantages similar to series hybrids. If the ICE is only used for sustaining stead-state speeds, it is essentially load leveled. Therefore, a properly sized APU running at wide open throttle (WOT) can be point tuned and capable of achieving low emissions and very high fuel economy because the APU directly propels the vehicle. Employing such a powertrain eliminates the inherent losses found in a series hybrid because fewer energy conversions are necessary. Consequently, the EM must provide the motive power during urban driving situations and acceleration when needed on highways. Lastly, APU recharging is possible using the regenerative capability of the EM controller. However, if the APU recharges the battery pack, fuel economy decreases.

Table 3 reveals the approximate power conversion efficiencies.

Table 3: Powertrain Efficiency Comparison

Power Conversion	Efficiency Series	Parallel
fuel - mechanical (combustion)	28%	25%
mechanical - electrical (generator)	90%	-----
electrical - chemical (battery in)	95%	-----
chemical - electrical (battery out)	95%	-----
electrical - mechanical (motor)	80%	-----
mechanical - transitional (drive-train)	85%	85%
Total	15%	21%

The parallel hybrid advantages are intriguing but a more complex powertrain is necessary since the output shafts from the EM and APU must be coupled. This complexity creates APU control problems.

Overcoming these disadvantages, however, creates an exceptional performing vehicle. Therefore, the UC Davis HEV Team opted to use the more efficient parallel configuration to achieve its design goals.

PARALLEL HYBRID DESIGN - AfterShock's parallel configuration utilizes a small APU and a large EM packaged into an integrated, mid-engine, rear drive powertrain. The APU is a modified Briggs & Stratton, Vanguard, ICE. The EM is an Unique Mobility SR180P, 32 kW continuous duty motor. The two powerplants are coupled, with a single shaft, into a 5-speed, Chevrolet Sprint transaxle. An oversized friction clutch allows for the increased torque (up to 176 N-m) produced by both powerplants. Cable actuated shifting utilizing a standard shift pattern yields precise shifting.

Located between the APU and EM is a Warner SF825 electromagnetic clutch which engages and disengages the APU depending on the driving situation. The EM drives the transaxle input shaft via a 36 mm, kevlar, Gates Polychain GT cog belt. The belt matches the EM rpm range to the APU with a 1.56:1 reduction. See Figure 1.

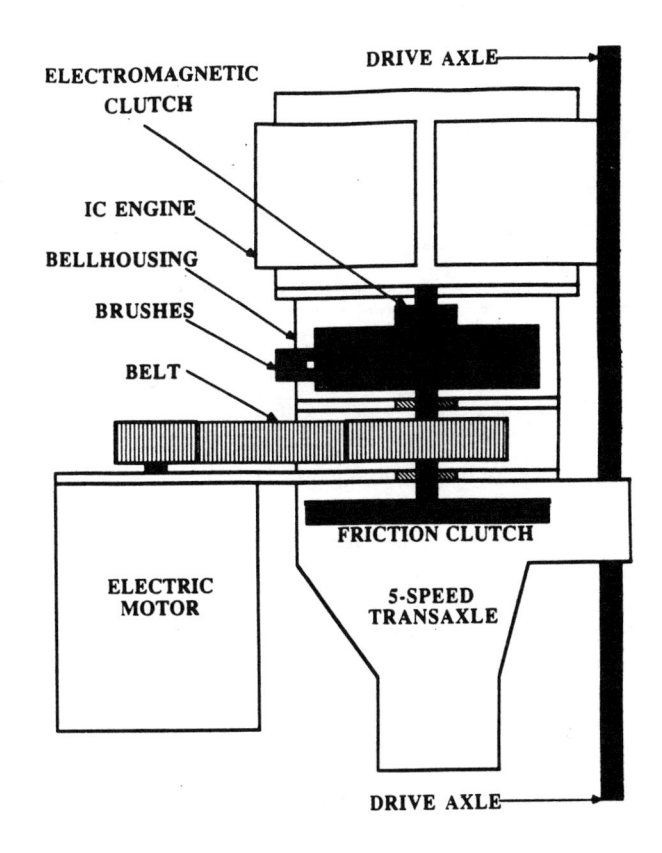

Figure 1: Parallel Configuration

AfterShock's powertrain is distinctively different than the powertrain found in Ground FX. Unlike the first generation HEV, AfterShock's powertrain is mounted transversely, increasing drivetrain efficiency due to the elimination of the bevel gear needed in a longitudinal transaxle. A reduction in powertrain size and weight were also achieved without loss of reliability or durability. The electromagnetic clutch and corresponding housing were reduced in size decreasing the weight by 23 kg compared to Ground FX.[1]

In addition to the weight savings and the increased efficiency, the transverse powertrain enlarged the available occupant space within AfterShock, reduced cabin noise, improved APU cooling, and powertrain access.

The complete powertrain weighs 173 kg including the EM controller, has a peak torque output of 176 N-m, and 60 kW of power at the transaxle input shaft.

POWERTRAIN CONTROL STRATEGY - AfterShock employs three primary driving schemes, HEV1, HEV2, and ZEV mode. In HEV1 mode, the EM provides the additional power required for acceleration and climbing while the APU (on continuously) supplies the power needed for constant speed cruising. When the ESS reaches 85% depth-of-discharge (DOD) the EM no longer provides additional power and the car is propelled by the APU only. The remaining 15% of the ESS is used to power accessories. In HEV2 mode, the APU engages above 72 kph and disengages below 56 kph. As in HEV1 mode, when the ESS reaches 85% DOD the EM no longer provides power and the car is propelled by the APU only.

In both HEV1 and HEV2 modes AfterShock is a charge-depleting vehicle. Charging is conducted while the vehicle is at rest from an auxiliary power source.

In ZEV mode, the EM provides the motive power during all driving situations. The onboard battery monitoring system constantly informs the driver of the remaining vehicle range. If charging is not possible, an HEV mode can be activated if the ESS is above 85% DOD.

AfterShock operates like a conventional automobile with an additional switch determining the mode of operation. Currently, the car is equipped with no complex computer control. In HEV1 mode, a dual stage accelerator pedal dictates whether or not the EM is engaged. HEV1 mode is a completely passive system. The HEV2 mode, however, is not a passive system at the present time. Engagement of the APU is done manually by moving the selector switch from the ZEV mode to the APU mode setting at the proper vehicle speed. (The mode switch has four positions, OFF, ZEV, HEV1, and APU. In APU mode, only the ICE propels AfterShock.) A micro-controller will be used in the future to ensure a passive HEV2 mode.

The two HEV modes allow greater flexibility for testing. The mode producing the optimum performance (lowest emissions and greatest fuel economy) would be preferred in a production HEV.

ELECTRIC MOTOR/CONTROLLER - AfterShock's propulsion system uses a Unique Mobility SR180P EM. This permanent magnetic, brushless DC, 32 kW continuous, 45 kW peak, motor best suits the vehicle's needs.

A Unique Mobility CR20-300 controls the EM. The 200 V, solid-state controller has a 300 A current limit resulting in a 45 kW intermittent capability for the EM/controller combination.

During initial EM evaluations, the Solectria brushless DC motor was compared to the SR180P. Table 4 outlines the characteristics of each EM. Since the Solectria is considerably smaller, two are used in the comparison.

Since the UC Davis HEV design concept dictates the EM provide vehicle acceleration, peak torque is critical for acceptable performance. The SR180P clearly satisfies this requirement. The Solectria, dual EM scheme does allow for increased EM/controller efficiency but is unable to provide the necessary power for acceptable acceleration times. The CR20-300 controller also has an appealing, speed sensitive, regeneration function.[1] Regenerative braking extends the vehicle's ZEV range up to 10%.

Table 4: EM Comparison

Electric Motor	UNIQ SR180P	Solectria BRLS11/BRLS240H
Motor Weight (kg)	24	29
Controller Weight (kg)	22	14
Continuous Power (kW)	32	16.4
Continuous Torque (N-m@rpm)	45@6,600	29@6,000
Intermittent Stall Torque (N-m)	90	63
Peak Power (kW)	45	30
Maximum Speed (rpm)	7,500	7,000
Peak Efficiency (%)	93	94

MAIN BATTERY PACK - The UC Davis HEV Team conducted an extensive battery investigation to determine the proper ESS for AfterShock. High power density, high energy density, long life, and affordability were deemed the most important battery characteristics. Table 5 summarizes possible battery options.

Table 5: Battery Options

Manufacturer Model	SAFT STM 5.100	Electrosource Horizon	Sonnenschein Gel 27	Ovonic	Eagle-Pitcher
Type	NiCd	Pb-Acid	Pb-Acid	NiMH	NiFe
Volts/Module	6	12	12	13.2	6
Weight/Module (kg)	13	27	29	18	24.5
Ah (@C/3)	96	112	72	90	200
Power Density (W/kg)	250	450 (est.)	100	250	150
Energy Density (Wh/kg)	45.7	49.7	29.7	66	49
Cycle Life	2000	1000 (est)	200	2000	850
Cost/Module	$700	$1100	$140	$3000	$1000
Cost/kWh (over lifetime)	$.59	$.80	$.81	$1.26	$.98

Since high power density is important, the Sonnenschein lead-acid and the Eagle-Pitcher batteries were undesirable candidates. Ovonic's NiMH battery, believed to be the optimum choice in the future by the UC Davis HEV Team, was too costly and unavailable to incorporate into AfterShock.

The two remaining choices were the Electrosource Horizon and SAFT STM5.100 batteries. The Horizon is a sealed lead-acid battery while the SAFT is a flooded Nickel-Cadmium battery. Although the Horizon has a lower cost/module, its lifetime is much shorter than the SAFT battery. Assuming a modest 80 km/charge over 2000 cycles, a SAFT battery pack should last 160,000 km. Hence, the cost over lifetime is lower for the SAFT battery. Furthermore, SAFT claims the batteries are 100% recyclable.

Since the Horizon is a sealed battery no maintenance is necessary but the SAFT battery does require periodic watering.

The SAFT battery pack, however, is equipped with a low maintenance watering system. Only one reservoir requires inspection and refilling.

The limited availability of the rare earth metals needed for the NiCd's construction is also a concern. Limited resources prohibit the mass production necessary for a large electric vehicle (EV) or HEV fleet. However, their use in AfterShock is acceptable because the NiCd performance levels best represents upcoming battery technology like the NiMH battery, which is capable of supporting a large vehicle fleet.

Another appealing characteristic of the SAFT battery concerns DOD. SAFT claims the batteries can be regularly discharged 100% without any adverse effect on life span.

Therefore, the UC Davis HEV Team chose the SAFT STM 5.100 Nickel-Cadmium batteries. AfterShock contains a 26 module, main battery pack. Each module has a nominal voltage of 6 V. The module capacity rating is 100 Ah @ C/5, giving AfterShock 15.6 kWh of energy storage. The nominal pack voltage when all 26 modules are connected in series is 156 V. Compared to Ground FX, there is a 200% increase in energy storage with only a 30% increase in total battery weight. This increased energy storage results in 180% greater range.

BATTERY CHARGING - Many EV and HEV disadvantages are associated with battery charging. In order to reduce the negative effects of charging, the UC Davis HEV Team dictated a 6 hour charge time and a convenient onboard, affordable charger. After investigating different chargers a second Solectria BC2000, stacked onto an existing BC2000 from Ground FX, best suited the charging requirements. The BC2000 is 93% efficient, weighs 5 kg, and can be controlled by a 0 - 5 V signal.

Due to AfterShock's 15.6 kWh battery pack capacity, a large charger was deemed necessary. Chargers in the 4 kW range, however, are not readily available and are costly. Thus, it was necessary to stack two BC2000s in parallel. Solectria has approved and assisted in the assembly of the unique charger. The resultant charger is capable of charging at 4 kW with a peak charge current of 20 A. The charging system is mounted onboard for convenience and weighs 11 kg.

The recommended charging scheme for the STM 5.100 battery is a two-stage, constant current method. Stage one charges the battery at 28 A until the module voltage reaches 8 V, or 208 V for the entire pack. Stage two is a 5 A trickle charge with a duration equal to stage one. A full charge is accomplished in under 6 hours.

BATTERY ENCLOSURES - Battery enclosure safety is another critical issue concerning EVs and HEVs. AfterShock's enclosures use a low weight, polypropylene material from Battery Mat for electrical insulation, fire resistance (in case of arcing), electrolyte absorption, and temperature regulation.

Specific to the UC Davis battery pack, is the temperature regulation of the SAFT batteries. The batteries must remain below 45° C during continuous duty and under 60° C during peak duty, otherwise battery life can be reduced. To maintain proper battery temperature and evacuate possible harmful gases, a heat transfer analysis revealed the battery enclosures must have 30 cfm of ventilation. Air is forced across three sides of each module, maximizing surface area for heat transfer. This is accomplished with four ComAir Rotron, 36 cfm blowers.

AUXILIARY POWER UNIT - As stated previously, the UC Davis HEV parallel powertrain uses the APU to directly propel the vehicle while cruising. Thus, only a small ICE is necessary to maintain highway speeds. Ideally, the engine is sized to run at wide open throttle (WOT), the most efficient setting. Hence, it is load leveled and can be point tuned. Consequently, the engine is tuned to maximize fuel economy and minimize emissions. An extensive APU investigation, therefore, was necessary to determine the proper ICE selection.

After investigating the Chevrolet Sprint, BMW K75 and K111, the UC Davis HEV Team selected the Briggs & Stratton (B&S) ICE. The three cylinder Chevrolet Sprint offered reliability, electronic fuel injection (EFI), good emissions and fuel economy but was too big, too heavy, and too powerful for AfterShock (APU operation would be approximately 3/5 WOT. Thus, fuel economy is not maximized and emissions not minimized). The BMW K75 and K111 engines, unfortunately, have high torque curves and rpm ranges which made them unfeasible. However, the B&S, 15 kW, 570 cc, overhead valve, 2-cylinder, ICE met all design restrictions. Its low weight, near prefect rpm and torque characteristics, size, and simplicity dictated its use.

Although the B&S best suits AfterShock's needs, several modifications were necessary to optimize its performance. These modifications include increased compression ratio (CR), the incorporation of variable spark timing, and custom camshafts. Consequently, engine torque in the 2000-3000 rpm range is maximized, increasing fuel economy and reducing emissions.

Increasing the CR intensifies the burn, and is one method to increase engine efficiency. However, if the CR is too high gasses explode rather than burn and push against the piston as it completes the compression stroke. This is known as knocking and pinging. Furthermore, increasing the CR raises the combustion temperature producing more NOx emissions. The stock B&S has an 8.2:1 CR. In order to increase the CR, the combustion chamber was reduced resulting in a 9.5:1 CR. This CR increases efficiency but avoids the undesirable side-effects mentioned above.

The stock B&S also has a fixed timing ignition system. Since the APU operates at a constant load and rpm, timing adjustment is not crucial. Varying the spark timing, however, can eliminate knock and ping as well as increase fuel economy through advancement. Torque can be increased with a slightly retarded spark. The Electromotive EFI system, utilized in AfterShock, has an integrated distributorless ignition system. This system senses engine position allowing precise spark advance. The Electromotive system also incorporates a knock and ping sensor which senses vibrations in the block and retards timing accordingly. This allows the engine to operate as advanced as possible, increasing fuel economy.

Modifications to the cams affect the APU performance the most. The cams control the air intake and exhaust by dictating when the valves open and close. Fortunately the stock camshaft already produces power in the correct torque bandwidth, so drastic changes are not necessary. Due to the other modifications, however, the camshaft is no longer optimized for the APU. Thus, testing has determined the correct cam shape.[2]

EMISSIONS - Battling emissions is a complicated task. The three main types of emissions are hydrocarbons and carbon monoxide, which increase during rich or excess fuel conditions, and NOx which increase with burn temperature and lean burn conditions.

To reduce emissions the Electromotive fuel injection system with feedback, previously mentioned, has been adapted to the B&S ICE with a 3-way catalyst. A modern 3-way catalytic converter significantly reduces all three emission types if the air fuel ratio (AFR) is stoichiometric (AFR is 14.7:1). Since the APU is not cycled, the AFR remains relatively stoichiometric producing few harmful emissions.[2]

Even though the APU only operates during steady-state, the AFR fails to remain exactly at 14.7:1. Thus, feedback control is used to ensure the AFR remains near stoichiometric. The computer controlled EFI utilizes exhaust gas oxygen (EGO) feedback from an EGO sensor installed in the exhaust manifold before the catalytic converter. Using the EGO sensor, the EFI system determines whether the engine is operating lean or rich and compensates by adjusting the pulse width of the fuel injectors. This reduces tailpipe emissions.

Although it is important to control tailpipe emissions, sacrifices are unavoidable. The equipment adapted to the APU reducing emissions also inhibits the air flow which reduces overall engine efficiency. As stated above, the AFR must remain a constant 14.7:1 to reduce emissions. Unfortunately this is not where peak power and peak fuel efficiency occur. Peak power is achieved with a slightly richer mixture and peak fuel economy is achieved in a lean burn situation. Hence, for the vehicle to be truly optimized under certain load conditions and cruising speeds, the AFR must be adjusted to meet the needs of the vehicle. The Electromotive fuel injection has proportional EGO feedback and the target AFR can be set for any operating condition based on engine load and rpm.

The emissions results from the HEV Challenge are list below in Table 6.

Table 6: HEV Challenge Emissions Results

HC	0.204 g/mile
NMHC	0.178 g/mile
CO	2.417 g/mile
NOx	0.333 g/mile

APU TESTING - Although designing new components and modifying APU variables may increases efficiency in theory, testing is required. A Mustang engine dynamometer is used for this purpose. The dynamometer interfaces with a 386 PC reading rpm, torque, and power. Incremental load settings allow the creation of torque curves for any throttle setting. Fuel flow data and brake specific fuel consumption (BSFC) can also be calculated by measuring the amount of fuel consumed for a specific operating condition. Through extensive modifications and testing, the APU has been optimized for its specific use.

FUEL SELECTION - After careful consideration, the UC Davis HEV Team choose reformulated gasoline. Although methanol and ethanol are appealing, the infrastructure for these fuels is not as established as the gasoline infrastructure within the US. Therefore, using these fuels creates a less appealing vehicle to the general public. Since, the UC Davis HEV Team wants to fabricate a practical car, reformulated gasoline is the only choice. After the Challenge conclusion, however, AfterShock will be modified, becoming a flexible-fuel vehicle.

CHASSIS DEVELOPMENT

FRAME DESIGN - The frame is an instrumental component of any vehicle. The UC Davis HEV frame must emphasize light weight, superior ergonomics, allow for a transverse powertrain, and fit within the existing body design.

To accomplish these objectives several frame designs including a composite monocoque, aluminum monocoque, steel space-frame, and an aluminum space-frame were investigated. A composite monocoque, utilized by the GM Ultralite, minimizes vehicle weight by combining the exterior shell and vehicle sub-frame into a single load carrying member.[3] Unfortunately, composite monocoques pose significant engineering and manufacturing difficulties. Although the concept is appealing and believed to be the best solution by the UC Davis HEV Team, it was unfeasible due to limited resources. An aluminum monocoque also offers the composite monocoque advantages as well as its fabrication hurdles.

Aluminum and steel space-frames, however, are easy to manufacture. Since low weight is a priority (aluminum is approximately 1/3 the weight of steel), the UC Davis HEV Team designed a 6061 T6, aluminum frame but incorporated aluminum monocoque frame features. The frame uses 25 mm x 76 mm x 3.2 mm (1 inch x 3 inch x 0.125 inch) box tubing and aluminum sheet metal for load carrying members. The resultant battery enclosures, wheel-wells, fire-wall, and occupant floor-board allow significantly more powertrain and passenger compartment space as well as reduced weight, 23 kg, compared to the Ground FX frame. Thus, AfterShock's frame is an aluminum monocoque/space-frame (AMS) combination.

AfterShock's AMS frame, as mentioned, uses aluminum sheet metal as load carrying frame members. The 1.6 mm

(0.0625 inch) thick aluminum sheet metal floor-board adds significant torsional stiffness to the frame. Wheel-wells are constructed from 1.6 mm and 3.2 mm thick aluminum sheet metal creating sufficient space for the suspension and transverse powertrain.

The box tubes used for occupant roll-over protection are 33% stronger than the HEV Challenge Regulations require and allow ample driver visibility. Battery enclosures also incorporate 1.6 mm thick aluminum sheet metal for strength.

Vehicle trackwidth is 1320 mm in the front and 1270 mm in the rear. Wheelbase is 2360 mm. The rolling chassis weight is 290 kg.

Hand calculations, Finite Element Analysis, and independent consultants provided sufficient proof that the frame design is safe during normal driving conditions as well as accident and roll-over situations. See Figure 2.

BATTERY ENCLOSURES - The battery enclosures are located along the outside edges of the vehicle between the front and rear wheels. This location seems natural considering the battery enclosures do not intrude into interior space or increase frontal area. Nor do they create a weight distribution problem.

Other possible locations included the center console and nose. Center console placement interferes with interior space and could possibly increase vehicle trackwidth and frontal area. Nose placement causes significant weight distribution problems. Additionally, the side battery enclosures provide supplemental side impact protection due to the large buffer mass created by the batteries.

The battery enclosures utilize 1.6 mm thick aluminum sheet metal, reinforced with box tubing and 51 mm (2 inch), 6.4 mm (0.25 inch) thick, angle aluminum. AfterShock's battery enclosures are 48% stronger (bending), 24% lighter, and 18% more space efficient than the battery enclosures in Ground FX. Internal battery enclosure volume is 0.12 m^3.

SUSPENSION AND BRAKES - Suspension design and fabrication is a long, difficult task. Therefore, UC Davis HEV Team members concentrated on integrating an existing suspension with the frame. Due to the transverse powertrain configuration, the low weight suspension must incorporate a space-efficient design consisting of available and affordable replacement parts.

A wide range of late model automobile suspensions were investigated including many MacPherson strut and rear swing-arm designs. The swing-arm suspension offered space savings for powertrain placement but interfered with the battery enclosures. A MacPherson strut design offers sufficient space for the powertrain as well as straightforward adaptation. Thus, a 1985 Toyota MR2, four wheel, independent MacPherson strut suspension was selected. The MR2 suspension is designed for a mid-engine vehicle with a weight distribution and curb weight similar to AfterShock's characteristics.

Utilizing the basic MR2 strut geometry, many suspension components were redesigned and remanufactured to reduce weight and allow for frame integration. UC Davis HEV Team

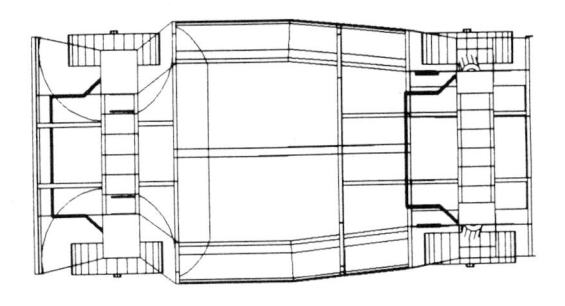

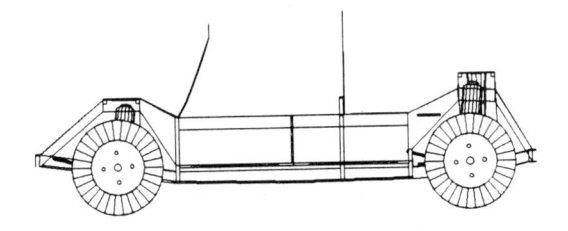

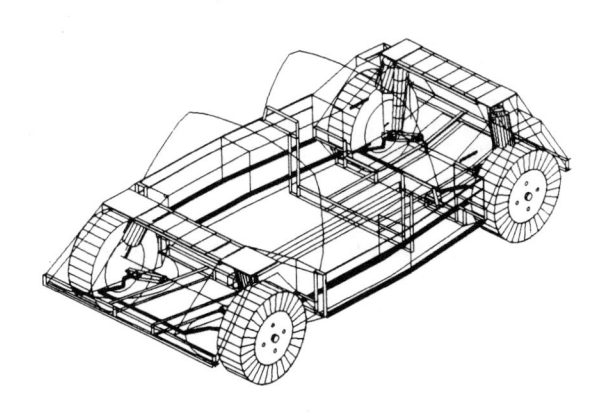

Figure 2: AMS Frame

members fabricated aluminum wheel hubs, aluminum lower A-arms, modified steering rods, shortened compression rods, and many other smaller components resulting in a 13 kg weight savings. Suspension travel, in compliance to the HEV Challenge Regulations, is 102 mm (4 inches). The brakes are a integral chassis component. The MR2 brakes were replaced due to their high weight. Aluminum CNC calipers were chosen for their low mass and high performance. Since the UC Davis HEV Team wanted to ensure proper braking power, the front wheels incorporate four piston calipers while the rear wheels utilize two piston calipers.

In order to reduce weight, aluminum and treated aluminum rotors were investigated. Due to past experience, plain aluminum rotors fail to offer sufficient durability. Anodized, Titanium Nitride (TiN) plated, and silicate treated (metal matrix composite) aluminum rotors were considered as possible options. Anodizing provides the required surface hardness but lacks long-term durability. TiN plating also provides the necessary hardness but the resultant surface friction coefficient fails to generate sufficient force for braking. An aluminum silicate rotor provides the desired weight savings and surface hardness but the material lacks manufacturability. Therefore, the UC Davis HEV Team choose 254 mm diameter, mild steel (9.5 mm thick on the front, 6.4 mm thick on the rear) rotors, despite the large weight penalty (9 kg) to ensure sufficient braking power.

Cast aluminum, mechanical parking brake calipers, produced by Wilwood, were adapted to the rear hub assemblies and actuated by the stock MR2 parking brake handle. AfterShock's brakes are 11 kg lighter than the stock MR2 brakes.

TIRES AND WHEELS - Rolling drag accounts for nearly 1/3 of the total drag for an aerodynamic vehicle. Thus, efficient vehicle design demands a reduction in rolling resistance. Reducing vehicle weight is one method to decrease rolling drag. Selecting the proper tire also helps reduce the effects of rolling resistance.

The UC Davis HEV Team researched several tire options from Goodyear, Michelin, Firestone, and Pirelli. Michelin and Pirelli were unwilling to provide sensitive tire characteristics as well as material support. Thus, the Firestone and Goodyear tire were the best options. The 350 kPa, recently DOT approved, 195/70R14 Firestone EVT (Electric Vehicle Tire) found on the Ford Ecostar has a respectable C_{rr} of 0.006. Goodyear produces the standard pressure, Invicta, found on many production vehicles. The Invicta, however, has a higher coefficient of rolling resistance. Therefore, AfterShock employs the Firestone EVT tires. The Firestone tires also weigh 8 kg lighter than the Invictas.[4]

Figure 3 compares the Firestone and Goodyear tires with a standard passenger car tire. Clearly, the Firestone EVT tire performance is superior.

Wheel selection also suffered intense weight scrutiny. Low weight wheels not only decrease total vehicle weight, but reduce rotational inertia which hamper acceleration. The cast aluminum rims selected weigh 5 kg each and are manufactured by Eagle. Standard, steel wheels weigh 11 kg.

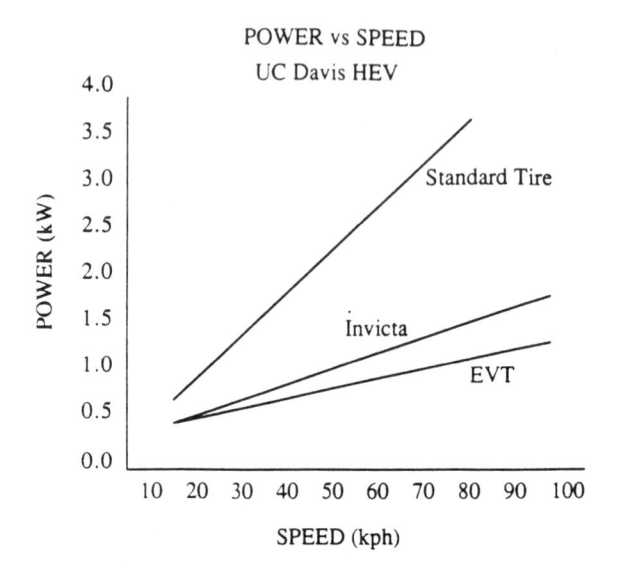

Figure 3: Tire Comparison

ERGONOMICS - The frame design optimized not only the battery enclosures and powertrain compartment but also the passenger compartment. The frame allows 1.0 m for easy occupant entry and exit. This large opening is necessary due to the non-standard door-jam height created by the battery enclosures.

The required roll-cage is concealed by integrating the body, chassis, and interior design. The concealed roll-cage enhances driver vision and an improves interior aesthetics.

Interior comfort is provided by Lear seats which translate 0.4 m (standard seat translation is 0.2 m) aiding vehicle entry and exit. AfterShock accommodates 1.6 m to 1.9 m occupants.

The dashboard further optimizes interior space. The large instrument panel houses all necessary gages for the dual powertrain with vehicle controls located within close proximity to the driver. Furthermore, AfterShock provides the "creature-comforts" found in production automobiles like cup holders.[5]

PASSENGER COMFORT - In order for alternatively powered vehicles to be commercially viable, standard features must include a climate control. Unfortunately, most climate control systems consume large quantities of power which hampers EV range. After examining conventional technologies, a thermoelectric system was selected. Thermoelectric devices (peltiers) are quite compact, light weight, and consume little power.

In addition to the appealing characteristics listed above, these thermoelectric devices are simple to construct and operate. When current is passed through the peltiers one side cools while the other side warms. Reversing the current creates the opposite effect. Two blowers force the cool or warm air into the passenger compartment while directing the unwanted air outside the vehicle.

The thermoelectric design incorporates two Melcor, 51 W (output) peltiers located between two aluminum heat exchangers. Air is forced over the heat exchangers by two ComAir Rotron, 36 cfm blowers. Total output is 102 W requiring 144 W of input power and weighing only 2 kg. The climate control produces 25 cfm of conditioned air over a 19° C to 43° C temperature range. Since conditioning the entire passenger compartment with 102 W is not feasible, the conditioned air is forced over the occupants neck and head. Military studies indicate sufficient passenger comfort when the head is kept at a comfortable temperature.

Noise reduction also aids passenger comfort. Noise is generated by the powertrain, tires, and airflow. At 60 mph, wind noise above 1000 Hz can reach 68 dB(A).[1] The EM drive belt specifications claim 110 dB(A) at 3700 Hz. To reduce noise levels to that in conventional vehicles, 19 mm, light weight foam from Scosche encapsulates the powertrain and insulates the entire passenger compartment.

MANUFACTURING - Manufacturability played an influential factor in all chassis design decisions. As mentioned earlier, composite and aluminum monocoques were considered unfeasible due to manufacturing problems.

During the design phase, special attention to frame member selection ensured standard, repeated sizes. No complex fixtures were required for the Tungsten Inert Gas (TIG) welded frame. TIG welding, however, produces heat affected zones. These zones suffer strength losses. Redundant frame members supplement the frame to avoid failure problems in the heat affected zones.

Heat treating the entire frame eliminates the zones and the need for the additional frame members. Heat treating facilities, however, were unavailable. The UC Davis HEV Team believes a further optimized frame is easily massed produced on an automated production line.

BODY DEVELOPMENT

BODY DESIGN - The body design is an important factor when producing a superior HEV. City driving dictates low vehicle inertia while highway driving dictates exceptional vehicle aerodynamics. Since good aerodynamics are key to an efficient HEV, the UC Davis HEV Team concentrated on the topics summarized in Table 7.

Table 7: Aerodynamic Goals

1. Minimal drag coefficient
2. Minimal frontal area
3. Complete under-belly
4. Low protuberance drag

A vehicle with these characteristics is aerodynamically correct but weight, manufacturability, and ergonomics must be considered as well.

After reviewing the UC Davis HEV Program resources, designing and fabricating an entirely new body was judged unfeasible. Hence, the basic body template from Ground FX served as the foundation for AfterShock's exterior shell. Testing revealed an aerodynamic body except for the under-belly and rear diffuser sections. AfterShock's body incorporates improvements to eliminate these undesirable characteristics while enhancing driver vision and ergonomics. Furthermore, special attention to manufacturing ensured the body panels fit properly. AfterShock's body also has the practical features found on conventional automobiles like a gasoline cap hatch, standard door handles, and a hatch for charging.

AERODYNAMICS - Equation 1, reveals the criticality of aerodynamics. At highway speeds aerodynamic drag accounts for approximately 65% of the primary drag forces.[1] AfterShock utilizes a distinct aerodynamic body to reduce these drag forces.

After reviewing several aerodynamic shapes, an airfoil was selected. Other aerodynamic shapes lacked the potential for styling or failed to comply with Challenge Regulations. The initial shape was based on a streamlined NACA 66-025 airfoil which maintains a decreasing pressure gradient over 60% of the chord length. The favorable pressure gradient permits laminar flow at higher velocities. To eliminate negative lift from ground effects, a 5% camber was introduced via a 0.7 NACA mean camber line.[6]

An aerodynamic body, however, does not decrease the drag coefficient sufficiently. Emphasis was placed on reducing protuberance drag from wheel-wells, tires, window recesses, drip rails, and external mirrors. Protuberance drag accounts for 28% of the total drag on vehicles. In order to reduce wheel-well drag, all tires are enclosed within the body. The faired tires, theoretically, decrease the drag coefficient by 0.084. Drip rails were integrated into the overlapping panels, effectively reducing this protuberance drag component to zero. Permanent windows were mounted from inside the vehicle. Hence, 1 mm separates the outer body and the windshield surfaces. [7]

Reducing external mirror drag was also investigated. Options included internally mounted mirrors, fiber-optic bundles, and video systems. These schemes, however, were not used because reducing external mirror size, utilizing an aerodynamic shape, and placing the mirror in an advantageous airflow position significantly minimizes mirror drag.

Other techniques to reduce the drag coefficient include boat-tailing (tapering the body) and decreasing the rear diffuser angle. The body tapers an average of 19 degrees, very close to the ideal 22 degree taper from the front tires to the tail-end of the vehicle. A small rear diffuser angle of 8 degrees also improves vehicle aerodynamics.[8]

Lastly, AfterShock incorporates a smooth under-belly. Under-belly drag potentially increases the vehicle drag coefficient by 0.015, roughly equal to skin friction drag.[7]

SIZING - The geometry factor in the force equation is the product of the vehicle shape and size. For bluff bodies, the standard area used for measurement is the frontal area. Reducing frontal area decreases the C_dA_f value. Frontal area reduction is limited by the occupant seating arrangement.

Several occupant seating arrangements were considered to reduce the frontal area (tandem, staggered, and standard). A standard seating arrangement was selected after impromptu interviews with the general public revealed their reluctance to non-standard seating positions. Thus, the standard seating arrangement forces a 1.8 m vehicle width. Overall length, dictated by the chassis length and HEV Challenge rules, is 4.2 m. Consequently, the frontal area is 1.77 m^2. See Figure 4.

BODY PANEL CONSTRUCTION - The main body panels are an advanced composite sandwich construction. This construction technique approximates the strength of stamped steel panels. A 6.35 mm (0.25 inch) expanded Nomex core placed between 4 plies of F-155 pre-preg E-glass ensures structural rigidity. McClellan Air Force Base Advanced Composites Laboratory had an instrumental role during body panel design and fabrication.

Several composite materials including carbon fiber, kevlar, E-glass/epoxy, and S-glass/epoxy composites were considered. All of these options offer excellent weight to strength ratios but some are quite expensive. Since panel stiffness is proportional to the panel shape; expensive, high modulus composites provide no significant benefits compared to E-glass. Thus, all body panels were constructed of an E-glass/epoxy composite. Total body weight is 95 kg.

VISIBILITY - To ensure proper visibility, AfterShock has a full compliment of large windows. The 1.4 m^2 windshield is due to the unique body shape. Each door has a side and upper window. The upper window is similar to a sun-roof but is used to detect high, forward objects like highway signs. Side mirrors, mounted on the doors, allow expansive rearward visibility.

Polycarbonate, acrylic, and safety glass were considered as possible windshield materials. A standard glass windshield was unfeasible because of the unique vehicle shape. Acrylic is easy to form but lacks impact resistance. Polycarbonate is more difficult to form but is very impact resistant.

Although the polycarbonate can be formed to the vehicle shape, its resistance to surface abrasions is poor. After investigating this problem, hardness treating could possibly increase its surface abrasion resistance but the treatment was unavailable.

Thus, for safety reasons, the front windshield is 6.4 mm (0.25 inch) polycarbonate. The windshield weighs 8 kg compared to a 17 kg glass windshield. The door and rear

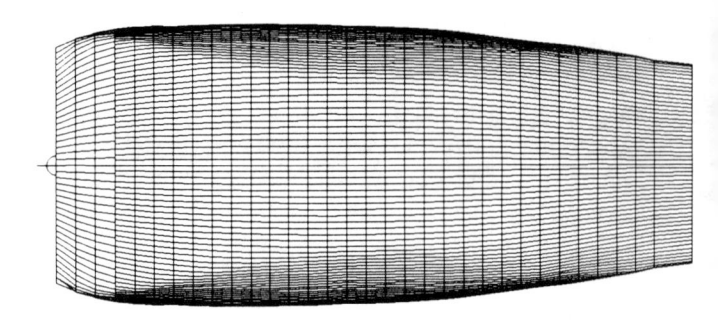

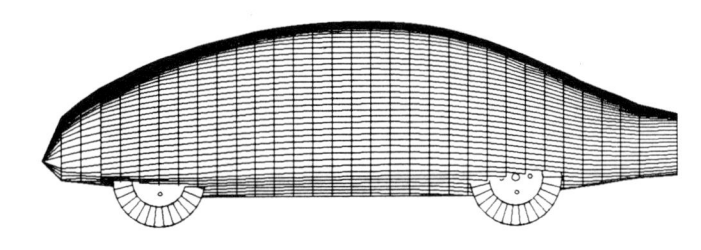

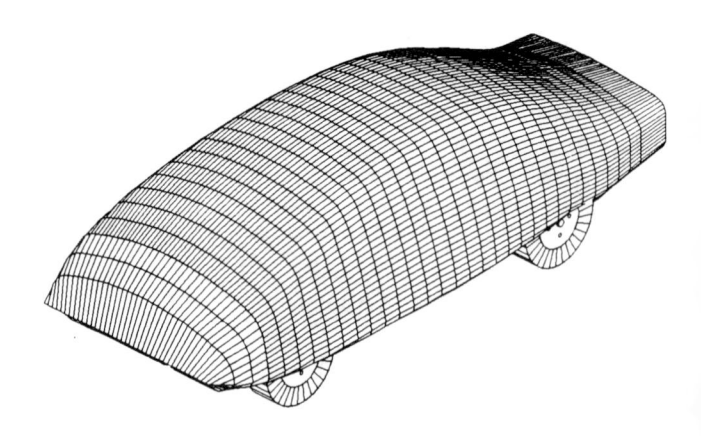

Figure 4: Body Design

windows are 3.2 mm (0.125 inch) acrylic because it is easier to form.

INTERNAL AIRFLOW - Internal airflow is necessary for powertrain compartment cooling. Studies show internal airflow fixtures are responsible for 2 - 5% of the overall vehicle drag.[7] Thus, the UC Davis HEV Team attempted to minimize this drag component.

Inlets are located in the vehicle's nose where a high pressure area is located. The air is routed through the passenger compartment to the engine compartment. An exhaust fan ensures the forced air travels completely through the vehicle cooling the powertrain compartment. The air is exhausted at the low pressure point at the vehicle's tail. Ducting in the passenger compartment is insulated to control noise.

MANUFACTURING - Standard composite techniques found in prototype and low volume production were used to construct AfterShock's body. The standard process has five stages: design, buck construction, tooling splashes, panel fabrication, and assembly. Design for manufacturing and continuous improvement strategies created a refined exterior shell for AfterShock.

The process described below was used due to limited resources. The UC Davis HEV Team is confident, however, large scale body production is possible using reaction injection molding (RIM) to create the light weight, strong body panels similar to those on the GM Impact.[9]

Design - As mentioned, testing revealed poor under-belly and diffuser aerodynamics on Ground FX. The poor aerodynamics were attributed to 25-30 mm panel gaps originating from the two bucks used to fabricate body panels. Hence, one buck was used for AfterShock to eliminate this problem.

To increase manufacturability, the underbelly, side transitions, and diffuser were designed as flat planes with large radius transitions simplifying buck and tool construction as well as panel lay-up. Importing chassis and wireframe body drawings in AudoCad r12 by Autodesk ensured the chassis fit within the new body design.

Buck - The existing Ground FX buck was modified to incorporate the new under-belly and diffuser. A 762 mm section was removed and additional masonite cross-sections were fastened to existing cross-sections. Polystyrene filled the void between the masonite. Two fiberglass layers provided strength and hardness. Body filler eliminated depressions or scallops creating a smooth surface. Finally, a polyester gel-coat added a hard surface coat to polish.

As a fore-thought to assembly, the back of the car was styled to taillights mounted into the buck to ensure a flush fit during assembly.

Tooling - Tooling was constructed from buck splashes. Splashes were fabricated from a 1 mm gel-coat backed with 6 mm chopped fiberglass applied at Leer West, Inc. Panel locations were marked by scribing the buck with a cutting tool. Seven tools were produced including two split tools.

Warped tooling in Ground FX created large gaps (10-25 mm) between panels. To prevent panel mismatch, plywood frames were bonded to the splashes prior to removing the tools from the buck. When possible, multiple panels were combined onto the same tools.

Body Lay-Up - To further ensure no gaps occurred between body panels, adjoining panels were simultaneously laid-up to the net part line. A non-porous Teflon sheet separated panels to prevent bonding while curing. The procedure simplifies assembly especially for the doors and hatch which mate to body frame members. Simultaneous fabrication also ensures panels match with minimal trimming and rework during assembly.

Assembly - Although a design may be excellent, poor execution during assembly results in mediocre aerodynamic performance. Tuft testing Ground FX revealed turbulence air flow was generated near panel gaps. AfterShock's 12 body panels compared to Ground FX's 20 panels significantly reduced this protuberance drag by eliminating possible turbulence generators.

To aid the assembly process, two body frame members were constructed. These members are mounted directly to the chassis and ensure proper alignment of other panels. The front frame supports the bumper, windshield, and doors. It is bolted to the chassis along the roll-cage, battery boxes, and wheel-wells. The hatch frame supports the hatch and tail panels. Other panels are indexed to these two permanent frame panels. Henceforth, the vehicle is easier to assemble and disassemble.

The gull-wing doors are attached to the chassis with custom aluminum hinges and are opened using gas struts. The rear hatch allows access to both the cargo area and powertrain. Routine maintenance is performed through the open hatch and removable fire-wall panels.

PRELIMINARY TEST RESULTS

ON-ROAD TESTING - In order to assess AfterShock's "real-world" applicability, on-road testing was conducted. Numerous coast-down tests reveal a 6.1 kW road load at 80 kph.

To evaluate AfterShock's reliability and road worthiness, the vehicle was driven from Los Angles to Davis on a single charge and 21.6 liters of gasoline. The official gasoline fuel economy for the journey was 3.0 liters per 100 km (77 mpg) and used nearly 12 kWh of electrical energy in HEV1 mode. AfterShock successfully climbed the Tehachapi Mountain Range (elevation: 1,900 meters) maintaining highway speeds and completed the 710 km trip in approximately 7.5 hours.

DYNAMOMETER TESTING - Dynamometer testing was utilized to evaluate the two HEV modes currently equipped in AfterShock. Although the data analysis at this time is incomplete, AfterShock demonstrated a 110 km ZEV range and used only 1.3 liters per 100 km (300 mpg) of fuel in HEV2 mode on the Federal Urban Driving Schedule (FUDS).

CONCLUSION

The UC Davis HEV Team has produced its second HEV named AfterShock. AfterShock demonstrates exceptional fuel economy, low emissions, and impressive ZEV range. It also illustrates the positive aspects associated with HEVs as well as some unavoidable vehicle compromises. Hence, AfterShock proves that with the correct vehicle configuration, control strategy, and material selection, a high performance, practical, and reliable HEV is possible.

REFERENCES

[1] Riley, Rebecca. "Hybrid Electric Vehicle Development at the University of California, Davis: The Design of Ground FX", SAE 940340.

[2] Heywood, John. *Internal Combustion Engine Fundamentals,* McGraw Hill, New York, 1988.

[3] Ashley, Steven. "GM's Ultralight is Racing Toward Greater Fuel Efficiency", *Mechanical Engineering*, May 1992, 64-67.

[4] Bosch Automotive Handbook, Ed. 2, 1986.

[5] Hartemann, Francois and Bernard Favre. "Human Factors for Display and Control", SAE 901149.

[6] Waters, D.M., "Thickness and Camber Effects on Bodies in Ground Proximity", *Advances in Road Vehicle Aerodynamics*, 1973.

[7] Carr, C.W. "Potential for Aerodynamic Drag Reduction in Car Design." Int. Journal of Vehicle Design, Special Publication SP3, 1983.

[8] Hucho, Wolf-Heinrich. *Aerodynamics of Road Vehicles,* Butterworths, London, 1987.

[9] Birch, Stuart. "Impact's Thermoset Body Panels", *Automotive Engineering,* April 1994.

ELANT III: A HYBRID ELECTRIC VEHICLE

Jun Lee, Jr/Sr EE   Mica Parks, Jr/Sr ME
University of California, Irvine

ABSTRACT

The ELANT III is a 1992 Ford Escort wagon modified to operate as a parallel hybrid vehicle with a battery powered electric motor and a gasoline powered internal combustion engine. The vehicle is designed as an environmentally friendly, two passenger commuter vehicle capable of operating in three modes. The first mode enables the ELANT III to operate electrically as a Zero Emissions Vehicle (ZEV) for short-range commuter travel. In the second mode, the car operates as an Ultra Low Emissions Vehicle (ULEV), powered by the gasoline engine tuned especially for highway speeds, This will provide a long distance travel capability. In addition, this mode also serves as an auxiliary measure when the electric drive is inoperable. The third mode allows both drives to be engaged simultaneously for extra passing power or dynamic braking (HEV mode), or for battery recharging.

1. INTRODUCTION

Taking its name from the University mascot, the "ELectric ANTeater" or ELANT III was designated to become a hybrid electric vehicle before the Ford Hybrid Electric Vehicle (HEV)Challenge was announced. The ELANT I electric vehicle was completed by University students in 1991 as part of an ongoing Electric Vehicle (EV) Project. Its performance and range were determined to be substandard. An AC motor powered vehicle was to be the next ELANT, but its projected range was similarly deficient. A hybrid electric vehicle was then planned to solve the range problem suffered by current pure electrics. Since the vehicle would have an internal combustion engine (ICE), reduced exhaust emissions would be a design requirement.

The ELANT III HEV power train became the design focus, since all vehicles in the UC Irvine EV Project were conversions from existing automobiles. There are two basic configurations for hybrid electric power trains: series and parallel. A series power train has the ICE drive a generator to charge a battery which in turn powers an electric motor. The electric motor is the only prime mover driving the wheels. In a parallel configuration, either the electric motor and/or the ICE drive the vehicle. Each configuration has benefits and disadvantages. It was UC Irvine team's decision to use a parallel configuration for reasons based on efficiency and simplicity. In a series HEV, less of the energy input goes towards useful work. In a parallel configuration, the ICE can efficiently direct power to the wheels through a standard automotive transmission when needed. For an equivalent power output, harmful exhaust emissions (such as carbon monoxide) from a parallel configuration could be slightly higher than from a series configuration, but the carbon dioxide emission of the former is significantly lower because the series system has higher fuel consumption.

2. PERFORMANCE CRITERIA

The HEV Challenge rules and regulations set the performance criteria which the ELANT III is designed to meet.

ACCELERATION - Vehicles should be capable of a 0 to 45 mi/hr acceleration in 15 seconds or less.

RANGE - Zero emission (ZEV) range should be 20 miles or more. ICE range should be 240 miles or more.

CURB WEIGHT - The Ford Escort wagon chassis is designed for a maximum weight of 3800 pounds. For carrying 400 pounds of passengers and cargo, the maximum curb weight of the unloaded vehicle should not exceed 3400 pounds.

FUEL ECONOMY - Entries into the Ford HEV Challenge will be judged heavily on fuel economy. Lower fuel usage of the ICE can also lower total emissions.

EXHAUST EMISSIONS - Harmful exhaust emissions such as reactive organic gases based on hydrocarbons (ROG or HC), carbon monoxide (CO), or oxides of nitrogen (NOx) are targeted to be much lower than current automotive emission standards. Ultra Low Emission standards (ULEV, as defined by the California Air Resources Board) are the targeted emission levels of 0.04, 1.7, and 0.2 grams per mile respectively.

BATTERIES - Current battery choices are limited by cost and thus restrict performance, but acceptable power and energy is nevertheless attainable with lead acid batteries. Aviation batteries, and batteries containing precious metals, are not allowed in the HEV Challenge. 400 volts is the maximum battery system voltage.

CHARGING - Battery charging time should be six hours or less. This is a typical overnight recharging time. Recharging power would come from a 230 VAC, 30 AMP outlet.

SAFETY - Safety rules for the HEV Challenge are of a race quality nature. Requirements include safety roll cages, 5-point racing harnesses (racing seat belts), and a special fire extinguishing system. Electrical safety is also notable. Proper insulation and circuit protection are required. Dynamic braking is allowed but may not be relied uppon. Emergency handling must be acceptable in order to prevent accidents. At the same time the car must be operable in fairly standard fashion and must not induce fatigue.

3. POWER TRAIN CONFIGURATION

The major task in converting the standard Escort station wagon into a hybrid vehicle was the design, fabrication, and installation of the dual drive system. Design requirements for the power train were:

1. 15 hp average (50 hp peak) power, 55ft-lb torque average.
2. Electric energy efficiency of 83% (motor and controller).
3. High fuel economy: no less than 40 mpg.
4. Mechanical transmission efficiency greater than 90%.
5. Similarity of driving control operation, regardless of drive mode (ZEV, ULEV, HEV).
6. Power train to weigh or 450 lbs or less and to have a width of less than 36 inches.
7. Components of ready availability and low cost.
8. Simple assembly, ease of maintenance.

The replacement power train was to fit the transverse mounted orientation and front wheel drive of the standard Ford Escort wagon. After calculating power and torque requirements desired for the hybrid vehicle and reviewing all affordable options, the UCI team decided on the use of rugged three-phase induction motor and a gasoline powered ICE.

The selection of an AC motor was somewhat influenced by the opportunity to test an electric dirve different from the DC series motor used with our ELANT I, a converted VW Rabbit. Similarly, 12 volt lead acid batteries were chosen for ELANT III after 6 volt units had been employed earlier.

The parallel-hybrid powertrain was designed with a unique merge unit comprising two electromagnetic clutches, one for each of the primary movers, and a chain for coupling the electric motor to the clutch on the main drive shaft. The car's original friction clutch was removed.

4. ELECTRIC DRIVE

4.1 Electric Motor

The ELANT III uses a 15 horsepower, 230V, 3-phase, class B, two-pole AC induction motor, donated by Electra-Gear of Anaheim, CA. The motor has parallel windings, NEMA 215M size frame and R type insulation capable of withstanding up to 360°F (200°C). Though our motor frame is rated at 15 horsepower, the R type insulation allows us to run the motor at higher than rated power for short periods of time. Thus we can draw up to 30 horsepower from the electric motor for up to two minutes, and more for shorter periods of time. The motor is forced air cooled.

4.2 Electric Motor Speed Control

In ZEV mode the ELANT III is driven by the 3-phase induction motor and controlled by a pulse-width modulated inverter acting as a variable frequency source. Figure 1 shows a general schematic for the motor controller and Figure 2 illustrates its position in the overall electric drive schematic.

Our Emerson Industrial Controls Eclipse II inverter/controller, donated by the manufacturer, is a high performance, general purpose, digital, adjustable frequency unit, usually operating on three-phase 50/60 Hz input power. The inverter produces an adjustable frequency, adjustable voltage, three phase output. This is usually achieved in the following manner.

Three phase, 60Hz, 220VAC power drawn from a line is rectified and filtered. The DC voltage is pulse-width modulated by the inverter to generate a variable voltage, variable frequency and approximately sinusoidal wave form.

Since the HEV will not be connected to an AC source except during battery charging, we removed the rectifier portion of the controller, supplying instead 312 VDC battery power directly to the controller.

A voltage/frequency curve is programmed into the Eclipse's onboard microprocessor. This allows the voltage to vary with frequency in order to maintain good magnetic flux density in the motor. The maximum voltage, where the V/Hz curve "levels off", is determined by the voltage characteristics of the batteries. Roughly speaking the ascending portion of the curve may be characterized as constant torque operation, the flat portion as constant horsepower operation.

As the electric vehicle is driven, the voltage available from the batteries will decrease. The voltage/frequency curve is set for the voltage still maintained at the 80 percent discharge level.

4.3 The Battery And Charger System

The battery selected is the Trojan DC-22NF Deep Cycle 12 Volt Pacer, a lead acid battery which weighs 30 pounds. It was necessary to use a wet battery to provide enough energy storage capacity. "Dry cell" batteries in the same price and weight range as the 22NF do not have nearly as much energy capacity. It has an Ampere hour rating of 56 at a 20 Amp current draw.

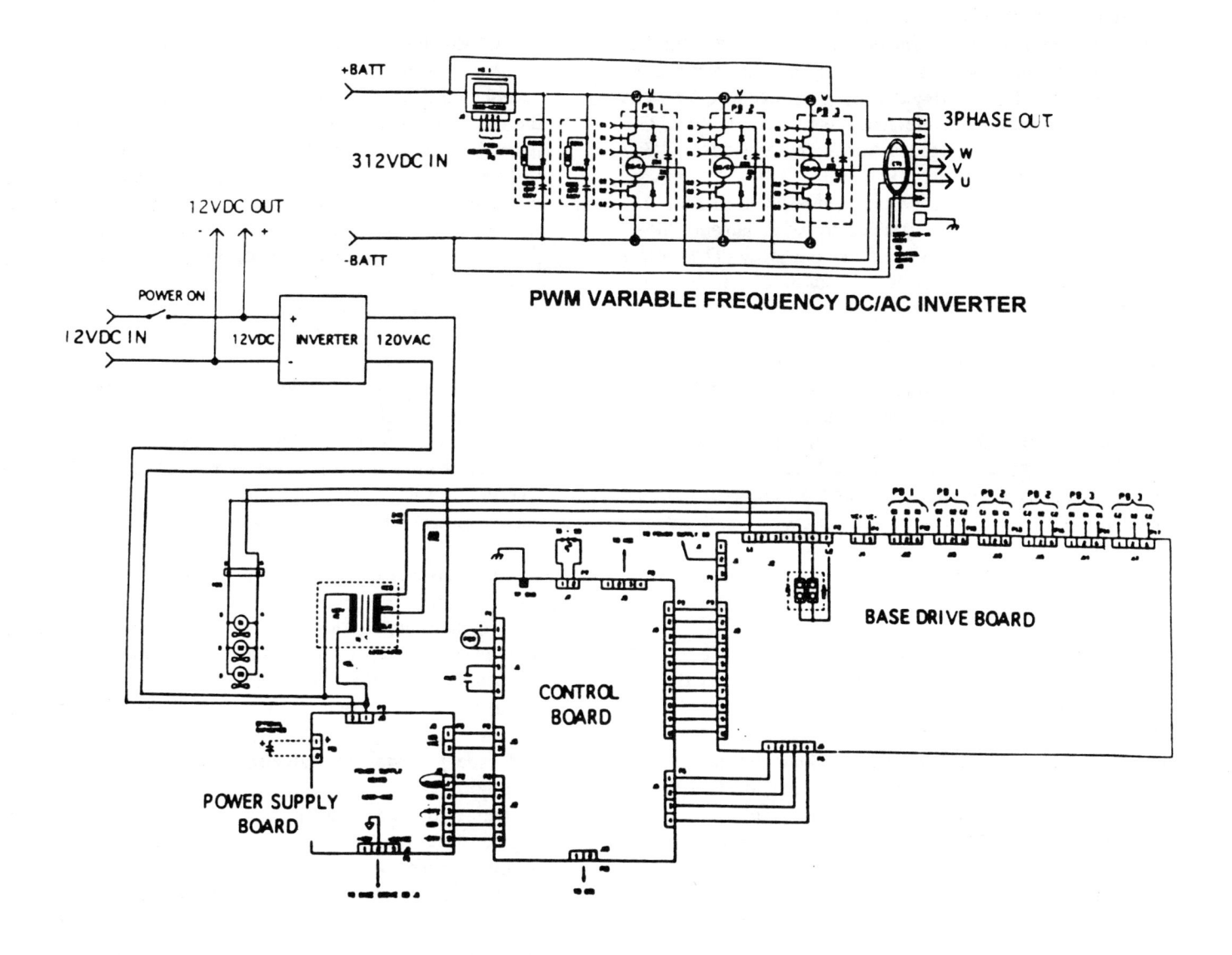

Figure 1: MOTOR CONTROLLER SCHEMATIC

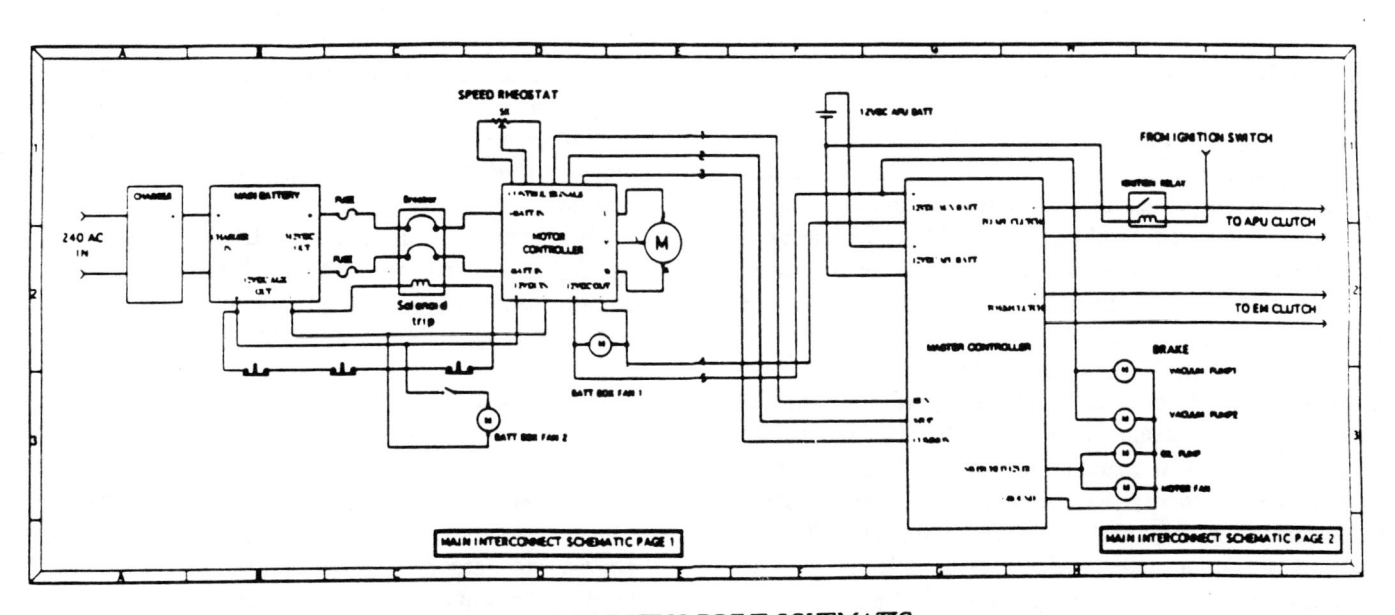

Figure 2: ELECTRIC DRIVE SCHEMATIC

The DC voltage required by the controller is between 290 and 356 volts. Using twenty-six 12-volt batteries connected in series gives a nominal voltage of 312 volts. Actual tests of the batteries being used show a voltage of 12.6 per battery when fully charged, and eventually a voltage of 11.8 volts when "discharged". Therefore the controller will see 327.6 volts DC with fully charged batteries and a voltage of 307 when the batteries are nearly exhausted.

The 26 drive batteries are connected in series and segmented by three fuses arranged to limit damage due to short circuits among batteries within the array. The battery box has a 29 battery capacity, leaving space for control system batteries which are not subjected to load swings.

For charging the batteries, a charger built with components from the rectifier section of the factory-supplied electric motor controller. To fully charge each 12 volt battery, at least 14 volts must be assigned to each. Therefore, at least 363 volt DC is required to charge 26 batteries in series, which would call for at least 269 volts AC. A step up isolation transformer with the proper tap setting provides the required voltage.

5. I.C.E. DRIVE

In ULEV and HEV modes the vehicle is powered by an internal combustion engine. The ICE is a used 1991 Geo Metro LSi 3 cylinder engine partially rebuilt by team members. This engine has a weight of about 140 pounds and a length of 15 inches from crankshaft pulley to the flywheel. The system will be described in an "energy transfer" order, starting with the fuel tank and ending with the exhaust gas emissions.

5.1 Fuel Source

Reformulated gasoline was selected as a fuel because the team foresees continuing use of the HEV for further tests and modifications which would be hampered by reliance on fuels less readily available in the University's immediate neighborhood.

Beginning with the fuel source, the standard Ford Escort 12 gallon fuel tank is retained. Since the ICE has an expected mileage greater than 40 mpg, less than six gallons of 87 octane unleaded gasoline fulfills HEV challenge range requirements. The Geo metro fuel pump is used to transfer fuel to the engine by means of the standard Ford Escort fuel lines.

The Geo metro intake manifold and throttle body assembly are used, as is the Geo metro throttle body fuel injection/controller computer. However, many of the features related to the exhaust emission are modified to reduce the amounts of pollutants in the engine exhaust. These features are described in the exhaust emissions section. The air cleaner is a custom made assembly using a K&N air filter which has a high air flow rate capability. This high flow-rate filter will allow better engine "breathing", which should reduce HC and CO emissions.

5.2 Engine

Almost no other mechanical features were modified on the ICE, however the engine head and block were partially disassembled/rebuilt for cleaning. The standard camshaft on the Geo Metro engine is used since it controls proper valve timing and lift. The standard Geo exhaust manifold is used. Custom made exhaust pipes are used to attach the exhaust manifold to the catalytic converter, and the standard Ford Escort muffler is also used.

The output from the Geo metro engine enters the merge unit through a lightened flywheel. The standard friction clutch is removed. Special engine mounts were fabricated to fit the altered engine configuration.

Engine cooling for the ICE has two modifications from the original Geo Metro cnfiguration. An after-market copper and brass radiator was used instead of the original aluminum/plastic radiator, and the mechanical water pump was replaced with a remote electrial pump for space considerations. Later this pump will be activated automatically when needed, thus conserving energy.

5.3 Emissions Control

The emissions system on the ICE employs features found on the Geo Metro and the Ford Escort. Evaporative emissions are controlled by the Geo Metro charcoal canister and Ford Escort fuel lines. The fuel filling nozzle employs a conically shaped receptacle as a seal which minimizes the amount of fuel vapors leaving the refueling nozzle during a refueling operation.

Most exhaust control components on the Geo Metro engine were retained and adjusted. Such items are the positive crankcase ventilation (PCV) valve as well as the exhaust recirculation (EGR) valve. Since exhaust back pressure is expected to be reduced, additional exhaust gas recirculation was required. The Geo Metro EGR valve is adjusted to permit more "burned" exhaust gas to be recirculated into the combustion chamber. Recirculating exhaust gases reduces NOx emissions and burns most unburned fuel vapor fumes present in the exhaust gas.

The Ford Escort catalytic converter is used since it is larger than the Geo Metro catalytic converter, thus a higher percentage of the exhaust gases volume will be converted into more environmentally friendly gases such as water vapor and nitrogen.

The preheater on the catalytic converter may reduce harmful emissions which occurs before the converter has attained its normal operating temperature. The preheater is to be viewed as an add-on experiment.

6. HYBRID DRIVE

The torque from the electric motor and the ICE is transmitted through a mechanical transmission device consisting of two separate components. These components are the standard five speed manual transmission from the Ford Escort wagon, and the "Merge Unit" which merges or selects power from the electric motor and/or gasoline engine to be directed to the transmission. Requirements for the transmission and merge unit combination are efficiency, mechanical simplicity, and the ability to utilize a relatively low speed power plant at high vehicle speeds. To fulfill these requirements, the existing 5-speed manual gear box was retained and two electromagnetic clutches were added. The clutches as well as a chain linking the electric motor to the main drive shaft comprise the merge unit. Control of the 5-speed transmission is through the gear shifter. Control of the clutches is described in the Master Controller section.

Advantages of using the Ford transmission are establishing direct mountings on the Escort chassis, use of existing gear shifting linkages, and utilization of the Ford Escort half shafts. The electric motor location is above and towards the front of the transmission.

6.1 Merge Unit

The merge unit (Figure 3) consists of two Mitsubishi Electric Electromagnetic (EM) clutches, one solid chrome-molybdenum steel shaft, two hollow, high-strength steel shafts, ball bearings, shaft seals, two steel sprockets, one silent chain, and a specially designed aluminum casing. The silent chain and sprocket used with the electric motor are located in a sealed compartment filled with gear oil for lubrication purposes. A Morse HV Silent chain was chosen for noise and strength reasons.

One of the clutches is attached to the ICE flywheel to transmit torque to the transmission input shaft in ULEV and HEV modes of operation. The clutch located in the bell housing is used to transmit electric motor torque in electric (ZEV and HEV) modes of operation. Both clutches will operate as a dry disc clutch would operate, i.e. with some "slippage" to place less stress on the merge unit shafts and give the driver more control. The clutch currents are controlled via a DC-DC convertor whose signal is derived from a potentiomemter linked to the clutch pedal.

6.2 Master Controller

The Master Controller enables the driver to select manually the HEV's mode of operation (a future modification wil allow an automatic selection option governed by gear shift position, vehicle speed and accelerator motion). At slow speeds where the electric motor is required for propulsion, the first, second, or reverse gears are used and the electric motor clutch can be activated, thus allowing the electric motor to propel the car. The ICE clutch will remain open. When driving at medium to higher cruising speeds with the ICE operating and the gear selector in third, fourth, or fifth gear, the ICE clutch can be activated and the electric motor clutch will remain open. At any operation, the clutch pedal will act as clutch operation control. Simultaneous operation of both drives is possible to allow the electric motor to assist the ICE in accelerating passing, to provide regenerative braking, or to recharge the battery during ICE operation.

The speed of the vehicle is controlled by means of the accelerator pedal. In electric modes of operation, the pedal position is translated into a variable voltage by means of a potentiometer. This control voltage is fed to the motor controller as a speed reference signal. The ICE can retain its mechanical linkage to the same accelerator pedal since fuel injection is inoperative unless the ICE is commanded to run.

6.3 Instrumentation

ELANT III retains the instrument panel provided with the Ford Escort. None of the existing components are removed or changed, but new displays will be added. These include: two tachometers, a motor temperature indicator, and battery meters (current, voltage, and energy).

7. SAFETY FEATURES

Throghout our design, fabrication, and assembly we paid special attention to possibilities of malfuctions and accidents. The following summarizes some of the many measures we took. Please also consult Figure 5.

7.1 Braking System

The braking system of ELANT uses a combination of Ford Escort wagon & Escort GT components. The standard front disc brakes from the Escort wagon were retained, but to increase braking effectiveness, an Escort GT rear disc brake assembly replaced the Escort wagon rear drum brake assembly. The brake master cylinder and brake booster from the Escort wagon were retained. The proportioning valve or brake bias valve has been set to permit more brake fluid pressure to act on the rear disc brake calipers. The brake booster receives vacuum assist from a vacuum pump-tank assembly powered by a 12 volt electric motor.

7.2 Body, Suspension, and Wheels

The existing body of the Escort wagon was not modified, but a rollcage designed especially for stock car racing was made to order by its manufacturer and installed by us. All standard Ford Escort seat belts were removed, and the front belts were replaced by Deist Safety racing-quality five-point harnesses. The replacement harnesses are fastened to the standard front anchors and to the roll cage.

The suspension was modified to handle the additional weight from the electric propulsion system, mostly the nearly 900 lbs of batteries. Suspension modifications include the replacement of all four coil springs and struts. All coil springs are stiffer than the standard Escort coil springs in order to cope with the additional weight as well as to reduce brake "dive" and body roll. Tokico twin tube struts have replaced the stock struts. The standard front and rear sway bars are retained.

The manual steering system of the Ford Escort was not modified. Although a power assist feature would be preferable, the additional weight from a power steering system was a drawback that discouraged implementation.

The tires on the Escort wagon have been replaced by 185/70/R13 Goodyear Invicta radial tires since they have a low rolling resistance and are capable of handling addition weight. The wheels were exchanged for American Eagle Series 240 wheels in a 13" x 5.5" size. These wheels were determined to be among the lightest wheels available.

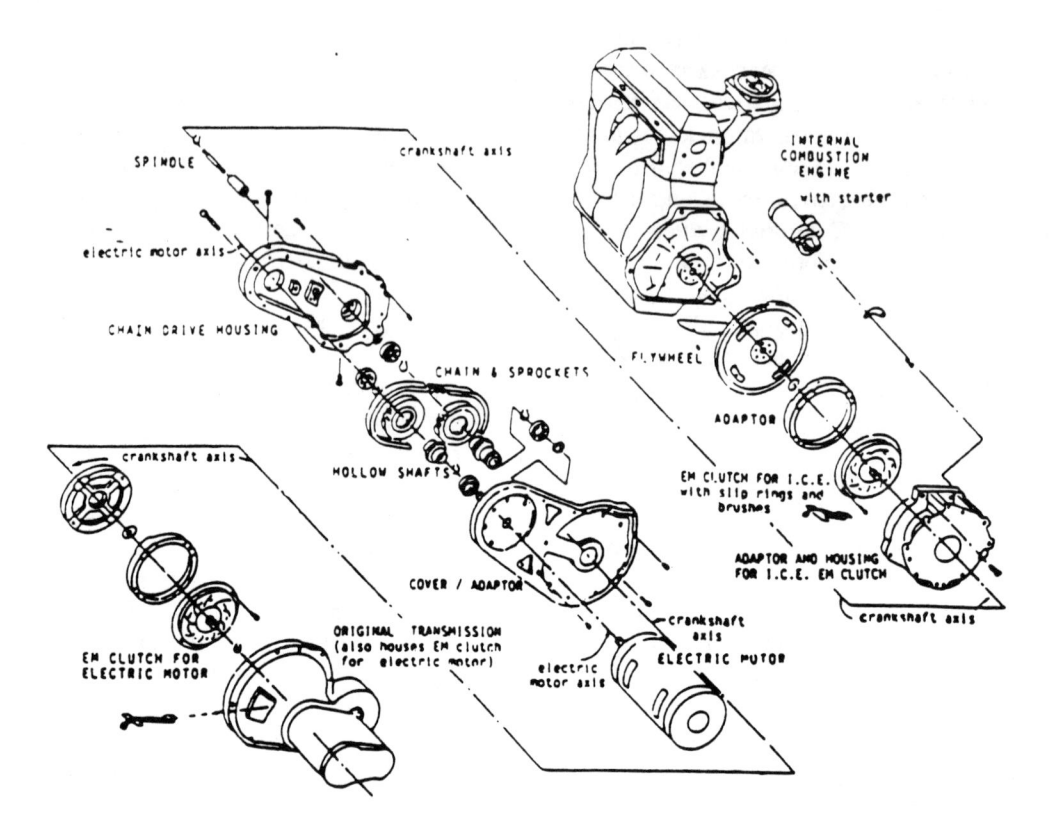

Figure 3: EXPLODED VIEW OF MERGE UNIT

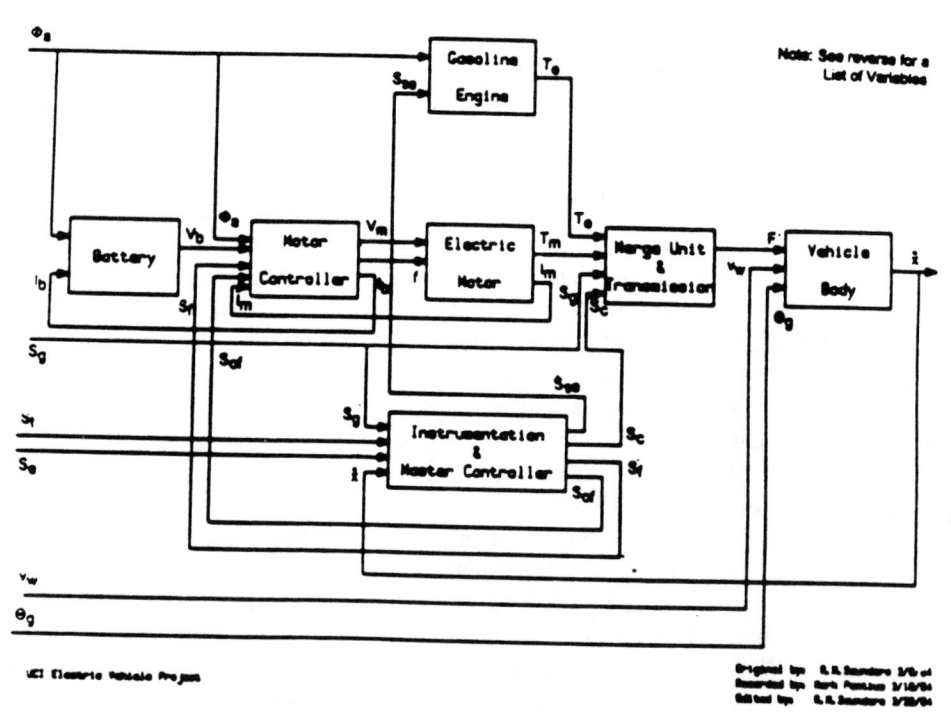

Figure 4: SIMULATION DIAGRAM FOR ELANT III

7.3 Fire Protection

A halon fire extinguisher designed for racing use was installed near, and accessible from, the driver's seat. This 1301-halon unit has a five pound capacity and its contents are harmless to the occupants. Discharge valves are located under the engine hood, in the passenger compartment near the front seats, and in the rear near the battery array.

The battery box is mechanically vented to the exterior of the vehicle to permit hydrogen developed during battery charging to escape. Industrial type safety stop buttons are mounted on the exterior of the car on both sides to permit quick isolation of the main battery array. The disconnect switch used is a high capacity, bakelite encased circuit breaker trpped by a solenoid or by hand.

7.4 Ergonomics, Manufacturability, and Maintainability

Ergonomics considerations heavily influenced safety related design aspects. The dashboard layout of instruments was left in its conventional configuration as much as possible. Operation of clutch, gear shift, and accelerator retained their traditional functions. Fail safe devices are incorporated in the motor speed controller, and the batteries are fused in subgroups. The battery disconnect system employs readily accessible and unmistakable palm push-buttons mounted both outside and inside the car.

The body was left intact, but the suspension was strengthened. The brakes were fortified (disc brakes in rear, and electric vacuum system for the power braking). The rollcage was tailor made by a stock-car racing supplier. We feel confident that we have a crashworthy vehicle. Only the battery box could leak some acid in a massive collision followed by rollover.

The UCI team's goal was not so much an entirely new vehicle, but the conversion of a standard commuter vehicle. Ideally such a vehicle would be available from a car manufacturer at somewhat reduced cost by leaving out the engine, including smaller components related to engine cooling, exhaust, and fuel supplies, while strengthening the suspension. Modifying such a vehicle given a ready made merge unit and control system could make conversion to HEV operation a profitable business for small, and large industry alike, especially if most additional parts and materials are readily obtainable.

One cautionary note must be sounded in vehicle conversion: When an existing car engine is removed and replaced by another, the tracing of the control wires from the dashboards and fuse boxes to the engines is a most difficult task. It is recommended that the harnesses and switch assemblies be exchanged in their entirety.

If maintainability is of great importance to a production vehicle which must win user acclaim, it is even more important in a test and research vehicle. All parts of the drive train, battery, and control systems were designed to be readily accessible. This is exemplified in the location chosen for the motor control unit (behind the driver's seat). Also, only conventional materials and parts were chosen for the vehicle modifications. Most components in the vehicle are "off the shelf" type components. Specially built items such as the merge unit were designed to be simple, inexpensive to build, yet strong enough to withstand the design loads. Power train components such as the induction motor and the Ford transmission or Geo Metro engine have been produced in large quantities in the last three years and have proven reliable. The battery selected is adaquate for the HEV Challenge, but batteries of higher specific energy (kW-hr/kg) are definitely needed to improve the performance of electric and hybrid-electric vehicles which rely on storage batteries.

8. PERFORMANCE

Vehicle performance was simulated in various ways. An overall systems diagram (Figure 4) was developed to clarify input/output and interface relationships. The computer program DIANE was found useful in establishing battery power requirements, and tests were carried out to determine true capabilities of the AC induction motor (e.g. full load efficiency of 87 percent) and the traction battery pack.

9. THE UCI HEV TEAM

The HEV Challenge came to the University of California at Irvine in 1992 when we were informed of our selection as one of the Ford conversion teams in response to a proposal spearheaded by a then junior/senior in mechanical engineering, Don Campbell. Because most of the interested participants were out of town on summer interships, we really did not get started until fall 1992. Since then the team membership has grown from 21 to 45, but each quarter we lose and gain individuals due to graduation and changes in study plans each quarter (three times a year). This has made planning a really difficult task for the project manager, Jun Lee (now a senior electrical engineering student). The UCI team struggled valiantly to complete the conversion of the Ford Escort wagon in time for the 1993 HEV Challenge, but assuming that it was not proper to complete the work at the proving grounds in Dearborn, decided to only send observers there. (Incidentally, the Challenges also coincide with our final examination and commencement dates.)

The work on electric vehicles at UCI is carried out as both a class project for credit (ECE 198, Electric Vehicle Project) and as an activity of the Electric Vehicle Association at UCI (EVA-UCI). The team is divided into five groups: 1) Battery Group, 2) Motor/controller Group, 3) Mechanical Group (internal combustion engine, body, and suspension), and 4) Instrumentation Group, and 5) Publicity/Fund Raising.

While an attempt was made to even out the work load, the battery group was most active in the early stages and the mechanical group heavily burdened toward the end. One of the most frustrating aspects of the project was the difficulty we experienced with machining parts of the merge unit. The problems stemmed from the use of machine tools which had to be shared with other classes and slow delivery times of those jobs we then had to farm out--all in all not an unfamiliar scene in industry, but a real eye opener for us students.

ACKNOWLEDGMENTS

For efforts in helping the students put the project and the car together we thank:

Jim Cook of Electra-Gear
Dave Eisenberg of UC Irvine (Fleet Services)
Bharat Patel of Ford Motor Co.
Kyle Petrich of Ford Motor Co.
Larry Marquiss of Tuttle-Click Ford
Theodore Sahli of Independent Ferrari Service of Costa Mesa
Robert M. Saunders and Keyue Smedley, professors at UCI
Special thanks to Professor Roland Schinzinger for his assistance throughout the entire project.

The team is most grateful for the support of sponsors listed on the right. They funded the entire 1992/1994 project in the amount of $13,500 in cash and $21,500 in kind: (including the Escort Wagon valued at $10,000). Also included in the $35,000 are $8,500 in travel and vehicle transport costs.

Ford Motor Co.
Emerson Industrial Controls
Electragear Motors (Regal-Beloit)
Flour-Daniel
Toshiba America
Western Digital Corp.
Los Angeles Department of Water & Power
California Air Resources Board
Energy Scholars Program (Department of Energy)
Siemens
Johnson Controls
Robert Bosch Corporation
United Technologies Automotive
American Eagle Wheel Corporation
Saddleback Radiator
Onan Corp.
ALABC
Borg Warner

Vicor Express
Tuttle Click Ford
Goodyear Tire
Doug's Towing
E.I. Dupont

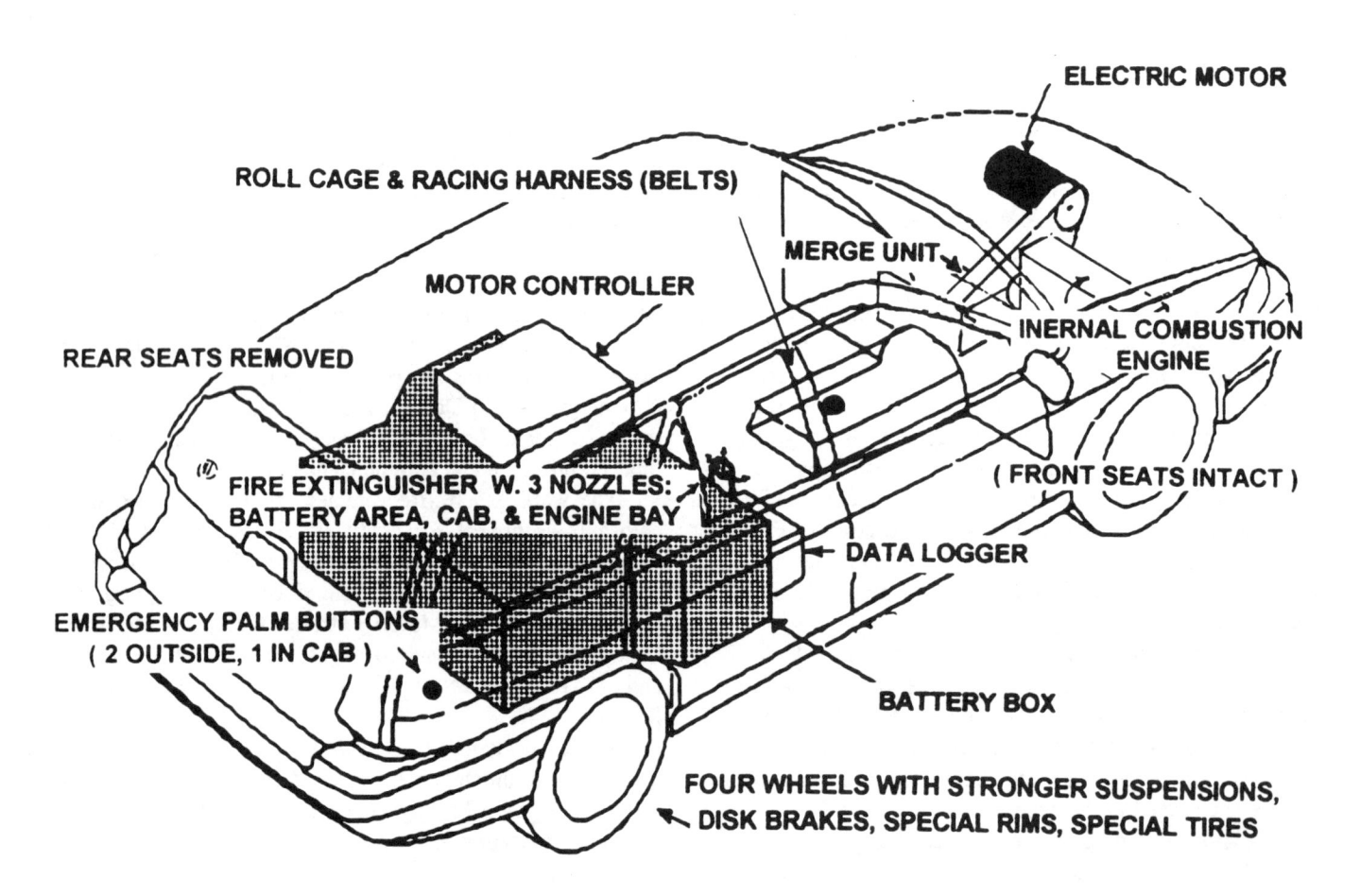

Figure 5: Modifications and Safety Measures

TABLE II: VEHICLE SPECIFICATIONS

VEHICLE TYPE: Front engine, front wheel drive, 2 passenger, experimental hybrid electric vehicle; converted Ford Escort wagon.

MOTOR
Type: 3-phase AC induction motor with R-rated insulation
Dimensions: 9" in outer diameter x 14.5" length
Power: 15 hp rated, 40 bhp @ 5000 rpm
Torque: 60 lb-ft @ 0 RPM
Redline: 6,000 RPM
Weight: 70 lbs
Controller: Emerson Eclipse 2A controller with PWM inverter

ENGINE
Type: GEO Metro/Suzuki G-10, Inline-3, aluminum block and head, Displacement: 997 cc
Compression ratio: 8.5:1
Emissions controls: catalytic converter with automatic preheater, feedback fuel-air-ratio control
Fuel Delivery system: Nippon Denso throttle body fuel injection
Valve gear: OHC, hydraulic lifters
Power: 50 hp @ 5700 RPM
Torque: 58 ft lb @ 3300 RPM
Redline: 6000 RPM
Weight: 140 lbs

DRIVETRAIN
Transmission/transaxle: Ford Escort 5-speed manual with powertrain merge unit
Clutch: dual electromagnetic powder
Final-drive ratio: 3.62:1

Gear	1	2	3	4	5
Ratio	3.24	1.84	1.29	0.97	0.73

Weight, transmission: 70 lbs
Weight: 70 lbs

CAR DIMENSIONS AND CAPACITIES
Wheelbase: 98.4 in
Track, F/R: 56.5/56.5 in
Length: 171.3 in
Width: 66.7 in
Height: 53.6 in
Ground clearance: 10.5 in

Curb weight: 3800 lb
Weight distribution: F/R 48/52
Fuel capacity: 12.0 gal
Oil capacity: 3.7 qt
Water capacity: 4.2 qt

CHASSIS/BODY
Type: unit construction
Body material: welded steel stampings

INTERIOR
SAE volume, front seat: 52.2 cu ft
Luggage space: 21.0 cu ft
Restraint systems: 5-point harness

SUSPENSION
Front: Strut type, independent front drive with upper strut mounted coil springs
Rear: strut type, independent twin trapezoidal links with trailing links and upper strut mounted coil springs

STEERING
Type: rack and pinion
Turns lock-to-lock: 4.3
Turning circle curb-to-curb: 31.5 ft

BRAKES
Front: 9.25x 0.87-in vented disc
Rear: 9.88x 0.35-in non-vented disc
Power assist: electric vacuum pump

WHEELS AND TIRES
Wheeels: 5.5 x 13 in
Wheel type: cast aluminum
Tires: Goodyear Invicta, P185/70/R 13
Inflation pressure: F/R 32/32

PERFORMANCE (PRELIMINARY)
Zero to 45 mph 14.0 sec
Top speed 75+ mph
ZEV mode 40+ miles
ULEV mode 400+ miles, 45+ mpg.

University of California, Santa Barbara
Final Technical Report

1994 Hybrid Electric Vehicle Challenge

May 9, 1994
Steve Barskey

ABSTRACT

The 1993 UCSB hybrid electric vehicle has been re-designed with improvements in performance, and economy in order to compete in the *1994 Hybrid Electric Vehicle Challenge*. The vehicle is powered by an ethanol (E100) powered internal combustion engine and two battery powered AC induction motors. The two drive systems are coupled in parallel via a road link, and the vehicle can operate in ethanol only, electric only, or in hybrid mode. The vehicle features efficient, clean use of renewable fuel; efficient, high voltage AC propulsion; regenerative braking; recyclable lead-acid batteries; recyclable steel frame; and a lightweight fiberglass body.

INTRODUCTION

More stringent vehicle emissions requirements and increasing environmental hazards force large automobile manufacturers to look for new technologies in automobile production that will minimize dangerous effects on the environment. One potential solution, the hybrid electric vehicle (HEV), incorporates both a conventional heat engine and an electric power plant. This type of vehicle offers the driver performance and range comparable to conventional gasoline powered automobiles, as well as a lack of emissions only possible with electric vehicles. With an HEV a driver does not have to sacrifice performance to enjoy the benefits of an electric vehicle.

To help promote the development of this relatively new technology, Ford Motor Company, the U.S. Department of Energy, and the Society of Automotive Engineers sponsored the first ever hybrid electric vehicle competition in 1993. Students from the University of California, Santa Barbara were invited to participate in the *1993 Hybrid Electric Vehicle Challenge*, along with students from 29 other schools across the United States. Schools chose between two entry classes, a conversion class and a ground-up class. Teams participating in the conversion class converted a Ford Escort station wagon into a hybrid electric vehicle. Teams competing in the ground-up class designed and built a hybrid vehicle from the ground up. Students from U.C. Santa Barbara participated in the ground-up class.

In June of 1994 the U.S. Department of Energy, Saturn Corporation, the Society of Automotive Engineers, and Natural Resources Canada will sponsor the *1994 Hybrid Electric Vehicle Challenge*. A new team of engineering students at U.C. Santa Barbara redesigned their existing ground-up class vehicle, and will compete in the 1994 Challenge. The modified vehicle includes an ethanol-powered internal-combustion engine operating in parallel with a pair of electric AC induction motors. The project is a ground-up effort to produce a practical, safe, efficient, commuter/service vehicle for the transportation of two passengers and a small amount of cargo.

DESIGN GOALS

The 1994 UCSB design team started with an HEV which was designed and built for the *1993 Hybrid Electric Vehicle Challenge*. The vehicle performed above average in many events, and poorly in others. Using the Challenge results as a basis, the 1994 design team determined several areas of the HEV which needed the most improvement. These areas became the design goals of the team, and guided the modification and construction throughout the year.

WEIGHT MINIMIZATION - One of the main goals of the 1994 design team was to reduce the overall weight of the vehicle. Originally UCSB's HEV was designed to be a lightweight commuter car, and each component was specifically selected for optimum performance with a light vehicle. When the vehicle was completed in 1993 it weighed almost 2800 lb, more than 50% heavier than predicted. Since each component was only expected to be used with an 1800 lb vehicle, the components were unnecessarily strained and the vehicle performance was crippled.

Rather than selecting entirely new components designed for use with a 2800 lb vehicle, the 1994 UCSB design team chose to reduce the weight of the existing vehicle to relieve the stress on each component. This could easily be accomplished by replacing the body, and modifying a few other areas of the car. Replacing the components would have involved purchasing a new internal combustion engine, motor controller, battery pack, and new electric motors. By reducing the weight of the existing vehicle, more time could be assigned to testing and optimizing vehicle performance rather than component selection and integration.

ECONOMY - UCSB's HEV placed below average in the economy event in the 1993 Challenge. One reason for this was because the vehicle was too heavy. The strain on the electric motors and ICE yielded meager vehicle efficiency and poor economy. Reducing the vehicle weight has remedied this problem.

Improper ICE tuning also affected the HEV economy. When the 1993 design team programmed the fuel injection system they did not have access to emissions testing equipment. The 1994 team utilized a chassis dynamometer and full emissions testing equipment to reprogram the fuel injection system. The fuel delivered to the injectors at any given load and engine rpm was optimized, ensuring efficient performance.

POWER TRAIN CONFIGURATION

The vehicle is a parallel configured hybrid featuring 4-wheel drive and optional electric-only operation. The parallel configuration was chosen because it serves to reduce component sizes since only a fraction of the total vehicle power is transmitted to the road by each propulsion system. In a series hybrid, all of the power flows through each component. Therefore, for the same vehicle power, systems are smaller and lighter in a parallel hybrid. The independent-axle parallel configuration eliminates the need to modulate between two propulsion systems with a differential. In addition, the losses of converting mechanical energy to electricity and back again are eliminated. The two power trains are mechanically independent, but can be coupled through the road if desired. This power train configuration also allows for easy design of a computer controlled, 4-wheel traction control system.

AUXILIARY POWER UNIT (APU) - The APU is composed of an engine from a 1992 Geo Metro, linked to a 5-speed manual transmission also from a Geo Metro. The engine and transmission are mounted in the front of the vehicle and directly drive the front wheels. Ethanol (E100) fuel was chosen for the APU due to its renewability, possibilities as a lower-emissions fuel, and to promote more development of its potential.

The Geo Metro engine displaces 1.0 liter (61 ci) and has three cylinders. When operating on gasoline it is rated as 36.5 kW (49hp). This engine was chosen to provide adequate power, but still provide high economy and low emissions. The necessary changes in fuel volume and ignition timing were considered in converting the engine to operate on ethanol fuel.

Compared to gasoline ethanol has a lower combustion rate, so it is necessary to deliver 30% more volume to the engine. The stock throttle body injector was used, and a fuel pressure valve was installed to deliver the fuel at much higher pressures. The stock engine control unit was replaced with a programmable engine control system from Motec Systems USA. This system allowed complete control and adjustment of the injection and ignition parameters. Chassis dynamometer and emissions testing allowed the parameters to adjusted for ethanol fuel. When running on ethanol the engine had a slightly larger maximum power output of 39 kW (52 hp).

Emissions Control Strategy - The emissions control strategy for the UCSB HEV is similar to that used in most current production small cars. Actively controlled fuel injection, including oxygen sensor feedback, exhaust gas recirculation, and a catalytic converter are used to control tailpipe emissions. A standard charcoal canister system is used to control evaporative emissions. The choice of a small engine also reduces overall emissions volume, and the use of ethanol fuel potentially could result in lower emissions. A chassis dynamometer and emissions testing equipment were also used to minimize dangerous exhaust emissions.

The catalyst substrate, which was manufactured and donated by Allied Signal Corporation, is formulated for use with ethanol and was prepared by Car Sound Exhaust of Rancho Cucamunga, CA. The stock exhaust gas recirculation and positive crankcase ventilation systems were retained from the Metro engine.

The stock Geo Metro evaporative emissions control system is used. It stores fuel vapors from the fuel tanks in a charcoal canister while the engine is not running. These vapors are then released to be burned in the engine when it is running under normal conditions. This prevents venting of such vapors to the atmosphere.

Fuel Consumption - The overall strategy for controlling energy consumption on the UCSB HEV was to connect both power sources to the road in as direct and efficient a manner as possible. Each system was then made as efficient as this design allowed. The car was designed to be small so that a small engine would be acceptable. A 5-speed manual transmission was selected because of its efficiency. There is no mechanical energy lost from a torque converter, which is found with an automatic transmission. A manual transmission also allows the driver to maintain the engine rpm in the most efficient range. The incorporation of a shift light would tell the driver exactly the optimum time to shift.

ELECTRIC MOTOR DRIVE TRAIN - The electric drive train consist of 12 deep-cycle lead-acid batteries; two 16.5 N·m AC induction motors; two 100 A motor controllers; and a double reduction, constant ratio chain drive system. The system was designed to provide low speed operation, adequate electric-only acceleration, additional power for hybrid mode acceleration, and power compensation as speed in hybrid mode.

AC Induction Motors - High voltage AC induction motors are used for superior efficiency, superior reliability, and low weight. The motors and controllers provide regenerative breaking and automatic differential action for the rear wheels. The controllers convert the 144 VDC input from the battery pack into three-phase AC to drive the motors, and determines the speed and torque output of the motors.

Chain Drive - A fixed 9.9 to 1 gear ration was chosen due to the exceptional torque characteristics of AC induction motors and the available 12,000 rpm of the particular motors used. This ratio is achieved through a double reduction of ANSI 40 roller chain passes. A double reduction was chosen for two reasons: the high ratio necessitated the use of two passes, and the use of an intermediate shaft located along the line of trailing arm rotation provides motor placement flexibility and zero chain stretch as the suspension moves.

Energy Storage - After many months of investigation, research, and deliberation, deep-cycle lead-acid batteries were chosen as the electrochemical storage means for the UCSB HEV. Although there are many newer battery technologies, no superior alternative emerged after looking at performance, price, and recyclability. The battery pack is designed to be easily upgradable when, and if, a higher performance, cost effective alternative is developed.

Electric Power Consumption - The strategy for the UCSB HEV electric power consumption was to develop as efficient a system as possible. The electric motors and controllers were chosen for their high efficiency. Regenerative braking also helps to conserve energy which would otherwise be wasted.

CONTROL STRATEGY

The control strategy for the UCSB HEV allows operation in any vehicle mode. In the hybrid mode power is supplied by both power trains, enabling a 4-wheel drive vehicle. In this mode, both the engine and the electric motors are providing power in parallel to the wheels. There is a zero emissions vehicle (ZEV) mode, in which only electric power is consumed. In the ZEV mode the APU does not run, since the parallel configuration does not require the APU to charge the batteries, and the car is driven through the rear wheels. There is also an ethanol only mode, in which only the engine is used to supply power. In this mode the engine drives the front wheels through the transmission, and the electric drive train is disconnected. The vehicle is front-wheel drive in this case.

STRUCTURE AND CHASSIS

The vehicle is built on a steel space frame, using existing front suspension components and a rear trailing arm suspension of original design.

FRAME - The tubular space frame has been designed of mild steel to allow for easy integration of the many systems of the hybrid electric vehicle. Frame stiffness was the primary design driver. The main structural group is a rectangular ladder frame base. This base has been designed to provide high stiffness while carrying most of the loads directly. Extending up from the ladder frame are triangular shock towers designed to hold the upper ends of the shocks and coils. These towers were designed to carry the full weight of the vehicle with the aid of struts and cross-bars extending between the towers. The front and rear roll structures define the remaining overall shape of the vehicle. Two side impact members on each side of the passenger compartment protect the passengers in the event of a side collision. The frame was analyzed using both conventional and finite element methods to ensure adequate stiffness without excess weight.

Where appropriate, the frame was modified to save weight and provide a more aerodynamic design. Several cross members in the front and rear of the vehicle, which were not providing any structural integrity to the frame, were removed to allow a narrower nose and tail. The forward roll bar was replaced with one that extends from the rear roll bar to the forward torque tube, allowing a narrower, more aerodynamic canopy. This is shown in Figure 1.

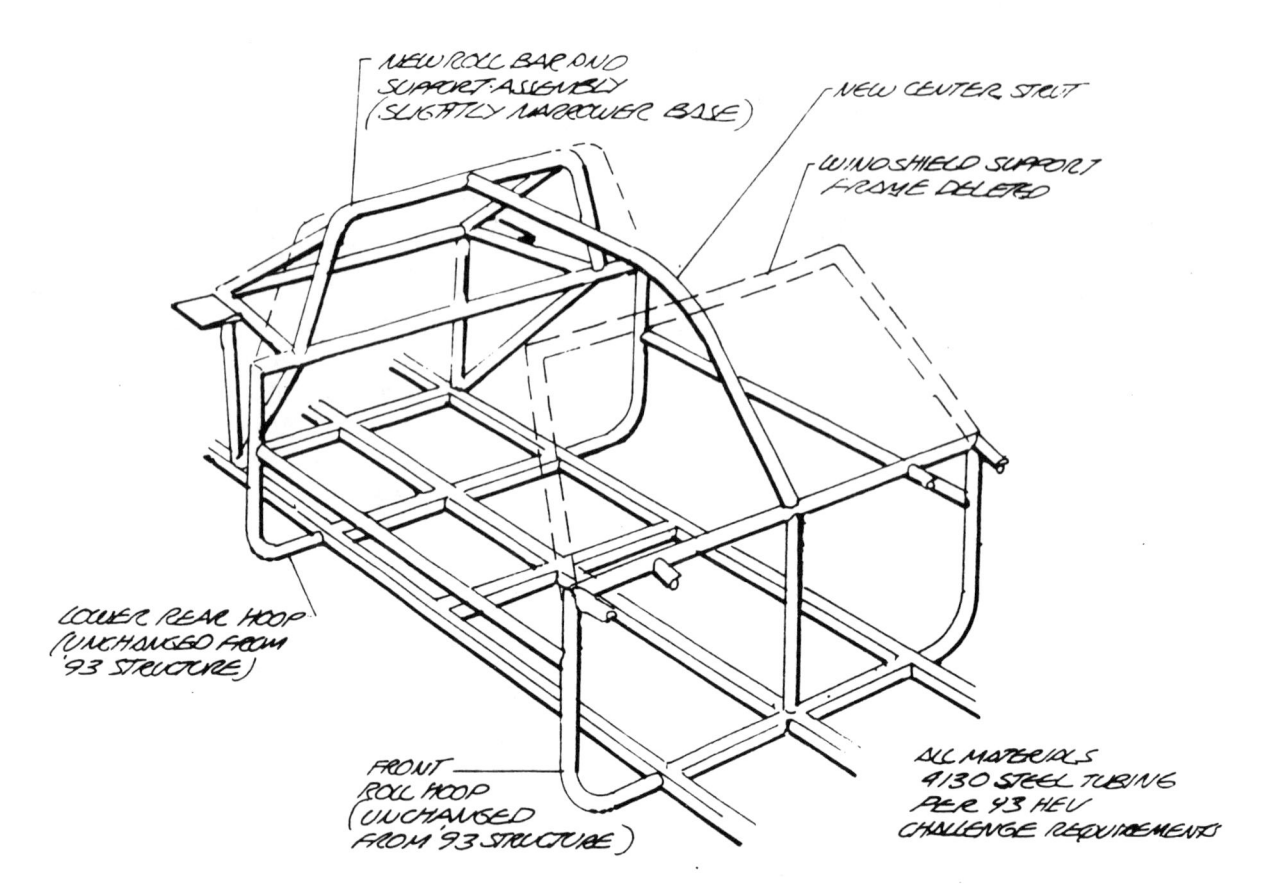

Figure 1. Roll cage modifications. *Removing the stock Mazda windshield and re-designing the forward roll bar allowed a narrower canopy, and thus a more aerodynamic body.*

CHASSIS - The vehicle utilizes brakes, steering, and front suspension from production vehicles. The hydraulic brake system uses Mazda 323 components, with the rear disc brakes from an Isuzu Impulse. The steering is a Mazda 323 rack and pinion unit. The front suspension is also a Mazda stock strut system. The rear suspension is an original trailing arm with coil-over strut design. The firewall is taken from the Mazda, but has been trimmed to save weight and space.

BODY AND ERGONOMICS

The vehicle is a two passenger coupe, with a storage area to secure a small amount of cargo. The styling concept is shown in Figure 2. The body was created from a foam/fiberglass sandwich construction. The interior and driver controls were designed for ease of driving, familiarity, and comfort.

BODY STYLING - The body shape was designed to have a minimal cross-sectional area, while keeping it as aerodynamic as possible. The open top roadster body from the 1993 HEV was replaced with a more practical, more aerodynamic coupe design. The coupe offers protection from rain and inclement weather, and offers improved aerodynamics and decreased wind drag, thus contributing to better economy. The canopy opens forward to allow the passenger plenty of room to climb in and out of the vehicle. The entire front and rear sections are each one piece, hinged to open away from the vehicle for ease of access to all vehicle systems.

BODY MATERIALS - The body is constructed of fiberglass with an epoxy matrix and a foam core. The fiberglass enables an ultra-light, stiff body that can be contoured around the space frame. The foam core will absorb energy on impact, helping to protect the passengers. The floorboard is made of a carbon-fiber honeycomb for superior strength and stiffness. The battery compartment and bulkheads are also constructed of fiberglass for strength, rigidity, light-weight, and non-flammability.

ERGONOMICS - The driver controls are familiar and have been carefully placed to allow operation by any driver. A custom dash board has been integrated with a conventional instrument cluster to facilitate familiarity and comfort. A stock steering wheel and conventional pedals were retained, as well as a stock gear shift. The cockpit displays were modified to reflect the hybrid nature of the vehicle and to provide the necessary information to the driver.

Figure 2. Body Design. *The body was designed to have a minimal cross sectional area, while keeping it as aerodynamic as possible. The entire front and rear sections are each one piece, hinged to open away from the vehicle for ease of access to all vehicle systems.*

MANUFACTURABILITY

The use of mild steel for the frame provided an easily machinable and formable medium of which to construct the frame. All machining and welding operations were performed by students at the UCSB Mechanical Engineering Department student machine shop. Components purchased and integrated into the vehicle were common, off-the-shelf components.

If the vehicle were put into production several changes would be made. The space frame design would be modified for compatibility with automated manufacturing processes. The rear trailing arms would be constructed of steel stampings rather than welded and machined tubes. The body panels would be produced in molds with sheet molding material. Finally, the fuel injection algorithm developed with the programmable system would be incorporated into a less expensive production system.

CONCLUSION

The design and construction of this HEV has been an educational experience for all involved. Many of the lessons learned and the experience gained would not be possible in the classroom. This project has provided an invaluable opportunity to sharpen skills in working with others, meeting deadlines, and communicating with companies.

The UCSB HEV was designed to demonstrate the practicality of hybrid vehicles in general, and specifically a parallel hybrid. This hybrid provides a practical, safe, efficient, and environmentally sound prototype vehicle which is a transition to lower—and zero-emissions vehicles of the future. It successfully demonstrates that vehicles of this nature need not compromise the styling, drivability, practicality, or performance common to current production automobiles.

The 1994 UCSB HEV team would like to acknowledge the invaluable assistance of our advisors, previous team members, sponsors, and the University, without which this project would not have been possible.

Cedarville College HEV Design and Development

Paul Armour, Nick Awabdy, Jeremy Barton, Andrew Bell, Nathan Bickel, Joseph Bower, Devin Cheek, Brian Cramer, Scott Crouthamel, Sara Denlinger, Kristine Deshetsky, Daniel Ernst, Scott Hartley, Brandon Kaercher, William Jariga, Matthew Lucas, Kevin McDermott, Greg Meek, Kendal Noller, Dale Radcliff, Andrew Shearer, Pamela Sorg, Adam Ulery, Mark Utz, Timothy Woodward

ABSTRACT

We converted the Saturn SL2 for the 1994 Hybrid Electric Vehicle (HEV) Challenge. We concentrated on reliability, safety, manufacturability, economy, and emissions in our design. We have replaced the Saturn's original powertrain with a series configuration that is powered by a permanent magnet, brushless DC motor. An internal combustion engine, powered by methanol (M85), drives a generator to produce electricity for the traction motor and battery recharge. When necessary, proven lead acid batteries will also provide power. This paper discusses component selection, control strategy, vehicle modification, emissions strategy, and drivetrain design.

POWERTRAIN CONFIGURATION

There are two primary configurations for an HEV, series and parallel. These configurations are named for the circuit analogues they resemble.

The parallel configuration couples the auxiliary power unit (APU) to the drivetrain in a conventional manner. An electric motor and battery pack are coupled to the drivetrain to assist the APU during hill-climbing and acceleration. In a parallel configuration, zero emission vehicle (ZEV) mode can be obtained by shutting off the APU and using the electric motor to supply motive force. The power supplied by the APU and traction motor is combined similar to the way two voltage sources combine in parallel to provide more current.

The parallel configuration is shown in Figure 1.

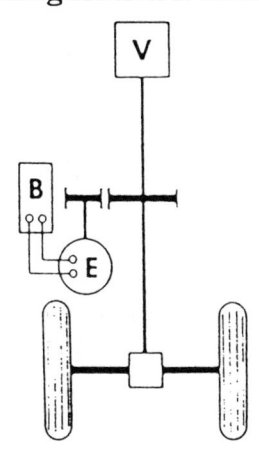

Figure 1. Structure of a parallel hybrid drive

The series configuration is not designed to couple the APU directly to the drivetrain, as in a conventional vehicle. Instead, the APU is connected to a generator that recharges batteries and powers the traction motor. In a series configuration, ZEV is achieved by shutting off the APU and operating solely on battery power. This configuration is called series because in some modes of operation power is transferred from the APU to the batteries and then to the electric motor. The power passes through each component in the vehicle, similar to the way current passes through each component in a series circuit. The series configuration is shown on the next page in Figure 2.

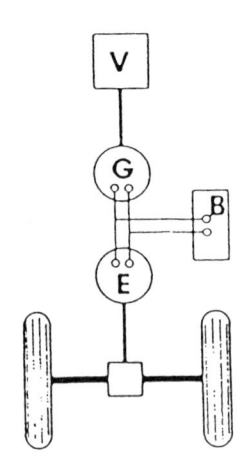

Structure of the series hybrid drive Figure 2.

We have chosen the series configuration, largely because we believe that it has the best abililty to supply both the ZEV and the ultra low emission vehicle (ULEV) modes. We do not want the vehicle to exceed ULEV standards in any mode. In the series HEV configuration, the APU supplies power to the car only when it is needed. When power is not needed to run the vehicle or to recharge the batteries, the APU is shut off. Operating the APU at a constant speed while generating power (load leveling) provides gains in both efficiency and emission control.

The series design is somewhat inefficient. When the APU is being used to charge the batteries, the retrieved by the traction motor for conversion back to mechanical energy. None of these conversions are 100% efficient, so we expect losses. Therefore, we plan to use the generator as the primary source chemical energy stored in the fuel must be converted to mechanical energy by the APU and then to electrical energy by the generator. This electrical energy is then stored in the batteries andof power for the vehicle. This will send the least amount of power to the batteries, so that their storage and retrieval losses will be minimized. The power normally spent by operating the APU at inefficient levels will be minimized by running the generator at the APU's optimum efficiency. This splits the generated power between the traction motor and the battery recharge.

AUXILIARY POWER UNIT

Engine - Selecting an APU for the Saturn was not an easy task. Next to the issue of series or parallel configuration, it was the most debated issue of the project. We considered many alternatives: the two stroke engine, rotary engine, IC engine, and the gas turbine. A chart containing some of the relevant characteristics is shown in figure 3 shown below.

Solution Selection Worksheet

Category Weights	5	5	4	4	4	3	Total
Solutions	Weight*	Power Output	Fuel Efficiency	Complexity*	Emissions	Risk*	
	5- least 4- 3- 2- 1- greatest	5- greatest 4- 3- 2- 1- least	5- greatest 4- 3- 2- 1- least	5- none 4- 3- modest 2- 1- most	5- lowest 4- 3- 2- 1- highest	5- lowest 4- 3- 2- 1- highest	Sum of all the Ratings
Turbine Generator Set	4	4	3.6	3.3	3.7	2.3	89.3
VTEC	3.1	4	3.7	2.9	3.9	3.5	88
Rotary Engine	3.7	3.3	2.9	3.4	2.4	2.9	78.5
Justy Engine	3.3	2.8	3	3.8	2.9	3.8	80.7
Honda Motorcycle Engine	4	3.8	3.1	3.3	2.5	3.3	84.5

*WEIGHT- Needed Fuel/APU Weight   *RISK- Time/Unknown/Conversion to Alcohol
*COMPLEXITY- Installation/Available Space

Figure 3

In spite of its high power density, we discarded the two stroke engine because historically, it has been uncivilized in the area of hazardous emissions. The unburned hydrocarbon output of the two stroke engine is generally five times greater than that of a comparable four stroke engine due to the fuel scavenging that takes place in the two stroke cycle (1). The two stroke engine mixes lubricating oil with the fuel, increasing unburned hydrocarbons even more. Finally, small displacement two stroke engines consume more fuel than four strokes. Recent developments in engine design have decreased fuel consumption, but the technology is not yet readily available in production engines.

The rotary engine also has high power density characteristics, but it, too, is plagued by high levels of harmful emissions. Again, recent developments have led to rotary engines with better emissions characteristics but these are unavailable or difficult to find. We were unable to identify a source for rotary engines in the 30-40 kW range.

Four stroke spark ignition (SI) engines have a number of pros and cons. Compared to the engines discussed above, they are heavy and large for equivalent power outputs. However, the four stroke SI engine has been widely used as a power source for over a hundred years. During this period of time, it has improved greatly in design. As a result, the four stroke engine is far more refined than its two stroke and rotary cousins. It now has good fuel efficiency and produces relatively low levels of harmful emissions. Burning alcohol in a four stroke engine allows the use of higher compression ratios and also increases the volumetric efficiency.

Gas turbines have power density values in the same range as two stroke and rotary engines. However, unlike the two stroke and rotary engines, the gas turbine has very low hazardous emission levels. Above 1950° F, fuel and air react to form significant amounts of nitrous oxides. The gas turbine operates below this temperature. The air/fuel ratio is much greater than stoichiometric, so combustion is virtually complete. This reduces the formation of carbon monoxide (CO) and hydrocarbons (HCs) (2). Use of a recuperator increases the fuel efficiency of the turbine by approximately 15%. Although turbines have been used in automobiles in the past, they are difficult to couple to the drivetrain because most of them operate at 90,000 rpm or above. Lubrication and gearing problems are difficult, though not impossible to solve in a conventional automotive application. However, in a series HEV configuration, the APU powers a generator and is not mechanically connected to the drivetrain. Most generators have no problem operating at high rotational velocities and can be downsized in both weight and volume when used at higher speeds. The aero industry regularly uses gas turbines in a variety of applications. These turbine powered generators are used in both portable and permanent installations, and have become so common that gas turbines can be purchased already coupled to a generator.

After weighing our options, we decided that the turbine engine was the best choice. We took steps to find a suitable turbine and were able to locate one, but could not arrange to get it for this year's competition. The risk involved in integrating the turbine into the vehicle was also considered to be very high. Along with the benefits of the turbine came many unknowns. Some of the problems we anticipated were heat and sound. The noise must be muffled below the 80 dB allowed, as stated in the competition rules. The flow rate of the exhaust is certainly much greater than that of a conventional four stroke SI engine and would use an exhaust pipe larger than the 1.5 in. diameter pipe required by the HEV challenge rules.

The second choice was the four stroke Honda VTEC engine. It has a displacement of 1500 cc and produces 90 hp at 6300 rpm. In the Civic, the VTEC has achieved 50-60 mpg and is designed to comply with current California emission standards. Due to increased volumetric efficiency when using alcohol, we estimate that the engine should now produce close to 100 hp at 6300 rpm (3).

Fuel Selection - We researched alternative fuels and found that the differences between E100 and M85 were not significant in relation to performance parameters. M85 has a higher density than E100 (.810 g/cm^3 to .791 g/cm^3) High density increases fuel economy in SI engines (4). This was one reason why we chose M85. Another reason was because the materials and processes used to produce the two fuels are different. E100 is produced from corn and sugar crops, while M85 is made mostly from natural gas. Considering the relative abundance and cost of raw materials, we decided that Methanol was a better alternative to gasoline (5).

ERGONOMICS

We made no major visible changes to the dash display. We wanted to keep the Saturn looking as close to stock as possible. We replaced the radio with an ammeter, energy meter, and a voltmeter. This placement allows optimum visual contact for both the driver and the test engineer.

Because the transmission with the pickup for the speedometer was removed, we designed a new way of measuring the vehicle speed. We calculated the speed by using a hall effect transistor in the traction motor. The signal is passed from the transistor through the CPU and converted to a signal that the stock speedometer can use. We added warning lights, which indicate high temperature in the battery box or APU, if the fan in the battery box is nonoperational, and if a fire is present in the battery box or under the hood.

MATHEMATICAL MODEL

Our main goal for the mathematical model of motion was to find the power required to reach a specified velocity within a specified period of time. These calculations were necessary for all competition events. It was crucial that the power calculations be derived for our Saturn vehicle. We found the mathematical model for the velocity of the car and related it to the coefficients of static rolling resistance and wind drag. We used the basic models of force and acceleration and then combined them to obtain the formulas needed to model velocity. Newton's third law indicates that the sum of the forces is equal to the mass of the object, times its acceleration.

$$[\sum F_x = ma] \quad -F_D = ma \quad (1)$$

We assumed that the drag force (F_D) was approximately equal to the sum of the coefficient of static rolling resistance (k_1) and the coefficient of wind resistance (k_2), times the square of velocity.

$$[Assuming] \quad F_D = k_1 + k_2 v^2 \quad (2)$$

According to theories of kinematics, we know that acceleration is equal to the derivative of velocity, with respect to time.

$$[Kinematics] \quad a = \frac{dv}{dt} \quad (3)$$

From the previous equations we compiled a general equation relating the derivative of velocity with the coefficients of static rolling resistance and wind drag. We did this to produce a useful formula for velocity.

$$-\frac{1}{m}(k_1 + k_2 v^2) = \frac{dv}{dt} \quad (4)$$

Finally, the integral of acceleration was taken to find an equation for velocity

$$\int_0^t -\frac{1}{m} dt = \int_{v_0}^v \frac{dv}{k_1 + k_2 v^2} = \int_{v_0}^v \frac{dv}{k_2 (\frac{k_1}{k_2} + v^2)} \quad (5)$$

$$v = \sqrt{\frac{k_1}{k_2}} \tan\left[-\left(\frac{\sqrt{k_1 k_2}}{m}\right) t + \arctan\left(\sqrt{\frac{k_2}{k_1}} v_0\right)\right] \quad (6)$$

Once we had determined the mathematical model for velocity, we found the coefficients of static rolling resistance and wind drag. In order to do this, we ran a Saturn coast-down experiment, recording (in 5 mph increments) the time it took to coast from 65-0. We then used this data in a computer program, which varied both k_1 and k_2 to minimize the error between the actual and calculated values of velocity, determining a theoretical curve for the experimental data. This program calculated the coefficients of static rolling resistance and wind drag for any road surface, any tire pressure, and any vehicle mass. (These two coefficients were then used to evaluate the power output required to achieve a particular velocity in a given amount of time.) Figure 4 on the next page summarizes these results.

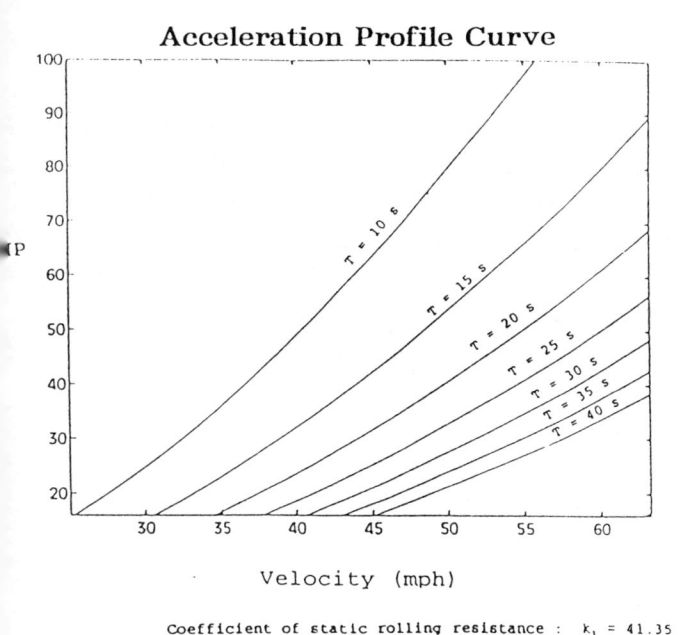

Acceleration Profile Curve

[Graph showing velocity (mph) vs P, with curves T = 10 s, 15 s, 20 s, 25 s, 30 s, 35 s, 40 s]

Figure 4
Coefficient of static rolling resistance : $k_1 = 41.35$
Coefficient of wind drag : $k_2 = 0.00705$

ELECTRIC MOTORS

Choosing the correct electric motor is vital in a series configuration. The motor is running every time the vehicle moves. The electric motor must endure and perform without error in every HEV Challenge event. We considered four types of motors: brushed DC, series wound; permanent magnet, brushed DC; permanent magnet, brushless DC; and AC induction.

Our primary selection criteria were efficiency and reliability. Because the majority of the HEV Challenge is performed at relatively constant speeds, we chose the permanent magnet, brushless motor for its efficiency.

The rotor is made of rare-earth magnets and the stators are coils imbedded in the motor housing. This eliminates friction and arcing with the rotor. Since heat is concentrated in the motor housing and not the rotor, the heat can be removed more easily which increases the efficiency of the motor. This motor provides two or three times more torque than a brushed DC motor, which is comparable in size and weight. This motor has a high power density and efficiency peak and it is best used at constant speeds. However, it has a few downfalls. First, it has three times the power circuitry of a brushed DC; so, the controller is larger and more complex. Furthermore, the sophistication of this motor makes it the most expensive.

We selected the Unique Mobility SR218/4.5FW motor and the CA20-300 controller, which is rated at 43 hp with a peak torque of 143 N-m from 0 to 2000 rpm and top rpm of 8000. Using a gear ratio of 4:1, the motor/controller system is approximately 95% efficient at 35-45 mph. From the above, we chose the Unique, SR218/4.5FW motor and the CA20-300 controller. The reason we chose the motor from Unique was because of their past reliability, high quality product, and customer service. The graph below in figure 5 shows the torque and efficiency curves of the Unique motor.

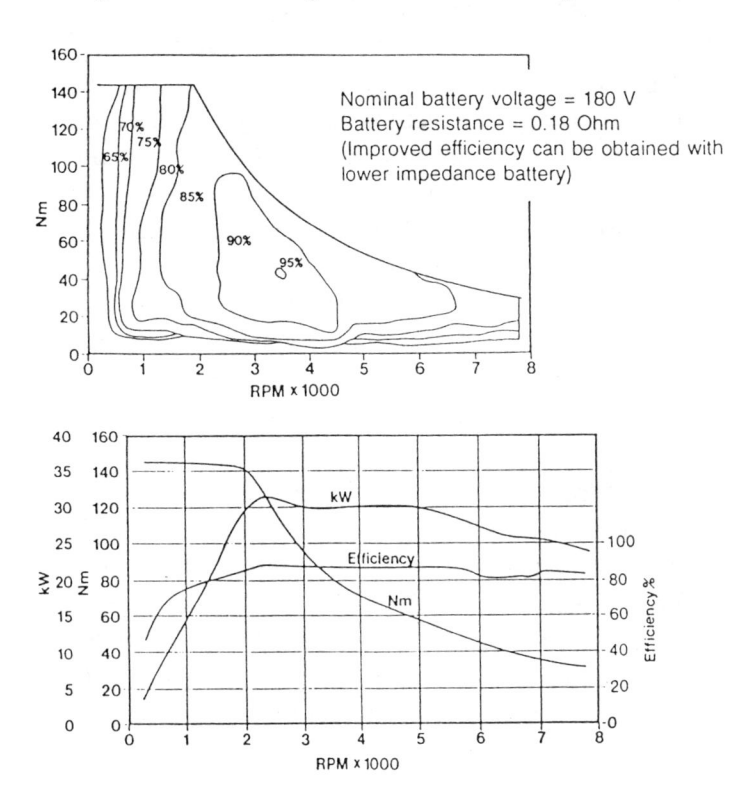

[Torque vs RPM contour plot with efficiency regions 65%, 70%, 75%, 80%, 85%, 90%, 95%]

Nominal battery voltage = 180 V
Battery resistance = 0.18 Ohm
(Improved efficiency can be obtained with lower impedance battery)

[kW, Nm, and Efficiency vs RPM plot]

Figure 5

GENERATOR DEVELOPMENT AND CONTROL

Our first goal was to develop a control strategy to share power between the generator and the batteries under certain conditions, such as acceleration and hill-climbing, while minimizing the amount of power extracted from the batteries. Our second goal was to develop a strategy to recharge the batteries while providing power to the motor.

Sundstrand donated a DC-9 aircraft alternator, rated for 40 kW at 6000 rpm. The generator is designed to output 270 Volts dc. We have modified the generator field current so that the generator voltage output is 200 Volts dc for the traction motor. The 200 Volts provides a bus to charge the batteries and power the motor. However, because of the lowered output voltage and a current limit of 150 Amps dc, the 40 kW generator is now rated at 30 kW. Therefore, with 95% efficiency, the generator is capable of outputting 28.5 kW. The generator can output 60 kW for 5 minutes and 80 kW for 5 seconds at 270 Volts, which is 43.2 kW for 5 minutes and 57.6 kW for 5 seconds at 200 Volts. Under normal cruising conditions, the generator output current is limited to 125 Amps dc. This 125 A is sufficient to charge the batteries and power the motor under most conditions. For acceleration and hill-climbing, a circuit built into the accelerator control circuit increases the current capability to 180 A. Our tests show that the generator can output 180 A for a short amount of time without overheating the rectifier diodes. However, to avoid overheating, a safety sensing circuit limits the current to 125 A when temperatures rise above rated levels in the rectifier. The generator operates at roughly 72% of its maximum output, causing the 40 kW generator to produce an optimum 28.5 kW.

When the vehicle is cruising at a normal, steady state speed of 45 mph, the generator provides all the power required by the motor with enough to charge the batteries. During acceleration and hill-climbing, the generator provides roughly 2/3 of the power, and the batteries provide the rest. The batteries are used for acceleration and hill-climbing only.

We obtained a generator control system from Sundstrand. They previously designed a similar type generator control unit (GCU) for an M1-tank, which, with slight modifications, yielded the design we decided upon. We tested and modified the GCU to better meet our requirements by adjusting the output resistance and the cooling flow to raise the output current to a maximum steady-state level of 125 Amps dc.

A full-wave bridge retifier rectifies the generator AC output to DC. A control unit senses the DC output and modifies the generator field current to produce a DC output of 200 Volts. During acceleration and hill-climbing, the control unit senses the increased demand for current and modifies the generator field current to sag the generator output voltage below the battery voltage. This allows the batteries and the generator to act as a combined power source, meeting motor demand.

Because the field current is only 5 Amps dc, we were able to build the controller entirely of light, inexpensive components.

DRIVETRAIN

We identified five options for the HEV drivetrain: gear reduction and differential, 6:1 differential, standard transmission, continuously variable transmission, and 4:1 differential.

A standard transmision would boost the acceleration time of the Saturn compared to a single speed configuration, because more torque could be transmitted in the lower gears. Although this option seemed favorable, the transmission did not fit after we put the APU in. The standard transmisison, one of the heaviest drivetrain configurations, would also make the vehicle overweight.

We have the least information on the continuously variable transmission (CVT), which makes it a less feasible option. Much like the standard transmission, the CVT is overweight and oversized in relation to the Saturn HEV.

We originally specified the gear reduction and differential option to use a single speed reducer. The input shaft and output shaft of a reducer are usually in line or 90^0, which makes integration difficult. A speed reducer also adds significant weight and cost.

We could accomplish the primary reduction with the use of pulleys or sprockets to take the place of the gear reducer. We ruled out high velocity chain as an option because of the complications of an enclosed chain drive with a special lubrication system for high speeds. This would add to cost and weight. We could accomplish the reduction through a Gates Polychain pulley system. The use of this high speed and high torque kelvar belt would not require special lubrication and would satisfy the weight requirements.

The results of the mule tests showed that a final drive ratio of approximately 4:1 was sufficient. Fom this, we decided to go with a straight 4:1 differential. We obtained a BMW 320i differential, which has a 4:1 ratio. This differential is the lightest and smallest of all options available. We also chose this design because it was simple and proven.

SUSPENSION

We decided to change the rear suspension in order to compensate for the batteries in the trunk. The stock Saturn coil springs were replaced with heavier duty springs which are rated at 3450 Newtons. These new shocks have the same damping ratio as the ones they replaced. This allows the overall handling and performance of the vehicle to remain the same.

BRAKING

Since safety and fuel economy were big considerations, we decided to alter the braking system. We designed the braking system to meet and exceed the braking requirements for the Saturn. We used special low drag drum brakes with larger aluminum drums in the rear. We reduced the frictional losses from the brake shoes rubbing by using a high-lift system.

MOUNTS

The mounts must support the weight of the components as well as withstand the forces the components exert on the mounts. The mounts must reduce vibrations from the APU and withstand the torque from the electric motor and APU. In order to account for unknowns, the materials we used incorporated a safety factor of no less than 2.5. We chose aluminum 6061-T6 because of its low weight and yield stress of 40 ksi. Physical properties of excellent weldability also enable efficient shaping of the mounts.

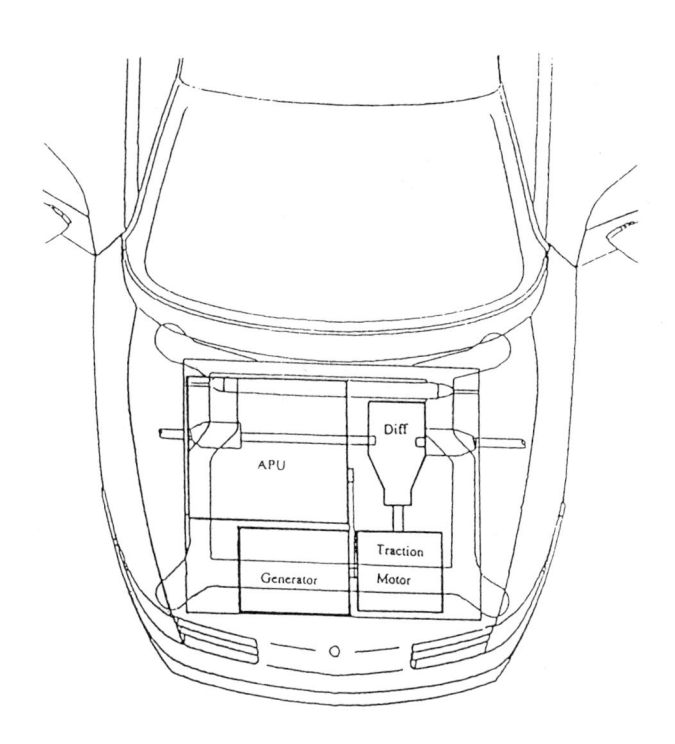

Figure 6

We used rubber bushings to reduce the vibrations from the APU, while solid-stop mounts on the differential and traction motor enable the mounts to survive the initial high torque of the motor.

We also reduced the vibration of the controller by using rubber isolation mounts. The traction motor and APU specifications did not indicate a sensitivity to vibrations, so extensive precautions were taken, only for the controller. The mounts for the APU and motor were designed to be strong enough to withstand a high impact collision. This safety consideration ensures that the vehicle's powertrain components will not enter the passenger compartment in the event of accident.

To simplify the analysis, we assumed the structural members to be simply supported beams. We determined a maximum allowable stress by using 2.5 as a safety factor for all the beams and stock forms of tubing. The highest calculated allowable stress was 48 ksi and the yield stress of the steel chosen is 50 ksi. All the other components have allowable stresses that are less then 40 ksi, which is the yield stress of aluminum 6061-T6. We determined that structural steel ASTM-A572 satisfied all of the structural requirements.

BATTERY SELECTION

We began the battery selection process based on a matrix-selection system, using energy density, power density, weight, size, volume, availability, and cost as selection criteria. The last two factors were weighted the most. Several batteries met the necessary requirements, but were eliminated because of cost and availability considerations. Among these were the lithium-based batteries, the nickel-iron batteries, and the nickel-metal hydride batteries. We reduced our battery choices to lead-acid (Pb-acid) batteries, nickel-cadmium (NiCad) cells, or sodium-sulfur cells. Pb-acid provides the advantages of low cost and ready availability, while NiCad has lower weight, higher power and energy density, and better durability. Sodium-sulfur also offers lower weight and higher energy density, but these advantages are offset by operating temperatures as high as 300 to 3500 C. We were concerned about maintaining that temperature while protecting other components. Analyzing these concerns and their possible solutions helped us narrow our selection to NiCad or Pb-acid.

We sized the batteries according to their power and energy density, and connected them in series. This allows the vehicle to operate at higher voltage and lower current, maximizing the power and energy available, and minimizing line losses. We decided on a total battery pack voltage of 180 Volts dc, since this was the rated voltage of the motor-controller unit. We compared the remaining choices and found a 1.2 Volt NiCad cell, each with a weight of 2.42 lbs and a capacity of 28 Ampere-hours (Ah). This battery yielded a total pack weight of 363 lbs with the capacity to drive the vehicle continuously for 20-25 minutes at 30 mph. However, the cost, $8,600, was beyond our budget. We also identified a 12V Pb-acid starter battery with a weight of 28.2 lbs and a capacity of 18 Ah. This battery yielded a total pack weight of 423 lbs with the capacity to drive the car for 13 minutes at 30 mph. The Pb-acid starter battery was finally selected on the basis of cost even though the NiCad battery pack would have left room for error and saved weight. Since these batteries were donated, cost was no longer a limiting factor.

	Lead Acid	Nickel Cadmium	Nickel Metal Hydride	Lithium Polymer	Sodium Sulfur
Power Density (W/kg)	280	190	150	120	120
Energy Density (W-hr/kg)	35	55	65	150	80
Cell Voltage	2.0	1.2	1.2	2.8-3.5	2.08
Availability	Excellent	Good	Poor	None	Fair
Cost ($/kW-hr)	70	600	600	?	600

Figure 7

Battery Selection Criteria

Enclosure - The batteries are contained in a fiberglass structure, as a single pack, located in the trunk of the HEV. We designed and manufactured this structure within the department. We selected fiberglass because of its strength, ability to withstand high temperature, acid resistance, fire resistance, and non-conductive characteristics. In our selection, we also considered ease of construction, availability, and cost.

The necessary air circulation is provided by fans. A 9 cfm fan and a 12 cfm fan operate constantly to exhaust any gas expelled from the batteries during charge and discharge.

MODE SELECTION AND ACTION

We used the 3-mode switch, required by the competition, in the control and operation strategy of the system. When the switch is turned to ZEV mode, it energizes a relay, allowing current draw from the batteries. It also signals the CPU to power down the APU and generator. When turned to the APU mode, it triggers the CPU to enter the APU control program. This program first determines if the APU is operating, and if not, it starts the APU, brings it to operating speed, activates the field windings, and energizes a relay allowing current draw from the generator. When the switch is turned to HEV mode, it takes the same actions as the APU mode, but also energizes the relay allowing current draw from the batteries.

FUEL CONVERSION

When converting the APU from gasoline to methanol, we took into consideration that gasoline contains almost two times more energy per unit volume than methanol (43.03 MJ/kg compared to 23.04 MJ/kg for M85). This means that in a conversion where equivalent fuel economies are achieved, twice as much fuel must be carried in order to travel the same distance. Similarly, the amount of fuel required for each combustion cycle doubles. The stock Honda fuel injectors were designed and sized to inject gasoline but could not handle the flow rates needed to run methanol. To correct this, we resized the injectors based on a stoichiometric fuel flow at the engine's maximum rated speed (6300 RPM). The calculations below yield the maximum flow rate required by the injectors.

Correct Air/Fuel ratio for M85 (By mass):

A/F for Gasoline = 14.6 parts air : 1 part Gasoline
A/F for M100   = 6.47 parts air : 1 part M100
A/F for M85   = 7.68 parts air : 1 part M85

Volume displaced in each cylinder per cycle:

1493 cc / 4 cylinders = 373.3 cc / cylinder

Mass of air displaced/Cyl. (Assuming a Vol. efficiency of .85):

Density of air @ 20°C (70°F), = 1.2 X 10^-6 kg/cc
Density of methanol = 785 X 10^-6 Kg/cc

M_{air} = (.85)*(1.2 X 10^-6)*(373.3 cc) = 380.8 X 10^-6 kg/Cyl.

Mass of Fuel injected / cylinder / cycle:

M_{fuel} = (380.8X10^-6 kg/Cyl/Cycle)/7.68 = .0496 grams/Cyl.

Volume of Fuel injected / cylinder / cycle:

V= (.0496 grams/Cyl/Cycle)/785X10^-6 = .063 cc/Cyl/Cycle

Injector Fuel flow at various engine speeds:

6300 RPM (Max rated engine speed.)
Flow = (.063cc/cycle/Cyl.) *(6300/2) = 199 cc/min.

1500 RPM (Idle)
Flow = (.063cc/cycle/Cyl.) *(1500/2) = 47.25 cc/min.

The fuel injectors must be capable of handling a volume flow rate of 50-200 cc/min. in order to run methanol.

Since the fuel injectors were resized for

different flow rates, the stock Engine Control Module (ECU) can no longer control them properly. We purchased a new ECU which allows individual mapping of the fuel injectors and spark timing vs. engine speed, temperature, and load. The new ECU is capable of closed loop control of the air/fuel ratio using a heated oxygen sensor in the exhaust manifold. This sensor gives a voltage reading that corresponds to the amount of oxygen in the exhaust. By monitoring the oxygen sensor output, we can modify the fuel injector pulse width to keep the fuel air ratio near stoichiometric. This allows methanol to be used in the engine, minimizes unburned hydrocarbons, and maximizes fuel economy.

EMISSIONS

Harmful emissions from the engine have historically been removed from the exhaust by either thermal or catalytic means (6). There are three basic types of catalytic converters: oxidation, reduction, and 3-way. The oxidation catalyst removes CO and HCs (unburned fuel), and the reduction catalyst, which removes NOx. The oxidation and reduction catalysts are used in series to remove all three pollutants from the exhaust. The 3-way catalyst removes all three pollutants when the engine is operated at or near the stoichiometric air/fuel ratio. The width of this stoichiometric window is narrow, about .1 air/fuel ratios for a catalyst with high mileage use, and varies with catalyst formulation and engine operating conditions. This window is too narrow to be controlled by an ordinary carburetor, though it can sometimes be achieved with sophisticated fuel injection systems (7). Fuel injection, with feedback control of the air/fuel ratio using an oxygen sensor, provides a range of air/fuel ratios narrow enough to use a 3-way catalyst effectively.

A major source of emissions is at start-up. Because the catalyst is cold at this point, emissions pass through without reaction. One solution is to heat the catalytic converter before startup. Heating the stock converter with high-resistance wires seemed to be the best choice, but research indicated that if the heating had to be done external to the substrate it would require far too much electrical energy to bring the catalyst to light-off temperature. Most converters are housed in a metal case and wrapped in fireproof insulation to prevent fires when they come into contact with dry grass. This design does not lend itself to external heating. When researching the problem, we discovered that an electrically heated catalyst (EHC) could solve the cold-start emissions problem. The problem with this was that EHCs were only available as prototypes. However, through much negotiating, we were able to obtain one, which is used in addition to the stock Saturn Catalytic Converter. In order to make the EHC more effective, an air pump is used to supply the oxygen needed for oxidation. This should drastically reduce startup emissions and pollutant levels to the low emission vehicle (LEV) and ultra low emission vehicle (ULEV) range. Testing of the system has not been completed.

The configuration layout is shown below in figure 8.

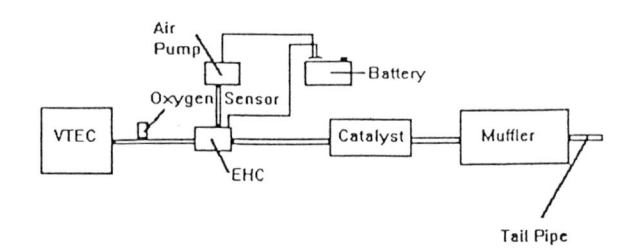

Figure 8

CONTROLS/INSTRUMENTATION

The engine speed is adjusted by a relay connected to a servomechanism that actuates the throttle. The mode switch notifies the CPU whether or not the driver wants the APU on. The driver turns on the APU by placing the mode switch in the APU position. The computer increases the speed of the APU to a set rpm, and then connects the load by energizing the generator field current. This action slows down the APU and causes the computer to take control. The computer raises the APU to its optimized speed by adjusting the throttle and then continues to monitor engine speed in order to maintain optimal performance.

CPU

The CPU is responsible for monitoring and controlling all necessary components of the vehicle. It is designed to optimize the vehicle's performance and provide feedback to the driver on the status of various parts of the vehicle. However, the computer's most important job is to control the generator and the APU.

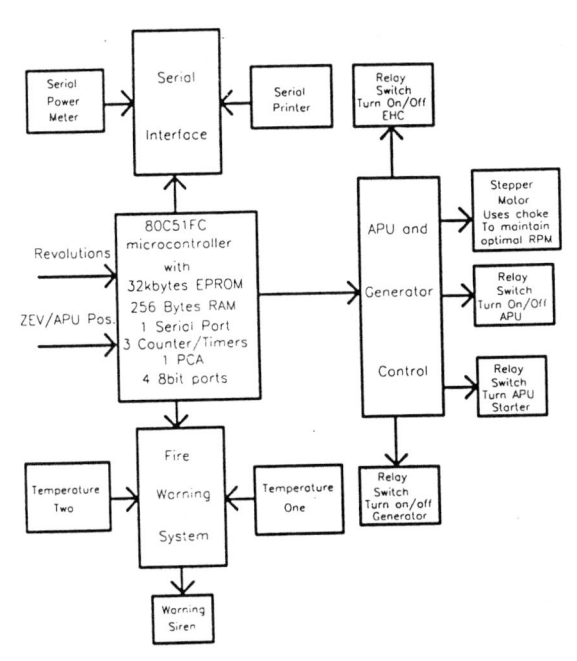

Figure 9

We chose the 8051 as the primary computer system because of its small number of parts. This allows the installation of two 8051 systems in the vehicle. The second system is used as a backup. Software is written to turn the APU and generator on or off, provide speed control for the APU, and act as a fire warning system.

The 8051 is able to input serial, analogue or digital information. It can then process it and export it as either analogue or digital output. Information used as input includes the position of the mode switch, voltage and current. The computer transmits signals which start the APU, adjust its speed, turn on the generator, and feed information to a serial printer for testing. The computer is able to process and modify this information in order to obtain the desired output.

The computer also processes math-related functions. These functions include the timing of devices, unit conversions of data, counting rpms, and control of external sources (APU/GCU).

FIRE SUPPRESSION SYSTEM

The primary function of the system is to prevent a fire from spreading and harming the occupants of the car. The system also isolates the fire to one compartment minimizing the damage to the other systems. After looking at current documentation and available systems, we determined that a remotely operated system using halon 1211 is the optimal design. This decision was based on our analysis of the effects of weight, cost, performance, and simplicity.

In the event of a fire, diode thermostats in the engine compartment and battery box trigger an alarm light in the dashboard. This alerts the driver or navigator to activate the fire extinguisher. The operator has the option of sending all of the halon at once, or sending a partial amount saving some halon in case of a re-flash. Halon 1211 extinguishes fires by reducing the amount of oxygen in the compartment to below oxidation level, thereby suffocating the fire. The halon is compressed by dry nitrogen which lowers the temperature in the compartment to below ignition temperature.

CONCLUSION:

The Cedarville HEV design demonstrates the practicality of hybrid vehicles, particularly a series hybrid. The design, development, and fabrication of this vehicle was a learning experience for all. This hybrid shows that low to zero emissions vehicles can have the performance, reliability, and safety that present vehicles provide.

Finally, the Cedarville College HEV team would like to thank all of the assistance that our sponsors, external consultants, advisors, and college provided. Without this assistance, the successful completion of this vehicle would not be possible.

REFERENCES

(1) --------. Internal Combustion Engine Fundamentals, McGraw Hill: 1988, pg. 881.

(2) Mackay, Robin. "Hybrid Vehicle Gas Turbines." SAE Paper 930044, 1992, pg. 5.

(3) --------. "Properties, Performance and Emissions of Medium Concentration Methanol-gasoline blends in a Single-Cylinder, Spark Ignition Engine." SAE Paper 881679, 1988, pg.5.

(4) Automotive Handbook. 1986, pg. 210.

(5) "Economics and Security Issues of Methanol Supply." Fuel Methanol - A decade of Progress. SAE Paper 872062, 1987, pg.6,

(6) --------. Internal Combustion Engine Fundamentals. McGraw Hill: 1988, pg. 655.

(7) --------. Internal Combustion Engine Fundamentals. McGraw Hill: 1988, pg. 655.

Colorado School of Mines'
Hybrid Electric Vehicle "Drive to the Future"

Christopher G. Braun
Assistant Professor
Member IEEE, SAE

Gustave W. Schlesier
Student Team Leader
Member IEEE, SAE

David R. Munoz
Associate Professor
Associate member SAE

Engineering Division
Colorado School of Mines
Golden, CO 80401

ABSTRACT

The Colorado School of Mines Hybrid Electric vehicle is a converted 1992 Ford Escort LX station wagon with a gross vehicle weight (GVW) of 1480 kg. The power train consisted of an electric motor coupled to the wheels through a continuously variable transmission. An auxiliary power unit was installed to provide extended range. The major components are: a 32 MJ (8.7 kW-hr) battery bank made up of 300 NiCd cells; a 32 kW (continuous) permanent magnet DC electric motor; a liquid cooled, pulsed width modulated (PWM) electric motor controller with regeneration; a 24 kW alternator; and a 620 cm^3, 20 kW Kawasaki V-twin engine modified to use neat ethanol.

The car is designed to operate in two modes; zero emission vehicle (ZEV) mode using batteries only and hybrid electric vehicle (HEV) mode using a combination engine/alternator and batteries. Safety features include a four-point roll cage, a single point ground, an electrical kill switch and an on-board halon fire suppression.

INTRODUCTION

This paper describes the Colorado School of Mines (CSM) entry into the 1994 Saturn/DOE Hybrid Electric Vehicle (HEV) Challenge. The project provided an unparalleled educational opportunity in engineering design and development for undergraduate students and faculty alike. The vehicle entered into the competition is a converted 1992 Ford Escort LX Wagon. The major modifications we made this year were a substantial upgrade of the power train, installing a new underbody battery tunnel and rebuilding the dashboard/interior.

The goal of the Colorado School of Mines Hybrid Electric Vehicle (CSM/HEV) team was to design, construct and test a working vehicle powered by an electric motor and a small internal combustion engine/alternator combination as a range extender.

Because this is an undergraduate student project, the role of the faculty advisors is to help the students understand the system issues, help procure resources and to ensure that the project is run safely. In turn, the students are required to do the work. This includes all aspects of the project from fund raising and public relations to system planning and design implementation.

The students are organized into an overall team structure with sub-teams in each major functional area. Our students range from Sophomore to Seniors working on the CSM HEV as part of their Senior Capstone project. Formal communication between the sub-teams was conducted in the weekly integration meeting which brought all student leaders and faculty advisors together. Generally, the weekdays were used to design, organize and prepare for construction; most construction took place on weekends.

Each sub-team, while working with their faculty advisor, developed the design specifications for their sub-systems. The sub-team brought their decisions and analysis to the weekly integration committee meeting for coordination and information sharing. Individual components were selected and purchased after exhaustive searches of commercial manufacturers. Decisions were made by the sub-teams after considering tradeoffs in performance, cost, ease of implementation and manufacturer support.

OVERALL DESIGN APPROACH

The operation of a hybrid electric vehicle (HEV) requires coordination of many different sub-systems and competing design strategies to achieve a fully functional vehicle. At the most basic level, a hybrid-electric vehicle must contain an electric drive motor, an auxiliary power unit (APU), some type of on-board energy storage, typically batteries, and a means to couple the power to the wheels.

The advantages and disadvantages of series and parallel power train configurations are shown in Table I. These different designs relate how the power is transferred to the wheels. Based upon the goal of higher efficiency, a series HEV configuration was selected. This entails that the entire motive power go through an electric drive motor.

While our performance at the 1993 HEV Challenge competition was reasonable, we felt that our vehicle was underpowered. Our two main goals were to reduced the weight of the vehicle and to upgrade the power train.

Series configured HEV	Parallel configured HEV
Key features: • All power to wheels from the electric motor(s). • APU is coupled to an alternator to charge batteries and power the electric motor. • Typically has a large energy store for range. **Pros:** • If the APU is optimized for constant speed operation, then a series configuration should be more efficient in converting chemical energy to motive energy. • Likewise, emissions from the optimized, constant-speed APU should also be substantially better than an engine designed to operate over a wide speed range. • Typically enough energy store to complete moderate trips without turning on the APU **Cons:** • Need a large, expensive electric motor and controller to achieve even mediocre acceleration and top speed performance. • Large motor/controller, large battery bank leads to a very heavy vehicle.	**Key features:** • Motive power comes directly from both the APU and the electric drive motor. • One end of design spectrum is to use the electric motor as an assist to reduce the power peaks. • The other extreme is a vehicle fully capable operating under just the electric motor, or just the APU, or both for superior performance. **Pros:** • Better acceleration, higher top speed. • Redundancy with two separate systems. • Power assist needs far less batteries -- less weight and lower cost. • A smaller electric motor and controller is easier to fit in vehicle and costs less. • Can use a standard (small) internal combustion engine. **Cons:** • Adding power mechanically is more difficult than electrically. Problems matching the torque-speed curves of the APU and electric motor. • Higher emissions than series. Possibly less efficient.

Table I. The relative merits of a series and parallel power train configurations.

SYSTEM CONFIGURATION

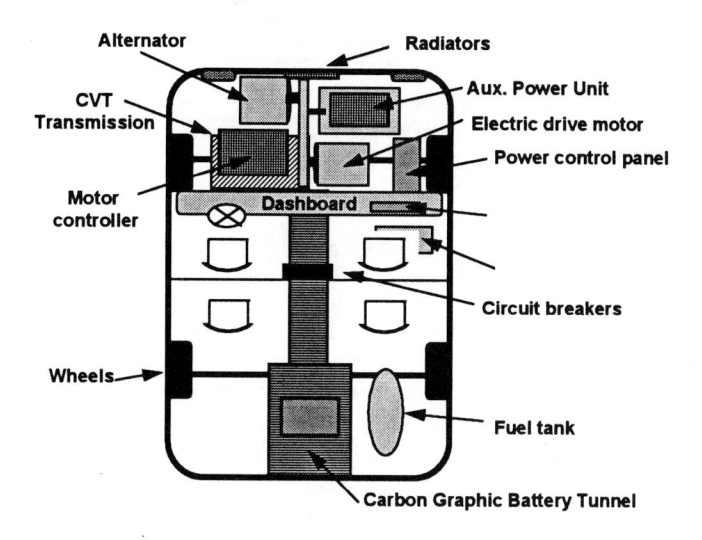

Figure 1 Shown is the physical layout of the CSM/HEV. The front of the vehicle is at the top of the drawing.

The CSM HEV design has two power sources: batteries and the APU. A block diagram of the vehicle power system is shown on Figure 1. When the car is in zero emissions mode (ZEV) the batteries are the only source of power for the electric motor and car systems. In the hybrid electric mode (HEV) mode the APU can be used to power the vehicle as well as charge the batteries.

The driver controls the mode in which the vehicle operates through the dashboard control panel. In case of emergency, both power sources can be disabled through high-current contactor as well as DC circuit breakers. In addition, there are two pull switches to disconnect the battery bank from the vehicle -- one located in-between the driver and passenger and one located outside of the vehicle.

Operation of the CSM HEV is simple. The driver must first ensure that the pull switches are connected. The vehicle's 12V power system is energized by closing that circuit breaker. Once power is supplied to the control and dashboard systems, a switch is turned on to enable the soft start for the electric motor. This soft start sequences the application of the battery voltage to the motor controller such that large spikes of current are avoided.

At the end of the soft start sequence, the two main contactor closes and connects the batteries directly to the motor controller. The vehicle is in ZEV mode at this time. The driver operates the rest of the car systems (throttle, brake, signals, etc.) as any normal vehicle. Regeneration of vehicle kinetic energy into the batteries takes place as throttle position is lower than the current vehicle speed. The magnitude of regeneration is controlled by an adjustable dial.

The APU is started by engaging the APU enable switch on the dash and toggling the APU choke switch to on. The standard Escort steering column 3-position keyswitch is used to start and operate the APU. The alternator is connected to the propulsion power bus by closing the alternator circuit breaker. The APU throttle is engaged, either manually or by the APU speed controller, to set the power flow from the APU/alternator. Shutting down is simply done in the reverse manner. An emergency "Kill Switch" may be used to shut down the propulsion system -- the vehicle and instrument systems will remain on line.

ELECTRIC MOTOR AND CONTROLLER - We have substantially upgraded our electric motor and controller over last year's system. Our previous system consisted of a series-wound DC electric motor from Advanced DC Motors FB14001 with a MOSFET-based controller PMC-1221 from Curtis PMC. This system was limited to less than 22 kW peak due to the maximum average power capability of the controller. It had the advantage of being inexpensive, simple and reliable, but lacked sufficient power to give the vehicle good performance.

Our new drive system is a Unqiue SR-180P permanent magnet electric motor and a CR20-300 controller. The advantages of this system are several. First, it has a peak power rating of 60 kW and a continuous rating of 32 kW. The motor is relatively compact and lightweight, although the controller is somewhat bulky. Second, it has a wide, almost flat torque curve and is up to 91% efficient for both the motor and controller at maximum power (48.5 N-m, 6600 RPM). The controller is liquid cooled for extended operating range while the motor is air cooled. The controller also has the ability to regenerate to low speed. The disadvantage of this system is it's high cost due to low volume production.

The SR-180P is coupled directly to our continuously variable transmission. The controller is mounted above the motor and transmission and connected to the main power bus through a high-current contactor.

BATTERY SYSTEM - We stayed with our existing nickel-cadmium battery cells. These were selected because they have a better lifetime, power density and energy density than lead acid cells. The disadvantage are that they are very expensive, use highly toxic materials (cadmium) that requires special handling if a leak occurs, and require substantial maintenance because they are unsealed, wet cells.

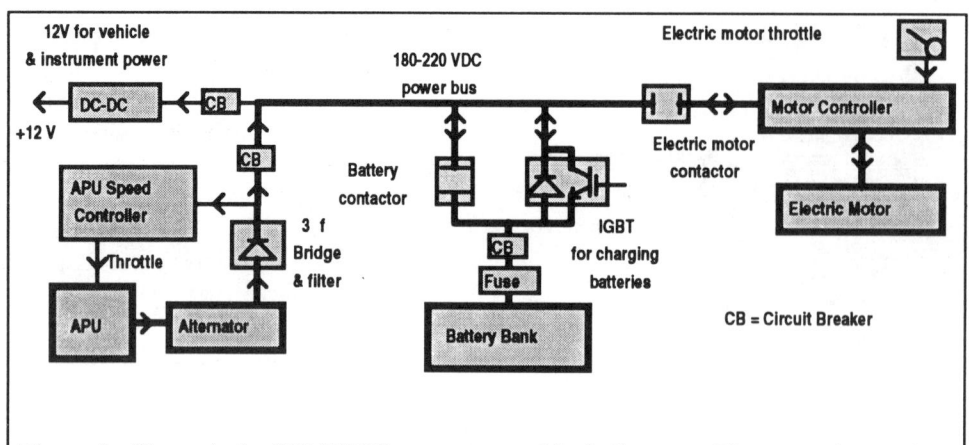

Figure 3. Shown is the CSM HEV power system block diagram. The power bus and all systems connected are electrically isolated from the vehicle for safety purposes.

The specifications of the batteries selected and battery system configuration are listed below:
- SAFT VP230KHB NiCd Sintered Plate Cells
    23 Amp-Hour, 1.26 VDC nominal
    36 W-hr/kg
    0.95 kg/ea.
- Two battery strings of each 150 cells in series
    210 VDC fully charged
    165 VDC discharged
    46 Amp-hour total
    32 MJ (8.7 kW-hr) energy storage

BATTERY TUNNEL - Our old battery enclosure consisted of an aluminum box located in the rear area of the Escort. It suffered from a number of drawbacks: reduced vehicle cargo and loss of rear passenger seating, high weight and difficulty servicing the batteries. A new battery tunnel system was designed and installed into the vehicle. This system has the advantage that it is lightweight (10 kg), the batteries can be easily pulled from the vehicle for maintenance and all the cargo and seating were reclaimed.

The tunnel consists of a long inverted T shaped enclosure make from carbon composite material. A full scale mold were fabricated and the tunnel was make in two halves. Each section consisted of inner and outer layers of carbon fiber cloth in-between a center foam core impregnated with epoxy. The two halves were mated together using the same material and techniques to form a hollow battery tunnel.

A center section of the vehicle floor and underbody from the bumper to the engine firewall was removed. The battery tunnel was fastened and bonded into this section. Several re-enforcing structural supports were welded underneath the tunnel. Because the tunnel extends down the center underbody of the vehicle, the rear axle was substantially modified. The center section of the axle was removed to allow the tunnel to be installed. A box-like system was built around the tunnel and attached. Space was allocated above the batteries along the sides of the tunnel and above the batteries at the end of the tunnel for electrical systems.

A carbon-composite battery tray was fabricated to hold the batteries together so they might be inserted and removed from the tunnel. A locking bar and plate cover the end of the tunnel. To get access to this cover, the rear bumper was modified to have a quick disconnect system. A battery cart was fabricated to hold the batteries for servicing.

ALTERNATE POWER UNIT - Our vehicle was designed with a total range of 500 km (300 miles) at 20 m/s (45 mph) and, a ZEV mode range of 42 km (25 miles) at 13.4 m/s (30 mph). Our analysis determined that we need 14 kW (19 HP) of on-board power to overcome the steady-state power losses of the vehicle while traveling at highway speeds. The power magnitude is the sum of the power required to overcome the wind, rolling resistance and propulsion system losses.

A 4-stroke, liquid cooled, electric start, carbureted, spark ignition engine (Kawasaki V-twin, model FD620cc) was selected as the APU. The engine weighs 41 kg (dry) and is sold in small John Deere tractors to provide shaft power for pumps. The engine is mounted under the hood using polymer engine mounts to reduce vibration and cabin noise.

This engine was substantially modified. Viton carburetor seals replaced the stock O-rings to allow the use of 100% ethanol. An electronic ignition timing control system from Haltech was added and the fuel injection system was rebuilt to better handle ethanol. The pistons, valves and heads were coated with ceramics to reduce heat losses in the engine. The engine is still able to burn gasoline by a simple remapping of the fuel-air mixture in the Haltech control system. A multi-spark discharge unit was added to ensure that a high quality burn takes place.

With these modifications, our APU power increased to about 23 kW (30 HP) at 3600 RPM from a stock 15 kW. Additionally, with the Haltech fuel injection system, the APU can be considered a multi-fuel power unit. We estimate that we need about 42 liters (9.5 gallons) of ethanol to meet our range requirements.

This fuel is contained in one satellite tank on the passenger side of the battery tunnel. It was fabricated from high-strength polyvinyl plastic and has a capacity of 12 gallons. All fuel lines were polypropylene sheathed in steel braiding for better wear under extreme conditions. The fuel pump is located at the bottom of the fuel tank. It is a positive displacement type with a maximum flow of 160 liters per hour. The pump draws a steady state current of 3 amps at 12 VDC.

CONTINUOUSLY VARIABLE TRANSMISSION - This is one of the most difficult modifications we made over last year's vehicle. Because we chose to use a permanent magnet motor from Unique, it was necessary to relook our coupling of the motor to the wheels. The promise of a continuously variable transmission (CVT) is that it can keep our electric motor near it's optimum RPM for efficiency and power. It also has the benefit of allowing us to regenerate power from the motor/controller down to lower vehicle speeds.

The selection of CVTs are very limited. We chose to use the Subaru Justy CVT. It had the advantage that it was smaller and lighter than the stock Escort five speed transmission it replaced. To adapt this transmission into our Escort we needed to modify the mounting points as well as have the stock axles re-machined to mate two different splines. The user control is simple -- a shift level for park, forward and reverse.

EMISSIONS - Ethanol was chosen as the operating fuel for several reasons. First, it is a renewable fuel, produced from distillation of corn or other organic matter. As a renewable fuel it does not add to the net carbon balance of the biosphere. Thus any emission carbon byproducts do not contribute to global warming as do fossil fuels. Secondly, it is an oxygenated fuel. This leads to lower CO and NO_x emission as compared to gasoline. Lastly, because of the higher autoignition temperature of ethanol we were able to increase the compression ratio from 8:1 (stock APU) to 11:1. This should lead to an increase of about 9% in the overall efficiency of the APU.

Our Haltech fuel injection system and multiple spark discharge ensure that a good, clean burn is achieved in the combustion chamber. A standard three way catalyst in the exhaust path eliminates most of the residual CO and NO_x.

ALTERNATOR - The alternator provides a means of converting shaft power from the APU to electric power needed for the electric motor or batteries. Based upon the power generation of the APU and requirements to propel the vehicle, we selected a modified Unique SR-180 HT high-efficiency, permanent magnet alternator. It is essentially the same as our drive motor except the winding configuration has been changed to produce 200 V out at 4000 RPM.

APU SPEED CONTROL - With the above permanent magnet alternator we must control the input shaft speed to control the output voltage. As a result, we acquired a Programmable Logic Controller (PLC) made by Eagle Pitcher to act as our throttle controller. In this system, the PLC monitors the voltage and current from the alternator. If they are not a desired, then a signal is sent to a small electric motor to rotate the throttle up or down.

The feedback variables and control strategy depends on the mode of operation of the vehicle. In "HEV" mode the current of the battery is monitored and a PID feedback loop is implemented to ensure that no net change is extracted from the battery. In this mode, the APU supplies the vehicle power except when a sudden acceleration requires additional power from the battery. Whatever energy was extracted from the battery is replaced by the APU/Speed controller as it adjusts to the new operating point. In the "Charge Mode" the APU is set to it's most efficient operating point above the current power demand need by the vehicle. The extra power charges the battery. A separate circuit controls the rate of charge and shuts down this mode when the batteries are fully charged.

USER INTERFACE AND DATA ACQUISITION - A critical design goal for our HEV was to provide the driver and/or the passenger with complete status and direct control of all the vehicle sub-systems. This year we built a custom dashboard. A mold was cut and a lightweight fiberglass shell was pulled from the mold. This was mounted to the stock dashboard frame. The fiberglass frame was coated with a rubberized compound. The standard Escort instrument cluster was kept essentially in place. The voltage and current meters and selector switches (for the many different systems) were placed in a small overhead console. The lower center part of the dashboard was used to mount the control switches and indicators, the watt-hour meter and temperature meter.

Two data acquisition systems were installed. The NREL data acquisition system kept track of vehicle velocity and battery and alternator currents and voltages. This data is used by the event organizers to determine how well each vehicle operated. Additionally, we installed our own data acquisition system to use for testing and optimization purposes. This system acquired up to 64 channels of analog, time varying data and sent it to a laptop computer for logging and real-time display.

BATTERY CHARGING - The usefulness of an (hybrid) electric vehicle is sometimes limited by the time it takes to charge the batteries. Typically, the user is able to wait overnight for charging the batteries so that a charger need only be of modest size. However, there are situations that require the vehicle to be quickly charged and in that case the charger must be rated for a much higher power level.

Our charger design focused on building a simple, inexpensive yet high-power charger capable of charging our batteries as quickly as possible. If we could obtain a power grid connection and charger rated sufficiently high, we could charge our NiCd batteries at 10C (460 amps peak) giving a fully charge in about 15 minutes. However, with our practical limitations we constructed our charger to give a full charge in less than an hour.

The charger circuit takes 1φ or 3φ 208 V input through 3φ, 20 kW isolation and autotransfomers into a diode bridge and capacitor filter. The autotransfomer wiper position, which sets the output voltage, is controlled by a motorized circuit. The output is connected to the vehicle through a standard Anderson connector. The output current is measured through a shunt resistor. The input power lines are run through a 3φ, 60 amp circuit breaker.

The rate of charge is controlled by an Eason Programmable Logic Controller (PLC). This PLC has a front panel keyboard and (small) display for user interaction. A program was written to provide the user with the choice of several charging modes including manual control of the autotransfomer, constant current or constant voltage or automatic three-step charge of the batteries. In this automatic charge mode, the PLC monitors the current and voltage and maintains a constant current I1 until the voltage V1 is reached, then it maintains current I2 until V2 is reached and then it maintains a constant trickle voltage limited to current I3.

This simple charger does not meet the competition specifications of IEEE 519 for total harmonic content. However, our approach is to examine the issues relating to quick charging and the cost for a unity power factor, low harmonic content power supply is beyond our means.

VEHICLE HANDLING - The suspension of the stock Escort was modified to accommodate the increased vehicle weight introduced by the batteries. In general, the goal was to use after market performance equipment for these modifications. Coil springs, stiffer than those on the stock Escort and fabricated by Suspension Techniques, were installed in the front and rear. These were designed to allow improved performance with the resulting weight distribution of the CSM HEV. In addition, stiff anti-sway bars were installed to augment the stock anti-sway bars in the front and replace those in the rear. For added stiffness, less compliant, polyurethane bushings replaced the rubber stock bushings. High performance gas shocks, supplied by Tokico were used in place of the lighter-duty stock shocks.

American Racing supplied the 0.356 m (14 inch) aluminum racing wheels and Goodyear supplied the P185/60R14 tires. The tires have a lower profile than those on the stock Escort [0.330 m (13 inch)] but were the same overall diameter and therefore provided improved handling of the heavy vehicle in a sharp turn. Lowering the mass of the vehicle by 80 kg from the previous year also improved the handling

CONCLUSION

The Colorado School of Mines hybrid electric vehicle has been substantially improved over last year. We have upgraded the power train by adding a higher power electric motor and a controller capable of regeneration. The stock Escort transmission was replaced by a continuously variable transmission. The rear passenger seating, rear cargo area were reclaimed and weight savings were realized by replacing the previous battery box with an underbody carbon composite battery tunnel. A new dashboard was fabricated and user-friendly instrument panel installed.

Our performance at the 1994 HEV Challenge was disappointing because of a failure of our CVT left our vehicle unable to shift. Even so, our vehicle did well enough for us to believe that our design approach is solid and that our vehicle has the potential to perform among the best.

ACKNOWLEDGMENTS

We gratefully acknowledge the support of our sponsors. We thank the Colorado Corn Administrative, the Public Service Company of Colorado and the Colorado School of Mines for their very generous monetary support. Also we would like to thank the many companies and individuals who donated money, in-kind support and advice. Last, but certainly not least, we thank all of the people who supported this HEV Challenge at GM/Saturn, the U.S. Department of Energy and SAE for their vision of the future and hard work to make this competition a reality.

Modeling the CSM Hybrid Electric Vehicle Using SIMULINK

C. Braun
Assistant Professor
Member IEEE, SAE

Dave Busse
Graduate Student
Member IEEE

Doug Cook
Graduate Student
Member IEEE

INTRODUCTION

Hybrid-electric vehicles (HEVs) are composed of many complex, interacting sub-systems. These systems include electric drive motors, battery packs and other energy storage elements, auxiliary power units, charging systems, supervisory controllers and so forth. Unlike conventional automobiles or similar systems, there is very little in established "common wisdom" in the best ways to configure an HEV system.

Likewise, because there is not a strong, established market for these sub-systems it is difficult to find exactly what one needs. Rather we are often forced to take what we can find and adapt that for use in our HEV system. Many of the technologies for these major sub-systems are rapidly changing and new approaches are constantly evolving. Because all EV/HEV systems are energy constrained, a major theme in all (hybrid) electric vehicle designs is efficiency. Any, even small, progress in improving system efficiency pays high rewards in overall system design.

Given these obstacles and goals in designing a hybrid-electric vehicle it is clear that we need tools to help understand the system implications various design changes. Specifically for our project, in the process of designing, building and improving the CSM hybrid electric vehicle, we needed to develop a simulation program that would help us understand how to optimize our system design.

There are several existing modeling programs designed for (hybrid) electric vehicles. Commonly available programs include SIMPLEV, developed by Idaho National Engineering Laboratory, MARVEL developed by Argonne National Laboratory. Additionally, there are commercially developed programs from ORTECH as well as Aeroenvironments. In the former, the programs are free but lack the flexibility to add elements in ways different than the designers allow. For the latter, these are proprietary, expensive programs that are neither open nor widely available.

One of our goals was to provide a simulation model that is easy to use and easy to add new elements. These elements may range from sub-system components that behave substantially differently than the standard building blocks to elements that embody sophisticated control strategies.

There are a wide range of suitable general-purpose programming languages as well as modeling/simulation tools. Examples of such tools include C++, MODSIM, Saber, Visual Basic, etc. We have found that MATLAB/SIMULINK is the best tool for our purposes. The advantages include:

- MATLAB is a standard, widely available tool; runs on range of computer platforms
- SIMULINK is a graphical, intuitive block-diagram programming language well suited for describing sets of differential equations
- Extensive set of special subject toolboxes including control
- Training material and third party books are extensive
- Student version and site licenses available

GENERAL MODEL

The mathematics of the model are simple in concept (the forces on the vehicle produce an acceleration), but can be complex when the details of the systems are taken into account. The following equations describe the major behavior of a hybrid electric vehicle.

(1) $m\ddot{x} = F_{propulsion}(\dot{x},t) - F_{drag}(\dot{x})$

where the forces are defined

(2a) $F_{drag}(\dot{x}) = mg(\sin(\theta) + f_{rolling}) + 0.5\rho A C_d (\dot{x} + V_{wind})^2$

and

(2b) $F_{propulsion}(\dot{x},t) = Torque_{motor}(\dot{x}, V_{battery\_out}, throttle)$
$Gear_{CVT}(\dot{x}, Torque_{motor}) Effic_{CVT}(\dot{x}, Torque_{motor})$

The current required by the motor controller causes a current to be pulled from the alternator and batteries

(3a) $I_{motor} = f_{controller}(Torque_{motor}, \dot{x}, V_{Battery\_out})$
(3b) $I_{Motor} = I_{ALT} + I_{Battery}$
(3c) $I_{ALT} = f(V_{ALT}, R_{ALT}, V_{Battery}, R_{Battery})$

The voltage on the alternator relates to the speed of the APU through the gear ratio

(4a) $V_{ALT} = K_E \Omega_{ALT}$

where

(4b) $\Omega_{ALT} = \Omega_{APU} G_{APU}$

with the APU speed determined by the torque produced by the APU

(5) $Torque_{APU}(\Omega_{APU}, throttle) =$
$(J_{APU} + J_{ALT} G_{APU})\frac{d\Omega_{APU}}{dT} + B\Omega_{APU}$

The battery output voltage is dependent on the no load voltage, the current, battery resistance, state of charge and battery temperature

(6) $V_{Battery\_out} = V_{Battery\_No\_load}(SOC, T_{Battery})$
$- I_{Battery} R_{Battery}(SOC, T_{Battery})$

with the state of charge of the batteries dependent on the current and rate of current

(7a) $SOC = 1 - \frac{\int_0^t f_{peukert}(I_{Battery}) dt}{Ah\ rating}$

with the function defined as

(7b) $f_{peukert}(I_{Battery})$ = Rate of battery capacity used

The difficulty here is that these equations are only a small part of the complexity. The limits of operation of each system must be taken into account along with the efficiency and dependence upon the environment (such as temperature, moisture, etc.). An example is that the a transmission is limited in the amount of torque it can transmit to the wheels. Every system in the vehicle may have limits that can affect the operation of the entire system.

Also not shown in these equations are the modes of operation and the feedback control strategies. Typically, the desired velocity profile is known (e.g., a FUDS cycle). Thus the system must solve or have a feedback loop to set the throttle/etc. to produce the correct amount of power such that the desired velocity is produced.

At the same time, there are several modes of operation for the APU - including off, supplying motive power, and charging the batteries. The amount of power from the APU is controlled by the throttle position. This throttle is in turn controlled by a feedback loop set to meet the objectives of that mode of operation. The model must incorporate all these control loops and modes of operation to simulate the overall performance of the vehicle.

The solution of this type of model is accomplished by time stepping through a numerical integration. As part of the process of building a reasonable model, we need to develop a test modules and process to check each sub-system and as well as check the overall vehicle. We can use an analytical solution of a simplistic case (i.e., constant power) to check the simulation. In addition, the results in some cases can be directly compared to the results from SIMPLEV for the same input systems.

SIMULINK MODEL

Shown below are the SIMULINK modeling block diagrams for the CSM HEV. These models are constructed graphically by linking the appropriate blocks and then defining the inner workings of each block. A collection of blocks can be grouped into a single, higher level icon. The top level block diagram is shown on Figure 1.

We have purposely constructed this model to be "local". That is, each block describes the complete behavior of a sub-system in the vehicle. All information relating to that sub-system is located within that block. Where information from one block is needed in

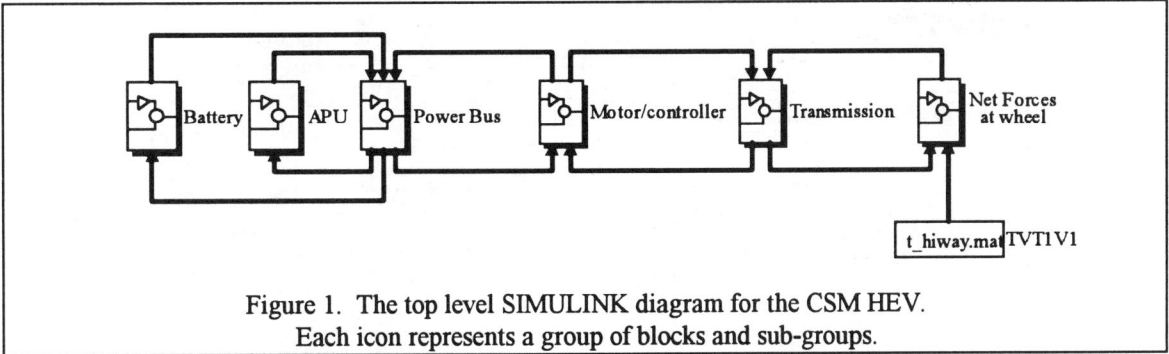

Figure 1. The top level SIMULINK diagram for the CSM HEV.
Each icon represents a group of blocks and sub-groups.

another location, the blocks pass the information through the connections shown.

The reasoning behind making each block local is that each sub-system can be replaced at any time without having to change the model in any other location. Thus, our APU model can be easily replaced by another APU model as long as it meets the connectivity standards that we defined. This modular organization then allows us to create a library of parts that describe different sub-systems or different operating strategies.

The connections shown between the functional groups on Figure 1 are multiplex "bus" connections. That is, each connection may represent any number of actual connections between the sub-systems. The box in the lower right is the input file specifying the desired velocity as a function of time.

This model calculates the required torque at the wheel and "requests" this from the transmission sub-system. The transmission sub-system applies the appropriate gearing, adds in the amount to overcome the transmission efficiency losses and sends this request to the motor sub-system. However, the transmission also limits this request to the torque it can handle.

The motor sub-system takes the transmission request and RPM and determines via a lookup table of the motor torque-speed curve if it can meet this request. If it can, then the request, plus the efficiency losses in the motor, are passed along to the battery/alternator. If the motor cannot accommodate this requested torque, then the request to the battery/alternator is limited to the maximum torque that the motor can produce plus losses.

This process continues through the battery and alternator. The actual battery and alternator powers are then passed back to the motor which then passes the actual motor torque to the transmission and then to the wheels. The actual force to the wheels will be exactly that needed to produce the desired velocity profile if there are no sub-systems that limit this force.

While this is a round about process, it takes into account the limitations and efficiency of each system. Shown on Figures 2 and 3 are the sub-groups for the "Net forces at wheel" and the "Motor/controller" groups, respectively.

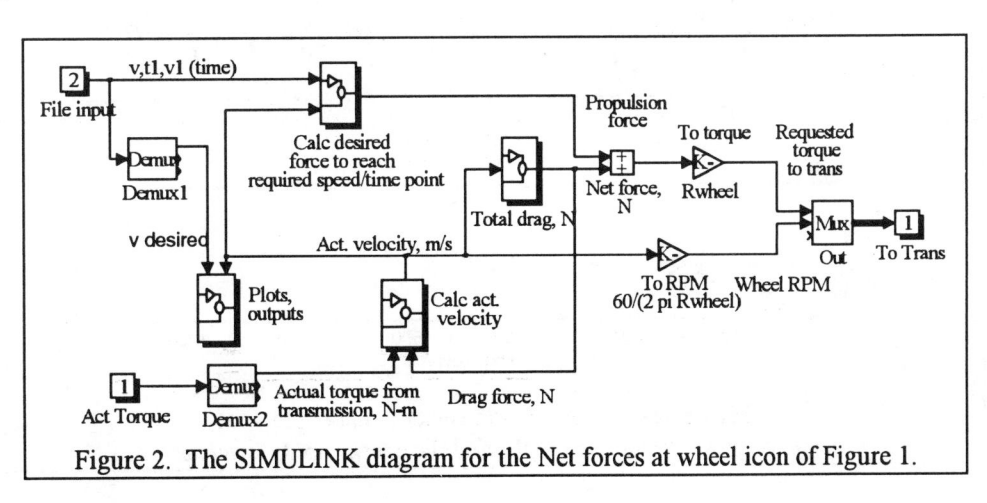

Figure 2. The SIMULINK diagram for the Net forces at wheel icon of Figure 1.

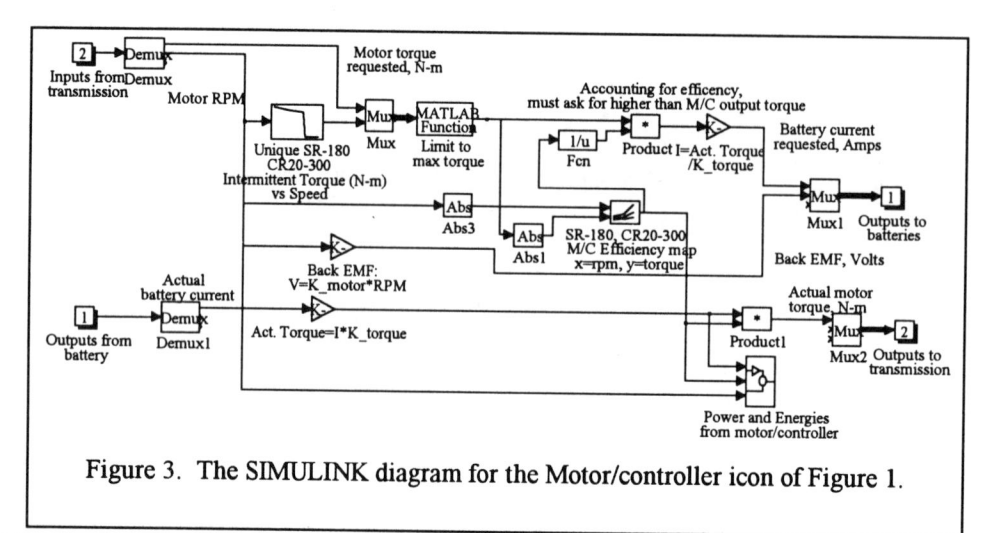

Figure 3. The SIMULINK diagram for the Motor/controller icon of Figure 1.

RESULTS

We have just started on the process of using this model as a design tool. It has taken awhile to refine this model sufficiently and gain confidence in the results. Many of the sub-systems are still being revised to more accurately capture the desired behavior. We expect that this process will continue as we develop a deeper understanding of our sub-systems.

This model has not yet been validated against the actual performance of the vehicle, although we have data showing that the results are in reasonable agreement. The validation process can be long and tedious and will wait until our model is more stable. For now we have a strong belief that this model can be used as a qualitative design and analysis tool to show what configurations work well together and indicate how to optimize vehicle performance.

Shown below on Figure 4a-c is the output from a run of the SIMULINK model. This run is one FUDS cycle and took about minutes on a 486, 66 MHz computer. The speed of calculation of the SIMULINK model is slower than the equivalent SIMPLEV calculations. Further work may lead to quicker calculations.

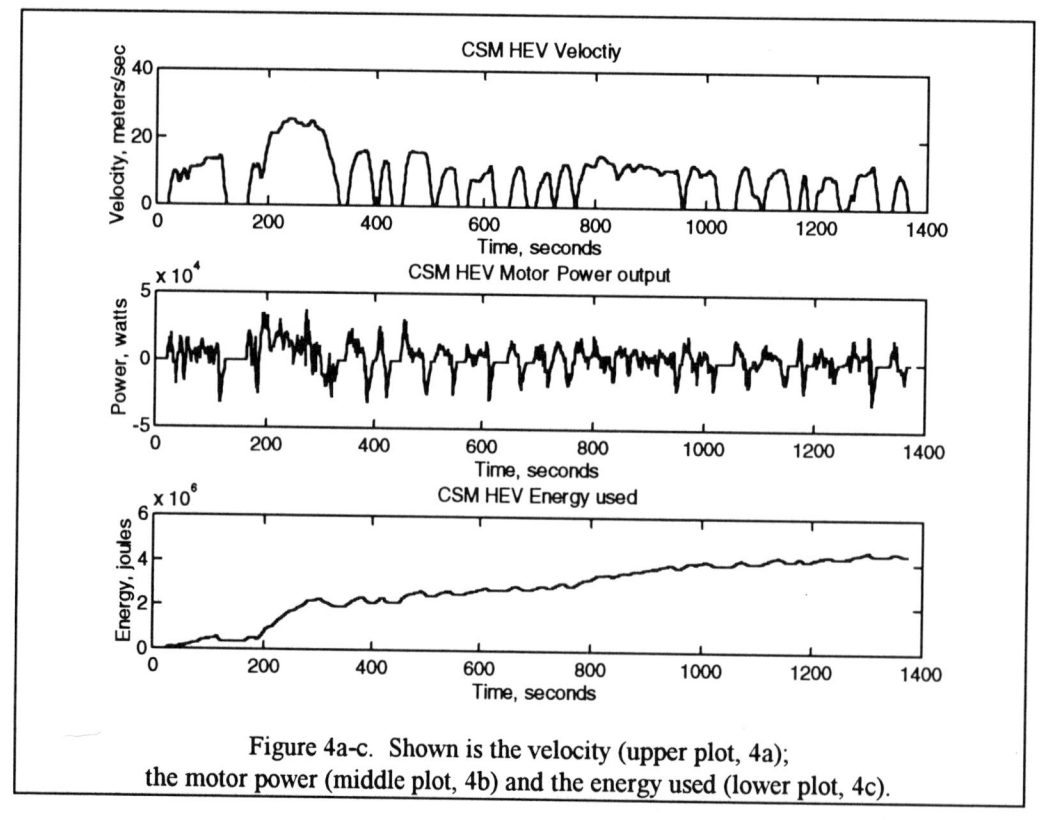

Figure 4a-c. Shown is the velocity (upper plot, 4a); the motor power (middle plot, 4b) and the energy used (lower plot, 4c).

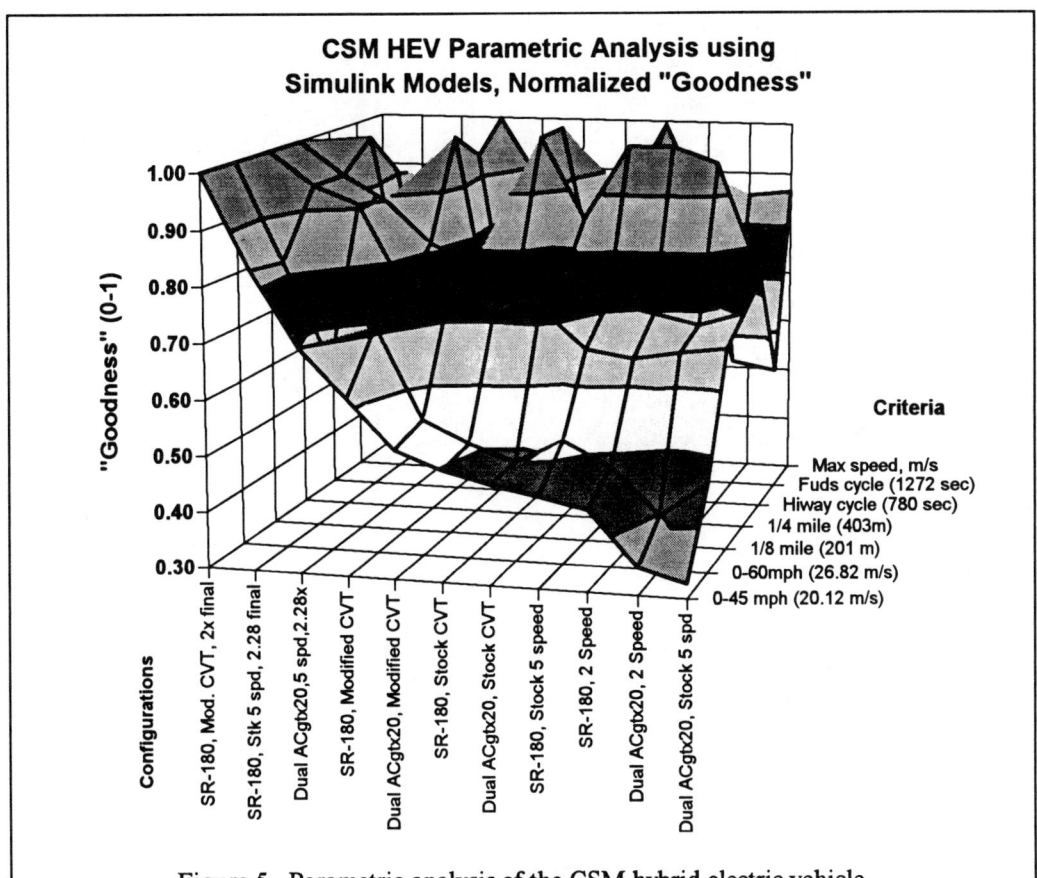

Figure 5. Parametric analysis of the CSM hybrid electric vehicle. Shown are some different possible power train configurations vs various performance criteria.

We have conducted some parametric analysis using this model. While the results are still preliminary, they do point out the tradeoffs in optimizing the vehicle power train. Shown on Figure 5 is a parametric map of the "goodness" (normalized performance) of several different power train configurations versus several different performance criteria.

The better power train configurations tend to be toward the left hand side as indicated by high scores for most of the criteria. One surprise was that for optimizing the current CSM HEV continuously variable transmission for higher power seems to lead to a slight decrease in efficiency. This was not expected since the optimized CVT operates in a higher RPM range than the stock CVT. This better matches the load to the peak power and efficiency point of the motor torque-speed curve. The loss in efficiency may be due to subsequent differences in low speed operation, rather than in high speed.

The overall efficiency numbers for the Solectria ACgtx motor (FUDS and Hiway cycle energy use) are suspect as the available efficiency curves had to be extrapolated to high and low speed operation.

This last point is a general criticism of any model such as this. Any comparisons of results can only be as valid as is our modeling of the performance and efficiency of each component. Considerable work is often required obtaining accurate and reliable data for the sub-systems.

CONCLUSION

This paper can only begin to discuss how the details of this SIMULINK model and the results we have obtained. It is our belief that this model will enable us to develop a better understanding of the operation of hybrid electric vehicles and can be effectively used in evaluating system and component performance, tradeoffs between approaches, and operating and control strategies.

Readers interested in this SIMULINK model should send an email request to "cbraun@mines.colorado.edu".

REFERENCES

MATLAB Reference Guide, The Mathworks, Inc. Natick, Mass. 1993.

SIMULINK Users Guide, The Mathworks, Inc. Natick, Mass. 1993.

L. Trigger, J. Paterson, and P. Drozdz, "Hybrid Vehicle Engine Size Optimization," Electric Vehicle Power Systems: Hybrids, Batteries, Fuel Cells, SAE SP-984, 1993.

T. D. Gillespie, <u>Fundamentals or Vehicle Dynamics</u>, SAE, USA, 1992.

Unique Mobility technical specifications and literature, Golden, Colorado, 1994.

Solectria Electronics Division technical specifications and literature, Arlington, Mass., 1993.

SAFT VP230KHB Battery Manual and Specifications.

The Concordia University 1993-94 Hybrid Electric Conversion Vehicle

Denis Dionatos, George Metrakos, Achilles Nikopoulos, John Theofanopoulos

ABSTRACT

Major modifications to the 1992-93 Ford HEV were performed on the electrical system and the internal combustion engine. An emphasis was placed on efficiency, emissions reduction and driveability. This vehicle's unique feature is that it can function within different driving modes that have been labelled as series, electric and cruise control. The gas-powered engine can either drive the vehicle at a constant speed of 72 km/h or perform on board charging with the aid of a high powered alternator. A permanent magnet brushless DC motor is used as the main electric drive unit, while the addition of a series wound electric motor satisfies the acceleration requirements. Extensive battery testing is summarised. Lead-acid batteries are used due to their cost versus performance effectiveness. A mathematical model using governing vehicle dynamics equations, a simulation and optimization process provided the final determining factors in our component selection and overall design.

INTRODUCTION

The Concordia HEV Team (CHEV) employs two electric motors and an internal combustion engine (ICE) to meet the competition performance requirements. Complementing the electric motor drive system, the ICE supplies an energy source for a driving range extension which can be applied in a series or a limited parallel configuration. A series setup uses a generator to convert mechanical into electrical energy. The limited-parallel drive configuration, also referred to as a cruise mode is established by directly coupling the ICE with the transaxle but only at certain fixed ICE speeds.

All of the three powerplants are connected to the drive shaft through a chain-driven extension shaft. Existing engine mounts were used to support the motors and to allow the transaxle, motors and attachments to move together as a combined unit. A drive compartment schematic is shown in Figure 1.

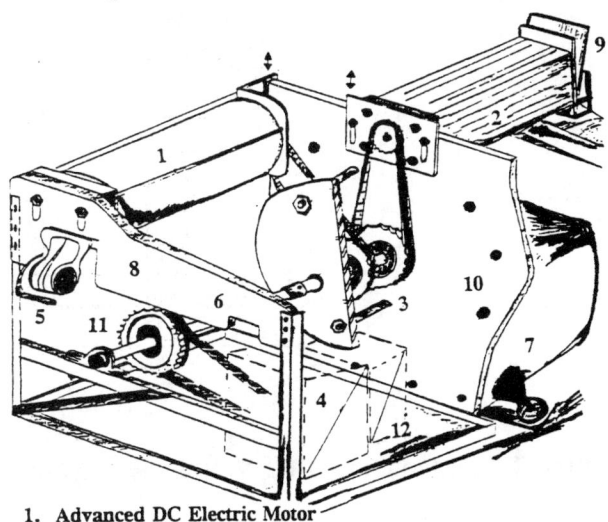

1. Advanced DC Electric Motor
2. Solectria Corp. Electric Motor
3. Drive Shaft and Sprockets
4. ICE Placement
5. Mount #1
6. Mount #2
7. Mount #3
8. Mounting Plate (Left)
9. Mounting Bracket (Right)
10. Trans-axle Cover Plate
11. Electro-Magnetic Clutch
12. Alternator

Figure 1 Drive Compartment Schematic

To promote fuel and electric efficiencies and to eliminate structural modifications the vehicle weight was kept within the outlined limits. An emphasis was also placed on manufacturability, safety, cost and ease of assembly for future vehicle mass productions. Furthermore, vehicle stability was also maintained through proper placement of the components. Enclosed within the interior, the shielded battery packs are securely attached to the vehicle structure, latitudinally with the weight distributed over the rear axle. The overall design shown in Figure 2 outlines the two main component locations on the vehicle, as follows:

- drivetrain in the front,
- energy storage system in the rear.

The illustration is explained in Table 1.

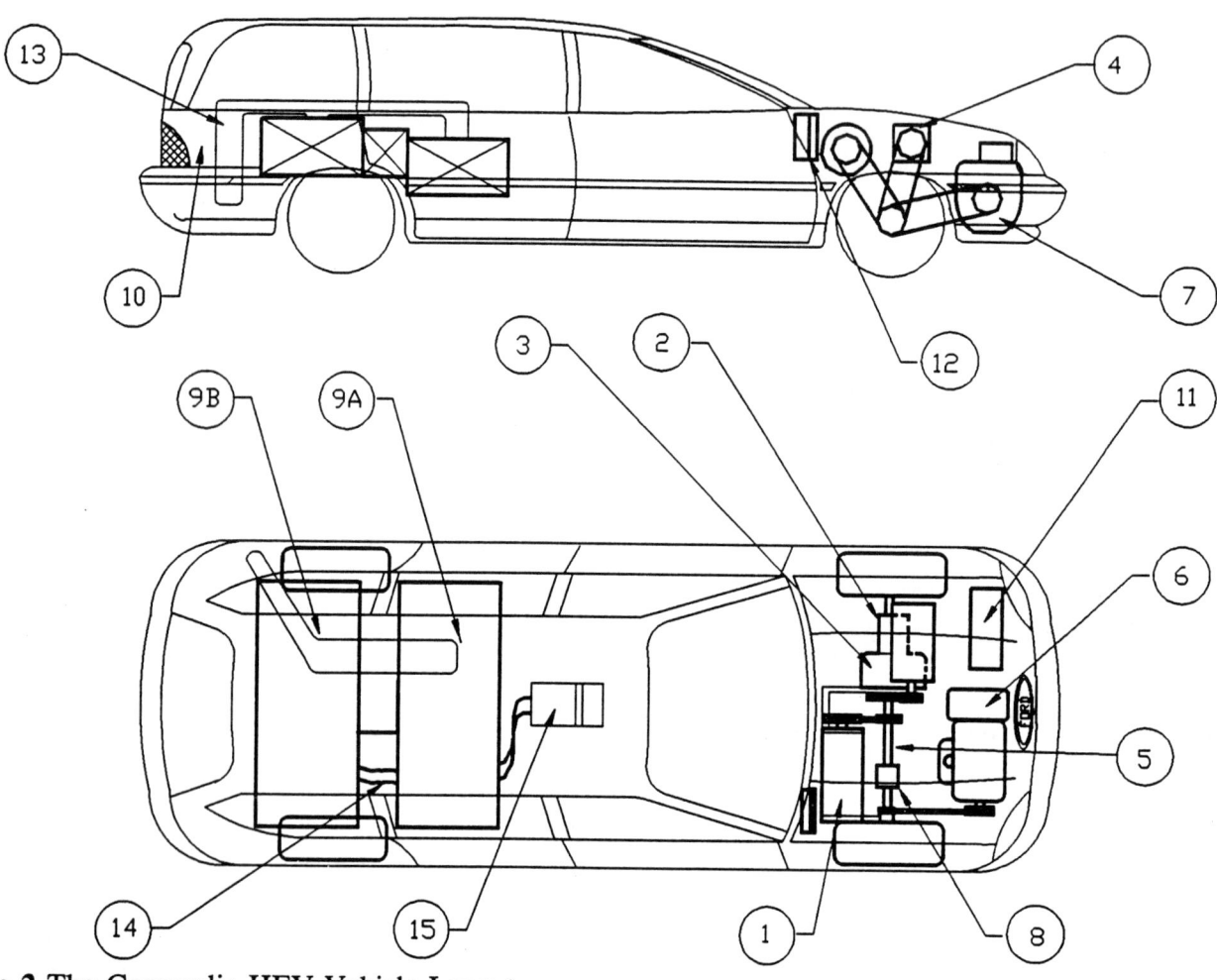

Figure 2 The Concordia HEV Vehicle Layout

Table 1 Component Description

1. Advanced DC Electric Motor
2. 5 Speed Transaxle
3. Clutch/Flywheel Assembly
4. Solectria Electric Motor
5. Extension Drive Shafts
6. Alternator Unit
7. Briggs & Stratton ICE
8. ICE Drive Clutch
9. Battery Trays A & B
10. Cargo Space
11. Solectria Motor Controller
12. Curtis Motor Controllers
13. Battery Exhaust Route
14. Wire Conduits
15. Battery Power Disconnect

A.I POWERTRAIN CONFIGURATION

A1.1 Motor and Controller Selection

Performance requirements, including acceleration and motor/controller efficiency along with a cost to power ratio were used in the selection of the electric drive system. The three objectives used were:

- acceleration from 0-72 km/h within 15 secs.,
- minimum overall electric motor and controller efficiency of 90%, which would be analyzed at cruising speeds of 72 km/h, in terms of power consumption, and
- minimum cost to power ratio ($/kW).

ELECTRIC MOTORS - AC induction motors were found to be the least expensive, next to series wound (SW) motors. However, when the complete AC drive system, including the controller and the inverter was reviewed, it was found to be a very expensive and heavy package. Also in terms of electric efficiency, brushless permanent magnet (PM) motors are capable of achieving higher peak efficiencies than AC induction systems. This is mainly due to the fact that AC induction systems are penalized for the DC to AC conversion performed by the on-board inverter. These systems were thus not considered as potential candidates for the CHEV.

A summary of the acceleration results, motor torque and speed, electrical efficiency, and power consumption for various electric drive units is found in Table 2. The acceleration requirement was satisfied by all the drive systems. The most effective system was the **Solectria/Advanced DC** configuration in combined terms of power consumption, response time and cost-power ratio.

Taking these requirements into consideration the brushless PM Solectria BRLS16 was chosen as the primary electric motor. In the CHEV this particular model was optimised for 156V rather than 120V thus increasing the vehicle's range with the use of extra batteries. The Advanced DC (A/DC) is a high output motor, which is small in size and weight. Since its is required only under hard acceleration its overall efficiency is not a critical factor. It also spins freely on its bearings when no field current is applied to it, thus causing no power loss. As in a typical SW-DC motor the torque in the armature will increase faster than the current applied and will effectively produce adequate torque to move the vehicle from rest [1].

A diagram of the electric motor torque vs. speed curves can be seen in Figure 3. The Solectria and the A/DC provide their maximum power at different speed conditions, therefore sprocket ratios were necessary to permit both motors to function at their optimum conditions. The procedure is discussed further in the simulation section.

CONTROLLERS - The **Curtis** Controller was selected as the control unit for the A/DC motor, since it used an efficient pulse width modulation (PWM) control strategy, operating at a frequency of 15 kHz. The supply voltage capabilities also match the motor requirements. In the CHEV the A/DC motor was matched with two of these controllers connected in parallel. The limits in performance are due to the maximum amperage of 500A the

Table 2 Summary of Drive System Simulations

Electric Motor	0-72 km/h Acceleration Time (s)	Motor Torque (N·m)	ω_m (rad/s)	η_e	Power Consumption (kW)	Cost to Power Ratio $/kW
UNIQ SR180LC	12.65	20	503.4	0.92	10.94	333
UNIQ SR218L	7.54	20	503.4	0.93	10.83	444
2-Solectria BRLS16	10.89	21.1	477.2	0.92	10.94	326
Advanced DC 203-06-4001	10.65	24.3	413.2	0.84	11.97	53
1-Solectria/Advanced DC	10.4	--	--	0.935	10.76	163

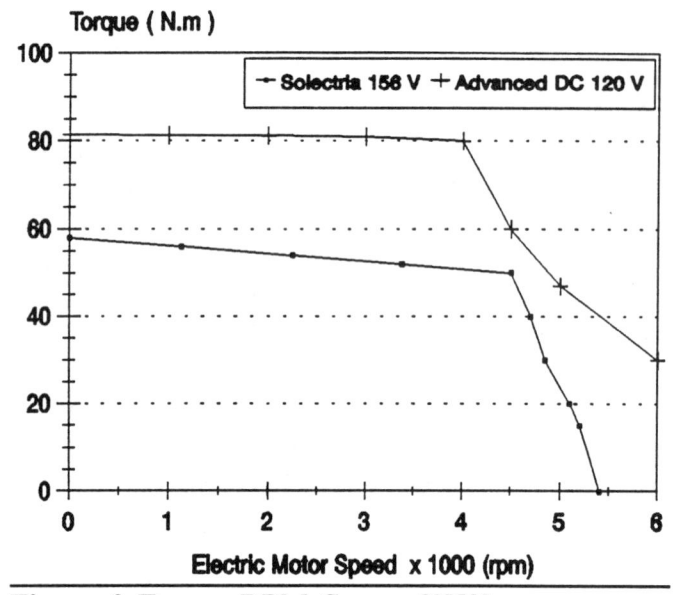

Figure 3 Torque-RPM Curves [2][3]

controllers can provide with the 120V DC they are connected to. The A/DC provides 60 kW of instantaneous power, limited by the controllers and not taking into account the controller efficiency. **Solectria Corporation** provides a custom controller for their motor which also has a built in regenerative braking circuit. The importance of this feature can be seen by the tests that were performed with and without regenerative braking. An average energy recuperation of 12.5% was obtained, (Figure 4), thus proving the importance of this feature.

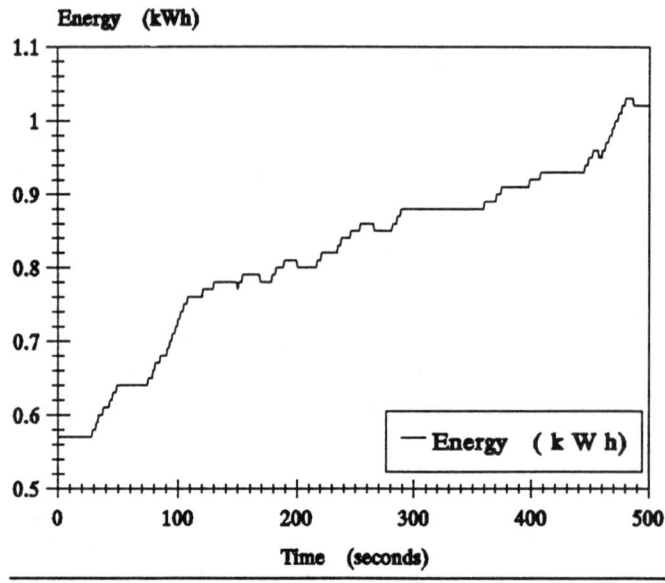

Figure 4 Energy Dissipated Versus Time - With Regenerative Braking

A1.2 APU Selection and Drive Configuration

The alternate power unit (APU) was selected based on the following criteria: specific fuel consumption, peak power, and physical dimensions.

In series mode the engine has to maintain the vehicle at a maximum speed of 64 km/h (flat surface), while in cruise mode the target speed is 72 km/h. The main advantage of this configuration is that the APU operates under a fixed speed and load environment, thus providing easy control of exhaust gas emissions. Using the overall efficiency outlined in equation 1, and the resistive forces, the power required to maintain the vehicle at each constant speed is determined (equation 2).

$$\eta_{so} = \eta_a \eta_e \eta_m = 0.82 \qquad (1)$$

where:
- η_{so} - series mode overall efficiency
- η_a - alternator efficiency (0.96)
- η_e - electric drive efficiency (0.93)
- η_m - drivetrain efficiency (0.92)
- η_{cco} - cruise control overall efficiency.

$$P_{APU_{mode}} = \frac{V_{HEV} \cdot (F_a + F_{ro})}{\eta_{mode}} \qquad (2)$$

$$P_{APU_{series}} = 8.2 \ kW$$

The overall efficiency for the cruise control configuration is $\eta_{cco} = 0.92$ and thus the power required to maintain the vehicle at a constant speed of 72 km/h on a flat surface is 10.2 kW. These greater power requirements took precedence in the sizing of the APU. A gasoline powered **Briggs and Stratton** (B&S) 13.5 kW Vanguard engine was selected based on its low specific fuel consumption as well as its low specific volume.

Evidently the APU has been sized generously in order to accommodate variances in load such as grade changes and face winds and yet still maintain the design cruising speed.

SERIES MODE - For the implementation of the series configuration, a customized alternator was built and installed directly onto the ICE shaft by **Fisher Electric Motor Technology Inc.** The design specifications were: 9kW net power output, 25.4 cm max. outer diameter, 11 cm max. thickness, an operating speed and voltage of 2700 rpm, and 170V. At 2700 rpm the B&S can achieve the 9kW of net power from the alternator (see Figure 11 section A3.1). When the alternator was tested it produced a maximum power and voltage of 8.5kW and 180 V respectively.

CRUISE MODE - In cruise mode, the APU transmits torque directly to the stock transmission through a chain driven electro-magnetic clutch. This configuration can be seen in Figure 5. The

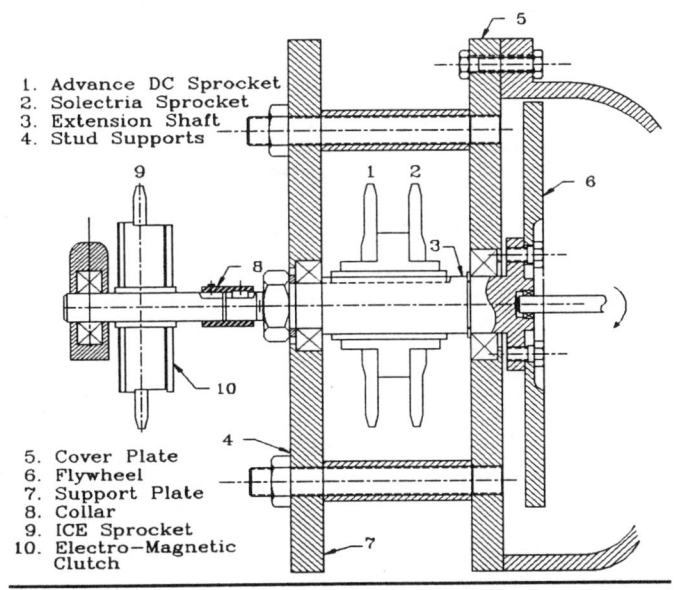

Figure 5 Component Assembly and Cruise Control Configuration

1. Advance DC Sprocket
2. Solectria Sprocket
3. Extension Shaft
4. Stud Supports
5. Cover Plate
6. Flywheel
7. Support Plate
8. Collar
9. ICE Sprocket
10. Electro-Magnetic Clutch

driver can select cruise mode by a switch in the driver's compartment. This is quite similar to a cruise control unit found in many passenger vehicles today (see Control Strategy). When this mode is not selected, the clutch is disengaged and the engine can run independent of the transmission, in series mode, or it can be turned off all together.

The APU-transmission speed reduction was selected such that a target speed of 72 km/h is reached in 4th gear. In equation 3 the calculations are shown for the 40 and 35 tooth sprockets.

$$\textit{Final Drive Ratio} = 3.619, \textit{ 4th Gear} = 0.972$$

$$\textit{Wheel Speed} = \frac{2700 rpm}{\frac{40}{35} \times 3.619 \times 0.972} = 674.52 rpm \quad (3)$$

$$\textit{Vehicle Speed} = 73.12 \textit{ km/h}.$$

DRIVE MODIFICATIONS - At certain times there will be up to three motors simultaneously driving the vehicle. These changes required a unique layout for the transmission system.

The original five-speed transaxle, differential and driveshaft were not changed. Simulations found this five-speed transmission to be necessary in order to meet the acceleration requirements. Extending the transaxle drive shaft allowed for the mounting of an external drive system. In terms of manufacturability and efficiency the direct coupling of an electric motor to the driveshaft is the best solution. The power of the chosen motors though did not allow for this option. Also their operating speed required a certain amount of gear reduction.

Cost, weight, lubrication and cooling precluded the use of gears to achieve the necessary gear reductions. It was determined that a simple set of motorcycle chains and sprockets as it was illustrated in Figure 1, was the best approach.

Chains and sprockets are readily available and easily interchangeable allowing for many different gear reductions to be easily obtained. In case of a break down, a change in the external drive system can be effected quickly so the vehicle could reenter the event and qualify for as many points as possible. A tensioning system was accomplished by allowing each motor to slide in bracket slots thus adequately tightening its respective chain. Covering the chains with a custom-designed shield reduces noise and protects the sprockets from contaminants.

ICE DRIVE CLUTCH - The electro-magnetic clutch is a main component of the vehicle since it will enable the vehicle configurations to be switched between cruise control and series at will. Tests proved that an oversized clutch is needed to avoid slippage caused primarily by the cyclic load from the ICE which averages out to 22 Nm. The **Pitts Industries** clutch model 28A-7 "high torque" rated at 88 Nm was chosen because of its high capacity and relatively small size. It works on a 12V DC source and requires only 4 Amps to maintain a magnetic connection. Installation and further tests proved that the clutch works admirably in its new environment with very little heat build-up or initial slippage.

A1.3 Fuel Type Selection

It was decided to operate the CHEV with

reformulated gas. The corrosive properties of methanol would add to the vehicle cost and conversion time, especially since the APU engine block is cast aluminum. Additional modifications to the fuel delivery system would be necessary and more testing would be required due to the availability of higher compression ratios. Thermal efficiency remains the same unless the compression ratio is changed.

As compared to gas, methanol is a heavier liquid and it would require twice the volume of methanol to travel a similar distance as with gas. The stoichiometric air-fuel ratio for methanol is 8.9 whereas for gas it is 15. Also, for a production vehicle the accessability of a fuel is of utmost importance. It was thus concluded that gas was the best option to meet competition requirements.

A1.4 Battery Selection

The battery selection process for the CHEV involved a compromise based on the batteries that best balance performance, cycle life, energy capacity and weight. It is important that the battery system:

- withstands continuous charging and discharging on a daily basis,
- functions under extreme working conditions such as high and low ambient temperatures,
- withstands vehicle vibrations.

Despite a considerable amount of research effort into the development of advanced batteries for electric vehicles, the only practical battery systems commercially available are still lead-acid and nickel-cadmium (nicad). A direct comparison between the **Eastern-Penn RV31, Saft Nife STM 1.130** and the **Marathon 44SP100** revealed that lead acid provides a comparable performance (Figure 6) at a fraction of the nicad system's cost.

The power and energy density requirements of a vehicle can be estimated through vehicle simulations. A total of five different commercially available lead-acid batteries were selected for constant testing. The weight of each 12V module was within our imposed maximum battery weight restriction of 30 kg, outlined in Table 3.

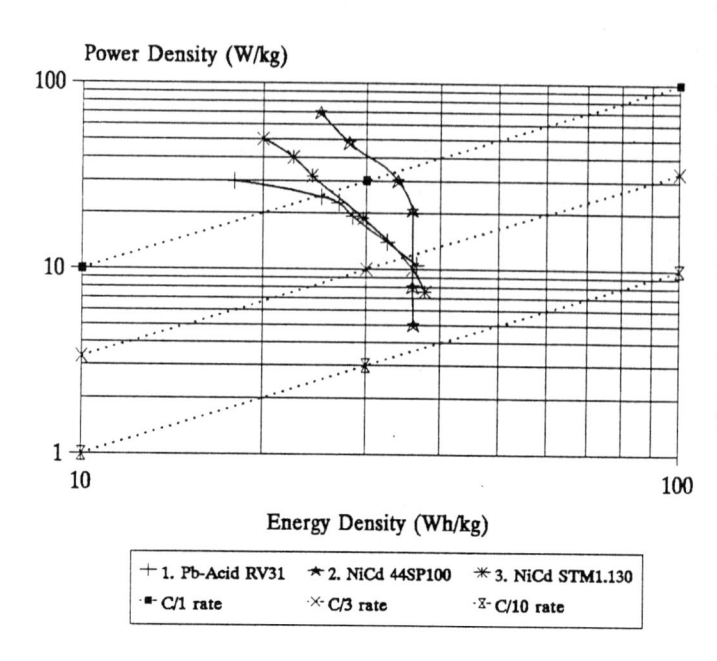

Figure 6 Ragone Curves - Battery Systems

Table 3 Battery Specifications

Battery Tested	Weight per 12V Module (kg)
Optima 800	18
Delco Voyageur 27FMF	24
Chloride 6EF78	26
Eastern-Penn RV27	24.6
Eastern-Penn RV31	28.5

The deep cycle open lead-acid **Eastern Penn RV31** batteries were selected for the CHEV. They are relatively maintenance free, aside from periodic watering. The battery casings are designed to operate as a "check-valve" allowing excess pressure to escape.

A preliminary calculation revealed that the CHEV would require a continuous draw current of 45 amps to maintain a constant velocity of 72 km/h with a bus voltage of 156V. Figure 7 depicts typical discharge curves for specific power vs. time. The East-Penn RV31, in terms of energy, out performed the other lead-acid batteries.

Although a continuous draw of 45 amps is not realistic in terms of a true driving cycle, it offers a good indication of the duration of the charge available from the batteries when subjected to a fast discharge.

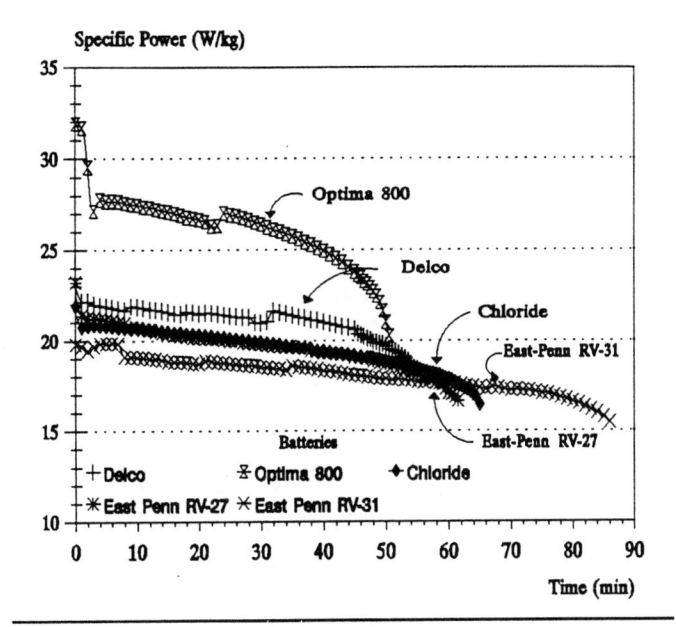

Figure 7 Energy Dissipated @ 45A Continuous Draw

A1.5 Power Conductor Selection

The CHEV power conductors were sized in accordance with the National Electrical Code. The electric drive system conductors were completely isolated from the vehicle chassis, and the 12V system was used for the vehicle instrumentation. High voltage conductors were selected with a 600V insulation rating, which corresponds to 3½ times the peak ground reference voltage of 173V. In Figure 8 one can see the location of the power conductors, high voltage areas, and fuses.

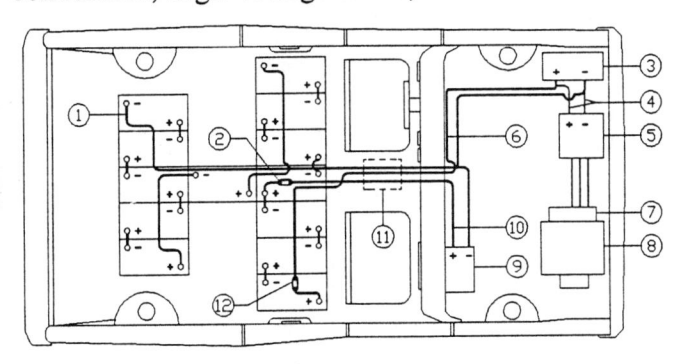

1. High Voltage Ground AWG I/0
2. 500 A Non-Time Delay Fuse
3. Solectria BRLS240H Controller
4. Rectifier Bridge Output AWG 10
5. Rectifier Bridge
6. Solectria Controller Conductor AWG 0/2
7. Fisher Alternator
8. Briggs and Stratton Engine
9. Curtis 1221B-74 Controllers
10. Curtis Controller Conductor AWG 0/2
11. Manual Battery - System Disconnect
12. 300 A Non-Time Delay Fuse

Figure 8 Power Conductor Locations

A.II CONTROL STRATEGY

The CHEV has two basic modes of operation identified as Series and Electric Vehicle (EV). Only one mode can be active at any instant of time, and the required mode is selected using the pushbutton pad located near the center of the dashboard. A third pushbutton enables the ICE to directly drive the vehicle at a constant velocity of 72km/h.

MODES OF OPERATION - Series is the primary mode of operation. In this configuration the vehicle is driven solely by the two electric motors. The ICE is also active and is used to recharge the battery pack, with the aid of the 9kW generator. In EV mode, the vehicle is running purely on electrical power. The ICE is shut off and is therefore not charging the main battery pack. A third mode, labelled as Cruise Control, is used to couple the ICE directly to the transmission, through a magnetic clutch. The ICE will then maintain the vehicle at a constant velocity of 72km/h. The control panel in Figure 9 contains all the pushbutton switches and indicator LEDs required for switching between operation modes.

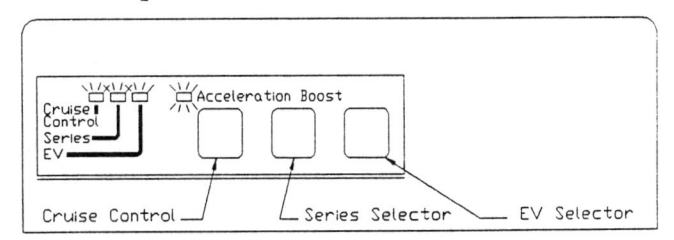

Figure 9 Vehicle Control Panel

CHANGING VEHICLE MODES - To ensure proper vehicle operation during startup and mode switching, several guidelines must be followed. When the vehicle is initially activated, with the use of the ignition key, the default mode of operation will be EV. The occupant may at this point drive the vehicle in EV mode, which is the recommended mode for city driving. If Series mode is required, the occupant need only press the Series push-button. The on-board controller will then take care of turning on the desired accessories and cranking the engine. Once the engine is running and a suitable oil temperature is reached, the controller will open the ICE throttle to achieve the necessary power output. The switch from EV to Series may be done on the move, without requiring that the vehicle come to a full stop. While on the highway, a cruising speed of 72km/h may be desired. This can be achieved most efficiently by enabling the vehicle's cruise control mode (the third pushbutton). In order to activate this mode, the vehicle must already be travelling at a speed approximately equal to 72km/h, and must be in Series mode. In the event that the speed of the vehicle is below 72km/h, the microcontroller will drop the engine speed to match the shaft speed as closely as possible before engaging the clutch. Once the clutch is engaged, the engine will attempt to reach the optimal speed setting. In this configuration, the electric motors are no longer required to propel the vehicle thus conserving battery power. The electric motors however are still online and can help the ICE if a steep grade is encountered.

HIGH ACCELERATION MODE - During driving conditions when acceleration is required, a pushbutton located on the stick shift, allows the driver to engage the A/DC motor. To facilitate this feature, the pushbutton acts as an input to the microcontroller and does not operate the motor directly. Using this approach, the driver is not required to let go of the gas pedal in order to reset the motor control potentiometer to zero. Instead the microcontroller uses a relay to momentarily short the potentiometer, thus making this process quite seamless.

ELECTRICAL CIRCUIT COMPONENTS - The control unit is located on the vehicle floor, just behind the firewall, between the driver and passenger. All control systems, such as charger relays, clutch relay and engine kill relay are controlled from this unit.

The entire circuit, illustrated in the Appendix is composed of a 68HC05 **Motorola** Microcontroller, a relay driver chip, an 11 channel A/D converter and 5 mechanical relays. The use of a 4MHz microcontroller gives this system a great deal of flexibility, and requires much less power than what would be required for a comparable logic gate system. To ensure that the system operates effectively under various conditions, a bandpass filter and notch filter were placed in line with the microcontroller supply. These filters eliminate all noise coming from the engine alternator, cooling fans, fuel pump and other high current sources and have proven to increase dramatically the reliability of the microcontroller unit.

A.III FUEL DELIVERY AND EMISSIONS CONTROL STRATEGY

A3.1 Fuel Delivery System

An electronic fuel injection system was developed for the ICE thus replacing the original gravity-feed carburettor. The **HALTECH** F7 fuel injection system is used, which is governed by a speed density configuration. The advantages of the HALTECH system over carburation are many. There is better fuel atomization due to high pressure delivery, leading to higher combustion efficiencies and controlled fuel enrichment during warm-up stages. This produces faster starts and precision tuning throughout the operational speed range. Tuning is possible by adjusting the injector open time up to 70 μs for varying load and speed conditions. Targeted stoichiometric conditions can thus be achieved.

SETUP - The fuel delivery layout is illlustrated in Figure 10. The fuel pump is submerged in the

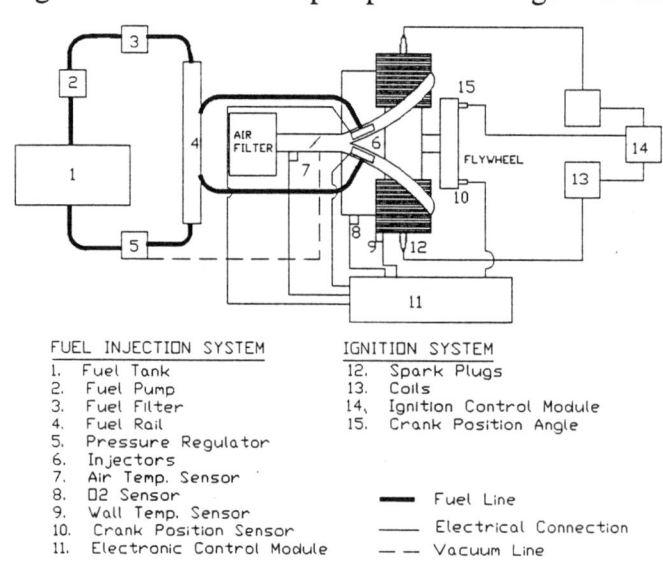

FUEL INJECTION SYSTEM
1. Fuel Tank
2. Fuel Pump
3. Fuel Filter
4. Fuel Rail
5. Pressure Regulator
6. Injectors
7. Air Temp. Sensor
8. O2 Sensor
9. Wall Temp. Sensor
10. Crank Position Sensor
11. Electronic Control Module

IGNITION SYSTEM
12. Spark Plugs
13. Coils
14. Ignition Control Module
15. Crank Position Angle

— Fuel Line
— Electrical Connection
— — Vacuum Line

Figure 10 Fuel Delivery Configuration

stock Escort fuel tank and it is used to deliver the

fuel through the filter to the injectors. A vacuum pressure regulator maintains a constant 40 psi pressure throughout the running system. The injectors used are from a 1.8 ℓ Mazda V6, since their flowrates closely match the requirements of the 520 cc B&S ICE.

As it is shown in Figure 11, peak torque for the injected engine occurs at 2700 rpm. This is also the point where the lowest brake specific fuel consumption occurs. It is for this reason that this speed has been selected as the operational speed in parallel mode.

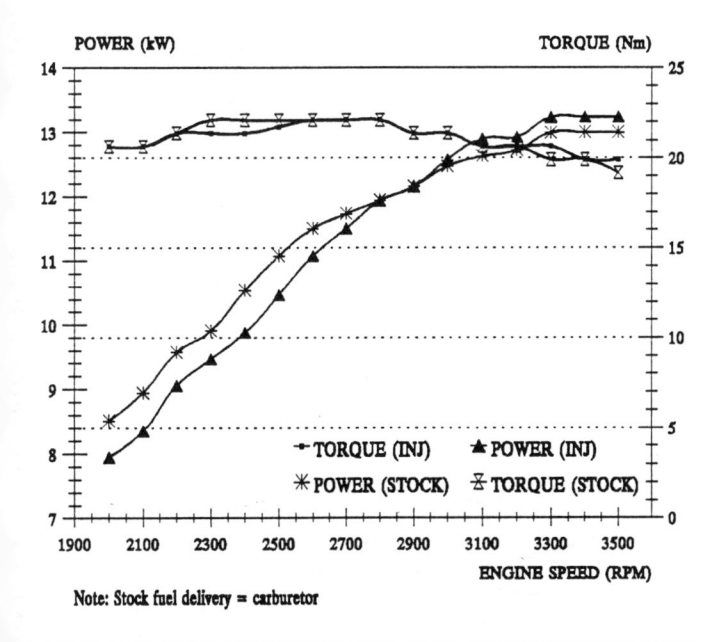

Figure 11 Briggs & Stratton (13.5kW) Engine Performance Characteristics

IGNITION SYSTEM - A **FIREPOWER** electronic ignition system was used to replace the stock magneto ignition system. The advantage to such hardware is the capability to set the required spark advance at any speed, in order to achieve maximum brake torque timing. This system provides an initial spark timing of 20° before top dead centre (BTDC) with the ability to advance the timing in steps of 2.8° at each 1000 rpm speed interval. The maximum total advance above initial timing is 19.8°. With the stock magneto system, an identical spark advance is used during all speeds.

A3.2 Emission Control Strategy

A three-way catalytic converter (TWCC) was selected as the main component of the emissions control strategy. The main advantage to such a selection is that the TWCC can achieve NO_x reduction as well CO and HC oxidation within one catalyst bed [4]. To achieve the highest conversion efficiencies for the aforementioned pollutants, the ICE should be operating at very near stoichiometric conditions. The implementation of an oxygen sensor situated in the exhaust manifold, permit the operator to precisely determine the fuel delivery parameters necessary to achieve the required stoichiometric conditions.

The TWCC provided by **Johnson Matthey** is selected with a 5.14:1 ratio of platinum to rhodium and a volume of 400 cm^3. The oxygen sensor is a HALTECH AF30 unit compatible with the HALTECH F7 electronic control module.

A.IV FUEL AND ELECTRIC POWER EFFICIENCY CONSIDERATIONS

ELECTRIC - The combined electrical efficiency map of the Advanced DC motor and Curtis Controller was strongly dependent on torque (Figure 12). The electrical efficiency of the Solectria BRLS16 motor and controller was strongly dependent on the torque and speed conditions. This is graphically illustrated in Figure 13. As it is shown by operating the motors at their respective torques of 35-45 Nm, the most efficient electric drive can be accomplished.

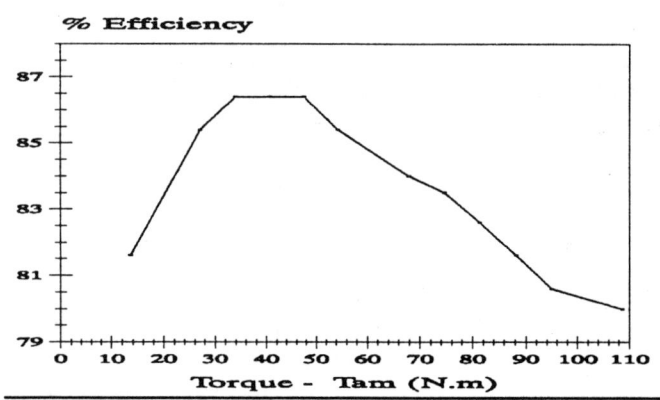

Figure 12 Advanced DC & Curtis Controller Efficiency Map

FUEL - To effectively incorporate the fuel control strategy certain steps were undertaken. The catalyst was placed at 56.5 cm from the ICE exhaust ports.

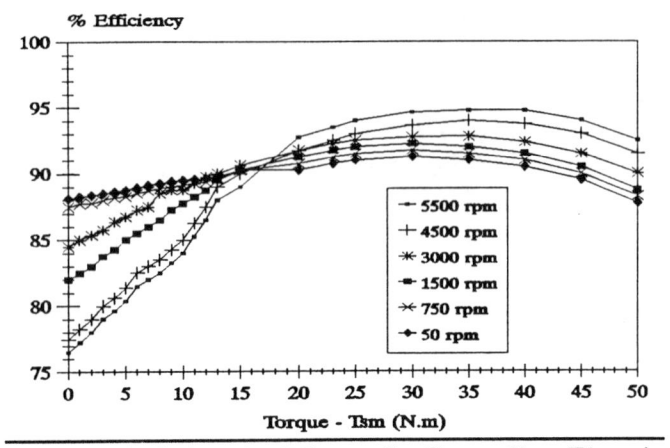

Figure 13 Solectria Motor & Controller Electric Efficiency Map

This effectively limits the power lost due to back pressure and also minimizes any catalyst deterioration. Optimum tuning of the engine is being determined with experimental test results such as emissions, power and fuel consumption. Figure 14 shows performance results for a stock 12kW B&S carburetted engine. Final results for the 18 kW engine are not available.

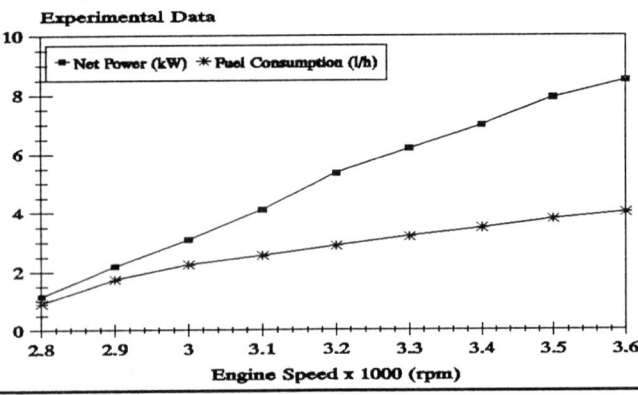

Figure 14 Power and Fuel Consumption vs. Engine Speed (Test - 12kW Engine)

A.V WEIGHT DISTRIBUTION

In order to get a breakdown of the HEV weight, the original unladen vehicle weight of the Ford Escort Station Wagon (1050 Kg) 59%/41%, front-to-rear bias and 52%/48% left-to-rear bias was used as a starting point. Tables 4 and 5 summarise the current rear and front weight distribution of the Concordia HEV.

The current vehicle weight at 1702 kg is below the allowed GVW+15% rating. Therefore, vehicle structural integrity and handling have not been compromised from the point of view of overall loading. Nevertheless, front/rear and left/right weight distributions have been calculated. The front/rear weight bias is calculated in equations 4-6. Calculation of the left/right weight bias is shown in equations 7 to 9.

$$FRONT\ WEIGHT\% = \frac{833}{833+869} *100 = 48.95\% \quad (4)$$

$$REAR\ WEIGHT\% = \frac{869}{833++869} *100 = 51.05\% \quad (5)$$

$$Front/Rear\ \%\ Weight\ Bias = \frac{48.95\%}{51.05\%} \quad (6)$$

$$LEFTside\ WEIGHT\% = \frac{421.5+422.5}{1702} *100 = 49.23\% \quad (7)$$

$$RIGHTside\ WEIGHT\% = \frac{417.5+446.5}{1702} *100 = 50.76\% \quad (8)$$

$$LEFTside/RIGHTside\ \%\ Weight\ Bias = \frac{49.23\%}{50.76\%} \quad (9)$$

Table 4 Frontal Weight Distribution

ITEMS	FRONT WEIGHT (Kg)	LEFT WEIGHT (Kg)	RIGHT WEIGHT (Kg)
Total Frontal Weight of Standard Ford Escort Wagon	620	295.5	324.5
Removal of Engine and Cooling System	-143	-53	-90
Removal of Existing Exhaust System (Weight at Front)	-13	0	-13
Weight of Passenger and Driver	160	80	80
Solectria Electric Motor BRLS16	29	29	0
Solectria Controller BRLS240H	7	7	0
Advanced DC Electric Motor 203-06-4001	49	0	49
Curtis Controller 1221-B74	10	10	0
ICE and Support Structure	50	10	40
New Exhaust System (Front)	4	2	2
Fisher Alternator	10	5	5
Cover Plate and Mounting Brackets	50	30	20
TOTAL ANTICIPATED FRONTAL WEIGHT	833	415.5	417.5

Table 5 Rear Weight Distribution

ITEMS	REAR WEIGHT (Kg)	LEFT WEIGHT (Kg)	RIGHT WEIGHT (Kg)
Total Rear Weight of Standard Ford Escort Wagon	430	202	228
Weight of First Battery Tray	228	114	114
Weight of Second Battery Tray	158	80	78
Wiring	4	2	2
Cargo Ballast Box with 20Kg	23	11.5	11.5
Rollbar	20	10	10
Exhaust System (rear)	6	3	3
TOTAL ANTICIPATED REAR WEIGHT	869	422.5	446.5

A.VI SUSPENSION AND BRAKING MODIFICATIONS

With the significant weight contribution of batteries in the rear of the vehicle, minor modifications to the braking system and rear suspension were inevitable. The objective was to retain the vehicle's original level, comfort, braking and handling. The overall front weight increase was not substantial thus modifications to the front suspension were not necessary.

SUSPENSION - To better estimate the suspension modifications required, the basic suspension model of Figure 15, has been used to simulate the behaviour of the system. Using this model, a force balance can be written for each mass:

$$m_u \ddot{y}_u + D_s(\dot{y}_u - \dot{y}_s) + K_u(y_u - y_g) + K_s(y_u - y_s) = 0$$
$$m_s \ddot{y}_s + D_s(\dot{y}_s - \dot{y}_u) + K_s(y_s - y_u) = 0 \quad (10)$$

The dynamics of the rear suspension system was simulated in Figure 16 for a quarter car model subject to a radical change in road profile of 5 cm. Original stiffness and damping values were linearized and approximated from vehicle specifications. The simulation displays the

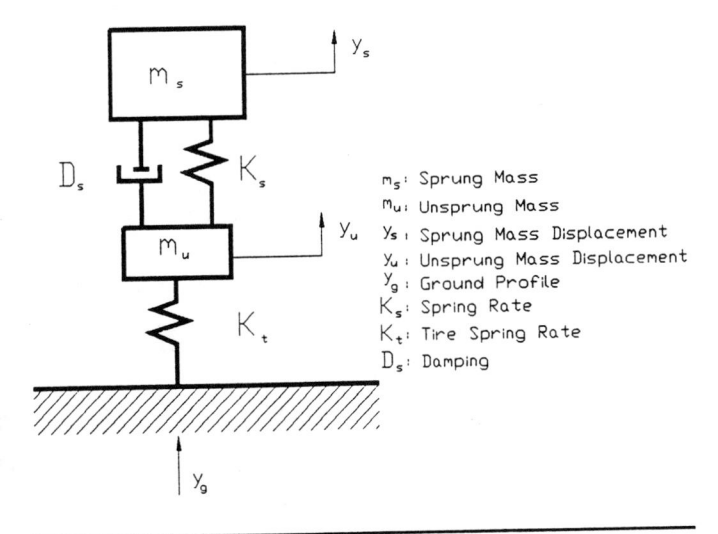

Figure 15 Basic Suspension Model

displacement of the sprung mass from its original position. The response of the second curve indicates an increase in oscillations when the additional weight of the converted vehicle is introduced. The third response is almost identical to the original system. This was accomplished by increasing the damping and stiffness by 50%. From these results, a rough estimate was provided as to the extent of modifications required to restore the comfort and driveability of the vehicle.

Modification of the rear suspension involved replacing the original springs with cargo springs and an addition of rubber spacers within the springs. Consequently, the level of the vehicle was raised by 5 cm, returning it to the approximate level before conversion. Road tests were performed with the modified suspension and good comfort levels and handling were noted. This compared favourably, relative to similar tests that were performed on the original vehicle.

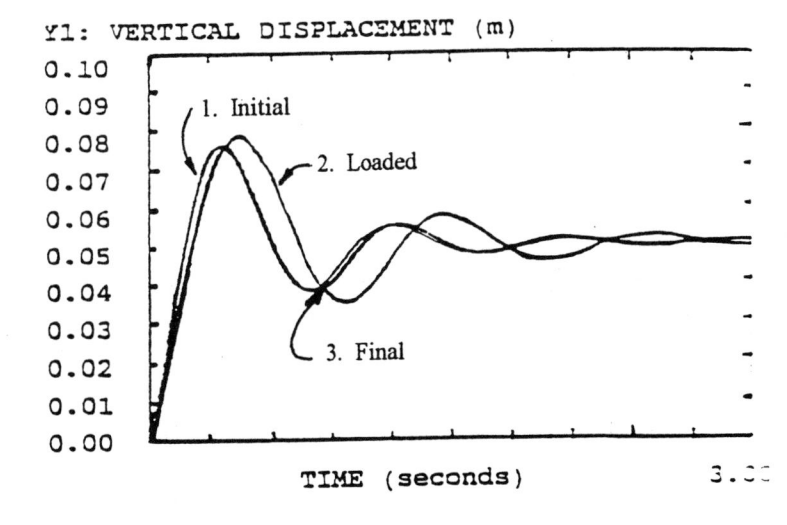

Fig. 16 Simulation of Suspension Dynamics

BRAKING - Initial tests on the braking system, revealed that the vehicle weight increase and the absence of engine vacuum, resulted in difficulties when trying to stop from speeds beyond 40 km/h.

An electric vacuum pump was installed for increased braking capabilities, beyond levels achieved by the standard brakes. Along with regenerative braking the vehicle now has a very reliable braking system, which even outperforms the original set-up. Conforming to the competition rules the CHEV incorporates both a primary and secondary braking system. An illustration of the electric vacuum pump and its associated components, supplied by **Solar Car Corporation**, is shown in Figure 17.

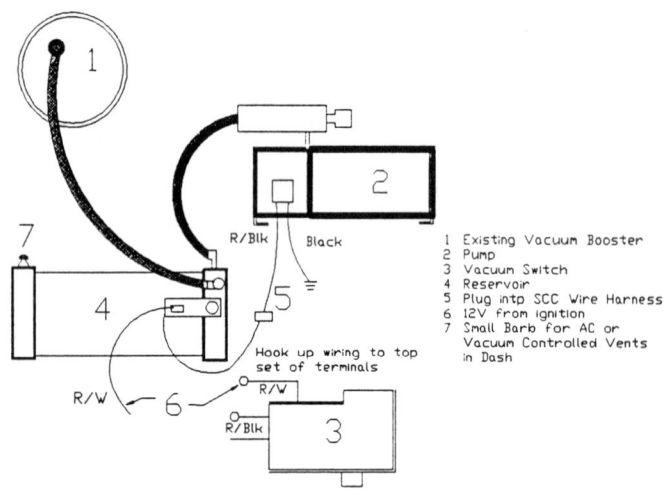

Fig. 17 Electric Vacuum Pump

Legend:
1. Existing Vacuum Booster
2. Pump
3. Vacuum Switch
4. Reservoir
5. Plug into SCC Wire Harness
6. 12V from ignition
7. Small Barb for AC or Vacuum Controlled Vents in Dash

A.VII MANUFACTURABILITY AND ERGONOMICS

The CHEV was converted into a two passenger vehicle having ample comfort and aesthetic appeal. By reducing vehicle modifications and employing workable simple solutions the conversion process was quick and effective. The vehicle is a finished product and great care was taken to ensure that controls and other items requiring maintenance and inspection are easily accessible.

There are no visible protrusions thus preserving the original vehicle aerodynamics and styling. The mounting of the controllers and motors was performed in an orderly fashion, paying careful attention to the placement of brackets and wiring. A roll cage, five point seat belt harness and a fire extinguishing system were added for safety. Additional gages were added to provide the driver with information on the status of the vehicle electrical systems. This is an important consideration when driving in EV or Series mode.

Each newly installed part in the vehicle was sized and manufactured accordingly. The use of steel was limited to high stress and load areas, but aluminum and fibre reinforced plastic were used elsewhere to reduce weight. Using alternatives to steel substantially reduced weight, cost and the overall manufacturing time.

The major components in the automobile were purchased "off the shelf", except for the customised alternator. The additional components required to mount these parts to the car were all built in the student machine shop. In light of this, duplication of the CHEV would be quite simple once detailed drawings of all components are documented.

A.VIII SIMULATION AND OPTIMIZATION

SIMULATION - A detailed mathematical model was established using governing vehicle dynamic equations. A simulation program was developed and used to optimise the design parameters, such as mass of the energy storage system, electric motor operating voltage and electric drive final gear ratio. The effects, these parameters had on the objective functions, namely range, acceleration, battery cycle life and cost were investigated.

A good correlation between the experimental and simulated acceleration time was established and is illustrated in Figure 18. In terms of range, the simulation projected 74.6 km on a flat surface, when travelling at a speed of 72 km/h, whereas during experimental tests a range of 71.2 km was obtained. However, it should be noted that the road surface on the test track had several banks (leading to a small increase in power consumption), which could account for this deviation of 4.8%.

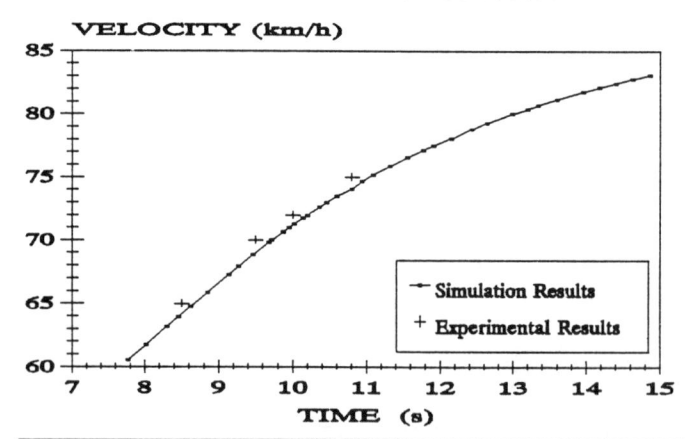

Figure 18 Acceleration Time Versus HEV Velocity

The time response to accelerate the CHEV to 72 km/h, which incorporated the Solectria and Advanced DC electric motors with a five speed transaxle, was 10.4 seconds. Since, under steady-state conditions the Solectria electric motor would only be active, it would provide the entire torque to maintain the vehicle at a constant speed of 72 km/h. By providing the entire load, the Solectria motor achieved an electrical efficiency of 93.5 %. The corresponding power consumption of this system was calculated to be 10.76 kW. This design configuration provided a very effective compromise between performance and cost, therefore, it was selected as the drive system for the developed CHEV in question. The system response is illustrated in Figure 19.

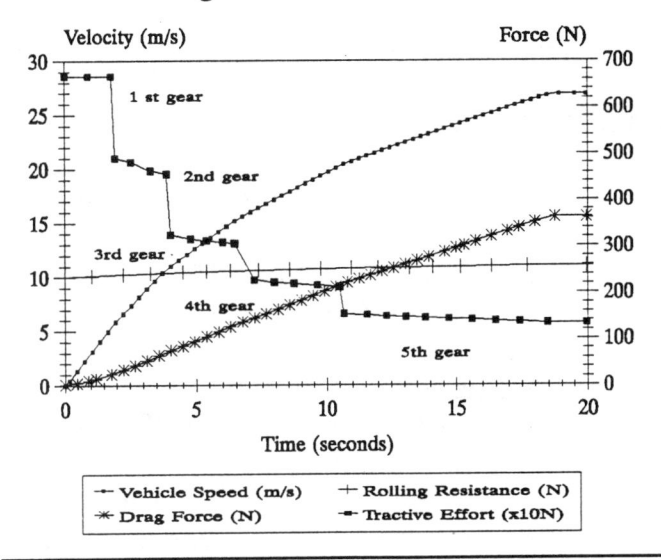

Figure 19 System Response - 5 Speed Transaxle Solectria and Advanced DC Drive System

OPTIMIZATION - In order to perform an optimization procedure, the feasible region of the entire system must be defined, by reasonably selecting design constraints. The lower and upper limits, that define the constraints for the electric drive system optimization process, are represented in equations (11) through (14), (also see nomenclature) along with nominal values for each design parameter.

Battery Mass:
$$275 kg \leq M_B \leq 425 kg \; ; \; M_{Bn} = 350 kg \quad (11)$$

Electric drive final gear ratio:
$$1.0 \leq N_g \leq 3.1 \; ; \; N_{gn} = 2.04 \quad (12)$$

Operating voltage of the electric drive system:
$$96 \, Volts \leq V_{OS} \leq 168 \, Volts \; ; \; V_{OSn} = 132 \, Volts \quad (13)$$

Energy density of the storage system (C/3 basis):
$$27 \frac{Wh}{kg} \leq S_E \leq 161 \frac{Wh}{kg} \; ; \; S_{En} = 33 \frac{Wh}{kg} \quad (14)$$

The nominal design parameters and these design constraints were utilized to performed a graphical optimization, which essentially outlines the sensitivity of the engineering system. In order to better understand the engineering system, which is being analyzed, a sensitivity analysis was initially performed with the aid of the established simulation. By varying the design parameters within the constraints formulated previously, their impact on the objective functions could be investigated. Each objective function was obtained by varying one design parameter, while holding the remaining nominal values constant. Results are shown for the varying of the electric drive final gear ratio (Figure 20).

An emphasis was placed on acceleration time and range. The following parameters were thus used as inputs to the simulation program:

- Battery mass: M_B = 400 kg,
- Electric drive final gear ratio: N_g = 2.05,
- Nominal operating voltage: V_{OS} = 168 V,
- Specific energy density of the storage system: S_E = 33 Wh/kg.

The anticipated performance of the vehicle is as follows:

- 9.31 seconds (0 to 72 km/h),
- 1600 $ battery pack cost,
- 11.9 kWh battery pack energy,
- 93 km range (full charge basis) travelling at 72 km/h on a flat surface, and
- 55 900 km obtainable range from the energy storage system during its entire life.

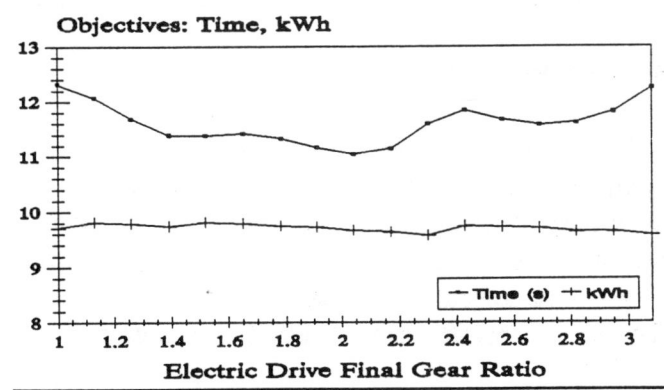

Figure 20 Acceleration Time and kWh Vs. Electric Drive Final Gear Ratio

CONCLUSION

As it was detailed in this report, designing a practical, efficient hybrid electric vehicle which meets today's performance standards involves considerable time and preparation. The objective of the design was to maintain a low overall vehicle weight while capitalizing on storage space. This facilitated the task of placing the batteries without significantly compromising interior space, comfort and styling.

A control strategy was developed that would not compromise the integrity of each drive system's autonomy. The control scheme was imperative in granting the driver a means of controlling the various operation modes. The fuel control strategy is geared towards the minimisation of harmful pollutants while maintaining adequate fuel economy. In conjunction with the ICE efficiency the electric drive efficiency was also determined to be optimal for this CHEV.

Final assessment of the weight distribution disclosed an excellent front/rear and left/right bias. The effects of the added weight were countered by minor suspension and braking system modifications.

The use of a valid mathematical model in conjunction with a computer simulation provided an overview of what is involved and where complications may arise. Together, they form a useful tool for projecting the performance of a particular system as well as the impact of varying certain design parameters.

It has been established that the CHEV would be very competitive with today's hybrids in range, acceleration and efficiency in all operating modes.

ACKNOWLEDGMENTS

The authors of this report would like to thank the balance of the CHEV team members for their hard work and dedication. We also appreciate the help and support extended by Sadi A. Arbid, Petros Frantzeskakis, and our faculty advisor Dr. T. Krepec and our sponsors. Team Sponsors :

Concordia University Dept. of Mechanical Engineering
Environment Canada, Mobile Service Emissions Division
Natural Resources Canada - CANMET
Hydro-Québec
Quebec Government
Transport Canada Vehicle Test Center - CEVA
Federal Muffler
Nutek - Firepower
Michelin
NGK
Power Surge Batteries

NOMENCLATURE

η_{mode} - overall efficiency for an operation mode,

F_a - Aerodynamic drag force, N
F_{ro} - Aerodynamic drag force, N
M_B - Battery pack mass, kg
M_{Bn} - Nominal value of M_B, kg
N_g - Electric drive final gear ratio (Solectria)
N_{gn} - Nominal value of N_g
$P_{APUmode}$ - APU Power in an operating mode, kW
S_E - Specific energy of the storage system, Wh/kg
S_{En} - Nominal value of S_E, Wh/kg
T_{am} - Torque developed by the Advanced DC motor, Nm
T_{am} - Torque developed by the Solectria motor, Nm
V_{HEV} - HEV Velocity, m/s
V_{OS} - Operating voltage of the electric drive system, V
V_{OSn} - Nominal value of V_{OS}, V.

REFERENCES

1. Ted Lucas and Fred Reiss, <u>How To Convert To An Electric Car</u>, Crown Publishers Inc., NY, New York, 1975.
2. Solectria Electronics Division, Pamphlet and Instruction Manual.
3. Advanced DC Motors Inc., Motor Technical Specifications.
4. Heywood, J. B., <u>Internal Combustion Engine Fundamentals</u>, McGraw-Hill, 1988, pg. 655.

APPENDIX

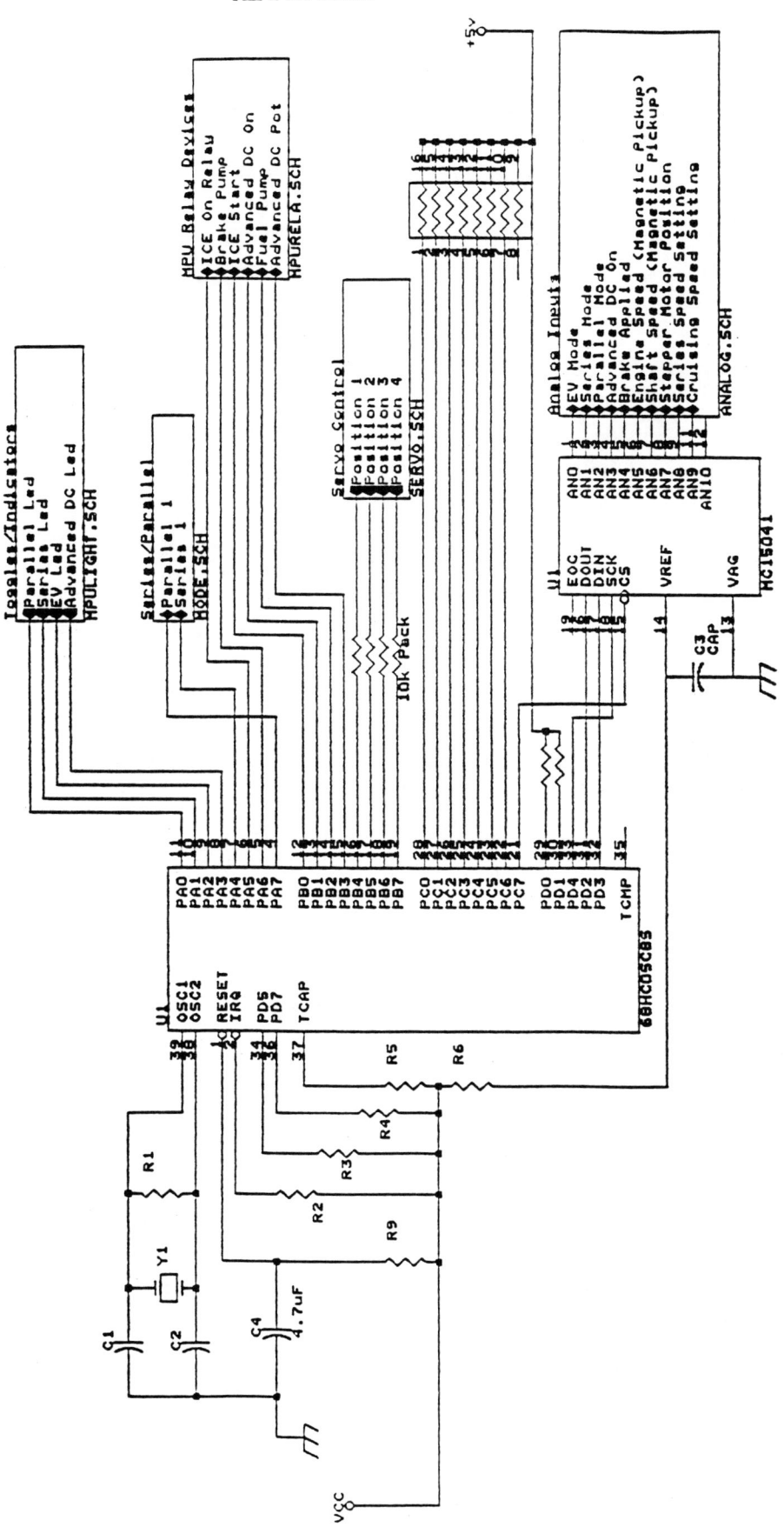

Fig. 21 Circuit Diagram

The Vortex - Design and Construction of a Two-Passenger Hybrid Electric Vehicle

Robert L. Holden
James A. Stevens
William T. Briggs
Abhinov Singh
Tara J. Wilson

Jose J. Garcia
Mariano S. Garcia
William D. Kolosi
Kevin Newman

Cornell Hybrid Electric Vehicle Team

1.0 ABSTRACT

Engineering students at Cornell University have designed, constructed, and tested a second generation, front-wheel drive Hybrid Electric Vehicle (HEV). The vehicle, called the *Vortex*, has been designed primarily for the 1994 Tour de Sol and the 1994 GM HEV Challenge Competitions. The *Vortex* uses a dual electric motor system coupled via a modular gearbox to a limited-slip differential to achieve high motor efficiencies (92%) and an excellent theoretical acceleration (0 to 72 km/h (45 mph) in 7.9 seconds). The vehicle's light weight (1100 kg fully loaded) and aerodynamic body (estimated C_d of 0.30) allows a range of 100 km (63 mi.) at 64 km/h (40 mph) from a 168V sealed-lead-acid battery pack. Titanium is used extensively throughout the suspension and frame to keep chassis weight under 220 kg. The fuel injected alternate power unit (APU) uses M85 as a fuel and outputs 15 kW while only weighing 50 kg. The APU is used in a series configuration so that the *Vortex* can be converted to a purely electric vehicle with as little difficulty as possible. A data acquisition system and digital display are combined to graphically provide the driver with all pertinent vehicle information.

2.0 TEAM STRUCTURE

Our team consists of two advisors (Professor Robert Thomas (EE) and Assistant Professor Richard Warkentin (ME)) and 43 engineering students from five different disciplines. To minimize development time, the students are divided into seven subgroups: APU, Business, Ergonomics, Fairing, Frame, Powertrain, and Suspension. Each group is responsible for developing their section of the vehicle. By using concurrent engineering and Total Quality Management techniques, most conflicts are resolved early in the design process.

3.0 VEHICLE POWERTRAIN

The Powertrain group is responsible for the transmission of power from the batteries and APU to the wheels. This includes the design or selection of batteries, motors and controllers, mechanical transmission, battery charger, wiring for all high power components, and an electronic management system which oversees the performance of all aspects of the powertrain.

Reliability, efficiency and modularity have been the guiding principles behind our design. Reliability is critical to everyday use, such as the typical commute between home and work. An efficient vehicle minimizes the depletion of natural resources while further decreasing operating costs. Finally, by making both mechanical and electrical systems as modular as possible, the vehicle can be altered to suit a wide variety of user needs.

3.1 MOTOR SYSTEM - Our first task was the selection of a suitable motor and controller system. After evaluating the performance of last year's vehicle, we determined target values for vehicle mass, coefficient of drag, rolling resistance and other parameters (See Table 1). Using these values, we calculated the minimum cruising power required for a vehicle speed of 88 km/h (55 mph) as follows:

Aerodynamic drag force:

$$F_{drag} = \tfrac{1}{2} \rho V^2 A C_D = 171.85 N \qquad (1)$$

Rolling resistive force:

$$F_{roll} = \left(mg - \tfrac{1}{2}\rho V^2 A C_l\right) C_{rr} = 110.24 N \qquad (2)$$

$$C_{rr} = C_{tire} + .00972 \frac{V}{44.704} = .0102 \qquad (3)$$

Total power required to cruise:

$$P_{cruise} = \frac{F_{total} V}{\eta_{trans}} = 8.67 kW \qquad (4)$$

Table 1: Target Vehicle Parameters

m	(mass)	= 1100 kg
V	(speed)	= 88 km/h
A	(frontal area)	= 1.75 m^2
C_d	(coefficient of drag)	= 0.27
C_L	(coefficient of lift)	= 0.5
C_{tire}	(coefficient of rolling resistance for tire)	= .008
η_{trans}	(transmission efficiency)	= 80%

A modified acceleration simulation program written for an automotive engineering class at Cornell was used to estimate vehicle performance. The inputs to the program include the vehicle parameters listed in Table 1, as well as information on the range event course. The program performs incremental analysis, calculating power and speed for each section of the course as well as aerodynamic and rolling losses. Using a single gear reduction which minimized acceleration time and allowed a top vehicle speed of at least 88 km/h, the theoretical 8.67 kW motor took 24 seconds to complete the 100 meter run and finished with a vehicle speed of only 30 km/h (19 mph).

Based on acceleration alone, a motor of this size is not suitable for use in a typical commuter vehicle, so we began the search for a high power motor. We discovered the Unique Mobility SR218P/(2) which could provide 63 kW with a peak efficiency of 92%. Along with its high efficiency, the Unique brushless DC permanent magnet motors have extremely high power to weight ratios due to high energy "rare earth" (Neodymium Iron Boron) magnets, and high pole count design. Initially, this appeared to be our ideal motor-it gave us an acceleration time of 7.95 seconds and achieved a speed of 79 km/h (50 mph) by the end of the acceleration run.

In order to determine the operating efficiency of the brushless DC motor, the torque versus motor RPM versus efficiency curves were examined. By overlaying curves of

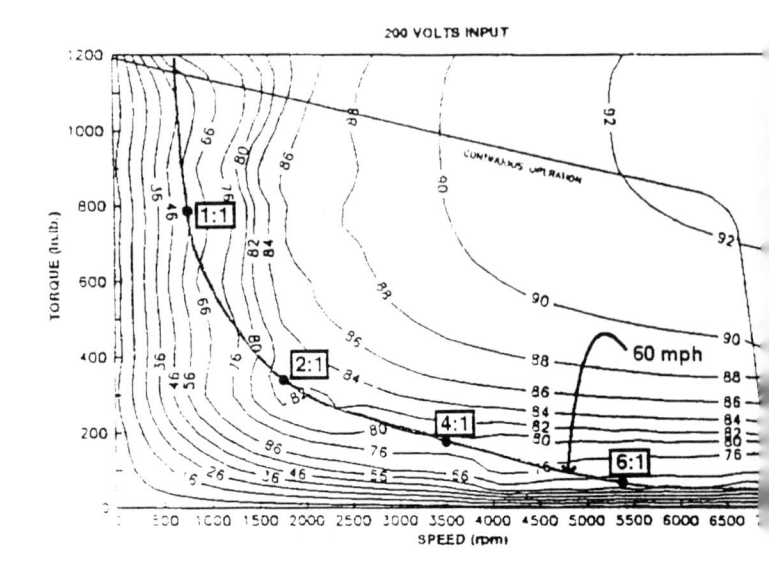

Figure 2
SR218P/(2) CR20-300 System Efficiency Map

constant gear reduction and constant cruising power at different vehicle speeds (Figure 1), we found that the optimal gear reduction for acceleration is 9.28:1. However, our peak motor and drive efficiency would be less than 30% (Figure 2). The required gear reduction for a peak efficiency of 82% at a vehicle speed of 96 km/h (60 mph) was approximately 2:1, which increased our acceleration time to 11.9 seconds. Similar calculations for other single motor systems led us to the conclusion that no single motor can provide both reasonable acceleration and high efficiency.

3.2 TWO MOTOR SYSTEM - By combining a 15.8 kW Unique DR156S motor with a 45 kW Advanced DC L91-4003, both high cruising efficiency and high on-demand acceleration were obtained. The Unique is used for cruising and normal acceleration, while the Advanced DC is used only

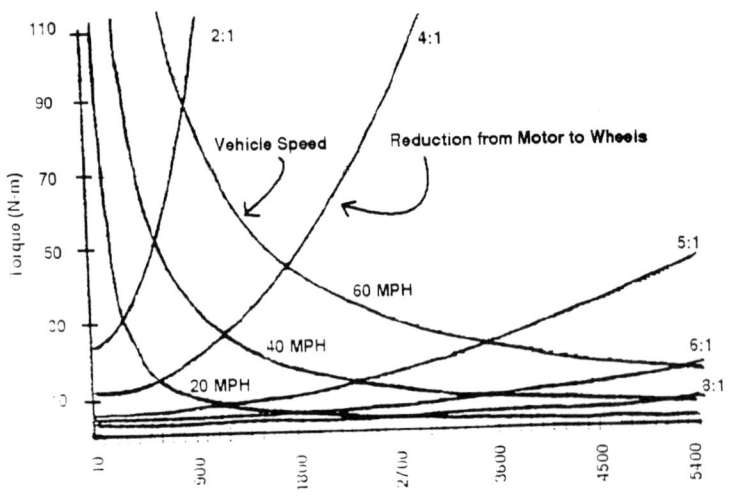

Figure 1
General Efficiency Map- RPM versus Torque

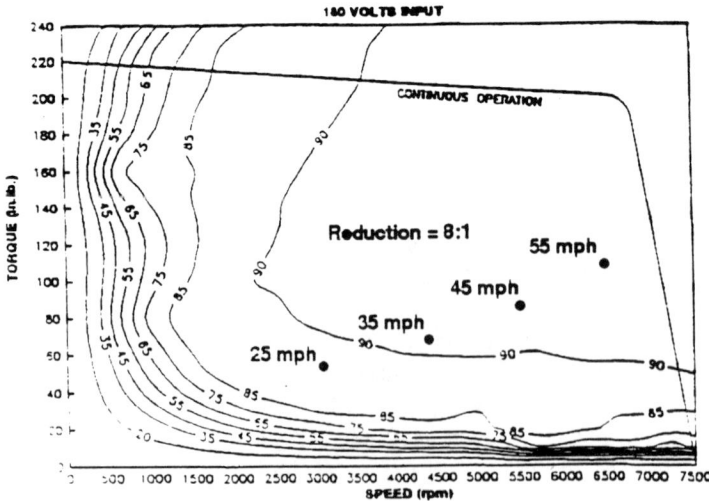

Figure 3
DR156S/CR20-150 System Efficiency Map

during high load situations. With this configuration, our theoretical cruising motor efficiency is over 90% (Figure 3) and our theoretical acceleration time is 7.8 seconds.

Due to the tedious nature of the calculations required to obtain these results, a complete vehicle simulation and optimization software package known as *BlizzOracle* (named in a friendly reference to last year's car) was developed. *BlizzOracle* takes as inputs: motor power and efficiency curves (for any number of motors), a range event profile, vehicle parameters, battery pack specifications, and vehicle performance requirements (such as top speed). It then outputs an ideal gear reduction for each motor and subsequent vehicle performance (top speed, acceleration time, range, etc.). This simulation enabled us to experiment with various motor and gear reduction combinations which would have been impossible to do by hand.

3.3 TWO INPUT GEAR BOX - Transmission of power from the motors to the road is accomplished through a custom two-input gearbox (Figure 4), mated to an 11 pound limited slip differential taken from a Honda TRX300FW FOURTRAX 4X4. This gearbox can also accommodate a belt or chain drive system. Changeover from gear drive to belt or chain drive simply requires an exchange of gears for pulleys or sprockets. Due to the modularity of this design, it will be possible to compare the efficiencies and drive characteristics of each system, allowing us to choose the best system for our car.

The output shaft of the gearbox is designed to be dimensionally equivalent to the output shaft of the Unique motor. This enables a direct coupling of the Unique motor to the differential without the additional weight of the gearbox and the Advanced DC. Other motors could easily be used in a similar fashion because of the versatility of the design.

As well as being modular, the gearbox is highly manufacturable. We began by CNC machining a positive mold of the gearbox, enlarged by 12 thousandths of an inch per inch to account for shrinkage. We then cast the gearbox in both aluminum and magnesium at a high volume casting facility in Ohio. The castings were hardened, machine finished, assembled, and tested in order to determine the advantages and disadvantages of each material.

3.4 CONTROL SYSTEMS - Figure 5 shows a block diagram of the motor and controller configuration. The DC permanent magnet motor is powered by a Unique CR20-150 controller which features regenerative braking, closed-loop speed control, and a user friendly interface. The Advanced DC is powered by the MHOG20-80 which was designed and built by the Powertrain Team. Under normal cruising conditions, the Unique motor runs continuously and the Advanced DC is isolated from the drivetrain by an overrunning clutch located within the gearbox (Figure 4). When the VCS detects that the vehicle cannot maintain a desired cruising speed with the Unique (due to a change in road conditions such as an increase in grade), the Advanced

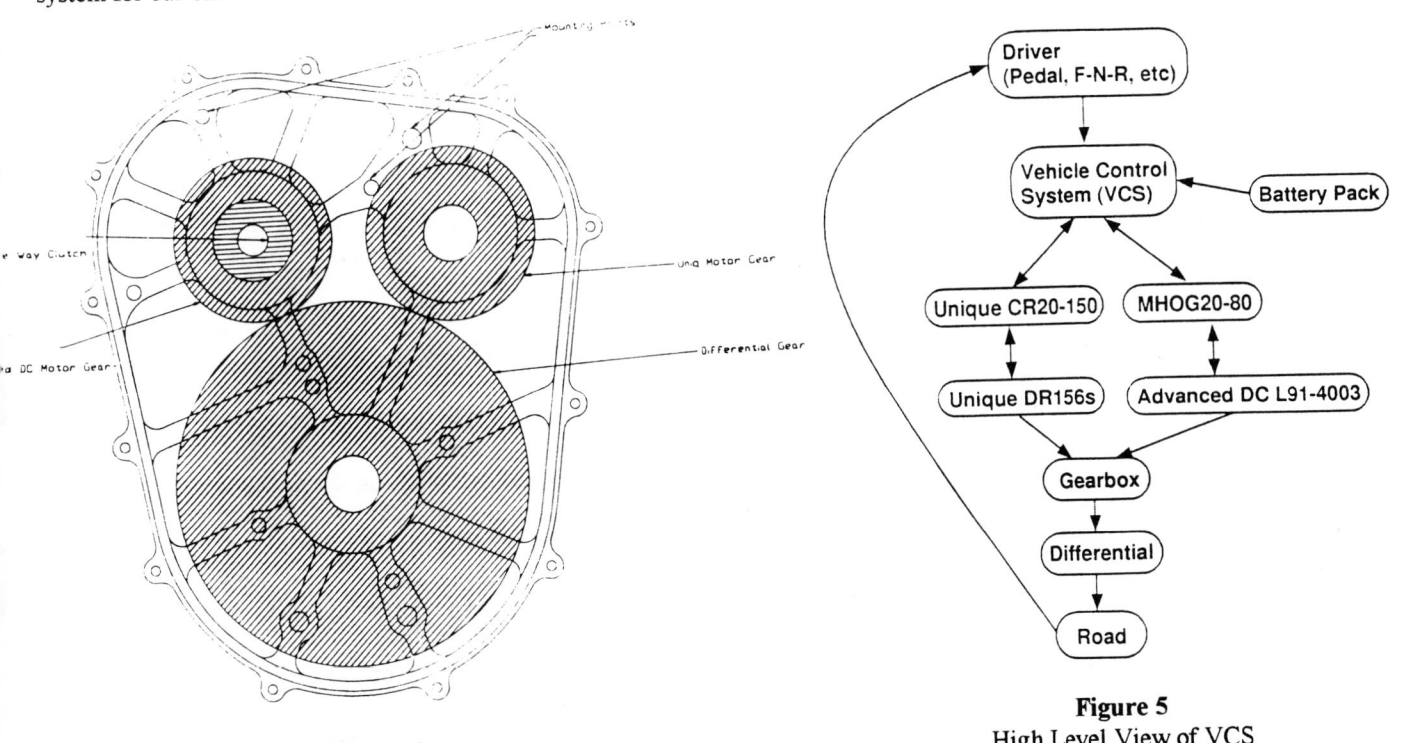

Figure 4
Gearbox Schematic

Figure 5
High Level View of VCS

DC speed is slowly ramped up to avoid shock loading the overrunning clutch. When the Advanced DC is no longer needed, it disengages, and the Unique provides all of the required power. By minimizing the use of the Advanced DC, this control algorithm maximizes overall efficiency, while increasing the life of drivetrain components.

The MHOG20-80 is a high-voltage, high-current PWM DC motor controller based around the Intel 87C196KB 16 bit embedded microcontroller. The MHOG20-80 replaces the Curtis 1221B controller commonly used to power Advanced DC motors. The Curtis controller cannot handle our high battery pack and charging voltages, nor can it provide the continuous high current levels required to achieve maximum power output from the Advanced DC. The MHOG20-80 can withstand pack voltages up to 200 Volts and can pass 800 Amps peak when cold and 500 Amps continuous under normal operating conditions. The power stage of this controller is mounted on an aluminum gold-chromate-coated bonded-fin forced-air heat sink of low thermal resistance (.022 K/W) assisted by twin 230 cfm axial 12 Volt DC fans.

Modularity has been the key design criterion for our powertrain. As a result, the control stage of the MHOG20-80 has been duplicated and is the basis of the VCS as well as the APU Control System. In order to make this possible, we designed our own printed circuit boards which can be used for any of these applications. Our battery charger is based on this PC board platform with a slightly modified power stage and reprogrammed control stage. Recently, burp-mode battery chargers (chargers which discharge packs at high rates for extremely short periods of time in an attempt to increase the pack's charge acceptance) have come to the forefront of electric vehicle technology. Earlier this year, a vehicle using this technology broke the world record for distance traveled by an electric vehicle in 24 hours. Any significant breakthrough in charging technology promises to bring the goal of practical electric vehicles much closer to fruition. By integrating these techniques into our charger, we expect to improve our overall energy efficiency, and provide our car with a competitive edge.

3.5 BATTERY SELECTION - By entering the range event track profile supplied by the challenge organizers into our simulation program, we determined that the energy required to complete a single 3.5 km lap at an average speed of 72 km/h is approximately 0.37 kW-hrs. Since the minimum ZEV range required by the competition is 40 km, the minimum energy supplied by the battery pack would be 4.26 kW-hrs. However, such a low ZEV range increases our dependence on the Alternate Power Unit (APU) which dramatically decreases our overall system efficiency. This led us to begin our search for a high energy battery pack.

Using the 85.1 Amp-hr (@ 80% DOD, C/3) Trojan 27TMH Deep Cycle Marine battery from last year's vehicle as an initial reference, we determined we would require 50 Trojans (at a total of 600 volts) to go the 480 km required in the range event. Although this eliminates the need for an APU, it is obviously not a practical solution.

Since our motor controllers cannot operate in excess of 200 Volts, we initially selected 192 Volts as our maximum pack voltage. At 192 Volts, the Trojans can provide 16.34 kW-hrs which yields an approximate ZEV range of 150 km. However, lead-acid batteries require a slight over-voltage condition during charging, roughly 2.4 Volts per battery. This would make our charging voltage for a 16 battery pack 230.4 Volts which is in excess of the maximum operating voltages of our controllers. The 200 Volt charging limit forced us to choose 168 Volts as our final pack voltage. While the range provided by lead acid batteries is acceptable, we investigated alternate battery compositions in an attempt to improve upon our design

Nickel Metal Hydride batteries have the highest specific energy and power presently available, and are less hazardous to the environment than other nickel-based batteries. However, due to their recent emergence on the market, they are prohibitively expensive and difficult to obtain.

Recent advances in lead-acid technology have led to the development of the Horizon ELSI1205 battery, which uses a unique fiberglass mesh coated with nearly pure lead to increase overall battery performance. In fact, the Horizons boast performance very near that of nickel based batteries at only 30% of their cost. However, Horizon is currently facing production problems and will not be able to provide batteries in time for the HEV Challenge.

After exhaustive research, we found that only the GNB EVolyte outperformed the Trojans without exceeding our battery budget. 14 GNB Evolytes will provide 16.52 kW-hrs at a 3 hour discharge rate. The EVolyte is a sealed, valve-regulated, maintenance free, deep cycle lead-acid battery intended for electric vehicle and traction applications. This pack should give us a range of approximately 80 miles during the Range Event.

4.0 ALTERNATE POWER UNIT

Like the 1993 Cornell Hybrid Electric Vehicle, the *Vortex* utilizes a series configuration for the APU and electric motors. We made this decision because a conversion from an series HEV to a pure EV is greatly simplified. The conversion is a simple task since the APU is only electrically connected to the motors and batteries. However in a parallel system, removal of the APU requires a complete redesign of the drivetrain.

The main purpose of the APU is to provide enough power to keep the vehicle moving and charge the batteries once steady state driving conditions are achieved. To accomplish this task, the *Vortex* utilizes a 725 cc two cylinder

Kohler engine and a Fisher Electric alternator connected in series to the motors and batteries. This configuration is advantageous due to light weight and increased flexibility. The system adheres to our main goal of building an environmentally friendly vehicle by using a clean burning fuel, M85 (85% Methanol, 15% Gasoline), instead of gasoline.

4.1 M85 CONVERSION - The conversion to methanol required that we address four main issues:

1. The stoichiometric air/fuel ratio for M85 is lower than gasoline.
2. Methanol has a higher octane rating than gasoline.
3. Methanol is corrosive to aluminum and some types of rubbers.
4. Methanol has a lower vapor pressure than gasoline.

While the stoichiometric air/fuel ratio for gasoline is about 14.7, the ratio for M85 is about 7.1. This required that we almost double the amount of fuel injected into the intake manifold. In order to compensate for this difference with carburetion, we enlarged the carburetor's main and acceleration jets to increase the fuel flow. In order to compensate for the difference with fuel injection, we used Bosch fuel injectors with a flow rating of 1.63 gm/s and adjusted the duration of the injection to 5.5 ms for idling and low loads to 12 ms for high loads.

Methanol has a 97 octane rating versus gasoline's range from 87 to 93 octane. To take advantage of the higher octane rating, we increased the engine's compression ratio from 9.0:1 to its theoretical optimum value of 12.1:1. This was accomplished by milling 0.092" from the face of both cylinder heads.

Once the compression ratio of the engine was raised, we protected the rubber and aluminum components from methanol corrosion. Methanol compatible fuel lines and O-rings were installed. We also obtained a fuel cell equipped with an experimental suppressant foam which is designed to resist methanol corrosion.

The final issue to address in the engine conversion was methanol's lower vapor pressure (6-14 psi vs. 14 psi for gasoline) which makes cold starting difficult. We solved this problem by programming the electronic fuel injection system to increase the spray duration by 30% at start up.

4.2 ELECTRONIC FUEL INJECTION - The programmable electronic fuel injection system (EFI) continuously varies the spark advance and electronically adjusts the fuel metering system throughout the engine's entire rpm and power range. Air and coolant temperature sensors detect a cold-start condition, while throttle position and manifold pressure sensors determine the load upon the engine. Using these sensors, we created three dimensional spark advance maps for a range of engine speeds and manifold pressures.

The installation of the EFI system first required the replacement of the fuel delivery system. This was achieved by replacing the mechanical fuel pump with an electronic fuel pump which was designed for alternative fuels. We manufactured our own throttle body to replace the carburetor. It was designed to accommodate a computer controlled stepper motor which actuates the throttle plate.

One of the most challenging problems we encountered with the installation was the spark timing trigger. In our direct-fire engine, the signal to the ignition coils is triggered by a flywheel mounted magnet located 22° before top dead center (BTDC). While this was enough advance for the engine's original system, it did not provide enough time for the FI system to process the information and deliver a spark. The FI system expects a signal in a window between 60° and 110°. Because the magnet is permanently attached to the flywheel, the coils could not be repositioned without creating a difficult packaging problem. Therefore, we removed the coils and mounted two infrared optical sensors in their place. We attached a signal tab to the flywheel to break the light beam at 80° BTDC. Since the EFI system requires a square wave input, we designed a circuit to perform the required wave transformation.

A heated exhaust oxygen sensor is used to determined the air fuel ratio of the exhaust. This information is employed as part of a closed loop to constantly maintain the engine at a set air fuel ratio by adjusting the injection duration. Using a predetermined range of $\lambda = 0.9 - 1.1$, we are able to keep the engine at a setting that optimizes the power output and emissions.

4.3 PLASMA COATING - To further reduce emissions and increase efficiency, the cylinder heads, valves, and piston crowns were coated with a plasma thermal barrier. The thermal barrier applied to these surfaces reflects heat back into the combustion chamber and reduces their temperature as well. This translates into more power, improved efficiency, and reduced emissions. With the methanol conversion and plasma coated heads, we observed a 11.4% increase in power and a 7.4% increase in efficiency with the fuel injected engine. The coating also protects these surfaces from the corrosive effects of methanol.

4.4 CATALYTIC CONVERTER SYSTEM - Basic exhaust control of a small engine can be achieved by adjusting the operating parameters such as air/fuel ratio, ignition timing, valve timing, and compression ratio. These adjustments can only reduce emissions to a certain point. After this point it is necessary to employ other strategies for reducing engine emissions such as exhaust gas recirculation, catalysts, and air injection. We are using a three way catalyst system from Engelhard (Model M85-CEX119) which is designed for use with methanol fueled engines. The catalyst will simultaneously reduce levels of hydrocarbons, carbon monoxide, and nitrogen oxides. It will be effective for all engine operations except cold-starting.

Because there is a lag time of 20-30 seconds before the main catalyst becomes active (called lighting off), we investigated methods of decreasing our cold-start emissions. Such methods include electrically heated catalysts (EHC), hydrocarbon storage devices, exhaust gas ignitors, and energy storage devices. Our research showed that electrically heated catalysts provided the best and simplest method for achieving the low cold-start emissions. We are using a electrically heated catalyst from Corning, Inc. followed by a light-off catalyst and finally the main catalyst from Engelhard. By preheating the EHC for 5-10 seconds before starting, the cold-start emissions will be reduced. After the engine starts, the light-off converter will activate, and finally the main converter will reduce engine emissions during normal vehicle operation.

4.5 FISHER ALTERNATOR - Using the data collected from last year's vehicle, we specified the alternator parameters with the assumption that we would obtain at least 15 kW (20 hp) of usable power from the Kohler at 3200 rpm. Fisher claims its alternators to be approximately 90% efficient indicating a minimum of 13.5 kW of electric power output. Using calculations of projected aerodynamic drag and drivetrain losses, 8.67 kW will be demanded by the motors at a cruising speed of 88 km/h. Based on the projected current/voltage characteristics of the battery pack (20 and 80% D.O.D.), the output voltage of the alternator windings at 3200 rpm was specified to ensure both proper loading of the Kohler and to avoid over-voltaging other power components in the car. The maximum-load voltage of the alternator is 176V and its no-load voltage is 214V. With these values and no other means of power control, over-voltages would occur if the dynamic load of the motors dropped significantly and the battery pack's state of charge was more than approximately 60%. Therefore, a control system governs the throttle position based on voltage and current present on the battery harness was developed. In addition to controlling the throttle on the Kohler, the system uses a fast-acting SCR rectifier bridge to protect against voltage surges caused by the motor load changing faster than the throttle can compensate for.

The APU control system measures the voltage on and the current through the battery pack as well as the temperature of the alternator and a general "trouble" signal from the injection/ignition system. The system controls a stepper motor directly coupled to the throttle shaft, the starter relay, the ignition and injection enable signals, and the control signal to a three phase controlled bridge. It allows the batteries to be discharged to an adjustable lower voltage set point before the APU is engaged. When running, the APU will output maximum power until an upper harness voltage is reached. The output power is then be reduced to maintain this voltage until a minimum charging current through the batteries is reached. The ignition is then turned off and the batteries are allowed to discharge. Alternator temperature is monitored and the output power reduced if over-heating occurs. Because of the dynamic nature of the motors as a load, the system does not only rely on throttle control to protect against over-voltage conditions. Instead, the APU system quickly "removes" the alternator from the wiring harness via the controlled bridge for a brief period of time. This will provide the time necessary for the engine/ alternator system to adjust to load transients.

4.6 AUTOMATIC THROTTLE CONTROL SYSTEM - The purpose of the Automatic Throttle Control System (ATC) is to allow the vehicle control system (VCS) to regulate the power and voltage output by the APU without the need for manual adjustment. When the APU is operating, the speed and torque requirements vary depending on load conditions. When the battery pack is significantly discharged and the vehicle is accelerating the APU supplies most of the power to keep the system voltage at acceptable levels. Alternatively, when the batteries are nearly charged and the vehicle is stopped, the APU power requirements are minimal.

The ATC consists of a throttle assembly on the combustion engine, a high precision stepper motor, a throttle position sensor, and a closed loop feedback control system which is integrated into the VCS. The throttle assembly controls the airflow to the combustion engine, and therefore the power output of the engine. Proper fuel mixture under all conditions is maintained by the electronic fuel injection system.

The stepper motor is linked directly to the butterfly valve in the throttle assembly. The stepper motor may place the valve in any position between fully open throttle (vertical position) and fully closed throttle (horizontal position). Its full range of motion is 90° and is designed for 1.8° steps. The motor can maintain desired positions under torques up to 0.40 N-m. The motor and butterfly valve configuration was selected because it provides very accurate throttle control with very few additional components over conventional manual throttle control, thus maximizing system performance with minimal weight.

The throttle position sensor's role is twofold. It provides airflow readings to the fuel injection throttle position information to the ATC. The ATC needs the position information so it does not allow the stepper motor to position the throttle outside of its intended range of motion. (i.e. 0° - 90°).

5.0 SUSPENSION

In the spirit of a ground-up vehicle and in keeping with good automotive engineering practice, we designed and built our own suspension, steering, and braking systems to match the target performance specifications of our car. Ignoring this aspect of design or using stock parts from other cars could lead to poor and/or unsafe handling characteristics. In addition, suspension engineering allows us to minimize weight and experiment with alternative designs that could have applications in the future EV market.

5.1 GEOMETRY - We use a unequal length A-arm suspension for the front and rear wheels because of its manufacturability, ease of integration into the frame and because it provides the best handling characteristics of any suspension design except active control. Figure 6 demonstrates the use of computer design for our "points" analysis where we choose tie rod and ball joint locations in order to minimize camber change and roll center migration of the front and rear suspensions in bump and droop.[1]

The A-Arms are constructed from round Chromoly tubing and are designed to withstand a 3g bump load with a 1g brake. They are held to the frame by threaded rod-ends which screw into plugs welded into the A-Arm tubes. Rod ends are used because, while they give a harsh ride and do not damp out vibration, they allow for some degree of suspension alignment (this is essential for a prototype design). Our principal load bearing rod ends are rated at 6900 kg radial static load capacity.

5.2 SHOCKS AND SPRINGS - In order to minimize packaging problems and allow for easy adjustment, we implemented an overhead front shock/spring design which allows us to move the shocks out of the way of the driveline, steering, and anti-roll linkages and still obtain a rising rate from the front spring. This involves triangulating the upper front A-arms to mount the shock/spring nearly at horizontal,

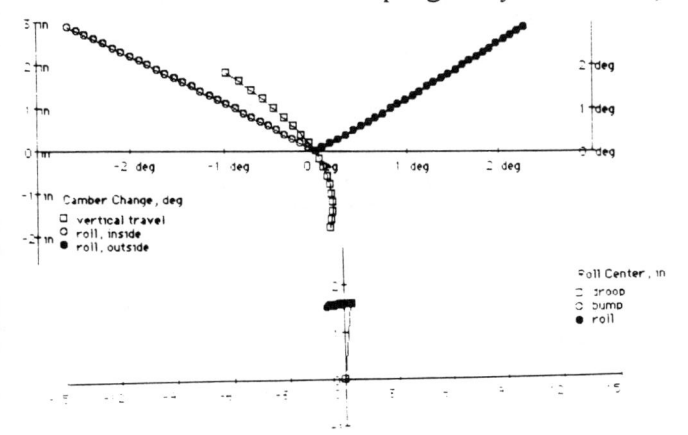

Figure 6
Camber Change and Roll Center Migration Curves

---

[1] Staniforth, Allan. <u>Competition Car Suspension</u>. Haynes Publishing, Somerset, England, 1991.

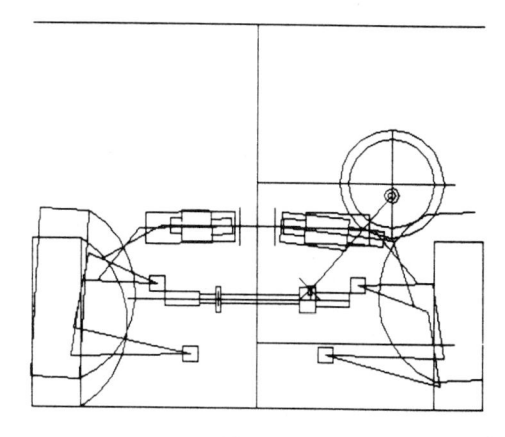

Figure 7
Front Suspension, Wheels and Steering (Front View)

just below hood level (Figure 7). However, this results in a small weight penalty (0.5 kg of steel bracketing) and a very small (<1%) rise in cg height.

AFCO aluminum gas shocks were used in the front and rear. This choice allows the suspension about 7 inches of total vertical travel. Gas shocks allow the shock to be mounted upside down in order to slightly reduce unsprung weight, and also allow the front shocks to be horizontal. Springs were designed to the shortest possible length to reduce weight. The static deflection was minimized to reduce the amount of time spent coiling and uncoiling springs while still giving the car an acceptable ride. Wheel ride rate can be calculated with the following equation:

$$RR = \frac{K_s K_t}{K_s + K_t} \quad (5)$$

where RR is the ride rate, and K_s and K_t are the spring constants of the tire and wheel, respectively. Our car has spring rates of 34,200 N/m (rear) and 27,200 N/m (front), and a tire rate of 201,000 N/m at 303 kPa (44 psi) which translates into wheel rates of 29,300 N/m (R) and 24,000 N/m (F). To calculate the suspension natural frequencies, we use[2]:

$$f_n = 0.159 \sqrt{\frac{RR}{W/g}} \quad (6)$$

where f_n is the suspension natural frequency (Hz) and W/g is the mass of one quarter of the car. Our car has suspension natural frequencies of 1.7 Hz (R) and 1.5 Hz (F) which is slightly higher than the optimal 1 Hz design. This means the car will ride stiffly but handle well. The two frequencies are slightly unequal in order to decouple pitching motions of the front and rear.

5.3 WHEELS, TIRES, UPRIGHTS & HUBS - Goodyear Tire And Rubber Corp. donated P175-65 R14 Invicta tires to our project. The tires have a rolling resistance coefficient of

---

[2] Gillespie, Thomas. <u>Fundamentals Of Vehicle Dynamics</u>. SAE 1992

0.0078 at 44 psi. Aluminum alloy rims were donated by American Racing Equipment. We chose 14" x 6" rims as a tradeoff between good handling and low rolling resistance, and to allow packaging of a 10" brake rotor in all four wheels. Inside the rims, we designed our uprights with a minimal kingpin angle (8.5°) and short stub-axle lengths (2.5 in) to minimize moment loads. The uprights have zero camber at ride height and 3° of caster to provide self-alignment. Packaging inside the front rim accounts for the design of the upright, the braking system, and the hub/wheel bearing assembly. The steering knuckle is aligned at a 5° Ackerman angle with an imaginary line connecting the outer suspension points.

We designed and built our own rear hub units out of 6061-T6 aluminum alloy and used the front hub units from a Ford Escort GT (since we could not manufacture splines or CV joints in-house). We constructed our own hybrid halfshafts using Ford outboard CV's coupled to hollow driveshafts, which mate with Honda inboard CV's.

Our vehicle's target weight is 1100 kg (wet), with approximately 3% forward weight bias. Anti-geometry is accomplished by tilting the suspension mounting members relative to the car frame. We use 20% anti-dive and 55% anti-squat geometry (these numbers are somewhat subjective). Our front and rear anti-roll bars give a total of 350 N-m per degree roll resistance, split evenly between front and rear. For other specifications, refer to the Spec Sheet at the end of this report.

5.4 STEERING

The vehicle uses rack-and-pinion steering, with an aluminum alloy steering rack that has four inches of rack travel. This amount of travel allows a curb-to-curb turn of 9.75 m (32 ft.) which makes the vehicle nimble enough to perform even the most demanding steering maneuvers. A steel steering column, cut to fifteen inches, with self-canceling turn signals, four-way flashers, five position tilt, and a standard GM wiring harness, was provided by Ididit, Inc. An intermediate shaft and needle bearing universal-joints, both splined at each end, are used to link the rack and column. The positioning of the rack and column required a double swivel U-joint at the column end and a single swivel joint at the rack end. The double swivel joint is able to operate at angles up to 60°, whereas the single swivel joint is limited to a 30° angle. The link between the U-joints was provided by a hollow, steel shaft with a 3/4 inch outer diameter and .065 inch wall thickness. The two inch splined ends were cut off of a solid 3/4 inch shaft and welded onto the hollow shaft in order to reduce weight.

5.5 BRAKE MASTER CYLINDER SYSTEM

The *Vortex* is equipped with four-wheel disc brakes with dual-acting four-piston calipers at each wheel. Each caliper includes self-retracting pistons, a feature which reduces rotor drag. Using our target wet vehicle weight of 1100 kg undergoing a weight transfer of 0.8g (worst case) results in a 7175 N load applied at the front axle. By choosing the standard piston diameters of 1.75 in (4.45 cm) for the front and 1.375 in (3.5 cm) for rear, the driver will be able to stop the car with a reasonable pedal force.

We installed a dual master cylinder assembly operating split front and rear brake systems. At the 1100 kg wet vehicle weight, we calculated the need for 2.22 cm diameter master cylinders with a 6.25:1 brake pedal ratio to allow for a pedal force of 334 N (75 lb.) to stop the car at a 0.65 g deceleration. This calculation led to the selection of a 7/8" diameter master cylinder. A 334 N pedal force is an acceptable expectation for normal vehicle operation when considering that an average human can exert up to 1400 N (300 lb.) on the brake pedal in a panic situation[3]. The front/rear braking bias can be adjusted via a balance bar in the brake pedal assembly to account for uneven weight distribution.

5.6 WEIGHT REDUCTION

In our effort to build an electric vehicle from the ground up, vehicle weight plays a vital role in any design choices. This is the rationale behind our extensive use of alternate materials in suspension, frame and in the vehicle as a whole.

We decided to investigate the use of titanium as a suspension material to further reduce the weight of the *Vortex*. Titanium offers a better strength-to-weight ratio than traditional steel, and it has better fatigue resistance than aluminum, when used properly. 3-2.5 Ti alloy has an ultimate strength of 100 ksi, as opposed to 60 ksi for 4130 steel, while weighing about 40% less. Its principal disadvantages are that it is more difficult to machine than steel, and must be purge-welded to preserve its purity. Furthermore, it is more prone to seizing and galling than steel, and it is more notch-sensitive.

By using titanium A-arms and uprights, we expect an unsprung weight savings of about 2.5 kg per wheel just. Our anti-roll bars are also constructed out of titanium; here, care must be taken in design since the modulus of rigidity of titanium is about half that of steel. Typical anti-roll design equations for steel will not work unless they are adjusted for this change. To weld our titanium anti-roll bars and A-arms, we constructed a purge chamber out of aluminum and Lexan, and kept a stream of argon running through the chamber while welding.

We also are experimenting with titanium in other areas. Titanium bolts are used that meet or exceed SAE grade 5 specifications and thus make them legal for use in the Tour de Sol competition. Recently, we have been in contact with a supplier of titanium coil springs, which, if used, promise further weight reductions for our car.

---

[3] Puhn, Fred. <u>Brake Handbook</u>. Los Angeles: Price, Stern, Sloan, 1985: 90-91

As an alternative to steel rotors, metallic coatings are available that make aluminum rotors feasible in vehicle applications such as ours. If possible, we plan to use aluminum rotors, with an expected reduction in unsprung, rotating mass of about 1 kg per wheel.

6.0 FRAME

In an effort to improve packaging, the space frame for the *Vortex* was the last design to be finalized and the first to be constructed. By understanding how all of the systems in the car work together, we were able to integrate mounting points, thermal management systems, battery storage, and passenger positions into a strong and lightweight frame design. The frame is divided into three sections, each of which serve a different function. The front bay houses the powertrain and the front suspension with maintenance and accessibility as primary design considerations. The main section of the frame provides an aesthetically pleasing passenger compartment, as well as storage for the battery pack. The APU, the M85 fuel cell, and the rear suspension are located in the rear bay.

6.1 BASIC STRUCTURE - The design process for the frame began with an in-depth examination of many different materials, construction methods, and geometries. We decided to use a spaceframe design due to its flexibility and ease of construction. The spaceframe is defined by the torsional box (t-box), which is the main source of structural rigidity in the frame. The torsional box serves as the "backbone" of the frame, and is combined with another classic chassis geometry, the "ladder." This combination grew from the necessity to package the batteries in a safe and practical location, while maintaining a comfortable, ergonomically sound environment. The ladder structure provides the space for two people, the alternate power unit, and the fuel cell, and the backbone houses the batteries. The cross sectional area of the torsional box has been reduced by 40% over last year's design.

The design change both improves passenger seating and decreases frontal area at the same time. The drawback of this design change is the reduction in the moment of inertia. Therefore, the t-box can not be the only source of torsional rigidity. To overcome this problem, our design uses the strength of the "ladder" and the roll cage to supplement the t-box. Cross members and a front bulkhead provide additional support. The bulkhead provides lateral strength, as well as a means to seal the passenger compartment from the front bay.

6.2 FINITE ELEMENT ANALYSIS - The CAD package PATRAN - P/3 from PDA Engineering was used to perform our finite element analyses. In order to test the structural integrity of the frame, four finite element analysis "jobs" were created: *pure_bend*, *tilt*, *roll_it*, and *torsion*. The first test, *pure_bend*, was designed to determine the maximum midspan deflection of the frame. This load case simulated the mass of the vehicle, 1100 kg, as if it were all acting at the center of gravity. This is a worst case scenario since the mass is actually distributed across the entire frame. The maximum deflection allowed in the *pure_bend* test was 0.76 cm. The second test, *tilt*, placed 907 kg on opposite corners of the frame while holding the other corners fixed. This test was intended to model typical worst case road forces. The figures were derived from the maximum cornering, bump, and steering forces that the suspension will theoretically be subjected to. The maximum deflection achieved in *tilt* was 3.2 cm which corresponds to a stiffness of approximately 1465 N-m / degree. *Roll_it*, the third test, was designed to simulate a 2.5g-rollover in which essentially two and a half times the mass (2750 kg) of the vehicle impacts the roll cage. The maximum deflection in the *roll_it* test was 8.55 cm.

Three impact attitudes were explored, including downward load, side crash, and impact at 45° angle relative to an axis parallel to the ground. The final test, *torsion*, did not model a realistic scenario, but rather exaggerated the limits of the frame. In this test, 1000 kg was loaded on one corner of the frame while the other three corners were held fixed. The goal of the test was to examine the stresses and strains on each member and to determine which frame members could be eliminated. Changes made to the design based on the torsion test results were reverified by again performing all other tests.

Several aspects of the frame design were scrutinized by the above tests. Choosing from a list of nine common sizes of 4130 Chromoly steel tubing and ten sizes of titanium tubing, various material combinations were evaluated. The primary goal was to find the lightest combination of materials without compromising from the structural integrity of the frame. Several options were tested, including mainframe triangulation, rear bay triangulation, front bay triangulation, roll cage supports, bulkhead height, and the optimum number of backbone members. By testing several different geometries and materials, we have optimized our frame for both weight and performance.

6.3 MATERIAL SELECTION - Based on the above results, the frame is made primarily of cold drawn seamless 4130N aircraft quality alloy steel. Its physical properties for the range of tube sizes used are the following: UTS = 724 kPa, S_y = 662 kPa, ρ = 4730 kg/m^3. The five square tube sizes used are the following:

O.D.(cm/in.)	Wall(cm/in.)
2.54 / 1.00	0.0889 / 0.035
1.91 / 0.75	0.0889 / 0.035
1.27 / 0.50	0.0889 / 0.035
2.54 / 1.00	0.1651 / 0.065
5.08 / 2.00	0.2108 / 0.083

The 2.54 cm-OD stock is the most widely used in both wall thicknesses listed. These two create the majority of the frame's strength. The smaller tubes are used to mount various components where structural members do not extend. Since these mounts are non-critical members, smaller stock is used to reduce weight. The frame was constructed by using the Tungsten Inert Gas (TIG) welding method due to its high precision and weld strength..

6.4 ROLL CAGE MATERIAL - The roll cage on the 1994 *Vortex* is made out of titanium, rather than 4130 or 1020 steel. We have determined that, for this application, a titanium roll cage has a performance superior to that of a steel roll bar-and-hoop assembly. The selected titanium alloy is 3%-Vanadium 2.5%-Aluminum; the size used is 3.49 cm OD (1.375") with 0.099 cm (0.040") wall. Its physical properties are the following: UTS = 828 kPa, S_y = 724 kPa, ρ = 2300 kg/m^3. To compare steel and titanium, three options were evaluated: 1) 4130 steel roll cage, 2) titanium roll cage, 3) 4130 steel roll hoop and roll bar assembly.

The first two options were geometrically identical. The roll hoop and roll bar both extend above the passengers heads and are connected by roof height cross members, creating a "roll cage". The roll bar is supported in the aft direction by braces of the same material. The third option is geometrically different. It is the design of the 1993 Cornell Hybrid Electric Vehicle, in which the roll hoop only extends to a height slightly above lap level.

The roll bar and hoop are constructed from seamless, closed sections of material. Steel roll bars and hoops would be welded to the frame, but since titanium cannot be easily welded to steel, a titanium roll bar and hoop the would be fastened to the steel frame with several aircraft grade steel bolts at each joint. When tested under the *roll_it* job, the 4130 steel cage had a maximum deflection of 5.89 cm and titanium had a deflection of 8.55 cm.

The roll bar and hoop assembly specified in the HEV Challenge rules was then modeled on the computer and tested. Using 4130 steel (3.5 cm od, 0.24 cm wall) as the roll bar and hoop material, we found that the assembly did not provide adequate driver or passenger protection. In this simulation, with the same roll loads as previously modeled, the plane connecting the roll bar and hoop deflected below the level of the driver's head.

Therefore, we decided to use the titanium roll cage configuration. The deflection of the titanium is minimal enough to not come in contact with passengers bodies in a roll over and will provide superior protect versus the rule specified roll bar and hoop configuration.

6.5 WEIGHT SAVINGS - We used 11.5 m of titanium tubing to construct our roll cage, which has a mass of 21 kg. The equivalent roll cage of 4130 steel would have had a mass of 82 kg, 61 kg more than the titanium option. The use of the thin walled Chromoly tubing also reduced the frame weight. Holes were drilled in all non-structural frame members, such as the seat mounts, in an effort to save less than 0.5 kg. A lightweight composite board, CIBA Composites' Fiberlam™, replaced last year's fiberglass and epoxy wrapping of the torsional box. Each square meter of Fiberlam has a mass of 2.9 kg, while the fiberglass wrap on the '93 car had a mass of 4.2 kg/m^2. The change in composite material saves a total of 7.0 kg. The final weight of the frame is 70 kg.

7.0 ERGONOMICS

The ergonomic goal of the *Vortex* is to provide for the interaction of the occupants and the vehicle, including packaging the occupants safely and comfortably. Easy access to controls and instrumentation is also a necessity for vehicle operation. Our design objectives for the *Vortex* were to ensure ample visibility for even the shortest driver (155 cm), to locate all controls within easy reach of a seat-belted driver, and to construct a stylish, well-finished interior.

7.1 OCCUPANT PACKAGING - Because of the diversity of size, shape and opinion among the drivers of our vehicle, it was unclear as to the best method for positioning such components as the seats or steering wheel. In order to devise a method, we sought to learn how industry approaches this same task. The first discovery was an SAE paper titled "The Best Function for the Seat of a Passenger Car."[4] This article described "comfort angles" which correspond to angles between limbs on the driver's body and are accepted by industry as guidelines for choosing seating positions. We used these angles to position the 50% male within the car in the optimum position. However, performance and packaging concerns as well as competition safety and visibility requirements forced us to deviate somewhat from the optimum comfort positioning of the driver.

An important benefit of the comfort angles is the determination of an angle at which to mount the driver's seat. Our seats had a back/seat angle of 95°; we tilted the back to make a 25° angle with the vertical. This was also based on advice from Designworks/USA, a vehicle design studio. Tilting the seat allowed us to keep the roof height as low as possible while still meeting the visibility requirements, and maintaining a high degree of comfort.

We chose Jaz Products' polyethylene plastic seats which weigh 11 pounds (5 kg) each. They were the lightest and most cost-efficient seats available. Positions of components including the steering wheel and brake pedal were determined using a full-size model of the interior.

---

[4] Halderwanger, Hans Gunter and Albert Weichenreider. "The Best Function for the Seat of a Passenger Car." SAE Technical Paper 850484. 1985

AutoCAD™ was used to manipulate driver geometries when choosing comfort positions. Our drawings rapidly began to include many other car components, until eventually the whole car was being packaged along with the drivers. This proved to be a necessary tool during some crucial design stages, when the only accurate visualization available was the CAD image.

7.2 DASHBOARD - The main goals in designing the interior were to achieve near-professional quality in fit-and-finish of the dashboard and the door panels. Our design includes a two-tone leather trimmed dashboard complete with color LCD instrumentation. In an effort to reduce weight, the dashboard substructure was constructed from Styrofoam panels. Before applying the leather cover, an epoxy resin was applied to the foam in order to harden the surface.

The dashboard was constructed in three distinct modules. The driver side module concealed the brake master cylinder and steering column, while the center module housed the Unique motor controller. The third module contained the vehicle control system, the high-power safety disconnect switch, and the data acquisition system. Removable upper panels constructed of polyester fiber composites allow easy access to each of these areas.

7.3 INSTRUMENTATION - There were two main concerns which played a key role in the design of our instrumentation. The first was the capability of evaluating the performance of our car with data recorded during testing. The second was the ability to easily present essential information to the driver. These requirements led us to a combination of an analog and digital dashboard.

Four traditional analog gauges, a speedometer, battery voltmeter, ammeter, and charge indicator are placed directly in front of the driver. The vehicle speed is displayed on a VDO electric meter. The current is displayed on a modified Faria gauge. This required recalibration of a gauge which was originally scaled between -60 and 60 amps to a required range of -800 to 800 amps. This was accomplished by changing the shunt resistance to a lower value. To display the voltage we stepped down our battery voltage from 168 V to a range appropriate for our automotive voltmeter. We then generated a new display for the gauge with the numbering calibrated to our system.

The digital dashboard incorporates the latest high resolution display technologies to present a graphical version of the analog gauges. This screen is adaptable and reconfigurable. Valuable supplemental information is presented via a color thin-film transistor (TFT) active-matrix liquid crystal display (LCD). The general layout of the digital portion of the dashboard is shown in Figure 8.

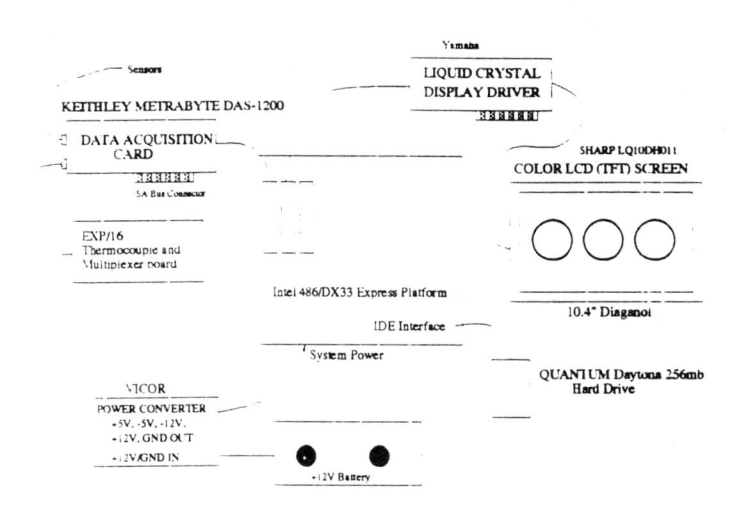

Figure 8
Digital Dashboard General Layout

The digital dashboard was developed to be not only a valuable data display, but also aesthetically pleasing. The choice to use traditional analog gauges complemented by a computerized dashboard allows us to have a system which can not only display information which would otherwise be unavailable, but also generate gauges which indicate changes in driving conditions for better performance from the car, such as an optimum speed for maximum range.

7.4 DATA ACQUISITION - The selection of a data-acquisition card was made by examining what measurements we wanted to record from the vehicle, keeping in mind that our goal was to be able to determine the overall vehicle efficiency. This allows us to determine what specific systems in the car could be changed to improve the overall efficiency and performance of our vehicle. The readings we deemed important are battery voltage, battery current, a battery temperature and the battery compartment temperature. The motor controller provides information about the electric motors. An analog output provides us with the speed of the motor which we scale to determine the actual speed of the car. The temperature, voltage and current through the motor controller are also measured. For the APU, we measure RPM, fuel level, and record oxygen-sensor outputs. We also use a single-axis accelerometer at four points on the suspension to determine the exact dynamic loads the *Vortex* encounters during driving and one tri-axial accelerometer is located in the center of the dashboard. Additionally there were some accessory inputs; turn indicators, lights and wipers.

The decision was made to use a Keithley Metrabyte DAS-1200 50/100 kilosamples/sec analog and digital I/O board, together with the EXP-16 expansion board. These boards plug into a standard 8-bit slot on an IBM PC bus.

Sensors were connected to the data acquisition card via the STA-U universal screw terminal connection board. This setup allows us to sample all the data we want to record with room for future improvements, while still having a low power consumption (2.3W total).

The DAS-1200 connects into the bus on our Intel Express 486/DX33 motherboard. The motherboard is placed in a dedicated area of the dashboard which is accessible via the removal of a detachable panel.

The information from our data acquisition system is stored on a Quantum Daytona 256 megabyte hard drive. This hard drive has a standard IDE interface, but weighs 0.11 kg and has dimensions of 7.0 cm x 10 cm x 1.25 cm. Additionally it has energy efficient modes, and typical power draws of 0.2W standby, 1.1W idle, 1.7W active, 2.2W operating. These characteristics makes it ideal for our application.

7.5 LCD DISPLAY - To present the information to the driver, we use a color liquid crystal display. A LCD has the lowest power requirements of any programmable pixellated display. It is also very narrow in depth; our display is 0.99" thick, including the backlight. We selected a thin film transistor (TFT) active-matrix display, which has slightly higher power requirements than competing dual-scan passive-matrix LCDs. However, the dual-scan passive displays lack the brightness, color vividness, and update speed that an active-matrix screen possesses. More importantly, passive-matrix LCDs have a limited viewing angle, so a viewer must be directly in front of the display to observe the image correctly. Additionally, dual-scans are subject to ghosting or the "submarine" effect. This is most commonly observed when moving a mouse on a laptop computer that uses one of these type displays - the mouse cursor will momentarily disappear or submerge while the mouse is moved. Obviously these characteristics would have negative results in a vehicle where information must be relayed with clarity and speed.

These problems do not occur with an active-matrix screen because the thin film transistor layer addresses each of the pixels of the display, resulting in a much brighter display and one where constantly moving images (such as a gauge needle) are presented without any negative effects.

The display we selected to use was the Sharp LQ10D011 TFT-LCD unit with a 640 x 480 resolution, 10.4" diagonal, and viewing angles in the range of ±45° horizontally, and -30°/+10° vertically. Working with our Sharp color display, we designed our own software using C++ to generate a display that can easily relay information to the driver or the passenger.

7.6 DISPLAY SOFTWARE - An integral aspect of the digital dashboard is the software subsystem used to present information to the driver. It must combine data acquisition, signal processing, and high speed graphics into a visually appealing and informative display.

Early in the design process, we decided that instead of having a hard-coded dashboard layout, the system would consist of an executable program which would read in a dashboard layout file at runtime. With this approach, the dashboard could be easily modified without rebuilding the driver software. Another advantage with this design is that different drivers could opt for different instrument layouts, with varying informational content, according to their preferences or needs.

There were a number of design considerations made before implementing the system. The most important display feature had to be the ability to display a wide variety of indicators. In addition to bar graphs and dial gauges, the system had to support pictorial icons for various status indicators, text for numeric displays, and static images for miscellaneous graphics. It had to be flexible enough to allow the designer to easily incorporate new indicator types into the framework if none of the existing indicators sufficed. All indicator types also had to be independent of the data acquisition code. This would ensure that a dial gauge could be easily replaced by a bar graph without changing any code, while still retaining a proper display. Should the need arise, it also allows us the flexibility of easily changing the underlying data acquisition system, with only minimal changes to the software.

The basic idea behind the instrumentation display is that in a typical indicator there are very few moving elements. For the most part, the image is static. For example, in a speedometer, the only moving part is the needle. Everything else, like the numbers and marks, remain fixed. Furthermore, by inspecting a large number of indicators, it is evident that the number of different types of gauges is quite small. There is the needle gauge in circular or semicircular varieties, bar graphs, numeric displays, and perhaps one or two others. In our display model, the designer is responsible for creating the background image. After that, the computer needs only to know what kind of indicator this is and where to draw the moving element.

The creation of this modular instrumentation system demanded the creation of a number of secondary computer-based tools. Dashboard layouts are merely collections of indicators where the designer positions these indicators as desired on the screen. Each indicator is assigned a variety of attributes, the most important of which are data acquisition source, signal processing function, and update frequency. These layouts are created with the Dashboard Maker utility. This utility allows the user to load, modify, and save existing dashboards, as well as create new ones.

The indicators used to create these dashboard layouts are created by means of an Indicator Maker. This utility loads a graphics image and allows the designer to crop portions of the image and save them as indicators. This program allows the designer to create a large library of reusable indicators. The indicators are saved as device-dependent bitmaps in the native Borland C++ format. This allows the graphics to be drawn on the screen very quickly.

The graphics images used are created by means of any paint program. The designer can create a graphics file with the pictures of many indicators. These pictures can then be converted into indicators using the Indicator Maker tool. The graphics format used by the display subsystem is the standard VGA 640x480 16-color format. However, this is not hard-coded and could be easily changed.

8.0 FAIRING

Designing and building a body for an automobile is a compromise between aerodynamics, aesthetics and manufacturability. In the case of alternate energy vehicles, a great potential exists for cars that are not only aerodynamically efficient, but also appeal to conventional market tastes. With these two objectives, we set about to design the *Vortex*'s fairing to be an aerodynamically efficient yet visually appealing vehicle that could be manufactured without any specialized equipment.

8.1 DESIGN AND CONSTRUCTION - To create our initial conceptual design of the *Vortex* fairing, we started with a number of scale models. These models were constructed by laminating two inch thick blocks of insulation foam. Initial cuts were made with a hot wire to obtain a rough shape of the body. Final shaping was done with hand tools. Once the skin form was decided upon, the foam model was coated with epoxy resin to produce a hard surface.

A computer model of the foam model was created by scanning the foam model using a XYZ coordinate measuring machine. Data from the digitizing machine was then imported into a CAD system and scaled to fit the frame design. Once the body model was fit correctly to the frame, a continuous surface model was created using a NURBS (Non-Uniform Rational B-Splines) based three dimensional modeling system. After fine tuning the body to better accommodate passenger needs, a final body design was determined.

Before beginning full scale construction of the body, we needed to decide what body material to use. To construct a single body, we narrowed the list of choices down to two: composite lay-up and vacuum thermoforming. We chose vacuum thermoforming of plastic for two reasons. Our goal was to achieve a high level of manufacturability without sacrificing environmental friendliness. Thermoforming allows us to do both. The Lexan panels can not only be mass produced, but are completely recyclable. Making the panels from composites involves an extensive amount of manufacturing time and a large number of hazardous chemicals that are not needed in vacuum thermoforming.

To construct the tooling needed for vacuum thermoforming, we first transferred the surface model of the car into CNC modeling software. Our intention was to generate NC tool paths for the entire body. Unfortunately, we were not able to secure any large CNC facilities and were forced to resort to using hand machining methods. Before any tool construction began, we divided the surface into appropriate body panels that are very similar to production vehicles. A total of thirteen different panels (including tools for the windows) were created. Further design modifications were introduced at this point to take into consideration the needs of thermoforming and the limitation of our hand held tools.

After the actual panels were formed, they were post processed to remove any extra plastic. Since the plastic panels have limited structural strength, an appropriate mounting system was designed and built from box tubing. All the panels were then affixed to this subframe using quick release mounting clips.

8.2 VEHICLE AERODYNAMICS - A significant amount of time was spent optimizing our design to be aerodynamically efficient. Most of the optimization was done by incorporating good design features in the initial foam designs. We tried to avoid any sharp corners and lines in favor of rounded surfaces and curves. Instead of designing completely flat sides of the car, we decided to taper the sides in to help move the turbulent transition point further towards the rear of the vehicle. By making the windshield and A-pillar cladding completely flush, we removed a large drag inducer. The door was designed to be further back from the A-pillar than in conventional designs to further reduce chances of the air flow becoming turbulent in that area.

A considerable amount of the drag produced by a vehicle is produced by the rotation of the wheels when the car is moving. Covering the rear wheel with body fascia is an easy method of reducing this source of drag. Covering the bottom of the vehicle with a continuous smooth panel not only reduces drag, but also eliminates the need for a large front airdam to divert air flow from below the vehicle.

In addition to the above mentioned "major" modifications, a through overview of the entire body surface was performed to smooth out any "rough" areas such as gaps between panels, etc. The result of these modifications is an efficient aerodynamic design.

8.3 AESTHETICS AND DESIGN - Consumers are generally not interested in a vehicle that they do not find aesthetically pleasing, so a considerable amount of time was spent researching how consumers define a "good looking" vehicle. By studying various prototypes and showcars, we developed a understanding of what appealed to the general consumer. This was further verified by conducting informal marketing analysis when we displayed our design to the public. To aid in the design our "ideal" body shape, we enrolled the assistance of a professional design studio. With the help of the automotive designers, we were able to arrive at a design that is both visually appealing and aerodynamically efficient.

We used extensive computer modeling for the final production of the panels, even though the initial models were fabricated using conventional hand tools. As evidenced by the vehicle's low C_d of 0.30, a balance between aerodynamic efficiency and an aesthetically pleasing form was achieved. A computer rendered surface of this year's fairing can be seen in Figure 9.

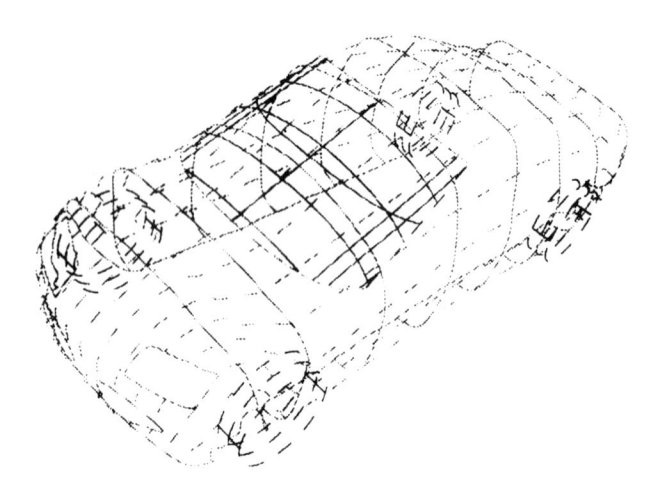

Figure 9
AutoSurf™ rendering of the *Vortex* Fairing

9.0 SUMMARY

In the state of California and possibly some states on the eastern coast of the United States, laws go into effect in 1998 which require that 2% of all vehicles sold be zero emissions vehicles. However, battery technology has not reached a point where purely electric vehicles can perform as well as internal combustion vehicles. Therefore, hybrid electric vehicles, such as The *Vortex*, are an excellent means of achieving a cleaner environment without sacrificing performance.

The 1994 *Vortex* includes many innovative designs in various aspects of automotive engineering, including:

- 168 Volt, two motor drive system with limited slip differential and custom modular gearbox
- Custom, high-amperage motor controller
- Sealed lead acid batteries
- Series configuration, fuel-injected M85 APU
- Plasma coated heads, heated catalyst
- Fisher alternator with Automatic Throttle Control
- High performance suspension geometry
- Aluminum gas shocks, overhead front design
- Titanium springs, plasma-coated aluminum rotors
- Low rolling resistance tires
- Titanium a-arms and anti roll bars
- Torsional box spaceframe
- Titanium roll cage
- Four wheel disc brakes with self-retracting pistons
- Digital dashboard and data-acquisition system
- "Cab forward" fairing design
- Thermoformed Lexan body panels

Designed and built by Cornell Engineering students, the 1994 *Vortex* is a practical solution for tomorrow's problems that can be driven today

10.0 BIBLIOGRAPHY

1) "Car Styling" Volume 96, Car Styling Publishing Co. Tokyo Japan. September 1993.

2) Janicki, Edward, Cars Detroit Never Built: Fifty Years of American Experimental Cars. Sterling Publishing Co., Inc. New York. 1990.

3) Gillespie, Thomas. Fundamentals Of Vehicle Dynamics. SAE 1992

4) Halderwanger, Hans Gunter and Albert Weichenreider. "The Best Function for the Seat of a Passenger Car." SAE Technical Paper 850484. 1985

5) Janicki, Gregory, Cars Europe Never Built: Fifty Years of Experimental Cars. Sterling Publishing Co., Inc. New York. 1992.

6) Juvinall, Robert, and Marshek, Kurt, Fundamentals of Machine Component Design. John Wiley and Sons Inc. New York, 1991.

7) Ogata, Katsuhiko, System Dynamics. Prentice Hall, Englewood Cliffs, NJ. 1992.

8) Puhn, Fred. Brake Handbook. Los Angeles: Price, Stern, Sloan, 1985: 90-91

9) Staniforth, Allan. Competition Car Suspension. Haynes Publishing, Somerset, England, 1991.

10) White, Frank, Heat and Mass Transfer. Addison Wesley Publishing Co. New York. 1991.

Cornell University HEV Specification Sheet

Vehicle Type:
Front-wheel-drive, 2-passenger, 2-door coupe, series hybrid electric vehicle

Sound System:
JVC 3-disc in-dash CD Changer, Infinity 6-speaker system

Alternate Power Unit (APU):
Manufacturer:	Kohler
Type:	V-twin, aluminum block, plasma-coated aluminum heads
Fuel:	M-85 (85% Methanol, 15% regular gasoline)
Fuel Capacity:	12 gallons (45 L)
Weight:	120 lb. (55 kg)
Displacement:	44 in^3 (725 cc)
Bore x Stroke:	3.27 in. x 2.64 in. (83 mm x 67 mm)
Compression Ratio:	12.1:1
Engine Control:	Haltech Prog. EFI system, CUHEV BOS EC System
Emissions Controls:	Electrically preheated catalytic converter for cold starts. Standard 3-way catalytic converter for normal operation. Evaporative emissions recovery system
Valve Train:	Single cam, pushrods, Overhead Plasma-Coated Valves, 2 valves/cylinder
Power:	24 hp @ 3200 rpm (18 kW @ 3200 rpm)
Torque:	40 ft.-lbs. @ 2500 rpm (54 N-m @ 2500 rpm)
Alternator:	Fisher Electric, 13 kW @ 3200 rpm
Fuel Consumption:	0.89 lbs/hp-hr (0.49 kg/kW-hr)
Redline:	4300 rpm
Hybrid Range:	350 miles (560 km)

Powertrain:
Electric Motors:
	Unique Mobility DR156S	Advanced DC L91-4003
Type:	Brushless Perm. Magnet DC	Brushed DC
Weight:	18 lb. (8.2 kg)	64 lb. (29 kg)
Controller:	Unique CR20-150	CUHEV MHOG-20-80
Thermal Management:	CUHEV, 72-fin radial heat sink Forced air convection	CUHEV, 8 fin axial sink Forced air convection
Peak Power	21 hp (15.8 kW)	60 hp (45.0 kW)
Peak Efficiency:	92% (motor/controller system)	80% (M/C system)
Redline:	8500 rpm	5600 rpm
Final Drive Ratio:	7.6:1	8.0:1

Driveline:
Gearbox:	CUHEV, 2 input, 1 output, single reduction for each motor
Differential:	Honda, limited slip
Halfshafts:	Hybrid, Honda inboard CV, 4130 Chromoly Shaft, GKN outboard CV

Batteries: GNB Evolyte 1123
Nominal Pack Voltage:	168 V
Cycle Life:	750 cycles @ 80% DOD
Energy Capacity:	16.5 kW-hr. @ 3 hr. rate
Energy Density:	39.2 W-hr./kg
Power Density:	170 W/kg
Weight:	924 lb. (420 kg)
Charger:	American Monarch Charger, 208 V, 30 A input
Charge Time	4.5 hours from 80% DOD

Performance:
ZEV Range:	63 miles @ 40 mph (100 km @ 64 km/h)
Acceleration (0-100 m):	7.9 sec. @ 45 mph (72 km/h)

Dimensions:
Wheelbase:	109 in. (277 cm)
Track (F/R):	54 in. / 52 in. (137 cm / 132 cm)
Overall Length:	154 in. (391 cm)
Overall Width:	60 in. (152 cm)
Height:	50 in. (127 cm)
Frontal Area:	19 ft^2 (1.75 m^2)
C_d (estimate):	0.30
Ground Clearance:	6 in. (15 cm)
Weight (wet, w/ psgrs.):	2400 lb. (1100 kg)
Weight Distribution (F/R):	53%/47%

Frame / Chassis:
Materials:	4130 Chromoly Steel, 3 Al-2.5 V Titanium
Construction Type:	Tubular space frame, torsional box backbone, full roll cage
Weight:	150 lb. (70 kg)

Fairing / Body:
Material (Body/Window):	GE Lexan Plastic 0.125" / GE Marguard Lexan Plastic 0.25"
Construction Type:	Vacuum thermoformed over CUHEV male tools
Weight:	85 lb. (38 kg)

Interior:
Seats:	Jaz Plastic bucket type, with cushion, 11 lb/ea. (5.0 kg/ea.)
Seat Adjustment:	Driver: Fore and aft, height. Passenger: Fixed
Restraint System:	Five-point racing harness, quick release
Luggage volume:	4.4 ft^3 (0.12 m^3)
Analog Instrumentation:	Speed, Voltage, Current, Kilowatt-hours, Odometer
Digital Instrumentation:	Range, Efficiency, Temperatures, Speeds, Vehicle Status

Suspension:
Front:
	CUHEV, independent, unequal length A-arms, overhead coil springs and aluminum gas-charged shocks, anti-roll bar
Toe/Camber/Caster:	-0.5°/0°/3° (at ride height)
Anti Dive Geometry:	20%
Spring Rate:	155 lb/in (27,200 N/m)

Rear:
	CUHEV, independent, unequal length A-arms, coil springs and aluminum gas-charged shocks, toe-control link, anti-roll bar
Anti Squat Geometry:	55%
Spring Rate:	195 lb/in (34,200 N/m)

Steering:
Type:	Rack and Pinion, Non power assist
Turns, lock to lock:	2.5
Turning circle:	32 ft. (9.75 m) (curb-to-curb)

Brakes:
Front:	CUHEV, 10" x 0.20" vented steel disk, hydraulic, no assist. Unique Mobility Regenerative Braking
Rear:	CUHEV, 10" x 0.20" vented steel disk, hydraulic, no assist. Independent caliper emergency brake

Wheels & Tires:
Wheel size:	American Racing 14.0" x 6.0"
Wheel type:	Forged Aluminum
Tire Make and Size:	Goodyear Invicta, low rolling resistance, P165-65 R 14
Inflation Pressure:	44 psi (303 kPa)

GMI HEV Student Design Team
Professor Colin Jordan, Advisor
Professor Mark Thompson, Advisor

Joe Bray	Jeff Harthett
Ryan Brown	Will Huber
Tim Beyer	Charles Marshall
Al Bsharah	Brad Mauch
Thad Conlisk	Ryan Milburn
Michael DaGama	Jerome Motley
Mike Dvorscak	Jack Phillips
Arthur Ekland	David Pray
Brady Erickson	John Rothermal
Sean Falkowski	Andrew Salas
Russ Faust	Jim Salviski
Terry Feldpoush	Duane Schuler
Jim Flanigan	Steve Smith
Sherry Fields	Ryan Suhre
Jason Forcier	Anthony Thomas
Dave Foster	Sean Wood
Jeff Grabowski	Roy Van Wynsberghe

ABSTRACT

This paper presents the design and specifications of the GMI Engineering & Management Institute's entry in the 1994 Hybrid Electric Vehicle Challenge to be held in Milford, Michigan from June 14-20, 1994. A greater concern for environmental issues such as air pollution and the use of fossil fuels has caused a push for more efficient vehicles. This competition provides an excellent opportunity for research into the fields of alternate fuels and electric power for automotive applications. For our competition, a 1994 Saturn was converted to run on an electric motor and an Auxiliary Power Unit (APU) fueled by ethanol. This new conversion class focuses on a power-assist hybrid vehicle with shorter Zero-Emissions-Vehicle (ZEV) range.

INTRODUCTION

The Hybrid Electric Vehicle (HEV) is designed to reduce the use of fossil fuels and to increase emission efficiency. The student design team at General Motors Engineering and Management Institute (GMI) placed the highest priority on designing the GMI HEV as being the hybrid electric vehicle of the future. Major emphasis was placed on safety, reliability, durability, drivability and low cost. With these items in mind the GMI HEV Team embarked upon this project with good sound disciplined engineering. Keeping in line with our main objective the GMI HEV Team has designed a vehicle, which we feel is energy efficient, with low emissions and provides the consumer with adequate performance.

APU Group Summary

Many engines were considered for utilization as the APU, such as the Geo Metro 1.0L, Kohler CH-20, Suzuki Bandit engine, Kawasaki FD620 (carburetor and fuel injector version), and Orbital engine or Sterling engine. Upon careful thought and deliberation, we decided to use the fuel injected version of the Kawasaki FD620. The various problems with our other potential APU candidates are as follows. The manufactures of the orbital and Sterling engines, which were our first choices, were unable to provide an engine satisfying our design requirements. The Geo Metro 1.0 L was too large for our compartment requirements. The main problem with the Suzuki Bandit engine was that it did not reach its maximum horsepower until high RPMs were achieved.

One of the primary reasons for choosing the Kawasaki FD620 was that it was compact and lightweight because of its V-2 design and aluminum engine block. Another plus in the design was that it has an electric starter (note this may not be needed if the generator can be used to start the engine). (See Figure 1) This engine is also water cooled, thus the location is versatile within the engine compartment, since an air inlet is not required for cooling. A water-cooled engine also allows us to have a functional heater on the vehicle.

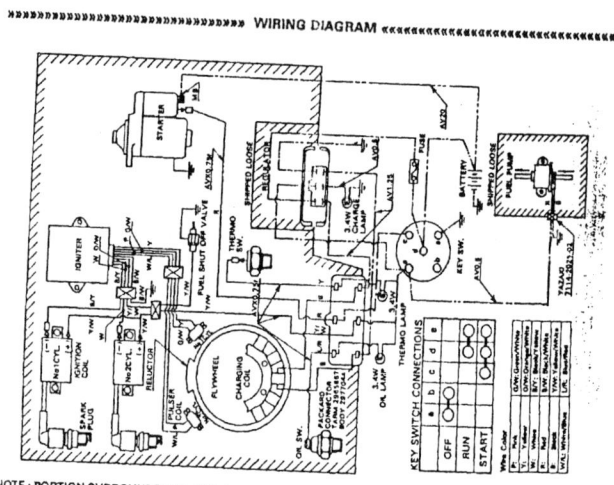

Figure 1

This engine is a four stroke which has some advantages over two stroke engines. One advantage is that two stroke engines usually are more efficient at high RPMs while the engine we chose was designed to be most efficient at low RPMs. Low RPMs aid in the reduction of engine noise level at cruising speeds. In addition, four stroke engines generally have better emissions than two stroke engines.

Another benefit for this engine was that it already had throttle body fuel injection, so setting up the fuel injection system for ethanol could be as easy as getting an injector with a higher flow rate. With the Kohler or the carburetor version of the Kawasaki engine, the fuel injectors would have to be retrofitted.

The throttle body injection unit was tested for flow, and it was determined that the existing fuel injector could be used if a larger pressure regulator was retrofitted to the engine. Doubling the fuel pressure effectively doubles the flow rate for the Kawasaki throttle body injection unit.

Ethanol was chosen over methanol after discussion with people knowledgeable in the field. Ethanol is much less corrosive and probably won't require flushing of the fuel system as would methanol. Overall, ethanol is a much safer fuel to work with. A programmable fuel injection system is to be supplied by Fuel Management Systems (FMS) in Mundelein, Illinois. An oxygen sensor has been installed to allow for the system to run in closed loop mode. The FMS LS-12 system allows for very precise fuel control. Oxygen sensor trigger voltage and jump-up and jump-down functions can be adjusted to allow for the lowest emission possible using ethanol.

The use of a programmable fuel injection system and the fact that the engine will operate at a constant RPM should allow the engine to produce minimum emissions. The use of a properly selected catalytic converter should further reduce emissions close to nothing. Utilizing this criteria, we decided to use a catalytic converter from a 1993 Ford Taurus.

Currently, the engine is being set up on an electric dynamometer to be calibrated. A special coupler was purchased to connect the engine to the dynamometer. The coupler uses rubber isolators to dampen vibrations that came from the V-2 engine. Engine damage has resulted in the past from not using such a coupler. The engine will soon be calibrated on the dynamometer and then installed in the HEV.

The generator must be relatively small, light weight, efficient, and compatible with the engine. The candidates considered by the generator group were the Magnetek generator, the Delco Remy DN50, or the Fisher generator. The Fisher generator was chosen because it met all of our design requirements. Fisher was also willing to design to our specific RPM and voltage requirements, thus we did not have to send the generator out to be rewound in order to meet our goals nor did we have to compromise performance by using an "off the shelf" item.

From the enclosed horsepower curve, it was determined that the engine would generate 14 kW @ 3200 RPM. Figuring a 15% efficiency loss, the maximum power alternator we could use would be 12 kW. The amount of effort to maintain 45 MPH and 60 MPH on a 0% grade was figured to be 7.5 kW and 13.5 kW respectively. Thus the alternator should be able to power the vehicle effectively with enough excess power at times to charge the battery.

The electric motor requires 230 V AC to run our engine at 3200 RPM. The alternator has a rated efficiency of 90-95% and was superior to the other generators when output per pound and output per unit volume was examined. It also offers an integral cooling fan as well as direct mounting to the engine output shaft or flywheel as shown in the figure. This eliminates the need for a coupling shaft and bearings which decrease efficiency and take up space. A 12 kW alternator operating near these specifications would be approximately 10" in diameter, 5" in length and weigh around 25 pounds.

Energy Storage Team Summary

The primary focus of the GMI HEV Energy Storage Team (EST) was to research available batteries for the GMI HEV power unit and to design the battery enclosure per 1994 Hybrid Electric Vehicle Challenge Rules and Regulations. Keeping in line with rules and regulations the GMI EST developed a list of project requirements. These requirements were as follows:

- Sufficient batteries to handle the motor requirements of 312 Volts and 106 Amps of maximum current draw.
- The availability of production batteries to meet the requirements stated above, which were relatively inexpensive.
- Size and weight requirements to keep in line with HEV rules and regulations.
- High cycle charging which was relatively fast. (See Figures 2 & 3)
- Reliability and durability.
- All pertinent safety items such as having a recyclable sealed battery enclosure, while operating at a low temperature with no harmful exhaust gases.
- Design of battery enclosure which was non-flammable, impact resistant, ventilated, and self-contained.

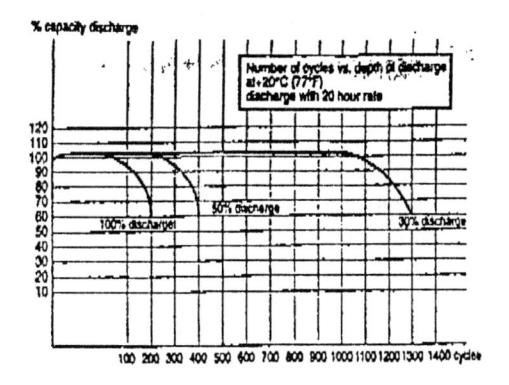

Figure 2

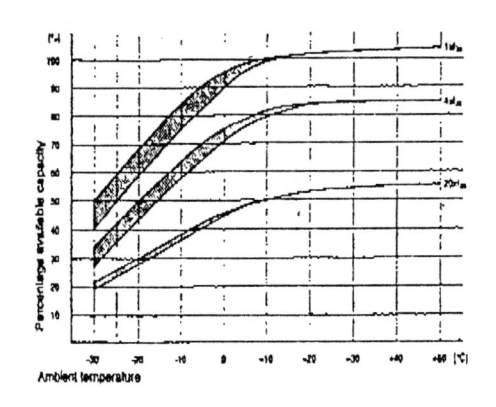

Figure 3

Initial research included studying various technical reports concerning electric vehicles. This was done to prevent repeating previous mistakes as well as to inform the group of developing trends of electric vehicle energy storage. This includes such topics as energy requirements, available technology, safety considerations, desired characteristics and potential suppliers. Using these resources a supplier contact form was developed to determine the availability and performance of various types and brands of batteries. The manufacturer information was then examined to determine which batteries satisfied the intended project requirements.

Culmination of these research efforts lead to the assumption that the only practical battery systems commercially available are still lead-acid and nickel-cadmium (nicad). After discussion with suppliers and further research, the team determined that sealed lead/acid batteries would be the best alternative for our application. Among the various lead/acid batteries, the individual characteristics needed to be analyzed. This was done through product literature as well as technical support from the manufacturers. The following batteries were analyzed.

- East Penn 8TU1
- East Penn 8GU1
- Concorde RG390E
- Trojan DC-9R
- Johnson Controls UPS 12-95

This list was reduced to only those batteries which would meet the design intent of the vehicle. The East Penn 8TU1 was eliminated due to the fact that it was a non-sealed battery. The Concorde battery was eliminated because it was intended for intermittent use only. The research efforts lead to the East Penn 8GU1 being selected as the preferred energy storage device for the GMI HEV. Installation of the batteries was the next item to be analyzed. Originally the EST was to use 2/0 gauge wire to connect the batteries with special connectors provided by the manufacturer. However, due to the configuration of the battery modules it was decided to use copper bar stock to connect the batteries. These connectors were fabricated using copper tubing which pressed flat then shaped to the needed specifications. However, there was still the need to use the 2/0 gauge wire for various length jumpers. These jumpers were used to make specific connections from battery module to battery module.

The next action item for the EST was to design the battery enclosure. This design was dictated by the rules and regulations for the Hybrid Electric Vehicle Challenge. To achieve the safety requirements, it was decided that the box should be constructed using a frame of 1" x 1" x 3/16" angle iron and a skin of .022" thick galvanized steel. The enclosure will house 24 batteries, which are arranged in two layers of 12 batteries. These batteries will be wired in series to achieve the necessary 312 Volts to drive our electric motor/controller. Spacing in the battery box was achieved using 3/4" insulation strips to provide adequate room for ventilation of any harmful exhaust gases present. This insulation also provides our energy storage unit with the proper means for absorbing shock during the competition. The exhaust outlets will be positioned on the top side of the battery box cover vented with aid of an exhaust fans through the rear trunk floor.

Safety Group Summary

The Safety group was formed to oversee safety matters relating to the preparation and performance of the vehicle. Adherence to all pertinent SAE HEV Challenge rules is the first priority of the group. During the actual competitions the safety group will assume responsibility for procurement of necessary safety and emergency items as well as crew safety training.

The principle identified areas for component procurement include driver's safety gear, cockpit safety equipment, fire protection equipment, and pit emergency safety equipment.

For driver and passenger safety, it is required by SAE that each occupant wear a fire-resistant suit, a safety helmet, and fire resistant socks. A single-layer Pyrotect Nomex standard-size suit, a Snell type SA92 helmet, and Nomex socks will be used in the competition and will meet all safety requirements.

On board remote fire protection is required for the APU area and the battery box. A three nozzle, 11 pound remote Phoenix Fire Suppression System will also be available for use in the competition. This system was chosen because it uses 1301 Halon which is non-toxic, thereby allowing the third nozzle to be used in the passenger compartment. Competitive systems use 1211 Halon that can cause brain damage. According to Phoenix, an 11 pound system is a typical requirement for the type of fire protection desired in the competition.

In addition to the safety equipment needed in the car, some pit area safety equipment will also be required. This additional equipment includes portable hand-held fire extinguishers, a chemical spill cleanup kit, and a portable eyewash station. Every precaution has been taken to insure the safety of the driver and everyone else at the competition.

Electric Motor/Controller Group

The motor that GMI decided upon was a VCM 703 donated by Magnetek. (See Figure 4 & 5) This is a 30 hp (22kW) AC induction motor that runs at a base speed of 1750 rpm. This motor offers several advantages for this application over equivalent motors, based on the extensive feedback equipment installed on the motor.

Figure 4

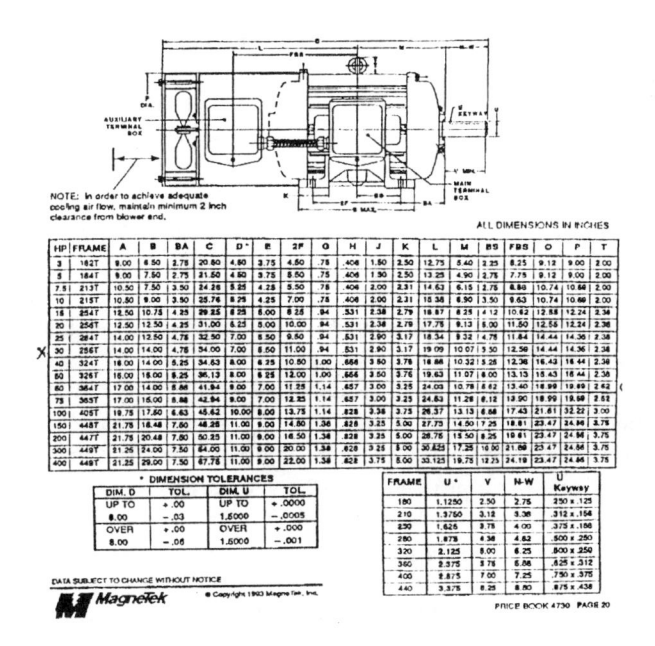

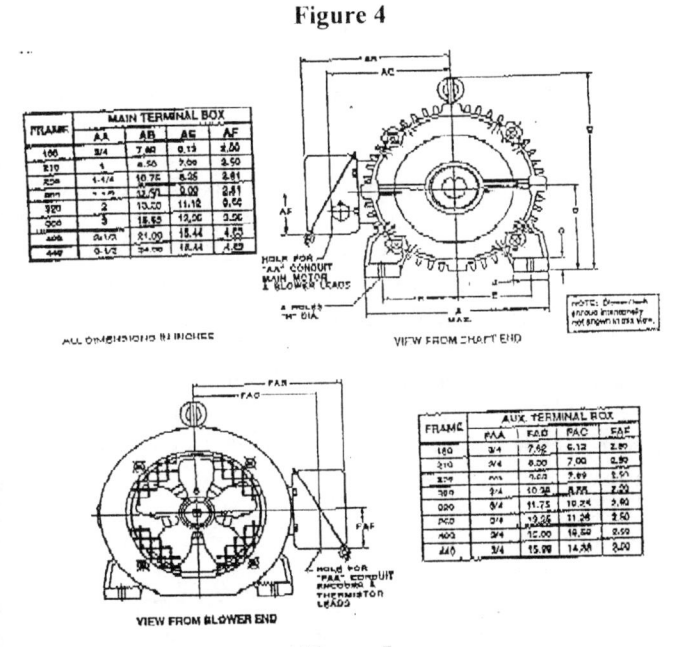

Figure 5

The matched controller for this system is a VCD 703 vector control drive, also donated by Magnetek. This controller is a sine-coded pulsewidth modulated AC motor controller. It has over 100 programmable functions that are programmed through the use of a keypad from the passenger compartment in the vehicle. The controller uniquely controls the torque output through the use of a special adapter card. This allows the vehicle to be driven at a constant torque rather than a constant rpm. The controller also has many safety features including overtorque, overspeed, overcurrent, and undervoltage protection to name a few, that are a great advantage for development work as well as vehicle operation.

The controller operation is based on vector control logic. This logic determines the slip frequency through the output of a pulse generator. The magnetizing current and resultant air gap flux are held constant by maintaining a constant Volts/Hertz ratio. With the rotor speed and flux held constant, the primary flow current directly controls the torque-producing current and the phase angle between the magnetizing current and the torque-producing current.

The motor and controller, as a set, produces optimum performance through the mathematical model developed by Magnetek. The parameters for operation were developed and then programmed through the use of the keypad in the passenger compartment. The keypad also displays useful output information as well as error messages. This was used for development purposes and is also being used during vehicle operation.

The controller is powered by twenty-four 13 volt batteries supplied by East Penn. For testing purposes, power to the controller was obtained by a cable setup using voltage from an external source. This was done to avoid unnecessary wear on the batteries while the systems were being tested. On both the negative and positive side of the batteries are 12 volt actuated relays developed by Kilovac and used on the General Motors Impact. Two more relays are used for switching the battery charging circuit and a for switching the APU output. This fourth one was mounted in the engine compartment rather than in the electrical box in the rear to prevent a live wire to run the length of the vehicle when the relay was tripped. These relays offer the only acceptable ratings for our application. (See Figure 6)

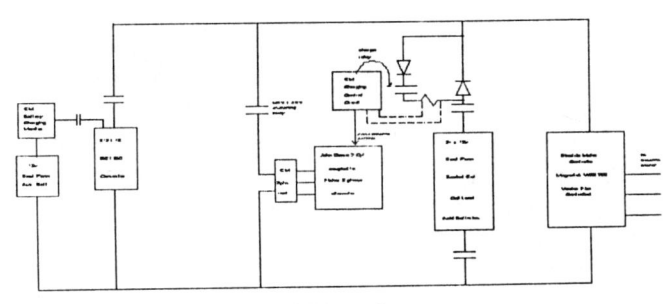

Figure 6

We developed a digital circuit for monitoring the charge rate of our main and auxiliary batteries. (See Figure 7) Through this we sample the charge current through a shunt, and switch off the charging circuit based on charge levels.

We also built our own rectification circuitry for our Fisher alternator, meeting the high frequency output of this alternator.

We charge our auxiliary battery from a Vicor DC/DC converter that converts the 312 volts from our main pack to 13 volts for our auxiliary systems.

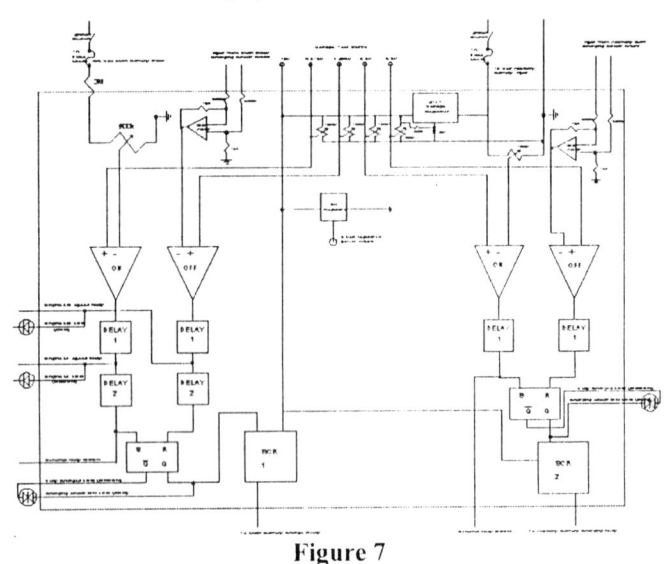

Figure 7

Modifying the existing Saturn wiring was an important issue that we tackled early. We also used an array of VDO gauges in our instrumentation section of our design. We our monitoring the APU current, voltage, the battery current voltage, and the auxiliary battery current, voltage.

We have made certain to meet all SAE requirements for switching ability, overcurrent protection, and systems rating. Overall, we have tried to achieve a high performance, efficient design.

Transmission Group Summary

The underlying goal of the competition is to develop an energy efficient vehicle with extremely low emissions. With this goal in mind it was decided to remove as many of the accessories that will draw a substantial amount of continuous energy. The first component to be removed was the power steering pump. This component was replaced with a non-power assisted unit.

The next step is the selection of a transmission design that is most compatible with our needs. The type of transmission that we selected to use is a manually shifted, front wheel drive transmission from a 1986 Honda Accord LXI. This type of gear box provides the best solution to efficiency and speed multiplication problems by using the following ratios:

1. 3.166:1
2. 1.857:1
3. 1.259:1
4. .935:1
5. .794:1

These gear ratios yield a final overall reduction of 4.18:1.

The transmission is supported by a 2" square tubulat frame that is welded to the existing saturn frame. The transmission is bolted to the tubular frame. Transmitting the power from the transmission to the wheels was the next item analyzed. This was accomplished by modifying the existing halfshafts. This was done by splicing together the wheel end of the stock Saturn shaft and the transmission end of the stock Honda shaft. The input shaft had to be lengthened to accommodate the pulleys for the drive system. round stock was used and turned down ot the proper diameter. The extended shaft is supported opposite the transmission end by a standard pillowblock. 11 gauge steel sheet was used as the drive system cover. This was done to eliminate potential hazards.

With the selection of the proper programmable controller for the AC motor, we are able to supply a high starting torque from zero up through 1800 RPM. This gave us reassurance to eliminate a clutch mechanism and direct drive the transmission. With the elimination of the clutch, shifts can be accomplished simply by reducing power output of the motor and manually shifting the transmission.

The transmission will be linked to a belt drive system. The belt drive system uses 4 belts with a 1.5:1 speed increase to the transmission.

Emissions / Fuel Systems

Fuel delivery and emissions control strategies in our Kawasaki FD620 APU are similar to that used in most current small cars. Actively controlled fuel injection, including oxygen sensor feedback and a catalytic converter are used to control tailpipe emissions. (See Figure 8) The choice of a small engine reduces overall emissions volume and the use of ethanol fuel shows potential to lower emissions even further. Unfortunately, the time did not permit emissions testing to be done of the engine before competition.

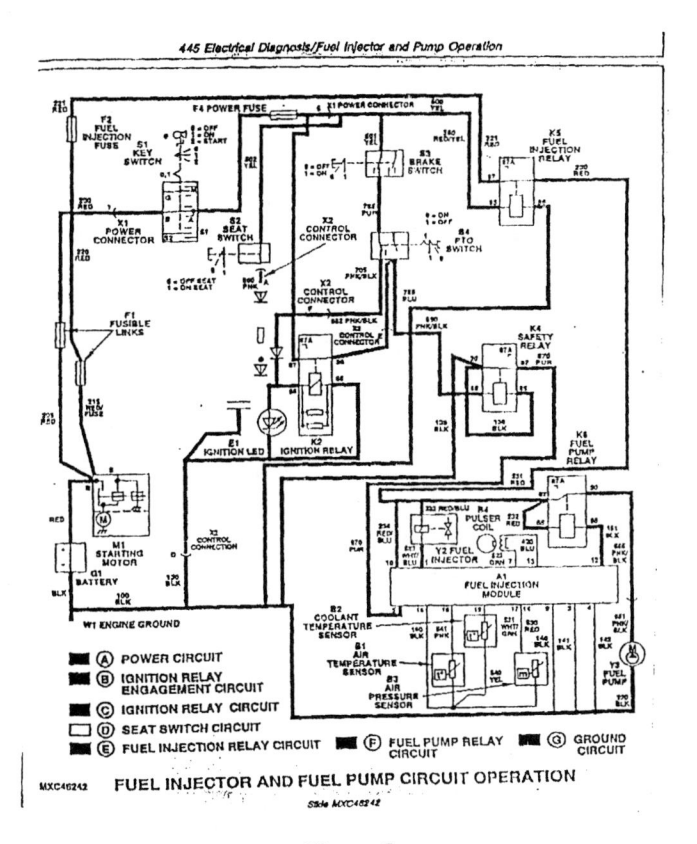

Figure 8

The catalyst substrate was manufactured and donated by Ford Motor Company. It is a three way catalyst formulated for use with ethanol. Ford also donated the nose cones, housing, and insulation for the converter.

The one brick catalyst was chosen because its reduced size better fit the size of the APU. This combined with the mounting of the converter close to the exhaust ports on the APU will result in a quicker light off time and better efficiency for the converter.

The exhaust header for our V-2 along with the exhaust piping and muffler configuration were designed and built by the HEV team. The standard production muffler supplied by Saturn was used in an effort to save time and money.

A programmable fuel injection system, supplied by Fuel Management Systems (FMS) in Mundelein, Illinois, was used to facilitate the switch from gasoline to ethanol. A programmable controller was needed to change calibrations from gasoline to ethanol. FMS coolant, air temperature, and Hall effect crank position sensors were also installed to better communicate with the controller.

Oxygen sensors are used to allow system feedback on exhaust output. By measuring oxygen in the exhaust stream the controller can optimize fuel delivery to the APU. This allows for higher converter efficiencies and is referred to as close loop fuel control.

The Kawasaki throttle body fuel injection system that come stock with our APU was utilized. Although V-2 engines suffer from inherent mixture distribution problems, the Kawasaki injector utilizes a special deflector tip to alleviate this problem.

The stock fuel tank and fuel level sender that come with our Saturn were kept for fuel storage and delivery. For more durable applications of ethanol, more corrosion resistant materials, such as stainless steel, should be considered for use in the fuel delivery system.

The fuel regulator, located at the injector, had to be replaced with a regulator rated at a higher PSI. In order to get the proper flow characteristics from the ethanol that our fuel management system required, the fuel needed to be delivered to the injector at a higher pressure, thus the reason for a more robust regulator.

Chassis & Accessories

Given the relative nature of the Chassis/Accessories Group and the Motor/Controller Group the GMI HEV Team decided to have these groups work in conjunction to facilitate installation of new chassis components. This joint effort has produced fabrication and installation of new cross members and brackets for mounting the electric motor and manual transmission. The original front cross member was removed to allow more room for the APU Group to mount the engine/generator. A new square tubing cross member was installed lower and more rearward than the original.

A Honda Prelude 5-speed manual transmission was installed and coupled to the engine with a V-belt drive system utilizing four belts. Proper belt tension will be achieved by shimming the electric motor as required. The Honda Prelude inside and Saturn oustside Half-shafts were mated and installed to complete the mechanical side of the drivetrain installation. The Honda shifter will be modified and used for gear selection. In fifth gear with our current 1.5:1 pulley ratio, our cruising speed should be about 65 MPH.

After weighing our HEV, we found it to be overweight based on the current rules. Our battery and motor selection limited our chances of meeting the weight requirements, however. Our current weight is 3150 without passengers and without approximately 300 lbs of additional components. We have ordered new springs based on a total weight of 3460 lbs without passengers, with 1620 lbs on the front wheels and 1840 lbs on the rear wheels. We will also be installing a rear swaybar to help handle our increased weight.

University of Illinois Hybrid Electric Vehicle: An Electric Vehicle for Today

Timothy Roethemeyer, Brandon Masterson, Marc Stiller, and the UIUC HEV Team
University of Illinois at Urbana-Champaign

ABSTRACT

Hybrid electric vehicles offer a next step in controlling fossil fuel emissions. A hybrid vehicle combines electric traction with an internal combustion engine to gain emissions advantages while maintaining long range. A team of 250 students at the University of Illinois, Urbana, has produced a practical hybrid vehicle using advanced industrial-grade off-the-shelf technology. This car has 30 mile zero emission range, coupled with over 400 miles of ultra-low emission range. This performance is accomplished through design innovations such as: preheated catalyst, tuned-port fuel injection, removable battery pack, high efficiency ac traction system and distributed microcontroller network. The estimated cost increase is $2,000 for mass production, compared to the Ford Escort on which the conversion is based.

INTRODUCTION

With legislation in California and other states requiring two percent of new automobile sales to be zero-emission vehicles in 1998, many automobile manufacturers are scrambling to produce marketable vehicles that satisfy these requirements. The University of Illinois at Urbana-Champaign has established a Hybrid Electric Vehicle (HEV) Program to address emissions concerns through hybrid architectures. The UIUC Team holds the philosophy that an economically viable hybrid electric vehicle that meets or exceeds the performance characteristics of a conventional car is attainable using off-the-shelf industrial technology, coupled with innovative solutions for design optimization, and backed by thorough engineering. Although an HEV is not a true zero emission car, it offers a realistic stepping stone toward fully electric vehicles, while providing better short-term emissions reduction. For example, a 90% reduction in internal combustion emissions can be obtained in hybrid operation compared to an equivalent conventional vehicle. Given the fact that HEVs do not compromise operating performance, it is expected that hybrids are more marketable than purely electric vehicles. Deep emissions reductions, combined with improved marketability, mean that hybrids are likely to have a greater effect on pollution than anticipated zero-emission cars.

The UIUC Team goal is to produce a practical vehicle that is within economic reach of the public and one that meets the intent of the recent zero emission mandates. This philosophy guided the team through the design stages, influencing decisions such as the choice of electric drive motor and the selection of the internal combustion engine. However, design decisions are ultimately backed by engineering principles, and the following design categories are discussed: power requirements, regeneration and braking strategy, battery packaging, suspension and structural modifications, hybrid control strategy, emissions, climate control, and driver interface. Of course, safety, manufacturability, reliability, and serviceability permeate all of these categories.

DESIGN CONSIDERATIONS

The UIUC vehicle utilizes a series HEV architecture, which combines an electric drive with a small engine-generator auxiliary power unit (APU). The electric drive provides excellent dynamic performance without emissions, while the APU provides long range with substantial emissions reduction. The ICE has no mechanical connection with the drive axles. The engine's operation can be optimized as necessary independent of the vehicle operating state for emissions, efficiency, or other requirements. To minimize emissions, the vehicle can be operated in a zero-emissions, all-electric mode for daily commuting and switched to hybrid mode for extended travel.

The design was implemented through conversion of a 1992 Ford Escort Station Wagon. This vehicle was refitted with an ac electric traction system, a 22 HP water-cooled engine-generator set, and a unique lead-acid battery package. See General Layout of HEV Components in Appendix.

Power Requirements

The electric drive system determines dynamic performance, and its power level was selected to match acceleration capabilities of the stock Escort. The system has a peak rating of 62kW (83 HP), and should be able to accelerate slightly faster than the stock car's 13.2 0-60 time. Figure 1 shows the measured road power requirements for the Escort. The drive system continuous rating is 22kW (30HP), and allows extended operation at speeds in excess of 110 km/hr. The APU is sized to provide cruising power on the highway. At 104 km/hr (65 mph), the Escort uses about 18 kW. An APU capable of 18kW output will allow freeway cruising at the legal speed limit, with range limited only by fuel tank volume. Thus, APU specifications are based on steady state vehicle performance in this architecture, while transients such as acceleration determine traction motor specifications.

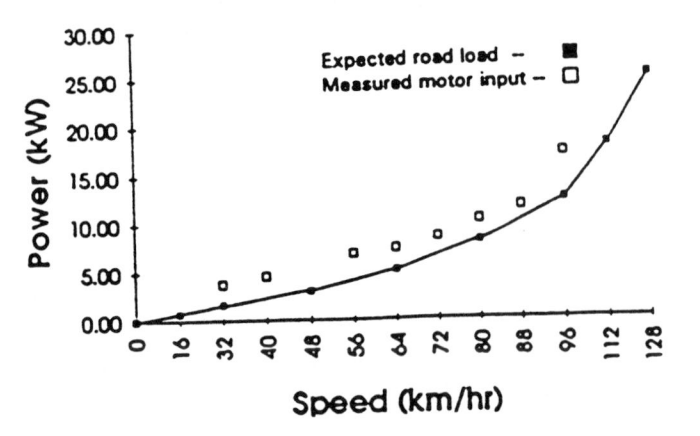

Fig. 1 Road Power vs. Speed

Battery Selection and Packaging

An extensive search for high-performance, rechargeable batteries revealed that few types are readily available as production parts. Lead-acid, nickel-cadmium, nickel-metal-hydride and other technologies were compared. As discussed in [1], lead-acid batteries offer the best compromise between availability, low cost, easy maintenance, recyclability, specific energy and power ratings, cycle life, and other environmental considerations. An HEV normally has a smaller battery pack than an all-electric vehicle, but it has similar power requirements. Specific power density is especially important in an HEV. Thus, Johnson Controls UPS 12-95 batteries were selected for use in the vehicle. This is a sealed glass-mat, lead acid battery intended for interruptable power supply applications. It shows excellent power density, good energy density, requires no maintenance, and will not disperse dangerous acids if opened.

Twenty-six sealed lead acid batteries weighing 11.8 kg each (totaling about 307 kg) provide a total of 8.58 kW·hr of energy at a three hour discharge rate. The batteries are placed on their sides and connected in series with a split ground that yields equal positive and negative DC busses with a nominal voltage 156 VDC per rail for a total battery nominal voltage rating of 312 VDC. Pressure plates are placed between individual batteries to distribute pressure on the center of the battery walls. The pressure prevents the electrolyte matting from separation with the plate structure and increases battery capacity by approximately ten percent. Foam springs maintain constant pressure and provide an insulation barrier to help maintain an equal cell temperature profile.

Battery Pack

All batteries are contained in an aluminum enclosure which is mounted between the rear floorpan and the rear suspension crossmember. The battery mass shifted the weight distribution from 60%/40% front/rear to 45%/55%. The effect on handling is discussed in Suspension Modification. The placement of the battery box is space-efficient and allows for maximum cargo space because it is in an area used for only the spare tire mounting and the fuel tank in the original vehicle. The spare tire is relocated behind the rear wheel well and mounted vertically. In a severe accident, the box is designed to separate from the vehicle and travel underneath the car. The battery pack is supported in front, rear, and side so the pack will not experience a pendulum affect.

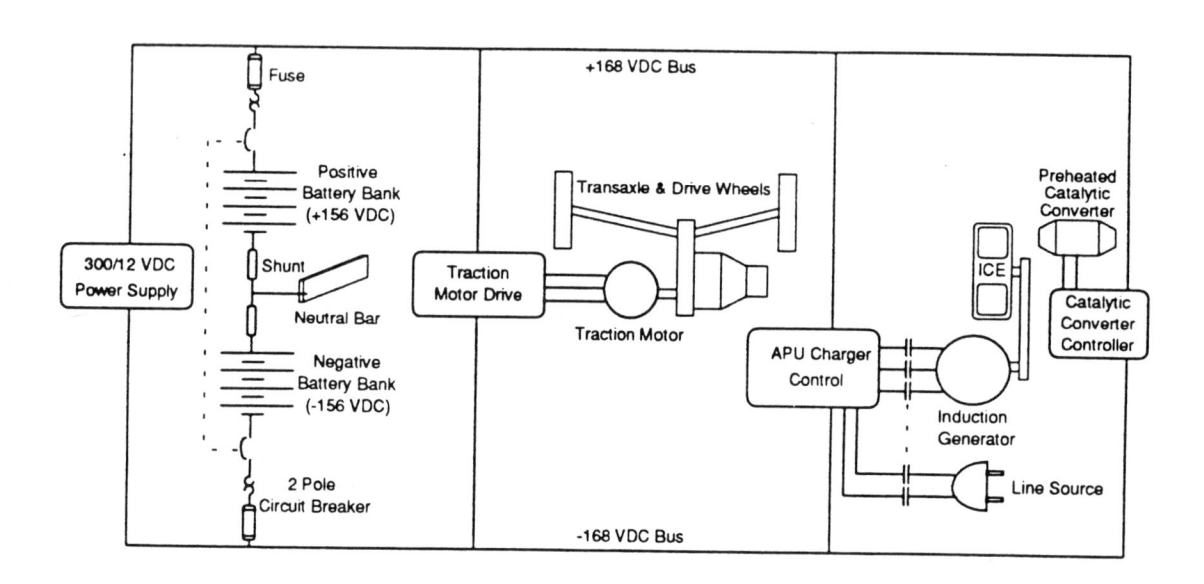

Figure 2 General HEV System Schematic

Power System Design

Overall electrical system architecture is illustrated in Fig. 2. Overcurrent protection is provided by two 250 VDC rated fuses sized at 250 amps. One fuse is installed in each bus for total overcurrent protection. A maintenance disconnect switch function is provided by a 250 amp, 250 VDC Class, 2-pole molded case circuit breaker and is wired to the output of the battery pack immediately before the fuses. The circuit breaker is opened remotely through a shunt trip. The shunt trip is powered indirectly from the main battery pack via an onboard 312/12 VDC switchmode power supply. To assure shunt trip operation in the advent of 12 VDC supply failure, a capacitive energy storage circuit has been implemented. Shunt trip operating locations are in the vehicle dashboard and underhood, with the underhood switch operable from outside the vehicle. Both the circuit breaker and the fuses are mounted within the main traction drive inverter enclosure for improved volumetric efficiency. The vehicle high voltage power system is grounded to the chassis at a single point through a silver plated copper bar. All electrical items added by the UIUC Team are connected to this ground bar, which is in turn connected to the vehicle chassis through a single #8 AWG conductor. This configuration prevents any chassis conducted currents from occurring under normal operating conditions other than those designed into the stock Ford Escort 12 VDC electrical system. It should also be noted that all loads connected to the high voltage DC system are operated between the positive and negative supply busses; there are no loads connected between bus and ground. Thus, only during a bus to ground fault will current flow through the high voltage power system ground. This arrangement permits the use of a hall effect based ground fault current sensing with milli-ampere sensitivity.

Hybrid Control Strategy

In order to emulate the look, feel and operation of the original car, a number of functions must be controlled automatically. For this purpose the vehicle features a distributed microcontroller system consisting of four Motorola 68HC11-based microprocessors. These microcontrollers are configured to monitor battery state of charge, control APU while in operation, control commanded traction motor torque, and control the on-board LCD. Information is shared between the microcontrollers via an RS-485 network. Control reliability is ensured with the distributed control architecture by minimizing the affects of controller failures.

In addition to exporting the battery state of charge data to the network for transmission and display in the instrument cluster, the state of charge calculator will also send messages to the APU controller to turn on and turn off the APU. Turn on and turn off thresholds are set at 40% and 80% respectively. The APU is also turned on when vehicle speed exceeds 45 MPH continuously for 3 minutes or instantaneously once the vehicle reaches 55 MPH. Once the APU controller is given the signal to start, it first turns on the catalyst preheater, then starts the engine and finally loads the generator.

There are three basic modes of operation: Zero Emission Vehicle (ZEV), Automatic Hybrid Electric Vehicle (AHEV) and Manual Hybrid Electric Vehicle (MHEV). In the ZEV mode, the APU will not start and the vehicle operates on battery power alone. This mode would be used primarily by the commuter, as she would know if the vehicle range is adequate for her daily requirements. In the AHEV mode, the APU operates automatically based on battery SOC and vehicle speed as described above. This mode would be used during normal driving other than commuting. The MHEV mode is primarily for use in emergencies or out of the ordinary driving conditions where the APU must be operated.

Instrumentation

The on-board LCD is located in the dashboard and serves to monitor vehicle parameters that cannot be accommodated in the instrument cluster, such as motor winding temperature and battery current and voltage. Also displayed are menus for the presettable cruise control, component temperatures and individual battery voltages. The LCD display obtains this information by interrogating the RS-485 network for periodic updates. In addition, the LCD display is used to provide complete alarm annunciation and system self-diagnosis for a large variety of foreseen vehicle failures. A keypad is provided to allow the motorist to change displayed parameters, adjust cruise control settings and acknowledge alarms.

After evaluating commercially available energy meters, a decision was made to develop a suitable meter internally. The algorithm utilized for the UI-UC HEV is a modification of the TVA method [2], which is based on averaged discharge currents to compensate for the rated amp-hour capacity of the battery pack. As can be seen in Fig. 3, lead-acid battery capacity is a strong function of temperature. Consequently, the algorithm used to calculate battery capacity has been modified to compensate for temperature derating. The primary advantage of the modified TVA method is high accuracy (median accuracy is approximately 5%) with a relatively simplistic processing algorithm.

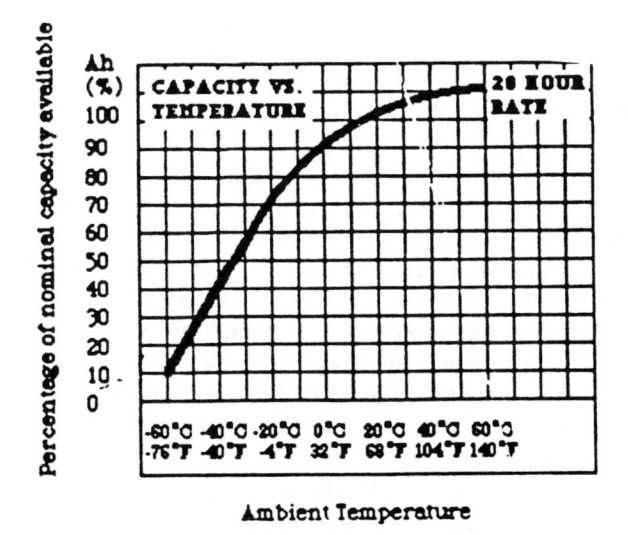

Fig. 3 Battery Capacity vs. Temperature

Auxiliary Power Unit -- Engine

An exhaustive search was conducted of small internal combustion engine manufacturers for an engine which would meet the vehicle power requirements as well as maximize specific power and minimize volume, brake specific fuel consumption, emissions, and noise. The APU internal combustion engine is a 37 kg Kawasaki FD620D that is liquid-cooled, V-twin, aluminum block, and fuel injected. This engine has a displacement of 620 cc with a maximum rated power of 17 kW at 3300 rpm. Brake specific fuel consumption values are excellent at a nominal 280 g/kW·hr using gasoline. Testing shows that this rated value is at a rich condition and stoichiometric running results in values as low as 265 g/kW·hr. The engine already meets California's 1995 emissions standards for off-road vehicles and extensive testing shows that emissions at stoichiometric air fuel ratio are excellent.

The University of Illinois team modified the stock Kawasaki engine in several different ways. The stock engine has an uneven fuel distribution between cylinders. To further decrease emissions and increase fuel economy, a tuned port fuel injection system was designed, fabricated and installed. Through dynamometer testing, we have demonstrated an increase in power of 20%. In addition to a new intake system, a complete fuel injection computer was designed and built to control each cylinder separately. This computer accounts for changes in coolant and air temperature, air density, manifold pressure, exhaust oxygen composition, and fuel composition. The new computer allows the system to be variable fuel. The engine will run on any combination ethanol, methanol, or gasoline. This system has demonstrated an equivalent of 70 miles per gallon using gasoline.

The team designed an 11 gallon safety fuel cell system located directly under the battery pack. The fuel cell was fabricated by ATL.

Emissions

Ethanol is used as fuel for the APU primarily for emissions considerations: higher heat of vaporization implies a lower peak flame temperature, reducing nitrous oxide emissions. Additionally, recent advancements in membrane separation and distillation techniques may allow for more efficient production of ethanol, resulting in an effectively closed carbon cycle with CO_2 produced during combustion and recovered during plant photosynthesis. Also, ethanol is typically non-corrosive, indicating relatively simple conversion from gasoline operation to ethanol operation.

Dynamometer testing of rpm, manifold pressure, load, emissions, air to fuel ratio, and fuel consumption have been performed in an effort to determine an optimization between low emissions and high power output of the APU. Running the engine lean reduces CO emissions and decreases brake specific fuel consumption. Dynamometer testing has shown emission levels on the order of ULEV levels. CO levels were reduced to .08 g/mile. Unburned HC were reduced to .02g/mile and NOx levels were reduced to .10g/mile.

The exhaust system for the APU includes the exhaust manifold, the catalytic converter, a catalyst preheater to reduce cold-start emissions, and a muffler. The team designed and built a catalyst preheater system. This system incorporates a properly sized 3-way catalyst enclosed in a preheater chamber which heats the initial air charge to allow the catalyst to reach light off temperature more quickly. This is incorporated into the control strategy to allow for a 20 second preheat and a 3 second postheat. This system is well suited for a series hybrid electric vehicle because the driver will not experience any changes in vehicle operation. The exhaust system upstream of the catalyst is made of stainless steel to further increase the rate of catalyst warm-up. Evaporative emissions use the production vapor recovery system of the 1992 Escort. Further reductions of evaporative emissions are achieved through the higher latent heat of vaporization of the ethanol fuel.

Auxiliary Power Unit -- Generator

One of the basic premises for the APU sizing and selection is that the system, when activated operates at 100% rated output. Consistent with prior work with hybrid and electric vehicles[3], the induction and permanent magnet synchronous machines stand out as candidates for the generator. Both machines have good efficiency. The induction machine has excellent robustness and relatively low cost, while the synchronous machine has slightly higher efficiency. Both types of APU systems have been developed by the UIUC Team for use in the vehicle. The generator is coupled to the internal combustion engine via a Gates GT Gilmer pulley system.

Traction Motor Selection

Desirable characteristics for traction motors in passenger vehicles are high torque, low mass, and high efficiency for improved range. In a series HEV, traction motor efficiency is more critical since all the vehicle's energy is transferred through the motor to the wheels. Referring to Table II, an induction motor yields the best balance between efficiency, robustness, and cost. Although alternatives were evaluated, available PM synchronous machines suffer from very high cost and questionable reliability; standard DC machines suffer from high cost, mass, and maintenance when compared to a squirrel cage AC induction motor; switched reluctance machines offer substantial gains in efficiency and cost, but torque characteristics from currently available models are inadequate for this application.

Two constraints influence the size and rating of the motor. The motor should be able to accommodate continuous cruising power. In general, this constraint is not demanding, considering that the 1992 Escort station wagon typically uses about 18 kW at freeway speeds. It should provide high momentary power for acceleration or regenerative braking. This constraint is more limiting since typical over torque capability for a standard NEMA Design B motor is only 250%. This implies that the motor rating should be selected primarily based on peak requirements, which closely parallels how automotive gasoline engines are currently rated.

Performance ratings for the UIUC traction motor are given in Table I.

TABLE I
Traction Motor Performance Specifications

Parameter	60 Hz Base Ratings	180 Hz Ratings
Voltage	67 VAC L-L	200 VAC L-L
Current	88 A	264 A
Breakdown Torque	103 N·m	103 N·m
Power Factor	0.83	0.83
Power	7.5 kW	22.4 kW
Slip	3%	2%
η at Peak Torque	89%	90%
η at 10% Torque	74%	76%
Rated RPM	1800	5400
Cost	$ 600	
Mass	55 kg	

The traction motor is mated to the stock Ford Escort transaxle via a flexible coupling. When driven at maximum torque through the standard transmission, the HEV has a 0-100 km/hr acceleration time that is 5% better than the stock Ford Escort Wagon.

Variable Frequency Traction Motor Drive

The traction motor drive is based on a 98% efficient commercially available PWM variable frequency inverter. The drive is a state of the art component featuring 400 amp IGBT switching devices controlled by a 32 bit microprocessor. Judicious repackaging reduced the mass by approximately 41 kg while providing an enclosure that satisfies IP42 specifications for environmental protection for improved reliability. The entire enclosure is shock mounted to the vehicle firewall for vibration isolation. The rating of the traction drive was selected to permit the motor to develop its maximum over-torque capability. Since power electronic devices are sized based on current ratings, the drive was selected with a 200% overcurrent capability matching over-torque capacity of the traction motor.

Induction Motor Torque Control & Clutchless Shifting

The traction motor drive was originally designed for industrial applications requiring speed control, but it has been modified to operate as an induction motor torque control. Extensive testing of a prototype HEV with similar design architecture determined that the motor speed control alone will not yield a satisfactory driver interface. Thus, motor torque control has been implemented by developing an interfacing controller to accept driver inputs corresponding to accelerator and brake pedals position and generates motor slip commands to the traction drive. Also, pedal signals are conditioned through a buffer circuit to provide safe failure modes. This technique also uses machine rotor speed feedback using a variable reluctance magnetic pickup and current feedback via a Hall effect sensor implemented around a standard scalar V/f control algorithm. The method is similar to conventional techniques [4], and a block diagram is shown in Fig. 4. This control strategy is in direct contrast to more sophisticated indirect flux vector sensing torque control techniques that are typically utilized for induction motor control. For high performance applications such as induction servo motors, flux vector control is necessary. However, for a traction application such as the HEV, very high performance flux vector control is excessive particularly when vehicle reaction time constants are taken into consideration. In addition, for direct or indirect flux vector control to operate reliably, an extensive knowledge of the induction machine parameters is required in order to ensure error term cancellation for stable motor operation. Since these parameters often vary with time, there is some evidence to suggest that flux vector control may not be robust enough to ensure satisfactory motor control over the projected life of the vehicle. On the other hand, scalar V/f control will provide sufficient performance with the additional benefit of providing a more robust control due to its insensitivity to changes in machine parameters.

The original clutch and fly wheel were removed to reduce weight, and a clutchless shifting system was installed. The clutchless shifting involves a unique procedure made possible by the rapid response of the electric motor. A micro switch from a Porsche auto-stick clutchless shifting mechanism is incorporated into the shifter to determine when a shift is initiated by registering any force applied to the shifter. This force instantaneously drops the motor torque to zero allowing the gear to disengage. The intended gear is determined by shifter column position and force direction while in neutral. For example, if the shift lever is in the 3rd-4th column and there is a forward force on the lever, the motor speed is matched to the correct rpm for 3rd gear based on the present vehicle speed. As the gear is engaged, torque is returned to zero until force is removed from the shifter, resulting in smooth, clutchless shifting.

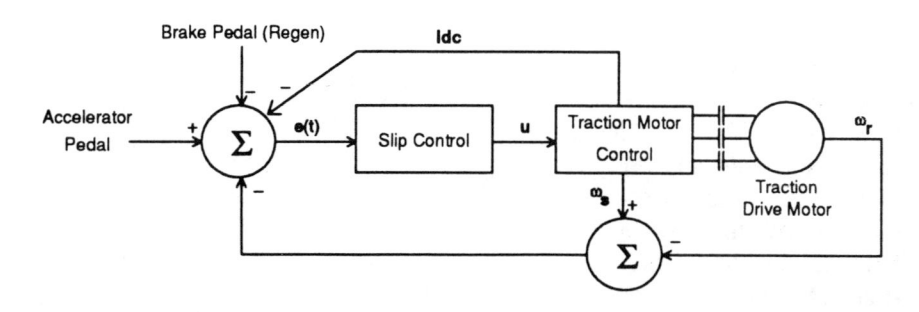

Fig. 4. Traction Motor Control Block Diagram

Regeneration and Braking Strategy

To increase vehicle efficiency and electric range, it is desirable to recover energy that would otherwise be dissipated as heat in the brakes. This energy can be recovered with regenerative braking. Simulations of the EPA city cycle show that a 24 percent range increase can be realized using this regeneration strategy.

The UIUC team developed a safe regeneration strategy that basically involves adding free-play to the brake linkage. As the brake pedal is depressed, the spring-loaded pedal arm activates a potentiometer, and braking is accomplished entirely through reverse torque on the motor, resulting in regeneration. The controller determines the torque to be applied by the motor based on potentiometer position and the current state of the transmission. The reverse torque applied is proportional to potentiometer position and inversely proportional to transmission gear ratio to achieve a constant ratio between wheel braking torque and pedal position. When the brake pedal is further depressed, mechanical braking is introduced at a deceleration rate greater than 0.3g, and will operate even in the case of a total regeneration system failure. This design prematurely locks-up the front wheels. Due to this situation, an Anti lock Regenerative Braking System (ARBS) has been developed. Sensors at the rear wheels and controls determine if the front wheels lock up before the rear. A situation like this is easily realized on a relatively slick road surface with regenerative brakes. Upon premature lockup of the front wheels, the ARBS decreases regeneration until the front and rear wheels turn at the same speed.

Structural Modifications

The rear structure of the vehicle requires modification to accommodate the battery pack. This involves modification to the bottom section of the tailgate, the rear seat mounts, the cross member to which the rear suspension attaches, and removal of the spare tire well. All chassis modifications result in greater moment of inertia at the modified area. Both the original and modified rear cross member were tested for structural integrity, and the results are shown in Fig. 5, Fig. 6, and Fig. 7.

From the figures, one can see that the testing shows that the modified cross member is stiffer than the original in the loading configurations tested. This is assumed to imply greater strength although yield testing was not performed due to the high cost of prototype fabrication. This is the most significant structural change.

Suspension Modifications

In order to match original ride characteristics, the natural frequencies of the front and rear of the vehicle were used as design parameters to be matched. With the changes to the mass of the front and rear of the vehicle, one can alter the respective spring rate to keep the natural frequency constant. Thus, with knowledge of both the original and modified vehicle mass distributions, one can calculate the new spring rates and make the necessary changes. Constant natural frequency implies that the ride height remains constant with no changes to spring free length.

However, with modified mass distributions, the front/rear cornering bias of the vehicle is altered as well. Specifically, the weight added to the rear of the vehicle increases the weight transfer at the rear end, resulting in a tendency toward an oversteering vehicle. To compensate for this, the diameter of the front anti-roll bar is increased until neutral to understeering characteristics were achieved. Moreover, tire pressures on the low rolling loss Goodyear Invicta GLR tires are biased higher to the rear to increase stability.

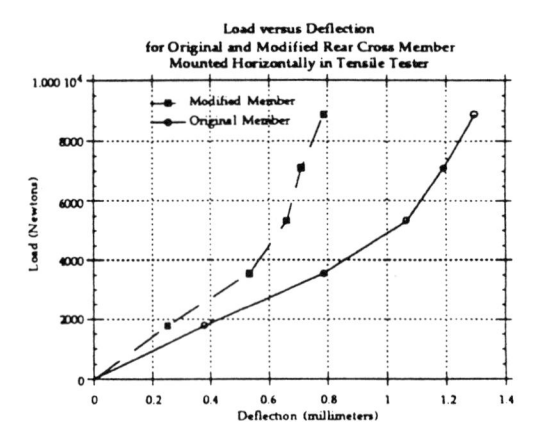

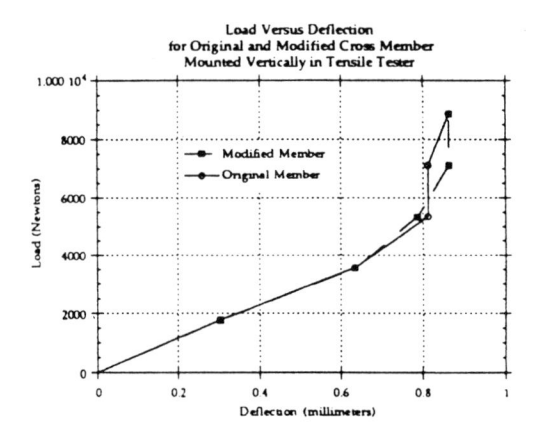

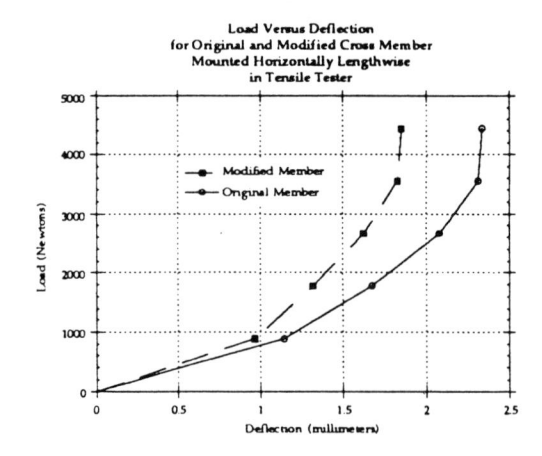

Figures 5, 6, and 7 Results of Structural Testing

Battery Charging

An external outlet is provided to permit battery recharging from an AC line. The DC outlet is installed in under the body mold trim in the front quarter panel. When the molding is lifted, the plug is revealed. The plug and fuel fill port are located on the same side of the vehicle for consistency. DC current is provided from a 220 VAC single phase line source by a regulated power supply system. This system is configured with three commercial power supplies, each rated at 150 V and 20 amps.

The power supplies are oversized to permit an increase in the main battery pack capacity without having to redesign the battery charging system. The three Sorenson DCS 150-25 power supplies are connected in series and controlled via a GPIB bus computer interface interconnected to a personal computer. The computer system permits constant current/constant voltage, or constant power charging algorithms, in addition to data logging of charging amps and voltage. The constant power algorithm enables the batteries to be charged aggressively without exceeding the 30 ampere limit imposed on line current draw by the HEV Challenge Rules. A kilowatt-hour meter is included to permit tracking of energy consumed during charging. The power supplies were not installed into the car to permit charging directly from an AC line input due to weight constraints. Charging is time limited to 6 hours, but in practice the batteries are recharged in 3 to 4 hours. Single phase true power factor correction may be added in the near future. Efficiency is rated at approximately 83% and claimed power factor is 0.68.

Climate Control

Approximately 80 percent of vehicles sold in the U.S. today have air conditioning, indicating that climate control is necessary for a marketable vehicle. Climate control system efficiency is crucial, because the power requirements for operation are significant (around six horsepower for the original refrigeration system) and energy storage is a fundamental limitation to electric mode operation. This problem is exacerbated by the drive motor producing insufficient heat to warm the passenger compartment. Climate control in the UIUC HEV is achieved through a vapor compression cycle which utilizes a reversing valve to toggle between refrigeration and heat pump operation modes. This system is illustrated in Fig. 8. This implies that only one system is needed for both refrigeration and heating.

Refrigerant 134a is used due to its environmentally safer composition. Extra evaporator volume is needed to accommodate the heat pump function, slightly increasing total system volume. The compressor is powered from the drive motor.

Manufacturability

Manufacturability is addressed throughout the component selection process by the choice of slightly modified off-the-shelf technology. Structural modifications are accomplished with the use of materials which match the original structure, allowing high volume manufacture with a simple change in die geometry at low cost. In production, all aluminum fabrications could easily be replaced by die castings resulting in a low marginal component cost. The entire drive train and APU system could be installed as one unit with the sub-frame consistent with modern assembly-line standards. The battery box could also be converted to a stamped and spot-welded assembly.

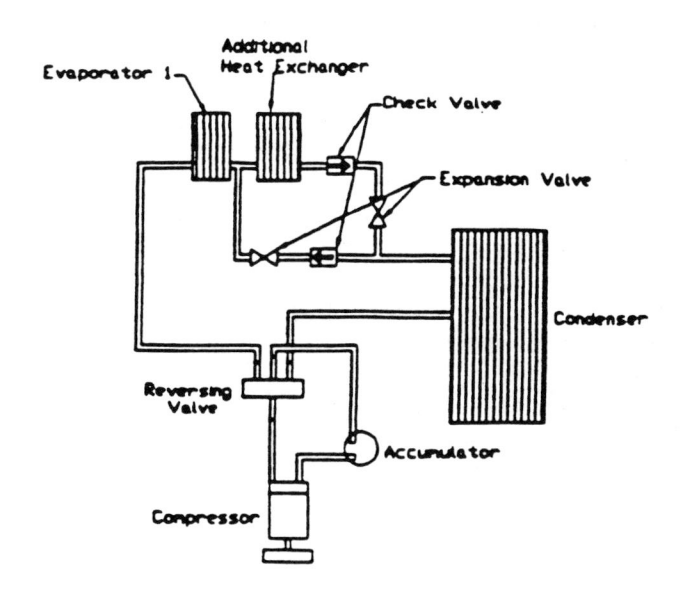

Figure 8 Climate Control System Schematic

Team Organization

The project has brought together over 250 graduate and undergraduate students in the last two years to carry out all aspects of production. Students are responsible for designing the architecture, developing selection criteria for components, modifying parts, writing codes, and fund raising. The project fosters student leadership and promotes "Green Engineering". The HEV program is now completely integrated into the University's curriculum through ECE and M&IE courses. Students are able to enroll in the HEV project through senior design classes, independent study, MS thesis and on a volunteer basis. Students have gained project management experience that normally requires years of work in industry.

An integrated, team-based concept is utilized to better prepare students with realistic training. Although there are two faculty advisers, the project is entirely student run. Two graduate students, one is a mechanical engineer and the other is an electrical engineer, form and supervise task teams to complete the development of the HEV. Tasks teams use concurrent engineering to assure that the prototype will be successful. An MBA student manages the fundraising and publicity efforts of the project. Most students admit that this program is the most rewarding experience they have had in college.

CONCLUSIONS

The manufacturability of the UIUC HEV is achieved, in part, using off-the-shelf technology. With this state-of-the-art, readily available technology, the University of Illinois

Hybrid Electric Vehicle Team believes that a marketable vehicle can be produced by 1998.

The final operational vehicle is the result of 24 months of design, analysis, iteration, and fabrication involving over 250 people at the University of Illinois. One objective of the design is to produce a vehicle capable of winning the Hybrid Electric Vehicle Challenge. However, the team also has an earnest desire to produce a safe, environmentally friendly vehicle.

Appendix

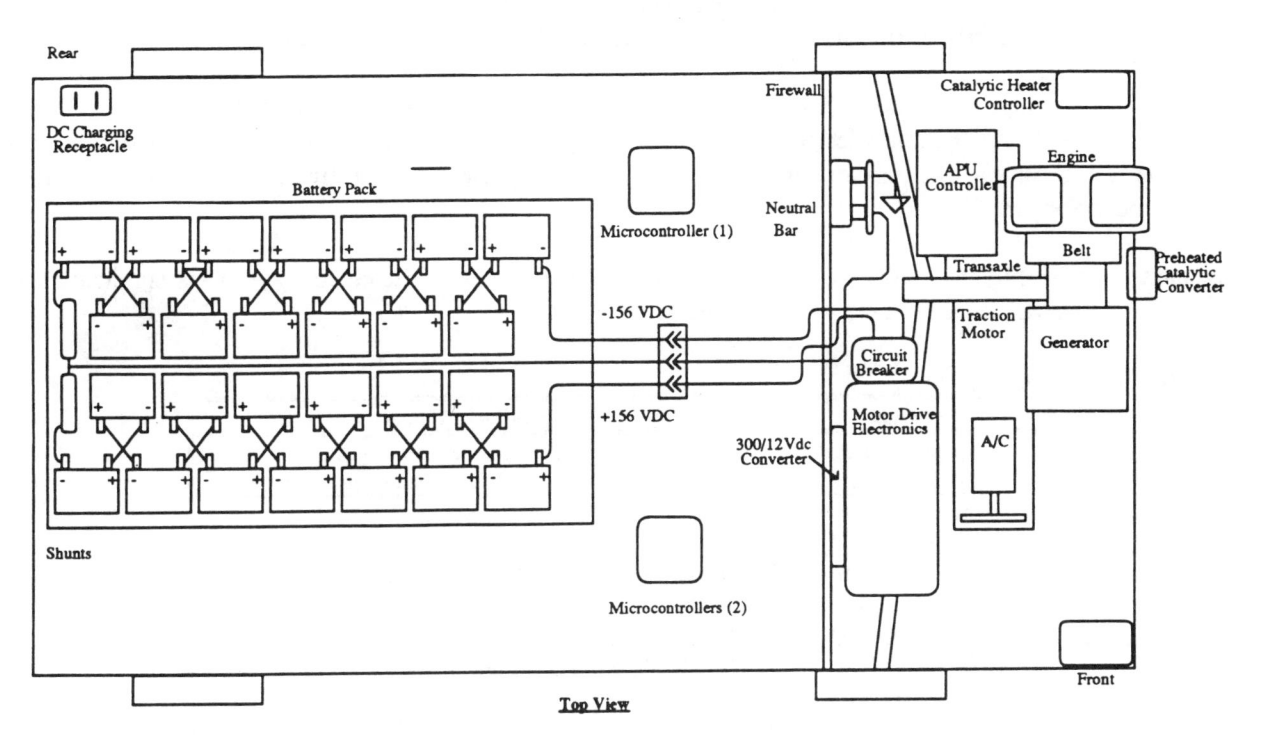

General Layout of HEV Components

General Layout of HEV Components

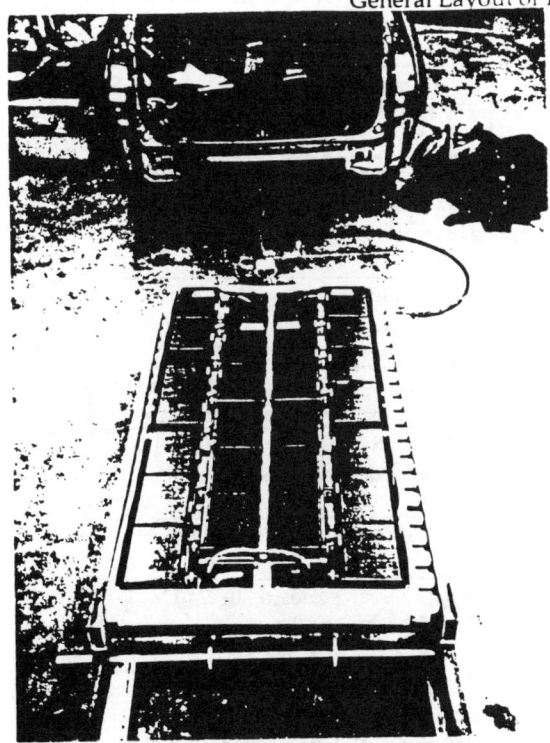

Battery Pack Open for Inspection

Stock Ergonomics

Summary of specifications and characteristics of University of Illinois College of Engineering Hybrid Electric Vehicle

Basic outline: The University of Illinois (UIUC) Hybrid Electric Vehicle was prepared for the Ford/Dept. of Energy/SAE 1993 Hybrid Electric Vehicle Challenge. It is a converted 1992 Ford Escort station wagon. The UIUC car was the only one in the Challenge to retain the complete five-passenger interior structure of the stock car.

System arrangement: Series hybrid, with full ac electric traction drive system. The electric system is supplemented by a small fueled engine-generator set.

Summary of major components:

System element	Description	Capability or Characteristics
Main drive motor	Three-phase ac squirrel-cage induction motor with forced-air cooling.	82 HP peak at 6500 RPM. 0-9000 RPM speed range.
Mechanical drive system	Stock Ford Escort arrangement, with elimination of clutch.	5 speed clutchless manual transmission, stock transaxle
Engine for E-G set	V-twin fuel-injected four-stroke water-cooled engine. Fuel adjustment for ethanol, gasoline, or any mixture.	26 HP at 3600 RPM.
Generator for E-G set	Three-phase ac squirrel-cage induction motor, or three-phase permanent magnet ac synchronous generator.	15 kW output continuous, 7200 RPM for induction machine. 18 kW continuous, 5000 RPM for synchronous machine.
Inverter for electric drive	Constant volts per hertz system with slip control.	100 A continuous output. IGBT inverter system.
Brakes	Dual regenerative and power-assist standard Escort brakes.	Regeneration for all normal braking action. Mechanical brakes are applied when sufficient pedal travel appears.
Batteries	26 sealed glass-mat lead-acid units from computer backup supply application, connected in series.	312 V nominal (split for ± 156 V), 8.8 kW·hr at 3 hour discharge rate, 307 kg.
Battery pack	Enclosed flat pack mounted externally in place of stock fuel tank and spare tire.	Fully enclosed plug-in pack system, with slide-in mount method. Dimensions: approx. 15 cm high, 80 cm wide, 145 cm long.
Displays	Stock dashboard. Stock display (from Mazda Protege) with LED graph used to show battery state-of-charge.	Separate LCD display added for diagnostics and testing.
12 V system	Dc-dc converter from main battery bus.	Up to 1200 W output, 12.0 V, fixed output as battery voltage decreases.
Electronics	Microcomputer network with multi-task operating system, for displays, diagnostics, and battery monitoring.	Complete vehicle control retained if computer network is nonfunctional.
Charging interface	Dc port in front fender. Electronic safety interlock.	Up to 20 A charge current at full dc input.
Charger	Switching power supply unit, garage mounted.	Input source: standard 208-240 V 30 A or 50 A outlet. Output: dc for battery pack, with interlock.
Interior	No modifications to interior space except roll cage required by Challenge Rules.	Complete five-passenger interior space retained. Cargo space, seats, dashboard, and accessories per stock Escort.

Performance characteristics:

Parameter	Value	Test basis
Acceleration	100 m in 9.1 s (0.25 g). 0-50 mph in 9.3 s.	Measured at 1993 HEV Challenge competition.
Top speed	95 mph (estimated)	92 mph achieved in early tests.
Range, electric	33 miles, 45 mph	Measured on highways near campus.
Range, hybrid	>400 miles, 45 mph	Limited only by fuel tank size. Batteries do not discharge in hybrid mode at 50 mph. Actual 100 mile test achieved.
Charging	90% of charge restored in 4 hours from 30 A, 208 V outlet. Full charge in approx. 6 hours.	Multiple charge cycles with full discharge.

Other characteristics:

    Curb weight   3240 lb.

    Date of first operation: June 2, 1993

REFERENCES

[1]   A.F. Burke, "Battery System Technologies - Overview", presentation at SAE TOPTEC, Dearborn, MI, 14 September 1992.

[2]   A.F. Burke, "Evaluation of State of Charge Indicator Approaches for EV's", SAE Technical Paper Series, 890816 (1989).

[3]   K. Rajashekara, "History of Electric Vehicles in General Motors", Conference Record, 1993 IEEE - Industry Applications Society Annual Meeting, pp. 447 - 454.

[4]   W. Leonhard, *Control of Electrical Drives*. New York, NY: Springer-Verlag, 1985, pp.204-214.

Jordan College Ecoscort

Scott Nichols, Henry J. Sarge, and James Giefer

Jordan College Energy Institute

This report presents detailed information on modifications to the Jordan College HEV Challenge entry.
Points covered are:
1) Power-train configuration
2) Control strategy for all modes of operation
3) Emissions control strategy
4) Fuel and Electrical power efficiency considerations
5) Vehicle design and modifications
6) Suspension design and modification
7) Vehicle manufacturability
and cost considerations

INTRODUCTION

The Hybrid Electric Vehicle Challenge was created in part to encourage a fresh new approach to automotive design. Unfortunately the words fresh and new in automotive design have become associated in recent years with the word 'expensive'. This does not have to be the case with hybrid electric vehicles. By combining existing components and technologies, a reliable vehicle can be created which will be affordable for the average consumer. The purpose of this paper is to outline the design of such a hybrid electric vehicle with a low
modification cost.

VEHICLE DESIGN

The single greatest change to any vehicle being modified for electric operation is the location and construction of the battery compartment. Therefore, the vehicle design began with the

BATTERY COMPARTMENT DESIGN

The main considerations for selecting battery locations were; driver and passenger safety, handling, and battery size and weight.

Batteries were selected to provide the most kilowatt hours for the lowest possible price while remaining within the competition weight requirements. The batteries used for the drive pack were 12 volt flooded lead acid, deep cycle cells. The size (34.3cm x 17.1cm x 29.2cm) and weight of (35 kg) was a major consideration in the design of the battery compartments and their positions.

After locating the vehicle's center of gravity, it was determined, that in order to maintain the original front/rear weight bias of 59% to 41% the batteries should be mounted in two locations. The larger pack, holding five batteries is situated under the rear seat. The smaller pack, with three batteries, is located in the engine compartment between the strut towers. The rear seat and stock fuel cell were removed to make room for the rear battery compartment. (see fig 1.)

The rear battery compartment is constructed of fiberglass-coated aluminum honeycomb, chosen for its non-conductibility, compressive strength, (4482 kPa), and tensile strength. (3206 kPa.)

This compartment is recessed into the floor of the vehicle. It is protected by an A-36 mild steel frame-work. This compartment is vented to the atmosphere through a 32 cfm brushless DC fan. The batteries are restrained within the box by a steel strap covered with a flexible nonconductive plastic coating connected to the frame with grade 5 steel bolts. In addition, restraint is provided by

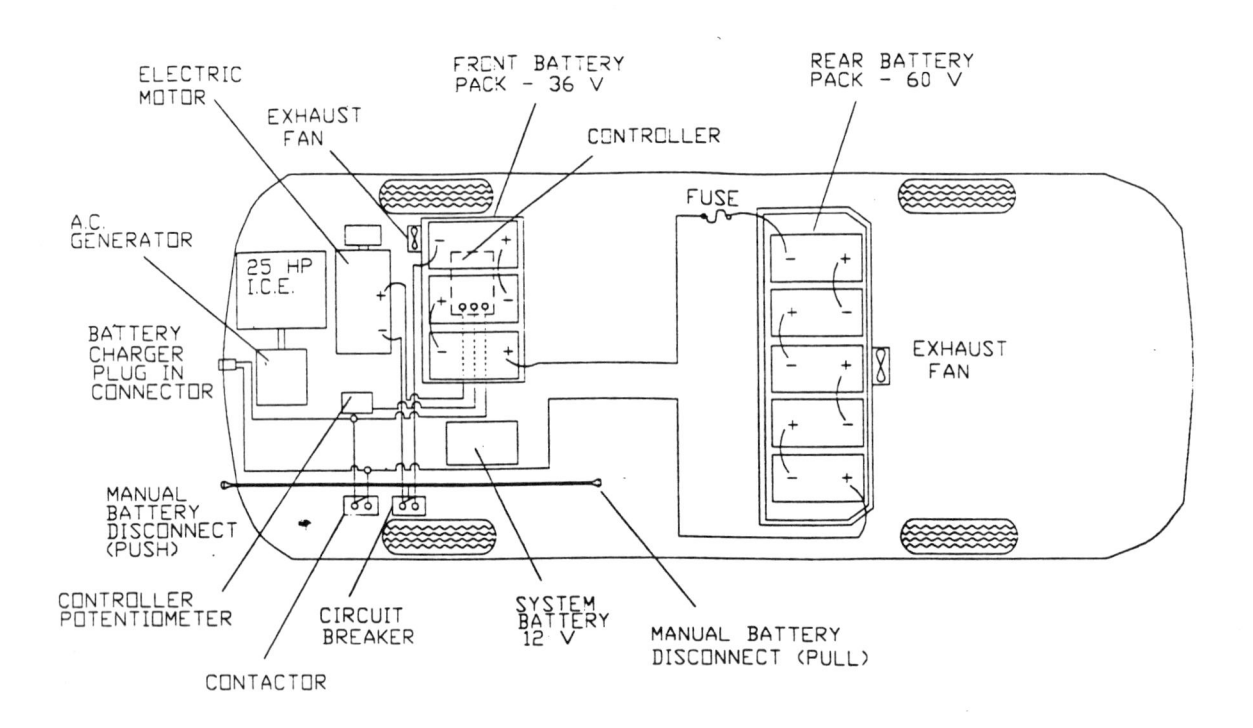

FIGURE 1. ELECTRIC DRIVE SYSTEM

a cover constructed of the same honeycomb material.

This cover is held by two nylon cargo straps rated at 34475 kPa ultimate strength.

The second battery box is located in the engine compartment, on the firewall, right side between the strut towers.

The front battery cover is constructed of 24 gauge sheet aluminum with a 4 mm non-conductive rubber liner. Aluminum sheet was chosen over honeycomb in this application due to size constraints in the engine compartment. Batteries are secured with a frame-work constructed from 38mm steel bolted to the front chassis rails. This compartment is vented to the atmosphere through a fifteen cfm brushless DC fan. Both fans derive their power from the traction batteries during charging or discharging through switches at the main circuit breaker and the battery charger plug-in connector.

For safety considerations, a study was made of electric vehicle crash test videos, which graphically demonstrated the result of improperly secured batteries during a collision. It was illustrated that the batteries had a tendency to become airborne, causing injury or death to passengers. To avoid this unfortunate incident the framework of the rear compartment was designed to prevent the batteries from exiting into the passenger area. (see figure 2.)

FUEL CELL MOUNTING MODIFICATIONS- Because the rear battery compartment was mounted in the location of the stock fuel tank a self-sealing plastic racing type fuel cell enclosed in a steel leak proof box was installed in the spare tire well area. Part of the surrounding sheet metal was removed to accommodate the replacement 45 liter fuel cell.

The fuel cell support structure was constructed from 38mm mild steel flat and angle stock and bolted to the chassis frame rails. This box has been further encased in a second layer of steel welded to the original floor pan sheetmetal.

The fuel cell is separated from the passenger area by an 24 gauge aluminum bulkhead.

APU MOUNTING MODIFICATIONS
The lower radiator support was modified to accommodate the APU. A section of box beam

connected to the positive side of the front battery pack through a 400 amp fuse. the front battery pack consists of three 12-volt lead-acid also connected in series aiding. The negative cable of the front battery pack is connected to the B-terminal of the controller through the ammeter shunt and a circuit breaker, rated at 125 VDC and 250 amps for 60s. The motor stator field is connected to the controller M- terminal.

The battery charger plug in connector is connected to the main battery pack through the circuit breaker and motor controller. (refer to fig 1.) This ensures that the vehicle may not be put into operation while the charger is plugged in.

4.)
This prevents electric motor feedback to the APU. When the APU is operating rotational energy is transferred through the armature shaft of the motor to the transaxle. The APU has an auxilliary drive-shaft which is connected to a modified AC generator allowing for battery recharging.

HYBRID MODE CONTROL STRATEGY

Competition rules required three modes of operation: APU only, electric only, and hybrid mode, which utilizes an efficient driving scenario

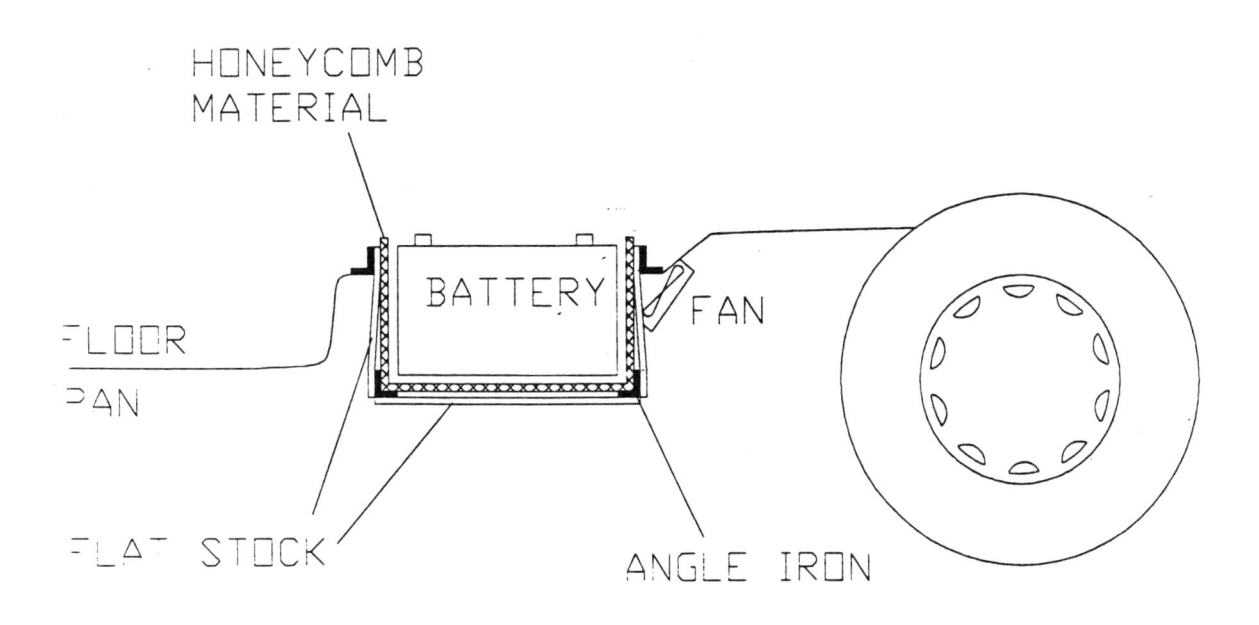

FIGURE 2. REAR BATTERY BOX LOCATION

POWER-TRAIN CONFIGURATION-
The power-train is a series-parallel hybrid configuration, utilizing a direct mechanical energy transfer combining a series-wound DC motor drive system and a twenty-five horsepower internal combustion engine linked through a synchronous belt. The electric motor is bolted to an adapter plate on the stock trans-axle, utilizing the stock clutch flywheel assembly. (see figure 3.)

The engine or APU is mounted on a plate forward of electric motor so that the shafts are parallel. The belt tension is adjusted by an idler pulley. The electric motor tail-shaft pulley utilizes a one-way clutch bearing assembly. (see figure

combining electric and APU drives. (see figure 5.)

Combined electric and mechanical power is needed when high acceleration or incline situations are encountered. A computer controlled power distribution system (logic controller) ensures smooth effective operation of the APU and the motor. In hybrid mode, the vehicle is started and continues under electric power up to a programmed setpoint (56 km/h). At this point the logic controller initiates start up of the IC engine and causes the motor controller to cut-out.

The vehicle is then driven by the APU, unless electric assist is needed. The operator can

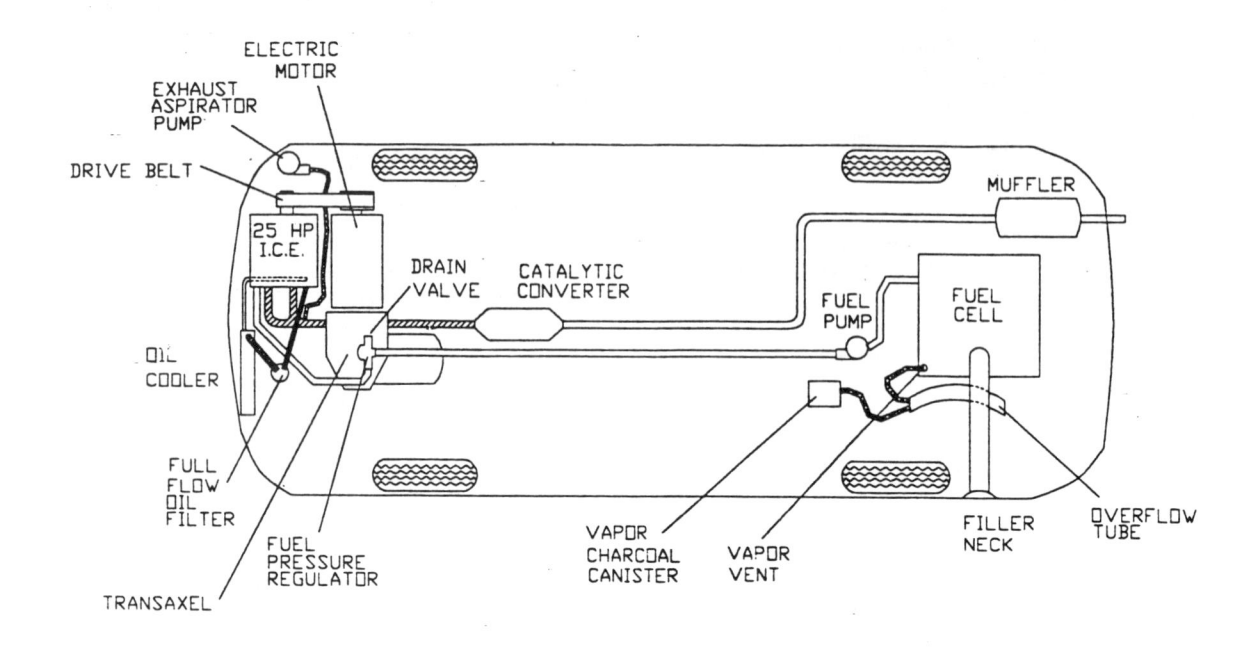

FIGURE 3. A.P.U. SYSTEM

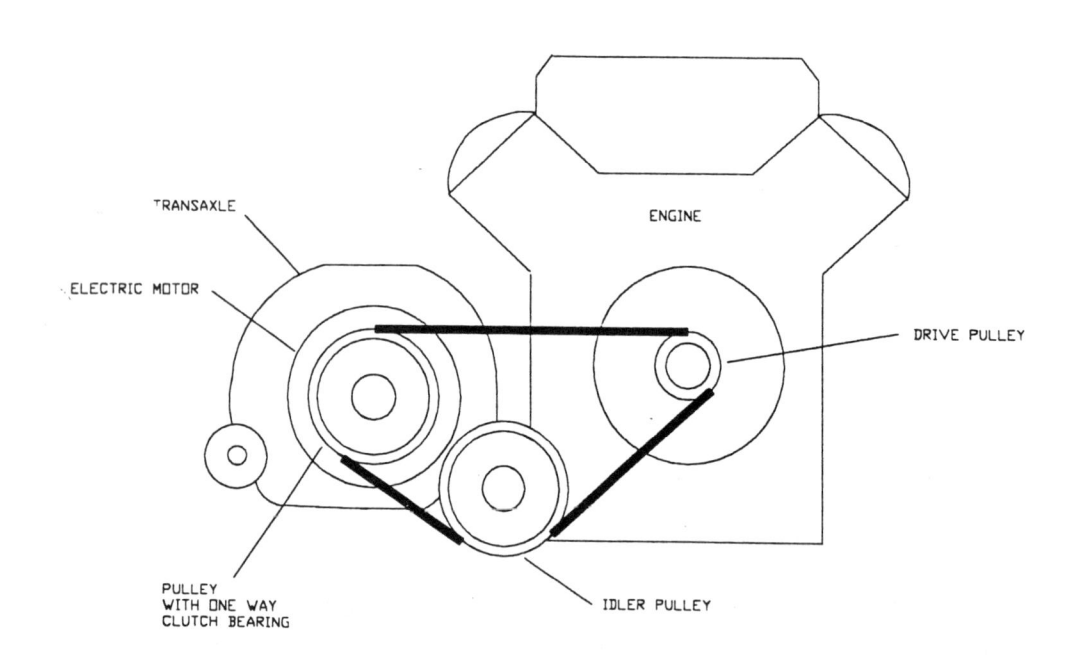

FIGURE 4. HYBRID DRIVE SYSTEM

engage electric assist by depressing the accelerator pedal fully. Electric assist will be activated automatically by the logic controller whenever vehicle speed is greater than 88 km/h.

Upon dropping below a programmed setpoint, (40 km/h), the logic controller shuts down the APU and reengages the electric drive system.

In addition, the APU also drives a small rectified AC generator with an output of 2200 watts. Generator output is reduced through the use of a variable resistance inversely proportional to pedal travel.

During heavy load conditions the generator field will receive decreasing excitation.

This will reduce loading on the IC engine during these periods. Also, during braking the generator field is fully excited, providing a regenerative braking effect.
(see figure 6.)

Some difficulty was experienced in getting the APU and electric drive sytems to work in sync with each other. After experimentation a pulley ratio of 2:1 was chosen because it produced the most low end torque from the APU. Idle speed of the APU was set at 2000 rpm. Because the governed speed is 3750 rpm, the top speed at the input shaft of the transaxle is only 1875 rpm. The operators had to learn to shift very quickly to maintain steady acceleration. Modification to the transaxle gear ratio would probably be required before untrained operators could sucessfully drive this vehicle.

EMISSIONS CONTROL STRATEGY

The emission reduction strategy calls for operation of the engine at its optimum rpm range for efficiency and clean-running. However, the type of small air-cooled industrial engine that was chosen for the APU has inherently high levels of unburned hydrocarbons in the exhaust. The engine used does meet the 1994 emissions standards for engines in it's category. However, these requirements are not as stringent as those of the auto industry, and some improvements were required.

Reformulated gasoline was chosen for a fuel, due to its availability, low cost, and because it's use does not require modifications to engines or the present infrastructure.

The stock charcoal canister was kept to prevent evaporative emissions.

The engine was equipped with a crankcase ventilation system, which directs crankcase fumes back to the intake manifold. The stock carburetor on the engine was utilized. To improve performance and reduce hydrocarbon emissions during starting an electric choke solenoid engages the choke only when the engine is cranking during cold starts, when the engine is warm, a temperature sensitive snap disk opens the choke circuit.

The engine's carburetor supplies a fuel mixture that is deliberately rich, to maintain lower combustion temperatures, and therefore lower NOX emissions. These rich mixtures significantly increase the levels of unburned hydrocarbons in the exhaust. The strategy selected to minimize emissions of both pollutants was to keep the intake mixture rich and further reduce temperatures by using a large (305mm X 305 mm) oil cooler and increasing the cooling fan airflow to preserve the low NOX levels. To decrease HC levels, the exhaust header was thoroghly insulated and atmospheric air was injected into the exhaust manifold by an electric pump. The combination of heat and oxygen reduced the HC levels in preliminary tests and insured high enough temperatures to light the three way catalytic convertor, that completes the emisssions reduction system. (see figure 3.)
Further testing and adjustment will be required to acheive the maximum reduction of emissions for the engine.

FUEL AND ELECTRICAL EFFICIENCY

The vehicle electrical efficiency was proven to be quite adequate in last years HEV challenge, therefore no effort has been expended on improvements in this area. The latest Test data verified by Argonne National Laboratory personnel on May 7, 1994 shows an expenditure of .17 kW/hr per kilometer in ZEV mode. Projected available energy is 7.4 kW/h, derived by equation 1. Where total battery energy is 9.3 kW/h and energy available from deep-cycle batteries is 80% of total.

EQ (1). $\quad 9.3 \text{kW/h} \times .80 = 7.4 \text{kW/h}$

Using the available energy, the total projected electric range is 44km, derived by equation 2. Where Kilowatt-hours per kilometer equals .17, and available energy equals 7.4 kW/h.

EQ (2). $\quad \dfrac{7.4 \text{kW/h}}{.17 \text{kW/h per km}} = 44 \text{km}$

Fuel efficiency was shown to be approximately 11 km/l, in these same tests. Although it may be difficult in a vehicle weighing 1758 kg, further testing and adjustments are underway in an attemp to improve fuel economy to a goal of 13 km/l.

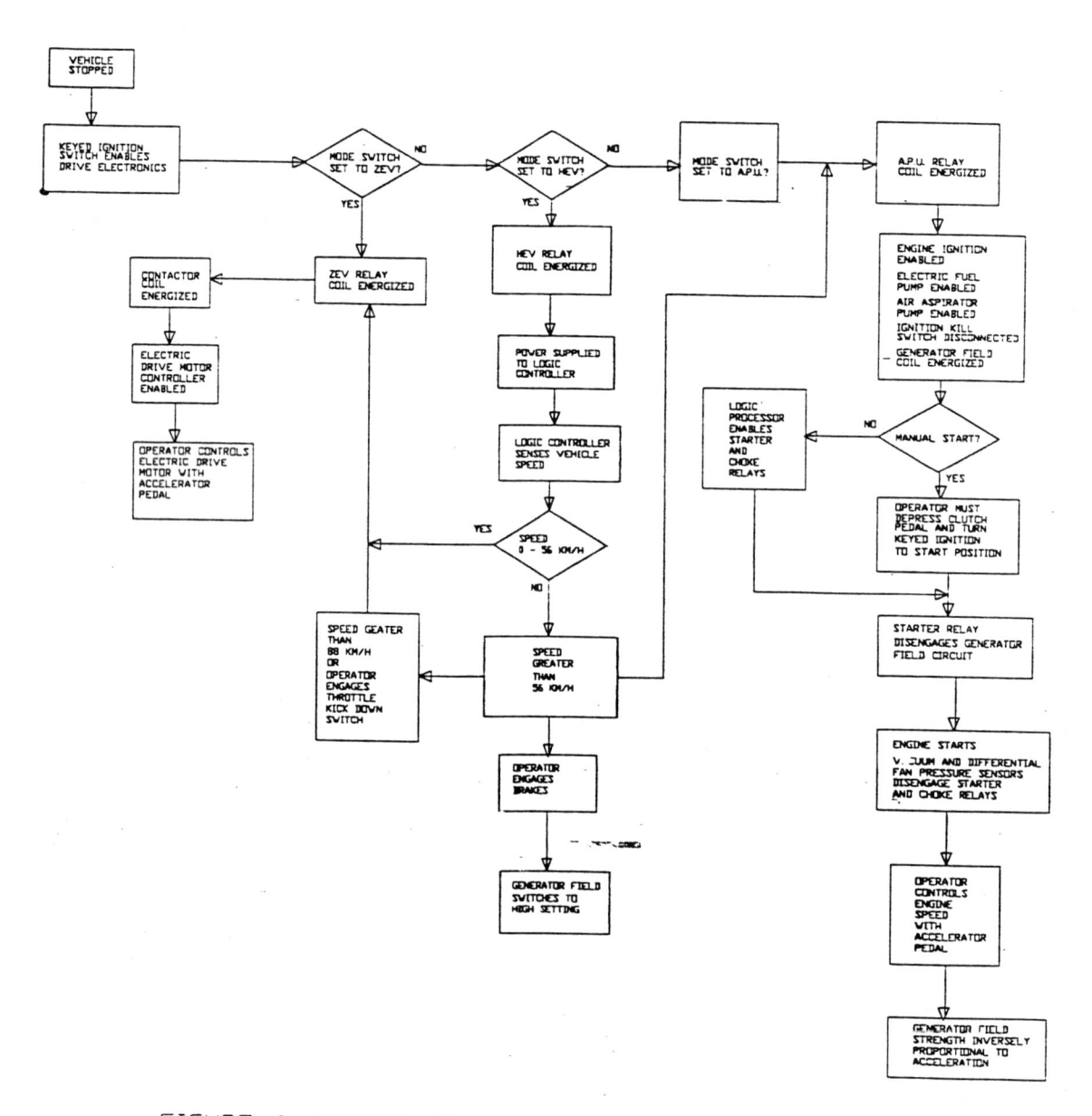

FIGURE 6. HYBRID CONTROL LOGIC FLOWCHART

Projected range in HEV mode is 495km, derived from equation 3. Where fuel efficiency is 11km/l, and total fuel capacity is 45 liters.

EQ (3).   45l x 11km/l = 495km

SAFETY CONSIDERATIONS

A roll cage constructed of 4.4 cm diameter steel tubing with a 3.5 mm wall thickness
was installed. A remotely triggered 2.26 kg halon 1211 fire suppression system was installed on the floor in front of the main traction pack. This system is triggered through a T-handle on the passenger side of the center console.

An additional heat-triggered system using a .90 kg halon 1301 cylinder is located in the luggage area.

A six point racing harness replaces the stock driver and passenger restraints.

VEHICLE MANUFACTURABILITY AND COST CONSIDERATIONS

This vehicle is manufacturable, however, it is simply not marketable. Range and performance are always limited in conversions due to the weight of the vehicle. Marketable electric vehicles are designed and built around the batteries, and weight is kept to a minimum, to extend range. If the components used in this conversion were to be installed in a light-weight chassis performance and range would be greatly improved.

As conversions go, however, this vehicle is relatively inexpensive and practical. Total cost of the conversion was approximately $11,000, not including the original Escort station wagon.

In conclusion, the best use for this vehicle may be as a test bed for new electric vehicle components and to demonstrate to the public, that a hybrid electric vehicle is a practical alternative to a typical IC powered automobile.

It may also demonstrate to automobile manufacturers, that reliable, low cost components are available, that could be used to produce an electric vehicle at a price consumers can afford, and would be willing to buy as a second car for short distance commuting.

The Lawrence Technological University's "Response II"

C.W. Schwartz, N. Brancik, Faculty, Sandra Smith, Felicia Ford, Paul Sharp, Tony Deshaw,

ABSTRACT

The powertrain configuration of the *Response II* contains a one liter, three cylinder, 53 hp engine from a Geo Metro and a 42 hp DC Uniq electric motor, driving through a five-speed manual transmission. The control strategy has been written for a Texas Instruments programmable logic controller (PLC). The emissions are reduced by the use of catalytic preheater and a dual substrate catalytic converter and air injection. The powertrain mechanical efficiency was improved over the *Response I* through the addition of an electromagnetic clutch to disengage the motor when it is not in use and the use of a manual transmission. Aircraft materials such as aluminum honeycomb and carbon fiber were combined with structural adhesives to construct a lightweight, rigid frame. Regenerative braking and 'opportunistic recharging' will be utilized to extend the range on battery. The suspension from a 1994 Probe was selected to simplify installation and provide superior ride and handling. The choice of materials was influenced by the need to reduce weight and the desire to use state of the art materials and fabrication methods. The body styling and ergonomics were assisted by the use ALIAS and ARIES computer software. The vehicle is considered to be very manufacturable in spite of the innovative materials and construction techniques. In addition, virtually the entire vehicle is readily recyclable, including the carbon fiber body shell.

THE DECISION TO CREATE AN ALL-NEW ENTRY FOR the 1994 HEV Challenge stemmed from the desire to incorporate the features of *Response I* that resulted in the Best Engineering Design Award, while improving the range, efficiency, emissions, power management, vehicle structure, suspension design, body styling, ergonomics, and achieving a significant weight reduction. The design of the *Response II* expands on the philosophy of the *Response I* with the goal of producing a "showroom-ready" practical prototype which would appeal to the buying public. The increased use of computers was instrumental in the design of the frame and body of the *Response II*. The frame was designed and analyzed with the help of a powerful modelling program, ARIES. Another sophisticated program, ALIAS, was used in the design of the body shell. The powertrain configuration was refined in several areas to improve performance at the competition. The addition of a PLC is hoped to make the use of the engine and motor more efficient. A new battery pack was selected for use in *Response II* in order to increase the range in both zero emissions vehicle (ZEV) and HEV modes.

OVERVIEW OF THE *RESPONSE II*

As stated earlier, the objectives of the *Response II* design team was to combine the sound features of the *Response I* with the improvements deemed necessary. With this in mind the Chassis Design Group sought to reduce the weight of the frame and optimize the placement of frame members through the use of ARIES. Another target for the Chassis Design Group was to make the traction battery pack easily accesible and removable.

VEHICLE STRUCTURAL DESIGN

The *Response II* combines front and rear structural aluminum frames with a monocoque midsection. The midsection is constructed from aluminum angle, aluminum honeycomb, and the existing cowl and windshield opening from a Geo Storm. The frame was designed to provide storage for the batteries, the required roll hoop and roll bar, attachment of the suspension, and attachment of the Geo Storm body components.

FRONT AND REAR FRAME SECTIONS - The

front frame is a weldment of 6061-T6 aluminum channel and tubing. The front frame was designed to provide for the placement of the engine, transmission and motor, and provide mounting points for the Probe suspension. The front frame also incorporates a 300 mm crush zone. See Appendix A for calculations of dynamic deformation.

The rear frame of 7.62 cm (3 in) channel provides attachement for the suspension cradle, roll bar braces, fuel tank, cargo box, motor controller, and power control center. The frame design conceived using the ARIES modelling program can be seen in Figure 1.

MIDSECTION - The traction battery box and tray are a vital part of the chassis structure. The battery box encloses the batteries, plus connectors and the ventilation system. The battery tray was designed to withstand a load of three times the 479 kg. weight of the batteries and was constructed of 6061-T6 aluminum angles and aluminum honeycomb. The passenger compartment floor, sides and top of the battery box were also made of the 2.54 cm (1 inch) thick honeycomb panels, reinforced with aluminum angles at the corners.

Roll Bar and Roll Hoop. The roll bar and roll hoop are constructed of 4140 chrome molybdenum steel tubing and have been integrated into the aluminum frame structure. The roll bar and roll hoop both have a diameter of 3.8 cm, with a nominal wall thickness of 0.24 cm. The Permabond joints were reinforced with mechanical fasteners to prevent peel failure. The adhesive also served to prevent a galvanic couple between the aluminum and steel. The placement of the roll bar and roll hoop was adjusted to allow the plane described by the tops of the roll bar and roll hoop to clear the occupant's helmets by two inches. This was accomplished through the use of ALIAS, as shown in Figure 2.

VEHICLE MANUFACTURABILITY

One of the highest priorities in designing Response II was to ensure that it could be fabricated using the University's facilities and could be constructed by the students involved in the project. With this in mind the primary manufacturing processes that were used were welding, riveting, and adhesive bonding.

ADHESIVE BONDING - A special structural adhesive containing rubber particles was used because of its high strength to weight ratio and moderate flexibility since honeycomb composite prohibited the welding to join the honeycomb to the other structural components. To ensure consistent bonding, the choice of adhesive was constrained by the following: cold cure, non-toxic, a working time of 10 minutes or greater, and the time to handling of 30 minutes or less. A two part acrylic supplied by Permabond was determined to be the best solution. Tests of the integrity of the bond were conducted using a tensile test machine.

MECHANICAL FASTENING - Rivets were used to supplement the bond, but primarily to prevent the failure of the adhesive through peel.

WELDING - The areas of the frame that limited the use of mechanical fasteners or adhesive bonding were welded using an inert gas welding process.

SUSPENSION DESIGN

An exhaustive review of suspensions in the weight class of the Response II led to the selection of the 1994 Probe suspension system. Although the suspension assemblies were not the lightest available, the low shock strut tower heights, coupled with the use of pre-assembled saddles greatly simplified the attachment to the Response II aluminum frame. The usable suspension travel in jounce and rebound exceeds the minimum 100 mm requirement stated in the HEV Rules and Regulations.

STEERING SYSTEM - The steering column, from a 1994 Dodge Neon, was selected for the ease with which it could be incoporated into the frame. The steering system is used without the benefit of power assist, in order not to burden the powertrain with the load of a hydraulic power steering pump.

HALF SHAFTS - The design of the half shafts and the constant velocity (CV) joints took into account two criteria: the ability of the shafts to transmit combined torques of the APU and motor and analysis of axial and angular shaft displacements. Completed shaft assemblys were proof tested to 600 lb-ft.

POWERTRAIN CONFIGURATION

The *Response II* powertrain has a parallel hybrid configuration. The parallel configuration was chosen for the flexibility it affords and because of the efficiency it possesses. The Powertrain Design Group kept the goal of producing a highly manufacturable vehicle in mind. The powertrain configuration of the *Response II* contains a one liter, three cylinder, 53 hp engine from a Geo Metro and a 42 hp DC Uniq electric motor, driving through a five-speed manual transmission. The control

ALTERNATIVE POWER UNIT (APU) - the APU is a 1990 GEO Metro engine. It drives through a 1987 Muncie Getrag 282 (MG282) five speed manual transaxle which was used in the 1987 Chevrolet Cavalier. A bridge, onsisting of two adapter plates and a cylindrical spacer adapts the APU to the transmission. Within the bridge is an electromagnetic clutch and pulley drive that engages and disengages the electric motor from the intermediate shaft. The intermediate shaft is splined to both the APU clutch and the transaxle. The electro-magnetic clutch and pulley drive system in combination with the motor

Figure 1. HEV Frame Design using ARIES Software.

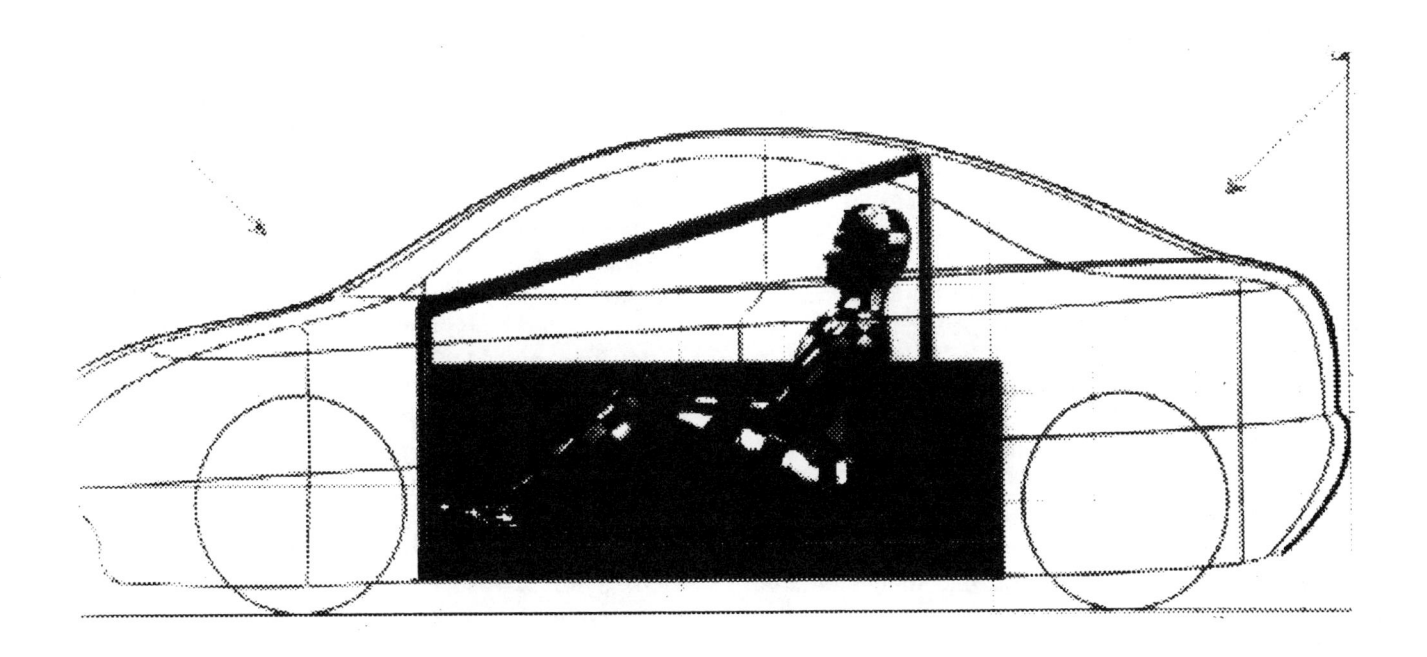

Figure 2. Placement of Roll Bar and Roll Hoop using ALIAS Software Package.

controller and programmable logic controller (PLC) will provide the driver with options of three (3) drive modes: APU (combustion engine as source of power); ZEV (zero-emission vehicle, with the Electric motor as source of power); and HEV (hybrid, both APU and electric motor power the vehicle).

APU MODE - The engine is the source of power in this mode. The electromagnetic clutch is disengaged by the PLC. The hydraulic clutch actuator is the "engaged" position, but a slide bar system allows the APU clutch to be engaged or disengaged the cable-clutch pedal assembly.

ZEV MODE - The electromagnetic clutch is the operating clutch. Its engagement is controlled by a switch under the clutch pedal (in = disengaged, out = engaged). The APU clutch will be held disengaged by the hydraulic actuator signalled by the PLC upon switching to ZEV mode.

HEV mode - The hydraulic clutch actuator is in the "engaged" position. The clutch pedal controls engagement or disengagement of both the electromagnetic clutch (via a switch in the clutch mechanicsm) and APU clutch (cable operated with clutch pedal). The PLC renders the pedal switch inoperable when the motor is recharging the batteries.

Reformulated fuel was chosen for the APU to ensure market-readiness of the vehicle.

The output shaft of the transaxle is connected to the 1994 Probe CS front hubs.

ELECTRICAL SYSTEMS

BATTERIES, FUSES, AND POWER CONTROL - The Response II uses absorbent glass mat, valve regulated lead acid 12-EVB-1180 batteries manufactured by GNB Industrial Battery Inc. Sixteen batteries, rated at 100 amphour at an eight hour discharge rate are configured in series for a total of 196 volts nominal. This 192 Volt system is fused for 300 amps at both the positive and negative terminals before providing power to the motor controller. A soft start system has been configured to supply negative voltage to the controller and which ramps the positive voltage to the controller for approximately 8 seconds before supplying full pack voltage. The accompanying schematic (Figure 3) shows the Albright 250 amp switching contactors used to transfer power, fuse state of charge indicators, battery pack ventialtion, fan control, watt hour meter shunt, and meter terminal strip.

MOTOR CONTROLLER AND MOTOR - The SR180/P 18 pole brushless DC motor and the CR 20-300 300 amp motor controller were manufactured by Uniq Mobility Corporation. The motor has been optimized for clockwise rotation and both cooling systems have been modified. Current regeneration to the main traction battery pack is controlled from the PLC during vehicle braking, Hybrid APU deceleration, and battery low state of charge in the Hybrid mode. Combined motor and controller weight is 100 lbs.

PROGRAMMABLE LOGIC CONTROLLER - The vehicle control system is provided by a Texas Instruments (Siemens) TI-435 Programmable Logic Controller, chosen because of its compact size, high noise immunity (2000 volts for 1 micro sec), and Boolean Algebraic graphical programming language. All inputs are software filtered to produce time average calculations. Analog and digital input and output modules are easily replacable for service. A Facts Engineering ASCII module programmed in basic is used to read the data from the RS232 link from the competition-provided watt hour meter. All control algorithms are contained in the controller. The control scheme for the HYBRID mode of operation is shown in Figure 4. Should the electric drive systems fail, the default condiditon allows the driver to continue under a "limp home" mode powered by the APU only.

WIRING HARNESS/LIGHTING SYSTEMS - The design of the wiring harness incorporated high mounted neon brake light as well as high intensity LED running lights. Newly designed Hewlett Packard transparent substrate high intensity LED's (Lum. 20 mills, 1000 mcandela) with an expanded viewing angle (60 deg. horizontal, 30 deg. vertical pattern) are also being utilized with a circuit design that incorporates a 12 v DC modulated 100 Hz square wave to enhance intensity and keep the substrate section cooler.

DASHBOARD DISPLAY - The dashboard display consists of a Futaba vacuum fluorescent display (VFD) for indicating vehicle speed, motor and engine RPM, engine oil pressure, and water temperature. A liquid crystal display for the 0.1 second rally clock and the vehicle odometer are used with battery backup to retain the displayed information during the ignition cycle or main battery supply removal. A heads-up display will duplicate vehicle speed, and turn signal indicators. If installed, the virtual image of the display will appear about one meter in front of the vehicle and provied a reduction in time for the driver to read the data from the instrument panel. Other telltales and gauges include engine vacuum, batter current and regeneration, mode indicators, battery state of charge indicator and a pre-catalyst indicator. The watt hour meter provided by the competition sponsors will be displayed as well. The accompanying schematic (Figure 5) is typical of the circuit design utilized for the VFD display.

BATTERY CHARGER - Two K & W lightweight (10 lbs each) battery chargers with an efficiency of 85% are being used for the out-of-vehicle charge system. The charging system is protected from the line and grounded

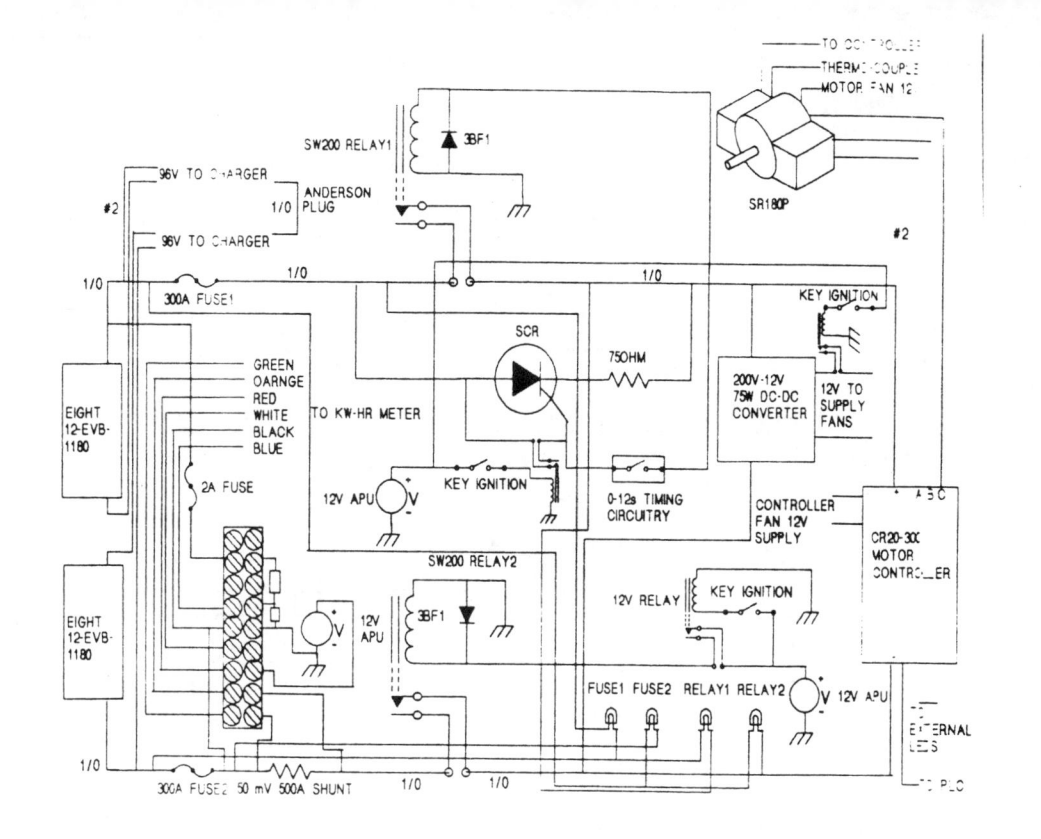

Figure 3. Soft Start System Schematic

HYBRID Strategy

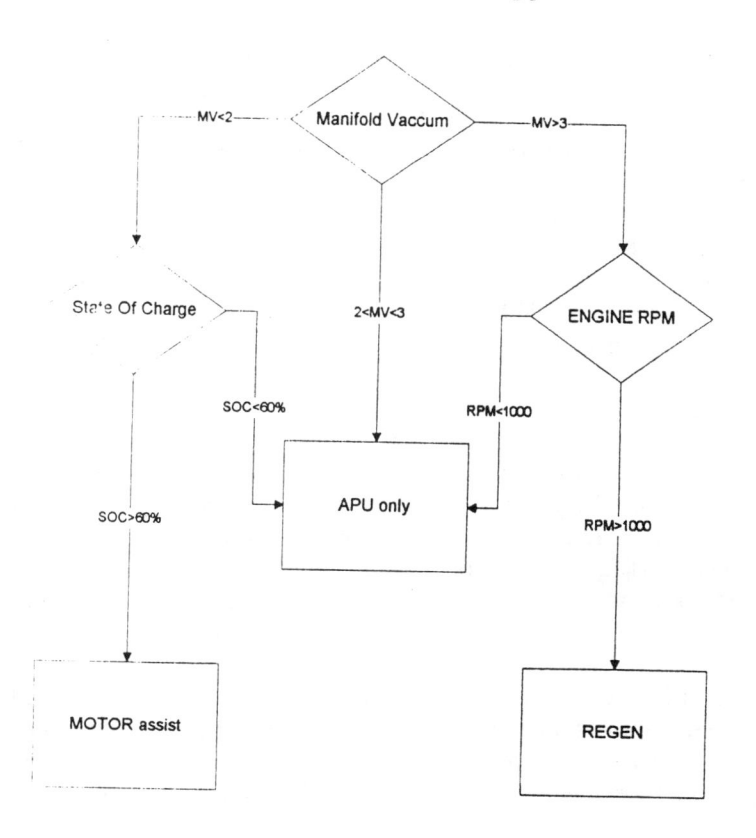

Figure 4: Hybrid Mode PLC Strategy

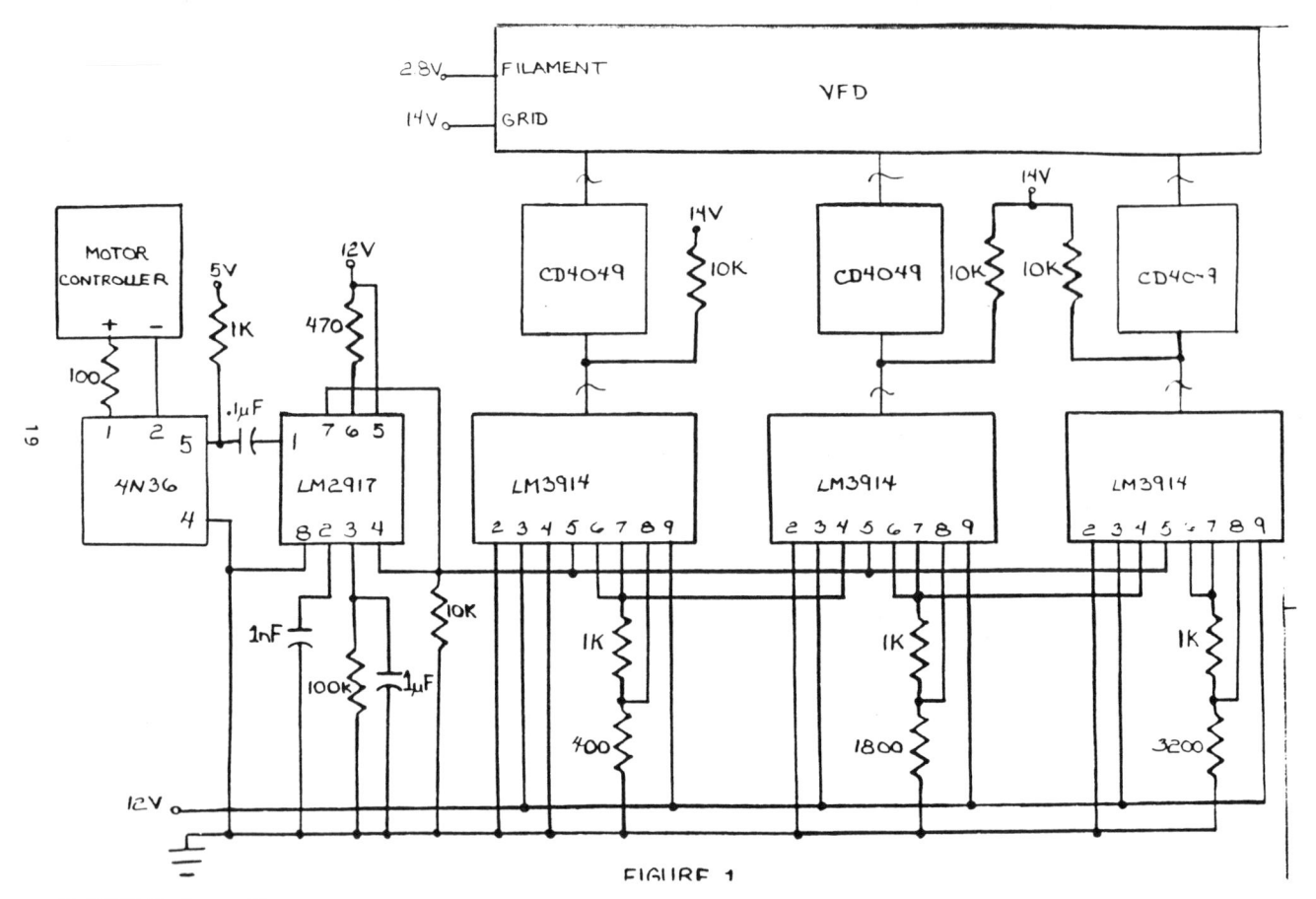

Figure 5. VFD Schematic

evacuation fan is powered by a Cicor 250 Volt DC-12 Volt DC convertor at 75 watts and is operated by a small 110 VAC to 12 VDC power source within the battery charger during the charging cycle to prevent hydrogen buildup. The Vicor also operates the electric traction motor cooling fan.
• Two traction battery disconnects are incorporated, a large mushroom push switch is located between the passenger and drive to act as a safety shutoff should an emergency occur as well as a manual anderson type shorting plug that is used for charging can be accessed through the electric fuel door in the event of a rollover or as a safety disconnect.
• A Dow Corning catalytic pre-heater is employed during the first eight seconds of APU starting, controlled by the PLC in APU and hybrid modes.

BODY

The body for *Response II* was designed with the use of a solid modelling software called ALIAS. The goals of the body team were to find the best balance between body styling and interior ergonomics and the engineering requirements defined by the Chassis and Powertrain Design Groups. In addition, the body team sought to achieve an ultra lightweight body using advanced composite technology. With these goals in mind, a local automotive prototype company, CDI/Modern Engineering, was asked to provide the facilities with which to design and produce the *Response II*. With their enthusiastic cooperation and assistance, CDI/Modern Engineering became a valuable resource.

BODY STYLING - The styling of the *Response II* was developed using the 3-dimensional modelling program that was supplied by CDI, called ALIAS. The first step in styling the *Response II* was to determine what structural elements could be used from a 1991 GEO Storm. After much debate, it was decided that the cowl, windshield surround and door hardware could be carried over into the *Response II*. A digitized file of the Storm, as well as the chassis and powertrain designs, were then imported into the ALIAS program. A study of body styles was conducted to find the most aerodynamic yet highly appealing design. This body design incorporated the Mazda MX-3 headlights, the Ford Probe taillights, and the GEO Storm windshield. The final results of the study can be seen in the design of the *Response II*.

ERGONOMICS - It was decided that the *Response II* needed to be significantly improved over the *Response I* in terms of ergonomics. With this in mind, a detailed study of automotive ergonomics was conducted. The packaging of the interior was designed to accomodate from 5th percentile to a 95th percentile person. ALIAS was also used to complete this part of the body. A sample of the ALIAS work can be seen in Figure 6. In addition, a wide range of people were asked to sit in the *Response I* and voice their opinion about the comfort of the car. One of the main improvements was the use of conventionally hinged doors instead of the gull

Figure 6. Ergonomics Studies using ALIAS.

wing doors of the *Response I*. This change greatly improved ingress and egress from the vehicle. A second important change involved the improvement of the visibility from the car. A driver can easily see 12 feet in front of the bumper compared to the 22 feet the *Response I* achieved. With all these changes, our team feels that the *Response II* gives the driver an atmosphere close to that of a production car.

BODY CONSTRUCTION - The body of the *Response II* was constructed using hand lay-up of carbon fiber and resin. In the process used by the body team, the Geo Storm body was used for the armature to hold the clay. A numerically controlled five-axis mill using the student generated data from the ALIAS program milled a full sized clay model. The clay was then sculpted to the final shape by hand and panel separation lines were placed in the clay. Two layers of Green Seal and three coats of wax were used as a parting agent between the clay model and the female molds. This step can be seen in Figure 7. Female molds were created using a surface coat, fiberglass with laminating resin, and a frozen epoxy tooling foam as an intermediate core material. The final body panels were constructed in a six step process. First, two sandable surface coats were applied by brush and allowed to air dry until the surface was tacked-up. One layer of 3K Carbon fiber with resin was carefully placed over the surface coat and stubbed by brush into place. One 0.317 cm (0.125 inch) layer of low density

Figure 7. Construction of Female Splashes

(98.6 g/cm^3, 3.5 lb/in^3) cellular foam called Klegecell R-55, used as a strengthening core, was placed over top of the first layer of carbon fiber cloth. A second layer of carbon fiber with resin was placed on top of the Klegecell. The entire panel was vacuum bagged for 24 hours, with a minimum of 20 inches of Mercury pressure. The final step involved preparing the part for painting. The larger sections of body panels were reinforced to stabilize them prior to attaching hinges and latch mechanisms on doors. The completed body panels can be seen in Figure 8. After removal from the mold the panels were attached to the chassis using fabricated aluminum brackets. The doors, hood and lift gate were removed from completed body mold and the hinges and latch mechanisms attached. The windshield surround, cowl, wiper mounting and hinge posts from the Geo Storm were then mechanically fastened and bonded to the front transverse frame member. Side intrusion beams were installed between the roll bars, across the door above lap height. All lamp assemblies were installed (head lights, tail lights/turn signals) and wired. Then the installation of instrument panel and chassis wiring harness was completed. The weight of body shell minus reinforcement and mounting hardware is 60 pounds. The completed body including attachment hardware, all windows

Figure 8. Completed Body Panels.

including windshield, and wiper motors is 180 pounds.

Seats are from a Geo Metro, weigh 31 pounds each and are fastened to the honeycomb floor. Testing of the honeycomb floor with reinforcements showed that a 5/16 inch bolt would withstand a force of 4100 pounds. The strength permitted the direct connection of the seats and 5-point harness attachments to the honeycomb.

The University of Maryland at College Park Methanol Hybrid Electric Vehicle

Jordan Wilkerson, Fred Householder, Mark Caggiano
Dr. David Holloway, Faculty Advisor

ABSTRACT

The University of Maryland is one of twelve universities converting a 1991 Saturn SL-2 sedan into a power-assist Hybrid Electric Vehicle (HEV) to compete in the Department of Energy's second annual HEV Challenge in June 1994. The primary objectives of a power-assist hybrid are to reduce the size of the heat engine, significantly improve the fuel economy, and reduce the toxic emissions. To accomplish these objectives, a Geo Metro 3-cylinder, 1.0 liter internal combustion engine is used as the primary power source and is coupled in parallel through an over-running clutch to a Unique Mobility 21.0 hp brushless DC electric motor. The transmission is a Subaru Electronic Continuously Variable Transmission (ECVT) which allows the engine and electric motor to run at a single, optimum RPM. The liquid fuel for the ICE is methanol (M85) and the electric motor receives power from 128 NiCd cells at 153.6 V. These 21 Ahr Saft Nife STH cells are specifically developed for hybrid applications. The University of Maryland's goal is to accomplish these tasks, reduce engine size, increase fuel economy, and reduce emissions, while maintaining conventional levels of performance and comfort in order to revolutionize the manner in which a Hybrid Electric Vehicle is designed and built.

POWER TRAIN CONFIGURATION

In order to convert the stock Saturn SL-2 sedan into a parallel hybrid electric vehicle, the original vehicle layout is mimicked: all of the controllers are behind the dash and the entire drivetrain is packaged in the engine compartment, which leaves the trunk space for battery storage (Fig.1). This configuration allows full utilization of passenger space and provides for adaptability to mass production.

Fuel Supply

Methanol (M85) is the chosen fuel for Maryland's hybrid vehicle for several reasons. Methanol produces slightly reduced emission levels than ethanol and the infrastructure for sale and distribution is currently being implemented; moreover, the University has extensive experience using M85 in previous Department of Energy

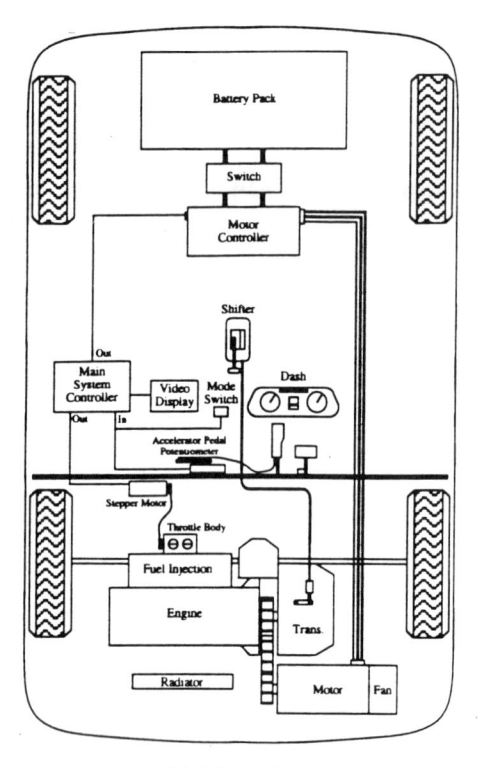

Figure 1. General Vehicle Component Layout

competitions.

To use M85, the Saturn's stock fuel system is entirely replaced. The original polyethylene fuel tank is methanol compatible; however, the fuel will eventually permeate through the single-ply walls; consequently, a new custom fabricated stainless steel tank is installed. The tank is made with 0.050 in. type-304 stainless, holds 13.2 gallons, and is mounted in front of the rear suspension with the original mounting straps. The five taps on the tank match the diameters of those on the original tank (fuel fill, fill vent, fuel supply, fuel return and vapor return). All fuel lines are methanol compatible, teflon coated hoses except the fuel supply line, which is high pressure, steel braided hose. A three-way valve is inserted in the fuel return line immediately after the fuel rail to enable the fuel tank to be drained when a vacuum pump is attached.

A self-priming, methanol compatible, Bosch fuel pump provides 134 lb/hr at 70 psi to "pencil stream" type, Seimens fuel injectors on a custom fuel rail. These 24 lb/hr methanol injectors allow the fuel to be accurately aimed and generate excellent atomization (50-80 μm particle size). The injectors have been regulated to 55 psi, which improves the atomization over a standard 40 psi gasoline system but is less than the 60 psi vapor pressure of methanol. The fuel rail also includes a schrader valve (used during testing only) and a pressure regulator set at 55 psi. An inertia switch will turn the fuel pump off when it is activated during a collision.

The Heat Engine

Several modifications have been made to the original Geo engine to increase the efficiency and power output. All intake and exhaust ports are ported and polished for smoother air flow. The combustion chamber, exhaust ports, intake and exhaust valves, and pistons have been coated with a ceramic coating from Swain Tech Coatings to increase thermal efficiency. To reduce rotational losses, the crankshaft is polished and balanced, and the crank and camshaft are coated with an anti-friction coating. Brass valve seats replace the stock seats to allow better sealing around the valve and combustion chamber. The cylinder walls are bored to 0.040 in. over stock and the connecting rods are cut and honed to reduce the stroke. Finally, the cylinder deck height is lowered to give a 13.5:1 compression ratio; accordingly, the pistons were fly cut to avoid collision with the valves.

The stock cam has been evaluated and is sufficient for the new engine modifications. To increase the volumetric efficiency of the stock Geo engine, Custom intake and exhaust manifolds were computer designed using Dynomation and Engine Expert software packages. The fabricated manifolds are thus optimized to produce peak torque and efficiency at 3300 RPM. Dynamometer testing has verified the modifications to the ICE have increased efficiency and power over its entire operating range.

Engine Emission Control Strategy

Since the internal combustion engine is modified to reduce emissions and improve fuel economy using methanol, a new engine management system controls all engine functions necessary to surpass 1994 Federal Light-Duty Vehicle emission standards.

The Engine Control Unit (ECU) is an Emtech E6 system. The E6 system is comprised of a fully programmable ECU and all necessary sensors for complete fuel and ignition control, including closed loop feedback. The E6 system uses the following I/O units for operation of the ICE.

ECU Inputs
- Manifold Absolute Pressure (MAP) Sensor
- Tachometer (RPM) Adaptor
- Throttle Position Sensor
- Coolant Temperature Sensor
- Air Temperature Sensor
- Heated Exhaust Oxygen (O_2) Sensor
- Battery Voltage

ECU Outputs
- Fuel Injectors
- Ignition Driver
- Exhaust Gas Recirculation (EGR) Solenoid
- Idle Air Control (IAC) Motor

The E6 system interface is MS-DOS software to allow adjustment of the engine operating parameters through an RS232 port connected directly to the ECU. All adjustable functions of the system may be changed while the engine is operational, either in the vehicle or on a test dynamometer, for real time emulation. The University of Maryland HEV uses the versatility of the E6 system to calibrate all engine operating conditions for the newly modified, methanol fueled ICE. Adjustable features of the ECU software include:

- Fuel and Ignition Control
    (11 speed & 32 load ranges)
- EGR Control (6 speed & 32 load ranges)
- Coolant Correction for Fuel, Ignition, and EGR
- Air Temperature Correction for Fuel and Ignition
- Closed Loop Control (O_2)
- Idle Speed Control
- Cold Start Control for Fuel
- Battery Voltage Correction for Fuel

Testing calibrations of the ECU adjustable features operate the engine near stoichiometric combustion conditions at all times since all adjustments were made in reference to the oxygen sensor. The exhaust gas recirculation adjustments are based on GM Corsica EGR operating parameters (an M85 test vehicle used for previous DOE competitions).

In order to reduce emissions even further, a three-way Electronically Heated Catalyst (EHC) is used during cold starts. The EHC is programmed for time control after the engine has been started to reduce effects on the starting battery voltage. The catalyst is placed nearest the engine exhaust manifold to reduce "light off" time when the emission conversion efficiency is low. The EHC uses 100 A for 15 seconds during electric heating via insulated 2 gauge welding cable as specified. A manufacturer supplied controller and circuit breaker allow for programmable adjustment of the EHC's electric heating.

Battery Pack

The main energy storage device chosen for the Maryland HEV is Nickel-Cadmium (NiCd) STH 210B's made by Saft Nife. These NiCd batteries exhibit higher power density, more favorable charge/discharge characteristics, lesser voltage depression with respect to discharge rate, and longer cycle life at 80% Depth of Discharge (DOD) than lead acid batteries. These NiCd batteries do not suffer from the charge memory phenomena common to the smaller rechargeable NiCd's found in many rechargeable appliances, nor do they suffer large voltage loss with higher DOD. Overall, the STH batteries are the best commercially available battery for this HEV's application and strategy.

Current strategy is to operate the batteries between 85% and 65% State of Charge (SOC) during HEV mode. These limiting values are set due to the inefficiencies of the battery charge and discharge differences (Fig.2&3). This will also nearly eliminate the required maintenance on the batteries, except for normal periodic checks.

The following is a list of battery specifications.

Weight per Cell:	2.3 lbs
Voltage per Cell:	1.2 V (nominal)
Cell Capacity:	21 Ahr
Electrolyte Reserve:	3.66 in^3
Total Number of Cells:	128
Total Pack Weight:	310 lb
Total Pack Voltage:	153.6 V (nominal)
Total Pack Energy:	3.23 kWhr
Temperature Range:	75-110°F

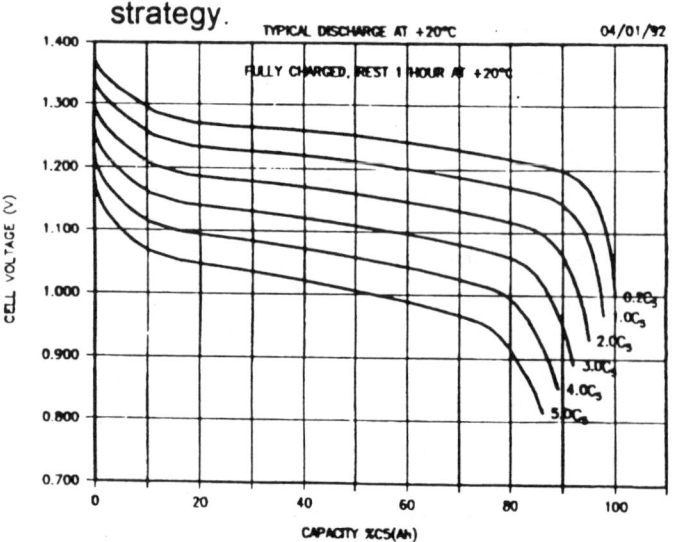

Figure 2. NiCd Discharge Curves

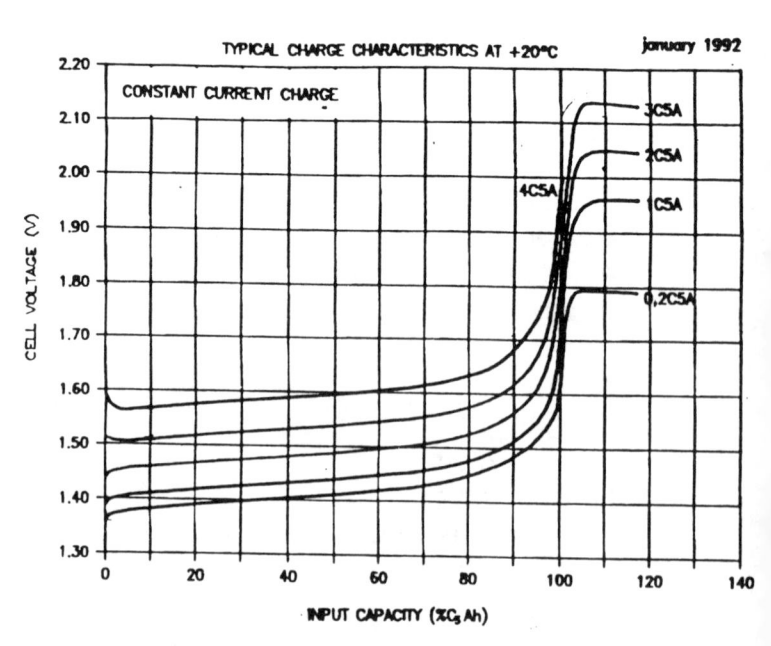

Figure 3. NiCd Charge Curves

Battery Storage

The batteries are configured in the battery box in 5 columns of 16 cells and 4 columns of 12 cells, with a 0.4-inch gap between each column for ventilation. The ventilation system utilizes two 3-inch, brushless fans exhausting a total of 27 SCFM. To assist the fans, ram air is channelled from under the vehicle into the box through four nozzles that help maintain an even airflow distribution. The fans then force the air to an exhaust port in the rear of the vehicle opposite of the ICE exhaust pipe.

The battery box is made of Hyzod, which is a transparent polycarbonate that does not react with the battery emission or electrolyte, has a high strength-to-weight ratio, is a poor electrical conductor, and resists heat up to 270°F. Its transparency allows for quick and easy general inspections of the batteries. Vehicle impact studies of a Solectria Force electric car show that a 30 mph collision translates to a 32g deceleration; consequently, the 0.25 in. thickness of the Hyzod and the number of mounting bolts (18) have been chosen to resist a 32g impact.

Electric Motor/Controller

The electric motor is Unique Mobility DR156s, AC permanent magnet (Neodymium Iron Boron - NdFeB) synchronous motor and is rated at 21.0 nominal hp. At a nominal voltage of the NiCd battery pack (153.6 V), the motor's maximum power is 28.8 hp. The continuous torque is nearly constant over the entire RPM range and is approximately 17 ft-lbf; peak continuous stall torque is 25 ft-lbf and peak intermittent stall torque is approximately 30 ft-lbf. This 24-pole motor offers 4 quadrant operation, regeneration to zero speed, closed loop speed control, and high power density. Almost half of the operating range of the electric motor is at least 90% efficient. The NdFeB magnets provide a very high power density; however, they are very sensitive to heat. In order to keep the electric motor temperature below 325°F, it is cooled by an Ametek "Windjammer" blower supplying 37 SCFM.

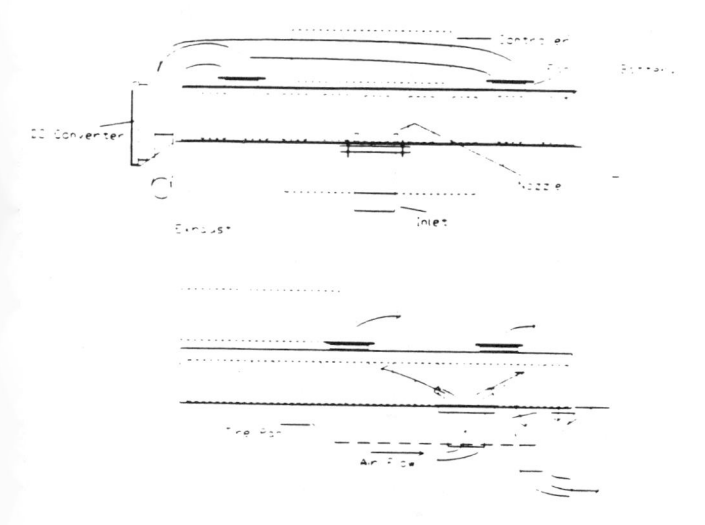

Figure 4. Battery Box and Ventilation

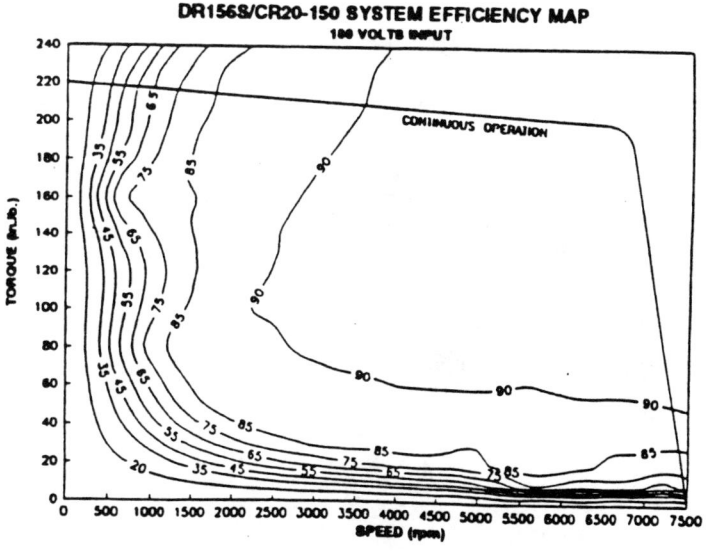

Figure 5. Electric Motor Efficiency Curves

The matched motor controller (CR20-50) consists of a power module containing the electronic switches, control module with microprocessors, and necessary software. Since the motor uses AC, the controller converts from DC for interface with the battery pack. The controller is limited to 200 volts and has a maximum current rating of 225 A.

Transmission

The transmission is a front-wheel drive Electronic Continuously Variable Transmission (ECVT) from a Subaru Justy. This transmission is capable of an infinite number of gear ratios from 0.50 to 2.50:1 with a 5.83:1 final drive ratio, and uses a steel segmented belt in compression between two V-shaped pulleys to transfer torque. The ECVT is an automatic transmission that uses internal, differential hydraulic pressures to vary the ratio between the primary and secondary pulleys which changes the ratio between the input and output shafts (Fig.6).

When the accelerator is depressed, the transmission is engaged and the primary pulley (input from ICE/electric motor) begins to rotate. As the engine and the primary pulley increase speed, the pressure in the pitot tube increases. When the pitot pressure is greater than the shift control valve pressure, a ratio change between the primary and secondary pulley (output to drive axle) is initiated and the gear ratio increases. As the pitot tube pressure approaches the shift control valve pressure, the ratio change is stabilized and the transmission remains at the same gear ratio. This pressure equalization during transient operation allows the ICE to operate at its most efficient RPM. The transmission shift cam determines what engine speed to shift the transmission. By modifying this shift cam, the point at which the pressures are matched can be adjusted; thereby, adjusting ICE RPM. The most energy efficient point in the ICE is approximately 3300 RPM; therefore, the shift cam has been modified to hold the ICE at this RPM. When the transmission is in Drive (D) mode, the modified shift cam allows the ICE to approach 3300 RPM before beginning to change the internal gear ratio.

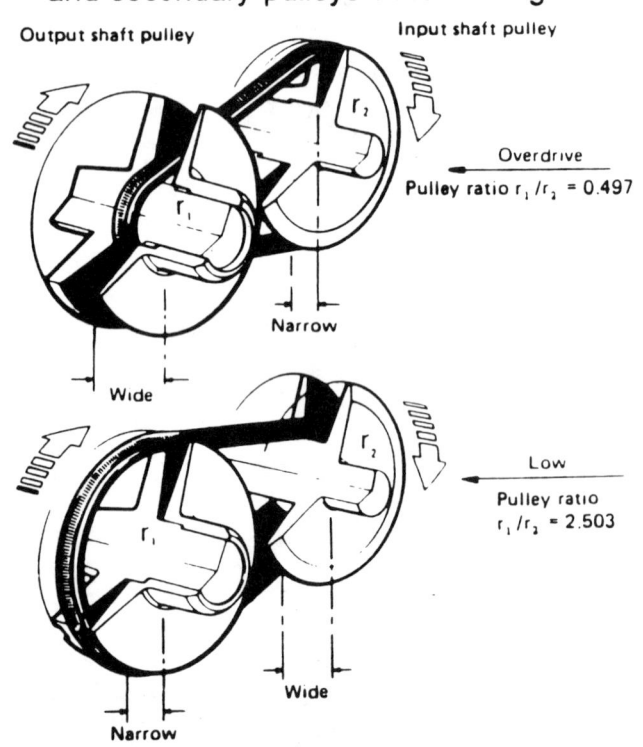

Figure 6. Belt and Pulley Mechanism

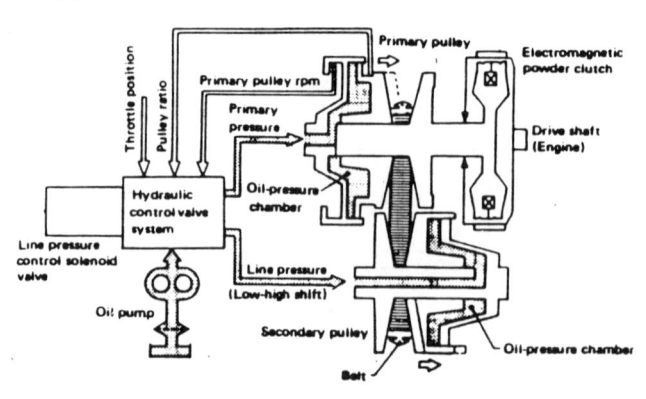

Figure 7. Transmission Layout

In addition to the Drive mode of the transmission, a Drive Sporty (Ds) mode can also be selected which allows for a higher level of engine performance by keeping the ICE/electric motor at 4800 RPM. The Maryland hybrid vehicle design does not allow the driver to select Ds mode directly; instead, the lower gear ratio is reserved for regenerative braking. When the vehicle control parameters are met for the electric motor to be in regenerative mode, the vehicle controller will signal an after-market Ds solenoid to shift the transmission to the lower Ds gear ratio. This allows the electric motor to spin at a higher RPM than the normal Drive configuration which increases the rate of energy conversion and pushes the motor into a more efficient operating range.

The ECVT's electromagnetic powder clutch provides positive engagement of the transmission without the pump loses and transfer inefficiencies associated with fluid torque converters. The clutch engages with a 12 volt signal through a one-second ramp-up circuit. It also allows the transmission to operate like an automatic transmission by disengaging at vehicle speeds below 3 MPH.

Coupling Assembly

The coupling assembly has been designed to combine the ICE and the electric motor in parallel before transmitting their combined power into the transmission. To do this, a Lovejoy CS-16 elastomeric coupling made of natural rubber is bolted to the original Geo manual flywheel wheel which will help dampen the excessive torque spikes from the three-cylinder ICE. The electric motor coupling interface is a Gates sprocket bolted to a Dana AL-2 overrunning clutch. An ECVT adapter plate and spacer ring bolt to the electromagnetic clutch and provide the spline connect for the ECVT shift control valve oil pump.

The inner race of the overrunning clutch is bolted to the ICE, and the outer race is bolted to the transmission side of the coupling assembly. The electric motor attaches to the outer race on the Gates sprocket through a kevlar reinforced belt; therefore, the electric motor can drive the transmission independent of the ICE since the outer race RPM can be higher than the inner race. When the inner race matches the speed of the outer race, the overrunning clutch engages for HEV acceleration. If the electric motor is not

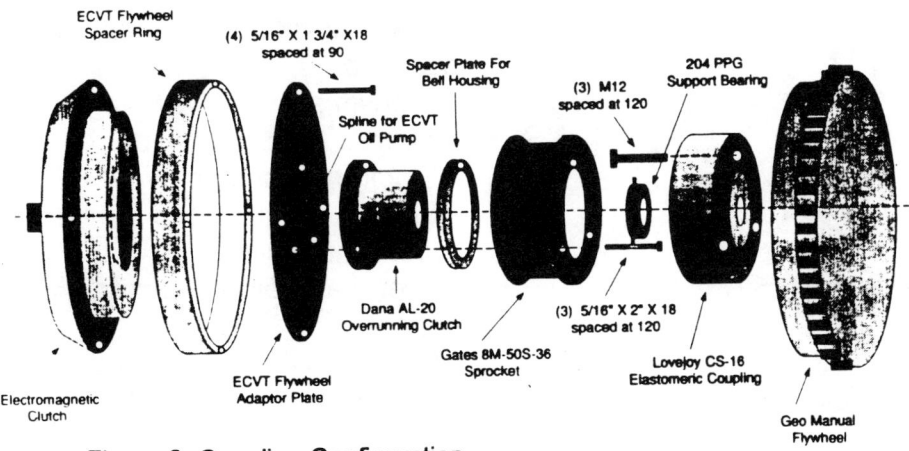

Figure 8. Coupling Configuration

supplying power to the drive train, it will be spun by the ICE with very little resistive losses.

The entire parallel drive coupling assembly is housed between the ICE and transmission and is supported by a bearing in a custom aluminum bell housing. This housing also maintains the proper distance and alignment between the ICE and the ECVT, provides protection to and from the drive components, and serves as a base for mounting other equipment such as the starter motor. Specifically, the engine and transmission are bolted to flanges on each side of the bell housing with the coupling assembly contained inside. The electric motor attaches to a structurally-reinforced, adjustable support arm mounted to the transmission side flange, with the electric motor belt fed through a slot cut into the bell housing.

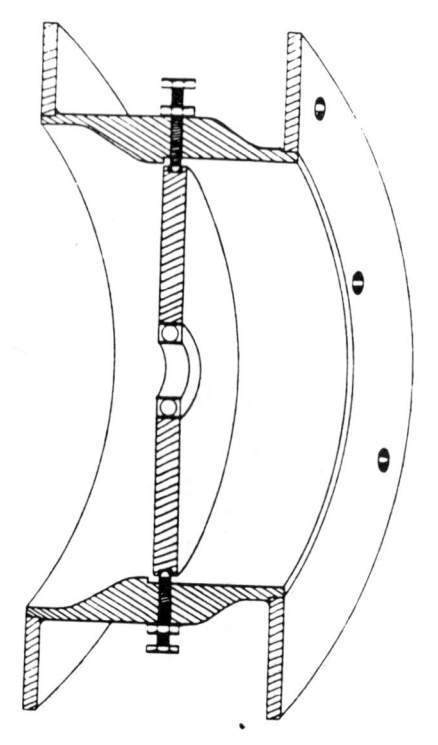

Figure 9. Bell Housing with Support Bearing

CONTROLLER CONFIGURATION

The control of the Maryland hybrid vehicle is delegated by two processors: an Allen-Bradley SLC-500 PLC (Programmable Logic Computer) and a 33 MHz Octagon 386 computer. Working together, the 386 will generate a real-time control strategy using its inputs, while the PLC is responsible for controlling outputs and ensuring all safety requirements are met.

Programmable Logic Computer

The PLC consists of a rack, a 16 point TTL discrete input card, a 16 point TTL output card, two 4-point analog input cards, and a 16 point relay card. Three discrete inputs and three discrete outputs have been reserved for communication with the Octagon 386 controller. These channels will be used for command and report registers between the two controllers. The PLC is primarily responsible for the following critical I/O signals:

- Accelerator pedal potentiometer.
- Discrete pulses to the ICE throttle stepper motor controller.
- ICE throttle feedback potentiometer.
- ICE starter motor.
- Electric motor cooling fan.
- Transmission line pressure control solenoid.
- Transmission D/Ds solenoid.
- Subaru Shifter position.
- Electromagnetic powder clutch.
- Electric motor speed reference and regeneration limit.
- Thermocouples: electric motor and batteries.
- Ignition switch.
- Operation mode switch.

386 Microprocessor

The 386 system is composed of six Octagon control cards, a 16-key keypad, a Sharp LCD, and communication libraries. The following is a very brief explanation of each primary component and its purpose.

- Microprocessor Card: Contains the 386SX processor, EPROM, SSD's, LPT1, and two RS232 ports (COM1, COM2). It has an on-board EPROM eraser and programmer and is based on standard XT (8-bit) architecture. It provides the vehicle's main processing capabilities and makes all major control decisions. It receives complete operating information from the ICE.

- SVGA Video Display Adapter: Drives a Sharp flat panel display in the center dash console and can be connected to a CRT for software development.

- A/D Converter: Contains 16 A/D channels and 8 digital I/O channels. This card gathers all of the analog signals including accelerator pedal, battery temperature, and vehicle speed.

- 4 Channel DAC: Capable of producing voltage for output analog signals to the motor controller and the PLC.

- Timer/Counter Card: Contains two clocks and three counter/timers capable of generating and reading frequencies. This card controls the OEM Speedometer.

- Multipurpose Card: Contains two serial ports (COM3 and COM4), LPT2, and a general purpose I/O. By disabling the IRQ for Com2 on the Microprocessor card, the Octagon 386 system can have three interrupt control COM ports. This card provides connections to the battery sensor and the engine management system.

- LCD: A passive 320 X 240 matrix used to output vehicle information to the operator. It also provides, with the key pad, an interface to the controllers of the vehicle to help determine operating characteristics.

- Key Pad: Allows the operator to have limited control of feedback from the system. Items like *Charge Rate*, *Silence Alarm*, and *Choose Display* are typical selections available through this interface.

- ASYNC Libraries: These Libraries provide an easy method of communicating with asynchronous ports of the controllers and are completely interrupt driven.

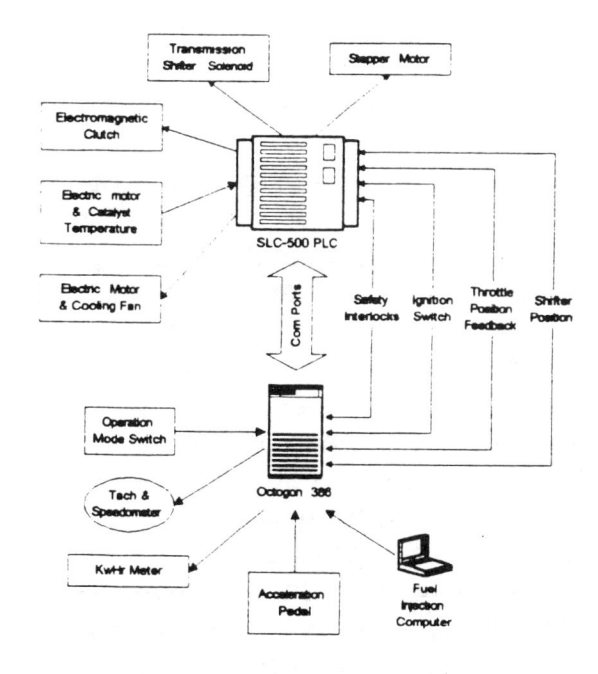

Figure 10. Controller Configuration

Vehicle Electrical System

The battery interconnects and the cable from the battery pack to the controller are 2/0 welding cable. The have an ampacity of 375 amps and a 600 volt breakdown voltage. The resistive losses through the cable are insignificant due to the short lengths. The factors of safety over the expected current and voltage are 1.67 and 2.67 respectively.

The 6-gauge cable connecting the electric motor and controller is rated at 75 amps and 600 volts. It is specifically designed for this application by the manufacturer and provides a factor of safety of 2 for the expected current and 2.67 for the expected voltage.

Several components in the vehicle require a steady voltage that won't fluctuate; consequently the original 12-volt starting system is not sufficient. These components, such as the computers, receive their power from DC/DC converters. Three Vicor converters are used: a 24-volt/2-amp converter powers the 386 computer; a 48-volt/8-amp converter powers the PLC; and a 48-volt/16-amp converter powers the throttle stepper motor. These Vicor converters are extremely light weight and are more than 90% efficient.

The cable and wires connecting various components throughout the vehicle are bundled together for structural protection and organization. For ease in maintenance and troubleshooting, each bundle and its branches are designated using a distinct numbering system. Great lengths have been taken to severely limit the amount of floating interference received by either computer system. Both computers have been caged in custom electric shielding boxes. Shielded, twisted pair wires and electrical shielding tape are used primarily for control signals. All of the wiring connectors are rated to 94 V and are made with AMP's self-extinguishing, fire-resistant thermoplastic.

The engine compartment is expected to reach 275°F; consequently, all wiring and shielding have been gauged accordingly. Flex loom tubing is wrapped around all existing bundles for further protection and aesthetic uniformity.

OPERATIONAL MODE CONTROL STRATEGY

The Saturn hybrid has four modes of operation. ZEV-only (Zero Emission Vehicle/electric), ICE-only, HEV, and Charge.

HEV Mode

A parallel hybrid drivetrain is extremely difficult to optimize. Ideally, a neural network will discretely optimize each data point due to its speed and learning advantages over more conventional control decision makers. Since the Controller (combined efforts of the 386 and PLC) does not have the artificial intelligence to take all inputs, discretely calculate the most optimum output, and send that output before the next data point is received, it takes a few critical inputs, checks the safety flags (battery state of charge, accelerator pedal position, etc...) and "load levels" the ICE with an assist from the electric motor. This forces a traditional "power assist" hybrid and still incurs loses. Since the electric motor is generally in a 90% plus efficiency range, the primary fluctuating dynamic loses come from the ICE. Aside from the fixed efficiency improvements (Fuel injection, engine coatings, etc...), load leveling enables the controller to operate the inefficient ICE over a more efficient range through most of the driving cycle. Since the ECVT will hold the ICE at 3300 RPM (most efficient operating speed) the primary variable is the ICE torque.

Assuming all initial conditions are met, the controller uses manifold air pressure (MAP) to determine if more or less power is required to maintain the best fuel consumption rate at the given load requirement. If more power is required, the electric motor will assist the ICE with enough power to bring the engine load down to a better fuel consumption rate. Similarly, if the MAP indicates less power is required, the electric motor is used as a generator which increases the load on the ICE, bringing it up to a more efficient fuel consumption rate.

If, for example, the battery SOC is near, above or below the chosen limiting values, the ability of the electric motor to supply or receive power will be reduced proportionally or eliminated.

The throttle position is set to require all of the ICE power at 69% displacement. If the throttle is displace beyond 69% both the electric motor and ICE will propel the vehicle. The maximum acceleration is 0-60 mph in 8.2 seconds

ICE-only Mode
When the mode switch is in this position, no load leveling can take place.

ZEV-only Mode
When the ZEV position is selected, only the electric part of the drivetrain is used. Logically, this can be used as a "limp-home" mode since it has a limited amount of available power.

Recharge Mode
The battery pack can be directly recharged with the ICE when the transmission is in neutral or park and is activated when the ignition switch is in the accessory position. This can only occur if the battery SOC is lower than the maximum discharge voltage and automatically disengages when that SOC is reached.

VEHICLE STRUCTURE AND HANDLING MODIFICATIONS

Powertrain Installation
The engine, electric motor, coupling assembly, and transmission are all designed and constructed to be assembled as a single unit through the bell housing, as previously mentioned. This enables independent mounting of the powertrain to the vehicle subframe so that any vibration does not create non-uniform displacement stresses. Mounting of the powertrain is through four points on a reinforced Saturn subframe; two on the passenger side of the engine, one to the rear of the transmission, and one on the front driver side of the transmission. All mounting points are isolated through rubber compound engine mounts.

Designing and manufacturing custom splines and halfshafts for this vehicle is costly and time consuming considering Saturn wheel hubs must match the Subaru transmission differential. Consequently, Saturn halfshafts were cut and matched with Subaru ECVT halfshafts. To fabricate new halfshafts, each shaft is cut to the appropriate length and then inserted into a 1030 steel sleeve to be welded. The outboard portions of each side (Fig.11) are fabricated to be equal lengths, eliminating torque steer during acceleration. Equal lengths are permitted by the use of a support bearing mounted to the rear of the engine.

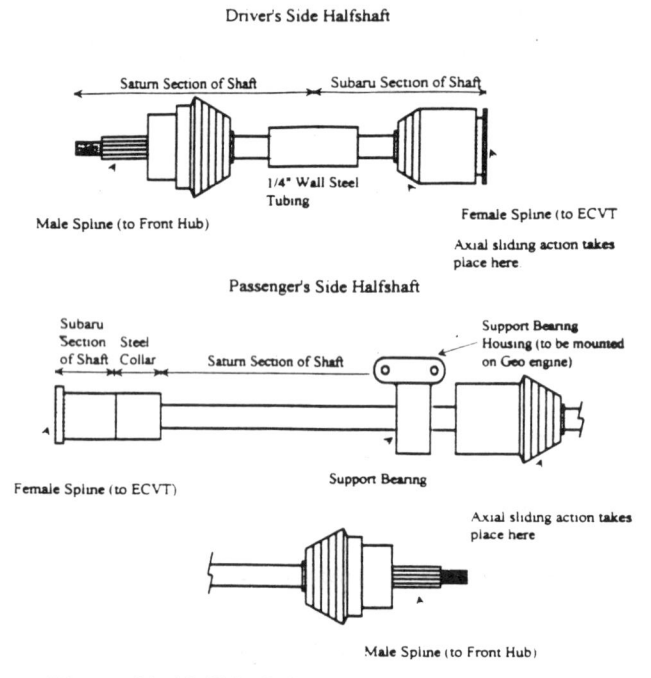

Figure 11. Halfshaft Assembly

Brake System

The original Saturn brake system is designed to safely stop 20% more weight than the Gross Vehicle Weight (GVW). Since the hybrid Saturn weighs less than the original GVW, the wheel cylinders, drums, discs, and calipers have not been modified. An electric GM vacuum pump from a 1985-87 Cadillac has been added to the existing brake vacuum system to assist the braking while the ICE is not running. This pump performs at 21 inHg; this corresponds with the original Saturn brake system which operates at 10.5 - 24 inHg. This passive system is controlled by a vacuum switch that will activate the auxiliary pump anytime the ICE vacuum is less than 10.5 inHg.

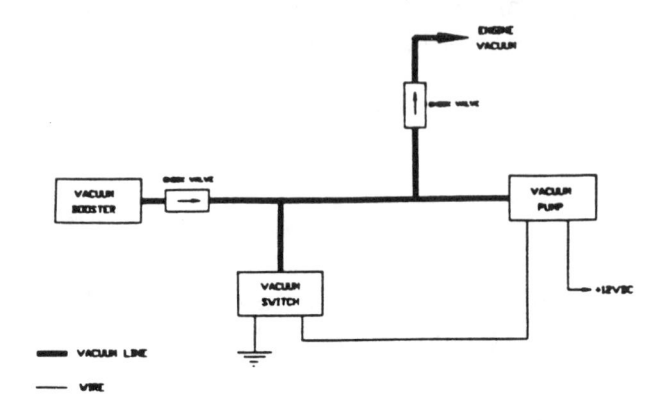

Figure 12. Hybrid Brake System

Wheels/Tires

Low rolling resistant P185/75R14 tires have been developed by Kelly Springfield Tires especially for this vehicle. They have a large cross section with a narrow footprint and a relative shallow non-skid, and a singly-ply polyester carcass is designed for minimum sidewall bending stiffness. An "S" speed rating allows the tires to be stamped for 44 psi maximum inflation, increasing performance over standard 35 psi tires. A rubber inner liner sacrifices permeability to gain elasticity to reduce rolling resistance; however, the dominant factor is a proprietary low rolling resistance tread compound credited for a 30% reduction in the rolling resistance coefficient.

Vehicle Suspension

In order to maintain the original Saturn ride height and handling, the front springs are compressed and heat treated while the rear springs are replaced with springs of a higher spring rate designed and manufactured by Coil Spring Specialties. The adjustments are necessitated because of the significant weight reduction in the front and increase in the rear. However, weight distribution has been equalized to approximately 50/50 so that the ride and handling qualities are slightly improved over the original Saturn suspension.

VEHICLE MANUFACTURABILITY

Throughout the design function and component integration of the University of Maryland's Hybrid Electric Vehicle, manufacturability, cost competitiveness, and driveability were items of consistently high priority in order to reach our goals of producing a revolutionary HEV. Nearly all of the vehicle's components are commercially available and are used in conjunction with easily produced integration parts fabricated with the University's own facilities. Common materials are used in the HEV for all major components, including steel, aluminum, and Hyzod. This combination allows for rapid adaptability to assembly line manufacturing techniques, which will reduce the vehicle unit mass production cost immediately.

Consumer receptivity to the HEV is also a great concern. When a consumer considers a vehicle purchase, cost, comfort, safety, performance, and maintenance evolve into the buyer's immediate concerns. Accordingly, the University of Maryland's HEV is designed to address each of these issues directly and competitively. The price of an HEV has been reduced in this design by purchasing as many "off the shelf" items as possible so that when an HEV similar to our design begins mass production, availability and inflated cost is of little consequence. Comfort is retained in the HEV by maintaining all existing passenger space and retaining a useable amount of trunk space. Ranking first on the priority list, however, is the safety of the passengers. Since the original GVW of the Saturn is not exceeded, structure modifications may not be needed; accordingly, reinforcement precautions have been taken since the Saturn design is not optimized for placement of all the necessary components of the HEV. Added precautions are also included in the vehicle's safety, such as an inertia switch to stop fuel delivery in a collision, throttle return in case of power loss, and battery box intrusion beam installation. Performance of the HEV is emphasized through the selection of highly efficient, high power components so that the vehicle provides its own mobility security during normal driving. Maintenance of the vehicle is convenient due to pre-planned design criteria and numerous installation modifications.

Finally, the driveability of the University of Maryland HEV is versatile and user-friendly enough that nearly anyone with a basic understanding of an HEV should quickly become familiar with the vehicle's operation. This user will quickly be able to operate the vehicle in a safe and relatively efficient manner.

University of Maryland Hybrid Electric Vehicle Sponsors

Maryland State Energy Administration
University of Maryland College of Engineering
University of Maryland Department of Mechanical Engineering
University of Maryland Department of Electrical Engineering
University of Maryland Engineering Alumni

Corporate Sponsors

Allen Bradley
American Methanol Institute
Ametek
Arias Industries
Auto Meter Competition Instruments
Battery Warehouse
Branch Electric
Capitol Cable and Technology, Inc.
Chuck's Used Auto Parts
College Park Jiffy Lube
Coil Spring Specialties
Corning, Inc.
Dana Corporation
Douglas Battery Company
Emitec - Gmbh
Engine Management Technologies
Fluke
Gates Rubber Company
General Electric
Hayes-Ligon Corporation
Jeff's Muffler Shop
Johns Hopkins Applied Physics Laboratories
Johnson Mathey, PLC
Kelly Springfield Tires
Marlboro Auto Parts
Saturn of Marlow Heights
Octagon Systems Corporation
Precision Engine Machine
Saft-Nife Inc.
Siemens Automotive
Solarex
South Atlantic Controls
Subaru of America
Swain Tech Coatings
Unique Mobility, Inc.
VICOR
Warner Electric
Walker Manufacturing

The 1994 Spartan Charge, Michigan State University's Hybrid Electric Vehicle

The MSU Hybrid Electric Vehicle Team
Faculty Advisor: Elias G. Strangas
Department of Electrical Engineering, Michigan State University
East Lansing, MI 48824

Abstract

The objectives of the the work described in this paper were to analyze the strengths and weaknesses of the hybrid electric vehicle that was designed and used in the 1993 HEV competition, to evaluate alternatives, and to redesign those components of the vehicle that would result in the greatest benefits in terms of safety, range, efficiency, emissions and ergonomics. The work involved redesigning almost all the components of the vehicle, as well as extensive testing of all alternatives.

1 Introduction

The Michigan State University's Hybrid Electric Vehicle participated very successfully in the 1993 competition, winning five out of eight engineering design prizes and finishing third overall in the competition. This success was to a large extent a problem for this year's team. On one hand, the team wanted to preserve the characteristics of the car without risking a complete redesign, while the need for certain improvements, some of them drastic, was evident from the beginning. At the same time, the car was requested to participate in many shows and competitions, leaving very little time for the students to work on it at MSU.

As a result most of the time was spent on simulation and component optimization, as well as on a 'mule', a modified production car equipped with the same electrical and computer systems as the Hybrid Electric Vehicle.

Given these externally imposed limitations and the fact that a decision to participate in this year's competition was not taken until late September 1993, the student team was organized and set its goals early on. A relatively loose organization was decided, which was modified as time went on and the nature of the projects changed. The technical objectives set forth were:

- Keep the same chassis and body shell form, while making it possible to access, inspect and repair most parts without removing the shell,

- Improve the brake system, so that braking can be accomplished with much less pedal effort while utilizing the existing Anti-lock Brake System,

- Modify suspension to improve traction and minimize real wheel wear and improve ride height,

- Redesign the alternator and its control, so that it would be impervious to magnetic dust and make it easier to control the rpm and output power of the engine,

- Redesign the engine to further decrease emissions and improve efficiency,

- Modify the control system to better monitor and control power generation, storage and usage,

- Further improve the ergonomics of the driver's instrumentation and control,

- Improve response and increase torque for the electric motor/inverter system,

- Design an intelligent battery charger for off-line charging,.

- Decrease vehicle weight.

2 Electrical Systems

The electrical power group was responsible for inverter/motor controller, electric motor, energy storage systems, high voltage generator, and low voltage system. The integration of these systems create the drive train and auxiliary power systems of the vehicle. It is important to note that the overall design is not a new system. The group's goals were to refine existing systems and upgrade various components to take advantage of recent technological developments.

2.1 Controller - Motor

The drive source is a three phase, two pole, AC induction motor with a rated output of $15HP$ at $190V$. The motor is the vehicle's only drive source. It is directly connected to the front axle via a 5 speed standard transmission and pressure clutch. The motor is controlled by a Vector Drive, Pulse Width Modulated Inverter. This is a standard factory inverter that normally operates in a closed loop system, receiving feedback from a shaft encoder. Past experience proved that a shaft encoder is not feasible, due to the extreme vibrations involved with the dynamics of vehicle operation. Therefore, the inverter was programmed to operate as an open loop system.

For the inverter to operate adequately, even with a rotor position sensor, the motor parameters have to be known to its controller. A series of tests were performed with the motor on a dynamometer to find the optimal parameters for both motors under consideration, so that efficiency and response were maximized.

The inverter operates using torque control rather than speed control. This mode was selected to provide higher start up torque and greater torque at low speeds, thus improving driveability. The inverter receives power from two sources, the batteries and the generator, coming together at the same node.

2.2 Energy Storage System

The batteries used in the vehicle are an array of prototype Nickel-Metal Hydride cells. These batteries were chosen for several reasons: they are considered to be a next generation battery, they have superior energy storage per volume, low internal impedance, and an excellent ability to be discharged deeply. In addition, their cycle life is superior to other cells and they have no memory.

254 cells at $1.43V$ per cell are connected in series providing $365V$s DC. Improvements from last year include the installation of thermocouples inside each battery box to monitor temperature of the cells during charging. The battery terminals are connected to a $600V$ rated current interrupting switch that is used as a safety switch to disconnect the high voltage DC from the inverter. These lines are protected with two $600V$ rated, $200A$ fuses. A $12V$ fan is powered by a $12V$ DC-DC converter, which is supplied on the cold side of the high voltage switch, therefore operational when the switch is on.

2.3 Battery Charger

To charge these batteries, a new system has been designed. Due to failure during last year's competition the new system was required to consistently achieve a full state of charge. The system consists of an isolation transformer, a controlled rectifier, a microcontroller, and current, voltage and temperature sensors.

The microprocessor controls the charger by monitoring voltage, current, and temperature simultaneously. This gives the charger autonomous control over the voltage applied across the battery terminals. The potential between the charge voltage and the battery voltage, as well as the internal impedance of the batteries cause current to flow into the cells. A single phase controlled rectifier is used to control the voltage applied to the battery terminals. The ability to control voltage enables the charger to operate until a preset maximum voltage is reached. At this point the charger shuts off the voltage to the batteries, allowing them to 'settle'. If the batteries settle below the preset voltage, the charger repeats the cycle until the voltage remains above a predetermined value. For safety purposes a seven contact high voltage connector is used to attach the charger to the vehicle. The contacts are diode protected to prevent exposure to high voltage when the charger paddle is not connected.

A $12V$ loop must be completed for the charger paddle to be energized with high voltage. This is possible by using two contacts on the vehicle. This $12V$ loop actuates a relay inside the charger allowing current to flow into the vehicle's batteries. This same contact supplies power to the battery fan during charging.

2.4 Generator and Controlled Rectifier

A second source of power is provided by an IC engine coupled to a generator. The generator is a 12 pole Nd-Fe-Bo permanent magnet AC machine, designed to provide $28A$ at $470V$ at 2900 rpm after full rectification. Protection is accomplished by an internal thermocouple to prevent any overheating and three $40A$, $500V$ fuses at the AC side.

Testing of the 1993 vehicle indicated that a change from high load to a no load condition, such as cresting a hill, while the IC engine/generator is operational,

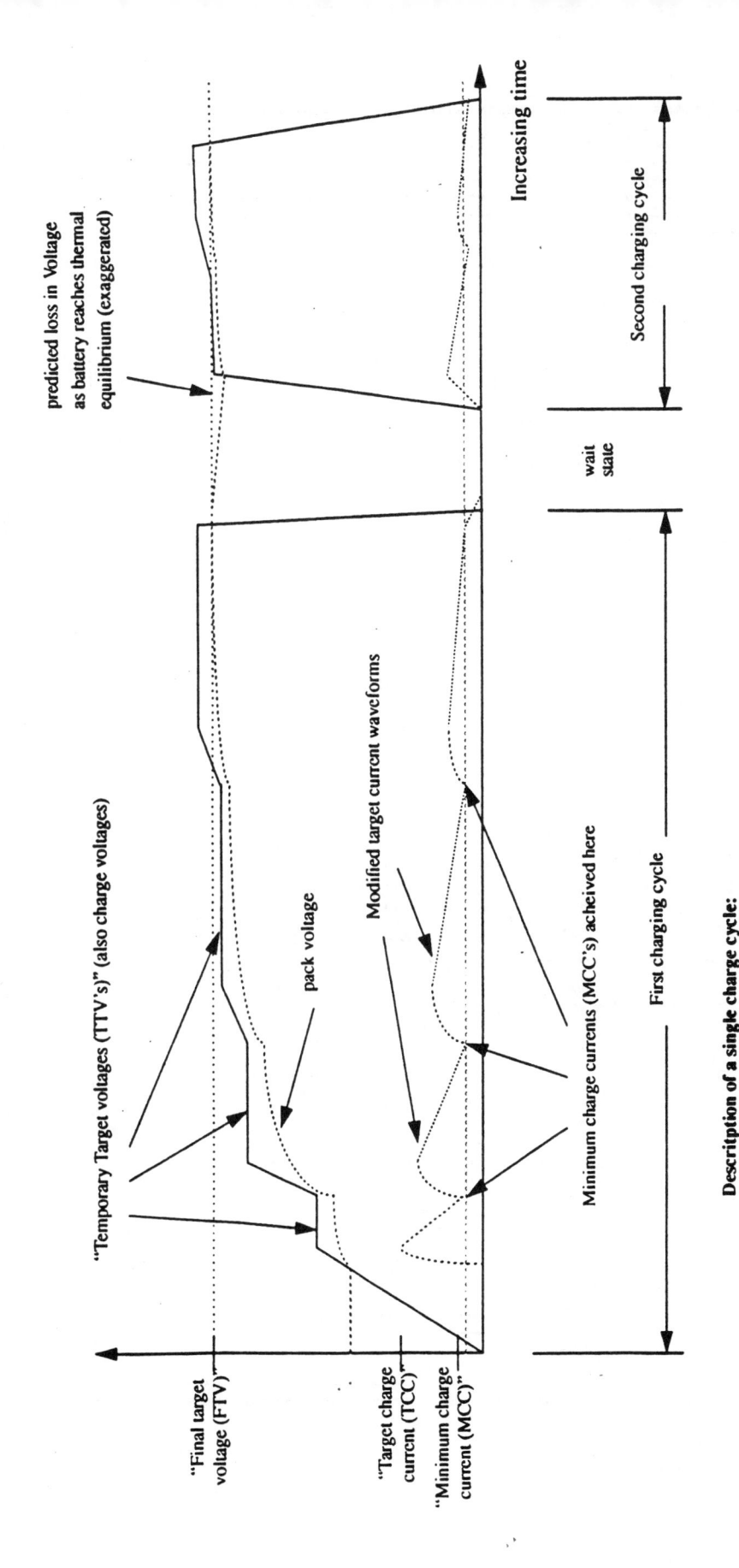

Figure 1: A Plan for Battery Charging

Description of a single charge cycle:
1 - Charge voltage rises gradually until "target charge current (TCC)".
2 - Charge voltage "levels-off" at "temporary charge voltage (TCV)" to allow pack voltage to "catch-up"
3 - Charge current reduces as pack voltage increases
4 - When current reduces to "minimum charge current (MCC)", voltage is raised again until TTC is reached.
5 - When "final target voltage (FTV)" is reached, charger halts for a time t, then the charger begins the next "charging cycle" (starts over at step 1).

197

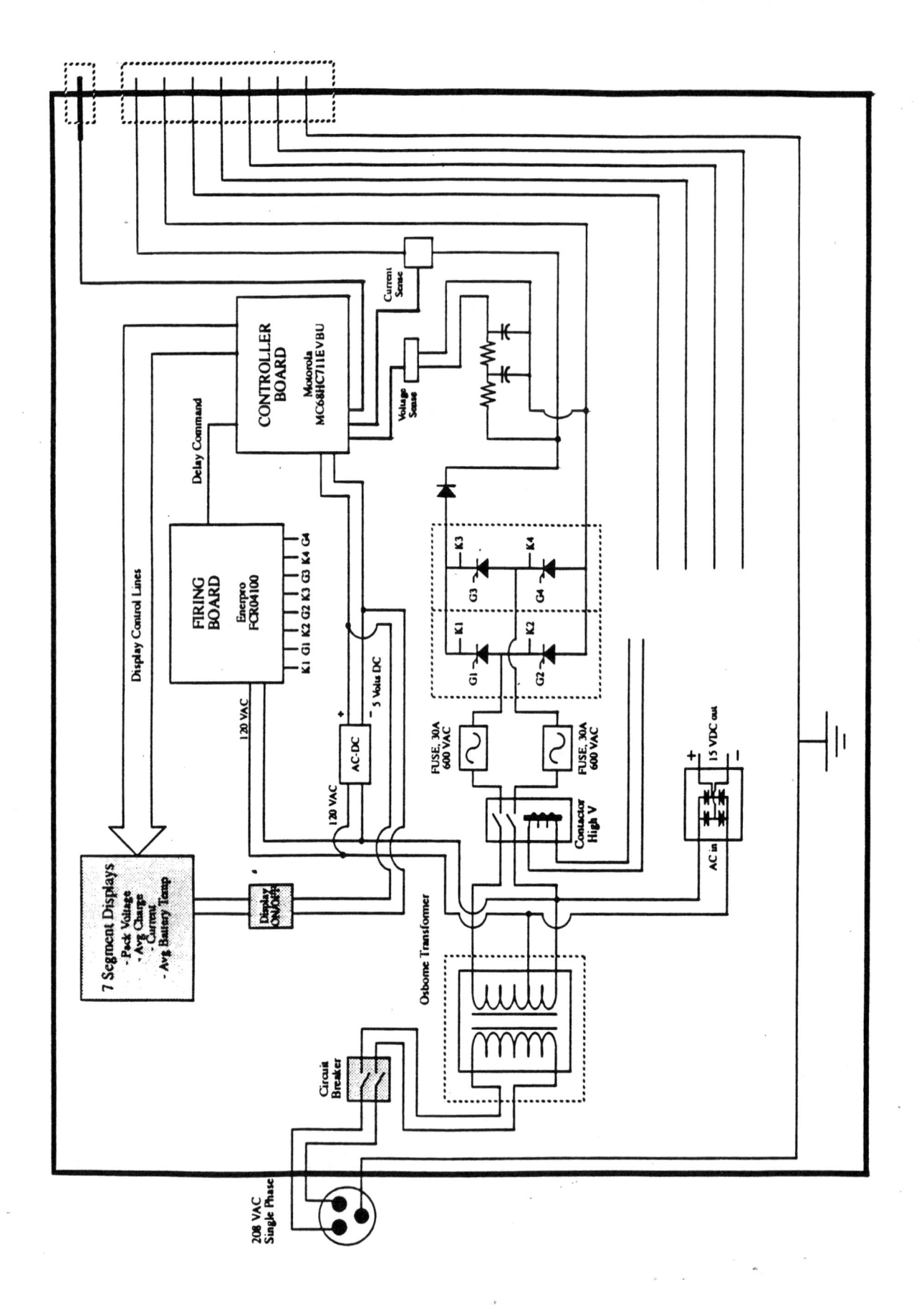

Figure 2: The Schematic Diagram for the Battery Charger

drastically changed the load of the generator, either decreasing its speed and emissions or, when the load was decreased, increasing the charging voltage beyond an acceptable level. The problem was solved by the addition of a three phase controlled rectifier, that increases its firing angle to maintain an acceptable output voltage and generator power. This rectifier is controlled by a firing board which is governed by the on-board CPU.

2.5 Low Voltage System

The low voltage, or 12 volt system, has been upgraded from last year's design. The 12 volt battery is directly connected to a fuse block which supplies overcurrent protected power to all 12 volt components. Stray currents are prevented from flowing through the chassis through the use of a common reference as a ground. Using only one point on the chassis as reference keeps the low voltage system and the chassis at equal voltage potential at all times. This reduces the chance of receiving a shock from contact with the chassis. The low voltage battery is used primarily to start the IC engine and run auxiliary systems such as brakes and lighting. The IC engine is equipped with an alternator to charge the battery while it is running. However, during ZEV mode the battery is charged by a 15 volt DC-DC converter that is powered by high voltage battery pack. The converter is wired to the master switch so as to draw power only when the vehicle is operational.

3 Controls

3.1 Main Control System

The controls of the HEV are an essential element of integrating a large number of subsystems. The need to monitor power flow and control various elements dictates the use of a micro-controller. The brain of the vehicle is a Motorola 68332 microprocessor. Several sensors allow the 68332 to monitor critical functions of the vehicle. The 68332 is given the task of operating several components to optimize the overall system. This must be accomplished by balancing optimal efficiency and adequate HEV performance.

Operating at 16.7MHz, the 68332 has many built-in functions to handle timing as well as a dense instruction set. This is attractive for systems where memory is limited and development time is crucial. The 68332 has the added attraction of having many developmental tools available. Code is written in C++ and cross compiled on a Unix workstation. This eliminates debugging assembly code during the developmental stages. The code can be downloaded for testing and modification. The final code is burned to an EPROM and boots at startup.

The software has two primary goals:

- The first is to control two major feedback loops which consist of rectifier power output and the APU rpm. The two loops work in conjunction to provide a good emissions profile without sacrificing performance.

- The second is to provide information to the driver about vehicle conditions and to make the control system as robust as possible.

The APU control loop consists of a throttle position sensor, a stepper motor, and a rpm sensor. The throttle position sensor is used to insure that the stepper motor is functioning properly and to locate the throttle during start up and shut down sequences. During startup, the computer engages a relay which in turn energizes the pre-heater controller. When the catalytic converter reaches operating temperature, its controller signals the computer, which actuates the starter motor. The starter motor is engaged until the motor reaches a minimum rpm. The stepper motor is stepped up or down to maintain a steady rpm.

The controlled rectifier uses a digital to analog channel which supplies a control signal to a firing board. The firing board in turn controls the delay angle of the thyristor bridge rectifier. By introducing a delay, current being drawn on the alternator decreases. This enables control over APU loading. This is especially useful under hard acceleration as putting a sudden load on the engine would cause an increase in emissions.

3.2 Dashboard

Inputs from various sensors help monitor the state of the vehicle at all times and relay this information through the dashboard.

The dashboard was designed to provide the driver with as much critical information as possible. Voltage and current sensors enable the CPU to control power generation and calculate energy stored in the batteries. Voltage is displayed to let the driver know when the electric motor is going to "stall". Motor rpm is displayed to assist the driver with optimal shift points. This is equivalent to a tachometer for a regular car. The driver is provided with a digital speedometer that has two functions. The speed

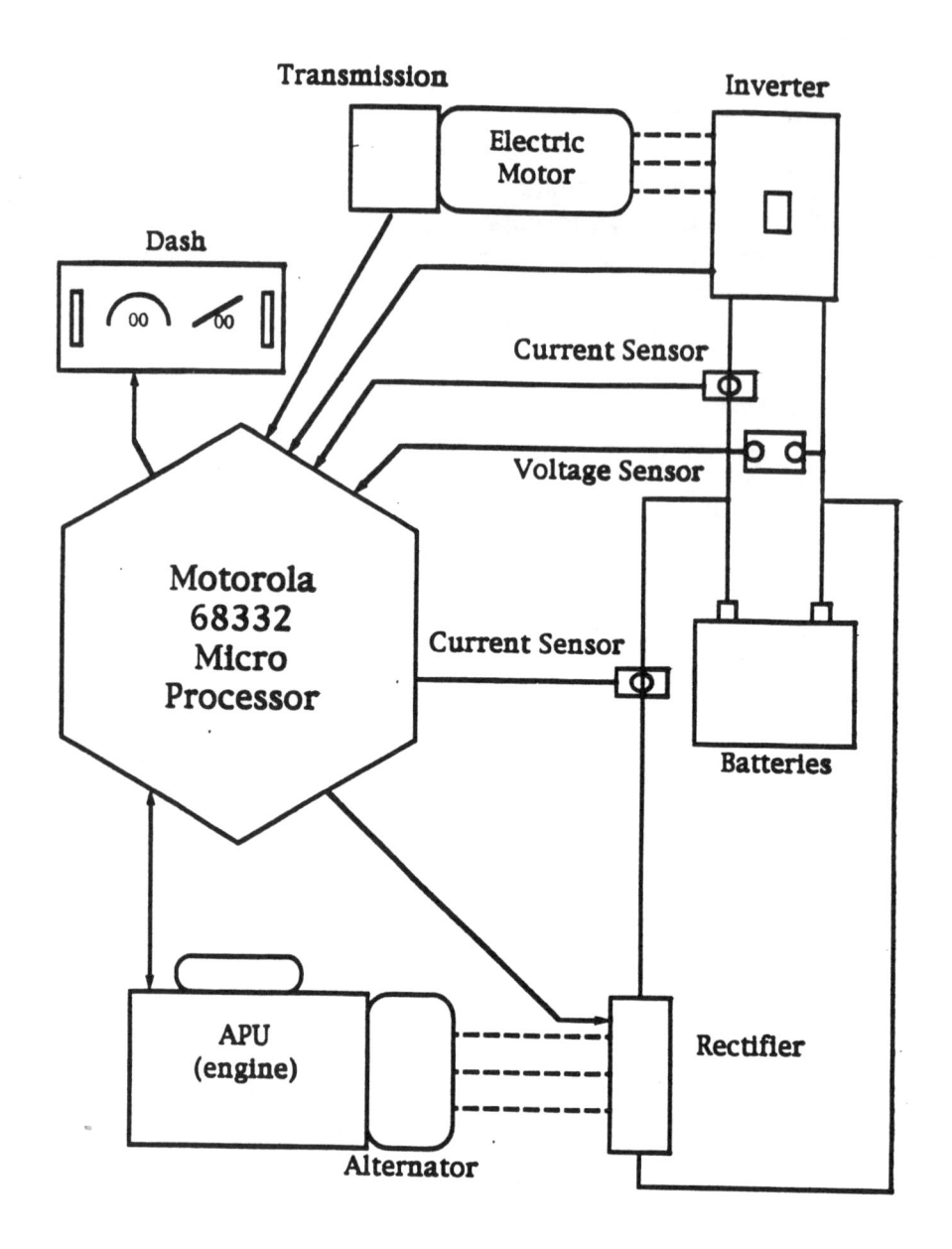

Figure 3: A Schematic of the Controller and Sensors

signal serves as a feedback value for emergency situations that may arise with components of the vehicle as well as providing driver information.

The hardware for the controls was designed to be as robust as possible. Pull down resistors are used on all circuits where power loss of one voltage source will shut down critical components rather than throw the system into an unstable state. The auxiliary CPU board houses 16 analog to digital channels, two digital to analog channels, and various status registers.

A secondary but important control loop is the cruise control. Speed can be controlled by the second digital to analog channel. The range even dictates that it would be advantageous for the driver as well as energy economy to maintain a constant speed. The circuit to engage and disengage the cruise control was given tremendous consideration to avoid potentially dangerous situations.

Control of the HEV is highly dependent upon the components of the vehicle. Sensors and system control inputs were selected for reliability as well as flexibility. Parameters can be altered, tested, or new components can be quickly added to the vehicle.

4 Suspension

The suspension consists of front coil over strut assembles and air ride springs coupled with shocks in the rear. The front suspension utilizes factory Geo Metro struts and control arms with heavier coil over springs. The rear uses Geo A-arms modified to accept air ride springs used by Ford's Mercury Mark VIII. The airspring rear suspension provides in-ride height adjustment and a reduction of 20 pounds of the vehicle weight from the coil spring suspension used last year. At the time of this writing, it has not been determined if benefits of an on-board air source would out weigh the added weight and complexity. A possible air source is the air injection pump used on the internal combustion engine. If it is determined that on-board adjustment is not necessary, air will come from an external source. The steering remains unchanged from last year. It consists of a manual rack and pinion system.

5 Engine

The vehicle team selected M85 methanol fuel after much research and many tests. The team decided to remain with the 1 liter 4 cycle Suzuki from a 1993 Geo engine. This is the same 3 cylinder all aluminum engine used in the 1993 entry. Implementation of M85 fuel required many changes to engine and its related components.

The most significant change was the addition of a port fuel injection (PFI) system. The factory engine was equipped with a throttle body injection system that was inadequate for the conversion. PFI was needed in order to facilitate cold starts and provide better atomization of the fuel. The team obtained a fully programmable engine control unit (ECU) to control the fabricated injection system. This ECU came with all of the necessary wiring for the PFI system. Bosch flex-fuel injectors were selected due to their resistance to the corrosive properties of methanol. An Afco high pressure fuel pump was utilized to provide the pressure required by PFI system. A Bosch high pressure regulator was installed to maintain proper operating pressure for the PFI. The regulator is mounted on a custom stainless steel fuel rail. Fuel lines and connectors were replaced with M85 compatibility materials. Finally, bosses were machined in the cylinder head for the three fuel injectors.

The additional octane of M85 over gasoline dictated an increase in compression ratio for better thermodynamic efficiency and reduced hydrocarbon emissions. The small combustion chambers of the Suzuki engine suggested that a piston change would be the best way to increase compression. A goal was established to attain a static compression ratio of approximately 12:1. To achieve the desired compression ratio, a dome piston was needed. The pistons were design and custom fabricated to specifications. A set of oversize pistons were obtained that offered enough material to produce custom pistons. The pistons were machined to match our engine's bore and dome size. Finally, the piston skirts were machined to match the factory pistons. As a result of machining the tops of the pistons, the distance from the top piston ring to the top edge of the piston was reduced. Research indicates that this area contributes to a large portion of the unburned hydrocarbons in an engine and reductions result in improved engine efficiency.

In order to increase efficiency, we designed a new camshaft. The camshaft is based on the Otto-Atkinson cycle. The profile of the Otto-Atkinson camshaft has a reduced intake duration with an early intake closing. This results in a reduction of normal pumping losses. A benefit of this style of camshaft is increased part load efficiency. While a normal Otto cycle engine may have a calculated efficiency as high as 45%, the Otto-Atkinson cycle has a calculated efficiency as high as 60-70%. One result is a decrease in fuel consumption. Another benefit of using this cycle is the normal decrease in nitrous oxide emissions. However, it reduces the effective compression ratio. Power is also reduced by 50-60 %. Once the camshaft was designed, a camshaft was sent out to be welded and ground to specifications. The camshaft will be installed in the engine and evaluated upon completion.

Along with these internal changes to the engine, many external changes were made as well. Extensive modifications were made to the exhaust system. The most important change involved the pre-heated catalyst. The pre-heated catalyst heats up in half the time and uses approximately half the current as that of last year's model. This significantly reduces cold start emissions and the drain on the twelve volt system. Welded to the back of the pre-heated catalyst is a light-off catalyst. A factory Geo catalyst is located in line after the light-off catalyst. Additionally, a thin-wall stainless steel tubing exhaust manifold was fabricated. This manifold allows positioning of the pre-heated catalyst closer to the exhaust ports of the engine, thus reducing heat loss. Thin-wall material was used in the manifold in order to reduce the amount of heat the manifold absorbed from exhaust

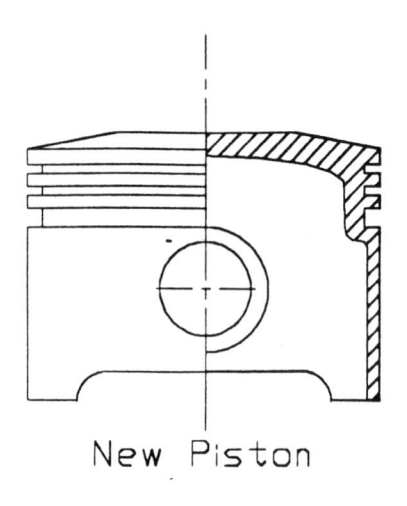

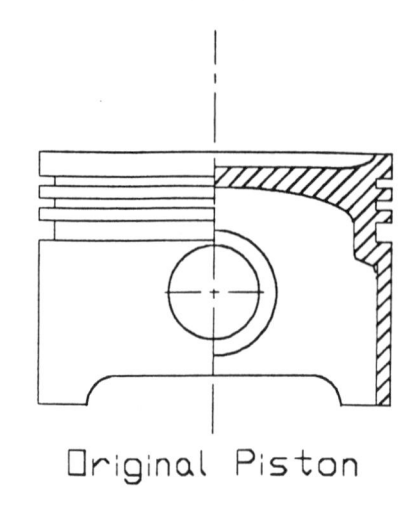

Figure 4: The Modified Engine Pistons

gases. The exhaust pipes are wrapped in an high-temperature exhaust wrap to further insulate the tubing before the catalysts.

A change that is underway involves air assisted atomization for the fuel injectors. Standard fuel injectors emit large droplets of fuel that can puddle on valves or in intake ports. This problem was witnessed at the engine research facility at MSU. To remedy this problem, an air injection system that facilitates fuel atomization is in the developmental stage. The system uses a vane type air pump that delivers approximately 30 psi to an air reservoir. Air regulated at approximately 10 psi is pumped into an injector shroud to facilitate fuel atomization. The system is expected to yield a fuel droplet size of about $10 \mu m$, which will prevent puddles and promote complete combustion.

The ignition system was modified by adding a multiple spark discharge ignition coupled with a high-energy coil to the ignition system to promote better burning of fuel. Optimal ignition timing and spark plug selection are being experimented with at this time.

Evaporative emissions systems were modified because of the higher latent heat of vaporization of M85 fuel. The capacity of the carbon canister was increased to meet the added demands on the system. Also, a number of different gas caps were investigated to minimize evaporative emissions.

Before and after each modification, tests were performed to determine the effectiveness of the change. Primary testing included emissions of CO and HC. During all testing procedures, the alternator was mounted on the engine and loaded to twelve kilowatts by a controlled rectifier and an adjustable resistor. Exhaust emissions and temperature were monitored while the engine was under the normal load. Exhaust temperature sensor were located in four positions: one position at each exhaust port and one positioned at the pre-heated catalyst. This setup was also used to fine tune the fuel curve.

6 Brakes

The brake system consists of a power assisted dual master cylinder, split front to rear, with 8.25 inch disc brakes in front and 8 inch drum brakes in the rear. This system includes a Bosch anti-locking brake (ABS) system. The front wheels are supplemented by regenerative braking controlled by the inverter. The power assist unit replaces a manual master cylinder resulting in reduced pedal effort and increased braking effectiveness. The master cylinder is an electric, hydraulic booster, unlike a conventional vacuum assist. This model was selected because there is no readily available source of vacuum on the vehicle. The booster is a complete system, including its control circuit, making it more compact and reliable than what could be accomplished by fabricating a vacuum control system with its many bulky components. The unit relies on pressurized fluid from an electrically actuated pump that assists in the application of the brakes. A reserve of pressure is stored in an accumulator which allows for several stops in the event of a power interruption.

The front discs and rear drums are factory Geo Metro components. The factory linings were replaced

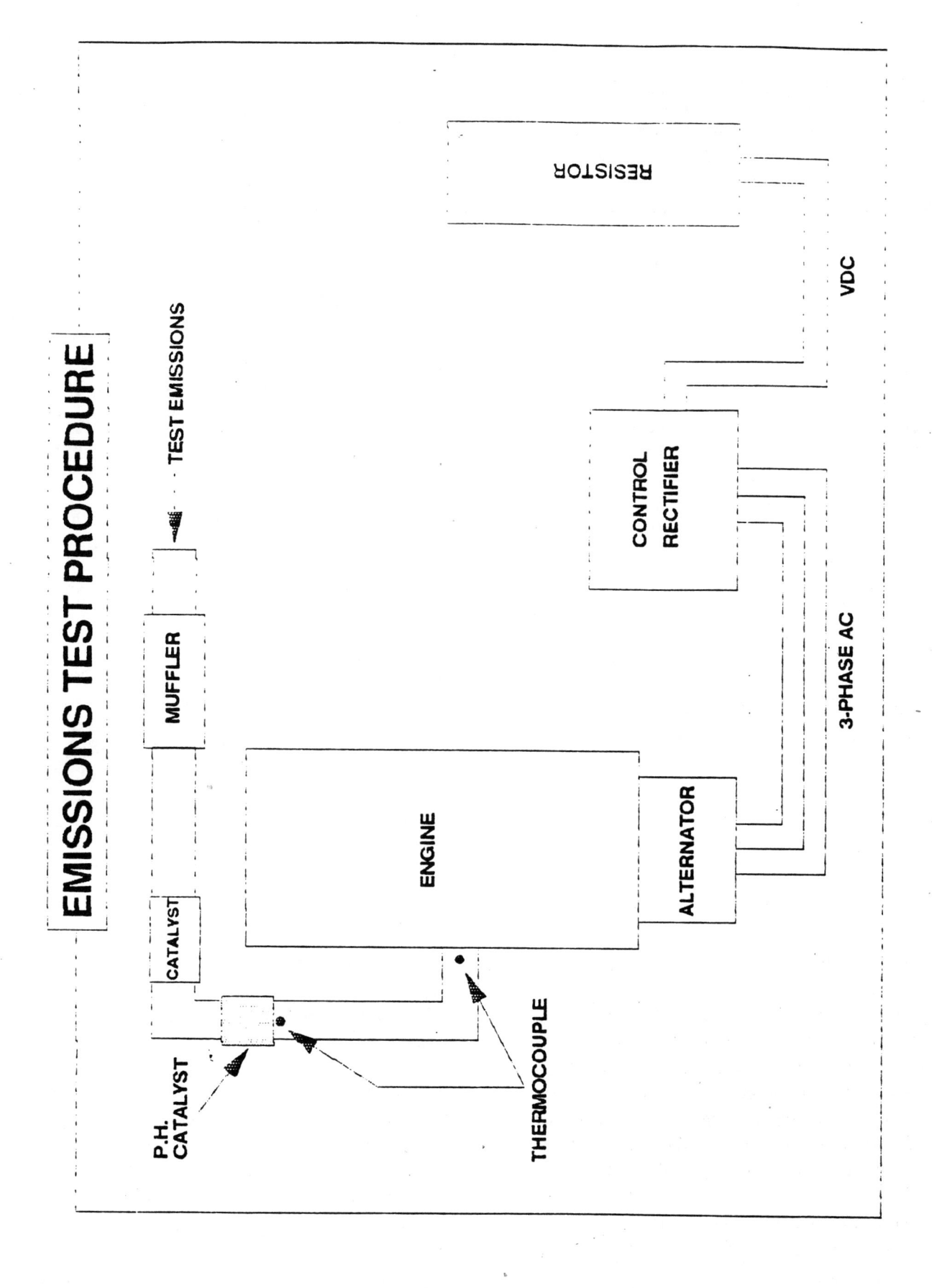

Figure 5: Emissions Test Procedure

with after market semi-metallic material, that provides a higher coefficient of friction and resistance to fade than factory materials. Sensors for the ABS are of the pulse wheel type. The front sensors are machined on the outside hubs front half-shafts. The rear sensors are integral with the wheel bearings. Front to rear bias is adjusted by an in-line proportioning valve. The team elected to take the 22 lb. weight penalty and make a statement about safety as electric vehicles demand high pressure tires and ABS is a necessity rather than a luxury.

7 Chassis and Body

Both the chassis and body were used in the 1993 HEV Challenge. Many improvements to the chassis and body were designed to decrease weight, provide better accessibility, and adapt to different components. Significant design changes have resulted in various modifications to the chassis and body. Utilizing a computer-aided design software package, I-DEAS (Integrated Design Engineering Analysis Software) by Structural Dynamics Research Corporation, was the most efficient way to develop the HEV chassis and body. I-DEAS is an integrated package of mechanical engineering software tools that provides a variety of applications for product design which was used to create a finite element model (figure 6) and a solid model (figure 7) of our vehicle. Utilizing the available computational capabilities of I-DEAS aided in the design of the structural systems.

7.1 Chassis

The chassis design was first constructed in the Finite Element Modeling package. A monocoque midsection, straddled by space-frames in the front and rear were roughed out on paper and then translated into a finite element model. Aluminum was used for the monocoque and was modeled using isotropic thin-shell elements, while the steel space frames were conveniently modeled by beam elements. Thicknesses and material properties of the thin shells could easily be governed, in addition to the beam element's cross-sections. A series of finite element strength analyses ensued after the creation of the preliminary model. Excessive high local stresses and deflections were reduced and material thicknesses were optimized in an iterative manner. This was especially helpful in optimizing the front and rear space frames. A collection of beam cross-sections were created. Larger cross sections were used where needed and tube geometry could be easily relocated and analyzed. Finite element modeling is flexible, in that separate sections, such as space frames, could be loaded and analyzed as separate entities.

After considerable design and analyses, a mid-chassis monocoque passenger compartment and front and rear bolt-on space frames were developed. A modular space frame/monocoque can prove to be very convenient for small collision repairs. In the event of a small crash, the space frame will deform to absorb the crash energy leaving the monocoque unharmed. The damaged space frame could then be removed and a new space frame installed with relative ease. The monocoque is made of recyclable sheet aluminum and is reinforced around the perimeter with an internal space-frame. The design incorporates large box sections along the sides which dramatically improve the torsional stiffness of the chassis. These boxes also double as battery housings. The monocoque contains five uniform battery packs. The batteries are inserted into the monocoque from the bottom. This enables the sides and top of the box sections which encase the batteries to be bent from one sheet of aluminum to ensure that the batteries are completely sealed and inaccessible from the passenger compartment. The space-frames are constructed of square steel tubing. The steel space frames contain all of the major propulsion components except the batteries. The space-frames bolt on and off. This feature enabled the space-frames to be taken off so that more people can work on the vehicle during it's development.

7.2 Body

The Spartan Charge team used the same carbon fiber body shell from the previous year (figure 8) only after a failed Kevlar body lay-up attempt. A fiberglass surface layer did not bond well to the Kevlar, thus time constraints forced us to modify and improve the old body. Making the body weather tight and sectioning the one piece body into three pieces were the two major projects. The old body was constructed from a computer surface model. The computer surface was used to generate a tool path for CNC machining. A male mold (plug) was constructed using plywood, and covered with a machinable epoxy tooling dough. The CNC machined plug was used to create a very accurate female mold (tool).

The high-temperature capable tool was wet layed up in a sandwich construction consisting of tooling dough in the middle of six layers of fiberglass. Three

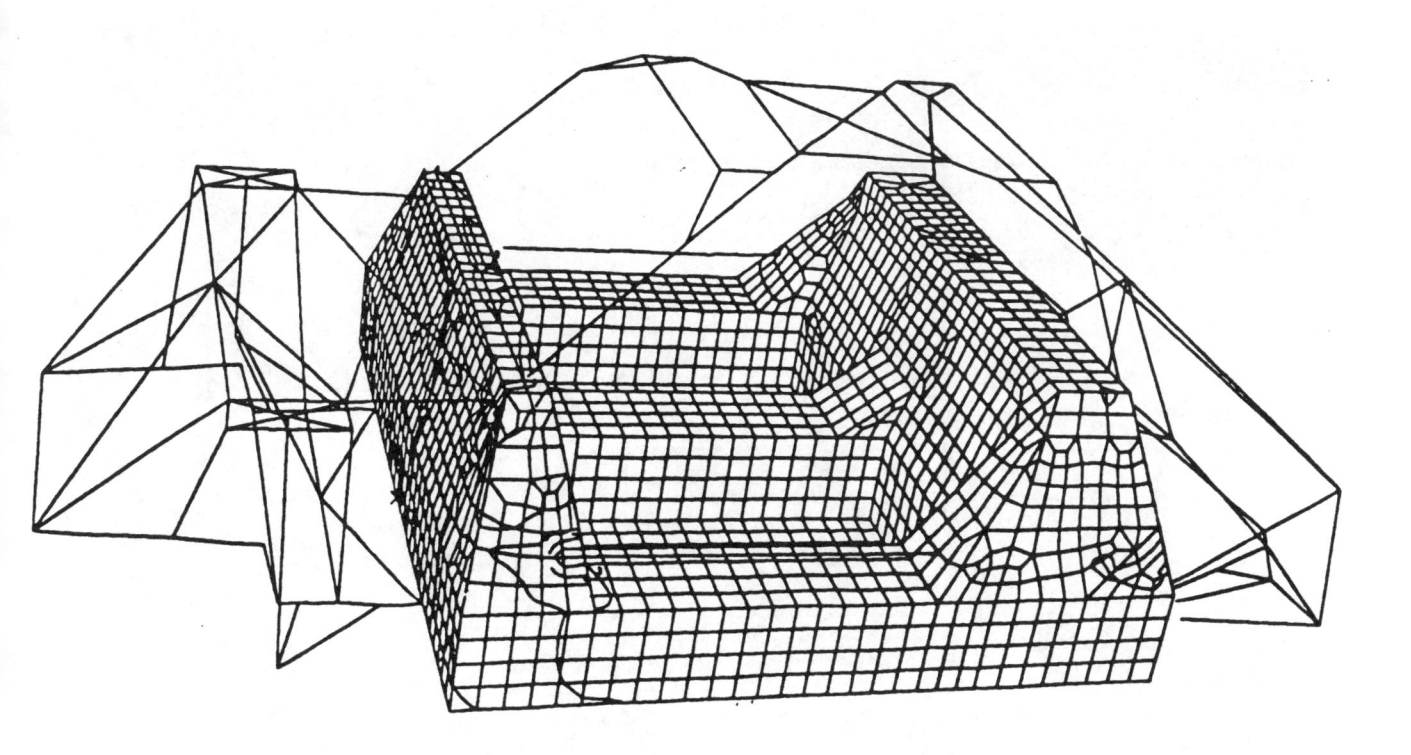

Figure 6: Finite Element model of the Chassis

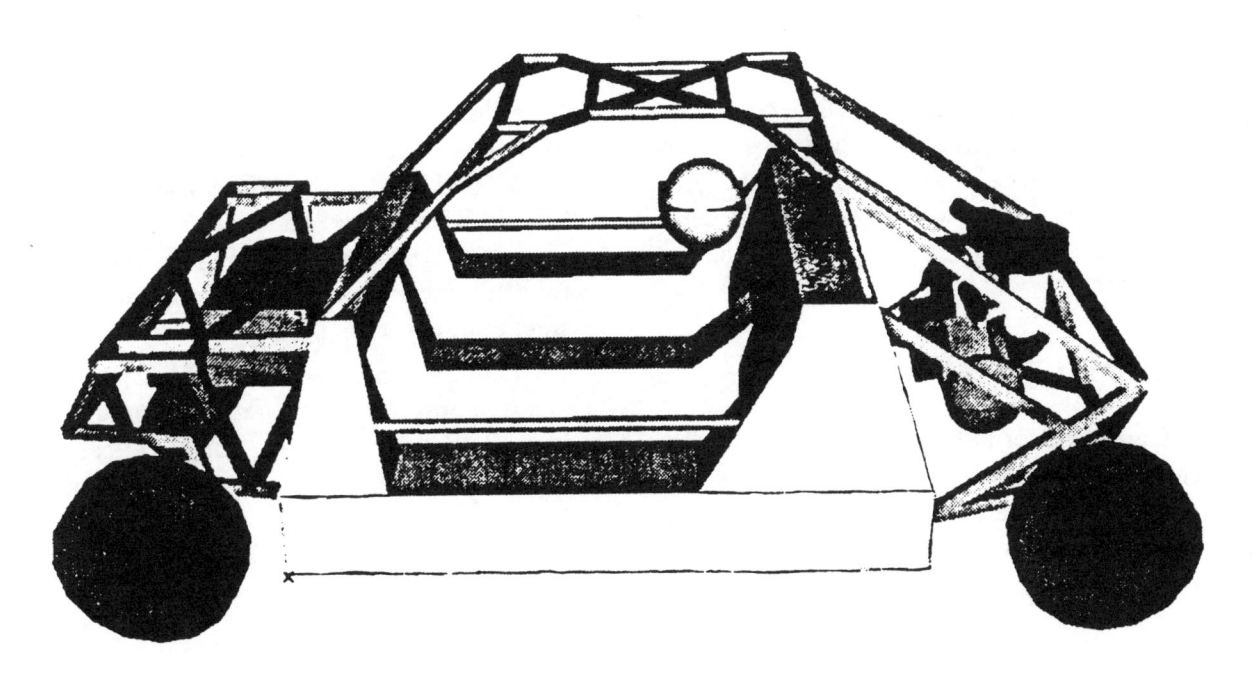

Figure 7: Computer Solid Model of the Chassis

pre-preg woven cloth carbon fiber layers and one fiberglass surface layer were layed up, vacuum bagged, and baked in the tool to fabricate the body. Sectioning the old body into three parts improves accessibility to the components. The body was cut into sections, then the middle section cut lines were traced into the female mold to make flanges for fastening and weather-stripping. Sheet wax was cut and pasted along the scribe lines in the tool to account for the weather-strip gap. Six layers of fiberglass cloth was wet-layed up over the wax to make the flanges. Epoxy was used to bond the flanges to the middle body section.

Accessibility and presentation drove the effort to

Figure 8: Panoramic View of MSU's Hybrid Electric Vehicle

mount the front and rear sections on slide rails. The sections roll out 20 inches and then pivot to open up like a clam shell. The rails were bent into an oval section from conduit and ride on steel ball-bearing rollers. 1/4 turn Dzus fasteners are used to fasten the sections together.

8 Conclusions

Through a team effort of MSU students from all disciplines, the design decisions for the 1993 MSU Hybrid Electric Vehicle were critically evaluated. Many alternatives were considered in view of the limited time that was available to the team to work with the vehicle. Changes were made in the engine configuration and fuel, suspension, brakes, generator, controller, drive motor and inverter, battery charger, and driver controls. All components have been successfully tested, promising a very efficient, non-polluting, safe, and driver-friendly vehicle.

The main success of the project has been in the creation and effective work of a team of students who tackled complex technical and logistical problems, of a scale much larger than expected in an undergraduate project, and the experience that these students acquired.

NEW YORK INSTITUTE OF TECHNOLOGY HYBRID ELECTRIC VEHICLE

BRIAN PESKIN, HEV PROGRAM
NEW YORK INSTITUTE OF TECHNOLOGY
OLD WESTBURY, NEW YORK

1 ABSTRACT

The NYIT hybrid is intended to be the closest immediate and practical replacement for the contemporary gas engine automobile. The NYIT hybrid provides full acceleration performance, can be driven continuously without any battery imposed limitations, does not require any exotic technology, does not need to be externally recharged, and can be easily adapted to virtually any fuel. This design will perform the same way as modern automobiles, and will adapt easily to current transmissions, fuel systems, electrical systems, mounting spaces, and automobile safety standards. The NYIT hybrid uses an engine that is about 1/4 the size of a comparable vehicle, and emissions can be reduced by a similar factor. The NYIT hybrid will adapt readily to the changing availability of fuels in the future.

The NYIT hybrid is based upon a collection of hybrid vehicle concepts that have been tried before, but this is believed to be the most efficient configuration.

The objectives of the NYIT Hybrid design are to successfully utilize a small, direct drive heat engine in conjunction with a small, elegantly designed, boost power system that requires little or no increase in total vehicle weight, and to identify a practical configuration that is transparent to the contemporary consumer, and can be easily implemented using existing technologies.

2 INTRODUCTION

The NYIT Hybrid utilizes a unique energy management system optimized for use in hybrid automobiles and other types of vehicles.

The fundamental concept of the design is to provide a source of continuous mechanical power with the ability to provide peak outputs much greater than the running average. The goal of the energy management system is to use a power source that is only slightly larger than the average power output required, and to store excess energy when available. The stored energy is then used for meeting the peak power requirements. The design consists of an energy storage system, an energy to mechanical power conversion element, a mechanical power to energy conversion element, and an auxiliary power unit.

The initial design application and configuration will be for a hybrid powered automobile that operates using a gasoline engine for the auxiliary power unit, an electrical generator for the mechanical work to energy conversion element, an electrical motor for the energy to mechanical work conversion element, and batteries for the energy storage system.

The unique aspect of the energy management system design is that the engine is directly coupled mechanically to the drive train for maximum efficiency, and the reserve energy system is not used for drive during low power operation. The reserve energy system batteries are charged during low power operation using the generator, and discharged during high power operation using the electric engines. In addition, energy recovered during braking can be stored to provide additional net system efficiency.

Two specific benefits are derived from this arrangement:

a   When electrical power is used for drive, this power is provided at much lower efficiency because energy is converted three times; Fuel to mechanical, mechanical to electrical, and electrical to mechanical. (Engine to batteries to electric motor) For this reason, the stored power is only used to meet the peak power requirements. The engine is used directly for low power operation with maximum efficiency.

b As the engine is sized only slightly larger than the total average power required, higher efficiency is achieved. The engine size is much smaller than the peak power available. In the hybrid automobile, a 20 HP engine can be used with an energy management system that provides 60-80 HP peaks. This configuration will allow normal acceleration and performance in a 2000 Lb vehicle with a 20 HP engine replacing a 60-80 HP engine. The very small engine will provide high fuel economy, and very low emissions.

The peak power system battery reserve is much smaller than the systems used on vehicles where electricity is the prime power source, and the NYIT Hybrid system does not handicap the design with high weight.

The NYIT Hybrid opposes current design philosophies where a hybrid or all electric vehicle contains over 1000 pounds of batteries, and tends to have a total vehicle weight of over 4000 pounds. These designs replace currently available vehicles with an addition of over 1000 pounds of weight. NYIT is focusing on the fact that even with a net increase in system efficiency, the additional weight can easily cancel the benefit, or provide only a small net improvement in efficiency. This is never going to practical for the consumer; these vehicles are far more costly to purchase and operate, and are offering only marginal benefits.

The objective of the NYIT Hybrid configuration is to successfully utilize a small, direct drive heat engine, in conjunction with a small, elegantly designed, boost power system that requires little or no increase in total vehicle weight. The battery bank and electrical motors are offset by a much smaller and lighter heat engine. The small heat engine is a prime candidate for natural gas, and would offer very low total emissions. The prime objective of the NYIT Hybrid is to identify a practical configuration that is transparent to the contemporary consumer, and can be easily implemented using existing technologies.

CONTENTS

1 ABSTRACT
2 INTRODUCTION
3 APPLICATIONS
4 DEFINITION OF TERMS
    4.1 Hybrid automobiles
    4.2 Average energy
    4.3 Generator
    4.4 Electrical motor
    4.5 Auxiliary power unit (APU)
    4.6 Energy storage system
    4.7 Power
    4.8 Peak power
    4.9 Efficiency
    4.10 Regenerative braking
5 IMPORTANT CONCEPTS
    5.1 Direct mechanical drive by the APU
    5.2 Energy management system
6 NYIT HYBRID CONFIGURATION
    6.1 ZEV mode
    6.2 HEV mode
7 EMISSIONS CONTROL STRATEGY
    7.1 Tailpipe
    7.2 Evaporative
8 VEHICLE STRUCTURE DESIGN
9 SUSPENSION DESIGN
10 CHOICE OF MATERIALS
11 VEHICLE MANUFACTURABILITY
12 BODY STYLING AND ERGONOMICS
13 SUMMERY
14 LIST OF DRAWINGS
    14.1 Average, peak, and, operating power
    14.2 NYIT Hybrid configuration
    14.3 Control system power integration

3 APPLICATIONS

The NYIT hybrid is a suitable configuration for any vehicle that requires a substantially higher peak power than average power. The greater the difference between peak and average power, the greater the benefits provided by the NYIT hybrid design. The vehicles include but are not limited to automobiles, vans, small and large trucks, buses, and utility vehicles such as golf carts and indoor use carts.

4 DEFINITION OF TERMS

4.1 HYBRID AUTOMOBILES

The term hybrid will be applied to automobiles and that are powered by a combination of a heat engine and an electric motor, generator, and battery system.

4.2 AVERAGE ENERGY

The average energy usage of the vehicle is the total energy consumed in a particular trip divided by the time of the trip. The term average energy will be used to describe all the energy used on the trip. Mechanical losses due to road resistance, air resistance, drive train friction, A/C systems, accessory systems, heat losses, and any other source, are included in the total energy used.

4.3 GENERATOR

The electrical generator is a device that converts mechanical energy to electrical energy. The electric generator includes but is not limited to alternating current, direct current, and motor generator types.

4.4 ELECTRICAL MOTOR

The electrical motor is a device that converts electrical energy to mechanical energy. The electric motor includes but is not limited to alternating current, direct current, and motor generator types.

4.5 AUXILIARY POWER UNIT (APU)

The APU or Auxiliary Power Unit is the gasoline engine used to provide mechanical power to the hybrid using internal combustion of any type of fuel. The APU includes two stroke, four stroke, and rotary engine types. The fuels include but are not limited to gasoline, ethanol, methanol, alcohol, propane, hydrogen, and natural gas.

4.6 ENERGY STORAGE SYSTEM

The energy storage system consists of the batteries used to store electrical energy. The energy storage system can consist of a single battery, or any number of separate batteries in a series, parallel, or combination arrangement. The battery types include but are not limited to lead/acid, nickel/cadmium, and carbon/zinc.

4.7 POWER

Power is defined as the work per unit time, and the units will be horsepower or HP. Power is the rate at which mechanical or electrical energy is transferred.

4.8 PEAK POWER

The peak power is the maximum rate at which energy is transferred. The transfer includes but is not limited to electrical to electrical, electrical to mechanical, mechanical to electrical, and mechanical to mechanical.

4.9 EFFICIENCY

Efficiency is the fraction or percentage of the energy put into the system compared to the energy effectively recovered by the system. For the hybrid, the efficiency will be the miles per gallon or MPG for gasoline. Each gallon provides the vehicle with a certain amount of energy, expressed in BTU/POUND of fuel, as compared to the total distance traveled per pound of fuel. The efficiency comparison for different fuels, liquid or gas, have to be adjusted for the different BTU/POUND values of each fuel. For the NYIT initial hybrid, efficiency will be evaluated by MPG, and gasoline will be the fuel.

4.10 REGENERATIVE BRAKING

Regenerative braking is the energy recovered by the energy management system from braking the vehicle. Regenerative braking includes but is not limited to the use of the generator, or motor generator, to provide electrical energy that can be stored for future use. The generator or motor generator can be attached mechanically to the wheels or to any part of the drive train of the vehicle where the generator action can be used to brake the vehicle.

5 IMPORTANT CONCEPTS

WHAT IS UNIQUE ?

The NYIT hybrid is a combination of design concepts incorporated in a specific arrangement. The NYIT hybrid is based upon a collection of separate hybrid vehicle concepts that have been tried before, but this is believed to be the most efficient configuration. A very important aspect of this design is the ability to achieve an extremely high efficiency vehicle with ultra low emissions using readily available contemporary components and technologies.

The fundamental concept of the design is to provide a source of continuous mechanical power with the ability to provide peak outputs much greater than the running average. The goal of the energy management system is to use an APU power source that is only slightly larger than the average power output required, and to store excess energy when available. The stored energy is then used for meeting the peak power requirements. The APU is the prime drive for the vehicle, and the electric power is considered less efficient, and is never used where APU power can be used. The design consists of an energy storage system, an energy to mechanical power conversion element, a mechanical power to energy conversion element, and an auxiliary power unit.

The NYIT hybrid design uses the APU, electric motors, electric generators, and batteries in a moderate manner. Nothing exotic is required to meet the system requirements. The small size of the APU would allow very low emissions using existing gas engine technologies. The same gas engine technology that provides over 40 MPG on the highway in contemporary automobiles can be used in the NYIT design, but the engine is about 1/4 the size, with a corresponding reduction in emissions and increase in efficiency. In addition, natural gas internal combustion engines are a good choice in the low HP ranges required, and would be similar to a forklift engine which can be operated in living spaces due to the cleanliness of the emissions.

The way in which the batteries are used is also of extreme importance. The batteries are for reserve power only, and do not have to be optimized for power density as is required in vehicles that use the batteries as a prime drive power source. A loss of battery capacity these designs results in a loss of vehicle range. The NYIT hybrid does not require the full capacity, or a full charge condition to operate normally. The NYIT hybrid will function with a loss of 25% or more of battery capacity, and the battery will charge and discharge more efficiently below full charge. The batteries will be used partially charged most of the time, and a loss of full battery capacity will not affect performance in any way. This will probably be true up to the point of failure of the batteries. It is also important to note that because the batteries are only being used for peak power storage, the size of the battery bank is much smaller and lighter.

The NYIT hybrid is defined by the following unique configuration, and includes these primary concepts:

a   The APU is the prime power source for the vehicle, and is used to drive the vehicle directly.

b   Electrical power is less efficient than APU power for mechanical work, and therefor must be conserved.

- The electrical motors are never used where APU power is available.

- The generator is never used at the same time as the electrical motors.

- The energy management system is the focal point of the system; the energy management system is the "fully optimized" integrated concept that combines standard system components to provide a higher level of efficiency.

5.1   DIRECT DRIVE BY THE APU (a)

The key concept in the NYIT hybrid is that the use of electrical energy stored in the batteries is at a lower efficiency than the use of direct mechanical energy from the APU. This is because the electrical energy is converted twice before use. The mechanical energy from the APU is converted to electrical energy by the generator, and stored in the batteries. The electrical energy is then converted to mechanical energy by the electrical motor. Energy losses occur at each stage. There is a loss in the generator, batteries, and the electrical motor.

The most efficient power in the NYIT hybrid is the mechanical energy from the APU. In the NYIT hybrid, the APU is used for all drive purposes up to the full power of the gas engine. The electrical energy is used for higher power requirements such as accelerating, or climbing a tall hill. The APU is sized such that there is sufficient power available to drive the vehicle at constant velocities. The electrical power is therefor not required for constant speed cruse. The APU is sized such that there is some excess power available which is used to recharge the batteries.

5.2   ENERGY MANAGEMENT SYSTEM (b)

During normal operation of the vehicle, the power requirements vary from low to high. In the NYIT hybrid, the batteries are charged during low power operation, and discharged during peak power operation. The APU is therefore used at close to full capacity at all times, and will allow very efficient operation for the gas engine. The gas engine is of low power, about 25% of the available peak, and will mostly be operated at or near full power which is usually more efficient in internal combustion engines.

The point is that a 2000 lb vehicle with a 20 HP engine would provide an extremely high MPG, but would have very poor acceleration performance. The NYIT hybrid utilizes the electrical battery energy reserve system to provide the peak power and performance of an 80 HP engine with a 20 HP engine. The NYIT hybrid concept works based upon the fact that the average usage of the 2000 lb vehicle is about 15 HP, and that the engine has power available to charge the reserve system.

The NYIT hybrid is specifically different from other designs in that the electrical system is never used unless necessary, and that the electrical generator is never used during operation of the electrical engines. The generator is powered by the drivetrain, and acts as a loss to the system. The generator is only used during low power operation of the APU when there is available extra power. For operation near the capacity of the APU, the generator is off. For operation at higher power than the capacity of the APU, the electric engines are used, and the generator is not used. Regenerative power is stored in the batteries for all engine braking conditions, and any energy recovered contributes to the total efficiency of the vehicle.

The NYIT hybrid has three operating modes:

a   Low power where the APU provides drive power and generator power. A variable charge system may be used to allow charging at lower rates when more power is required for drive, and at higher rates when there is extra capacity available. This architecture augments the near continuous power level operation of the APU, and will contribute to the net efficiency of the system.

* The APU is operating near full power which is efficient, and the excess power is saved.

b   Medium power where the APU provides drive

power, and the generator and the electrical motors are not used. In this mode, there in not sufficient power to allow charging. On a variable charge system, the medium power mode could be reduced to about 97% of the APU maximum power level.

- * The APU is operating at or near full power, which is efficient, and the electrical system is completely off, and causes no losses to the system.

c High power where the APU and the electrical motors are used, and the generator is not used.

- * The batteries are discharged for drive power, and the charging system is off, and causes no losses to the system.

Note: Regenerative braking is used in all modes where engine braking is available.

6 NYIT HYBRID CONFIGURATION

6.1 ZEV MODE

In ZEV operation mode, the NYIT Hybrid utilizes battery power alone for drive, and regenerative braking is used to provide some energy recovery during engine braking of the vehicle. The APU is not operated during ZEV mode.

The NYIT Hybrid is intended to be a true hybrid vehicle, and a direct functional replacement for contemporary automobiles. The NYIT Hybrid battery/electric motor boost system will allow a very small, low emissions, heat engine to power the vehicle while providing very high fuel economy.

ZEV operation is expected to be limited, and will serve to provide local operation of the hybrid in the event of APU failure, and to allow short distance operation of the hybrid with zero emissions. Local ZEV operation may be attractive for very heavy traffic areas, or tunnels.

The small boost-only battery bank is simpler, lighter, and utilizes currently available automotive batteries. The use of very large battery banks to provide direct vehicle drive defeats the purpose of a hybrid; the extra weight opposes efficiency. The NYIT Hybrid uses only 400 pounds of batteries, and this is largely offset by the reduction of the engine weight to 80 pounds for the APU. Otherwise, the NYIT Hybrid will be in the same class as contemporary vehicles, with much better environmental performance, but usually not operating in ZEV mode.

In ZEV mode, the NYIT Hybrid will provide 60 HP peak and up to 30 HP steady state using the battery reserve power.

6.2 HEV MODE

In HEV mode the APU provides the source of power, driving the wheels directly, and charging the batteries when any excess power is available. The batteries are recharged at a varying rate allowing low level recharging even when most of the APU power is used for drive, and to provide rapid charging when excess APU or regenerative braking power is available. The energy management system controls all of the system components to maximize energy usage, storage, and recovery. The NYIT Hybrid is very simply a gasoline powered automobile with a very small engine. The energy management system provides energy integration to meet peak power demands, and to allow normal acceleration and grade climbing abilities.

An important aspect of the NYIT Hybrid is transparent operation of the vehicle by contemporary drivers. The vehicle can be started and driven under any normal profile; city stop and go traffic, or steady state driving at any legal speed, for any distance. The NYIT hybrid is self charging in normal operation, and only the usual gas stops would be required. The vehicle does not require an external recharge during or after use. In addition, as the peak power levels are similar to a comparable gasoline vehicle, no special training or consideration is required to operate the NYIT Hybrid.

In HEV mode, the NYIT Hybrid will have a total of 80 HP peak available from the electric motor and APU parallel operation. Steady state operation of up to 20 HP is possible using APU power, and without depleting the battery reserve. Steady state operation of up to 50 HP is possible with the parallel configuration.

7 EMISSIONS CONTROL STRATEGY

7.1 TAILPIPE

The NYIT Hybrid APU is a 20 HP gasoline engine that utilizes a standard catalytic converter to meet emissions requirements, and no other modifications are required. The vehicle emissions will meet current standards for percentage pollutant content. It is important to remember that the measured emissions are a ratio of the tailpipe flow as taken on a small engine, and that the total emissions are a fraction of what would be generated by a modern vehicle with a larger engine and the same performance numbers.

7.2 EVAPORATIVE

The NYIT Hybrid uses a closed fuel system to prevent fugitive emissions. The fuel is pumped using an electric fuel pump that is in line between the tank and the APU. The tank vent line is fed back to the intake system to burn off any evaporative emissions.

8 VEHICLE STRUCTURE DESIGN

The NYIT Hybrid utilized a qualified NASCAR frame for the foundation of the ground-up vehicle. This frame assures a high degree of passenger safety on the prototype vehicle. Total vehicle weight is about 2000 lbs.

The NASCAR frame allowed convenient addition of the primary drive components of the NYIT Hybrid without concern for the basic integrity of the frame as related to passenger safety. Additional attention was paid to the battery storage containers to ensure that the acid was contained under failure conditions, and that the batteries were secure during collision and rollover. A fiberglass, acid resistant, enclosure is used to contain liquid from the batteries, and a steel structure secures the batteries to the vehicle frame. The batteries are located near the center of gravity of the front-rear axis to reduce forces in a rollover condition.

9 SUSPENSION DESIGN

The front axle uses a MacPherson strut, coil spring, independent suspension from a Volkswagen rabbit. The rear uses is a solid axle, coil spring suspension, also from a Volkswagen. The bearings and brakes are unmodified, and as the original vehicles were heavier than the NYIT Hybrid, these components provide an extra margin of safety in the prototype design.

The springs were set to provide a level vehicle profile when fully loaded, and the NYIT Hybrid has close to a 50-50% weight distribution allowing for an intrinsically stable design.

10 CHOICE OF MATERIALS

One of the objectives of the NYIT Hybrid program is to identify a working configuration using readily available contemporary technologies. All of the materials and components used on the NYIT hybrid were commercially available, and no special permits were required. Gasoline is used on the hybrid, and is contained in a rated vessel with proper safety vents. Lead-acid batteries are used on the hybrid, and these are a standard automotive type, and are suitable for recycling after use. Secondary containment provisions were made on the NYIT Hybrid vehicle to ensure that the battery acid is secured during a rupture of the battery, a collision, or a rollover condition.

Fiberglass was chosen for the body to reduce weight and to allow ready fabrication of the unique shape of the prototype vehicle. Aluminum is used extensively for the passenger compartment floor and firewalls. The aluminum provided the necessary strength with less weight than steel. Steel was used for all structural members such as engine supports, electric motor and battery sub-frames, and suspension supports.

The chassis is a welded tubular steel, NASCAR qualified frame that was purchased complete. This previously optimized design has a known record of safety and strength at a very low weight.

11 VEHICLE MANUFACTURABILITY

The NYIT Hybrid is a ground-up class vehicle, and is not designed for high volume production. The drivetrain components are coupled via timing type belts, and the drive system is an open frame configuration. The system layout is not fully optimized, but is oriented for ease of assembly and maintenance access. The body sections are removable, and the vehicle has no doors.

The NYIT Hybrid that will be competing in the 1994 HEV Challenge is a proof of concept vehicle to determine the net efficiency and environmental performance of the system configuration as compared to the components used. In the event that the NYIT Hybrid system performance exceeds an equivalent system based upon a gasoline engine in a modern vehicle, then the prototype will be a success.

The objectives of the NYIT Hybrid design are to successfully utilize a small, direct drive heat engine in conjunction with a small, elegantly designed, boost power system that requires little or no increase in total vehicle weight, and to identify a practical configuration that is transparent to the contemporary consumer, and can be easily implemented using existing technologies.

In keeping with these goals, the NYIT Hybrid system configuration is built on very standard building blocks. The APU is a small gasoline engine, and there are many readily available types of low HP heat engines that operate using a variety of fuels. The electric motors and generators can be AC or DC, and do not have to be of unusually high efficiency to meet the system requirements. A generator/motor could also be used. The motor controller is a high efficiency pulse width modulated type that was readily available. The batteries do not have to be of exceptionally high power density to function in the NYIT Hybrid system, and standard lead/acid type batteries can easily be used.

The ready availability of the system components required for the NYIT Hybrid, in combination with the great flexibility allowed in the component implementation, make the NYIT Hybrid configuration the potential automobile of tomorrow. Existing contemporary automobiles can be retrofit to meet the NYIT Hybrid system configuration, and these vehicles can be sold to the modern consumer without a reduction in vehicle performance. The NYIT Hybrid is an energy conservation philosophy that implements energy integration using a very basic approach. The design is extremely manufacturable in a modern plant environment, and the simplicity of the design is the greatest advantage.

The problem with many modern electric vehicle designs is the effort to provide complete ZEV operation although modern batteries do not have the power density required for an effective design. Modern ZEV vehicles have very poor acceleration, are overly heavy, and are very costly. In addition, most contemporary hybrid designs neglect the heat engine as the primary driving power source, and rely on the batteries and the electric motors more than is necessary. A far more practical approach is to fully optimize existing technologies, and to maintain an on board heat engine. The NYIT Hybrid allows for a very small heat engine, which can easily be adapted to natural gas providing exceptionally low total emissions, on a vehicle that can be manufactured today, and without compromising performance.

12 BODY STYLING AND ERGONOMICS

The NYIT Hybrid is styled as a two passenger, convertible sedan with an aerodynamic fiberglass body. The vehicle does not have doors, and the convertible top consists of a retractable roof cover that leaves the top chassis tubular frame in place when open. Passengers will have an open air feel with the top open, but the passenger compartment is still completely protected by the frame, and the roof cover will provide protection from rain. Entrance and egress will require climbing over the side frame structure, and the seats include a five point safety harness.

The NYIT Hybrid is a ground-up class vehicle, and the functional ergonomics are somewhat compromised by the focus on safety in the design. Once located in the driver/passenger seats, the environment is as similar as possible to a standard automobile. The view of the road, operator controls, and seating height are comparable to a contemporary sub-compact automobile. Operator controls specific to the hybrid design are minimal, and are clearly labeled. The primary operator selection will be for HEV or ZEV mode, where HEV is the expected standard, and battery charge information will be available in either mode. Additional controls and defeats that will be used to research the NYIT Hybrid under a wide variety of conditions are available, but these controls will usually not be used during normal driving.

The NYIT Hybrid will operate as a standard transmission vehicle with no driving restrictions in HEV mode, and with range restrictions in ZEV mode. Battery status indication will provide charge state information to the driver.

13 SUMMERY

The NYIT hybrid is intended to be the closest immediate and practical replacement for the current gas engine automobile. The NYIT hybrid provides full acceleration performance, can be driven continuously without any battery imposed limitations, does not require any exotic technology, does not need to be externally recharged, and can be easily adapted to virtually any fuel.

The NYIT hybrid is a suitable configuration for any vehicle that requires a substantially higher peak power than average power. The greater the difference between peak and average power, the greater the benefits provided by the NYIT hybrid design. The vehicles include but are not limited to automobiles, vans, small and large trucks, buses, and utility vehicles such as golf carts and indoor use carts.

The objectives of the NYIT Hybrid design are to successfully utilize a small, direct drive heat engine in conjunction with a small, elegantly designed, boost power system that requires little or no increase in total vehicle weight, and to identify a practical configuration that is transparent to the contemporary consumer, and can be easily implemented using existing technologies.

The NYIT Hybrid that will be competing in the 1994 HEV Challenge is a proof of concept vehicle to determine the net efficiency and environmental performance of the system configuration as compared to the components used. In the event that the NYIT Hybrid system performance exceeds an equivalent system based upon a gasoline engine in a modern vehicle, then the prototype will be a success.

14 LIST OF DRAWINGS

14.1 AVERAGE, PEAK, AND, OPERATING POWER

14.2 NYIT HYBRID CONFIGURATION

14.3 CONTROL SYSTEM POWER INTEGRATION

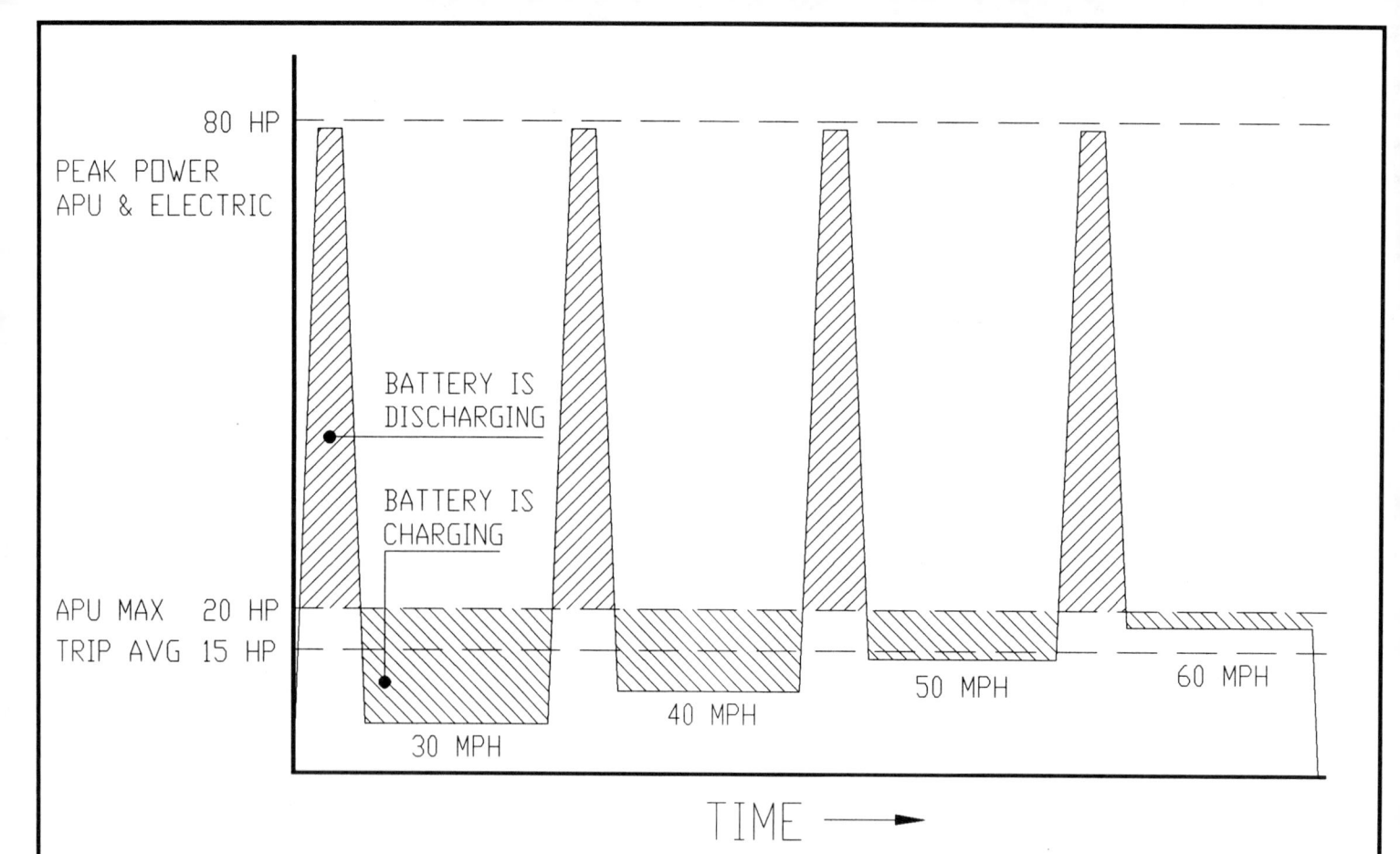

Shown is a graph of a short trip that demonstrates the application of the NYIT Hybrid as an energy intergration system. The trip consists of acceleration at full power to 30 MPH, a short cruise period, acceleration at full power to 40 MPH, a short cruise period, acceleration at full power to 50 MPH, a short cruise period, acceleration at full power to 60 MPH, a short cruise period, and stopping.

During acceleration the batteries are discharged, and the electric motors provide 60 HP of boost to the 20 HP APU. During cruise, the batteries are recharged using the excess APU power. The APU is running at full power at all times, but the vechicle uses power over a wide range, and up to 80 HP. This is the function of the energy management system, peak power much greater than the capacity of the APU is available. In addition, the system is optimized; the APU drives directly as much of the time as possible providing greater efficiency, and the batteries are only recharged when excess power is available.

NOTES:
- The trip average is below the operating power due to losses.
- The trip average will vary depending upon driving profile.
- The vechicle can operate at cruise without using the electrical boost, and is not limited in range by battery charge capacity.
- The batteries are recharged during operation, and the vechicle does not require external recharging after a trip.

14.1 AVERAGE, PEAK, AND OPERATING POWER

Block Diagram

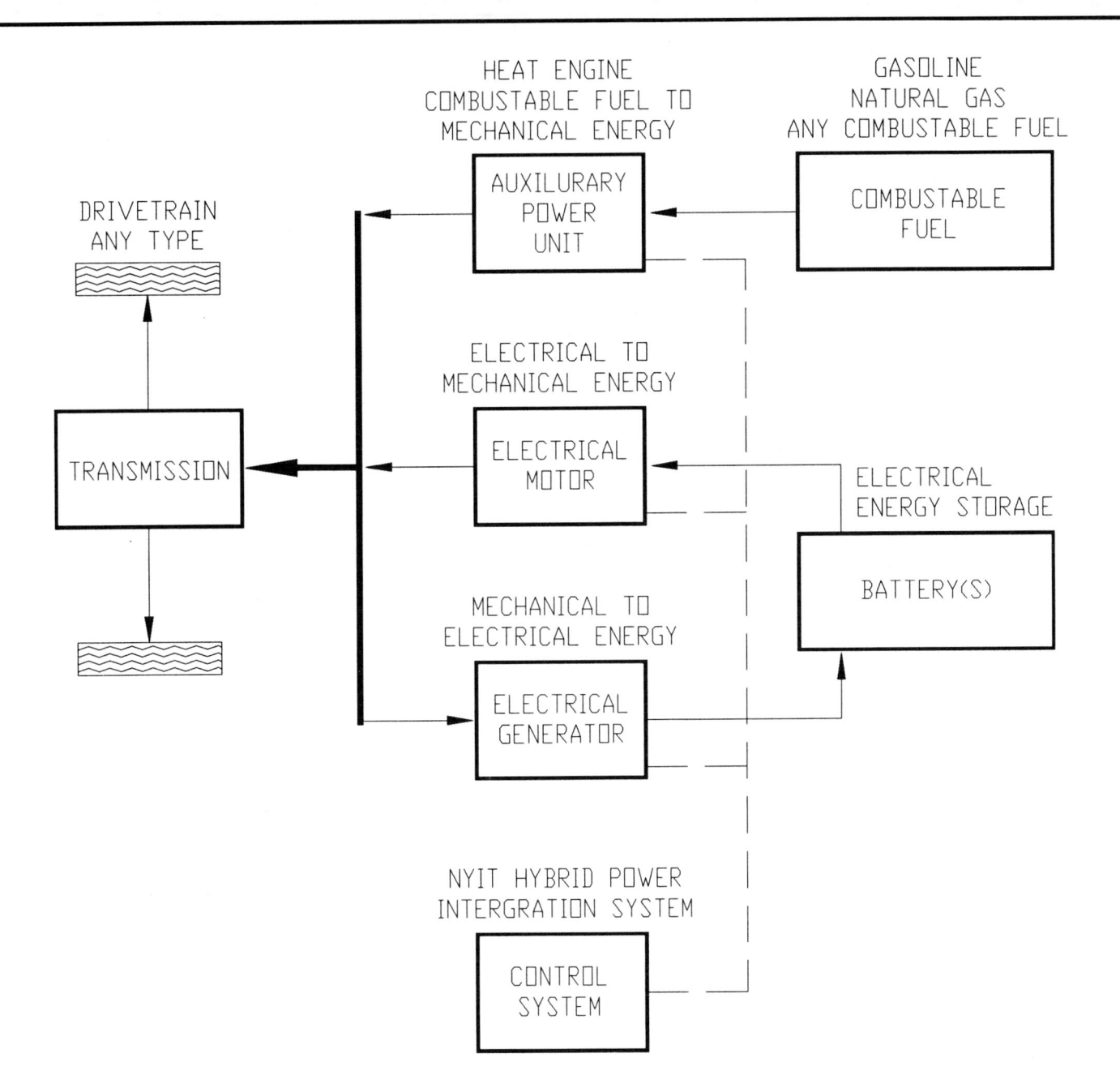

- **DRIVETRAIN ANY TYPE**
- **TRANSMISSION**
- **HEAT ENGINE — COMBUSTABLE FUEL TO MECHANICAL ENERGY**
  - AUXILURARY POWER UNIT
- **GASOLINE NATURAL GAS ANY COMBUSTABLE FUEL**
  - COMBUSTABLE FUEL
- **ELECTRICAL TO MECHANICAL ENERGY**
  - ELECTRICAL MOTOR
- **ELECTRICAL ENERGY STORAGE**
  - BATTERY(S)
- **MECHANICAL TO ELECTRICAL ENERGY**
  - ELECTRICAL GENERATOR
- **NYIT HYBRID POWER INTERGRATION SYSTEM**
  - CONTROL SYSTEM

This is the system block diagram for the NYIT Hybrid system. The APU, electrical motors, generator, batteries, and drivetrain, can be of any type. The configuration of the hybrid system has the following key features:

- The APU is used for direct drive up to the full power of the unit.
- The electrical motor is used for peak power operation.
- The generator is used only when excess APU power is available.
- The APU, electrical motor, and generator, are in parallel on the same drive system.
- Regenerative engine braking is available at all times.

14.2 NYIT HYBRID CONFIGURATION

SYSTEM ENERGY FLOW	MODE	APU	EM	GEN	POWER LEVEL
(diagram: Transmission ← APU ← Combustable Fuel; Electrical Motor; Electrical Generator → Battery(s)) ELECTRICAL MOTOR DOES NOT CAUSE LOSSES TO THE SYSTEM	LOW POWER EXCESS APU POWER AVAILABLE FOR CHARGING	DRIVE CHARGE	OFF	ON	100% — APU MAX — FULL CHARGE AVAILABLE
(diagram: Transmission ← APU ← Combustable Fuel; Electrical Motor; Battery(s); Electrical Generator) ELECTRICAL MOTOR AND GENERATOR DO NOT CAUSE LOSSES TO THE SYSTEM	MEDIUM POWER NO EXCESS APU POWER AVAILABLE FOR CHARGING	DRIVE	OFF	OFF	100% — APU MAX — NO CHARGE AVAILABLE
(diagram: Transmission ← APU ← Combustable Fuel; Electrical Motor ← Battery(s); Electrical Generator) ELECTRICAL GENERATOR DOES NOT CAUSE LOSSES TO THE SYSTEM	HIGH POWER BATTERY IS USED FOR FULL POWER DRIVE	DRIVE	ON	OFF	100% — APU MAX

The NYIT Hybrid has 3 distinct power modes, each fully optmized the bring maximum power to the wheels with maximum efficiency from the system components.

The NYIT Hybrid uses regenerative engine braking in all modes.

14.3 CONTROL SYSTEM POWER INTERGRATION

The Pennsylvania State University 1994 Hybrid Electric Vehicle

Brandon Vivian
SAE Student Branch V.P., HEV Team Leader

Douglas Buch
HEV Project Supervisor

Joel Anstrom
Mechanical Systems Design Head

Jeffrey Fry
Mechanical Design Engineer

Daniel Ellis
Microprocessor Control Head

Abstract

Emission free vehicles will provide personal transportation in the near future. To curb the reliance on fossil fuels and their environmental impact, a solution must be developed. Battery technology will make pure electric vehicles impractical for the near future. With today's technology the solution, to decreased fuel consumption and emissions, tends toward a hybrid vehicle design as an intermediate stepping stone to an emissions free vehicle. The Hybrid Electric Vehicle (HEV) is designed to decrease petroleum consumption and reduce emissions by featuring a hybrid propulsion system. This high-efficiency system contains an internal combustion (IC) engine in conjunction with an electric propulsion system. The IC engine could be fueled by either gas or alternative fuels. The IC engine in a HEV may assist the electric drive motors, power an on board in transit battery charging system, or a combination of both. This report details the vehicle engineering and strategy involved in Pennsylvania State University's entry in the 1994 HEV Challenge. This includes the power-train selection, vehicle and emissions control strategy, structure modifications, materials selection, and vehicle manufacturability.

1. Introduction

Performance and strategy of the 1993 Pennsylvania State University HEV have lead to vehicle design reconsideration and analysis for the 1994 HEV Challenge. The 1993 HEV employed an Auxiliary Power Unit (APU) in a primarily range extender role by powering a generator, but had the ability to drive the front wheels in a traction drive assist role while in the HEV mode. This ability necessitated the use of a reasonably powerful and sizable APU. The APU, a GEO Metro 1.0 Litre, the accompanying transmission, and the DC Brushless Motor employed as a generator occupied the engine compartment. Thus, the electric drive motors were located in the rear driving the rear wheels. The increased weight and traction drive redundancy accompanying this strategy was found to be unacceptable in performance and overall efficiency. The strategy used for the 1993 HEV was found unsuitable for a conversion of most production

2.6 Motor/Controller

2.6.1 Selection.

Motor qualities such as power-to-weight ratio, torque, reliability, motor efficiency over a wide range of operation, and low maintenance requirements are key selection points and are provided with the use of an AC induction motor; Solectria's ACgtx 20 accompanied with the AC 300 controller. The ACgtx20's high power-to-weight ratio contributes to weight reduction, an overall 1994 Penn State design goal. Appendix B lists complete specifications for the Solectria AC 300 controller and ACgtx 20 AC induction motor. To provide acceleration comparable to today's conventional vehicles high torque motors are necessary due to the Escort Conversion vehicle's weight. The ease of interfacing the controllers to the Penn State's microprocessor control scheme was a critical requirement. The Solectria AC 300 controllers provide this requirement with control complexity and number of available outputs.

2.6.2 Efficiency

To achieve the greatest range possible from the battery charge, the total efficiency would theoretically be 100%. To realistically achieve the greatest range, the highest efficiency must be realized. The ACgtx20's high motor efficiency throughout the operating range enables a high percentage of the theoretical possible range to be achieved.

3. Tailpipe and Evaporative Emissions

The Kohler CH25 engine used in the HEV presented unique problems with respect to tailpipe emissions. Although the Kohler CH25 engine meets 1994 California Emissions Requirements for small (under 25 horsepower) utility engines, the engine-out emissions required large reductions. Necessary reductions in engine-out HC+NO$_x$, and CO are shown in Table I along with approximate engine-out emissions in each category.

Table I: Calculated Engine-Out Emissions Normalized to the Emissions Event Testing Schedule

	CO (g/mi)	HC+NO$_x$ (g/mi)
50-point bracket	3.4	1.41
Kohler CH25	266.5	7.54
California 1999 Small Utility Engine Standards	96.9	3.10

Note that small utility engine emissions are regulated differently; HC and NO$_x$ emissions are combined.

To meet the 50-point bracket for the Ford conversion class emissions scoring schedule (0.41 g/mi THC, 3.4 g/mi CO, and 1.0 g/mi NO$_x$), a three-way catalyst was donated by Catalytic Exhaust Products. The catalyst capacity was chosen to handle the emissions from small utility engines. However, the three-way catalyst requires closed-loop control of relative air/fuel ratio, λ, to achieve high conversion efficiency. Specific requirements of the three-way catalyst were given as $\lambda=0.99 \pm 0.005$. Neither the closed-loop control nor the necessary λ were possible with the stock CH25 carburetor.

To enable use of the three-way catalyst, a closed-loop system was fitted to the Kohler engine. A throttle-body fuel injector was adapted to fit the CH25 engine to replace the stock carburetor. To achieve proper feedback, a speed-density system was used in conjunction with a λ-sensor mounted close to the exhaust ports.

Evaporative emissions remained as in the stock Escort with the exception of the carbon canister which was removed from the engine compartment in the interest of space.

4. Vehicle Design or Modifications Including Suspension and Structure

The suspension design and geometry the stock Ford Escort suspension with minor modifications. The suspension modifications were selected to accommodate the increased vehicle weight. Escort GT rear disc brakes replaced the drum brakes to improve stopping distance and reduce fade. The suspension stiffness and performance was enhanced by replacing existing components with KYB GR-2 Heavy Duty Gas struts and specially designed springs from Coil Springs Specialties.

The vehicle structure is stock Ford Escort with minor modifications in the rear. The frame rails aft of the rear-strut towers have been strengthened by the addition of 1/8 inch angle iron. The angle was welded to the inside of each frame rail to accommodate last year's drive train. The rear cross member between the rear-strut towers was also strengthened through the addition of 1/8 inch angle iron. The two pieces of angle were welded on the underside of the crossmember between the frame and the crossmember to increase the load-carrying capacity and the rigidity. Although these modifications were initially for last year's design, they have been left in place to increase the vehicle structure's strength to compensate for the 10%-15% increase in stock gross vehicle weight. The floor pan aft of the trailing arm mount was removed to allow for battery box and fuel tank placement. The floor pan was replaced with 18 gauge steel; equivalent to stock floor pan.

The stock gas tank has been replaced with a fuel cell. The battery container is located in the area which the stock gas tank previously occupied to locate the largest weight component close to the center of the vehicle. This meant the gas tank had to be placed aft of the rear suspension. Due to this fact, a fuel cell was utilized in the interest of safety. This is an eight gallon unit and 17"x17"x8" dimension with a sending unit used for fuel measurement.

The stock power brake system needs a constant vacuum to operate and this can not be provided in ZEV mode or even in HEV mode with a small APU. To solve this problem a small low power electric pump creates a vacuum, stored in a small canister, to provide the necessary vacuum for the power brakes in all operating modes.

Roll over protection was in accordance to section 19.7 in the 1994 Rules and Regulations and no other vehicle structural modifications were performed.

The Penn State HEV was developed to simulate a conventional vehicle driving environment. The stock automatic transmission shifter was employed as a transmission position selector switch and a key was retained to start (power up) the vehicle. These subtle details should help to bridge the gap between conventional and pure electric vehicles while providing a familiar driving environment.

5. Control Strategy

Vehicle control is achieved via two Little Giant microprocessors. The main microprocessor controls the performance, mode selection, and motor controllers. As in a regular vehicle, the driver has the option of down shifting and overdrive. This option allows the microprocessor to determine the batteries recharging schedule. A selection switch for ZEV, HEV, and APU modes will be utilized to enable or disenable the APU recharging system as required in the rules. Gas and brake pedal sensors allow for the control of the motor controllers. The gas pedal determines the speed of the motors via sending position and acceleration information of the pedal to the microprocessor which in turn determines the current that the motors draw. The brake sensor informs the microprocessor to the degree of regenerative braking. The other processor controls the APU recharging system. This processor receives information from the mode selector switch, motor controller, microprocessors' battery, Kohler starting battery, engine RPM, and performance selector. The outgoing information from the processor starts the motor, informs the throttle controller to accelerate the engine to the desired rpm, and turns on an Led on the dash to inform the driver that the recharging system is on. The performance selector determines the recharging strategy which will includes maximum vehicle efficiency, energy economy, or minimum recharge time for performance and city driving.

The main microprocessor will consist of a 12 MHz Z80 with 512K of RAM. Its primary function will be controlling the drive motors (One per front wheel) as efficiency as possible. To achieve this maximum efficiency, (and therefore range), the vehicle will utilize a ELECTRONIC DIFFERENTIAL. To employ this scheme, the processor will be aware of the load (stress) on each motor, speed of each motor, and the angle of the steering wheel. The software will run in a multitasking environment, with the high priority task being the differential (To minimize lag while turning). The next lower priority task will be managing the braking, with its goal being 100% regenerative, minimizing energy loss to heat(conventional brakes). Other Running tasks on this processor will consist of displaying vital statistics on a 80 x4 character display (On-Dash), watching for system faults, and controlling driving modes.

The secondary processor will consist of a 9 MHz Z80 with 256K of RAM. Its primary function will be controlling the APU and charging system. Again, its main goal will be to achieve maximum efficiency. It will monitor the drive batteries as well as the secondary power batteries. This processor will contain extensive data consisting of the battery charge/discharge rates, and data collected since the last recharge. Using this data, the processor will determine the optimum time and throttle to activate the APU.

Power for the processors will be supplied by a secondary, monitored, battery and isolation system. The purpose of this is to supply stable, non-rippled, surgeless power to the CPU's. This source will be coupled to the APU and the APU's battery. The APU will monitor this secondary source, charging it when necessary.

6. Choice of Materials

The design required a minimization of the vehicle weight. To save weight, many transmission parts were fabricated with aluminum and unnecessary material was removed from the structure where possible. Large pulleys made of surface treated aluminum were ordered from Gates and the side plates where milled from aluminum. Small gears and shafting were ordered in steel since the additional cost to weight reduction was not practical.

The engine compartment frame, which supports the two double belt transmissions, was also constructed of aluminum angle. The battery compartment was also designed and constructed out of fiberglass reinforced plastic.

7. Vehicle Manufacturability

The utilization of the Escort body and a division into subsystems provides for a highly manufacturable vehicle. The major components added to the vehicle are designed so as to be installed in sections. The transmissions can be installed or removed individually and the Kohler engine mounts on the support designed to hold itself and the transmissions. The mount for the engine compartment was designed for ease of installation by having some welded and some bolted together pieces. The micro-processor container was designed with concept of ease of installation and removal from the vehicle. All of the connections are made on the box with connectors. When the connectors and restraining hardware are removed that the unit can be removed from under the dashboard. The battery enclosure was a simple design that could be manufactured in two pieces and placed into the car. The enclosure was also designed for ease of manufacturing and installation.

8. Conclusion

The 1994 Penn State HEV has achieved the design goals through unique and practical engineering of vital components. The utilization of the designed double reduction transmission provided the necessary weight reduction while obtaining an excellent efficiency. Additionally, an electronic differential and a custom fuel injection for the Kohler Command 25 were integrated into the design. An engine compartment modularization concept allows ease of maintenance and installation of transmission and APU. Attention to weight reduction was evident in every designed aspect of the vehicle with the main concentrations in the drive system and battery compartment. Overall, the design resulted in a cost effective, practical, and manufacturable vehicle with available technology.

Appendix A: Double Reduction Transmission Illustration and Load Diagram

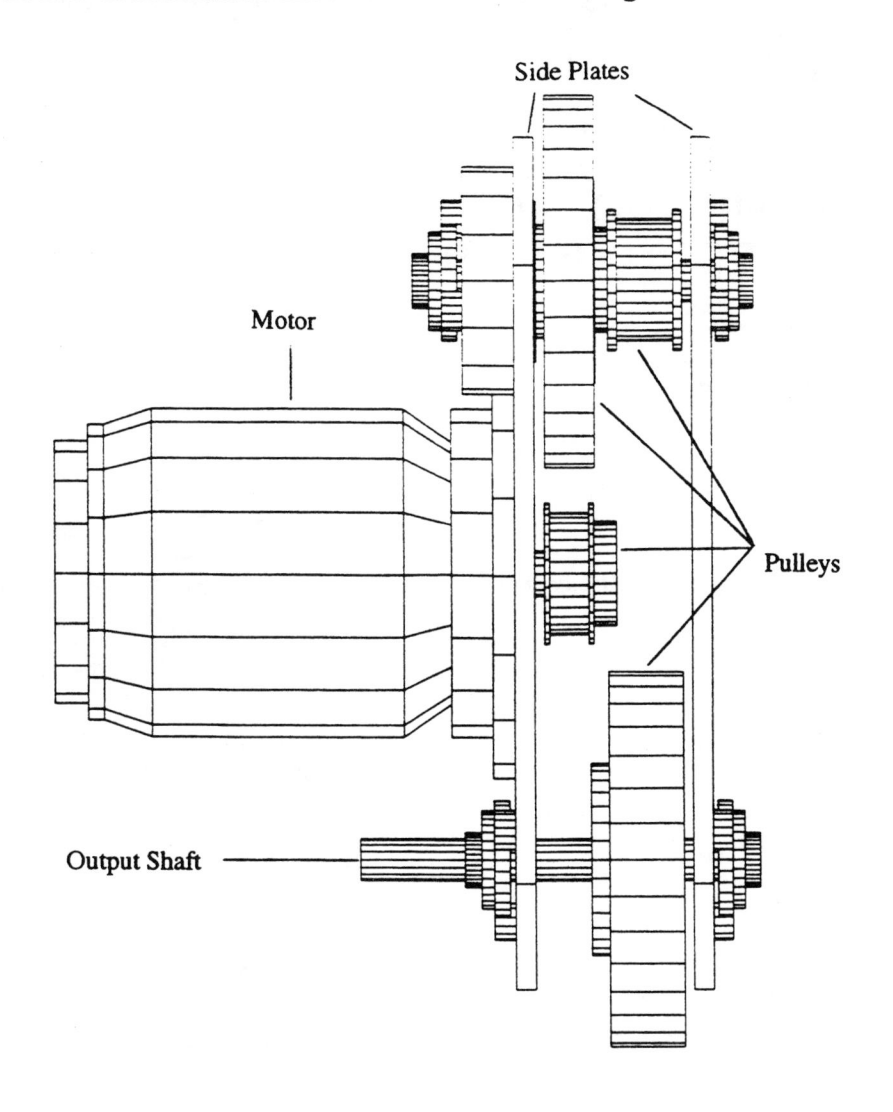

Double Reduction Transmission: Side View

PENN STATE SAE HEV BELT DRIVE BEARING LOAD DIAGRAM

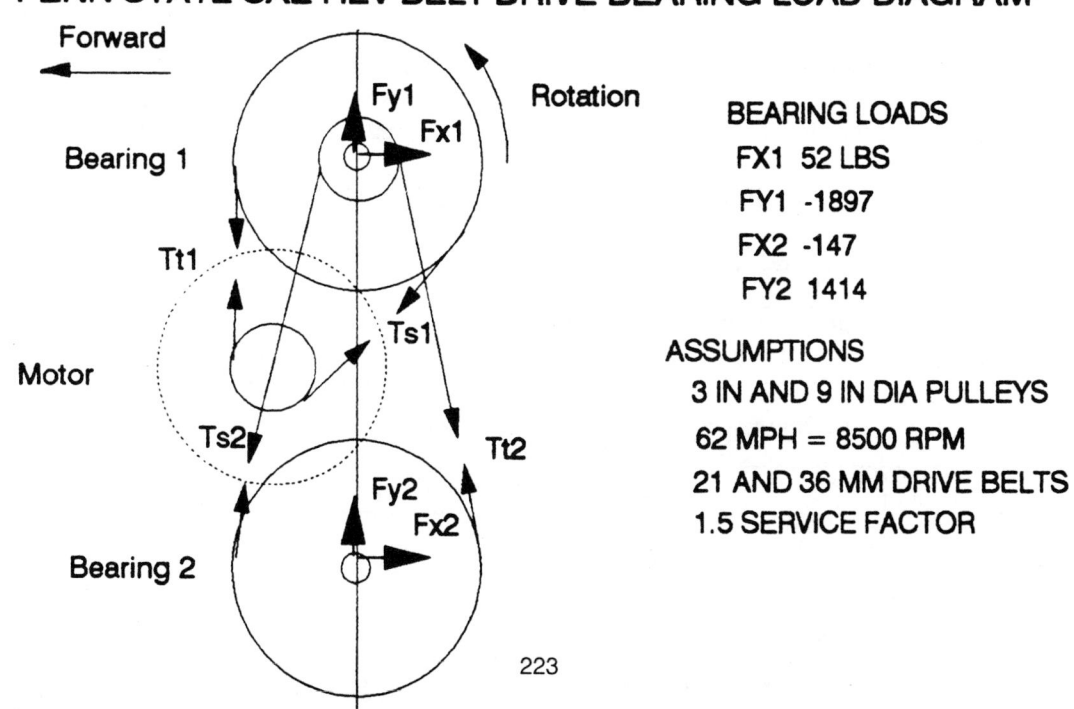

BEARING LOADS

FX1 52 LBS
FY1 -1897
FX2 -147
FY2 1414

ASSUMPTIONS

3 IN AND 9 IN DIA PULLEYS
62 MPH = 8500 RPM
21 AND 36 MM DRIVE BELTS
1.5 SERVICE FACTOR

Appendix B: Component Specifications

GNB

ELECTRIC VEHICLE BATTERY

HIGH ENERGY DENSITY - DEEP CYCLE - VALVE REGULATED BATTERY FOR ELECTRIC VEHICLES / TRACTION APPLICATIONS

12-EVB-1180

6 Cell, 12 Volt Electric Vehicle Battery

1180 Watt-Hours at 3-Hour Discharge Rate

Valve Regulated Lead Acid (VRLA) Technology

Oxygen Recombination Cycle

Can be Configured in Series and Parallel Strings to Match Any Vehicle System Voltage

INNOVATIVE FEATURES

VRLA Operation
- Never requires watering
- Spillproof
- Tested for leak tightness
- Operates at low internal pressure
- Minimal gas evolution on charge

Immobilized Electrolyte
- Installation upright or on the side
- Absorbent glass mat design
- Low resistance construction
- Survives freezing

Proprietary Positive Grid Alloy
- Deep cycle capability
- Low self-discharge rate
- Long life

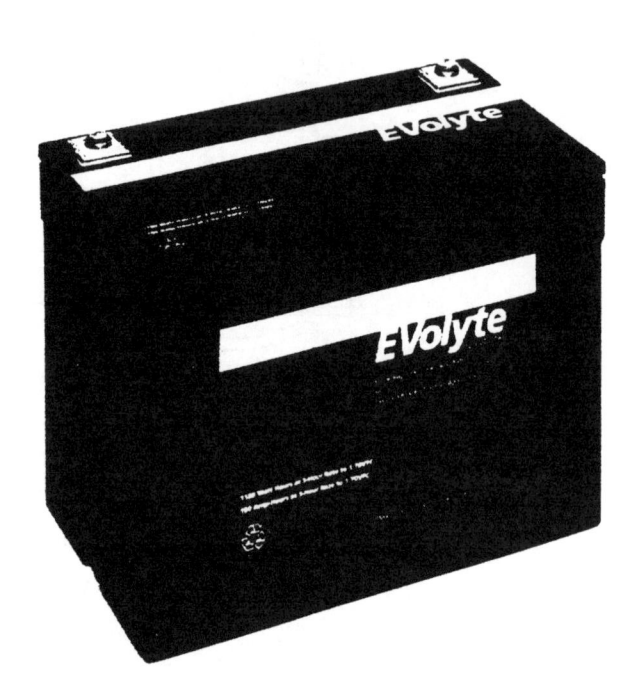

SPECIFICATIONS

Container and Cover - Polypropylene
Separators - Spun Glass, Microporous Matrix
Safety Vent - 3-8 psi, Self Resealing
Terminals - Threaded Heavy Duty Copper
Power Density - 170 W/kg
(@ 80% DOD)   435 W/liter

Positive Plate - Proprietary Low Antimony
Negative Plate - Lead Calcium
Self Discharge - 0.5-1.0% per week maximum
Estimated Cycle Life - 750 to 80% DOD
Energy Density - 39.2 Wh/kg
(@ C/3 Rate)   101 Wh/liter

PHYSICAL CHARACTERISTICS

Type	Length In	Length mm	Width In	Width mm	Height In	Height mm	Weight Lbs	Weight Kg
12-EVB-1180	12.1	308	6.9	175	8.7	221	66	30

ELECTRICAL PERFORMANCE (Preliminary)

Type	Cells Per Module	VDC Per Module	1	2	3	5	8
12-EVB-1180	6	12	75	90	100	105	110

Ah Capacity To 1.70 VPC @ Discharge Rate (Hrs)

Note: Design and/or specifications subject to change without notice. If questions arise, contact GNB for clarification.

SOLECTRIA
ELECTRIC / SOLAR-ASSISTED VEHICLE TECHNOLOGY

Dual-Drive Configurations
Peak Torque and Peak Power

Two AC200 Controllers

Motor Type	Nm	kW	HP
2 x AC12 (Δ)	2 x 35	28	36
2 x ACgtx20 (Δ)	2 x 45	28	36
2 x AC30 (Δ)	2 x 55	28	36

Two AC300 Controllers

2 x ACgtx20 (Δ)	2 x 55	42	56
2 x AC30 (Δ)	2 x 70	42	56

Two AC300-216V Controllers

2 x ACgtx20 (Δ)	2 x 50	56	68
2 x AC30 (Δ)	2 x 64	56	68

AC12   ACgtx20   AC30

INDUCTION MOTORS

Specifications	AC12	ACgtx20	AC30
Nominal power	4 kW	7 kW	8 kW
power w/AC200 (144 V)	14 kW	14 kW	14 kW
power w/AC300 (144 V)	--	21 kW	21 kW
power w/AC300-216V (216 V)	--	28 kW	28 kW
Nominal torque	20 Nm	20 Nm	30 Nm
torque w/AC200 (144 V)	35 Nm	45 Nm	55 Nm
torque w/AC300 (144 V)	--	55 Nm	70 Nm
Nominal speed	4,000 rpm	4,000 rpm	3,000 rpm
Maximum speed	12,000 rpm	12,000 rpm	9,500 rpm
Weight	51 lb.	66 lb.	93 lb.

Single-motor 144V systems range from 18-28 HP, dual motor systems 36-56 HP.
Single-motor 216V systems range from 18-34 HP, dual motor systems 56-68 HP.
Other system voltages and drive configurations are available; ask for assistance.

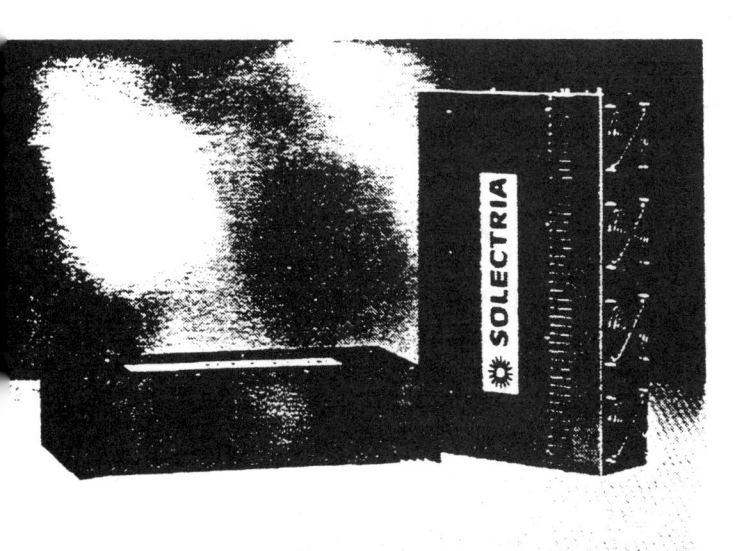

Applications
These motor and controller designs are particularly suited to commuter automobiles, trucks, buses and shuttles, industrial plant vehicles, airport service vehicles, and other applications where low-speed torque is an important requirement.

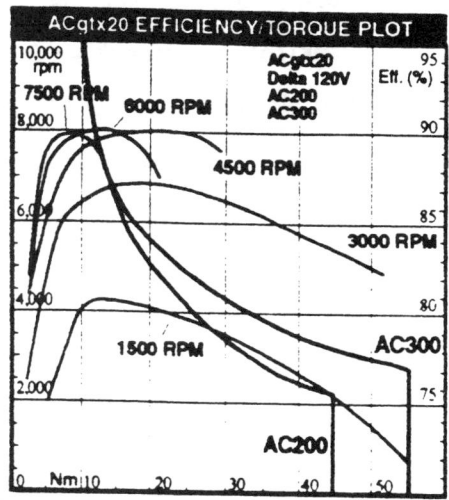

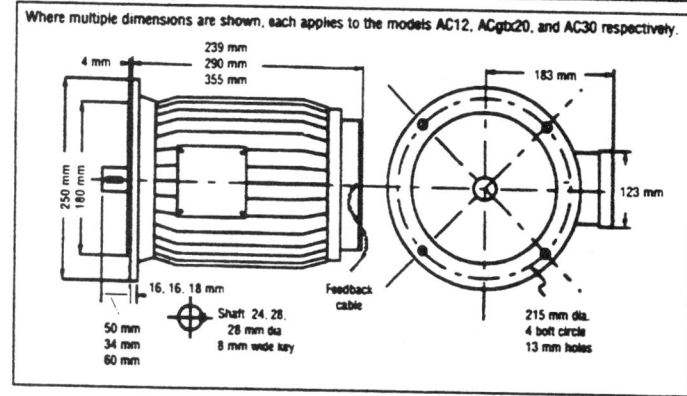

ACgtx20 EFFICIENCY/TORQUE PLOT — ACgtx20 Delta 120V AC200 AC300

Where multiple dimensions are shown, each applies to the models AC12, ACgtx20, and AC30 respectively.

AC200   AC300

AC INDUCTION MOTOR CONTROLLERS

Specifications	AC200	AC300	AC300-216V
Maximum power	14 kW	21 kW	28 kW
Nominal voltage	96-144 V	96-144 V	180-216 V
Safe operating range	70-170 V	70-170 V	150-260 V
Max. motor current (phase)	150 A	210 A	200 A
Max. battery current	100 A	160 A	140 A
Efficiency @ nominal power	98%	98%	98%
Efficiency @ full load	95%	96%	95%
Power for circuit electronics	12 W	12 W	12 W
Weight	19 lb.	20 lb.	20 lb.
Dimensions	17" x 9" x 4"	17" x 9" x 4"	17" x 9" x 4"
Operating temperature	-20 to +85°C	-20 to +85°C	-20 to +85°C

below by creating equation (2) using the general P_{rr} equation (C) in Fig. 2 and the constants of Fig. 1.

P_{rr} = CC_0(mass)(v) = 3.2 Hp @ 22.35 m/s at a GVW of 1100 Kg

2386.24 = (0.09706079)(1100)(22.35)   (Hp)

(2)   ∴ **P_{rr} = (0.09706079)(mass)(v)**   **(Hp)**

For the total power, the process is repeated using equation (A) from Fig. 2 and constants from Figure 1. The calculation below illustrates that the Ford supplied values for P_{d-air} and P_{rr} do not equal the value of P_{total}, thus a factor of 1.08 Hp is added to validate the equation. The factor's source is unknown and is assumed to be independent of velocity.

P_{total} = P_{d-air} + P_{rr} = 10.5 Hp @ 22.35 m/s at a GVW of 1100 Kg

10.5 = 6.22 + 3.2 + 1.08

(3)   ∴ **P_{total} = P_{d-air} + P_{rr} + 1.08**   **(Hp)**

Total power, P_{total}, needed to maintain the Escort at a constant speed can be calculated using equation (3). To accurately calculate the power required for the Escort in HEV configuration, the electric motor drive efficiencies must be considered. The efficiencies and the corrected equation (4) are listed in Fig. 3.

HEV Electric Drive Efficiencies:

  Motor Efficiency   = 91%
  Controller Efficiency = 98%

(4)   Power Required with Efficiencies = Power Required +[1 - (0.98)(0.91)] (Power Required)

Where: Power Required = P_{total} from equation (3)

Figure 3: Power Required with Efficiencies

equation (5), in Fig. 4, was utilized to relate the power requirement to an amp-hour (A.H.) battery rating. This rating can be used to estimate the electric vehicle's range without the use of the charging system.

---
(5)   A.H. Rating = (Electrical Power)/(Battery Pack Voltage)

Where: Electrical Power = Power Required with Efficiencies; equation (4) in Figure 3
Battery Pack Voltage = 144 Volts for 1993 Penn State HEV
A.H. Rating is at a one hour discharge rate

Figure 4: Amp-hour Rating, Equation (5)

A summary of the vehicle analysis for operation at various steady-state speeds can be seen in Table 1. This table provides required wattage ratings and a battery A.H. rating required to maintain a constant speed for one hour. Refer to Fig. 1 for horsepower to watt conversion factor. Also, the power required was calculated including 10% marginal factor.

The power required to maintain constant velocities has been calculated. This is significant since the generator must provide enough power to charge the batteries. This information represents the base lines for the generator's required power output at specific constant velocities.

Operational Speed Calculation

The v_d value supplied by GNB, vehicle's batteries manufacturer, stated not to exceed 170 Volts for the 144 Volt battery pack. That voltage supplies the quickest charge, a desired characteristic. To provided this v_d value, the generator's operational speed was calculated in the following manner.

The battery recharging DC voltage was related to the generator speed using equation (1).

(1) $v_d = 3V_s/\pi (1 - X_c I_d / V_s)$

Where: V_s = 28 Volts / 1000 rpm ≡ Generator(motor) peak line-to-line voltage
X_c = 32μH ≡ Motor Commutating Reactance line-to-line
π = 3.1416

The unknown in equation (1) was I_d. It was related to the DC voltage using the general power equation (2).

$$(2) \quad P = v_d I_d$$

Where: P = 14.9 kWatts ≡ Rated Motor Power or desired output

From equation (2), I_d = 87 Amps. This value is the current supplied to the batteries if the motor produces 14.9 kW and the rectified DC voltage is 170 Volts. The generator's operational speed to output these values is obtained by substituting I_d = 87 Amps back into equation (1) and solving for V_s. The value for V_s calculates to equal 178V. Since the DR 156s specifies a peak line-to-line voltage equal to 28 Volts / 1000 rpm, the corresponding speed for the calculated V_s value is 6357 rpm or ≈ 6400 rpm.

The generator's operational speed, to maximize charging capability, was 6400 rpm.

Table 1: Power Requirements at Constant Velocities

VELOCITY			POWER REQUIRED (Hp)				POWER REQUIRED W/ EFFICIENCIES	POWER REQUIRED W/ 10% ERROR FACTOR	A.H. RATING AT 1 HOUR RATE
mph	m/s	Pd-air	Prr	Ptotal	Watts		Watts	Watts	A.H.
5	2.2352	0.006	0.503	1.589	1185		1313.33	1444.66	10.03
10	4.4704	0.050	1.006	2.136	1593		1765.01	1941.52	13.48
15	6.7056	0.168	1.509	2.757	2056		2278.40	2506.24	17.40
20	8.9408	0.398	2.012	3.490	2603		2884.33	3172.76	22.03
25	11.176	0.778	2.515	4.373	3261		3613.65	3975.01	27.60
30	13.411	1.344	3.018	5.442	4058		4497.22	4946.94	34.35
35	15.646	2.134	3.521	6.735	5022		5565.88	6122.47	42.52
40	17.882	3.185	4.024	8.290	6182		6850.49	7535.54	52.33
45	20.117	4.536	4.527	10.143	7564		8381.89	9220.07	64.03
50	22.352	6.222	5.030	12.332	9196		10190.92	11210.02	77.85
55	24.587	8.281	5.533	14.894	11107		12308.45	13539.30	94.02
60	26.822	10.751	6.036	17.867	13324		14765.32	16241.85	112.79
65	29.058	13.669	6.539	21.288	15875		17592.37	19351.61	134.39
70	31.293	17.072	7.042	25.195	18788		20820.46	22902.51	159.05
75	33.528	20.998	7.545	29.624	22090		24480.44	26928.48	187.00
80	35.763	25.484	8.048	34.612	25810		28603.15	31463.46	218.50

The above values are for the standard escort at a GVW = 3812 lbs.
Shaded areas are values that are not allowed because of the 20 KWh limitation.

Appendix D: APU Requirements

The design calculations and assumptions made to determine the actual power requirement of the generator on the Kohler engine are outlined below. Calculations were performed according to the Kohler *Engine Application Guide* except where noted.

The basic formula for calculating actual power requirement is listed below as equation (1).

(1) $\quad P_{act} = P_{load}/(P_f \times P_t \times P_{trans} \times P_{alt})$
where
P_{act} = actual power requirement
P_{load} = maximum power of the load
P_f = power factor
P_t = temperature power factor
P_{trans} = transmission efficiency
P_{alt} = altitude power factor

The maximum power required by the generator for optimum charging has been previously calculated as 20 hp @ 6400 rpm and was substituted directly into equation (1).

The power factors used in equation (1) comply to the standards given in the *Engine Application Guide* except for the power factor, P_f, which is usually set at 80%. Our power factor was 90% because an 80% P_f causes the actual power requirement to exceed 25 hp, the maximum rated power of the engine. The temperature power factor, P_t, of 97% was derived by assuming a 107 °F maximum ambient air temperature for the engine compartment. The transmission power factor, 98%, was estimated from the reported 97%-99% efficiency of the Gates belt/pulley system used as the transmission. Neglected was the power factor for altitude change since we do not expect to run the car at high altitudes. Using these values, the actual power requirement is 23.4 hp.

Because the engine makes only about 23 hp at 3200 rpm, 0.5 times 6400 rpm, the nominal operating speed of the generator, the pulley diameter ratio must be changed from 2:1, as was used in the 1993 HEV Challenge. A better recommended setting for nominal engine speed is 3400 rpm where the engine produces roughly 24 hp. This setting allows for at least 200 rpm upward and some lower variation, depending on the generator load at lower speeds.

As a check, the torque output of the engine at 3400 rpm, roughly 37.5 ft-lb, exceeds the torque required to drive the generator, 36.1 ft-lb. The torque required to drive the generator was calculated as in equation (2).

(2) $\quad T = [(P_{act} \times 5252)/n] \times r_v$
where
P_{act} = 23.4 hp
n = 6400 rpm
r_v = 1.88

To be noted is the narrow margin for error in engine operating conditions. Thus, there is *no room for any compressor* for an HVAC system.

Seattle University - Pegasus

Dr. Jack D. Mattingly
Gary L. McMann
Michael A. Jackola
Shane D. Jackola

Lachlan Pope
Panaipon Uawithya
Xang K. Moua

Seattle University

ABSTRACT

Concerns, both environmental and economical, have facilitated the automotive industry's need to explore new and unconventional options in vehicle power. The Hybrid Electric Vehicle Challenge is an expedition into these new and exciting worlds of ideas. Seattle University has modified the vehicle (Mach 0.1) that was used in the 1993 Hybrid Electric Vehicle Competition. In an effort to make the vehicle more driveable and efficient, efforts have been put forth to enhance not only the functionality and efficiency, but the ergonomics and aesthetics as well. These efforts were culminated in the 1994 Seattle University Hybrid Electric Vehicle (Pegasus) which provides the desired functionality, efficiency, aesthetics and ergonomics. A design motto was used throughout the design process that consisted of design elegance through simple functionality. It was decided that complex components and/or complex integration of the components could escalate the problem, especially for a eight member team including two faculty advisors. Since the components that were utilized in the Mach 0.1 were excellent, it was decided that the work effort would be focused on enhancing the integration of these components into a more efficient, ergonomic, and functional vehicle. A computer simulation program (PEGASUS) used in the design process allowed these three goals to be attained by aiding in the establishment of a design configuration that was also capable of greater performance.

INTRODUCTION

BACKGROUND - The Hybrid Electric Vehicle (HEV) Challenge as launched in 1993 and continued in 1994 focuses on the challenging environmental and economical demands facing automobiles of the future. These demands are present with the ever growing pollution from automobile emissions and foreign dependency on oil. By seeking alternative power sources and methods of power transfer, the automobile industry will be in a better position to meet these future economical and environmental demands.

A hybrid electric vehicle utilizes two power sources in its operation. One source is entirely electrical in nature while a second source, an alternative power unit (APU), may be either a internal comubstion engine or fuel cell. A HEV can be configured in either a series or parallel configuration. The series configuration involves a design in which an electric motor supplies power to the road while the APU is used to charge the batteries or some other energy storage device. A parallel configuration is a design in which both power sources are connected to separate drivetrain(s) to supply power to the road. There are two different design approaches under the parallel configuration, split and combined. A split parallel design focuses on each power source being connected independently to separate drive trains. A combined parallel approach utilizes a coupling device to allow both power sources to contribute power to one drive mechanism to propel the vehicle. In addition, a hybrid electric vehicle can operate in three energy efficient modes. The first mode of operation denoted as the zero emissions vehicle (ZEV) mode utilizes the electric motor alone. A second mode referred to as the alternative power unit (APU) mode uses the alternative power unit alone. Finally, a third possibility is to utilize both the electric motor and the alternative power unit together under a hybrid electric vehicle (HEV) mode.

Seattle University's 1993 hybrid electric vehicle design team of 25 engineers finished extremely well in the Ford sponsored competition with honors such as "Best Overall Electrical Efficiency" and "4th Place Overall." The 1993 hybrid electric vehicle design called the Mach 0.1 combined selective research and design to comprise success. The efforts put forth by the 1993 team paved the way for innovation and improvements in the 1994 vehicle design.

DESCRIPTION OF 1993 HEV - For the 1993 competition, the Seattle University (SU) design team chose a split parallel drive system, as shown in Figure 1. A parallel configuration was determined to have a better ratio of fuel to road power efficiency than a series configuration since electrical power generation losses are avoided. In addition, a parallel configuration has the ability for more power since there are two sources that can supply power to the road simultaneously. Figure 1 shows the configuration of the 1993 vehicle.

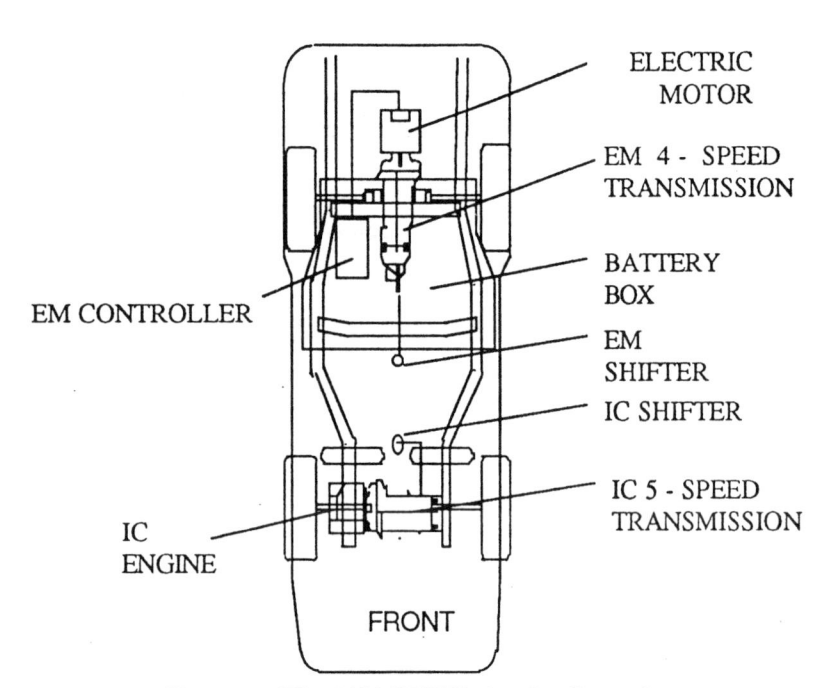

Figure 1 - The 1993 HEV Drive Configuration

The internal combustion engine drives the front wheels through a manual transmission, and the electric motor drives the rear wheels also through a manual transmission. Both systems are independent and may be operated separately (ZEV and APU modes) or simultaneously (HEV mode). The justification for this approach was to have four wheel drive capability and powertrain redundancy. If one power system failed the driver would be able to use the other.

DESIGN PROBLEMS OF 1993 HEV - Despite the success in last year's effort, the 1993 HEV Challenge and further vehicle testing exposed problem areas needing improvement for this years competition. The individual componentry selected for the vehicle were excellent, but time constraints on the 1993 team prevented proper integration of the components. Extensive testing and review of the 1993 vehicle and design documentation allowed us to identify the following problem areas:

Complicated Motor Integration - The operation of the vehicle extends over three modes: Zero Emission Vehicle mode (ZEV) on electric motor power only, Auxiliary Power Unit (APU) on internal combustion power only, and Hybrid Electric Vehicle (HEV) with the combination of power sources. The ability to control and "synchronize" both power sources in HEV mode is a complicated task. It requires complex mating of two power sources having different power, torque, and energy characteristics over different engine speed ranges. The 1993 design leaves the mating of these two engine curves in HEV mode to the drivers discretion lending to poor energy consumption and performance during vehicle operation.

Complicated Shifting Procedure - Because of the choice to use a split parallel system coupled with two manual transmissions, the 1993 design requires two shifters. This arrangement presents no problems in either ZEV or APU modes. In the case of HEV mode, the 1993 configuration requires a complicated shifting procedure to allow both drive systems to operate simultaneously and in harmony. Specifically, the driver must select the appropriate gears from two completely different transmissions with two entirely different gear ratios when driving the vehicle. This means that the driver must be educated and practiced in knowing which gears correspond to which vehicle speeds in each transmission. The driver must also have a complete understanding of how the drivetrain is functioning to avoid catastrophic failure of drivetrain components during operation.

Poor Driver Ergonomics - Ergonomic layout in the 1993 design was poor. The seats are moved all the way forward to make room for the batteries, leaving little or no leg room for the driver. Buttons and switches are in poor locations on the dash, allowing for the possibility for the driver to have inadvertent contact with them. The clutch is very stiff, resulting in difficult operation and extreme driver fatigue.

Problems with the Charging System - The charger purchased for the vehicle did not perform satisfactorily at the competition. Although it worked well during charges on campus at SU, it failed to perform with the charging station set up at the competition. In Detroit, a ground fault current detection device on the charger kept tripping the circuit preventing charging of the battery pack. To resolve this problem, two new chargers were purchased at the competition. The two chargers required the batteries to be re-configured into two groups each time the team charged the vehicles battery pack. This procedure required extensive re-wiring before and after each battery pack charging period.

Weight Distribution Problems - In the 1993 design, the front to rear weight distribution is 43%-57%, biased to the rear of the vehicle. This contributed to rear body sway in side to side transition maneuvers and emergency turn situations during testing.

Inadequate Clutches for the APU and Electric Motor - The clutches in the 1993 design for both power sources did not perform satisfactorly during testing. The APU clutch is inadequate for the mass of the vehicle. This clutch inadequacy greatly reduces the vehicles clutch life and in some cases prevents proper vehicle performance under extreme power request conditions. The clutch pedal pressure required is extremely high since the 1993 design actuates both clutch mechanisms sequentially. The electric motor clutch has hard spots on the friction surface initiating raised surface areas which reduce the overall clutch surface area.

Throttle Control - In the 1993 design the throttle was actuated using a cruise control vacuum servo system driven off the same potentiometer as the electric motor. This vacuum actuated APU throttle system performed poorly during testing with throttle progression being full on or full off in nature rather than linearly progressive. Testing of the system showed that full throttle pedal input yielded roughly 30% rotation of the throttle plate or dramatically reduced available APU power. This low power output could have also contributed to the heavy APU clutch wear observed during disassembly of the 1993 design.

Auxiliary Cooling Fans Power Control Problems - The DC/DC converters that run the cooling fans on the motor, controller, and the battery box are too small. In the 1993 design configuration includes one battery in the main pack isolated to run the auxiliary fan motors. This reduces the voltage available for use by the electric drive motor.

Battery Monitoring - The only monitoring of the batteries available in the 1993 design was the state of the pack as a whole. There was no ability to monitor the state of charge of each separate battery to determine the location of bad cells in the pack.

Data Gathering - The 1993 design provided no means to gather performance data on the vehicle during use for testing and validation purposes. There were no tests performed by the 1993 team and likewise there were no provisions into the 1993 design to allow for vehicle performance testing.

SCOPE and PURPOSE of 1994 HEV PROJECT - The purpose of the 1994 HEV project was to improve the 1993 design in as many ways as possible given available resources such as financial, technical, manpower, and funding. With a team of 6 engineers and a total budget of $20,000 (Competition costs included), this team produced a vehicle that was easier to drive and provided better performance and efficiency. Specifically, the goal of the 1994 design team was to improve the attractiveness of the vehicle for an ordinary driver by using only one shifter and providing more leg room, more convenient and easy to operate controls, and better instrumentation. We also wanted to obtain better handling, better acceleration, lower weight, increased electrical and APU efficiency, performance monitoring and data collection abilities, and improved power integration. With extensive documentation and heightened workmanship, we also planned to improve the ability to troubleshoot the vehicle in the event of a problem. As the rest of the report shows, Seattle University's 1994 Hybrid Electri Vehicle, which we call the **PEGUSUS**, meets all of our design goals

THE 1994 PEGASIS OVERVIEW

The 1994 design utilizes the main system components chosen by last years team for the 1993 configuration. For the 1994 design, the components were integrated using a dramatically different control strategy.

CONTROL SYSTEM OVERVIEW - To improve the performance, efficiency, and driver vehicle interface of the PEGASIS vehicle, a new and improved control strategy had to be designed and implemented. The new vehicle configuration utilizes the same UNIQ SR180P brushless DC electric motor with controller, the 1.0 litre Geo 3 cylinder internal combustion engine, and the battery pack consisting of (16) - 12-volt deep cycle DYNO 27M lead-acid batteries from the 1993 configuration. A markedly different design philosophy was taken in integrating the internal cumbustion engine (APU) power and electric motor (EM) power. This design involved a combined parallel drive system which synchronizes the belt-driven electric power and shaft-driven APU power onto the same input shaft leading to the stock Escort 5-speed manual transaxle. Utilizing these basic components, a control strategy was developed using a lap top computer and data acquisition card in conjunction with additional electronic circuitry to integrate the vehicles control system components. This system provides monitoring capabilities, real time system feedback via screen display, data collection capabilities, and performance and efficiency logic capabilities for electric motor (EM) power output. The layout of the vehicle's componentry is shown in Figure 2 below.

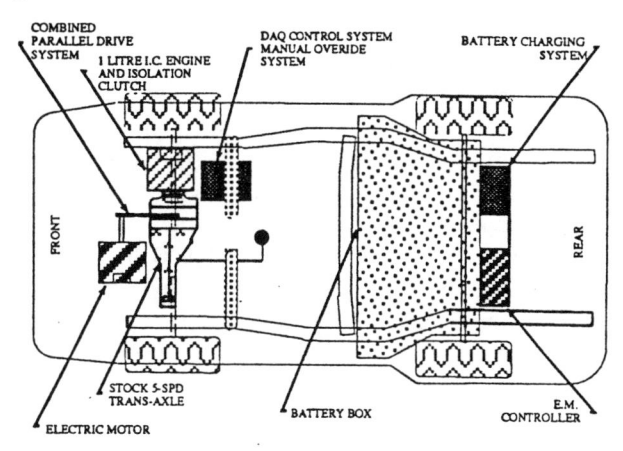

Figure 2 - 1994 PEGASIS HEV Configuration

OPERATION OF 1994 PEGASIS - The PEGASUS vehicle control strategy and layout has been designed so that the driver will operate the vehicle in a manner similar to that of the stock Ford Escort. Care was taken in the design to minimize the need for intensive driver training in operation of the vehicle. In addition, the vehicle control system is only active when the key to the vehicles ignition system is in the accessory position.

The vehicle has the ability to operate in Data Acquisition (DAQ) Operation or Manual Operation. In both DAQ and Manual Operation the vehicle has the ability to operate in three different modes. Those modes are ZEV, APU, and HEV. DAQ Control Operation turns over the control of the vehicle power sources to the computerized data acquisition system. Manual Control Operation provides for manual operation of the vehicle in the three modes in case the data-acquisition system fails. This system has been incorporated to act as a backup system in the event of a computer malfunction. If the data-acquisition system malfunctions, the driver will flip a toggle switch transferring operation of the vehicle to the manual control system. Whether in DAQ or Manual Operation, the vehicle may only be active in one of the three modes at any instant. The desired mode is selected using the three push-button switch array located on the dash to the right of the steering wheel.

MODES of OPERATION - The PEGASIS vehicle has been designed such that ZEV, APU or HEV mode can be activated at any instant in both DAQ and Manual Operation. The active mode is determined by the demographics in which the driver is driving and/or external driving conditions.

ZEV mode is activated when the vehicle is being driven in densely populated areas such as inner cities where lower vehicle emissions is desirable. In ZEV mode the vehicle is running solely on electric power while the APU is shut down.

In APU mode the internal combustion engine is the only

source of power. The electric motor is also running but is used to recharge the battery pack and does not apply any power to move the vehicle. This mode is primarily used for long distance driving at highway speeds and as a backup in the event of ZEV power failure.

HEV mode is activated when the driver needs additional performance out of the vehicle. In this mode the electric motor and internal combustion engine are simultaneously applying power to the same drive shaft in combined parallel fashion.

VEHICLE OPERATION - Due to the care taken in designing and building of the vehicle control system, most of the vehicle system control is done without input from the driver. Because the vehicle has additional control and operating features, the following guidelines need to be followed during vehicle operation.

After making sure the operation switch is in the DAQ operation position, the key to the vehicles ignition system must be rotated to the accessory position to activate the control system. Once this has been done, the driver may choose any one of the three modes of operation. To select a mode, the driver needs to take their foot off the gas pedal and push one of the three mode array buttons.

When APU or HEV mode is desired, the driver must let off the gas pedal and the ignition key needs to be rotated forward to activate the APU's starter. Once the APU is started, releasing the key will rotate the ignition to the accessory position thus enabling the driver to continue normal operation of the vehicle.

If ZEV mode is selected, the internal combustion engine is automatically turned off if switching from APU or HEV mode. The electric motor is then activated, if not already active if switching from HEV mode, so that normal vehicle operation can continue under electric only power.

ENGINEERING MODELING and DESIGN

What distinguishes back yard mechanics from engineering is the utilization of engineering principles to model and predict real world problems. The engineering approach was integrated into the design process of the PEGASUS HEV using engineering principles and computer modeling to predict the performance of different drivetrain configurations. The creation of a software program allowed us to analyze different drivetrain configurations and make sound engineering judgements in selecting the best solution to the HEV concept. Another benefit to this approach involved the selection of a Data Acquisition (DAQ)/laptop computer control system. Utilizing the DAQ/lap-top computer, we were able to use componentry such as potentiometers and mode selection switches, similar to the ones used in a HEV, to communicate with the computer and simulate vehicle operation. Many pieces of the software were used directly in the cars control software logic development. This design philosophy is summarized in Figure 3 below.

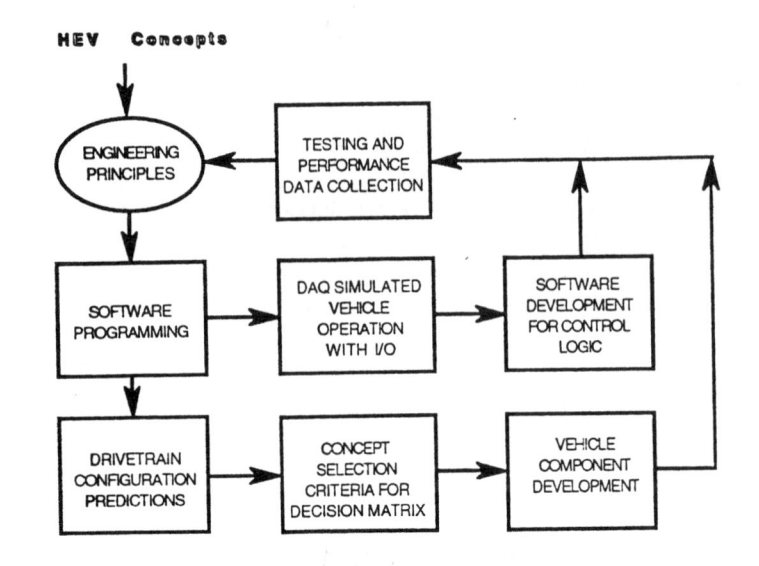

FIGURE 3- Engineering Design Philosophy

To accurately simulate the power requirements of a HEV, each force acting to restrict vehicle motion must be modeled separately and then combined for total road load. The dynamic analysis tools used in the PEGASUS modeling program are shown below.

DYNAMIC ANALYSIS of the HEV

The principal forces acting on a vehicle traveling up an incline are shown in Figure 4. These forces are:

- F_D Aerodynamic drag force
- F_P Propulsive force
- F_R Rolling resistance force
- N_F Front normal force
- N_R Rear normal force
- W Weight

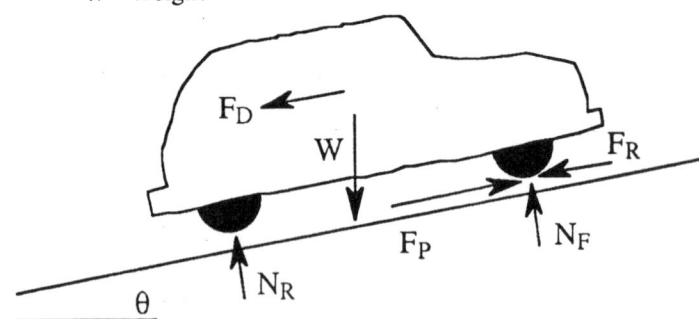

Figure 4 - Principle Forces on the Vehicle

The sum of those forces acting in the direction of travel equals the vehicle's mass times its acceleration or

$$F_P - F_D - F_R - W \sin \theta = \frac{W}{g} a \qquad (1)$$

where g is the acceleration of gravity. The aerodynamic drag force (F_D) can be expressed as

$$F_D = \frac{1}{2} \rho\, C_D\, A_F\, V_a^2 \qquad (2)$$

where ρ Density of air
  C_D Coefficient of drag for the vehicle
  A_F Frontal area of the vehicle
  V_a Velocity of the air relative to the vehicle
and the rolling resistance force (F_R) can be expressed as

$$F_R = W \cos \theta \left(K_0 + K_1 V + K_2 V^2 \right) \quad (3)$$

where K_0, K_1, and K_2 are rolling resistance coefficients and V is the vehicle velocity.

The different HEV configurations were simulated using equations (1), (2), and (3), along with the performance characteristics of 1.0 liter Geo (APU), the UNIQ 32kW electric motor/controller, and the different transmission/drive systems. The power characteristics of the APU and EM used in the PEGASUS program are shown in Figures 5 and 6, respectively. Using the computer models, a gear ratio of 1.25 was selected for the electric motor's belt drive system as optimum for HEV acceleration and range. The resulting input power and torque of the APU and EM to the 5-speed transmission are shown in Figure 7.

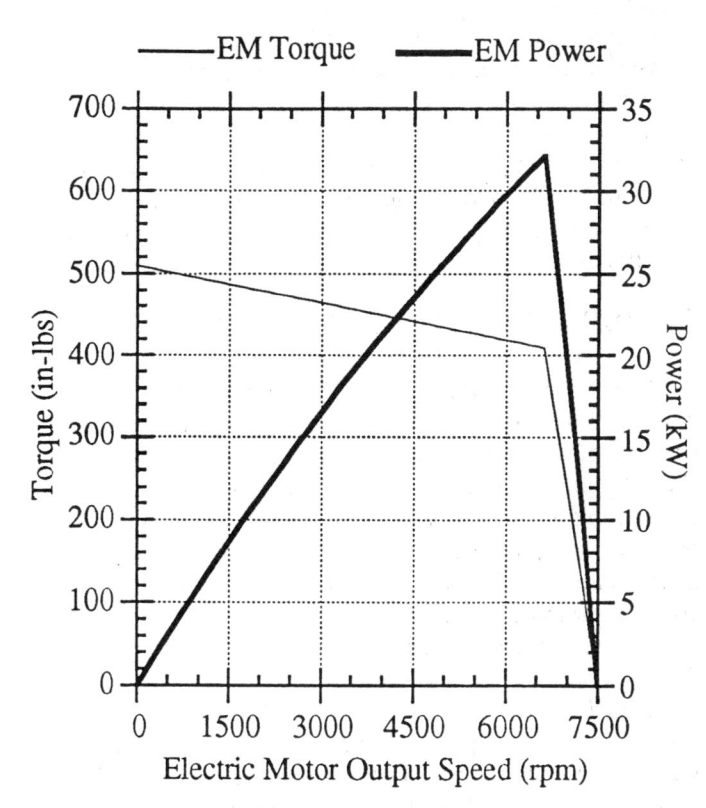

Figure 6 - Power Characteristics of Electric Motor

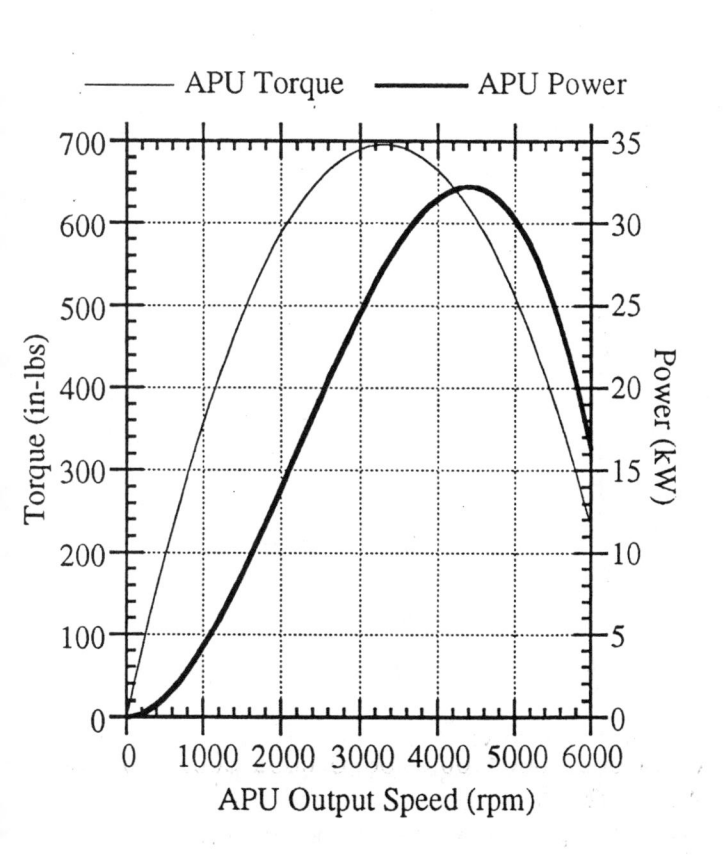

Figure 5 - Power Characteristics of APU

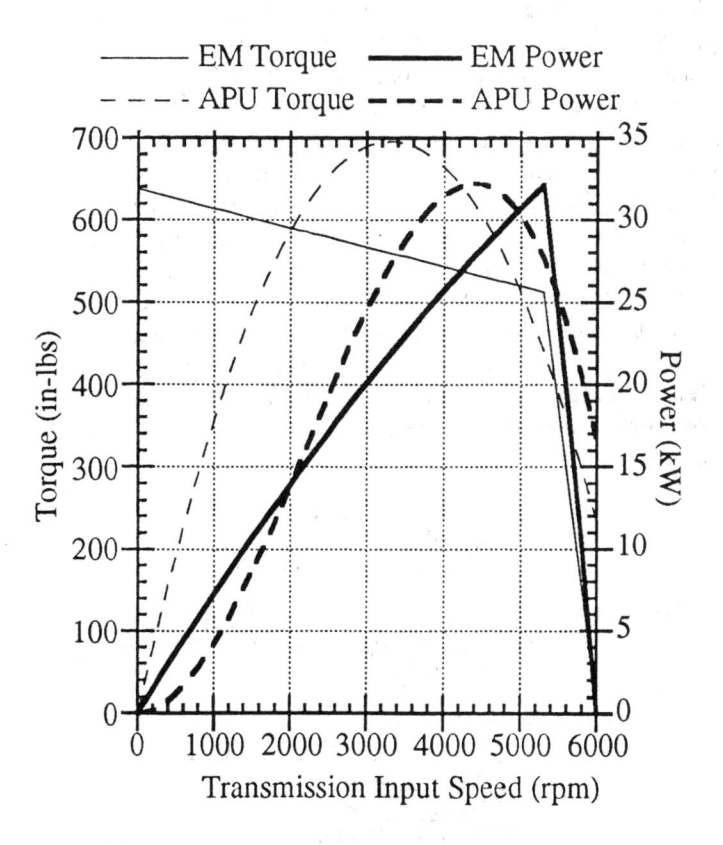

Figure 7 - Power Characteristics of EM & APU

The PEGASUS modeling program allowed us to visualize the effects of different drive system strategies by manipulating the vehicles power application abilities in the software. Since the vehicles road load characteristics remained relatively unaffected except for total vehicle weight changes, the results of each adjustment to the power application logic could be analyzed for significance in overall vehicle performance and efficiency. This proved to be a extremely effective tool in validating or dis-proving engineering predictions used in both drive system concept selection and control logic software development.

COMBINED PARALLEL SYSTEM DESIGN

The combined parallel drive system supports any of the three afore mentioned modes of operation. In HEV mode, the problem of integrating two distinctly different power curve characteristics onto one shaft to provide a broad smooth output power curve was resolved a combined parallel system design. In ZEV mode, only the EM would provide power to the transmission. In APU mode, only the IC engine would supply power to the transmission.

These components were needed to integrate the belt onto the input shaft to the transmission. A shaft extension was fabricated to transmit power through the bell housing extension. Secondly, a belt and related accessories were selected in a configuration compatible with the system. Third, bearings and support housings were design and selected to be placed strategically throughout the system accounting for the belt tension side load and thrust load from the mechanical clutch. To operate the vehicle in ZEV mode, the APU had to be physically isolated from the drive system to avoid compression rotation and inertial losses from the internal combustion engine. This was accomplished utilizing a freewheeling sprag type clutch. This clutch allows torque to be transmitted in one direction only, the direction of power application to the shaft extension. This is done with sprags that lock in one direction of rotation but over-run in the other. The sprags in the clutch are designed so that one horizontal (L1) is long enough to "jam" the inner and outer race together and allow torque to be transferred, while the other horizontal (L2) is shorter and does not allow contact, thus it over-runs (See Figure 10 below). The selected clutch utilized a keyway for torque transfer on the inner race allowing for a designed failure of the key prior to clutch or shaft failure.

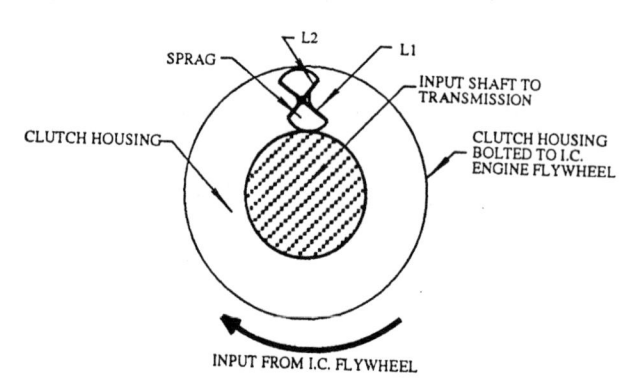

Figure 8- Sprag Overrunning Clutch

For shifting purposes in any mode of operation, the stock Escort manual clutch was selected since the stock 1.9 liter Ford engine had 108 Hp and the combined HEV horsepower is approximately 100 Hp.

ISOLATION CLUTCH DESIGN - The freewheeling sprag clutch met the requirements needed to isolate the APU from the combined parallel system during ZEV operation. This clutch was rated at a maximum continuous idling speed of 3200 RPM provided the operating environment involved oil lubrication and a projected 1000 hour clutch life. Since our design did not involve a sealed and oil lubricated operating environment, the clutch would have to be a self-contained, sealed, and grease lubricated. This type of clutch reduced the 1000 hour life as shown in the following equation provided by the manufacturer:

$$L = 0.7 (1000 (N_{max} / N)^{1.25})$$

Where: L = Clutch Life in hours
N_{max} = Maximum rated idling speed (RPM)
N = Operating speed (RPM)

A graphical representation of this equation is shown in Figure 9 below.

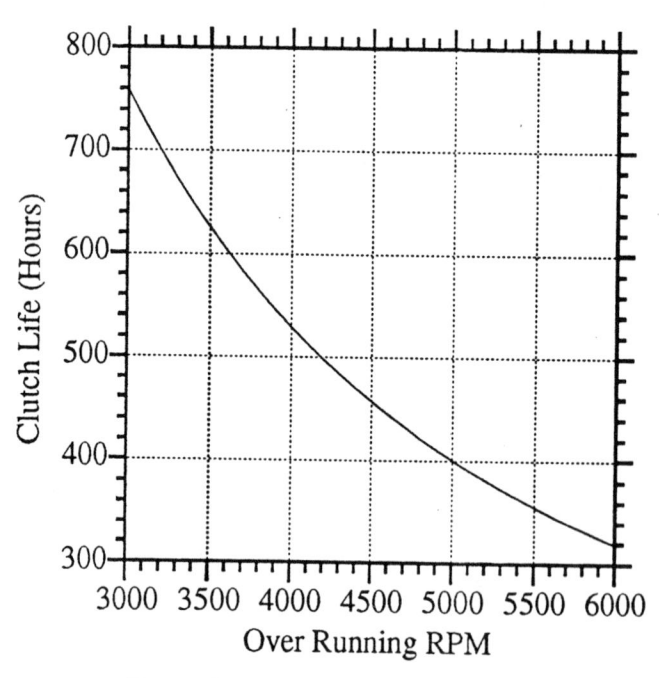

Figure 9 - Life of Over Running Clutch

The average overrunning clutch speed was estimated at 4000 RPM corresponding to a clutch life of approximately 500 hours. This is a very conservative estimate since the corresponding EM RPM from the belt drive ratio of 1.25 is 5000 RPM, at the top of the normal E.M. operating range. Since the clutch will only overrun in ZEV mode, at a typical urban speed of 30 MPH this corresponds to 15,000 ZEV miles.

Looking at the range proportion of ZEV (30 miles) to HEV (300+) miles, total vehicle life can be extrapolated to approximately 150,000 miles in terms of the overrunning clutch life. This was within the requirements of our design parameters. The peak torque rating for the clutch is manufacturer specified at 117 ft-lbs with a 12-13% safety factor. Since the shaft stress analysis utilized 2 times the APU's peak torque (116 ft-lbs) with a safety factor of 1.4 for worst case loading, the clutch provides a second protective mode of failure for the shaft extension after the clutch's inner shear key. This is important since the shaft extension is extremely complex and costly to replace.

BELT DRIVE COMPONENT DESIGN - The selected belt drive system was a Gates Poly Chain GT Drive. This Poly Chain GT belt utilizes tensile Kevlar fiber reinforcement and is temperature tolerant, corrosion and abrasion resistent, and light weight in construction. This belt occupies less space, has low maintenance, increased reliability, increased performance over conventional V-belts. Since this proposed drive system is not a conventional belt drive system, Gates sized the belt from performance and size limitations provided by us. To avoid bending moments on the shaft, a belt pull support bearing and bearing housing was indexed inside the driven sheave (see Figure 10). This sheave complies with Gates specifications and meets the space limitations imposed by combined parallel drive system design. A Driver and Driven sheaves for 62mm belts were machined down to accomodate the 21 mm belt sized by Gates. The driven sheave was bolted to the mechanical flywheel as shown in Figure 10 with the driver sheave keyed directly onto the EM.ouput shaft. The driver sheave on the electric motor side was selected to provide the desired gear ratio of 1.25 established by the PEGASUS modeling program. This belt drive system is the best compromise between space limitations, noise considerations, performance, and loading criterion. The Electric motor mount mounting system shown in Figure 10 was designed to attach to the bell housing extension with tightening mechanisms similar to those on a typical automotive alternator belt system.

BEARING DESIGN - The bearing selected to accommodate the 400lb belt side load specified by Gates was a double sealed double axis ball bearing. This bearing was geometrically and dynamically sized to be indexed and centered inside the belt sheave as shown on the Figure 10 to eliminate bending moments on the shaft extension. An additional bearing was needed to accommodate the 970lb thrust load caused by actuation of the mechanical clutch. The selected bearing was also a double sealed double axis ball bearing located within the APU flywheel. Another bearing was used as a pilot between the shaft extension and the transmission input shaft as shown in Figure 10. The thrust and pilot bearings are also used to aid in keeping the shaft extension, APU crankshaft, and transmission input shaft concentric.

SHAFT EXTENSION - The geometry of the shaft extension design was governed by other components of the combined parallel drive system. The Overrunning clutch with it's inner race keyway limited the shaft diameter to 15mm or 0.5906 inches. The shaft extension also required several steps in diameter to accommodate assembly of the thrust bearing, freewheeling clutch, radial load bearing, and finally the belt sheave (see Figure 10). A stress analysis was performed under these physical constraints utilizing a peak torque rating of 2 times the IC engines peak output torque of 58 ft-lbs. This was used to encompass any torque spikes associated with inertial loadings caused by rapid mechanical clutch release. This stress analysis involved stress concentration factors from the keyway, and the increases in diameter between sections of the shaft extension such as where the it flairs out to accomodate the driven sheave. The stress concentration for a sled runner type keyway was listed at 1.38, and a stress concentration factor for the step from the radial load bearing diameter to the belt sheave diameter of about 1.8, erring on the side of safety. Since this step in the shaft extension has a higher stress concentration factor as well as an out of plane stress to the "skin" of the shaft from the press fit of the belt pull support bearing, it was determined to be the governing case. A Mohr's circle analysis determined the maximum shear stress of 62,030 psi to be in-plane and be the limiting factor. Using the distortion energy criteria, this corresponds to a maximum shear stress of .577 times the maximum normal stress. This yielded maximum normal stress of 107,504 psi. Normalized and Tempered 4340 Aircraft Quality steel with a yield point of 150 ksi yielded a worst case factor of safety of 1.40. The best case factor of safety is 3.12 using only the IC engines peak torque. The clutch shear key has been designed to use 7075-T6 aluminum for torque induces shear failure before that of the shaft. The aluminum will not score the complex and costly to replace shaft extension allowing repair by a new aluminum key.

Bolt patterns were then designed for the freewheeling sprag clutch housing and the driven sheave attachment to the shaft extension as shown in Figure 10. These patterns were determined using the maximum applied torque as producing a negligable shear stress on the 6-bolt patterns.

The clutch housing was designed to take the inertial loadings from the rotation as well as the the freewheeling sprage clutch's outside diameter press fit (see Figure 10). The inertial loading was calculated using an order of magnitude approximation for the maximum radial stress at the outermost section of the housing. It was determined to be in the vicinity of 1500 psi and assumed to be negligible. The radial and tangential stresses in the housing were determined using the equations for an internal pressure from the press fit of the overrunning clutch. A Mohr's circle evaluation determined the minimum outside diameter needed to produce a favorable stress factor of safety of 2 for a 4340 steel housing.

BELT PULL SUPPORT HOUSING - The radial load bearing housing is a integral portion of the bell housing extension. A stress analysis was done to account for the 400 pound side load from the belt as well as for the press fit from the bearing. A stress analysis was performed similar to that of the clutch housing showing it to fail in shear. A factor of safety of about 3 resulted using 7075-T6 aluminum.

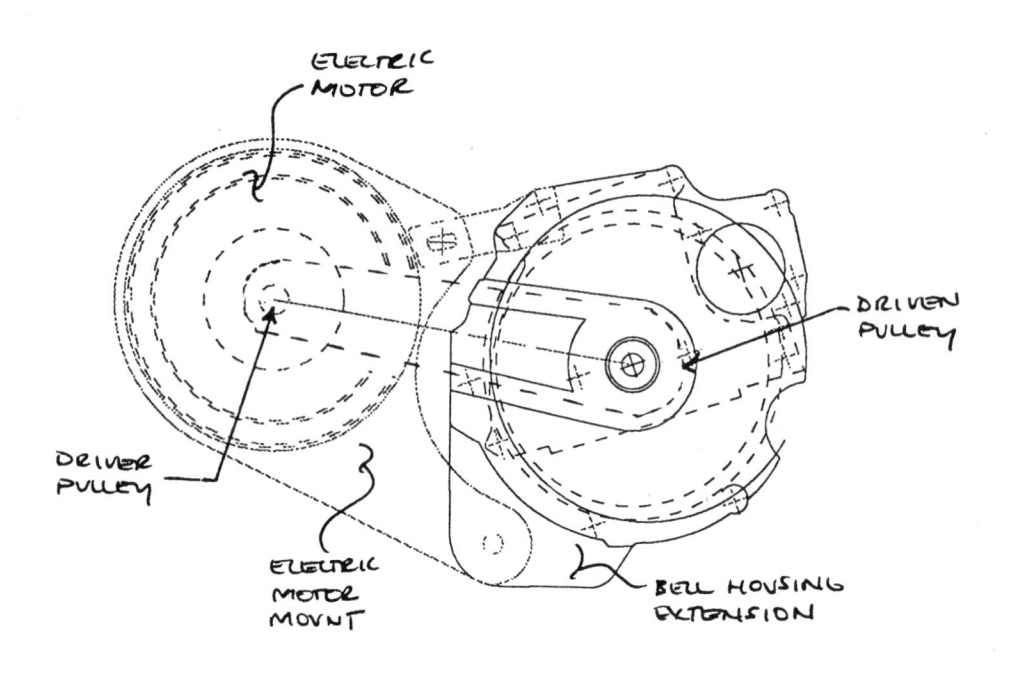

Bell Housing End View

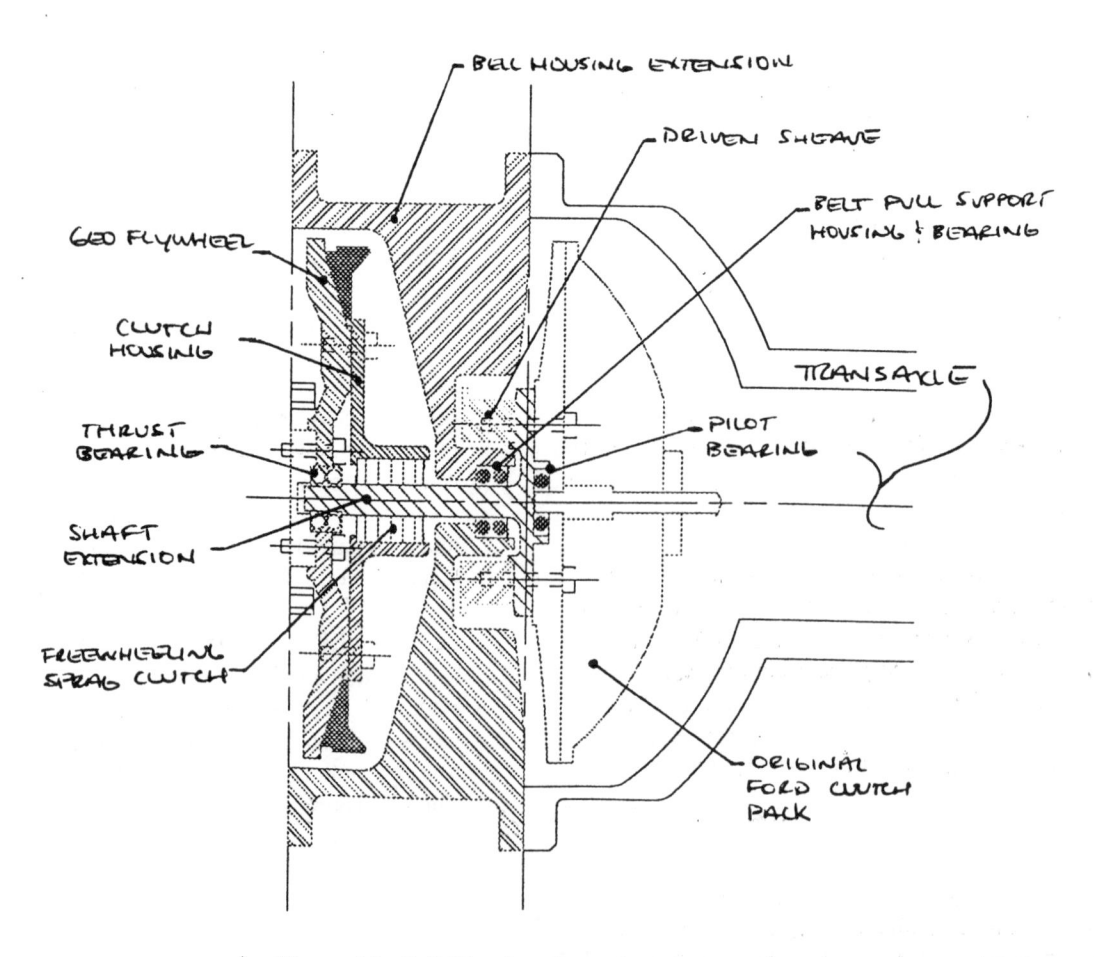

Figure 10 - Bell Housing Cross Section

DAQ CONTROL and MONITORING SYSTEM

The brain of the 1994 HEV control system is a National Instruments, Data-Acquisition (DAQ) Card 700. The DAQ-700 is a PCMCIA (Personal Computer Memory Card Interface Adaptor) card that slips into the PCMCIA port of an AST Bravo Notebook Computer. The DAQ-700 card reads both analog and digital inputs from peripheral sensors supplied to it by external electronic circuitry. The DAQ-700 card, used in the 1994 HEV control system and operating inside the Bravo lap-top computer, is mounted in front of the passenger seat (see Figure 11). The DAQ-700 card receives input data from electronic circuits. These circuits are broken down into systems for controlling the vehicle in APU, HEV, and ZEV modes of operation.

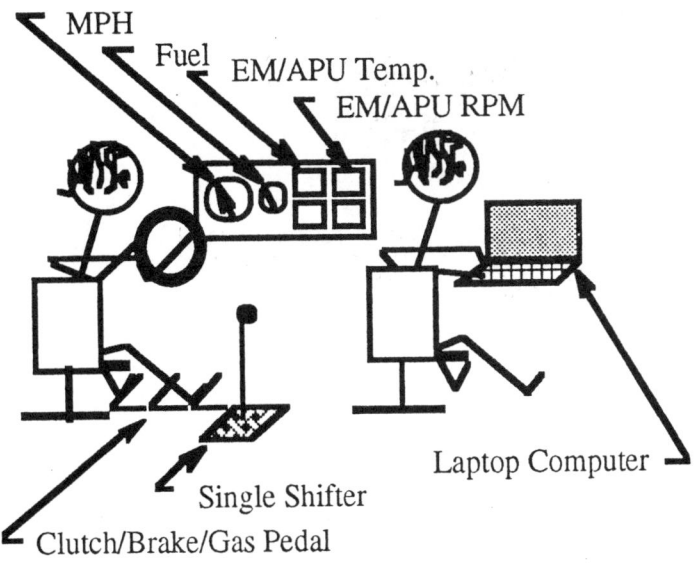

Figure 11 - Lap-Top Computer Location

ELECTRICAL HARDWARE DESIGN - The electrical control system was designed so that the vehicle would operate under DAQ operation as well as manual override operation. Automated control of the electric motor and internal combustion engine will occur using the DAQ-700 card. In the event of an unforeseen computer malfunction, the control of the electric motor and internal combustion engine will be done in a manual override fashion. Under manual override, the performance and operation of the vehicle will rely on the ability of the driver rather than the automated DAQ control system. This design approach was taken in order to ensure that the vehicle would be able to complete the competition if the DAQ control system cannot be troubleshooted and fixed. Therefore the electrical circuitry incorporates components that will enable the vehicle to operate in both DAQ and manual override operation.

THROTTLE CONTROL - This system, illustrated in Figure 12 showing mechanical and electrical hardware, ensures that the IC and EM throttles operate in APU, HEV, and ZEV modes. The IC throttle is controlled by a cable linkage from the gas pedal to the fuel injector throttle body. The fuel injector throttle is actuated from 0-100% open over the first 50% of the gas pedal travel. The electric motor throttle is controlled by the electric motor controller desired speed input. The speed input is a 0-10 volt analog signal, 0 volts representing zero desired speed and 10 volts representing full desired speed. The 0-10 volt analog signal originates at a potentiometer connected to the vehicles gas pedal by a second throttle cable. The potentiometer outputs a 0 volt signal when the gas pedal is not being pressed, and increases to ten volts as the gas pedal is pressed to the floor. When the driver is operating the vehicle in DAQ Operation, the 0-10 volt signal is routed to the DAQ-700 card to control the electric motor speed using a software program to control the vehicle for maximum performance. Under manual operation, the voltage signal is routed to the electric motor controller desired speed input directly from the potentiometer while bypassing the DAQ. The routing of the signal is determined by the DAQ/Manual Operation switch located on the vehicle control panel.

MODE SELECTION - Figure 13 is the circuitry used to control modes selection. This circuit ensures that the peripherals needed for vehicle operation are activated and deactivated in all three modes while under DAQ and Manual Override Operation. The mode select switch is a three button switch that enables the driver to choose only one mode of operation at a time. The mode switch incorporates an interlock with lockout feature. This feature ensures that no two buttons will be activated at once. The mode select system sends a digital signal to the DAQ-700 card notifying the software program which mode the driver has selected. When the driver chooses a mode, a five volt signal is sent to the DAQ-700 card, representing a digital one, notifying it which mode has been selected. The mode select circuit activates and deactivates the IC engine fuel pump and battery pack circuit depending on which of the three modes the driver has selected.

EM and IC ENGINE RPM's - The EM RPM is taken from the hall effect clock output on the electric motor controller. The hall effect clock is a TTL signal level (0- +5volts DC) which pulses at the motor rotational frequency. As shown in Figure 16, the hall effect clock signal is used as an input to a frequency to voltage convertor. The frequency to voltage convertor changes the frequnecy signal, representing motor rotational speed, into an analog voltage level between zero and ten volts. Zero volts represents zero RPM and ten volts represents 7000 RPM for the electric motor speed. The resulting analog voltage level is sent to the DAQ-700 card as an RPM for the electric motor.

The APU RPM is sensed directly off of the internal combusiton (IC) engines distributer. The IC engines distributer pulses a voltage level every time the ignition fires. Since the pulses can sometime be in the kilovolt range, which would damage the DAQ if inputed directly, a filter circuit was designed to suppress the high voltage spikes and produce a TTL (0 - +5 volt dc) level frequency signal. This signal is used as an input to another frequency to voltage converter which translates the varying frequency of the IC ignition into an analog DC voltage level. The anolog signal is sent to the DAQ-700 card to represent the IC engine RPM.

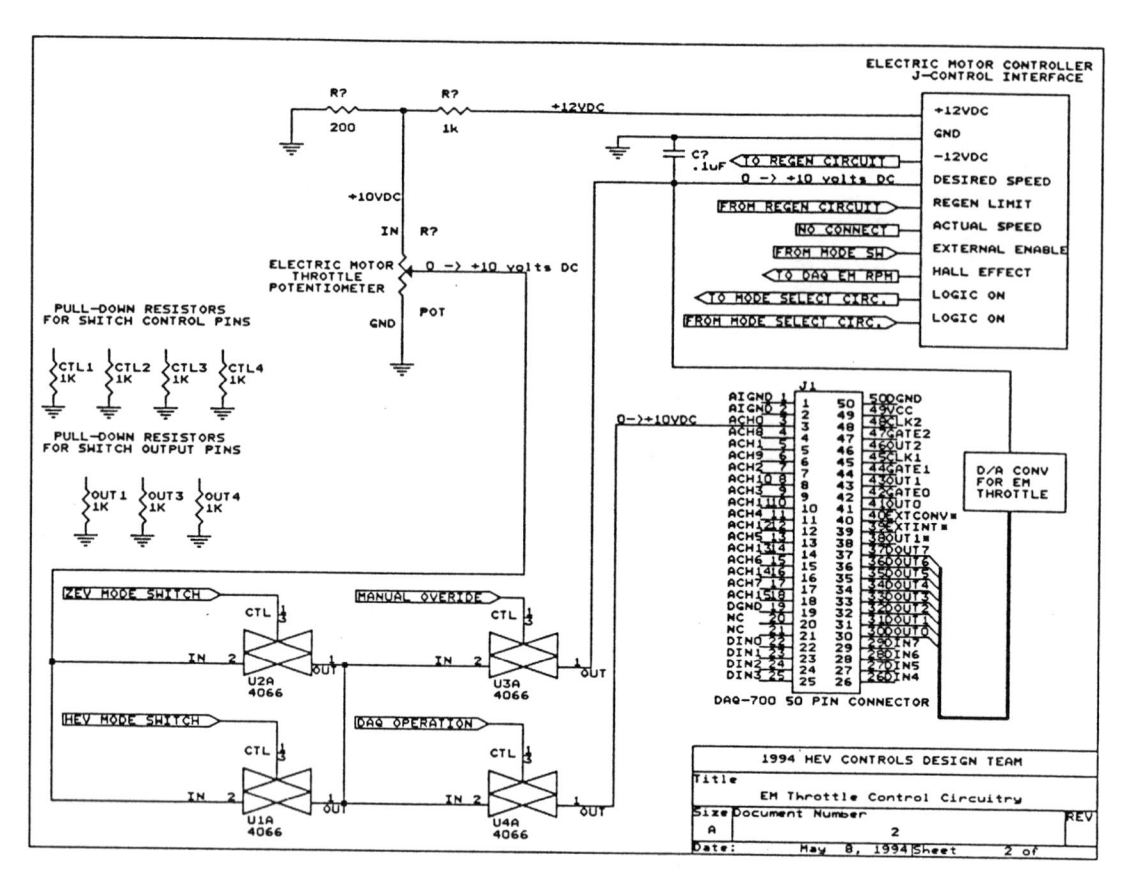

Figure 12 - EM and IC Engine Throttle Control

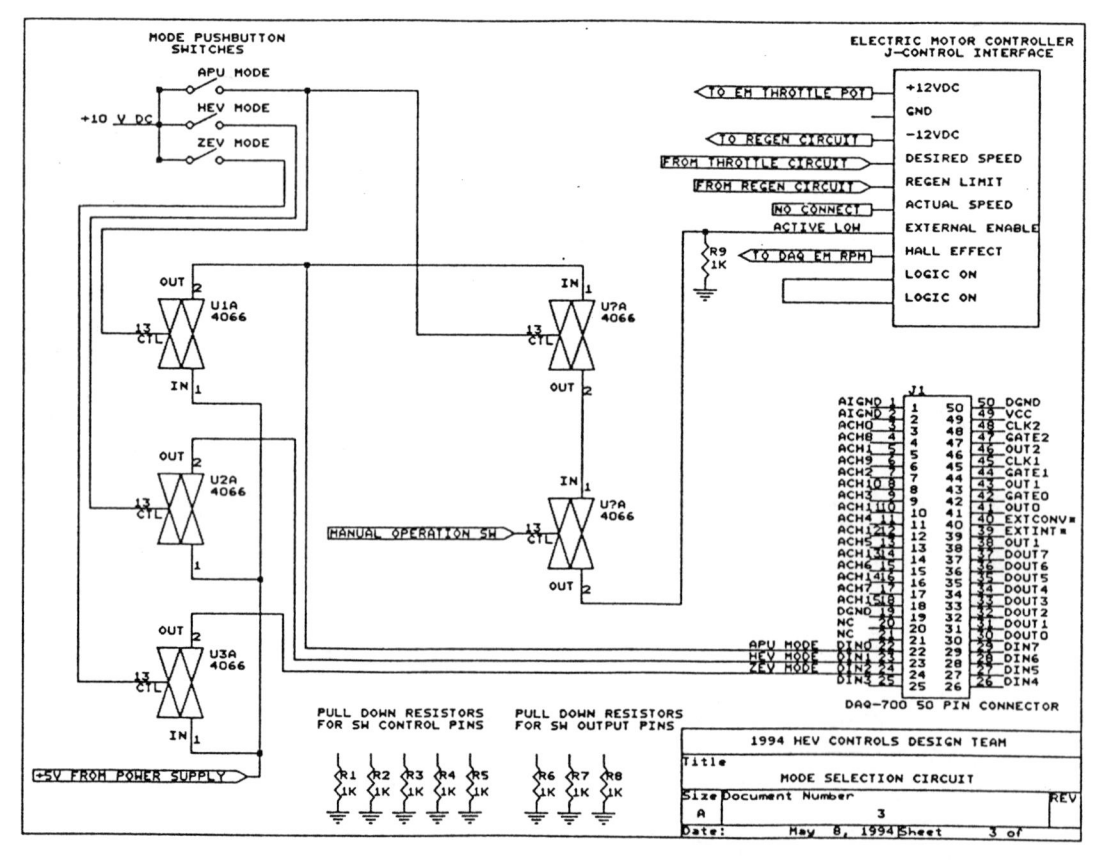

Figure 13 - Mode Select Schematic

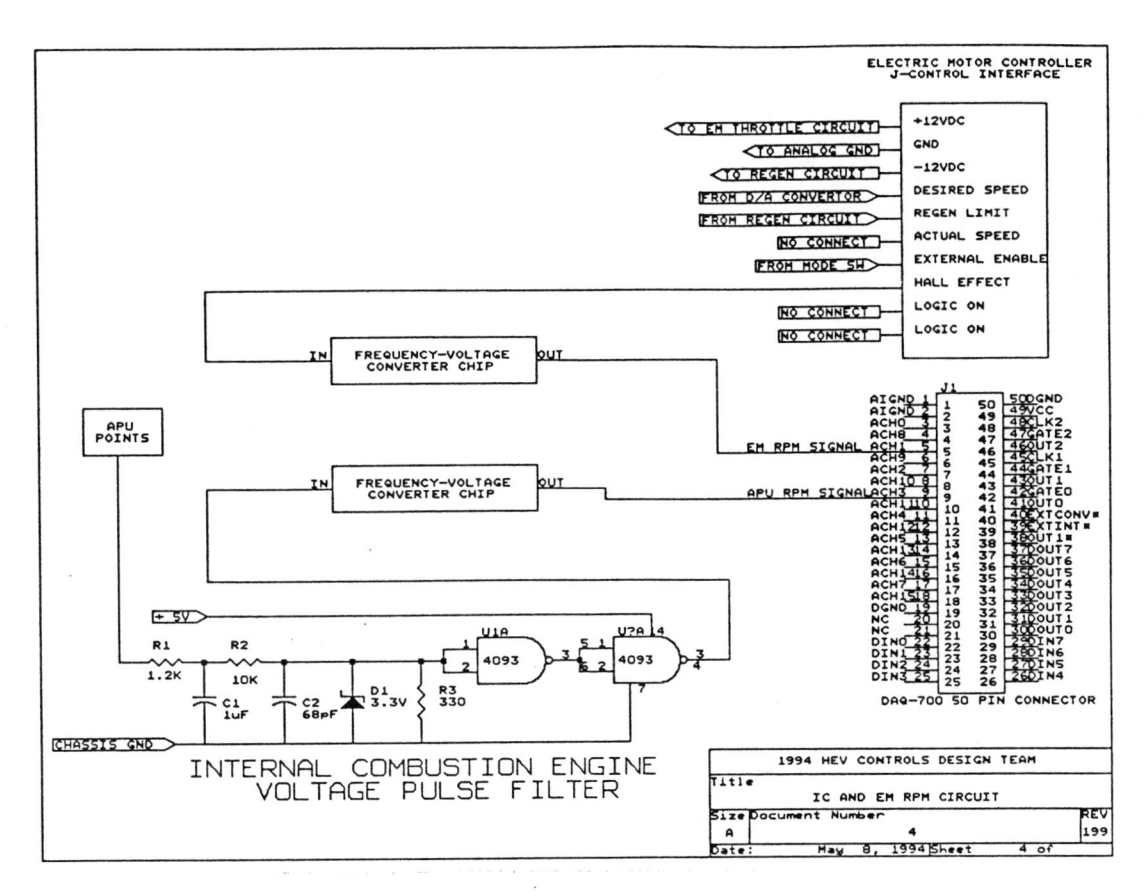

Figure 14- EM and IC Engine RPM Circuit

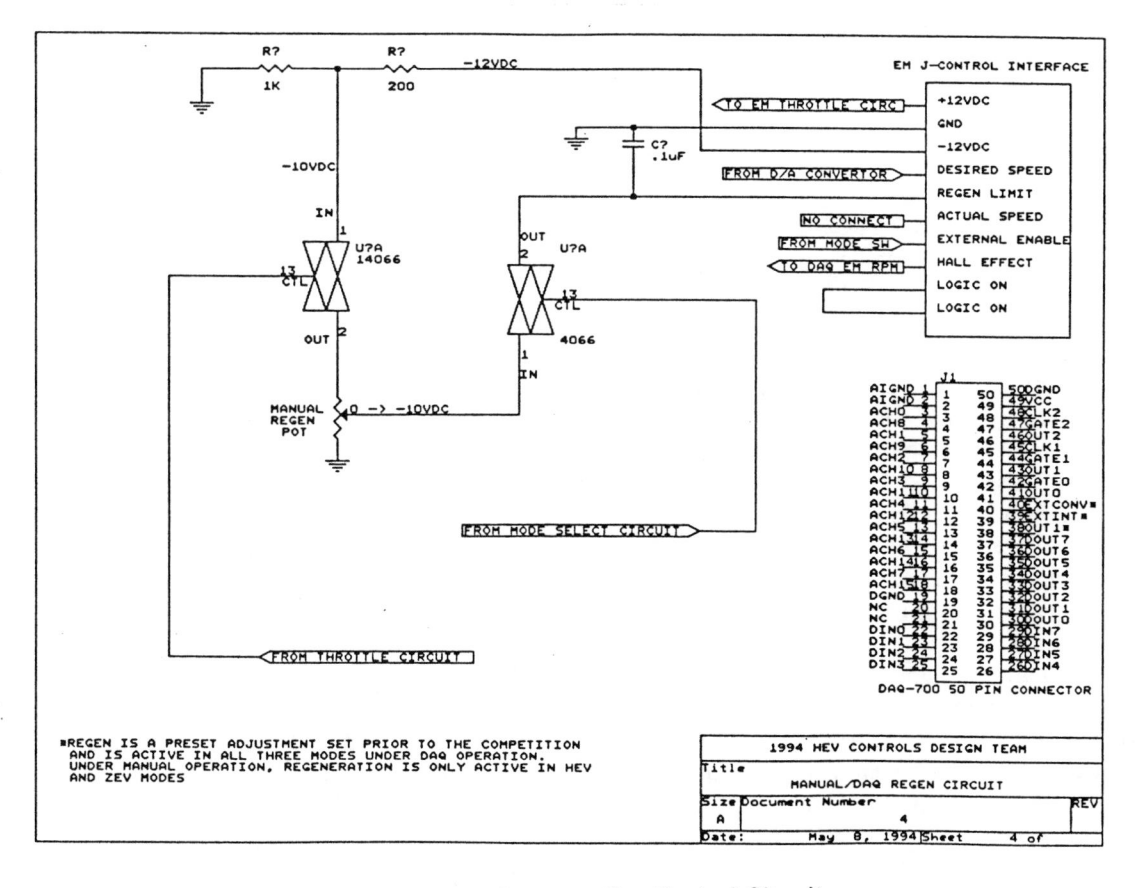

Figure 15 - Regeneration Control Circuit

EM and IC ENGINE TEMPERATURES - The electric motor temperature is sensed by applying a voltage level to the thermistor located in the electric motor. The thermistor resistance drops as the motor temperature increases. This results in a lower voltage output to the DAQ-700 card as the electric motor temperature rises. The voltage signal sent to the DAQ-700 card, representing electric motor temperature, ranges from 165 millivolts at 150 degrees centigrade to 5 volts at 25 degrees centigrade.

Internal combustion engine temperature is sensed using the 1.0 litre GEO engine temperature sending unit. The sending unit is a thermistor that operates in the same way the electric motor thermistor does. The analog signal sent to the DAQ-700 card, representing internal combustion engine temperature, operates throughout the same range set up for the electric motor.

BATTERY PACK REGENERATION - Shown in Figure 15 is the circuitry used to provide for the ability to recharge the battery pack when the vehicle is operating in any of the three modes while under DAQ Operation and in HEV and ZEV modes under Manual Override Operation. The amount of battery pack regeneration, between 0 and 100% of actual electric motor speed, is a preset value. Regeneration is activated only when the driver lets up completely on the vehicle gas pedal.

BATTERY PACK ENABLE - Illustrated in Figure 16, the Battery Pack enable circuit ensures that the battery pack is enabled when the vehicle is in all three modes under DAQ Control Operation. Under DAQ Operation the battery pack is enabled in APU mode for battery pack regeneration. In DAQ and Manual Override Operation, this circuit enables the battery pack in HEV and ZEV modes for electric motor driving and regeneration of the battery pack. The battery pack ventilation, electric motor controller, and electric motor fans are also enabled and disabled with this circuit.

BATTERY PACK CHARGE MONITORING - This system has two objectives: 1) To determine when the battery pack has received a full charge and shut off the battery charger automatically during charging, and 2) display the charge status of the individual batteries after charging in order to identify that all the batteries have received a full charge. The first of the objectives is accomplished by using a micro controller. The micro controller toggles through all sixteen batteries and shuts off the battery charger when all sixteen batteries are identified as having a full charge. A full charge is indicated when all the battery's have a voltage level of 14.65 volts. The second objective is accomplished using LED bar graph displays shown in Figure 17. One bar graph display is used to show the voltage level of each battery. They have been calibrated to display a battery voltage ranging from ten to thirteen volts. If a battery is unable to receive a full charge, this will be indicated by fewer LED's lit. The purpose of this is to allow the driver to visually identify the charge status of the individual batteries so that it can be removed from the pack and replaced with a new battery.

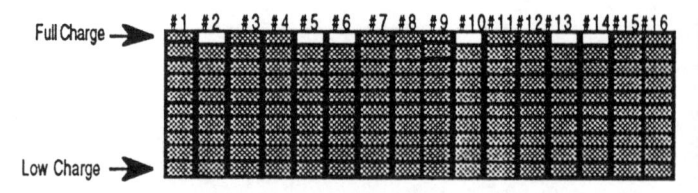

Figure 17- Battery Pack Charge Status

DAQ SOFTWARE DESIGN - The DAQ-700 Card is controlled by a customized software program written using the National Instruments Data Acquisition (NIDAQ) Driver software in conjunction with Microsoft QuickBASIC software. The NIDAQ Driver software controls the DAQ-700 card through instructions written in a QuickBasic program. Activated at vehicle start-up and running continuously during vehicle operation, the program instructs the DAQ-700 card to read input data from peripheral devices and store it in memory so the QuickBasic program can analyze and manipulate it for use. Analysis and manipulation is performed in order to display the data on the lap-top screen for dynamic performance monitoring, write the data to the lap-top computer hard drive for off-line performance analysis, and send an output to the digital to analog converter for controlling the electric motor speed during HEV and ZEV operating modes.

DAQ INPUT DATA - Input data is supplied to the DAQ-700 card in both analog and digital forms by electronic circuits. Table 1 shows the input signals names, their type, and their levels.

INPUT SIGNAL NAME	TYPE	RANGE
Gas Pedal Input	Anaolog	0 - +10 volts
Electric Motor RPM	Anaolog	0 - +10 volts
Electric Motor Temperature	Anaolog	+5 - +.17 volts
IC Engine RPM	Anaolog	0 - +10 volts
IC Engine Temperature	Anaolog	+5 - +.17 volts
APU Mode Selected	Digital	logic 0 or 1
HEV Mode Selected	Digital	logic 0 or 1
ZEV Mode Selected	Digital	logic 0 or 1

Table 1 DAQ Input Signals

The QuickBASIC program instructs the DAQ-700 card to read these inputs and store their immediate values to determine the appropriate response for display purposes on the lap-top computer screen, writing performance data to the hard drive, and ouput a digital signal to a digital to analog (D/A) convertor for controlling the electric motor speed.

LAP-TOP COMPUTER DISPLAY - Due to the fact that PEGASIS is a prototype HEV, the screen of the lap-top computer has been utilized for providing instantaneous performance monitoring capabilities. The performance conditions being displayed for monitoring by the vehicle passenger are displayed in the figure below.

Figure 16

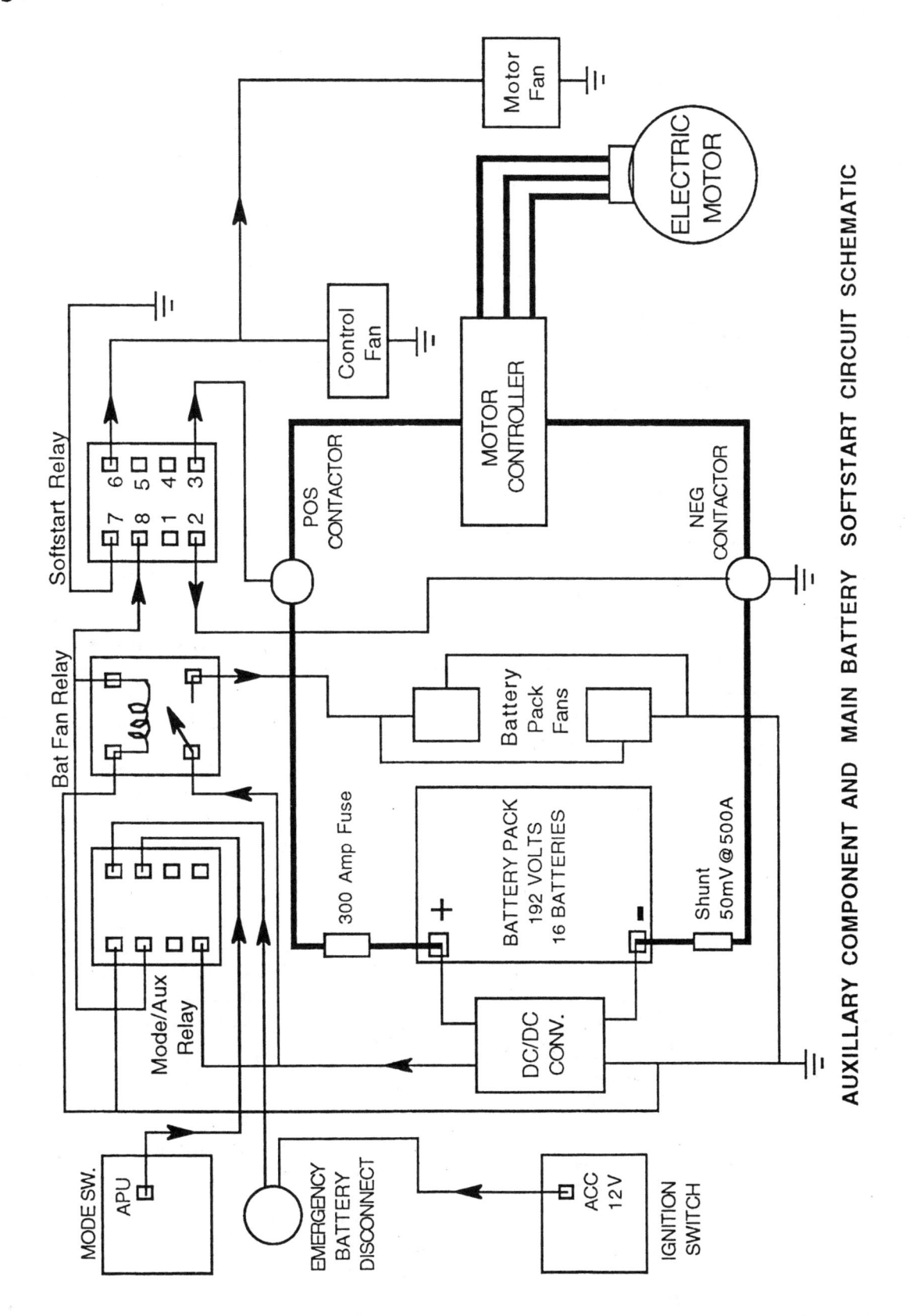

AUXILLARY COMPONENT AND MAIN BATTERY SOFTSTART CIRCUIT SCHEMATIC

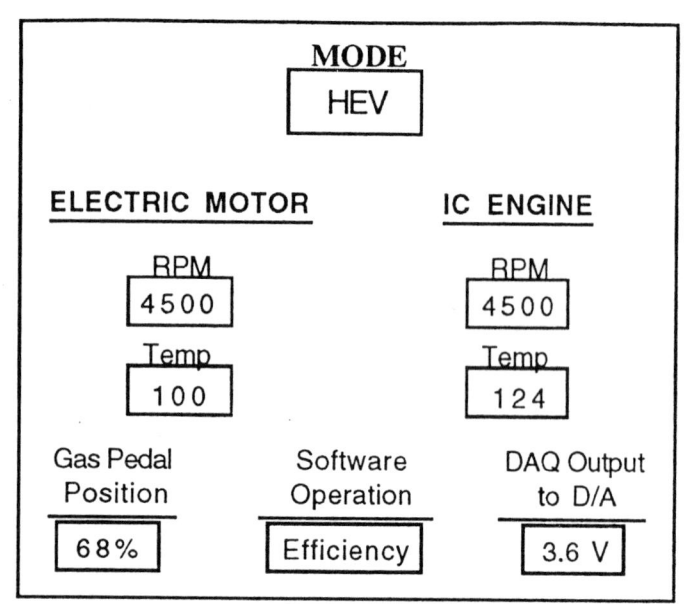

```
        MODE
        ┌─────┐
        │ HEV │
        └─────┘

ELECTRIC MOTOR        IC ENGINE

    RPM                  RPM
   ┌─────┐             ┌─────┐
   │4500 │             │4500 │
   └─────┘             └─────┘
   Temp                 Temp
   ┌─────┐             ┌─────┐
   │ 100 │             │ 124 │
   └─────┘             └─────┘

Gas Pedal    Software      DAQ Output
Position     Operation      to D/A
┌─────┐     ┌──────────┐   ┌─────┐
│ 68% │     │Efficiency│   │3.6 V│
└─────┘     └──────────┘   └─────┘
```

Figure 18: Lap-Top Display

The EM and IC Engine RPM's are displayed so the vehicles passenger, during testing, can monitor how the two power sources are operating with respect to one another. EM and IC Engine temperatures are displayed so that the passenger can monitor the operating temperatures to make sure they stay within safe operating limits. The gas pedal throttle position input is displayed, along with the EM desired speed output, in order to identify whether or not the software program is reading the input of the gas pedal position and output the appropriate digital signal to the D/A convertor for proper EM control.

INPUT DATA STORING - The software program includes instructions that store the input data it is taking in from the external monitoring devices and store this data on the hard drive of the computer. The program will instantaneously download the input and output data values to the hard drive every 10 seconds so that analysis of the vehicles performance can be reviewed after a dynamic test run has been completed.

OUTPUT CONTROL of EM SPEED - The DAQ-700 card has the capability of sending an 8-bit digital output signal from its 8-bit digital output port. The digital output port is being used solely for the purpose of controlling the EM speed in APU, HEV, and ZEV modes of operation. The software program determines the output based on the gas pedal input signal, mode select input signal, and software parameters determined for optimal performance. Since the EM desired speed input requires an analog voltage signal from 0-10 volts for controlling the speed of the EM, and the DAQ can only send a digital output, an external digital to analog convertor is used for the conversion. With the 8-bit output port, the analog signal can be broken up into 2 to the 8th, or 256, different analog voltage levels. This provides for the conversion of the digital signal into analog voltage levels stepping up and down at .038 volt increments. Table 2 shows the digital to analog relationships.

| DAQ Digital Output | Analog |
MSB LSB	Equivalent
1111 1111	10.00 volts
1111 1110	9.96 volts
1111 1101	9.92 volts
═══════════	═══════════
═══════════	═══════════
0000 0010	0.08 volts
0000 0001	0.04 volts
0000 0000	0.00 volts

Table 2 Digital to Analog Relationship

The output of the DAQ varies depending on which mode the driver has selected. When the program reads the mode select input and determines that the driver has selected APU mode, the DAQ sends a zero volt digital signal to the D/A regardless of the gas pedal input. In APU mode, when the IC engine is the sole power source, the EM will not be applying power to the vehicle. When the mode select input is showing HEV mode, the output of the DAQ is a scale of the gas pedal input signal. The output signal is scaled between a zero and ten volt digital signal relationship when the gas pedal input varies between five and ten volts. The scaling of the input to the output has been done to make the EM act much like a turbo charger during HEV operation. The EM will only supply power when the gas pedal has been pushed passed its 50% position. This is being done to increase the battery pack efficiency while the vehicle is in HEV mode. When in ZEV mode, the output will be a mirror of the gas pedal position input signal. When the program reads an input voltage level from the gas pedal input, it will output the exact voltage level in digital form. This will ensure that the EM reacts the same way the IC engine does when it is the only source of power.

CONCLUSIONS

A tremendous amount of time and effort was spent researching 1993 design documentation and prepararing the 1993 configuration testing. This effort provided the 1994 team with an excellent HEV concept knowledge base, a clear problem definition, and rigid scope of work. The process of making the 1993 vehicle configuration test ready, as well as the actual testing process, allowed each student to become intimately familiar with the problems associated with the design. This first hand experience reduced the problem definiton to a scope of work feasable in the time available and within manpower and budget limitations imposed on the team.

It focused the team on the projects critical issues and accelerated developement of engineering design tools such as the PEGASUS modeling program by bridging the gap between academics and real world problem solving. The creation of the PEGASUS modeling program was the key design tool used to aid in making engineering design decisions on both the mechanical and electrical fronts. The PEGASUS modeling program allowed quick and easy comparison of design concepts accelerating the team into detailed design of the individual components necessary to meet the selected design concept requirements.

The PEGASUS modeling program layed out the ground work for the control ysstem software logic development. Many communication problems in interfacing the lap top computer with hardware peripherarls werre ironed out by simulated drivng with the program. Base code from the program accelerated software contol logic development.

All of these efforts resulted in a design process that transitions smoothly from problem definition to concept generation, concept selection, conceptual design, fabrication, and lastly testing and validation. Testing and validation will carry on through the competition in June of 1994.

Over half of the problems outlined in the Background section of this report were resolved directly by selecting combined parallel drive system concept. The elimination of the rear drivetrain by combining both the internal combustion and electric power into the same stock 5-spd transmission resolved several issues. With the elimination of the rear transmission and clutch, the clutch pedal pressure was reduced to normal pedal pressure and shifting procedure to that of a stock vehicle. The 3 batteries behind the front seats were moved to the rear of the vehicle in the space opened by eliminating the rear drivetrain creating much needed leg room. This also reduced the vehicles overall weight and rearward weight bias allowing better performance and handling characteristics. The problem of clutch inadequacy was resolved with the re-introduction of the stock escort manual clutch.

The selection of the DAQ / Lap top computer control system created a means to govern the application of power and overall control philosophy of individual components in the vehicle. This control system also provides the means to collect, monitor, and store vehicle performance data. This control system provides the engineers with a powerful tool in studying the effects of design changes on vehicle performance.

Complete documentation of the design process has created good schematics for troubleshooting, maintenance, and repair of the vehicle. The 1994 team hopes that these improvements and control system analysis tools will help future engineering students make the HEV concept a production reality.

ACKNOWLEDGEMENTS & RECOMMENDATIONS

At the time of publication of this document, sufficient validation through testing has not been performed on the vehicle. Testing and validation of the vehicle will be performed on the vehicle during the remaining 3 weeks prior to and during the competition. It is the recommendation of the 1994 PEGASUS design team that Seattle University utilize the tremendous test and validation capabilities of this vehicles control system. This vehicle can be used in both the electrical and mechanical engineering curriculum by students and faculty for confirming engineering principles and theory. The 1994 PEGASUS HEV provides Seattle University with a whole host of analysis tools needed to validate engineering theories that otherwise can only be taken at the instructors word. In other words, the student can learn the strengths and limitations of engineering theory.

Team PEGASUS would like to thank the Seattle University Engineering Design Center for all their cooperation, Team Advisor Dr. Jack D. Mattingly for his expertise, patience, and guidance to help this team achieve insurmmountable goals. We would also like to thank our school machinist Blaine Shaeffer for his guidance, excellent craftmanship, and motivation. In addition, we would like to thank Dr. John Bean of the English Department for his guidance and insight.

References.

1. Thomas D. Gillespie, "Fundementals of Vehicle Dynamics", SAE, 1992

2. Joseph E. Shigley / Charles R. Mischke, "Mechanical Engineering Design (5th edition)", McGraw-Hill, 1989

3. Theodore Baumeister, "Mark's Mechanical Engineer's Handbook (6th edition)", McGraw-Hill, 1958

4. "The 1993 Ford Hybrid Electric Vehicle Challenge", SAE SP-980, 1994

Leland Stanford Junior University
"The Winds of Freedom II"

Stanford Hybrid Automobile Research Project
Shawn Sarbacker   Joel Miller
Bailey White   Phillip Chen

ABSTRACT

SHARP, the Stanford Hybrid Automobile Research Project, has built The Winds of Freedom, a prototype hybrid electric vehicle (HEV) conversion of a 1992 Ford Escort. The project is unusual in that it has focused on designing and building custom components rather than off-the-shelf pieces to emphasize student learning and explore new technology. The vehicle is a series type, electric/gasoline hybrid. It principally operates as a purely-electric vehicle, powered by an electric motor and batteries for 60-80 miles. Subsequently, a 13 kW auxiliary power unit or APU (consisting of an internal combustion engine coupled to an alternator) produces electricity to drive the electric motor and recharge the batteries. The series system allows for full performance capability even in purely-electric operation and non-polluting driving for most of users' needs. However, the APU provides The Winds of Freedom with a projected range that exceeds conventional gasoline vehicles. This design's advantages include efficient and clean operation of the APU.

PROJECT SYNOPSIS

In his speech at the Stanford Manufacturing Partnerships conference in May 1994, Jack Smith, CEO of General Motors, stated that he felt hybrid vehicles were the direction the automotive industry would take in designing more environmentally–friendly cars. Hybrid vehicles are expected to be the design of choice in meeting increasing air quality laws. He went on to sight the Stanford Hybrid Automobile Research Project (SHARP) as one that created a valuable link between academia and industry in exploring this new technology.

SHARP has given students a strong relationship with industry and allowed them to learn about working on real engineering projects. SHARP has designed a hybrid vehicle that addresses the needs of the public. Our primary goals in building an HEV are to create a car that: 1) maximizes benefits for the environment and 2) produces a vehicle that would fit in with the existing infrastructure. Based upon these goals, we have converted an existing vehicle into a series hybrid configuration.

By converting an existing vehicle, we are able to demonstrate the immediate viability of this technology to the public. The warm public reception we received at the San Francisco International Auto Show has shown that such a conversion is very attractive. Such a conversion demonstrates that environmentally friendly cars can have the functionality of today's vehicles.

Our series hybrid design, long ZEV range, large passenger space, and regenerative braking ability help form a very environmental and practical car. California Air Resources Board estimates that 95% of daily driving covers less than 60 miles. Our car is designed to go this distance in ZEV mode, emitting zero tailpipe pollutants. Only when drivers travel more than 60 miles in a day does our computer-activated generator supply power. The car has maintained all passenger space to preserve consumer usefulness.

Throughout the project, SHARP tried to probe the limits of technology. Because of this focus, many engineering decisions valued performance enhancement over cost. We believe that once the capabilities of hybrid vehicle technology are demonstrated, less expensive methods of implementing it can be developed. For example, AC induction motors can be produced inexpensively in mass since they require no rare metals.

VEHICLE CONFIGURATION

ELECTRICAL CONFIGURATION —Because the Winds of Freedom is a series hybrid vehicle, the car's wheels are driven solely by the electric drive motor. The electricity for this motor is supplied first by a 108V, 15 kWh Ni-Cd battery pack and subsequently enhanced by the 13 kW APU.

All high-power electrical components in the vehicle are wired, in parallel, to a central power bus. This power bus routes electricity from all sources, to all destinations. The components located on this power bus are the drive motor, the APU, the traction battery pack, the battery rechargers, and the DC-DC converter (which provides power to a 12V accessory bus). For example, if the vehicle is fully charged, during normal operation, the batteries supply energy to the power bus while the traction motor and DC-DC converter draw from it. During regenerative braking, the traction motor supplies energy to the bus, and the DC-DC converter and batteries draw from it. When the APU is operating it supplies 13 kW to the bus. The traction motor would draw 12 kW, the DC-DC converter about 0.25 kW, and the remaining 0.75 kW would go to the batteries.

System Evaluation — While the car's electrical configuration is quite satisfactory, a holistic, production-oriented design would lead to significant integration of components, especially controllers and rechargers: For example, the motor controller, alternator controller and chargers all contain inverters/rectifiers which could be integrated into one unit, significantly more compact and efficient.

PHYSICAL CONFIGURATION — SHARP's overriding goal in component placement was to preserve as many of the ergonomic advantages of the Escort as possible. Preserving seating for five, full rear cargo space, and maximum leg and foot-room were prime considerations. Items which would not be included in a production vehicle, such as a roll cage and a halon fire extinguishing system, were not placed with any less consideration for space preservation. Refer to Figure 1 for a schematic diagram of the location of main components.

SHARP's original goal was to fit all components except the batteries in the original engine compartment, and place the batteries in locations which would not interfere with passenger or cargo space. The primary location for batteries is beneath the cargo area floor. Thirteen of the eighteen batteries in the car are in a fully enclosed box in this area. These batteries are arranged in three rows: the front row containing three batteries and the other two with five batteries each. By tightly locating components, we were able to place the additional five batteries in the front driver's (hereafter left) side of the engine compartment. These five batteries are in two layers: three on top, and two beneath. Because these five batteries account for 90 kg, this placement greatly improves the weight distribution compared to putting all 18 in the rear.

The differential unit was removed from the original Escort transmission, the final gear was replaced with a sprocket and High-Velocity (HV) chain, and the entire assembly was re-cased. We selected an HV chain because of its high strength and high efficiency. The case was designed to be able to accept the original half-axles and be located where the differential had been in the original Escort. The chain was added because it enabled the motor to be mounted transversely above the right half-axle. As the traction motor has an eight inch diameter, if a final gear reduction had been used instead of a chain, the motor would have interfered with the half-axle. The chain was also used because it was significantly more cost effective in a single-vehicle application. The motor is attached through a 4:1 planetary gear reduction to the pinion gear of the chain system.

The traction motor power inverter is located directly above the motor, and sits in a flat plane. The main contactor is between the motor and the firewall. Beneath and slightly forward of the motor is the one gallon oil tank which holds the oil to cool and lubricate the motor, gear reduction, chain reduction, and differential.

Figure 1, Physical Layout

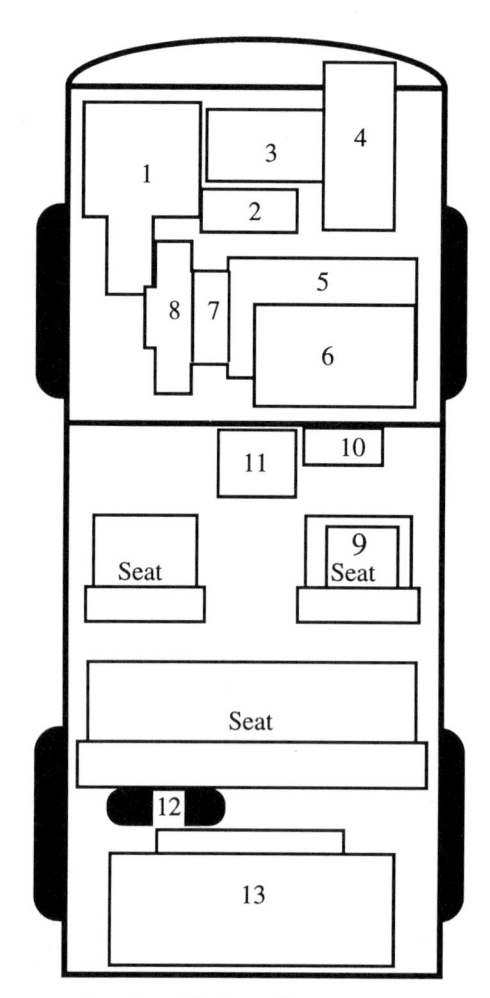

1: Front Battery Box
2: Alternator Controller
3: Alternator
4: Combustion Engine
5: Electric Motor
6: Motor Power Electronics
7: 4:1 Planetary Reduction
8: Chain Reduction/Differential
9: Motor Controller
10: Vehicle Controller
11: Rechargers
12: Spare Tire
13: Rear Battery Box

The engine's fuel injection system controller is mounted outside the right frame rail, in front of the wheel well, and to the right of the engine. As the alternator (which forms the other half of the APU) is coupled directly to the crankshaft, it lies to the left side of the engine.

The left half of the original radiator area is used for the engine's radiator and the traction motor's oil cooler. An electric fan is mounted to draw air through the radiators. The alternator controller sits in a vertical plane between the alternator and the differential/chain case. Outside the left frame rail lie the two DC-DC converters, the 12V emergency/accessory battery, and the fuse blocks.

The traction motor electronic controller is stationed under the passenger seat, and the main vehicle controller is mounted under the passenger's side of the dashboard. The recharger units are under the center of the dashboard. This setup allows the main battery cables to run along the left side of the car, under the driver's seat, along the center console, and to the rechargers, main contactor, traction motor power electronics, and alternator power electronics all in one line. The cabling from the front battery box is able to go directly to the DC-DC converters, run along the left frame rail to the rear of the engine compartment, and meet the rear battery cables in the center of the firewall. Routing these cables together reduces the space they occupy and minimizes electromagnetic radiation.

System Evaluation — Working within the original Escort design significantly constrained the physical configuration of The Winds of Freedom and forced SHARP to locate certain components in less than ideal locations. For example, the front row of the rear battery box intrudes four inches into the cargo area because it had to be located above an already-existing suspension arm. In addition, the motorcycle engine's air intake and exhaust system are routed rather circuitously because the engine was approximately two inches taller than the original engine compartment. However, working within these constraints allowed SHARP to achieve its goal of demonstrating that an existing vehicle could be retrofitted with a hybrid-electric drive system.

The car's physical configuration within the Escort's constraints is acceptable. We were able to locate most of the high power components in the car in the same order that they attach to the high-power electrical system, thereby creating a fairly straight wire route from the rear battery box to the center of the engine compartment. We were able to locate the motorcycle engine far enough forward to secure significant natural cooling, and were able to maintain enough frontal area to place all the radiators in the coolest area of the engine compartment. We were also able to stagger components and preserve "crush zones" in the engine compartment such that the compartment has sufficient room to buckle in the event of a frontal impact.

VEHICLE SYSTEMS

The car's systems are divided into five main groups: the traction drive system (Motor, transmission, differential), the battery system (batteries, rechargers), the alternate power unit or APU (combustion engine and alternator), the vehicle control systems, and the vehicle structure itself.

TRACTION DRIVE SYSTEM: COMPONENTS 5,6,7,8,9

SPECIFICATIONS — The SHARP drive system consists of a student-designed and built, second-generation 56 kW continuous 3-phase AC induction motor and controller.. These drive the wheels via a 4:1 planetary gear reduction, a 3.5:1 High Velocity chain reduction, and the original Escort differential gears.

DRIVE MOTOR — AC induction motors are very attractive for hybrid vehicle applications because of their relatively low cost, high specific power, high efficiency and rugged durability. They offer wide power ranges that are superior to DC motors, and maintain high efficiency throughout that range.

Last year, the SHARP motor development team designed and fabricated an AC Induction drive motor system. However, due to manufacturability problems leading to a significantly out-of-specification air gap, the motor produced only about 30% of its design performance. Additionally, the motor demonstrated insufficient cooling of the rotor and stator, leading to over temperature shutdowns and an overall loss of efficiency.

This year SHARP chose to redesign and build a second-generation 75 hp (continuos-rated) AC induction motor. The motor development team found that no commercially available motor was competitive with the performance projected for the redesigned motor system. Additionally, our motor design team felt that they had the technical competence and ability to bring the old motor design up to its full performance specifications.

The motor was redesigned by a team of SHARP students, with the FMC corporation providing limited technical support in the area of electromagnetics. The system was designed to meet the requirements and specifications decided on by SHARP in Tables 1, 2, and 3 attached. The new motor is 8.5 in. in diameter and 11 in. long. Figure 2 shows the redesigned motor performance, with Figure 3 showing the SHARP vehicle acceleration curve when powered by our new AC induction motor.

Figure 2, Performance Curves for the Second-Generation SHARP AC Induction Motor

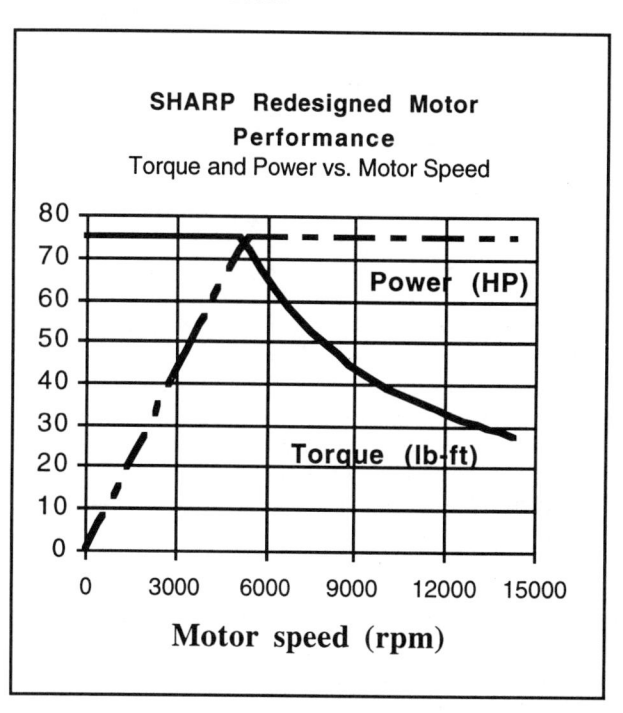

Power and Torque outputs are shown for the second generation SHARP redesigned AC induction drive motor

Figure 3, SHARP Vehicle Acceleration

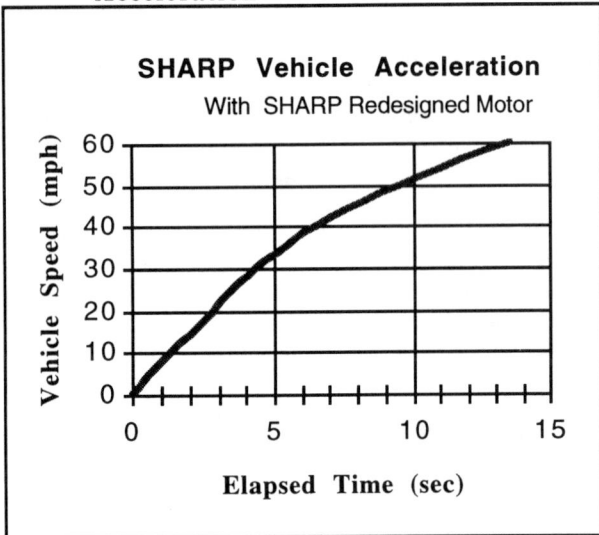

**SHARP Vehicle Acceleration
With SHARP Redesigned Motor**

[Graph: Vehicle Speed (mph) vs Elapsed Time (sec), showing acceleration curve reaching 60 mph at approximately 14 seconds]

SHARP vehicle acceleration is virtually identical to original Ford Escort with 75 hp AC induction drive system

MOTOR DESIGN PROCESS

The drive motor development team focused on three areas: design for manufacturability, cooling system redesign, and electromagnetic redesign. The redesign of the individual motor systems followed a "bottom-up" approach. Fixed motor system design parameters were identified, such as use of the existing rotor hardware, and the interfacing and supporting systems were developed around them. Design issues were addressed on a critical-path basis, beginning with stator lamination redesign effort, due to its long procurement lead time.

RECONFIGURING THE MOTOR FOR IMPROVED COST AND MANUFACTURABILITY

Based upon our voiced concerns over the identified issues leading to oversized air gap problems with last year's SHARP motor design, consulting engineers from FMC corporation suggested investigating a four pole configuration for the motor in place of the existing eight pole configuration. This investigation was driven by our goal of motor system cost reduction and addressing manufacturability issues. The use of a four pole design, due to its inherently lower switching frequency, allows the use of thicker stator laminations. By going to a four pole motor design, we could double the thickness of the stator laminations (from 0.007 inch to 0.014 inch each), without increasing eddy current losses in the laminates. By going to a larger lamination thickness, assembly difficulty and manufacturing costs are reduced in several ways. First, 50% fewer laminations have to be stamped for each 6 inch long stator assembly. The thicker lamination, with its improved stiffness characteristics, allows much more accurate stamping at a lower cost due to less required die "tweaking" for proper dimensioning. Also, with the reduced number of laminations per stator and increased handling ease of the thicker laminations, the laminations are much easier to assemble accurately into a stator stack. This results in shortened manufacturing times as well as better control of lamination alignment during stacking. Furthermore, the motor with the four pole design is much easier to wind as the motor coils in a four pole design have 50% fewer jumper connections on the endturns of the stator windings. These factors lead to the manufactured cost of a four pole configuration to be approximately 30% less than an eight pole design.

For evaluation of the performance capability of the four pole motor configuration, FMC conducted simulations of a new four pole motor design (see Appendix B). From evaluation of these simulations, we found that a four pole design would be capable of fully meeting SHARP vehicle performance specifications.

The four pole design, however, does require additional backiron material be added the stator to avoid magnetic saturation. Due to the reduction of the number of poles from eight to four, each individual pole must produce twice as much flux, in order to keep the same performance output as an "equivalent" eight pole design. Working with FMC engineer Craig Joseph, we determined that an additional 0.55 inches of backiron (radially) would need to be added in order to accommodate the additional magnetic flux. This increase in stator diameter would necessitate the redesign and fabrication of a new housing system, with a commensurate increase in motor housing OD. Investigation of the SHARP vehicle packaging was able to verify that, with some vehicle modification, the new motor diameter could be accommodated.

Based on the overarching goals of increased manufacturability and cost reduction of the motor system, while meeting specified performance goals, the decision was made to go with the four pole motor configuration. This required the unexpected burden of redesigning and fabricating new motor housings, since we had previously thought the existing housings could be modified to accept an improved stator design. This added effort was accepted, and project resources will be reallocated as needed to accommodate the housing redesign effort.

STATOR ASSEMBLY DESIGN – Since a new stator assembly was being designed and fabricated, we had the opportunity to address stator stack assembly techniques. For lower-performance motor applications, with associated loose-tolerance air gaps, stator stack dimensional tolerance is not highly critical. For these stator systems quick stator assembly is accomplished by running a weld seam axially along the length of the stator stack at several circumferential positions. This method was used last year for stator stack assembly. Unfortunately, this technique often induces a great deal of stack distortion due to thermal warping during the weld process. This phenomenon caused the welded stator stack last year to distort to over 0.050 inches out cylindrically (on a nominal 0.008 inches air gap!).

Through discussions with FMC and M/G Electric (our stator assembly and winding vendor), a new stator assembly system was identified which could meet stator stack tolerance requirements. This system will be discussed in the following section.

SIMULTANEOUS ENGINEERING OF STATOR SYSTEM – In an effort to improve manufacturability of the stator system, reduce costs, and improve fabrication lead times, a large amount of effort was dedicated to simultaneous engineering of the stator system. All parties involved in the design, fabrication, assembly, and use of the stator system were involved in the initial development of the lamination and stator design. Figure 4 shows the parties involved in the simultaneous engineering process for stator system development, and their individual contributions to the process.

Figure 4, Simultaneous Engineering of the Stator System

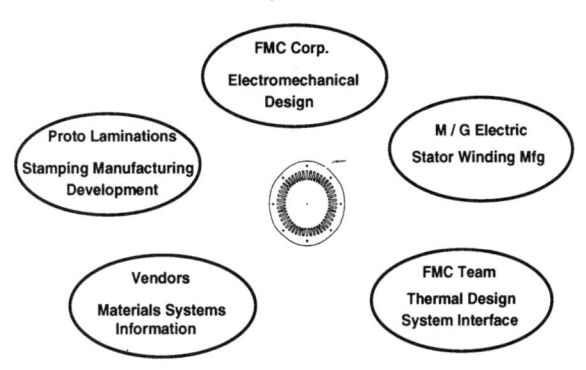

All parties involved in the design, fabrication, and end use of the stator system were actively involved up-front in the design and iterative refinement of our stator system design.

Several critical design features were identified through concurrent discussions with each party involved in the simultaneous engineering process. Each party had features of the lamination and stator stack design that were critical to their particular process. In many cases, although many of these features were critical for an individual party, they did not significantly affect the other parties. Therefore, many features could be optimized based on individual requirements or preferences. Using this arrangement, the individual parties all expressed their concerns up front and prior to unalterable commitment to any single design. Approximately three complete iterations of the lamination design were completed before release of the final lamination design, with each party giving their individual "blessing" of the part before final design release.

Several stator lamination design improvements resulted due to the simultaneous design approach. The basic stator lamination was released to the ME210 team for thermal analysis, Proto Laminations for stamping fabrication input, M/G Electric for stack assembly and winding concerns, as well as vendor input for magnet wire and insulation system expertise. Each party added valuable insight into the overall design of the system.

M/G electric was able assist in the development of a novel 8-bolt assembly system for the stator, eliminating the need for the distorting welding assembly process. This system promises to provide a dimensionally-correct stator stack, without expensive honing operations to bring the part into conformance. M/G electric also identified slot profile issues which would have lead to difficulties for them in the winding of the motor. These problems were then eliminated from the lamination design without compromising the design performance. Further, M/G Electric pointed out the necessity of the topstick system for retaining the windings during assembly. FMC had hoped to eliminate that feature for incremental cost and magnetic flux improvements, and had done so on their initial design. However, M/G electric showed that the incremental improvements identified by FMC would not justify the manufacturing complexity and variability that the elimination of the topstick feature would cause. The use of the topstick was therefore reinstated.

Proto Laminations identified several design features to allow them to more effectively perform the stamping fabrication of the individual laminations. These improvements included increasing the aspect ratio of the topstick retaining feature in the stator tooth. For clean stamping of sheet metal, a feature depth of at least the thickness of the stock being processed is required. The tooth detail was therefore improved to accommodate this concern. Proto Laminations also requested that the default fillet radius be increased to "minimum radius 0.020 inches" from the default of 0.005" to easily accommodate the wire thickness of the wire Electric Discharge Machining (EDM) tool used for cutting the stamping dies. This feature dimension did not adversely affect lamination performance, as the 0.005" was simply a default standard used by FMC. This concern was therefore easily accommodated. Finally, Proto Laminations also was able to discuss the tolerancing conventions used by FMC in their drafting system, so that all parties were seeing "eye-to-eye" in terms of intended final part tolerancing. This communication eliminated unnecessary fabrication costs and post-fabrication tolerancing disputes due to miscommunication of actual component requirements.

HEAT TRANSFER ANALYSIS OF STATOR COOLING

Electric motors, while extremely efficient in transferring electrical power to mechanical power, are not 100% efficient. Thus, in operation, inherent power losses within the motor system lead to loss of power in the form of rejected heat energy. This heat energy is lost via two main mechanisms: coulombic heating of the windings and eddy current losses in the backiron. Since this finite heat rejection will tend to increase the temperature of the motor structures--particularly the high voltage windings--it is critical to provide sufficient cooling to avoid thermal damage to the motor components. It is the goal of this thermal analysis to verify cooling performance of the three cooling schemes by analytically determining motor structure temperatures under critical operational conditions. Once completed, this data allows evaluation of the cooling method options, and identification of the best option.

An initial "pseudo-2D" heat transfer analysis was performed to benchmark the existing stator cooling scheme, determine the general cooling requirements, and rapidly evaluate cooling system alternatives (see Appendix C for development of heat transfer analysis). Figure 5 shows the Resistance Model used to determine stator winding temperatures.

RESISTANCE MODEL OF THE STATOR HEAT TRANSFER – This resistance heat transfer model was used to evaluate maximum stator winding temperatures during motor operation based on varying cooling system parameters.

This analysis gave us insight into the stator cooling system behavior. Two key points were observed from the results of this analysis. First, the maximum steady state temperature reached in the stator windings for standard OD oil cooling was approximately 160°C, which is well below the allowable 220°C maximum winding temperature. Further, we see from Figure 6 that the steady state winding temperatures level off as the convection coefficient of the OD cooling oil (h-value) increases above the 500 W/(m2 C) range.

Figure 5, Resistance Model of Stator Heat Transfer

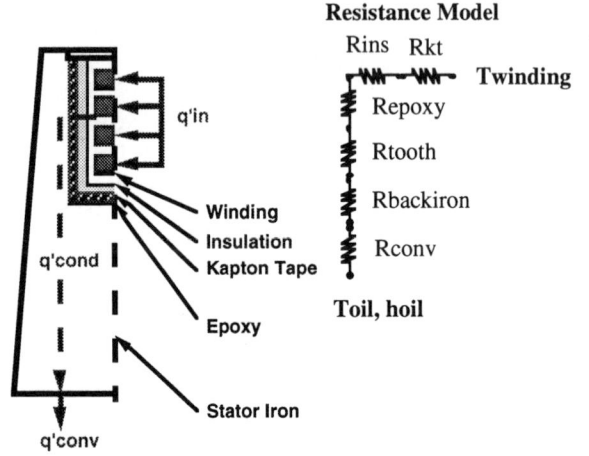

This model solves the heat transfer energy balance equations explicitly for the winding temperatures, based on assumed heat rejection rates, material conductivities, geometric features, cooling oil temperature and convective heat transfer coefficient (h-value).

Figure 6, Calculated Stator Winding Temperatures

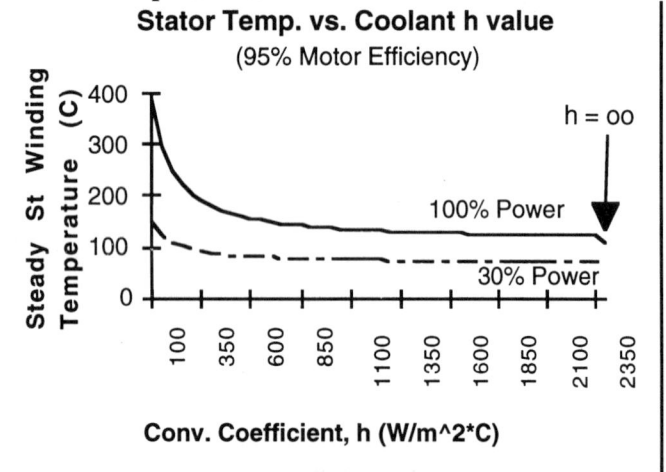

The "pseudo-2D" resistance model heat transfer analysis determined stator winding temperatures based on varying stator OD convection coefficient values.

Thus, this initial heat transfer analysis was able to quickly show that:

1) For our specified stator cooling requirements (1.0 kW steady state) stator cooling passages are not necessary, since winding temperatures could be maintained well below the maximum allowable conditions with stator OD cooling, and

2) External fins are not required, since h-value of the coolant oil flow on the stator OD without fins is approximately 1000 W/(m^2 C), which is well above the point of where the winding temperatures have leveled off based on h-value increases. Therefore, the stator fins would not provide a significant stator winding cooling benefit.

These results were concurrently validated using a sophisticated Finite Element Analysis package with full 2D heat transfer analysis capabilities (FEA is discussed in the following section). Based on the results of the "pseudo 2D" analysis, and agreement of the Finite Element Model, the Standard Stator OD Cooling Method design was selected, eliminating expensive stator cooling passages and external fins from consideration

FINITE ELEMENT MODELING OF THE STATOR- Concurrently with the "pseudo-2D" analysis, a 2-D Finite Element Model in IDEAS was constructed and used for verification of the "pseudo-2D" analysis results. This model was able to more accurately take into account 2D heat transfer effects, such as choked heat transfer in the stator tooth and heat transfer out the bottom of the slot, which could only be approximated in the analytical model.

Our team created and solved a full 2-D Finite Element Model (FEM) using Ideas Finite Element software written by Structural Dynamics Research Corporation. This model consisted of a stator tooth and backiron section. It was possible to use this simplified model because of the symmetry that exists in both the stator and the heat generating elements (windings). The model consists of 80 quadrilateral shell elements and 120 associated nodes. The windings, Kapton tape, mechanical insulating tape, epoxy, and stator lamination are all represented in the model. The heat generated by the windings is represented by 16 nodal heat sources. The heat is removed through convection to the cooling oil which is represented by edge convection off of the elements that represent the outer edge of the stator.

Figure 7, Finite Element Heat Transfer Results

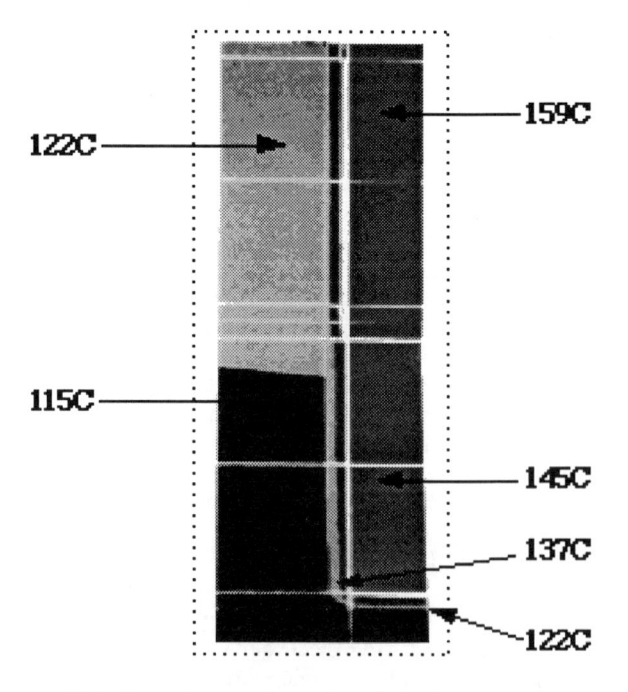

This figure is a contour plot of the heat transfer results around the windings, the critical section of

the stator model. Various temperatures are shown to illustrate the gradient through the various material layers.

The results of this analysis correlate well with the "pseudo-2D" analysis shown earlier based upon classical heat transfer equations. We were therefore comfortable with accepting the results of these analyses for the selection of stator cooling system design options.

ROTOR COOLING ANALYSIS – During the winter quarter, discussions with FMC identified possible problems with peak rotor temperatures exceeding specifications (200C). In order to investigate this issue and propose possible solutions, we created a resistance heat transfer model for the rotor so that peak rotor temperatures could be calculated. Figure 8 shows the model derived for the rotor system.

The resistance heat transfer model below was used to evaluate maximum rotor bar temperatures during motor operation based on varying cooling system parameters.

Figure 8, Resistance Model for Heat Transfer Analysis of the Rotor Cooling

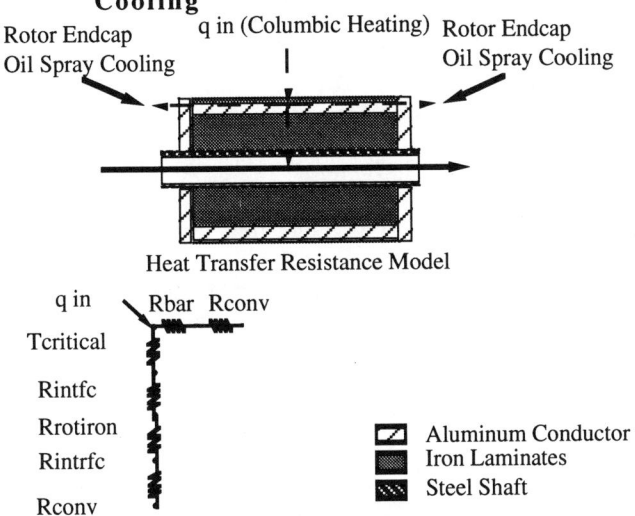

This resistance heat transfer model was used to validate cooling system design for our redesigned motor.

This resistance heat transfer model was implemented on an Excel spreadsheet, so that peak rotor temperatures could be quickly obtained based on varying motor and cooling system parameters. Shaft cooling was assumed to be equivalent to obtained data from last year's team, with an h value of 2000 (W/m^2 C). Figure 9 shows the calculated values for rotor bar temperatures at maximum (100%) power output, and at a typical vehicle cruise (30%)

The resistance model was used to calculate rotor temperatures. From this analysis we found that rotor endcap cooling was required to meet maximum rotor temperature specifications.

Figure 9, Calculated Rotor Bar Temperatures

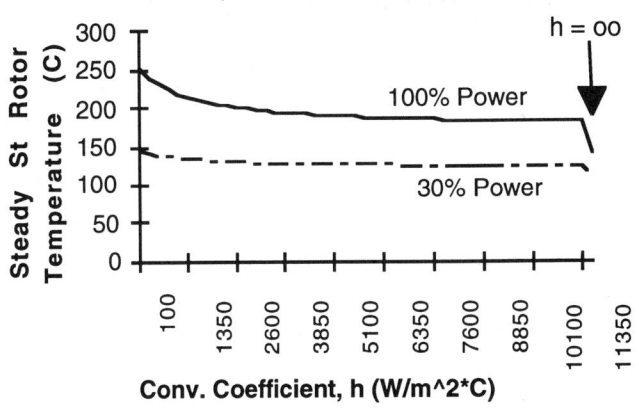

The results of rotor heat transfer modeling showed temperature levels well within the 200°C maximum temperature specification.

We found from this analysis that meeting rotor maximum temperature specifications at peak power conditions necessitated the use of rotor endcap spray cooling. Based on model assumptions, such as liberal use of interface resistances as well as conservative thermal conductivities and convection coefficients, we feel that this model is fairly conservative. Therefore, we feel that although the maximum rotor temperatures calculated slightly exceed the maximum temperature specification, the system will meet maximum temperature specifications with the addition of rotor endcap spray cooling.

We also noted that the rotor peak temperature is fairly insensitive to h value of the impinging cooling oil. However, due to the extremely high relative velocity of the rotating endcap surface to the impinging oil, h values of the cooling oil as expected to be as high as 100,000 (W/m^2 C).

HOUSING REDESIGN

DESIGN ALTERNATIVES-The increased diameter of the stator and increased cooling requirements necessitated the redesign of the motor housing. In the redesign process we strove to incorporate features to assist with housing fabrication, motor assembly, rotor alignment, and motor integration into the SHARP vehicle.

To expand the number of design solutions from which we had to choose, we conducted a morphological analysis of the motor housing. This analysis laid out the many individual design options for the housing system, and allowed the visual coordination of required design functions with their complementary solution options.

MOTOR HOUSING DESIGN FOR MANUFACTUR-ABILITY-Design for Manufacturability is often overlooked or ignored during the design process of a project. Even in our project it was tempting to worry about manufacturing 'later'. Yet most manufacturing issues must be addressed in the early phases of the design process in order to be implemented economically. Often a little time spent at the beginning of a project can considerably reduce manufacture time during prototyping and production, as was demonstrated in our stator assembly design process.

Redesigning the housing gave us the opportunity to incorporate features to ease manufacturing, testing and serviceability. The team will work closely with the machinist fabricating the housing to better understand the housing fabrication process and to develop the best possible design. Figure 10 and the following paragraphs describe features added to the motor housing.

Figure 10, Housing Redesign

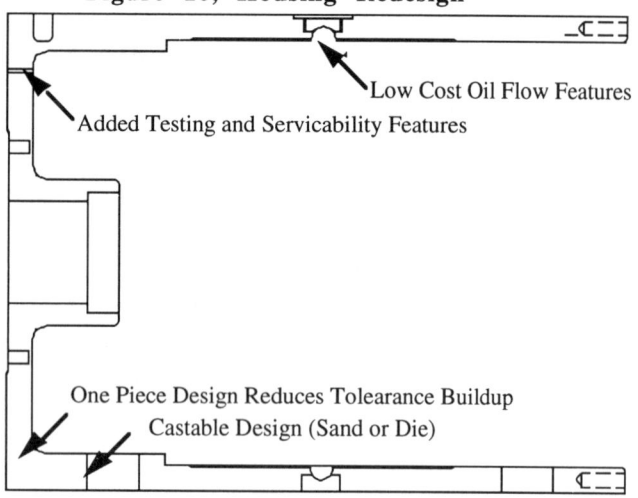

- Low Cost Oil Flow Features
- Added Testing and Servicability Features
- One Piece Design Reduces Tolearance Buildup
- Castable Design (Sand or Die)

The redesign of our motor housing led to significant improvements in manufacturability and cost

<u>Reduced tolerance buildup between rotor and stator</u> - The amount of air gap between the rotor and stator plays a big part in the electromagnetic efficiency of the motor design. An 0.008" air gap was designed to separate the rotor and stator. Minimizing the variation by reducing the tolerance buildup markedly improves efficiency of the motor. To accomplish this we went from a separate endcap and casing to a one-piece design. This removes variation caused by the endcap shifting relative to the casing. Concentricity is improved as well.

<u>Castable Design Features</u> - All features in the housing were altered to allow die or sand casting in a high volume production environment.

<u>Motor Installation Eyelet Holes</u> - Tapped holes were added to the outside surface of the housing. A chain hooked into eyelets mounted into these holes can be used to easily lower the motor into the engine compartment of the hybrid vehicle with a engine lift or chain fall.

<u>Test Probe Access Holes</u> - Holes in the endcap will allow for temperature probes to be inserted into the interior of the housing. Temperature readings will then be taken of the internal components of the motor during the testing phase of the design process.

MOTOR CONTROLLER

The new FMC/Stanford motor controller utilizes state-of-the-art AC induction, vector controller technology. It replaces our previous controller and is designed to meet 5 criteria: 1) Improve the acceleration profile to make the HEV 'feel' more like an internal-combustion vehicle 2) Improve the deceleration profile (regenerative braking) to make the HEV feel' more like a vehicle with power brakes 3) Reduce or eliminate the jerky motion of the motor at low speeds 4) Reduce the physical size of the controller and number of components 5) Maintain compatibility with existing wiring harness and inverter.

The motor controller system was upgraded in the spring of 1994 with design support from FMC Corporation of San Jose. Stanford's previous controller was designed using analog devices such as operational amplifiers, frequency-to-voltage converters, and a pulse width modulation generator while the new controller is based on an AC vector processor.

The heart of the controller consists of an inverter that uses IPMs: intelligent power modules rated for 400 Amps per phase. IPMs consist of a integrated package of two IGBTs, diodes and thermal protection circuitry. The use of IPMs in the design has decreased the overall size, and increased the reliability of the inverter. These devices carry their own thermal and electrical protection, substituting for MOSFETs or IGBTs, traditionally the Achilles' heel of AC induction power electronics. In addition to drive circuitry required by the IPMs, the inverter package also houses the main control circuitry. A speed signal from the motor through a magnetic pickup is used to close the control loop. An input at the accelerator pedal by the driver through a potentiometer, commands a required motor torque by varying the slip frequency.

The controller and motor are capable of alternator operation and hence allow regenerative braking. Simulations indicate that in a SFUDS cycle, this regenerative braking increases overall efficiency by 15%.

Having released the accelerator pedal, the first inch travel of the brake pedal provides direct regenerative braking. Maximum regeneration is achieved at the end of this first inch of travel. If this is not sufficient, the brake pedal can be depressed further and the vehicle's original hydraulic brake system is activated to support regenerative braking.

Air cooling would be inadequate to fully utilize the motor's capabilities. An integrated cooling and lubrication system utilizes turbine oil to lubricate all bearing surfaces and to cool the rotor and stator. Oil ports direct oil through the stator and rotor shaft, as well as through the controller.

TRANSAXLE — SHARP also designed the final reduction/differential unit (hereafter: final drive). The differential gears themselves were removed from the Escort transaxle to be used in our unit. Using the Escort gear package allowed us to keep the existing half-axles, which we saw no reason to replace. The bottom half of the final drive was manufactured to place the differential in exactly the same location as in the original Escort, and mounts were designed to attach to SHARP's lower mounting platform. The final drive reduction, a High-Velocity chain, extends vertically, where its pinion gear is concentric with the traction motor and planetary gear shafts. The final drive case is directly coupled and bolted to the planetary gear system, which in turn is bolted to the traction motor. These four components (motor, planetary gears, chain reduction, and differential) compose the car's unified traction drive system. The unified system, rigidly integrated, is then vibration isolated from, and mounted to, the vehicle structure.

<u>System Evaluation</u> — Though the system has not yet been sufficiently tested to permit a conclusive evaluation, preliminary testing indicates that this system is extremely

successful. In production quantities, the HV chain would almost certainly be replaced by gears. In addition, a more holistic design of the entire system would have lightened the system by some 15-25 kg by eliminating unnecessary adapter plates and a fully integrated casing design. The system is able to generate a large amount of power and do so very efficiently and in a compact space.

BATTERY SYSTEM: COMPONENTS 1,11,13

SPECIFICATIONS — The battery pack consists of 18 Saft STM 5.140 (6V) Nickel-Cadmium batteries. This is a nominal 108V, 140 A-hr pack which should provide 60-70 miles of range at a constant 55 mph and have a listed life cycle of 2000 cycles at 80% DOD. The pack weighs 300 kg when fully hydrated, and occupies 0.176 m^3. The batteries are recharged by two interconnected 3 kW rechargers.

BATTERIES — Batteries were evaluated on the criteria shown in Table 5, in the Appendix. Energy density is energy storage capability per unit weight (wh/kg); power density is power release and charging capability per unit weight (w/kg). Voltage density is a measure of how many volts per unit weight without a significant reduction in energy density (for example, since Ni-Cd batteries are chemically limited to 1.2V per cell, achieving a high system voltage with a realistic amount of weight calls for so many small cells that the weight of cell casing begins to approach the weight of the cell material itself: hence, Ni-Cd's have a poor voltage density). Temperature is a compilation of efficient operating temperature and charging temperature. The remainder of the criteria should be fairly self-explanatory.

Although contacts with a Nickel Metal Hydride (NMH) battery company led to a tentative agreement to provide Stanford with a test set, the company was not able to produce enough batteries for us before the HEV challenge. Due to their extreme cost and minimal availability, we were not able to procure Nickel Iron batteries. Based on availability and state of development, we purchased Nickel Cadmium batteries. We reject Ni-Cd batteries for use in a production vehicle for two reasons: the serious recyclability and potential health problems related to cadmium, and the scarcity of this element. NMH batteries offer similar performance and could be used in a production vehicle.

Our battery pack was designed to meet or exceed the following specifications:
- more than 55 mile ZEV range
- 60 kW required for acceleration and regenerative braking
- High enough running voltage to allow sufficient motor power and system efficiency
- Recharge fully in 6 hours
- Weigh under 350 kg to meet GVWR requirement.

RECHARGERS — The batteries are recharged by two interconnected, fully automatic 3 kW recharging units. These units are automatically controlled for quick and efficient charging. When charging begins, both units are active (charging at constant voltage and decreasing current) until the batteries reach their nominal 108V. Subsequently, one recharger turns off, while the other charger slowly reduces its output until the batteries are fully charged at 20% over their nominal voltage.

System Evaluation — Ni-Cd batteries offer the advantage of a well-developed technology. These batteries are well documented, and fairly commonplace both for household and electric vehicle applications. Although projected, the same is not yet true for Nickel Metal Hydride batteries, which offer similar performance. Nickel-cadmium bus voltage does not drop appreciably until over 80% depth of discharge. This advantage allows us to maintain 55 mph over most of the batteries range. Also, Nickel-Cadmium batteries have been shown to have long life cycles. Life spans of 2000 cycles (80% DOD) and user lives of 10 years or more are very common. While damage can quickly occur if excessive overcharging occurs (> 10% overcharge), this problem can be easily overcome with carefully designed rechargers.

The battery pack offers a large electric-only range which benefits the environment. However, despite weighing over 300 kg, this battery pack's nominal voltage is only 108V. This low voltage creates performance compromises, since most electrical components in the vehicle are current limited, not power limited. In addition, lower voltage necessitates higher amperage and therefore higher losses in wiring, fuses, contactors, and controllers. This problem would be reduced through the use of Nickel Metal Hydride batteries.

ALTERNATE POWER UNIT (APU): COMPONENTS 2,3,4

SPECIFICATIONS — The APU is powered by a Honda 250cc liquid-cooled V-twin 4-stroke gasoline engine, with grouped throttle-body computer controlled fuel injection. This engine is coupled to a 13 kW 3-phase AC induction alternator with controller.

ENGINE — The decision to build a series hybrid vehicle drove our APU strategy. We decided that the APU should provide sufficient power to drive the car on a flat road at highway speeds (13 kW, as determined by coast-down tests), and 25% extra to slowly recharge the batteries. This yielded an APU power requirement of 17 kW, and assuming 90-95% alternator efficiency, we concluded that we needed an engine which would supply a constant 18 kW.

Since our strategy called for our engine to run a constant speed at its most efficient point, we were left to research engines and find one whose emissions were low and efficiency was high at 18 kW.

We researched two stroke, four stroke, rotary, and gas turbine engines. The decision matrix based upon this research appears in Table 4, in the Appendix. As can be seen in the table, the two stroke engine had a definite power to weight ratio and size advantage over a four stroke, but fully developed two-strokes suffered emissions and efficiency handicaps. Although some information on recent two-stroke engine developments indicated that they could be considered a viable choice in the future, these developments have not yet led to fully developed, proven systems.

Rotary engines' high power to weight ratios and extremely compact size was also attractive, but we could not find any rotaries in the size range we required. One, slightly under powered, was made in England to power remotely controlled drone (target) aircraft. It's reliability was suspect due to its intended application, and its $9000 price tag and 3 month production lead time made it an unfeasible option.

Sealing problems remained a question, as did emissions, efficiency, and cold start difficulties.

Gas turbines were excluded due to high cost and limited development in small power versions as would be required for a hybrid electric vehicle.

Having decided upon a four stroke engine, power requirements of 18 kW (23 hp) excluded all car engines as being oversized. Three separate motorcycle engines met our requirements and seemed essentially equivalent for our application. The Honda VTR250 was selected because of the higher availability of engines, parts, and information regarding the engine.

Subsequent Engine Modifications — Having selected the engine, we made the following modifications to enhance its performance for our application:

1) Since they were unneeded, the transmission, clutch, starter motor, alternator, and flywheel were removed to conserve weight and reduce parasitic losses. This forced us to modify the engine case to both seal now-open oil passages and drive the alternator directly off the crankshaft. A flexible chain coupling helps to isolate the vibration inherent in the engine from the alternator.

2) A tuned exhaust system was created to minimize back-pressure losses and further decrease fuel consumption at 18 kW output.

3) To eliminate evaporative emissions and enhance engine performance, efficiency, and control, the carburation system was replaced by dual throttle body fuel injection and computer control system. This system allowed us to find the optimal operating point for the engine at 18 kW, while minimizing fuel consumption and emissions. This system also allowed us to tune the fuel system for efficiency and emissions control, rather than for power and drivability as in a standard motorcycle application.

4) To alleviate the temperature problems created by running this engine at constant high speed a larger radiator was installed to increase the cooling capacity. The radiator is roughly sized for a one liter engine and has an electric fan which pulls air through the radiator at all times when the engine is up to temperature. An oil cooler was also considered and prototyped, but was discarded in favor of increasing the water cooling capacity in order to reduce system complexity.

ALTERNATOR — Our first major alternator decision was whether or not the alternator needed to be controllable. As our overriding reason for designing a series hybrid was to allow for constant RPM operation of the engine, our alternator necessarily had to be controllable. Operating the engine at constant RPM necessitates a constant load be placed on the engine by the alternator. To accomplish this, the current and the voltage output from the alternator may vary, but their product must remain constant. Since the power bus voltage rapidly fluctuates with changes in motor current draw and regenerative braking, the alternator controller needed to compensate by varying the output current. In addition, we back-drive the alternator to start the engine, thereby eliminating the need for the engine's starter motor. This operation also necessitates alternator control.

Having concluded that we needed a controllable alternator, our decision on type matched our decision for type of traction motor. Again, due to weight and efficiency considerations AC induction seemed the logical choice. Further supporting this decision is the fact that the optimal speed for our engine was 8500-9500 RPM. Using an AC induction motor eliminated the need for a speed reduction between engine and alternator, as would have been required with a DC generator.

Due to factory shortages, we were forced to initially purchase a 13 kW alternator rather than the planned 17 kW alternator. Financing constraints made us continue with this smaller, suboptimal alternator.

System Evaluation — Though the engine has performed much to our expectations, some aspects of the system have surprised us. The engine's size has been somewhat of a liability. Due mostly to time constraints, we have not reduced the size of the engine crankcase to reflect the components removed. As such, and compounded with the bulky cooling jackets necessitated by a liquid cooled engine, fitting this engine within our crowded engine compartment has taken considerable efforts. Given more time, the size difficulty could be overcome. Except for those two difficulties, the engine has exceeded our expectations. It was readily adaptable to fuel injection, and removal of unneeded parts such as the transmission, clutch, and flywheel was not as difficult as expected. The power output of this engine is as we expected it to be, and its particularly low fuel consumption has exceeded our expectations (BSFC appears to be 4.6-4.7). General knowledge concerning this engine has been readily available, as have replacement parts. The engine has an excellent reputation and has proved exceptionally adaptable. This engine seems to be an excellent choice for our application.

The basic operation of this alternator has been acceptable for our prototype application. Some aspects of the alternator, however are less than optimal. Our current alternator generates less power than our initial designs called for. This lack of power is due to funding constraints. Also, the alternator, capable of less than 1/4 the power, is nearly the same size as our traction drive motor. By scaling our traction drive motor to meet our alternator requirements, we would trim 2-3 inches from the overall length and about 2 inches from the diameter of the alternator. Some of the discrepancy is because the alternator is air cooled, and therefore has large cooling fins which surround the case.

Using an AC induction alternator was a good decision, not only because it increased APU system efficiency by 4-7%, but because it eliminated the need for a speed reduction between engine and alternator.

Coupling the alternator rotor to the crankshaft has proven to be a more difficult problem to solve than was expected. Our work to lighten the rotating mass in the engine has exacerbated the harsh angular vibration experienced at the coupling to the alternator. Since a substantial portion of the rotating mass is in the alternator rotor, the coupling experiences high peak torque at each power stroke. The keyway in the alternator failed last year, and has been replaced with a splined joint. We are also considering adding back the flywheel to the engine to help smooth out the power delivery.

EMISSIONS

The hybrid vehicle concept is based largely upon its reduced emissions over normal automobiles. While a hybrid provides several unique methods of reducing emissions, the APU emissions of the hybrid vehicle must still be reduced.

SYSTEM OVERVIEW — Overall emissions are reduced in a hybrid concept by several means: decreasing total fuel consumption, increasing efficiency, constant speed operation, and maintaining the air-fuel ratio close to stochiometric. In the SHARP design, fuel consumption is reduced by using a small but powerful 250cc motorcycle engine. The engine is run at close to its minimum specific fuel consumption point, approximately 8500 rpm. To ensure continuous operation at peak efficiency, the engine's speed is held fixed at its optimum point. In addition, the engine, as indicated above, was retrofitted with a computer-controlled fuel injection system. This system allows for significantly faster and more accurate control of air-fuel mixture and emissions.

<u>Evaporative emissions</u> — Evaporative emissions in the crankcase are eliminated with the installation of the fuel injection system. Evaporative emissions from the fuel tank are controlled by the stock pressure control valve which routes fuel vapor back to the fuel tank. This system, included on the original Escort, was not modified.

<u>Exhaust emissions</u> — Exhaust gases are the primary source of emissions, and require one or more catalytic converters in order to eliminate HC, CO, and NOx. Since a disproportionate share of the exhaust emissions occur in the first few minutes of engine operation, before the catalytic converters reach light-off temperature, our design incorporates an electrically-heated catalytic converter (EHC). This system heats the catalytic converter before engine start (since the engine is started by the vehicle controller, it is easy to heat the catalytic converter 10-15 seconds before ignition). Preheating the catalytic converter eliminates the large amount of emissions normally released before a car's catalytic converter can reach operating temperature. Eliminating these emissions is extremely significant since a study by the California Air Resources Board estimates that these emissions account for 70-90% of all emissions released during a normal round-trip. A second down-stream catalytic converter provides most of the emissions reduction during normal engine operation.

Emissions are further reduced with the use of a fuel injection system. With the close-looped system (by use of an oxygen sensor), we are able to achieve near stochiometric air-fuel mixtures.

Final tuning of the engine is scheduled at a facility with exhaust monitoring systems. This final tuning will allow us to come closer to the team goal of ULEV status.

CONTROL SYSTEMS

As The Winds of Freedom is a prototype vehicle, many systems have individual control units. In addition to the overall vehicle controller, the AC induction motor and AC induction alternator, fuel injection system, and rechargers, each have a controller. The vehicle controller is discussed below. Each system controller is discussed in that system's description.

OVERALL VEHICLE CONTROLLER

SPECIFICATIONS — The vehicle controller is constructed from the following components.
- Microcontroller: Motorola MC 68HC16 - 16MHz 16-bit microcontroller with built-in Analog-Digital converter, general purpose timer, serial I/O unit.
- Sensor input: built-in A/D converter, general purpose timer.
- Data Output: Computer terminal / LCD connected through serial unit.
- Control Output: through D/A converter (Burr-Brown PCM56P)

SYSTEM OVERVIEW — Utilizing a Motorola 68HC16 microcontroller, the vehicle controller serves three distinct functions in the car. First, it carries the overall vehicle control strategy, monitoring battery state of charge and determining when to activate the APU. Second, it controls the APU output and ensures constant speed operation of the engine when active. Third, it monitors all vehicle systems, activates cooling systems, controls driver display, and activates warning lights when necessary.

The first function of the controller entails monitoring battery state of charge and activating the APU when necessary. The controller reads the operation mode (ZEV, Zero Emission Vehicle, Batteries Only; HEV, Automatic Hybrid Operation; or APU on) from a manual dashboard switch. ZEV mode and APU on mode override the HEV mode operational strategy. ZEV and APU on modes are required for the HEV challenge competition and for testing of this prototype vehicle. This switch would not be included in a production vehicle. In HEV mode, the controller monitors the battery level by integrating the current and voltage coming out of the battery. At 80% Depth of Discharge (DOD), the controller activates the APU (see Appendix, Figure 13). At about 85% DOD, the batteries' performance will begin to decrease sharply. However, 80% was selected because at that DOD, there is still enough energy left in the batteries to compensate for peak demands.

When the controller determines that APU activation is required, it proceeds through the strategy illustrated in controller strategy figure 12 (see appendix).

Having followed this flowchart, the engine begins constant RPM operation, the second controller function. To maintain constant RPM entails keeping a constant load on the engine. This is equivalent to keeping a constant power output of the alternator. Power is the product of the current and voltage output. Since the main power bus voltage fluctuates with acceleration and regenerative braking, the vehicle controller must fluctuate the alternator output voltage inversely to maintain constant power. The controller outputs a -6V to +6V to the alternator controller which varies the output current accordingly. This control method allows the engine to run constantly at its cleanest, most efficient speed.

The controller's third function, a system watchdog, is perhaps the controller's most vital function. The controller monitors critical temperatures in the motor, both battery boxes, the engine, and the alternator. It also monitors battery state of charge, fuel level, currents, and voltages. Recently installed, a speedometer has been successfully connected to the controller which uses the motor RPM to calculate the exact speed of the car. An odometer has also been implemented. The controller activates warning lights when it detects a system malfunction, and will, in potentially dangerous situations shut systems down automatically. The controller is normally powered by the DC-DC converter, but a separate 12V battery is included to ensure controller operation in the event of a battery pack failure.

BASIC CONTROL STRATEGY — The product of the main bus current and voltage is integrated to determine the total energy to and from the battery to determine the remaining battery level, and when running in hybrid mode, when to turn on or off the APU.

The controller controls the APU by varying the throttle position and alternator load on the engine. The RPM of the APU sub-system is closely monitored both for strategy and for ensuring proper operation. Stalling, over-heating and other malfunctions of the APU sub-system are dealt with by the controller, and the APU is shut down in an emergency.

When the APU needs to be turned on in either the APU only or HEV mode, the controller goes into the start-up cycle. First, the alternator is used as a motor to start the engine up to a set RPM, after which the engine will be run at a certain speed until the engine has warmed up. When the engine has reached the desired temperature, it is sped up by increasing the throttle, while the alternator load is carefully adjusted to control the RPM of the APU.

When the optimal point is reached, the controller goes into APU on mode. For optimal efficiency, the engine is kept running at constant speed and constant load. When the battery is fully recharged, or in any other situation when the APU needs to be shut down (apart from an emergency), the engine speed is decreased to idle after which the fuel injection will be cut off.

System Evaluation — The vehicle control system, is satisfactory for a prototype vehicle. Its functions provide exceptional vehicle safety and system monitoring. The control algorithm for engine control seems quite satisfactory. This system has not yet been fully tested and calibrated due to time constraints. Time permitting, parameters will be changed to enhance performance.

In a production vehicle, the overall control system would be integrated with all other control systems in the car to eliminate redundancy. Set algorithms, such as regenerative braking operation, could be varied based upon battery state of charge to ensure optimally efficient operation. In addition, integration of low power control systems would simplify electrical insulation and electrical noise reduction.

VEHICLE STRUCTURE

STRUCTURAL MODIFICATIONS — The structural modifications to the car are intentionally minimal. The most significant modification was a removal of the cargo area floor between the frame rails to accommodate the battery box. The area removed mostly served to house and support the spare tire, which has since been relocated to the backside of the rear seat. This modification, for safety reasons, did not cut or alter any of the structural members in the car. The floor was replaced by a fiberglass/Nomex composite box which houses the 13 rear batteries. As this box is rigid, it is isolated from the more flexible Escort structure. Steel straps are enclosed in the fiberglass structure of the box, which are then bolted to the frame of the car through rubber vibration isolators.

We made small modifications in the engine compartment (cutting of sheet metal) to install various components such as the engine, DC-DC converters, and radiators. Various holes were drilled and sealed in the firewall to allow proper wire routing. We were able to rout most wires and hoses through holes which existed in the original Escort.

We modified the dashboard and instrument panel to incorporate appropriate indicator lights and gages, fire extinguishing system and power interrupt controls, and operational switches. In keeping with SHARP's goals, we were able to make most of these modifications while preserving the original Escort's ergonomic features.

SUSPENSION — Final curb weight is about 3200 lbs (700 lbs heavier than a production Escort). However, the weight distribution improved from the original 59% front/41% back to 51/49. To compensate for the change in weight and weight distribution, we replaced the original springs with stronger and stiffer ones. In addition, we replaced the fluid in the shock absorbers with a more viscous fluid, to achieve approximately the same damping rate and ride as in the original Escort.

System Evaluation — SHARP was successful in achieving its goal of minimal structural modification, even though many aspects of the Escort's original design (soft handling and a flexible body) made it less suited than other cars to adding 300 kg of batteries in the rear. Strengthening of and rigidly connecting the rear strut towers makes the car handle better with or without the batteries in back. We were pleased with the adaptability of the original instrument panel to a hybrid design. The original instrument panel was equipped with enough indicator lights and gages that no additional positions would be needed in a production vehicle. (Some additional displays, such as an oil pressure gage were incorporated in the car to allow us to monitor the operation of various operating systems.)

MANUFACTURABILITY

SHARP's design is very manufacturable. Our design uses custom components, but these could be mass manufactured to reduce costs. As indicated in the system evaluations, the controls and cooling systems should be integrated for a production vehicle. A gear drive could also replace the HV chain drive we use now. The Ni-Cd batteries should be replaced with NMH batteries. However, the overall system is very manufacturable. The car's reliance on AC motors makes the car inexpensive to mass produce. The year of design to make SHARP's traction motor easy to manufacture also facilitates production of this car. Finally the lack of major vehicle structure modifications demonstrates the easy adaptability of this car to an HEV.

Figure 11, The Winds of Freedom

The SHARP hybrid electric vehicle.

APPENDIX

REQUIREMENTS FOR MOTOR DESIGN:

Table 1, Physical Requirements

System	Specification	Reason
Motor	15,000 RPM Continuous	SHARP vehicle performance spec.
Motor	Compatible housing	Simple integration, cost
Motor	Mate with existing drive train	Simple integration, cost
Motor	Not made of exotic materials	Reduce production cost
Cooling	Keep rotor below 120C (220C peak)	Motor reliability, efficiency
Cooling	Keep stator below 120C (220C peak)	Motor reliability, efficiency
System	Cabin noise less than 50db	Passenger comfort
System	Weight under 100 lb	Meet GV weight specifications for Hybrid Vehicle Challenge

Table 2, Functional Specifications

System	Specification	Reason
Motor	22 ft-lb @ 11500 RPM	Sustain 6% grade at 55 mph
Motor	70 ft-lb @ 100 RPM	Climb a 30% grade for 10 seconds
Motor	75 ft-lb regenerative braking	Increase overall vehicle efficiency
Motor	100,000 mile operation	Vehicle reliability
Motor	Low electrical noise	System functionality
Motor	Interface with mounting system	Integration with vehicle system
Motor	Interface with drive train system	Integration with vehicle system
Cooling	1.8 KW from rotor	Motor efficiency, control rod temperature
Cooling	1.04 KW from stator and windings	Motor efficiency, control winding temp
Cooling	No winding insulator breakdown	Retain motor integrity
Vehicle	Simplified Federal Urban Driving Schedule (SFUDS)	Prove hybrid vehicle is viable alternative
Vehicle	Adequate torque	0 to 60 MPH in 13 seconds
Vehicle	Low mechanical noise	Driver and passenger satisfaction
Vehicle	Prevent electrical shock	Safety under normal operation / eliminate liability issue

Table 3, External Requirements

System	Specification	Reason
Motor	Accept bus voltage of 0 to 150 volts	SHARP vehicle battery and generator voltage range
System	Function on 30o horiz and vert incline	SHARP vehicle operating conditions
System	Function in temps. 10-100°F	SHARP vehicle operating conditions
System	Resist climatic effects (rain, snow)	SHARP vehicle operating conditions
System	Resist foreign object damage	SHARP vehicle operating conditions
System	Tolerate 2g horizontal & vertical accelerations	Maximum automotive operational loadings
System	Function up to 10,000 ft. elevation	Possible operating conditions
System	Compatibility with existing gear box	Simple integration into vehicle
System	Standard operator interfaces	Safe operation and ease of use
System	Variable torque capability	Vehicle performance requirements
Cooling	Easily accessible cooling system	Easy maintenance

SELECTION MATRICES

Table 4: Engine Selection Matrix

		Four Stroke			Two Stroke				Gas Turbine
Criterion	Weight	Car	Motor cycle	Wankel	Orbital	Ultra light	Motor cycle	Outboard	
Emissions	5	10	8	4	1	4	4	4	9
Cost	3	4	10	2	3	2	10	5	1
Power / Wt.	3	7	7	10	1	3	8	5	5
Availability	1	10	10	5	4	5	7	7	1
Fuel Cons.	4	7	10	5	1	4	8	5	5
Development	6	10	9	7	1	5	9	8	3
Size	4	1	7	10	2	6	8	6	6
Reliability	4	10	8	6	1	4	8	6	5
Total		225	<u>295</u>	187	43	126	231	173	155

Table 4: Battery Selection Matrix

Criterion	Weight	Pb-Acid	Ni-Cd	Ni-Metal-H	Ni-Fe	Na-S	Bi-polar*
Energy Density	5	1	7	9	8	8	10 (est.)
Power Density	4	1	9	9	8	6	10 (est.)
Voltage Density	2	10	1	8	8	5	10 (est.)
Temperature	2	9	9	9	9	1	10 (est.)
Availability	2	10	7	2	2	1	?
Projected Avail.	4	10	9	5	4	3	?
Recyclability	3	3	1	9	9	2	3
Cycle Life	2	6	10	10	9	5	6
Production Cost	4	10	8	9	7	5	?
Current Cost	1	10	4	3	2	2	?
Total		178	199	<u>225</u>	202	128	N/A

APPENDIX C: CONTROLLER STRATEGY FLOWCHARTS

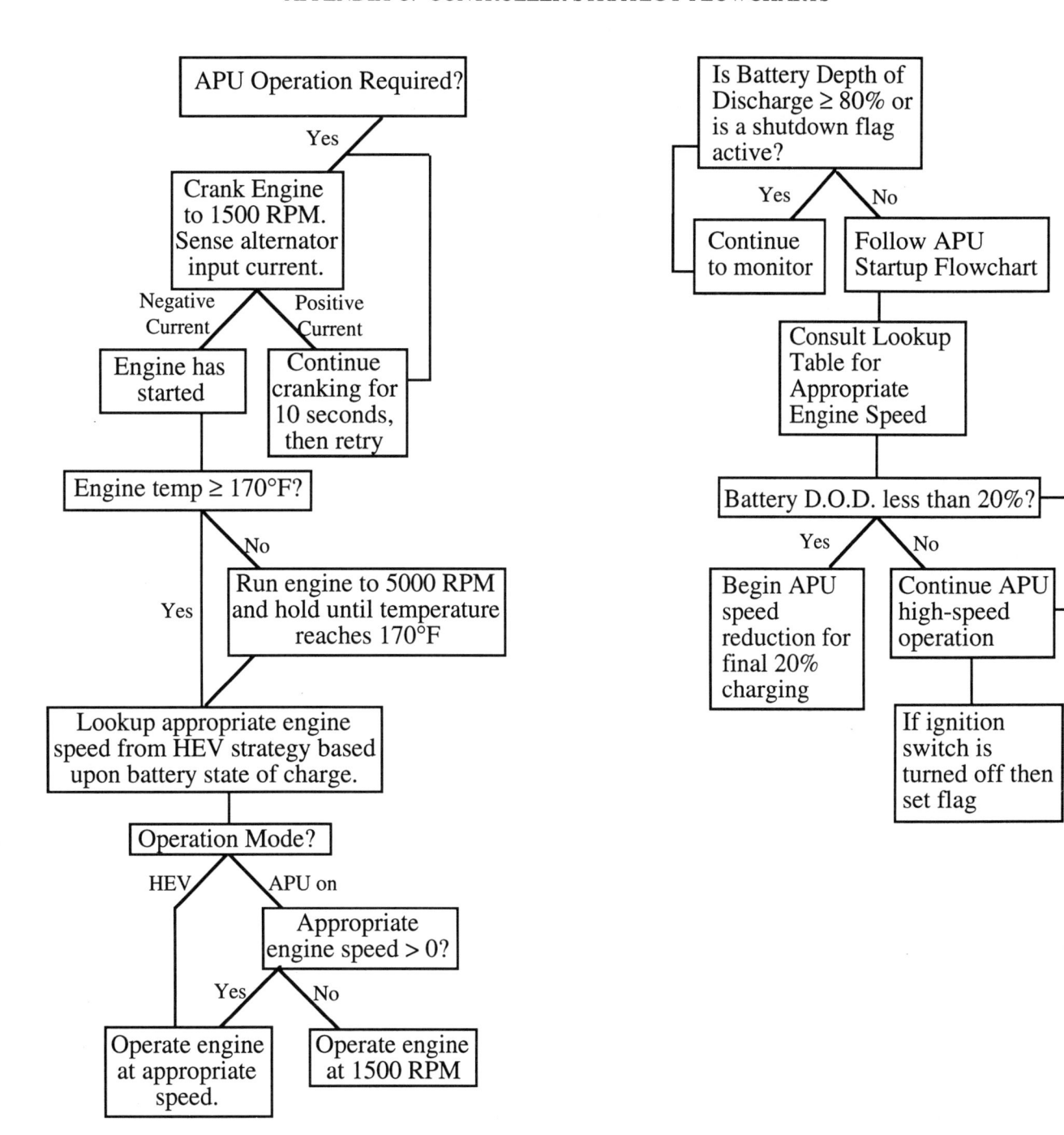

Figure 12: APU Startup Procedure

- APU Operation Required?
  - Yes → Crank Engine to 1500 RPM. Sense alternator input current.
    - Negative Current → Engine has started
    - Positive Current → Continue cranking for 10 seconds, then retry (loop back to Crank Engine)
- Engine temp ≥ 170°F?
  - No → Run engine to 5000 RPM and hold until temperature reaches 170°F
  - Yes → Lookup appropriate engine speed from HEV strategy based upon battery state of charge.
- Operation Mode?
  - HEV → (continues)
  - APU on → Appropriate engine speed > 0?
    - Yes → Operate engine at appropriate speed.
    - No → Operate engine at 1500 RPM

Figure 13: HEV Mode Operation

- Is Battery Depth of Discharge ≥ 80% or is a shutdown flag active?
  - Yes → Continue to monitor (loop back)
  - No → Follow APU Startup Flowchart
- Consult Lookup Table for Appropriate Engine Speed
- Battery D.O.D. less than 20%?
  - Yes → Begin APU speed reduction for final 20% charging
  - No → Continue APU high-speed operation
    - If ignition switch is turned off then set flag

263

THE UNIVERSITY OF TENNESSEE'S HYBRID ELECTRIC VEHICLE: THE 1994 HIGH POTENTIAL

Authors: Bryce Anderson, Ken Augustinovich, Kevin Barfield, Krista Conner, Ramon Gonzalez, Teddy Miller, Steven Mundy, Scott Sluder, John Taylor, Brent Thomas, and Nicole Vickery

Project Manager: Lori Snook

Faculty Advisors: Jeffrey Hodgson, John Snider, Fred Symonds

The University of Tennessee College of Engineering, Knoxville, Tennessee

ABSTRACT

Students from the University of Tennessee's College of Engineering have successfully designed and built a hybrid electric vehicle to compete in the "ground-up" division of the 1994 Hybrid Electric Vehicle Challenge. The 1994 "High Potential" is a two passenger, series hybrid electric vehicle which uses an alternative power unit (APU) fueled with reformulated gasoline to extend the range of the vehicle. This APU is an air-cooled 623 cc displacement, two-cylinder spark ignition engine equipped with port fuel injection. The emission control system features an electrically-heated three-way catalyst and exhaust oxygen feedback to the fuel metering system. The vehicle is powered by an air-cooled brushless permanent magnet DC motor and a 180 Volt pack of deep cycle lead-acid batteries having a total energy storage capacity of 7.46 kW-hr. The vehicle's energy management control system monitors the battery pack state of charge and activates the APU as required. The vehicle chassis is a steel space frame carrying a carbon fiber composite body. Simplicity of design, cost effectiveness, acceptable performance, and low environmental impact were the focus of this hybrid electric vehicle design effort. The 1994 High Potential has a pure electric range of 46 km and acceleration capability of 0 to 75 km/hr in 10 seconds.

INTRODUCTION

A serious effort exists in the United States and around the world to develop alternatives to conventional gasoline fueled vehicles. One promising alternative is the electric vehicle (EV). Electric vehicles are not new...they have been around for over a century (1)[*]. In fact, in the 1890's and 1900's electric vehicles were considered to be superior to vehicles

---

[*] Indicates reference at the end of the paper.

fueled with gasoline due to the noise and exhaust associated with the latter and three-quarters of the vehicles produced in 1900 were powered by electricity or steam (1). However, the course of history was changed with the introduction of assembly-line produced gasoline engines in 1904.

It now appears that we may be on the verge of coming full circle. Once again EVs are gaining attention due to the undesirable pollutants that are emitted from gasoline powered vehicles. In 1990, the California Air Resources Board (CARB) adopted more stringent tailpipe emission standards for light-duty vehicles, beginning with the 1994 model year (2). These regulations include a mandate to produce and offer for sale zero emission vehicles (ZEV). Each major manufacturer is required to produce ZEVs at a rate of 2% of total light duty vehicle California sales, beginning with the 1998 model year (2,3). This requirement increases to 5% of total California sales in 2001, and 10% of California sales in 2003, when all but the smallest vehicle manufacturers must produce ZEVs. Right now, "ZEV" means "EV". The only emissions associated with EVs are the emissions from the electricity generating systems, which in California are extremely "clean" (2). There are, however, several problems to overcome before electric vehicles can be re-introduced into the U.S. market. These problems include the relatively low power and energy densities associated with batteries, the lack of a recharging infrastructure, the lengthy recharging time, and public acceptance of EV performance.

One way to bridge the gap between traditional gasoline fueled vehicles and EVs is with the hybrid-electric vehicle (HEV). The official definition of a hybrid road vehicle is one whose specified operation requires propulsion energy to be available from two or more kinds or types of energy storage, sources, or converters; with at least one store on-board. An HEV is a hybrid vehicle in which at least one of the energy stores, sources, or convertors can deliver electric energy (4). The HEV offers several advantages over EVs as an interim vehicle. These include a longer range, no special infrastructure needs, a ZEV mode when needed, and backup power for weak batteries. Additionally, the California Air Resources Board is considering offering ZEV credit to manufacturers that can demonstrate alternative power unit emissions which are no more than the power plant emissions generated to provide electricity to charge the vehicle's battery pack (2).

As part of the research into Hybrid Electric Vehicles, students from the University of Tennessee's College of Engineering have designed and built an HEV, the 1994 High Potential, for the purpose of competing the 1994 HEV Challenge. This paper details the design, construction, and performance of this unique vehicle.

DESIGN DETAILS

Power-Train Configuration

OVERALL CONFIGURATION - The 1994 High Potential is a series Hybrid Electric Vehicle. The decision to use a series drivetrain configuration rather than a parallel setup was made for the following reasons:

1) Potentially higher overall efficiency
2) Lower weight
3) Greater ease of implementation
4) Potentially lower APU emissions

In the series configuration, there is no direct connection between the drive wheels and the alternative power unit (See Figure 1). The APU is directly connected to the generators which, electrically in parallel with the batteries, power the electric motor to turn the wheels through a transaxle. A discussion of each powertrain component follows.

MOTOR - The 1993 University of Tennessee HEV team decided that the vehicle should be capable of accelerating from 0 to 75 km/hr in less than 10 seconds. The required motor size was then approximated with the aid of computer simulations including the SIMPLEV modeling software developed at the Idaho National Engineering Laboratory and EMACC (a BASIC program developed in-house for acceleration simulations). After a thorough search, the team chose the Unique Mobility SR180 motor and controller. The SR180 is a permanent-magnet air-cooled motor which operates from a DC supply up to 200 Volts. However, maximum performance is achieved from a 180 Volt DC supply. The motor is rated at 50 hp (peak) and 25 hp

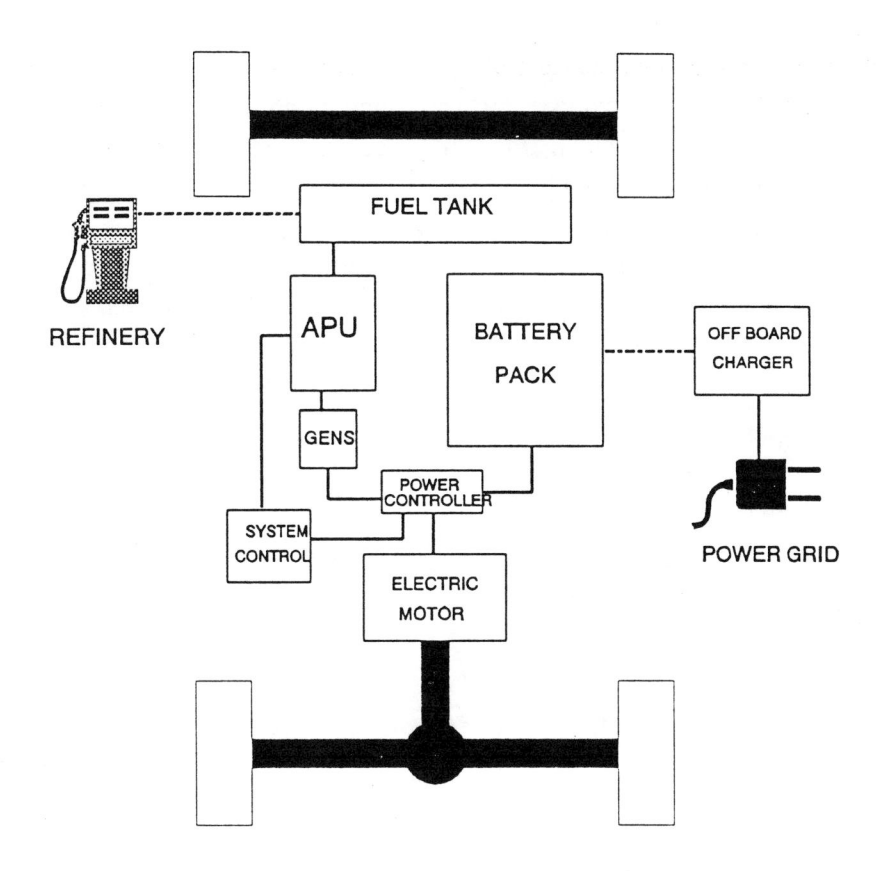

Figure 1 Series Drivetrain Configuration

(continuous) (see Figure 2). The SR180 is current limited by its controller to 275 amps.

GENERATORS - The electrical generating system requirements of the UT High Potential are delivery of 50-70 amps at 180 Volts, microprocessor controllability, high efficiency, and minimal weight. The 1994 UT HEV team conducted testing on two different generator systems to see which would best meet these requirements. One incorporated a single generator, which had been rewound to produce 180 volts. The other system consisted of two 90 volt Leece-Neville alternators. Mounting brackets were fashioned for both machines and testing was done using a General Electric DC dynamometer. The generator load was provided by a variable resistance load bank. The test results showed the Leece-Neville generator out-performed the other system in both current output and efficiency. Therefore, two Leece-Neville alternators are used on the 1994 High Potential. Table 1 compares the data obtained for the two generator systems, and Figure 3 shows the experimental data measured from the Leece-Neville generator system.

Table 1 Comparison of Experimental Generator Data

System	Max Output Current (amps)	Efficiency	Weight (lbs)
Leece-Neville	75	85%	56
Rewound Generator	60	72%	110

The field current in the Leece-Neville generators is regulated as a means to vary the output power. The on-board microprocessor determines the state of charge of the battery pack and sets the power needed from the generators by establishing (using a table look up method) a suitable combination of APU speed and generator field current settings. The field current is regulated by a student designed voltage controlled current source (VCCS) which takes an input voltage of one to four volts from the microprocessor to produce a generator field winding current of one to four amps.

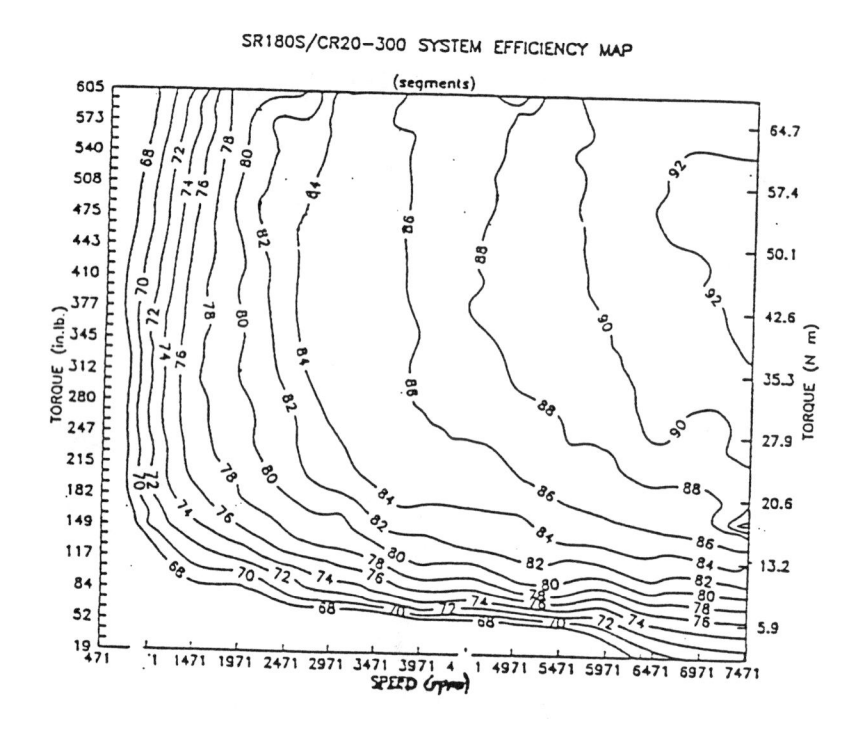

Figure 2 Electric Motor Characteristics

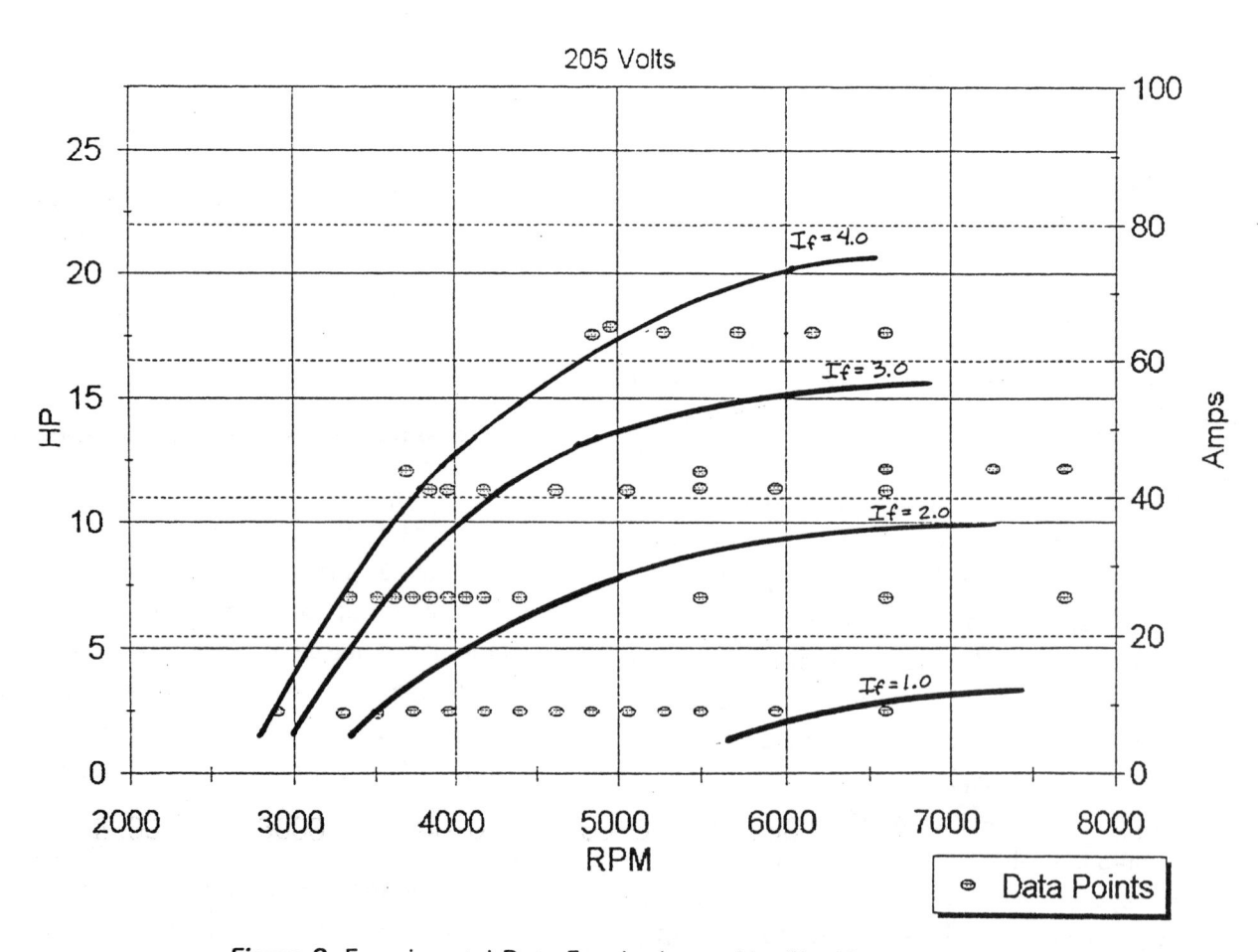

Figure 3 Experimental Data For the Leece-Neville Alternators

DC to DC convertors are used to transfer power from the 180 volt battery pack to the appropriate devices. Figure 4 depicts a block diagram of the major components of the electrical power system.

ALTERNATIVE POWER UNIT - The function of the Alternate Power Unit (APU) is to provide energy to drive the twin alternators which in turn provide power to the electric drive system. For this purpose, the APU can operate at a constant speed, eliminating the inefficiency associated with transient speed conditions. The selection of an APU was based on considerations of weight, volume, efficiency, and power output. The advantage of the series HEV is that the APU can be sized to meet the required <u>average</u> load and can operate at a fixed point that is independent of the instantaneous <u>vehicle</u> operating point. Additionally, the engine should be capable of producing its rated power at a relatively low speed. Based on these considerations a Kohler CH-20 industrial engine was selected as the APU. This engine is a lightweight, air-cooled engine which can be purchased as a unit.

In order to prepare the APU for the UT hybrid electric vehicle, several modifications were necessary. The original engine incorporated a magneto (fixed advance) ignition system with a carburetor to control the air-fuel ratio. In order to provide better control for the APU in this application, a new engine controller was fit to the engine. This controller provides closed-loop EGO control of the air-fuel ratio and enables the use of fuel injection as a fuel metering system. A Bosch fuel injector was fitted to the intake manifold of the engine at each intake port, the existing carburetor was utilized to throttle the engine, and the throttle shaft was fit with a servomotor to enable the system to control the engine speed. The APU was coupled to a small eddy-current dynamometer to permit

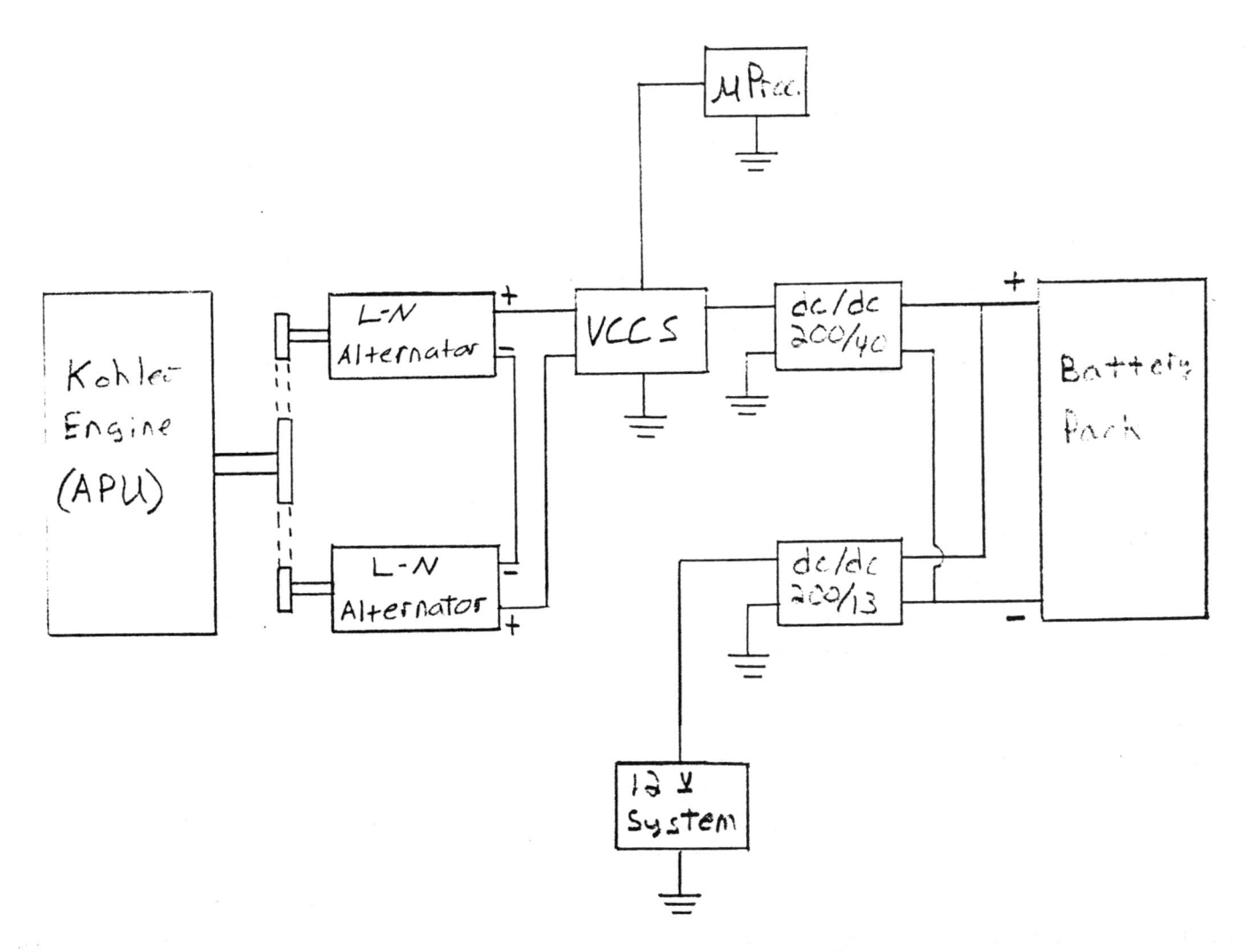

Figure 4 Schematic of the Electrical Power System

optimization of the engine spark advance and air-fuel ratio. The result is a brake specific fuel consumption of 0.299 kg/kW-hr (0.491 lbm/hp-hr) at the APU operating point of 16 BHp and 3100 RPM (5). Based upon modeling of the vehicle using SIMPLEV, this suggests a fuel economy during maximum charging of 17.86 km/l (42 mpg). Testing of the engine revealed a 13.5% power loss from that published by the manufacturer - a result of operating the engine at stoichiometric conditions as opposed to the slightly rich conditions for which the engine was originally designed.

Reformulated gasoline was chosen over M85 and E100 as the fuel for the engine due to its higher energy density. Although M85 and E100 normally provide a lower emissions potential than gasoline, the constant speed operation of the APU enables gasoline to provide lower than normal emissions without sacrificing space in the vehicle for a larger fuel tank. Use of either M85 or E100 would require a substantially larger fuel tank.

BATTERIES - Simulations were performed by the University of Tennessee's 1993 HEV team which showed that a battery pack energy capacity of 7.46 kWh was needed to meet the 33km (20 mile) (@ 50 miles per hour) ZEV range requirement (6). The 1994 ZEV range requirement is 25 miles at an average speed of 30 miles per hour. Tests consisting of running the car in ZEV mode on a chassis dynamometer were conducted to see if the same battery pack would meet the new, 1994 requirements. These tests showed that the average current draw went down sufficiently as the average speed decreased, that the same battery pack energy capacity rating of 7.46 kWh should allow the car to meet the 1994 ZEV range requirements.

Since the SR180 motor achieves its rated power ratings at 180 volts, the battery pack consists of 15 deep-cycle twelve volt batteries connected in series. The type of battery used in the battery pack is flooded electrolyte lead-acid. There are four main reasons for choosing this type of battery:

1. Lead-acid technology has been proven to work well through extensive testing and information is readily available (7).

2. Lead-acid batteries can handle slight overcharging without permanent damage (unlike some sealed types) (7).

3. Lead-acid batteries do not require a special environment such as adding heat (7).

4. Lead-acid batteries are economical, readily available, and there is already a recycling system in place to handle them (7).

In an article in the November 1992 edition of the IEEE periodical "Spectrum", lead-acid were predicted to be the dominate battery for the next few years until other types develop more fully.

Trojan Corporation brand 22NF wheelchair batteries were used on the University of Tennessee's 1993 HEV. Electrotek Concepts in Chattanooga, Tennessee ran tests on the 22NF to check its suitability for this project. Comparing their data and the data from the ZEV dynamometer tests that were conducted this year, the 22NF was found to meet the requirements for the 1994 competition. The 22NF is rated at 56Ah at the 20hr rate (41.4Ah at the 3hr rate). Electrotek's data shows that at a current draw of 26 amps the 22NF will discharge in 68.43 minutes. Our ZEV dynamometer test showed that at 30 miles per hour the average current draw was 25 amps and the depletion time was approximately 57 minutes. At 48.3 km/hr (30 mph), Electrotek's data gives a ZEV range of 55 km (34.2 miles) while our test gives a range of 45.9 km (28.5 miles) both of which meet the 40.25 km (25 mile requirement). One note to make is that our test was run using a 180 volt battery pack consisting of 15 used 22NF's while Electrotek was using a 24 volt pack made of two new ones.

Tests were also run on a single twelve volt 22NF to try to estimate the percent charge left in the main battery pack by reading the voltage and discharge current. A single twelve volt 22NF should simulate the 180 volt main pack because the current draw from each of the individual 12 volt batteries in the pack should be approximately the same. The tests involve draining the battery down while measuring the voltage, discharge current, and time. At first, specific gravity was also measured but this approach was abandoned because it the specific gravity would take to long to measure during the high current tests. At each data point the power, in kW, and the kilowatt-hour was calculated. The kilowatt-hour readings were "summed" by adding each individual value to all the preceding values. This integration gave the total kilowatt-hour output. The kilowatt-hour value corresponding to a battery voltage reading closest to 8.6 volts was set as the zero percent charge point. The electric motor has a low voltage limit of about 130 volts which is approximately 8.6 volts across each twelve volt battery in the main pack, explaining why 8.6 volts was chosen as the zero percent point. The 100% charge point was calculated using the initial voltage and current readings at time equal to 0+. The initial voltage was measured before the load was turned on and the initial current was calculated by dividing the initial voltage by the average resistance. The percent charge values were then calculated for each data point. The data was plotted on a graph of voltage versus current with each point being assigned its percent charge value. The lines of 10%, 20%, 30%, and etc. were found graphically. The end result is the graph in Figure 5. The information on this graph is entered into the microprocessor and allows it to approximate the percent charge of the main battery pack by reading the battery pack voltage and the discharge current.

The main battery pack is located in a tunnel between the driver and passenger seats as in Figure 6. The battery pack is lowered from underneath the car. To isolate the pack from the tunnel walls, a Lexan cover is secured over the top of the pack itself. This also serves to protect anyone who removes the pack from the car. The individual twelve volt batteries are separated in the tray by composite spacers held in by bolts or J-hooks as in Figure 7. Each twelve volt battery has dimensions of 9.5in long, 5in wide, and 8.812in high. Air is drawn up from the open-bottomed tray and vented out the top of the tunnel by a 12cfm fan.

HIGH-VOLTAGE WIRING - The high-voltage wiring system uses 2/0 welding cable. This neoprene covered cable has an ICEA current rating of 375 amps and a voltage rating of 600 volts making it more suitable than regular battery cables which cannot handle these high currents and voltages on a continuous basis. Battery connectors were chosen to allow for the shortest lengths of cable which will provide less weight and resistance. Colored battery terminal protectors and cable, red for positive polarity and black for negative, were used to insure that the

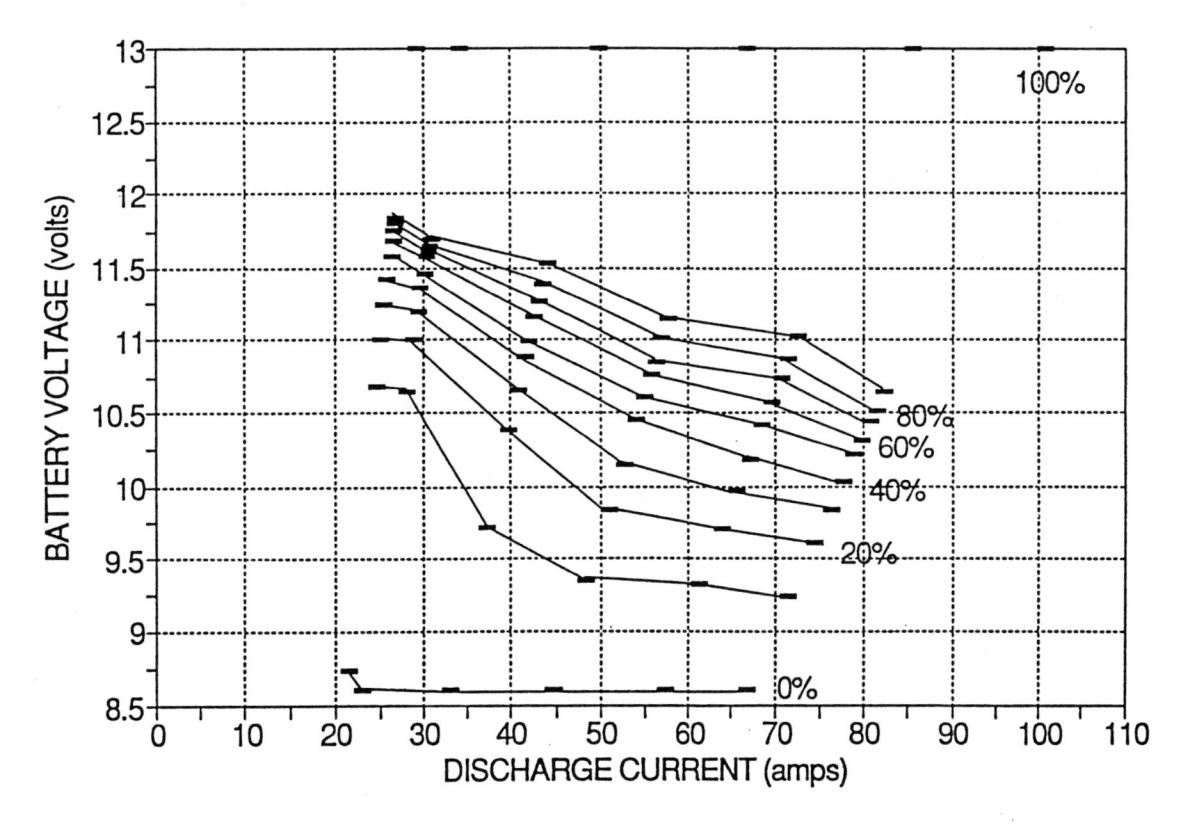

Figure 5 Experimental Battery Data

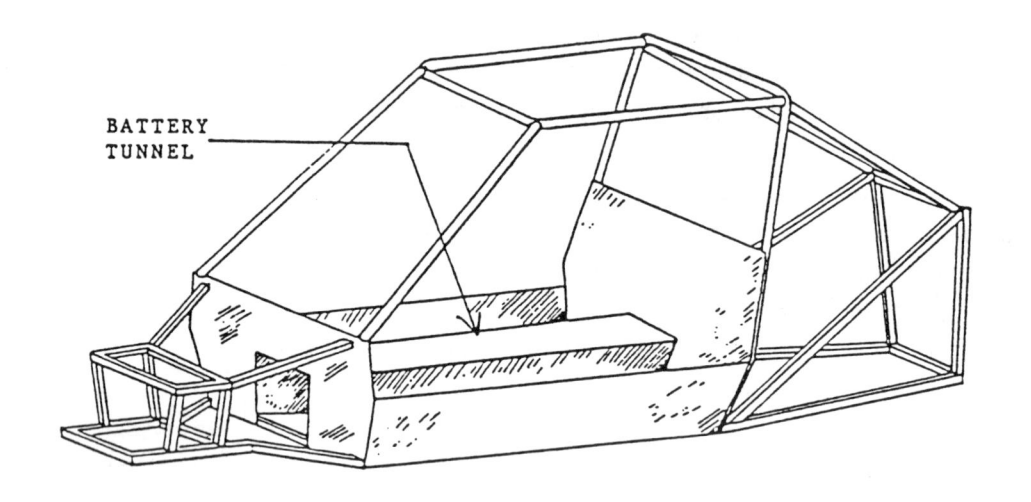

Figure 6 Chassis View Showing Battery Tunnel

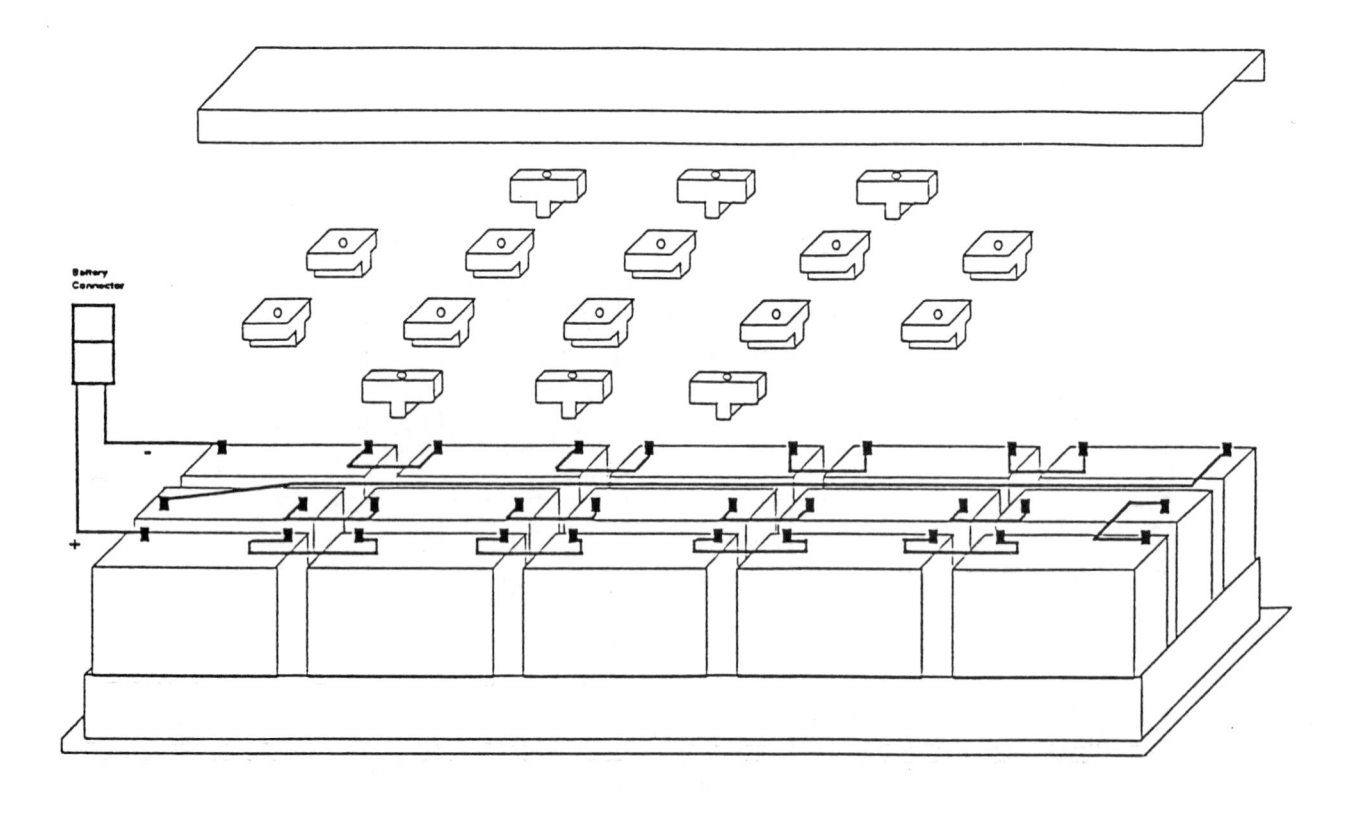

Figure 7 Battery Placement with Shield and Mounting Hardware

polarity throughout the high voltage system is always correct. Heavy wall copper lugs are used on all connections in the high voltage system requiring terminal lug connection. Each sub-circuit of the high voltage system is fused. The motor controller circuit has a 300 amp fuse while the generator circuit has a 100 amp fuse and the main battery circuit a 400 amp fuse. Current shunts are used in the ground legs of the motor, generator system, and battery pack. The shunts are used by the microprocessor control to monitor the operation of the electrical system. A manually actuated 400 amp General Electric switch is used to electrically disconnect the main battery pack from the rest of the electrical system. All connections for the high voltage system are made in an enclosure on the back firewall. The main disconnect switch is mounted in this enclosure such that it can be operated from inside the car beside the driver's seat.

OFF BOARD CHARGER - Out-of-vehicle charging of the battery pack is accomplished by the use of a Good-All Electric model C90-1Z 260-12 offboard battery charger. This charger is a 240 Volt DC charger, which can be adjusted to 216 Volts which makes it capable of charging the battery pack in a maximum of 6 hours with an output current of 12 amps. It has an isolation transformer located between the AC source and the charger and also has a DC output breaker installed. The charger requires 220 Volt AC single phase service and is built in accordance with NEMA PE-7 specifications with operating temperatures between 0 and 50 degrees Celsius.

TRANSAXLE AND TIRES - Based on acceleration simulations, a Volkswagon GTI transaxle was chosen with the first three speeds used in the acceleration run. In order to reduce the tractive power requirements, special low rolling resistance tires provided by Michelin are used. These tires have a coefficient of rolling resistance of 0.0067.

Control Strategy

The control system used on the 1994 High Potential is entirely passive, as required by the HEV Challenge rules. The center of the control system is a Motorola 68HC11 EVB. Hardware circuits are used to interface between the microprocessor and the vehicle, and a modular software program executes on the microprocessor. Several vehicle subsystems are interfaced to the 68HC11. Each system is described below.

Uniq motor: Since the Uniq motor is connected to battery ground, it is isolated from the control systems. Relays are used to enable the motor controller, potentiometers are used for the accelerator pedal and for the regenerative braking control, and an optoisolator chip is used to interface the tachometer to the microprocessor.

APU: The APU has its own engine control system so the only power outputs from the microprocessor to the APU involve relays to switch the electrical power on and off and to energize the starter. In addition, a throttle control signal is sent from the microprocessor to the throttle actuator. A tachometer signal from the APU is connected to the microprocessor for feedback control.

Generators: A student-built generator regulator controls the field current on the generators. A DC signal from the microprocessor controls the field current.

Watt-hour meter: The Watt-hour meter sends out a data line of serial output every second. This output is connected to the serial communications interface through an optoisolator chip.

Current Shunts: There are two current shunts for the main battery pack and the generator output. These signals are fed into the Analog to Digital converter on the 68HC11.

Dash: There are four lights on the (Saturn) dash used for giving the driver information. These are the battery overcharge light, the gear shift light, APU failure light, and the APU controller failure light. In addition, there is a tachometer gauge and a speedometer gauge on the dash.

The software for the 68HC11 is written in Motorola assembly language. The routines are written in a modular format to simply development, testing, and updating. Control of the APU throttle and the generator field current is accomplished in interrupt routines.

The control strategies for the three modes of operation are as follows:

ZEV mode: The microprocessor has a very limited role in ZEV mode. It simply reads in the system status and outputs the necessary information to the dashboard for the driver. This information includes the gear shift light and the battery overcharging light.

HEV mode: The controls are more involved for HEV mode. When the vehicle enters HEV mode, the microprocessor reads in the voltage and current of the main battery pack from the Watt-hour meter and current shunt, respectively, and determines the state of charge of the battery. If the battery is below the minimum state of charge set point, the microprocessor engages the APU/generator combination. A delay for the electrically heated catalyst warm-up prior to engine start is used. The microprocessor then reads the APU tach output and varies the generator field current setting and APU throttle setting to keep the APU speed within a certain range. Once the battery pack has been recharged, the APU and generator combination are turned off. In addition, the microprocessor warns the driver if a situation occurs that could damage the vehicle.

APU on mode: APU on mode is exactly like HEV mode, except that the APU and generator are turned on immediately when the mode is turned on and they stay on until the mode is turned off.

Emissions Control

To date, the most successful emissions control strategy for light duty vehicles involves the use of a three-way catalyst (TWC) in combination with exhaust-gas oxygen (EGO) sensor feedback control of the engine's air-fuel ratio. Feedback control is required because the three-way catalyst is most effective in reducing oxides of nitrogen (NOx) while simultaneously oxidizing carbon monoxide (CO) and unburned hydrocarbons (UHC) when the engine air-fuel ratio is essentially stoichiometric. This stoichiometry can be expressed in terms of the fuel-air equivalence ratio, ϕ, defined as $(F/A)/(F/A)_{stoich}$. Fuel rich mixtures ($\phi > 1$) result in low CO and UHC conversion, while fuel lean mixtures ($\phi < 1$) give low NOx conversion. This is shown in Figure 8.

Tailpipe emissions for the University of Tennessee's 1994 High Potential HEV are controlled by operating the APU in a closed-loop system with feedback from a Bosch heated exhaust gas oxygen (EGO) sensor. The heated oxygen sensor quickly provides an accurate assessment of the exhaust condition and allows the fuel metering system to make corrections sooner than a conventional EGO sensor (9). It also assures that the EGO sensor remains active even under extended engine idling conditions when unheated EGO sensors may cool off and become inactive.

An electrically heated, close-coupled, three way catalyst system from Corning is utilized on the UT vehicle to control engine-out emissions of CO, UHC and NOx. This system is designed to enable the catalytic converter to light off within five seconds of engine start, drastically reducing startup emissions. The electrically heated metal substrate requires four watt-hours of electrical energy to initiate the catalytic conversion, and this energy is provided from the vehicle's high voltage system. Tests performed by Corning indicate that this catalyst system is capable of exceeding even the stringent California 1997 ULEV standard (10). Since the APU operates at constant speed, the need for devices aimed at reducing power enrichment emissions is eliminated, resulting in an emissions control system that requires less energy for operation and contributes less to the overall weight of the vehicle.

Additionally, the vehicle is fit with a standard carbon canister in order to control evaporative emissions from the fuel tank. This device collects the vapors from the fuel system and burns them in the engine when the vehicle is in HEV mode.

Suspension Design

The suspension utilizes existing components with student designed geometry and modifications. The use of existing components saved the engineering effort of designing and building the structural members. The roll centers and camber gain curves were calculated using student written computer programs. The roll centers were designed at 3 inches in the front and 4 inches in the rear. The camber gain was designed to maintain the 1 degree negative camber up to the maximum body roll of 5 degrees. The car has relatively stiff springs and sway bars to aid in vehicle handling in events such as the commuter challenge.

The front suspension uses unequal length A-arms with concentric spring/damper assemblies. The components are from a Triumph sportscar with

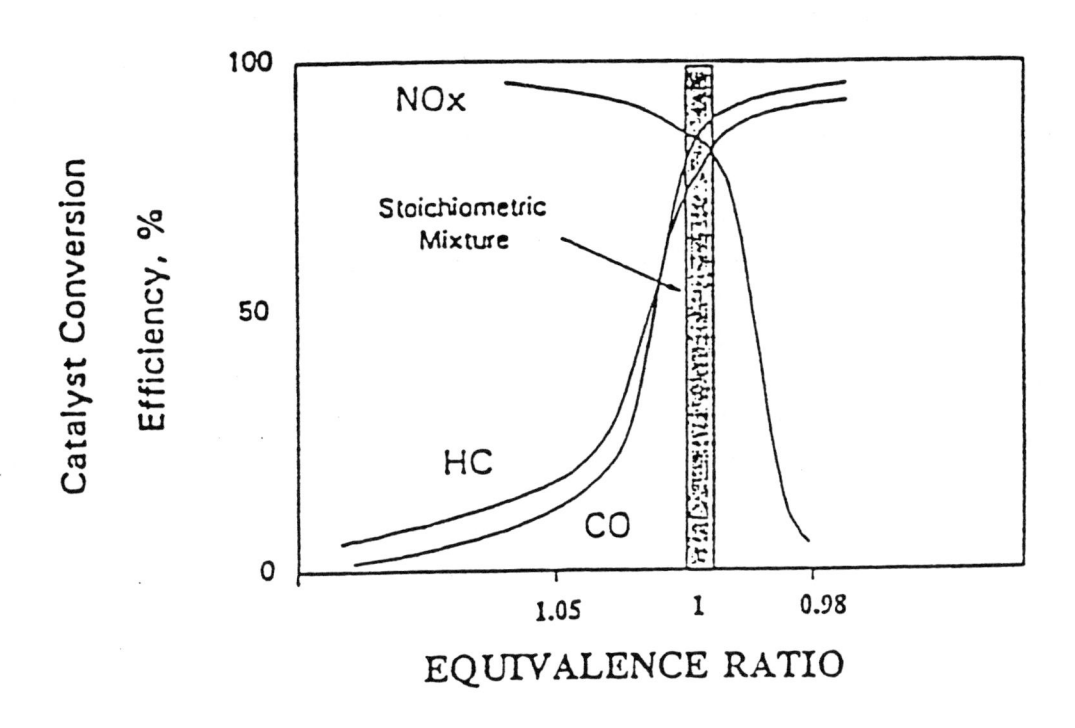

Figure 8 Catalyst Conversion Efficiency as a Function of Fuel/Air Equivalence Ratio (8)

approximately the same weight as the UT HEV. These components were selected because of their lightweight construction and availability.

The rear suspension configuration utilizes the front-wheel-drive McPherson strut assembly from a Ford Fiesta. This configuration was chosen as the best method of independently suspending the drive wheels. With this suspension setup initially having steering capability, it was necessary fix the tie rods rigidly to the chassis.

The braking of the H.E.V. is handled by disk brakes on all four wheels. The front brakes use 12" solid disks with cast-iron twin cylinder calipers from a Triumph GT-6 whereas the rear braking system incorporates 11" solid disks and floating calipers. These systems were chosen mainly because of their compatibility with the front and rear suspension components. Proper fore and aft braking balance was insured by using a dual master cylinder arrangement. Power boost was not felt to be necessary because the calculated maximum pedal force required was less than 100 lb.

Body Design

EXTERIOR BODY - The design of the exterior of the HEV is a compilation of various stylistic components gathered from existing cars on the market as well as other electric cars such as the GM Impact. The criteria of obtaining a low drag coefficient had the greatest impact on the decision for the overall design. The process of deciding on a design was time consuming since the exterior of the car needed to be stylish yet it still needed to fit the existing frame on the car.

The first steps in deciding on a design consisted of gathering data about existing cars and their associated drag coefficients. It was found that the GM Impact had the smallest drag coefficient on the market (0.19). Therefore, the design team looked at the design of this electric car and embellished upon it to come up with the final design for the Tennessee High Potential. Students on the team made their own models from clay and various styling characteristics were taken from each.

The next step in constructing the body was to

transform the chosen design into a life-sized model, or buck, from which the molds for the body parts would be made. The buck was constructed of wooden bulkheads, wooden stringers, and fiberglass (See Figure 9). Since the buck was only a skeleton, there needed to be a filler material to make the complex curves needed for a stylish exterior. The team, after consulting major car companies, decided to use Chavant brand CM-50 clay for this purpose. This clay was heated to a temperature where it became pliable, then it was molded onto the buck in several layers until it was 3.17 mm (1/8 in) above the bulkheads. After this arduous process was completed the final surface development began. This process included using various clay styling tools to develop the clay into the desired curves and smoothing the clay to the smoothest surface as possible.

After finishing the clay development, the surface had to be sprayed with a surface sealer, adhesion promoter, and vinyl ester in order for the molds to be built on the clay. This surface had to be waxed and buffed in order to obtain a perfect finish on the body parts. Before building the molds on the clay surface, the clay was sprayed with a gel coat which would become the inside surface of our female molds. The molds were constructed of fiberglass and wood.

The actual body parts were constructed of successive layers of carbon fiber, epoxy, composite core, then carbon fiber. These parts were formed in the female molds then vacuum formed using plastic sheeting and nylon polyamide/cotton absorbent layers between the part and the plastic sheeting in order to soak up excess epoxy, resulting in a light, strong body panel.

To attach the body parts to the frame various methods were used. Industrial strength Velcro was used to secure the body panels onto the aluminum/steel substructure.

The windows of the car were obtained from existing cars on the market such as the Chevrolet Camaro with eighth-inch Lexan forming the rear glass and one-fourth inch Lexan forming the side windows.

VEHICLE INTERIOR - The design of the interior of the car began in October of 1993. Major constraints were found in creating the interior since the battery tunnel utilized most of the interior space. Students

Figure 9 Picture of the Buck (Under Construction)

were able to implement their own designs in styrofoam to judge feasibility and ergonomics. Again, existing cars on the market were looked at in order to decide on a final design that was feasible for our car. The driver's gauges used in the car last year were sufficient for overall looks and ergonomics so the changes took place mainly on the battery tunnel, or center console, and the dash. The dash and console were constructed first of styrofoam and shaped so the interior was aesthetically pleasing. The next step was to construct the dash and console out of fiberglass and then have them covered with the chosen cloth/vinyl. Controls were placed in a manner such that driver and passenger could easily reach or read the ones utilized most often.

MATERIALS AND MANUFACTURABILITY - The design team for the exterior put great thought into the material to be used. Considerations were taken in recyclability, strength, weight, and availability from suppliers. The team first considered aluminum due to its recyclability, however, forming this material for a prototype vehicle was beyond the skills of the team. Plastics were considered for this characteristic too but the cost, lack of strength, and manufacturability of this material was not sufficient. The parts, if made from this material, would have to be made at a local plastics company which would cause problems if a part was damaged later on. The final decision of utilizing carbon fiber came about after researching into how racing boats are constructed. A local raceboat builder was consulted and the material was found to be superb in its strength, durability, and low weight. The manufacturability of the body parts from this material was found to be manageable for team members rather than contracting the work out to local companies.

REFERENCES

1. Traister, Robert J., All About Electric and Hybrid Cars, TAB Books, Inc., Blue Ridge Summit, Pennsylvania, 1982.
2. Cackette, Tom, "California's Zero Emission Vehicle Requirements and Implications for Hybrid Electric Vehicles", NIST Advanced Components for Electric and Hybrid Electric Vehicles Workshop, Gaithersburg, MD, October 1993.
3. Initial Statement of Proposed Rulemaking Amendments for the Low-Emission Vehicle Program, Mobile Source Division, Monitoring and Laboratory Division, CARB, El Monte, CA, 1992.
4. Innovations in Design: 1993 Ford Hybrid Electric Vehicle Challenge, Society of Automotive Engineers, Warrendale, PA, 1994.
5. Heywood, J., Internal Combustion Engine Fundamentals, McGraw-Hill, 1988.
6. Hodgson et al, "The University of Tennessee 1993 Ford Hybrid Electric Vehicle Challenge Final Design Report", University of Tennessee, Knoxville, 1993.
7. Linden, D., Handbook of Batteries and Fuel Cells, McGraw Hill.
8. Black, Frank M., "Control of Motor Vehicle Emissions - The U.S. Experience", Critical Reviews in Environmental Control, 21(5,6): 373-410, 1991.
9. Ribbens, William B., and Mansour, Norman P., Understanding Automotive Electronics, Howard W. Sams & Company, 1989.
10. Corning Incorporated, Environmental Products Division, E-293-01, 1994.

ACKNOWLEDGEMENTS

In addition to those organizations that made the 1994 Hybrid Electric Vehicle Challenge possible, the University of Tennessee HEV Team would like to thank the following organizations and individuals who made the UT effort possible:

Tennessee Valley Authority
Tennessee Valley Public Power Association
Lane-Magneto and Electric
Kohler Co.
Electrotek
Corning
Honda
Kawasaki
Analog Devices

Michelin Tire
Hallmark Electronics
Textron Aerostructures
East Tennessee Race Prep
British Cars, Ltd.
Trojan Battery
General Electric
East Tennessee Battery
Southern Armature Works, Inc.
SDRC

Special thanks are extended to the following individuals: Dr. William Snyder, Chancellor of the University of Tennessee, Dr. Jerry Stoneking, Dean of the College of Engineering, Janene Connelly,
Steve Hunley, Lanny Wallace, Danny Graham, Gary Hatmaker, Janene Jennings, and Linda Stooksbury

1994 HEV TEAM MEMBERS

Bryce Anderson, Ken Sugustinovich, Kevin Barfield, Monique Brown, Brad Caldwell, Anthony Carter, Krista Conner, Ross Cormia, Kenny DeHoff, Chester Duffield, Kenneth Frazier, Ramon Gonzalez, Clinton Groves, Kevin Humphreys, Woramol Khamkanist, Robert Lee, Sittichai Lertwattanarak, Michael Loope, Michelle MacFarlane, Jim McDonald, Sean McGowan, Allen Miller, Teddy Miller, Todd Morley, Steven Mundy, John Norton, Greg Peters, Scott Sluder, David Smith, Johnnetta Smith, Lori Snook, Stan Spence, Chris Spock, John Taylor, Brent Thomas, and Nicole Vickery

Texas Tech University
Hybrid Electric Vehicle Challenge
Final Report

Jeffrey Bratcher
Brent Crittenden
Jeff Earhart
Shawna Salyer
Roy Sanchez
Robert Tolentino
Scott Wilkes

Electrical Engineering

Davie Benner
Boyd Burnett
Ookyong Kim
Dale May
Casey Osborne
Gary Romero

Mechanical Engineering

Roland Shafner
Mechanical Engineering Technology

ABSTRACT

This paper describes the conversion of a Ford Escort station wagon to a Hybrid Electric Vehicle. The converted car is an electric vehicle with an auxillary power unit (APU). The APU is used for range extension and for low battery charge. The APU is an internal combustion engine running on ethanol. A special hybrid configuration minimizes emissions from the APU. This system was developed by Texas Tech University in response to the Hybrid Electric Vehicle Challenge sponsored jointly by the U.S. Department of Energy, Saturn, the Society of Automotive Engineers, and Natural Resources of Canada.

INTRODUCTION

The U.S. Department of Energy, Saturn, the Society of Automotive Engineers, and Natural Resources of Canada have organized a 1994 intercollegiate competition focusing on the advancement and use of practical hybrid vehicle technology. A hybrid electric vehicle (HEV) is defined as an electric vehicle with an alternative power unit (APU). The APU is normally an internal combustion engine (ICE). Texas Tech University has been selected as one of the universities in North America to compete in the HEV Challenge. The HEV Challenge is designed to provide an exciting and practical interdisciplinary exercise for college engineering students. The Texas Tech team includes students from the Electrical Engineering, Mechanical Engineering and Engineering Technology Departments.

The hybrid electric vehicle concept has been developed in order to reduce environmental problems caused by current forms of transportation. The internal combustion engine vehicle, which is the most popular form of transportation, produces emissions that are harmful to the environment. The refineries that provide the fuel for the vehicles also add to the pollutants in the air. The problem is a growing one that is nationwide. The hybrid electric vehicle is one solution to the problem.

The HEV addresses two major areas that have prevented wide acceptance of electric vehicles. The HEV can provide for long distance driving capability and for continual operation even with a low battery pack. The Texas Tech University team is converting a Ford Escort Station Wagon to a hybrid electric vehicle that will assure minimal pollution emissions from the vehicle. The TTU HEV uses two electric motors as the primary source of power. The electric motors do not emit pollutants to the air and are very energy efficient.

The alternative power unit for the Texas Tech HEV is a small ethanol powered internal combustion engine. Ethanol fuel has lower emissions than gasoline. Thus, this HEV should be able to satisfy the needs of the driving public while providing substantially reduced emissions.

POWER TRAIN CONFIGURATION

The Texas Tech Hybrid Electric Vehicle utilizes a parallel configuration which allows for battery recharging during cruising situations. The basic layout for the TTU HEV is shown in Figure 1. The interconnection between the electric motors and the IC engine is shown in Figure 2. The vehicle is powered by the electric motors during slow speeds and erratic situations, such as in-town driving. The auxillary power unit (APU) is an ethanol powered internal combustion engine (ICE). Ethanol fuel was selected due to the very low emissions it offers after combustion when compared to gasoline. The APU is used when the vehicle is driving in cruising situations, such as in highway driving or when the batteries are low.

Figure 1. HEV Layout

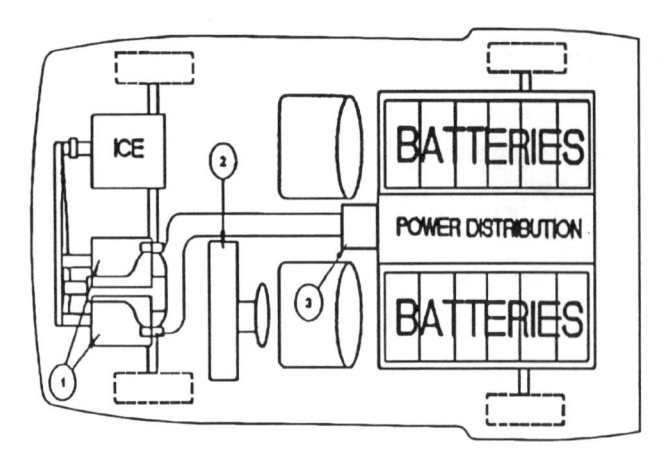

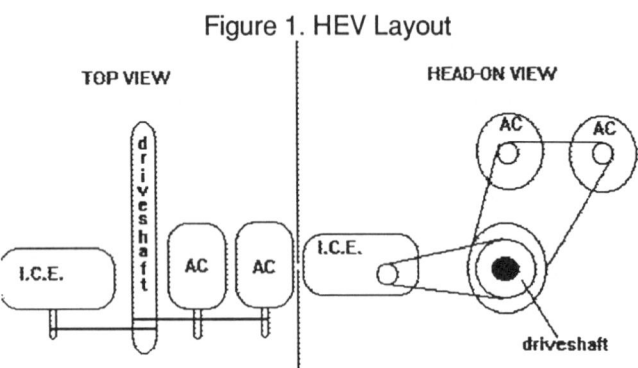

Figure 2. Parallel Motor Drive

The drive train utilizes a differential unit from a Datsun 280Z as the main transmission for the vehicle. This differential connects to the existing drive shafts and support members in the Escort Station wagon. The two electric motors and APU connect to the input shaft of the differential via two separate belt drives. The electric motors are permanently connected to the differential which allows them to rotate whenever the car is moving. This allows the motors to be used for recharging when the APU is driving the car. The APU is isolated from the differential via a sprag clutch and the clutch in the engine. At a vehicle speed of 60 mph, the input shaft to the differential rotates at 2500 rpm, the electric motors rotate at a peak of 5000 rpm, and the output shaft of the APU rotates at 2500 rpm, which requires a 2:1 reduction for the belt drive for the electric motors and a 1:1 drive for the APU.

AUXILLARY POWER UNIT - The APU is a 650 cc single cylinder, water cooled motorcycle engine powered by ethanol. The ethanol is fed to the engine via a fuel injection system that has been modified to use ethanol. The output from the engine is engaged by a mechanical sprag clutch. The sprag clutch allows the engine to be isolated when the APU is not in use.

An internal combustion engine creates power by burning an air-fuel mixture. The air-fuel mixture is important in determining the amount of power produced, as well as the efficiency of the engine. Air and fuel amounts vary with changing conditions, such as engine revolutions per minute. The throttle regulates the amount of air entering the engine, while the fuel delivery system regulates the amount of fuel to be mixed with the air.

A fuel injection system is an alternative fuel delivery system to a standard carburetor. The amount of air flow into the manifold can be determined if the air temperature, manifold pressure, and throttle position are known. Once the amount of air flow is known, the fuel injection system can measure the amount of fuel needed for the proper air-fuel mixture. The fuel is measured by a pulse generator in the control unit of the fuel injection system. The electric pulses turn the injector on and off, spraying fuel into the engine. By controlling the width of the pulse, the amount of fuel delivered can be controlled precisely.

The fuel injection system has several advantages over a carburetor, which include: air-fuel variability is reduced, fuel delivery is matched to specific operating conditions, stalling and engine run-on are prevented. All of these combined mean better engine efficiency, less fuel consumption, and fewer emissions produced. The fuel injection system used in the TTU HEV is the Total Engine Control (TEC) system made by Electromotive Inc. The system has been modified for use with the one cylinder engine in the TTU HEV.

ELECTRICAL SYSTEM - AC motors were selected for the electric drive due to cost and capability. The use of AC motors allows for the elimination of the transmission with its associated weight and volume. The electric motor system block diagram is shown in Figure 3.

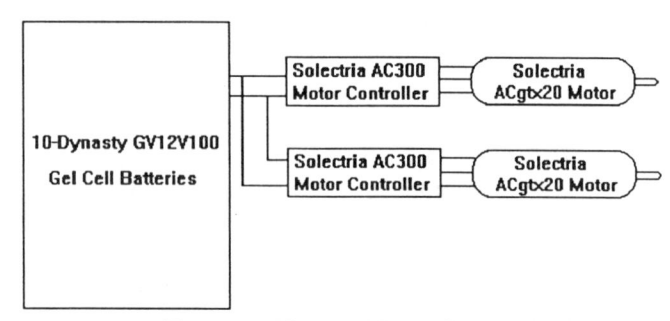

Figure 3. Electric Motor System

The vehicle is powered by two Solectria ACgtx20 three phase induction motors, each capable of delivering a nominally constant 28 horsepower at 4000 RPM. The motors are connected to the differential drivetrain via a two inch grooved belt. Each motor is driven by a separate pulse width modulation Solectria controller. Each controller is optimized for an input DC supply of 120 volts, and begins limiting current from the batteries when the voltage drops to less than 108 volts. The controllers are "vector" drive controllers, in that they minimize the slip of the motors by using speed sensing feedback, which constantly updates the controllers with the shaft speed of the motors. By doing this, the controllers are able to maximize the efficiency of the motors at any given speed. Throttle inputs are obtained from a single 5 kilo-ohm potentiometer, whose signal is sent to both controllers. Finally, each controller provides signal outputs which can be used to monitor the condition of the vehicle. Each controller provides speed, battery voltage, and current information, which can in turn be used to display the current condition of critical sections of the system for the driver.

CONTROLLER STRATEGY

One objective of the HEV is to reduce or eliminate emissions, especially in urban areas, and to reduce the dependency on foreign oil. Therefore, the controller strategy must be in line with these objectives to be practical. Since an IC engine is added to an electric vehicle to extend the range, the operation of the ICE must minimize emissions and be at the most efficient operation point, if possible. With the above objectives, the ICE should be operated with as constant a load as possible, which implies constant RPM and constant torque.

The TTU HEV is a parallel type HEV. It can operate in three different modes for maximum flexibility. The three modes are: electric drive only, ICE only, and a hybrid mode consisting of both power sources in a parallel configuration.

The AC motor controller is designed to operate the vehicle at a constant velocity, which is determined by the operator input through the accelerator pedal. The AC motor controller monitors the speed of the vehicle through an RPM sensor mounted on the AC motor. The AC motor controller adjusts the slip in the AC motor to increase or decrease torque to operate the vehicle at constant velocity. Therefore, if the AC motor controller detects a change in load through a change in velocity, the controller will adapt to the load change by adjusting the slip.

The TEC fuel injection system provides for reduced emissions in the ICE only mode. The transmisson of the ICE is also used to provide for a more constant load when the ICE is in use.

In the hybrid mode, the electric motors are used to enhance the output of the ICE. This allows the ICE to operate with a reduced demand and be smaller in size resulting in fewer emissions and a decrease in fuel consumption. The hybrid mode requires the system controller to match the output of the electric motors to the output of the ICE drive shaft. The system controller receives an input from a RPM sensor on the ICE drive shaft. The drive shaft input is used to calculate the required input to an AC motor controller unit. The analog output of the system controller is then routed to the AC motor controller.

If the ICE power output is larger than the power needed to operate the vehicle at the desired speed, the AC motors are operated as generators by providing a negative slip to convert the additional power into current to recharge the batteries. The amount of load the AC motors place on the ICE is variable proportional to the slip. Therefore, if a hill is encountered, the load is decreased, thus allowing more power from the ICE to climb the hill. With this control strategy the ICE can be operated in a small operating region. The better the response of the AC motor controller, the smaller the operating region.

INSTRUMENTATION - The function of instrumentation is to provide an interface between vehicle operator and vehicle function. Instrumentation provides feedback to the operator about the current status of various systems. The monitoring of miles per hour, revolutions per minute, and fuel consumption are the primary parameters and variables of concern to the driver.

With the dual system of the HEV, individual monitoring of the RPM for each motor system is required. Dual displays have been incorporated for the RPM monitoring of the HEV.

Battery usage is monitored by a kilowatt-hour meter. The instrument display is located in the center console of the vehicle. The meter shows the stored energy in the battery array. The display figure on the screen is indicated by a "plus" sign or a "minus" sign. Consumption will appear as a negative figure. When recharging, the display will indicate the restored energy. This instrument also displays the voltage and current in the battery array.

For the monitoring of the ethanol fuel, the existing fuel gauge that was in the original Ford system is used. Indicators and warning indicators are also utilized. Mode lamps are indicated by blue lights. Mode 1 is the electric mode and will automatically turn on when the system is in the electric car mode. Mode 2 signifies the ICE mode. During the operation of both systems at once, the hybrid mode, both mode lights are on. A check engine lamp is used for the safety of the vehicle. This display will warn the driver of possible problems. An oil pressure lamp is also provided.

With the exception of the kilowatt meter, located in the center console, all instruments and indicator lamps are located on a new panel located in the dashboard in front of the original Ford instrument panel. A diagram of the new panel is shown in Figure 4. The driver may see the complete system at a quick glance. The displays exhibit high visible numbers with good lighting. The displays may be seen both in direct sunlight and at night with no problem. The meters used are easily accessible and very user friendly. The lamps are also located in the panel. They also exhibit high visibility for driver convenience.

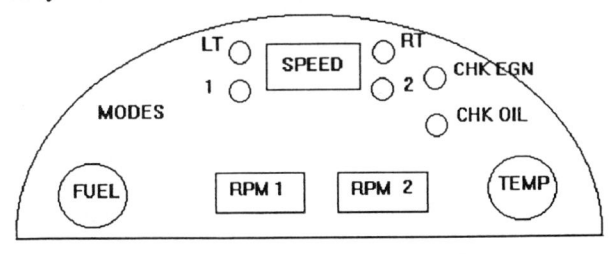

Figure 4. Instrumentation Panel

EMISSIONS CONTROL STRATEGY

One objective for minimizing emissions is to try to maintain a constant power output from the APU. When more power is needed to pass another vehicle or climb a hill, the electric motors will help the APU drive the car so that it still sees a constant load. The car requires 15 hp to drive the car 50 mph on a level road. The output of the APU is 47 hp at 3500 rpm which is well above the requirements for the vehicle.

The emissions from the use of ethanol are controlled by catalysts and engine tuning. The operating speed and load of the engine is controlled and kept constant which allows the engine to be tuned for that bandwidth. The engine is tuned to run at 3500 rpm with a lean mixture which produces very little emissions. The exhaust system for the APU is the stock Escort exhaust system that has been modified to contain an ethanol catalyst. The crankcase and fuel tank are vented to the carburetor intake to be combusted.

The ventilation system for the battery enclosure box serves to remove hazardous fumes from the vehicle and cool the batteries. A 160 cfm brush-less type fan draws the air through enclosure. The air flow pattern allows fresh incoming air to travel from the floor-board to the bottom of the enclosure and exit through the rear top of the enclosure and back through the floor-board.

FUEL AND ELECTRICAL POWER CONSUMPTION

Ethanol was chosen to power the APU to promote it as an alternative fuel to gasoline. Ethanol is a renewable fuel with a one year cycle where gasoline is a fossil fuel which takes thousands of years to produce. Ethanol can be produced from corn which is in excessive supply and no shortages are foreseen. The fuel produces very little emissions and has the highest heating value of any of the other alternative fuels.

One significant advantage of the AC motor controllers is in the area of regenerative braking. The controllers provide automatic regenerative braking, permitting recharging of the batteries during coasting, as well as in the hybrid mode.

BATTERY CHARGERS - Two battery charging systems are used. One system charges the batteries when the vehicle is operating. This charging mode occurs when the AC induction motors are acting as generators to charge the motor during hybrid operation. The second system is an off-board charger which recharges the batteries at the end of each day's events during competition.

The hybrid electric vehicle must also be capable of accepting a battery charge from an external off-board charging system. To provide for maximum life of the batteries, the off-board charger is designed to follow a safe charging cycle. Inherent features of the charger are voltage and current monitoring, overcharging protection, and faulty battery determination. The charger has a 220-volt input. This input will provide greater power than the 110-volt input, so the batteries can be charged quicker. The charging circuitry is a one-phase thyristor rectifier bridge. This design consists of a thyristor module, a B642-2T, and a triggering circuit, a PTR6000. The triggering circuitry controls the firing of the thyristors. A transformer provides isolation for the charging system. The entire battery array can be charged with this design.

VEHICLE STRUCTURE MODIFICATIONS

The structure of the Escort frame has not been modified in any way. The four components required for the hybrid drive train; a 1978 280-Z differential, two A.C. motors and one Kawasaki I.C. engine, are being supported by compact, simple structures. The foundation for these structures is the O.E.M. under carriage member and two rubber motor mounts fixed to this member. The differential utilizes the two motor mounts for support and two adapter plates for connection to the drive shafts. The adapter plates are necessary to connect with the O.E.M. tripod joint at the inside end of the drive shafts. By locating the two electric motors directly on top of the differential, the three shafts are parallel and when viewed from the front, they form a triangle, The electric motors are supported by means of an aluminum plate and a steel frame that bolts to the mounting holes of the differential. The motors are flush mounted to the plate which is then bolted to the steel frame. The I.C. engine is supported on the driver's side by a fabricated steel bracket that originates from the front differential support. On the passenger side of the engine, there is a steel structure attached to the O.E.M. motor mount that reaches down to grasp the engine at it's original mounting point. To minimize the forward and backward torque that will ensue from driving the car, a brace will be added to the top of the engine on the driver's side. This brace will attach to the existing bolt holes on the two motors and extend laterally towards the passenger's side of the engine bay. In conclusion, there are a total of five individual structures made from numerous parts that make up the support frame. These five components have been designed to be light and strong but also very serviceable.

BATTERY ENCLOSURE BOX-The battery box is located in the passenger compartment behind the front seats. It contains the ten batteries needed to power the car.

The battery containment enclosure was designed, constructed, and installed to achieve three goals: 1) supply the vehicle with electrical power, 2) secure ten gel-sealed batteries in all orientations, 3) ventilate the enclosure to insure that the interior of the vehicle was free from hazardous fumes produced from the batteries.

The main frame of the enclosure was constructed using aluminum 6061 T6, 2" x 2" x 3/16" angle. The base of the enclosure was designed so that each battery was separated by angle. Each angle can contain the force of one battery in case of impact to the vehicle. Nylon straps surround each battery to secure the batteries to the base in case of vehicle roll-over.

The exterior of the enclosure was constructed using 1/16" thick aluminum sheets. Silicone sealant was applied between the sheeting and the frame to insure an air-tight seal. The total weight of the batteries and the enclosure is 765 lbs. The fan was mounted to the exterior of the vehicle to reduce the noise from the interior.

SUSPENSION MODIFICATIONS

The suspension and braking systems of the Hybrid Electric Vehicle (HEV) is an affected area of the modified Ford Escort Wagon. Because of the increased weight added to the vehicle in the form of batteries, AC motors and components, it was determined that the suspension and braking systems needed to be improved so that the Ford Escort Wagon will be able to handle this increased load. These modifications required minimal losses in vehicle handling and no change in track width and ground clearance.

The current rear suspension was not rated for the additional load distribution. Therefore, the strut assemblies were removed and replaced with strut assemblies which could support higher side loads and with spring rates approximately equal to 300 lbf/in. which is 250 percent higher than the stock rating of 85.1 lbf/in. Two modifications to the aftermarket struts were necessary to accommodate the rear coil springs. Both modifications were due to a smaller outer and inner coil diameters of the 300 lbf/in spring. First, a shim was fabricated out of aluminum and installed in the lower strut spring cup to fill the gap caused by the outer diameter difference of 0.5 inches of the 300 lbf/in spring as compared to OEM. This shim is supposed to aid in keeping the spring's base from shifting within the cup. Secondly, the top spring cup was altered to hold the new spring. A new smaller centering section of 3.5 inches outer diameter was welded in place of the existing centering piece. Also, the existing rubber bushing for the top cup was cut and shortened and placed around the new centering section.

The front suspension also shared the problem of not being rated for the additional load. Therefore, the front spring/strut assemblies were replaced with springs from the Escort GT package (167.9 lbf/in.) with rates approximately thirteen percent higher than original factory equipment (151.3 lbf/in.). The struts were replaced with aftermarket replacements capable of enduring high vertical and side loads. No modifications were required to install the GT springs.

The added weight of approximately 1000 pounds force was a major concern to the handling characteristics of the Ford Escort Wagon. In order to improve these characteristics, stiffer aftermarket sway bars and struts with increased damping were installed. This was necessary for an attempt to pass the required maneuvering tests of the competition, which include the slalom, U-turn and lateral stability tests.

Low profile tires mounted on fifteen inch wheels from the Escort GT package improves both the braking and handling ability of the vehicle. Since the internal combustion engine (ICE) has been removed, a supply of vacuum for the brakes was needed. An electric pump was selected because of ease of mounting and power usage. The use of a large vacuum reservoir or purge canister with an adjustable vacuum measuring switch solved the problem of the electrical vacuum pump having to run continuously by providing a consistent mean vacuum of 15 inches of mercury.

Another modification of the braking system was the reversing of the 60/40 (front/rear % line pressure) proportioning valve to provide a 40/60 split diverting more braking force to the rear brakes. This modification was performed because most of the vehicles weight is in the rear and was not considered a drastic loss in front braking force.

CHOICE OF MATERIALS

BATTERIES-The batteries that are used to power the AC induction motors are the Dynasty GC12V100 gel-cell batteries. These batteries were chosen primarily because of their low weight per energy rating. Each battery weighs 65 lbs. The Dynasty GC12V100 batteries are deep cycle batteries with a rating of 90 amp hours rated at a 20 hour discharge rate.

WIRING-The wiring scheme for the HEV was designed in such a way as to conform with the rules and regulations (17.4.1-thru-17.13.4) set by the 1994 Hybrid Electrical Vehicle organizers. The various types and sizes of wire used in the wiring scheme were all sized in such a way that the current rating of the chosen wires surpasses the normal operating current expected in that wire by a factor of at least 25% and the insulation of the wire must have a breakdown voltage of at least two times the peak voltage in that conductor.

The battery array was wired with a 4/0 wire that is rated for 2000 volt and a current carrying capacity of 415 amps. The connectors used were blackburn compression connectors and are rated for 35 kilo-volts. Each connector in the battery array is concealed with a hood of heat shrink to prevent accidental contact. The battery array is completely sealed with a vapor proof strip, which will contain any gases emitted from the battery. The power and charging leads are routed out of the battery array through special explosion proof connectors (EY-100) which allow for a complete seal from the battery array to the disconnect switch. The battery array is completely enclosed by a aluminum housing.

The disconnect switch is a Square D 250 volts direct current breaker and has a current rating of 400 amps. The disconnect switch is mounted where it is accessible to the passenger as well as the driver. The disconnect switch has a shunt trip which allows for an emergency shutdown. This safety switch is mounted to be accessible from the passenger compartment as well as from outside of the vehicle. From the disconnect switch the wire is secured to the power distribution blocks. The power distribution blocks allow for a 4/0 wire on the primary side and a range from #8 to 2/0 wire on the secondary. The power leads for the AC motor controllers require a current of 150 amps each. Each positive lead for the AC motor controllers is fed through a 160 amp fuse. The wire used was #2-THWN and has a current rating of 190 amps. The wire from the AC motor controllers to the AC induction motor are fed through a 1" seal tight flexible metal conduit. Each phase requires 73.3 amps, so the wire used is #4 THWN with a current rating of 110 amps.

The instrumentation and various signal carrying lines are either a #18 or #22 gauge wire. These lines are fed through a molex (plastic sheathing) which allows for easy trouble shooting and access. The instrumentation is powered by a DC to DC converter which converts the 120 volts from the battery array to 12 volts for the instrumentation. The wire used to feed the DC to DC converter is a #10 and has a current rating of 55 amps. The wires are fed through a 1/2" seal tight flexible metal conduit. The battery used for the ICE is the original Kawasaki battery and is wired directly to the ignition switch for the Kawasaki engine. The wire used is the original wire for the Kawasaki engine.

All electrical devices, cable and terminations were selected and constructed so that safety was the number one priority.

ROLL BAR -The roll bar is made from seamless mild steel tubing with a 1.5 inch outer diameter and a wall thickness of 0.120 inches. The roll bar is no more than 3 inches from the ceiling of the automobile. The roll bar has 2 support bars (4-point roll cage) extending from the hoop at an angle of about 30 degrees from the vertical position. The roll cage has 1 inch of padding with a 3/8 inch inspection hole drilled into main hoop. Two steel mounting plates hold the main frame rails in compression with the frame of the car to avoid welding to the thin uni-body floor pan. The bar is bolted with 3/8-inch diameter hardened steel bolts using self-locking nuts. The required support arms were fabricated from material identical to the main hoop and were mounted to the shock towers using a steel plate integrated into the shock tower. Using the shock tower for the connection of the rear support arms provides for a strong roll bar and also helps maintain lateral stability in the rear suspension.

VEHICLE MANUFACTURABILITY

The goal of the Texas Tech HEV team is to develop a hybrid electric vehicle that is cost effective and realistic to produce on current assembly lines. To assure this is the case, all components used in the TTU HEV are readily available, off-the-self items. Changes to the basic Ford Escort were kept to a minimum while still assuring the vehicle would meet all requirements. The number of major components was kept to a minimum. One of the advantages of the parallel configuration is that an alternator for charging the batteries is not required.

CONCLUSIONS

A majority of the funding for the project was provided by the Texas Corn Growers Association and the State of Texas. The association donated $5,000 and two 55-gallon drums of ethanol to promote the use of the alcohol as a fuel. The State of Texas donated $10,000 toward the project. A number of smaller contributors also aided the project. Ford donated the Escort wagon with the company's approval of TTU's initial HEV proposal.

TEAM ORGANIZATION-Over 20 students and five faculty members from three departments within the College of Engineering have participated in the 1994 HEV Challenge. Students from Mechanical Engineering, Electrical Enginering and Engineering Technology have worked on the project through special task groups, or teams. These groups, composed of 4 to 6 students and a faculty advisor, worked to accomplish specific task assigned by the complete team. The task teams have met on a regular basis, usually more than once a week. The complete team has met on a weekly basis to discuss the status of the project, to resolve any problems and to make further assignments. Membership of the teams has changed considerably since the begining of the project due to graduation and new students entering the project.

The University of Texas at Arlington Hybrid Electric Vehicle - High Bred

by
Robert L. Woods, Faculty Adviser
Rob Lloyd, Team Captain

ABSTRACT

Presented in this paper are the design concepts and implementation of the Hybrid Electric Vehicle from the University of Texas at Arlington. The basic concept of the UTA vehicle is a high-performance, sporty, hybrid electric vehicle that has the aesthetics, performance, handling, and attractiveness of a true sports car combined with the utility and desirability of an electric vehicle.

The vehicle uses a brushless DC electric motor with a power rating of 16 horsepower continuous and 36 horsepower intermittent. The battery pack is 120 Ni/Cd cells with a maximum voltage of 170 volts. In the electric mode, a continuous speed of 55 MPH is expected with a intermittent speed of 85 MPH.

The combustion engine is a 600 cc motorcycle engine with turbo, electronic fuel injection, using M-85 fuel. The combustion engine delivers in excess of 100 horsepower and drives the differential directly with a parallel mechanical drive system from the electric motor, so that either can propel the vehicle. The combustion engine can be used to recharge the drive batteries while driving down the road.

The body is fiberglass with gull-wing doors for the two-seat configuration. The battery packs are located in the side pods on both sides of the vehicle. The tube frame is constructed with mild steel. The engine and the electric motor are both located in the rear of the vehicle.

INTRODUCTION

Due to the ongoing concerns of pollution control and the increasing restrictions placed on automobile emissions, the search for alternatives for the internal combustion engines has intensified. Electrically-powered cars are thought to be a viable replacement to current automobiles in the near future; however, today's battery technology limits the useful range of an electric car. Consumers are reluctant to purchase a car that has less than 100 miles range. With increased political pressure on auto makers, new alternatives are being researched to bridge the gap between internal combustion and electric power.

One alternative is the hybrid electric vehicle (HEV). A HEV is a vehicle that utilizes an electric motor drive system as its main power source for inner city driving while using a combustion engine as a supplemental energy source for highway operation and to extend the useful range of the car. To accelerate the research into such alternatives, the Department of Energy, Society of Automotive Engineers, Ford, Saturn, and others have teamed together to sponsor the Hybrid Electric Vehicle Challenge. This paper presents the University of Texas at Arlington's technical design of a hybrid electric vehicle named "High Bred".

OVERVIEW OF THE UTA VEHICLE

A "Electric Commuter Vehicle" often conjures images of an unstylish utility vehicle that is predominately functional and not particularly exciting or fun to drive. While our goal is to meet the emissions, fuel economy, and range standards, our basic concept is to build a two-seat, mid-engine, sports car that looks great, handles well, and has good acceleration.

Power train - There are two types of HEV power trains recognized for the competition. A series-type power train uses an electric motor as the main power source while using a combustion engine to drive a generator to recharge the batteries. The generator allows an increase in range for the vehicle but limits the acceleration and speed to whatever the electric motor can provide. In a parallel-type power train, both the combustion engine and the electric motor are connected mechanically to the drive train so that either (or both) can propel the car. This set-up allows the option of either driving in electric mode when low pollution is desired (such as in the inner city), or in combustion mode when good acceleration or long range is needed (such as entering or driving on a highway). The direct drive of the vehicle in combustion mode, increases the power to the ground and efficiency of the fuel used. Since the combustion engine can also be used to recharge the batteries while driving in combustion mode, this system has the advantages of both parallel and series systems. Therefore, UTA has chosen the parallel system.

The combustion engine is a 600 cc Honda motorcycle engine fitted with an electronic fuel injection system that was designed, built, and programmed by UTA students. It is a modified speed-density system. A turbocharger is installed in the tuned exhaust system. The engine is fueled with M85 (85% methanol, 15% gasoline). The engine has an integral 6-speed transmission that is shifted with a mechanical clutch and a sequential shifter. An Allied-Signal catalytic converter is used after the turbo to reduce emissions.

The electric motor is a Solectria brushless DC motor (BRLS16) with a BRLS240H 200 amp controller. The motor produces constant torque to 5000 RPM. The maximum continuous power is 16 HP with an intermittent power rating of 36 HP.

The electric motor is attached through the combustion engine case to the first shaft of the transmission by a chain. This is the clutch shaft from the engine. In this manner the electric motor is always connected to the rear wheels through the transmission. Thus, the electric motor has the same benefit of torque multiplication through the transmission as does the engine. The friction and inertia of the electric motor are small so very little loss in performance of the combustion engine is realized.

The speed ratio of the chain is selected so that the engine speed of 11,500 RPM corresponds to 5,000 RPM of the motor. These are the maximum power points of each. In order to accomplish this, sixth gear was removed and replaced with the chain sprocket. Fifth gear was then replaced with sixth gear so that maximum vehicle speeds could be obtained. Therefore, this 6-speed transmission is now actually a 5-speed with the top gear retained. An external oil pump is installed to provide lubrication to the transmission when running in the electric-only mode.

The chain drive configuration of the motorcycle is retained for the vehicle. A Zexel-Gleason "Torque Sensing" differential is used. The differential is mounted in a housing that has the chain sprocket attached to the housing, and the housing rotates with the differential.

Battery Pack - Various battery technologies were investigated for this vehicle [2]. These technologies include lead-acid, nickel-cadmium, and nickel-metal hydride. In order to maintain a lightweight sporty vehicle, a high power and high energy density battery is needed. The lead-acid batteries were less desirable since their weight would be about 700 pounds. Some technologies were very expensive while others were not available. However, the nickel-cadmium batteries were ideal for several reasons. The Ni/Cd batteries are high power and relatively light while maintaining a fast recharge time. Therefore, the SAFT Ni/Cd batteries were selected. Two battery packs of 60 cells each were connected in series and charged to 170 volts. The total battery packs weigh 420 pounds and contain about 3.5 kW at the voltage used.

A Solectria constant voltage, current limited charger is used to charge the main battery pack. Ni/Cd batteries should be charged with constant current. This is achieved since this charger is normally operated in the current limited mode so that constant current is given to the battery pack for most of the charging time. As the voltage rises, the charger switches to the constant voltage mode to top-off the batteries. Since the Solectria controller is limited to 170 volts, the batteries are only charged to 170 volts which represents 1.42 volts per cell on the Ni/Cd batteries. Ni/Cd batteries can be charged to 1.55 volts per cell; however, our experience is that when you try to take all of the cells to 1.55 volts per cell, some cells will start outgassing much sooner than others and therefore, will cause damage to those cells and reduce their life. It was decided that the batteries would be taken to 1.42 volts per cell in the hope that the outgassing would be minimized. We still observe some outgassing of a few cells even at the 1.42 volts per cell.

12 Volt Electrical System - A 12 volt service battery is used to power the combustion engine and to run all of the instrumentation and lighting systems. The combustion engine has a 30 amp alternator on board, and a DC-to-DC converter from the main battery pack is used to supplement the combustion engine alternator as well as to provide power when the combustion engine is not running.

Chassis - The vehicle structure is designed to integrate well with the suspension, drive train, and batteries while allowing passenger comfort. The chassis is a mild steel tubular space frame that allows easy access to components. The frame was analyzed with MSC/NASTRAN and ANSYS to insure structural integrity and driver safety under impact loads. The chassis has space for the two battery pods on the sides and uses the stiffness of these pods to add strength and rigidity to the chassis and to increase safety to side impact. See Figure 1 for the chassis layout.

The chassis can withstand a side-impact of an 8 g force at a speed of 13 mph, a front-impact an of 8 g force at a speed of 19 mph and a roll-over simulation of approximately 4 g's. The chassis has a torsional rigidity of 1,333 ft-lb/deg.

Body - In order to create a unique look for "High Bred" and to maintain a true sportscar appearance, a fiberglass body was designed and built. Figure 2 is a photo of the car in the configuration for the 1993 HEV Challenge. The front section of the body is one piece that rotates at the front bottom to allow access to the gas tank. The rear section is also one piece that rotates about the rear bottom to allow access to the cargo storage, engine, motor, and electronic controls. Two side covers are removable to allow access to the two battery packs place along either side of the driver compartment. Gull wing doors allow access for the driver and passenger. Plastic flex-hinges are used on the door. The T-top roof is steel. A fiberglass dash and instrument panel was constructed.

To manufacture this body, foam molds were poured and carved by hand to an approximate shape. The rough shape was then sanded to the final shape and finish and fiberglass molds were made. The fiberglass body parts were built from these molds.

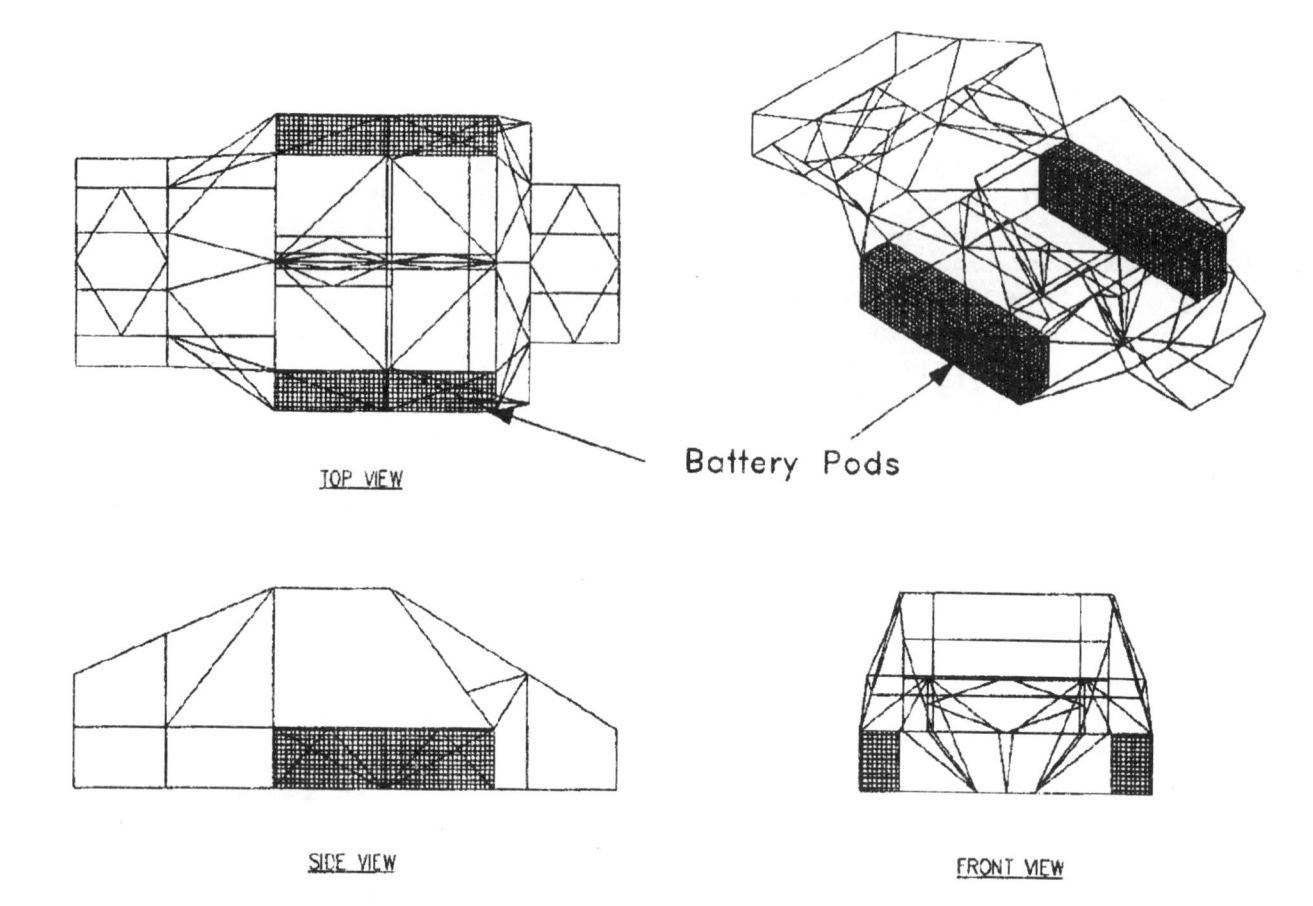

Figure 1. Tube Chassis of UTA Vehicle

Figure 2. The UTA "High Bred" Body Design

Suspension Design - To achieve sportscar handling, an independent front and rear double A-arm suspension is used. A computer analysis of the kinematics was used to optimize camber change, roll center position, roll center movement, scrub, damper length, motion ratio and wheel rate for static bump and rebound conditions. The bump and rebound travel is six inches.

The geometry of the suspension is established to provide shock clearance and good connection points to the frame. These connection points minimize the amount of stress transmitted to each frame member. The A-arms are made of 4130 steel.

OPERATIONAL MODES

The vehicle has four operational modes, ZEV or electric only mode, ICE or internal combustion engine only, DEDICATED HEV mode in which both electric and combustion modes are active, and PROGRAMMED HEV in which the electric motor and the combustion engine are switched by a computer that is using decisional control to determine which mode should be active. Currently, the first three modes are implemented on the vehicle.

ZEV Mode - In the Zero Emissions Vehicle (ZEV) mode, the electric motor is active full time. The throttle pedal operates a pot that gives a command to the electrical controller and a pot on the brake pedal commands regenerative braking in addition to hydraulic braking. The electric motor is driven through the 5-speed transmission of the motorcycle engine and is always connected to the rear wheels through the transmission. Shifting is accomplished by releasing the torque on the drive train and shifting the transmission without a clutch. The inertia of the electric motor is very low so that engagement transients are very small. Therefore, only two pedals are used, the throttle pedal and the brake pedal. The clutch for the internal combustion engine is held in the disengaged position so that no power is transmitted to the engine.

ICE Mode - In the internal combustion engine (ICE) mode, the auxiliary power plant is used to propel the vehicle. In this parallel configuration, the ICE powers the rear wheels directly through a 5-speed transmission. The normal mechanical clutch integral to the engine is used for shifting and take-off. The electric motor is always connected to the transmission shaft so the motor turns all the time in proportion to the engine speed; however, the motor has very low inertia and friction losses so very little power or performance is lost.

In the ICE mode, the electric motor can be used as a generator to recharge the main batteries. A pot on the dash can select any percentage of regeneration. This regeneration control is in parallel with the brake pot used for regen braking.

Dedicated HEV Mode - In the dedicated hybrid electric vehicle (Dedicated HEV) mode, both power plants are continuously engaged. The throttle pedal actuates the engine throttle as well as the electric motor controller pot. Thus, the two torques sum together in the transmission and the power output to the rear wheels is enhanced to over 140 HP. The clutch is used to shift the combustion engine and a relaxation in electric motor torque (by the throttle pedal) is given during the shift.

Programmed HEV Mode - In the programmed hybrid electric vehicle mode, there are several options for running the electric motor or the combustion engine. The pot on the throttle pedal is intercepted by the computer and reinterpreted to give a voltage command to the electric motor controller. One concept is to have the throttle pedal become a constant power control as it is displaced. Thus, in the combustion-only mode, the throttle will command about 100 HP per full stroke. In the electric-only mode, the throttle pedal will command full electric power at about 30% of full throttle stroke. If the combustion engine is engaged during electric mode, the throttle interpretation for the electric controller could go to zero or to full electric command at 100% throttle displacement (depending upon the state of charge, the throttle position, speed, and the amount of time the throttle and speed have been experienced). In this manner, the "hybrid" operation will be almost transparent to the driver as the electric and combustion modes are engaged and disengaged.

Fuzzy logic and neural networks can be used to shift the transmission, engage and disengage the combustion engine and electric motor [3]. For operation at slow speeds and with partial throttle depression, the electric mode can be used if the batteries are charged. If the speed and throttle depression exceed a specified amount for a specified time, then the combustion engine could be engaged and the electric motor command would be reduced or set to zero. Since an external oil pump is engaged in electric mode to lubricate the transmission, the engine as well will be fully lubricated if the combustion engine were suddenly engaged; therefore, the only adverse effect of allowing the computer to determine when the combustion engine will be engaged will be the water temperature. The engine could be pre-warmed before dedicated HEV mode if this were a problem.

If the speed is relatively high (e.g., driving on a freeway), the electric motor could be put into regeneration so the main batteries could be recharged while driving. Since the engine has excess power, the additional regeneration load on the engine might actually improve the engine specific fuel consumption and thus the overall efficiency. At higher speeds and throttle positions for long periods of time, the electric motor could be neutralized to improve the combustion power. At full throttle, both electric and combustion could be engaged for maximum power.

A driving simulator has been implemented on a computer with screen graphics to allow drivers to try different operational modes of acceleration and shifting. A mouse and the keys are used to control the throttle, brake, and shifter, and a display gives the responses from a dynamic simulation model of the vehicle [3].

SUMMARY AND CONCLUSIONS

The hybrid electric vehicle designed and built by UTA meets our expectations a high-performance sporty car and will have the impressive performance, range, and aesthetics that was our goal. The Ni-Cad batteries combined with the brushless D C motor and drive train will provide the weight, acceleration, cruise speed, top speed, and range that were desired. The combustion engine and drive train will provide exceptional acceleration, range, emissions, and the ability to recharge the batteries.

The tubular steel frame has the structural stiffness and factor of safety to maintain the integrity of the vehicle and suspension while providing for adequate factors of safety and crash worthiness. The suspension system will provide exceptional handling and will provide good ride qualities. The fiberglass body has the aesthetics and ergonomics that were desired and are light weight.

This exercise in engineering a new hybrid electric vehicle has introduced young engineers to the problems facing industry and our nation, and has brought great insight into the problems and compromises required in a major project such as this. UTA feels that it has met the challenge in an optimum manner based upon our concepts of the car.

ACKNOWLEDGMENTS

This paper is an accumulation of student projects written over a period of 2.5 years. The following students are responsible for the data and analysis: Khader AbuKhadijeh, Hector Amador, Terry Autrey, Soloman Aregay, Zak Barakat, Todd Baughman, Kevin Bevans, Bobby Brown, Doug Box, Anthony Bundrant, Bob Bundy, Kakanin

Bunnang, Mike Burns, Yean Neng Choong, Greg Cleveland, Scott Crowder, Kevin Culver, Rusty Davis, David Dillmore, Dang Dinh Nick Dringenberg, Doug Evans, Richard Fallas, Mark Fisher, David Foken, Mike Fortner, Tory Gallier, James Gorden, Boyce Hardin, Craig Henry, Kevin Hill, George Ipe, Ming Japutama, John King, Thun Kham, Pete Leboulluec, Kris Little, Rob Lloyd, Tim Lockhart, Richard Martinez, Ann Maslyk, Ben McCurley, John McFadden, Marvin Mitchell, Steve Nance, Thuan Nguyen, Chinedu Okeke, Steve Oliver, John Pullman, Kevin Rainey, Scott Rice, Keith Sauer, Bassem Srour, Steve Stahl, Philip Suder, Faiz Taqi, Zeb Tidwell, Maureen Traynor, Paul Tucker, Brent Warren, Peter Wells, James Whitt, Ray Wolshlager, and Jeff Zimmerer.

UTA is greatly indebted to the following sponsors and contributors that have helped with parts or cash: A. E. Petsche Wire Co., Allied Signal, Arendale Ford, Arlington Fastener, Aromat, Associated Fiberglass Engineers, Aurora Bearing, Austin, AvCom Aviation, Bowman Industries, Buz Post Auto Park, Dayco, Collmer Semiconductor, Foam Supply Inc., Ford Motor Co., Fluke, General Motors, Goodyear, Hayes Brake, Hiley Mazda, Hillard Mazda Auto Park, Hinderliter Heat Treating, Honda of America, J & H Machine, Norman Kamb, Marshall Electronics, Martin Sprocket and Gear, Mazda, Microtooling Systems, Mikuni, Morse, Motorola, Nohau, Ono Industries, Peterbilt Motors, Raychem, SAFT America, Seimans, Spectra Technologies, Solectria, Southco, State of Texas Energy Office, Texas Electric Utilities, Texas Section of SAE, Texstar, Thomas Industries, World Car Limited, and Wiseco. Numerous other companies have provided direct funds as major or associate sponsors of the HEV Challenge Event which has also been critical to our project.

REFERENCES

1. Woods, Robert L., et. al., "The University of Texas at Arlington - 'High Bred'", 1993 Ford Hybrid Electric Vehicle Challenge, SP-980, SAE 1993.

2. DeLuca, W. H., et. al., "Performance and Life Evaluation of Advance Battery Technologies for Electric Vehicle Applications", SAE paper #911634, 1991.

3. Fisher, Mark, "Implementation of Neural Networks with Integer Arithmetic", Ph.D. Dissertation, U. of Texas at Arlington, May 1994.

Hybrid Electric Sport Utility Vehicle Prototype Design and Development by a Student Team

Daniel J. Alpert and Scot B. Smith
The University of Tulsa

ABSTRACT

Several student teams at The University of Tulsa have contributed to the design, construction and development of a hybrid electric vehicle (HEV) prototype. This series configuration HEV has a reformulated gasoline powered internal combustion engine supplementing charge to recyclable lead-acid batteries. A flux vector variable frequency controller transfers this energy to an AC induction motor coupled to a transmission for a rear wheel drive system. The high efficiency, low emissions power design is packaged in a uniquely handcrafted fiberglass monocoque pickup truck body structure. Also a custom polyethylene lamination process added safety and ergonomic improvements to the dash and bumpers to uniformly style the exterior and interior.

INTRODUCTION

The students at the University of Tulsa are in the third year of developing a hybrid electric sport utility vehicle. After participating in the 1993 Ford Hybrid Electric Vehicle Challenge, the vehicle named the Hybrid Hurricane, is now improved for the 1994 HEV Challenge. The project design, development, vehicle operations, unique composite construction techniques, and custom charging system are outlined in this report.

The team technical goal is to provide safe, practical, sporty, affordable, alternative fuels transportation that uses state-of-the-art, off-the-shelf components in a composite structure. The team educational goal is to discover, combine, analyze and improve existing technologies while educating the public about hybrid electric vehicles and transportation technology.

BACKGROUND

The 1990 Clean Air Act Amendments were created to help some of the over 40 cities nationally that are currently not in compliance with the National Ambient Air Quality Standards for carbon monoxide and ozone. California, New York, and Massachusetts have since passed tougher legislation aimed at decreasing regional pollution problems. The California Air Resources Board Low Emissions Vehicle Program requires percentages of sales of automobiles with progressively lower emissions while also mandating the sales of vehicles with zero emissions beginning in 1998.

Vehicles are needed to satisfy these legislative demands, however, electric vehicles are the only commercially available vehicle with no tailpipe emissions. Consumer acceptance for electric vehicles is dependent upon performance, range and infrastructure like conventional vehicles, but battery technology has not advanced enough to allow the needed performance with the range of a conventional vehicle. Also an infrastructure has not been developed to support wide spread electric vehicle commuting or long distance travel.

The need for low and zero emission vehicles is best satisfied by our hybrid electric vehicle. The Hybrid Hurricane generates low emissions, with zero emissions capabilities, and the performance and range that meet or exceed that of conventional vehicles. Also, our vehicle is gasoline powered and uses a complete infrastructure of refueling stations with the option of plugging into a 120, 240 or 208 Volt electrical outlet.

The design necessary to build our HEV lead to the development of several customized techniques to create a conventionally operated urban commuter vehicle using commercially available technology in a practical, safe, highly efficient, low to zero emissions, sporty pickup truck package.

POWERTRAIN

Hybrid electric vehicles have the option of a series, parallel, or combination configuration. A series configuration was determined to have higher efficiency, lower emissions, and lower weight than the parallel or combination configuration. The next critical decisions were the motor, controller, battery, and APU selection, modification, and integration.

Initially DC motors were researched because of the simple speed control and low loss connection to the batteries. Then brushless motors were found with longer life and higher efficiencies, but higher maintenance costs might result if high temperatures

FIGURE 1. Powertrain components

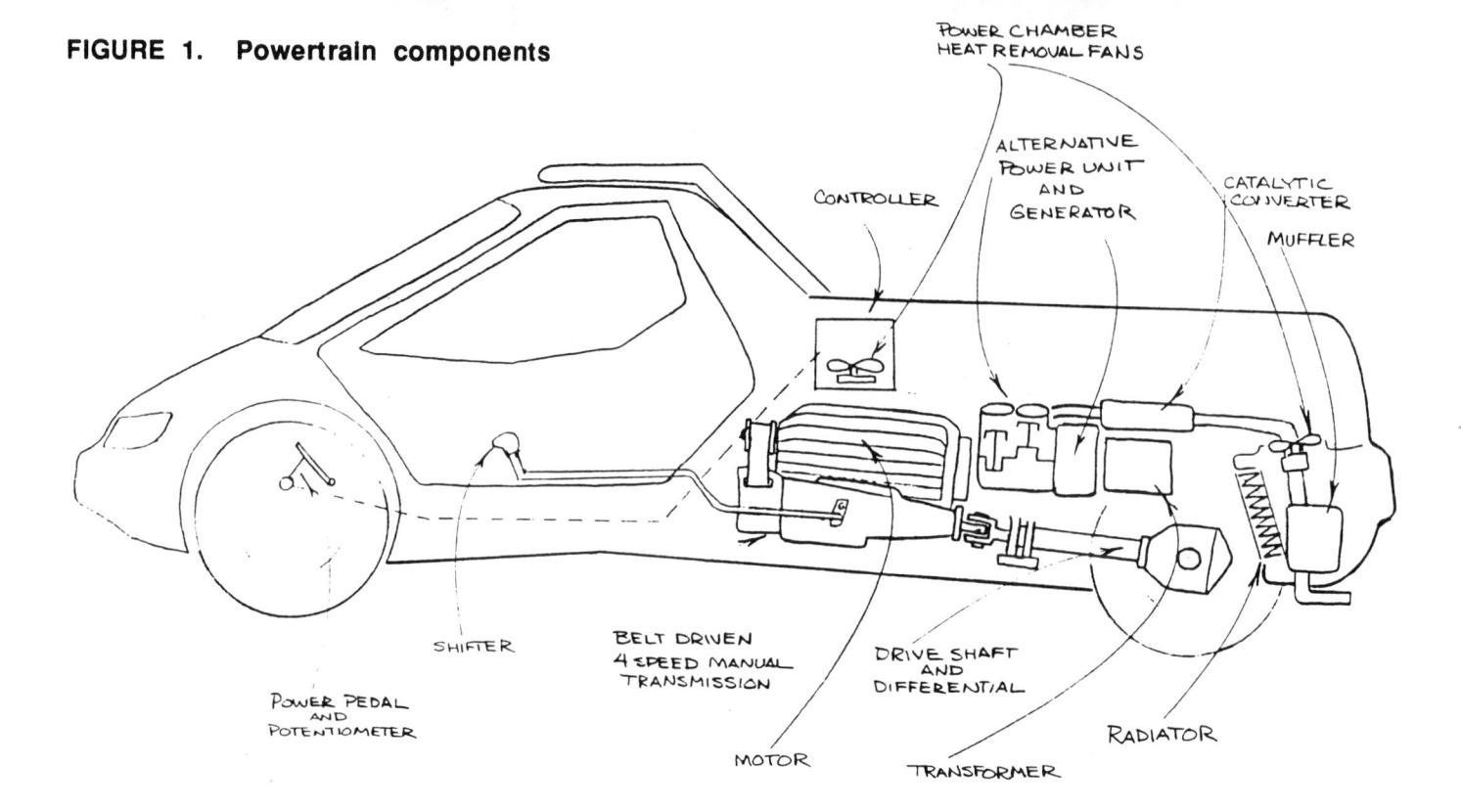

degauss the field magnets. Operational concerns became secondary when few DC motors, and even fewer high voltage DC controllers, could be found in the ratings necessary to propel our vehicle. Also, finances prohibited the purchase of several suitable components, or a custom made system.

An AC synchronous and induction systems comparison determined that the induction system was superior in efficiency and cost. Both had low power to weight ratio, but the synchronous system had the disadvantage of complex motor design and slip rings supplying the field current.

The decision to use AC induction was made based on power needs and system availability. Baldor Electric in Fort Smith, Arkansas eliminated the financial restrictions by donating the AC induction system that we selected. The 11.2 kW AC induction motor, with a variable frequency flux vector controller, was the system that best met our computer simulated power needs for torque, speed and climbing power. The controller was oversized to handle up to a 18.65 kW motor, thereby having reserve capacity during short term acceleration.

Computer simulation also aided in determining that a transmission was necessary to prevent a stalled start on an inclined path over 22 percent. A four speed manual transmission, controlled by an electronic clutch, is coupled to the motor by a toothed belt. The addition of a transmission also allowed the use of a lower power motor than would be necessary in a direct drive system.

The battery system to supply the necessary power was the next major decision. A comparison of Lead-Acid, Nickel-Cadmium, Nickel-Iron, Metal-Air, and Sodium-Sulfur batteries was conducted. Lead-Acid batteries were found to be the most affordable, available, and recyclable, with competitive life expectancy, safety while charging, pond low maintenance. Twenty-nine Exide DC-9, size U1, 12 Volt, deep cycle, marine/wheelchair, 100% recyclable, lead-acid batteries provide the 8.0 kWhrs needed by our design. Also, the durability, reliability and donation of these batteries met our design and budget restrictions.

The vehicle range was then dependent upon the efficiency of the alternative power unit (APU). A commercially available gasoline-powered generator was selected because of its ability to fit into the existing vehicle structure, ease of refueling, and efficient power generation. Unleaded or reformulated gasoline was selected over ethanol or methanol due to the higher energy density and superior infrastructure. A modified Honda ES6500 with a 4-stroke, O.H.C., 2 cylinder, liquid cooled, unleaded gasoline-powered engine running an electric generator provided the needed power with a maximum output of 6.5 kW.

The APU was modified for installation by relocating the mounting brackets, radiator, exhaust, and integrated with extended electronic controls, custom fuel tank and fuel pump. The factory brackets and rubber bushings were replaced with L-brackets and the bushings were modified to secure the engine to the floor of the vehicle bed. The radiator was installed at the rear of the vehicle utilizing the air flow from under the vehicle rather than losing efficiency over the distance to the vehicle front. See Figure 1. The cooling hoses were repositioned, lengthened and the diameter was increased to maintain pressure. The exhaust tubing was also lengthened and a catalytic converter and muffler were added. The muffler is mounted vertically within the rear corner body compartment and exhaust exits out beneath the rear of the vehicle.

Ignition/starting switch, temperature gauge, current warning and oil warning lights are all cabled to the cab for driver information and control. A high density polyethylene, 30 liter, fuel cell with a modified filler neck is situated in a vented compartment next to the APU.

FIGURE 2. Electrical System

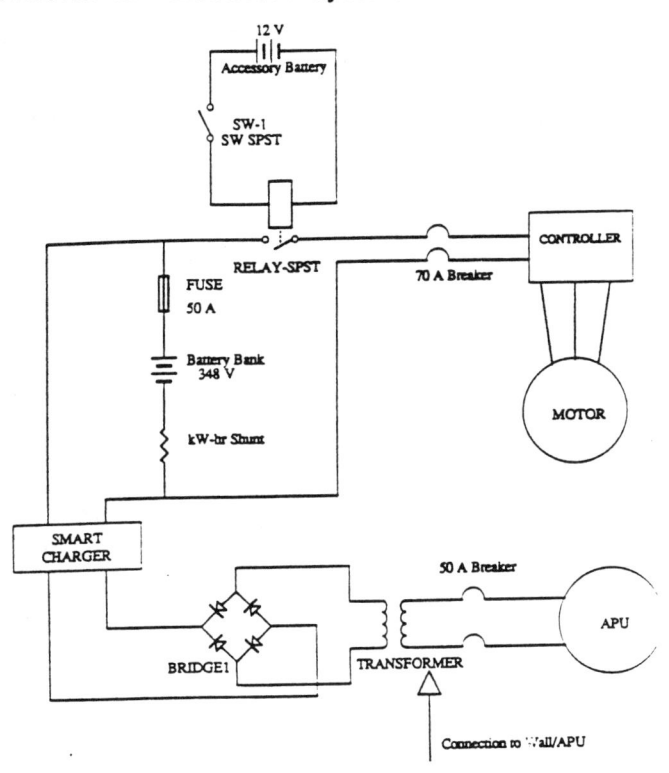

The racing specification fuel cell was chosen for its safety at high temperatures, high strength and high elastic modulus. A neck and filler cap were integrated into the side panel for easy access. Also a regulated electric fuel pump was added to deliver the fuel to the engine.

ELECTRICAL SYSTEM

The controller can operate at supply voltages between 220 and 385 volts. Initially it was decided to operate at 264 volts from two parallel battery banks with 22, 12 Volt batteries in each bank. This decision kept the voltage above the low cutoff and allowed room for higher voltages in charging. However, testing showed a 100 meter acceleration in over 13 seconds, a top speed of 65 km/h, and only a kW motor output. To improve these results a 348 Volt operating system with a redesigned controller interface and charging system was installed. See Figure 2.

A series of 29, 12 Volt batteries are now used provide up to an 80% increase in power over the original design. Operating at 348 Volts allows for more power without overloading the motor. A higher voltage was considered, but 348 Volt was more safely within the operating limits of the controller.

The controller was designed to accept a 240 Volt, three phase source, but we reconfigured the wiring to direct the DC current past the disabled three phase sensors and rectifier bridge. Then the AC cooling fans were replaced with DC brushless models of the same dimensions.

The controller, through an auto-tuning technique, determines the optimal values for slip, proportional-gain, windage and friction loss, and maximum current. These tuned parameters provide the best control strategy for the application of power to the motor. The controller receives a positive 0-10 volt signal from the accelerator pedal to control the variable torque output of the motor. A double throw/double pole switch is used to enable reverse mode. Also an electronic clutch has been installed to disable the motor during gear changes.

A switch in the dash panel allows for APU on/off control when the motor is not engaged. However the primary control for the APU is automatically controlled through a voltage sensing circuit in the smart charger that turns the APU on or off depending upon the status of the battery charge. Other conventional switches on the steering column control the 12 volt accessory system including headlights, wipers, horn and turn signals.

The APU is connected to an autotransformer that converts the 240 VAC signal to a 340 VAC that is then rectified before entering the custom charging circuit. See Figure 3. The charging circuit labeled smart charger senses the average battery voltage to enable the charger or prevent over charging. The APU 120 VAC output is then used to power the commercially available charger for the 12 volt accessory system. The 70 A breaker protects the controller and is the master cut off switch for use in emergencies or during maintenance. The 50 A breaker protects the APU.

The charging circuit is activated by connecting the components at a single group of connectors, called the electrical gas cap. This gas cap is designed as a master connection point between a power source and the charging system. See Figure 3. Consequently this same connection point is used for charging from the APU or from an isolated power grid source of 120, 240, or 208 VAC.

FIGURE 3. On/off board charger connections

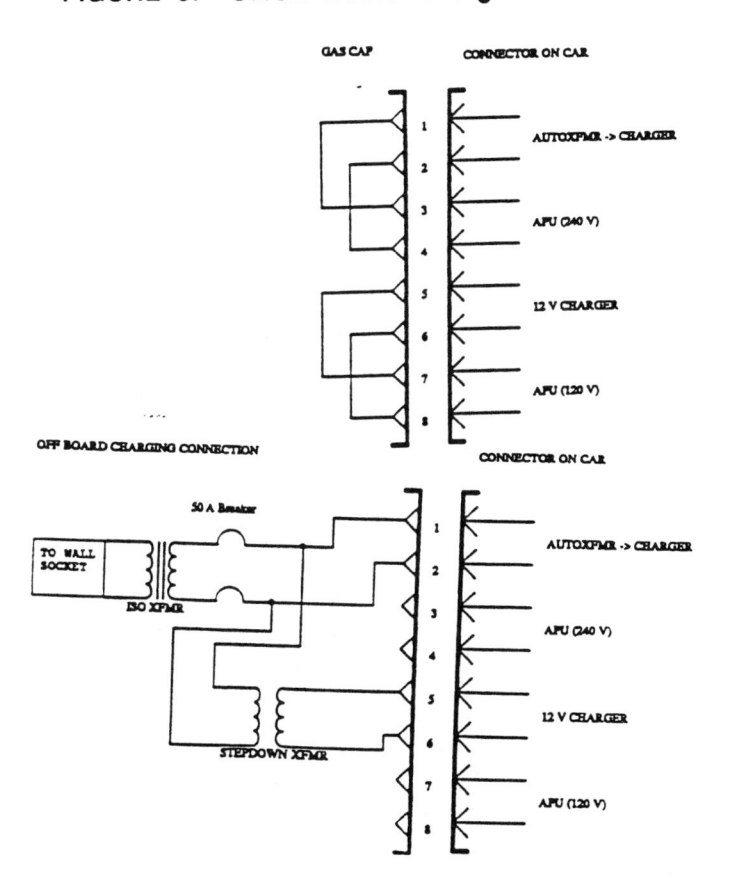

EFFICIENCY

The Hybrid Hurricane had the highest efficiency at the 1993 HEV Challenge with 52.76 miles per gallon of gasoline. The electrical efficiency is undetermined because of a dysfunctional kWhr meter at the Challenge and ongoing system improvements ever since.

EMISSIONS

A custom catalyst from Allied Signal Automotive Catalyst Company made our emissions reduction very efficient. At the 1993 HEV Challenge only 0.003 g/mi of Nitrogen Oxides, 16.068 g/mi of Carbon Monoxide, and 0.961 g/mi of nonmethane organic gases were emitted. After placing fourth in the emissions event at the Challenge we made no changes to our low emissions system.

Evaporative emissions from the fuel cell are carried through a liquid trapping filter directly into the intake manifold of the APU engine.

VEHICLE STRUCTURE

The vehicle structure is divided into the categories of body structure, interior components and exterior components.

BODY

The design criteria for the structure of the vehicle was determined to be the following:

1. Provide support and room for two people.
2. Provide a high strength to weight ratio
3. Provide storage room for the batteries.
4. Provide proper support and room for the motor, transmission, and suspension.
5. Have an aerodynamic appearance.

The construction of the Hybrid Hurricane began with the a PVC tubing "skeleton". The "skeleton" defined the shape of the final load bearing frame and served as a three-dimensional model from which the team could work. When the "skeleton" was complete, PVC foam was secured between the tubing to add structure. The foam was then coated with a mixture of resin and microspheres. The newly formed hard surface was then ground smooth. Next, layers of fiberglass and resin were placed on both sides of the surfaces. Another mixture of microspheres and resin was applied to the outside. This time a wax additive was included in the mixture. The outer surface was then ground smooth and finished off with paint. This produced a strong monocoque body structure.

Some places of the vehicle require additional compressive strength, especially at locations where mechanical fasteners are attached. At these locations, a hardpoint is required See Figure 4. A hardpoint is an area of the structure where the foam layer has been replaced with a microsphere and resin mixture. Hardpoints are made by removing a layer of glass and the foam underneath. This hollowed out volume is then filled with a resin and microsphere mixture. When the resinand microsphere mixture dry, the surface is ground smooth and a layer of glass is placed back on top.

The structural integrity is one of the main concerns of the body design. Calculations were made to alleviate the questions of bending moment, deflection,

FIGURE 4. Hardpoint construction

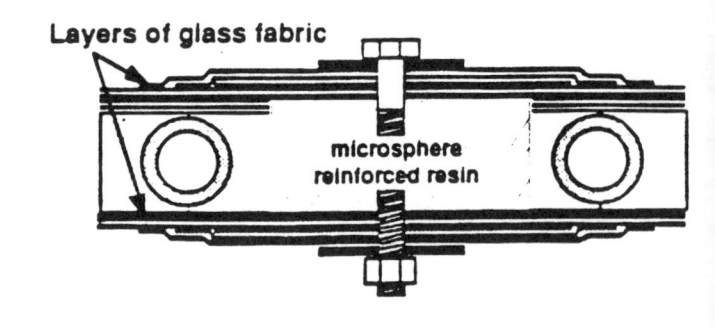

and torsion. The maximum bending moment was found to occur at from the front of the vehicle. A structural channel 2.1 meters long separates the motor and battery compartments, and reduces the bending stress in the exterior walls. With this channel, the body needed a total of four layers of glass arranged in a 0-90 degree orientation, with the axis running from the front to the back of the vehicle, so that it would provide sufficient strength for the bending stress calculated under a 2-G load. The total maximum deflection was found to be 1 mm.

The load that the front wheels transmit to the road was measured at 2.65 kN. Twice the measured loads were used to account for dynamic effect of the torsional loading. Torsional forces were calculated to simulate effects that the vehicle would encounter in everyday driving. The results of the calculations indicated that the torsional forces could be handled by adding two layers of glass at +/- 45 degrees to the 0-90 degree oriented layers.

Much thought went into the choice of the various materials throughout the vehicle. The vehicle is mainly constructed of a structural glass composite. The main advantages of the fiberglass are: ease of manufacturability, including mass production; good specific modulus; non conductive and non-corrosive to the batteries. The PVC tubing and structural foam were chosen because of their light weights since the fiberglass carries virtually all of the loading of the car. These materials also allow for the "skeleton" to be modified with little difficulty before the fiberglass is applied.

SAFETY

The Hybrid Hurricane design incorporates several safety features. See Figure 5. The high strength fuel cell contains a float to prevent fuel from exiting the cell. Testing showed that the fuel cell can tilt a total of 85° without leaking. This design feature reduces the danger of a fuel spill during a roll over.

Another feature is the installation of front and rear polyethylene foam bumpers. The bumpers serve to protect the passengers, and to protect the car. They are designed to absorb the force of a five mile per hour impact with no damage to the car. The dashboard is also made out of polyethylene except it is a foam of half the density as the bumpers. It also has the ability to absorb an impact. The dashboard made of foam helps reduce injuries if the occupants head strike the dashboard in a collision. Also the contoured dashboard design creates a large field of vision of the driver.

Steel side impact bars and a steel roll cage were

FIGURE 5. Safety Improvements

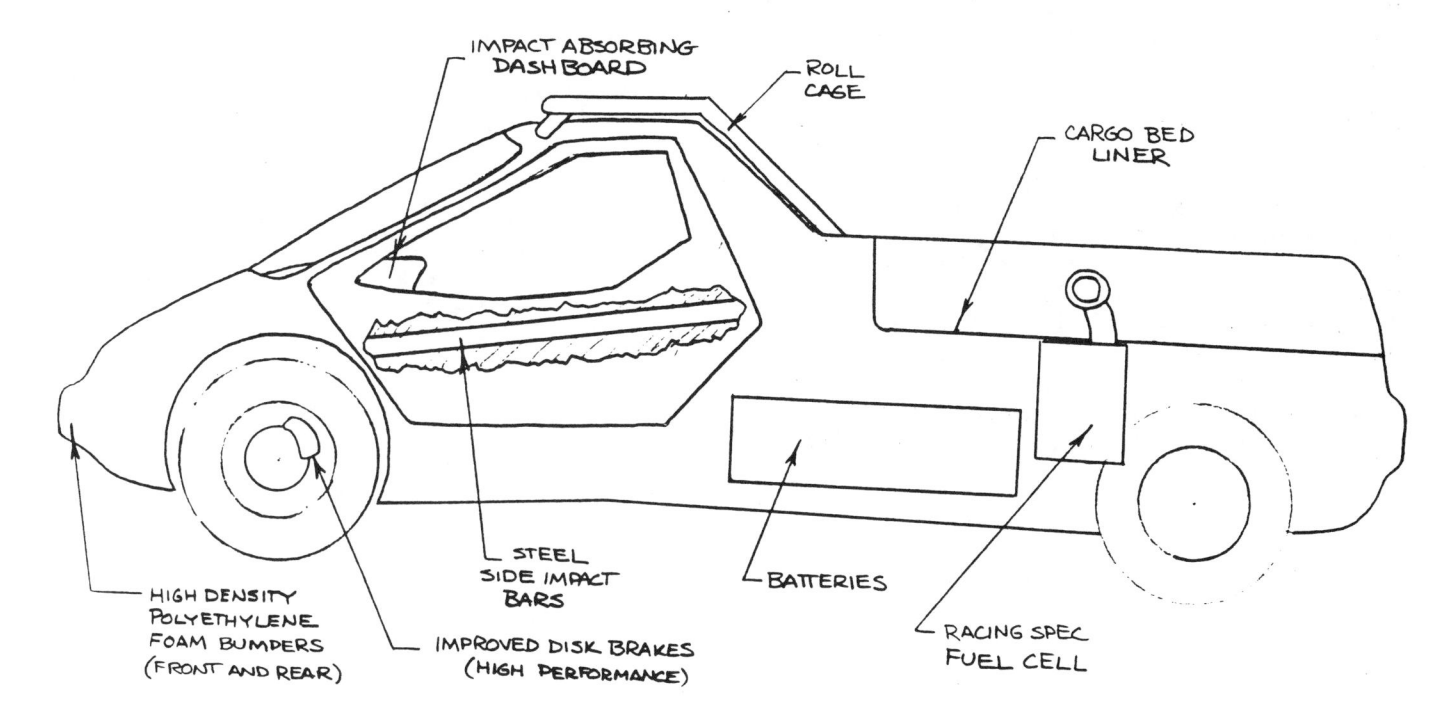

installed, in accordance with the Sports Car Club of America regulations for a vehicle over 2500 pounds. Also, a new set of disk brakes were installed to reduce the braking distance.

PRODUCTION VALUE

Although the methods described above could not be cost effective for mass production, the composites used in the vehicle are capable of high volume production. A process called resin transfer molding (RTM), makes it possible to mass produce composite automobiles. In this process, stampable dry glass fiber is performed into a 3-d geometry by an epoxy shaping die. Next, the perform is fitted with foam core. Finally, the perform is placed in a low pressure heated press, where resin is injected and allowed to cure. The finished fiberglass part can then be removed. Mass producing our vehicle would require four to seven parts: the monocoque chassis with wheel wells and the truck bed channel, the upper cab section, and finally the hinged parts such as the doors, tailgate, and hood. All of these parts could be fitted together to mass produce the hybrid electric sport utility vehicle.

MARKETABILITY

"In 1993 up to three times as many pickups were sold in Oklahoma as in 1989," wrote Rob Martindale in the Tulsa World (April 4, 1994). This trend has resulted in the sale of light trucks exceeding that of passenger vehicles in 10 states, including Oklahoma. These factors alone indicate a strong market in middle America for a light truck like we have built. The added feature of being a hybrid electric vehicle expands this market to include the coastal states with stricter emissions requirements and a greater need for low emission vehicles. California, New York, and Massachusetts in particular have set vehicle emission requirements more demanding than those regulations set by the federal government.

California's Low Emission Vehicle Program requires 50% of all vehicles sold in 1998 in California must be low emission vehicles. Our design qualifies as a low emission vehicle to further increase its demand. As an added bonus our vehicle is easily converted to be a zero emission, i.e. fully electric, vehicle to satisfy further legislative mandates. Whether through tax breaks or government rebates, or simply cost-effective mass production, the price of EV's must become competitive to satisfy the demand. Our design can be mass produced and prices brought down by using readily available components in a hybrid electric sport truck that meets the consumer demands for practical, efficient alternative fuels transportation.

SUMMARY

The University of Tulsa student teams have developed a truly unique HEV. The sport utility truck design makes it ideal for use as a commuter vehicle with the functionality of a delivery truck.

Goals were set early and attempts were made to reach them when occasionally new obstacle came into play.

The final result was:
- A stylish sport utility truck with a fully functioning bed.
- A powertrain that makes the truck perform at or above conventional levels.
- A suspension and balance that cause an increased handling ability.
- A monocoque composite body that is easily mass produced.
- A flexible charging system that requires minimal effort of the operator.

The University of Tulsa students involved in the project have learned a great deal from this project. The principles of electric vehicles and automobiles in general. In addition we learned how to create new processes when there is no technology available for meet the needs at hand. We learned how to design and then test new ideas. Most importantly, we learned how to work with other engineers as a team.

ACKNOWLEDGMENTS

The Hybrid Hurricane team would like to thank all of those who helped us with in-kind, monetary, and technical support for the project. Specifically, we would like to thank Baldor Electronics for the donation of the motor and controller, Exide for the batteries, and Ford Glass for a custom windshield. There are also numerous people and groups from the Tulsa area who helped the project. Finally we would like to thank the advisors to the project; Dr. John Henshaw, Dr. Robert Stratton, and Mr. John Orloski. Their advising was invaluable and greatly appreciated.

AMPhibian Evolution:
The 1994 United States Naval Academy Hybrid Electric Vehicle

Frank C. Madeka
Patrick L. Padgett
Steven M. Miner

Gary L. Hodges
Gregory W. Davis
Joseph L. Greeson

United States Naval Academy

OVERVIEW

The USNA midshipmen accepted the HEV Challenge as an extension of their commitment to serve their country -- in this case, to help America preserve its resources. The vehicle name, **AMPhibian**, was chosen by the midshipmen because, just as a real amphibian spends time both on land and in the water, the vehicle will operate using electrical energy from the battery system, and at other times using electrical energy that is derived from the gasoline powered generator. As a reminder that electricity will be the primary power source for the vehicle, the first three letters of **AMPhibian** were capitalized to represent the ampere, the basic unit of electric current. The name is also used in recognition and support of the military role provided by the Navy and Marine Corps amphibious team.

REVIEW OF VEHICLE DEVELOPMENT

For the 1993 HEV Challenge, the United States Naval Academy considered safety, time, and resources and came to the conclusion that the conversion vehicle was the best option for success. Consideration of resources included the significant limitations associated with of a four year, undergraduate, federal, service academy. By selecting the conversion approach, the **AMPhibian** team was able to concentrate on the development of the hybrid drive and control system without expending undue effort on the design of the basic vehicle (i.e., chassis, body, lighting, etc.).

1993 COST - For the 1993 HEV Challenge, the United States Naval Academy considered safety, time, and resources and came to the conclusion that the conversion vehicle was the best option for success. Cost was considered the major design goal. All design decisions were made only after the associated costs were analyzed. To help attain this goal, all components were to be based upon existing, readily available technology.

1993 POWER TRAIN - The **AMPhibian** is propelled using a series drive, range -- extender configuration. That is, the only component that is mechanically connected to the drive-train of the vehicle is the electric motor. This arrangement is depicted below.

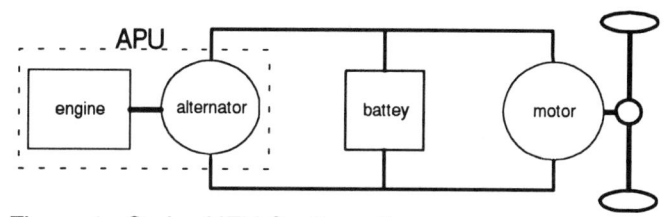

Figure 1. Series HEV Configuration

This arrangement was considered to be superior to the parallel drive arrangement, in which both the electric motor and the auxiliary power unit (APU) can propel the vehicle, for the following reasons. The series drive would require less structural changes to install, and thus provide a lower cost. The parallel drive system would also require a more sophisticated control system to minimize driveability problems such as those associated with the transition from electric vehicle (EV) mode to hybrid electric vehicle (HEV) mode. This would, again, result in higher cost, and, possibly, reliability problems due to the added complexity. The parallel drive is attractive because it has the potential to provide improved acceleration since both the APU and the electric motor are used to propel the vehicle. Additionally, when the battery is discharged, the parallel system cannot easily be used to recharge the system. Overall, the series drive was seen to be the *best* choice to meet the design goals.

The selection of the motor was driven by cost. Although AC systems were considered, cost and availability of AC controllers dictated a DC system since DC motor controllers are readily available and less costly. An estimate indicated that the motor would need to provide 9 kW to maintain a steady 80 kph. For accelerations, 0-72 kph in less than 15 seconds, the motor needed to provide an estimated 32 kW peak power for short duration. The motor selection was greatly simplified by the donation of a series wound, 15.6 kW @ 90 volts DC motor from General Electric. Although the steady state rating is less than the peak incurred during the acceleration, the motor should provide a peak power 2-3 times its steady state rating for short duration. To provide maximum torque, a high system voltage is required. Cost, size and the ready availability of a proven motor controller dictated the controller choice. A controller, rated at 120 VDC, was chosen, and thus this determined the system operating voltage.

The power train conversion to a series drive system required the removal of the standard Escort engine. Since the Escort has front-wheel drive, the transaxle was left intact so that a new axle would not need to be designed. The electric motor was attached directly to the existing bell-housing and flywheel. This arrangement also allows full use of the existing transmission, thus allowing for variable gear ratios. This was considered an advantage since it would allow the electric motor to be operated closer to its preferred operating speed over varying vehicle speeds.

1993 BATTERY SELECTION - USNA *AMPhibian* has two battery power systems. One system is at 12V and one at 120V. The 12V system is used to power the 12V lighting and accessories. The 120V main battery stack powers the electric motor and supplies power to recharge the 12V battery.

USNA *AMPhibian* battery selection was overwhelmingly driven by cost considerations. Key secondary considerations included: 1) the motor controller rating of 120V, and 2) the gross vehicle weight rating constraint. In general, an inexpensive, small, lightweight battery having high specific power and high specific energy was desired for use in the *AMPhibian*. With these key considerations, 120 volts, the maximum rating for the motor controller, was selected minimizing I^2R losses due to lower operating currents. This enabled an order-of-magnitude calculation of the costs of batteries having the desired characteristics. They were considered technically superior to those of conventional lead-acid batteries. This resulted in the following figure:

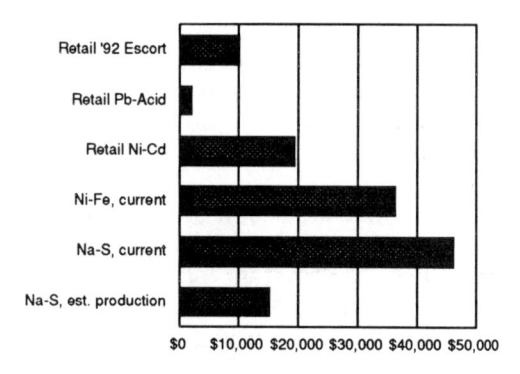

Figure 2. Battery Cost Comparison

These estimates far exceeded budget constraints. This lead the design team to limit battery selection considerations to only available off-the-shelf lead-acid batteries.

With the self-imposed limitation to lead-acid, the key design factor was to maximize battery capacity, and, therefore, ZEV capabilities, within the gross vehicle weight rating constraints, while considering battery volume as a tertiary consideration, assuming the majority of the interior passenger/cargo volume was usable for this purpose. This design objective lead to favoring multi-celled batteries. This was contrary to the practical considerations of battery maintenance & replacement. That is, a favorable battery stack composed of individual, replaceable cells to facilitate replacement of only bad cells as opposed to the replacement of entire multi-celled, batteries having only one bad cell. More then 50 batteries were considered, in terms of both direct and relative performance evaluations. A sample of the evaluations are shown.

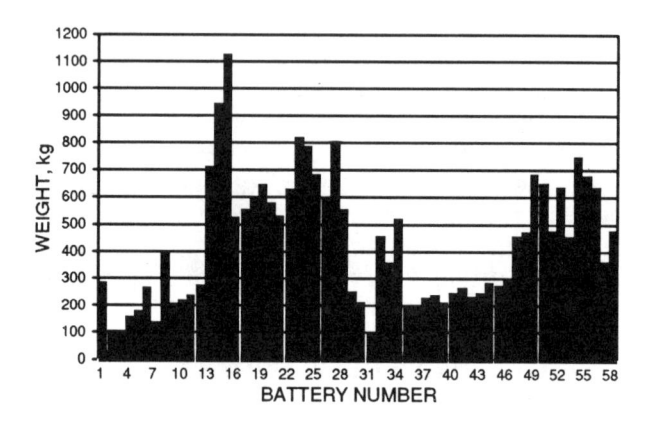

Figure 3. Total Battery Stack Weight.
120 VOLT ARRANGEMENT

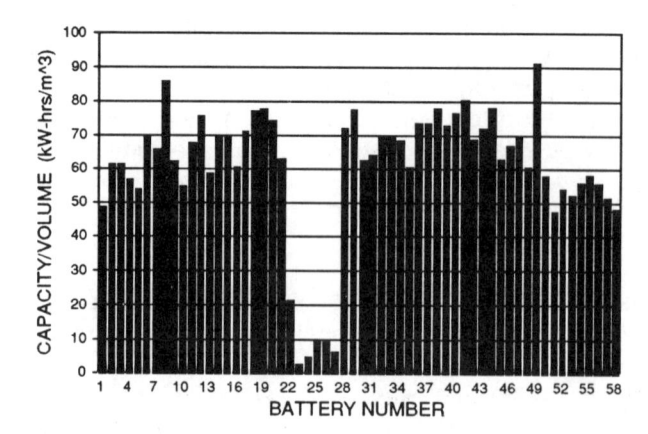

Figure 4. Battery Stack Capacity-to-Volume

Battery number 8 appeared to be the *best* choice when considering all evaluations. Battery number 8 is the Trojan 5SH(P) deep-cycle, wet-celled, 12V battery with

"L" type terminals. With the main battery selected, the 12V system needed to be defined and selected.

Several approaches were considered to power the 12V systems. This covered the extremes of using the existing 12V Escort battery, as is, or converting all 12V components to 120V. The other approaches considered included: utilizing one of the single 5SH(P) twelve volt batteries from the 120V stack, utilizing a higher amp-hr rated battery inserted into the 120V stack, a single, independent high amp-hr rated battery, and a large DC/DC converter, powered by the 120V stack, to meet all 12V demands including starting of the APU. The approach selected was a DC/DC converter, sized to handle the sustained accessory loads under moderate to heavy use, along with a small 12V lead-acid battery, in parallel, sized to accommodate the APU starting loads. A 30 amp DC/DC converter was selected to accommodate existing and future HVAC additions. To accommodate the 210 cranking amp APU starting load, the ultra light (4.58 kg), Pulsar Racing Battery, offered by GNB Incorporated, rated at 220 cranking amps was selected.

1993 AUXILIARY POWER UNIT (APU) - Based on estimated highway speed driving conditions (drag and rolling resistance) and drivetrain efficiency, original calculations indicated a minimum desired electric power availability of 10 kW. If the APU could deliver this power, the HEV should be able to sustain highway speeds, limited only by the amount of onboard fuel. However, this power capability alone would not allow for reasonable accelerations over this distance. Therefore, the APU must be capable of charging the batteries while at highway speeds so that if acceleration becomes necessary, the power may be drawn from both the batteries and the APU. The 1993 total calculated electrical requirement resulted in a specification of 12.5 kW (16.76 hp) output from the APU. Estimating the overall efficiency of the APU to be 80%, the engine then must be capable of mechanically developing 15.6 kW (20.9 hp). Since Briggs & Stratton Corporation donated a 13.4 kW (18 hp) "Vanguard" series engine, the cost was "right" and hence incorporated into the vehicle.

The Vanguard was integrated as follows. The fuel system includes the stock Escort fuel tank, an electric lift pump, a Puralator fuel filter & regulator, the Briggs & Stratton fuel filter that came with the engine, the existing diaphragm pump, and the existing carburetor. The engine intake uses the existing Escort air cleaner (cost) and fuel vapor canister along with the crank case vent and vapor recovery canister lines connected to be burned. The *AMPhibian* exhaust consisted of a stock Briggs & Stratton exhaust manifold, UEC 3-way catalytic converter, an electric air pump (i.e., a "small" 12v, 140 psi portable tire inflation pump), ceramic exhaust header wrap, and the existing Escort tail pipe & muffler. The Briggs & Stratton exhaust manifold and UEC converter were wrapped in the ceramic header wrap in an effort to aid quick light-off of the catalyst as compared to relatively expensive electric resistively heated catalysts.

AMPhibian experienced stalls under transient conditions. This was readily apparent when tested for emissions on a vehicle dynamometer during the 1993 HEV Challenge. The car was configured, as above, with the engine and exhaust warm/hot, the main battery partially drained, the throttle partially choked (immediate stall fix), and run to the FUD schedule cycle. *AMPhibian* was able to basically follow the cycle. Since the team "understood" the rules to mean fully complete the test or get no points (i.e., all or nothing) the engine was choked to minimize stalling problems. Test results were as follows:

Table 1. 1993 Emission Results

Component	AMPhibian (gm/mi)	TLEV Std (gm/mi)
NMOG	0.111	0.125
CO	36.8	3.4
NO_x	0.06	0.4

AMPhibian was ten times higher than allowed for CO. This can be explained, to some degree, due to *AMPhibian* running partially choked and stalling three times during the dynamometer test. Again, partially choked to minimize any stalling since this was the quickest fix at the time.

After approaching several major electrical machinery manufacturers with the generator requirements, and receiving estimates ranging from several thousand to fourteen thousand dollars for components that weigh as much as the DC drive motor or more, (e.g., 200 kg (440 lb)), a lower cost manufacturer was located. Fisher Technology, Inc. was contracted to build a custom 13.5 kW, 150 V, 3 phase alternator. The voltage was selected based on providing 144 V DC from a three phase bridge rectifier, the maximum recommended charging voltage for a 120 V battery stack. However, this voltage is well above the rated voltage for the Curtis 1221B motor controller. The controller manufacturer was contacted several times with several responses to voltage concerns. Responses ranged from do not exceed the 120 volts, 126 volts (fully charged 120 volt stack) and 160 volts (apparently built in protection). Based on schedule, the team decided to risk running above the controller rating by staying with the 144 volts. Reliability problems, to date, have not occurred with this condition.

RESULTING 1993 HEV CONFIGURATION - Considering the above, the resulting U. S. Naval Academy HEV, *AMPhibian*, configuration for the 1993 HEV Challenge can be found in the attached specifications.

REEVALUATION OF THE *AMPhibian* - Unlike the 1993 HEV Challenge, USNA now has a functioning vehicle to work with for the 1994 Challenge allowing the opportunity correlate estimates with measured data. This allows the design team to evaluate the "new" 1994 Challenge modified requirements including: ZEV range, HEV range, and vehicle weight specifications. Additionally, it is appropriate to reevaluate safety concerns and incorporate design changes. Similar to 1993, USNA team is constrained to very limited funds--COST is the key design driver.

LIMITED MEASURED DATA - Based upon extremely limited data, the 1993 configured *AMPhibian*'s electrical economy (ZEV) was determined to be 7.8 km/kW-hr at 48 km/h (30 mph). This electric economy is based on two runs along a 6.6 mile relatively flat course and subtracting 0.1 km/kW-hr to sustain minimal, required lighting (parking lights and head lamps). The 7.8 km/kW-hr value has some significant concerns including:

1) limited data
2) constant 30 mph (i.e., this is not a representative commuter economy since it contains no stops, no accelerations, no significant grade/environmental variations, etc.)
3) the data has no HVAC or other creature comforts (e.g., radio)
4) assumes nominal 120 volts

However limited, it still is actual performance data that can provide significant insight into the existing vehicle. For example, the data can be used to determine the average current draw for the *AMPhibian* traveling 25 miles in 50 minutes.

$$(7.8\frac{Km}{Kw \cdot hr} \times 120 volt \times \frac{1}{25mi} \times 50 \min)^{-1} = 51.6 amp$$

Likewise, this information can be used to evaluate potential vehicle modifications.

1994 ZEV/BATTERY SELECTION EVALUATION - The design team would like to reduce the size of the current battery stack to incorporate other desired features into the car. The USNA has, on-hand (i.e., no cost), two other 120 volt lead-acid battery stacks. Both stacks consist of ten, 12 volt batteries. One stack uses GNB Incorporated, Action Pack AP115's and the other Delco Remy, Voyager 105 (M27MF). Due to budget constraints, these are the only other batteries currently being considered for integration into the vehicle.

The task of evaluating batteries is complicated due to the general lack of published, comprehensive, technical battery performance data covering a wide range of current usage levels verified by an independent source. As such, HEV designers make do with what they have. For example, the available manufacturer provided data for the Trojan 5SH(P) covers a wide range of currents.

On the other hand, battery data for the two other possible batteries is presented in Table 2.

Table 2. Manufacturer Provided Battery Data

	GNB AP115	Delco Voyager 105
current, amp	time, hours	time, hours
5	20	
10	9	
15	5	5
20	3.4	3.6
25	2.67	2.7

This information is plotted in Figure 5.

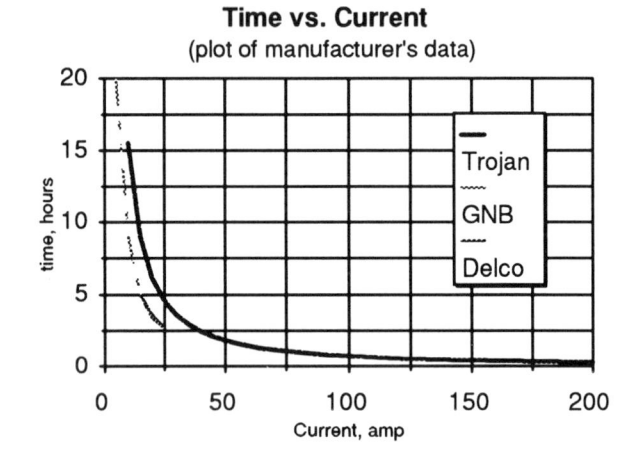

Time vs. Current
(plot of manufacturer's data)

Figure 5. Discharge Time vs. Current, Manufacture's Data

Using Figure 5, or a similar plot, like this, an HEV, or EV, designer can determine if a battery can deliver a desired current draw for the time required. For the 1994 Challenge ZEV range requirement of 40 km (25 mi) at 48 km/hr (30 mph), this means the vehicle must sustain the current draw for 50 minutes. Based on very limited data, *AMPhibian* will draw 51.6 amps. There is sufficient data for the Trojan battery to determine it can sustain the required load. The designer, must extrapolate the data for the other batteries to estimate their capabilities. One approach is to assume the battery characteristics are logarithmic. Making this assumption, the limited data can be extrapolated and is plotted in Figure 6.

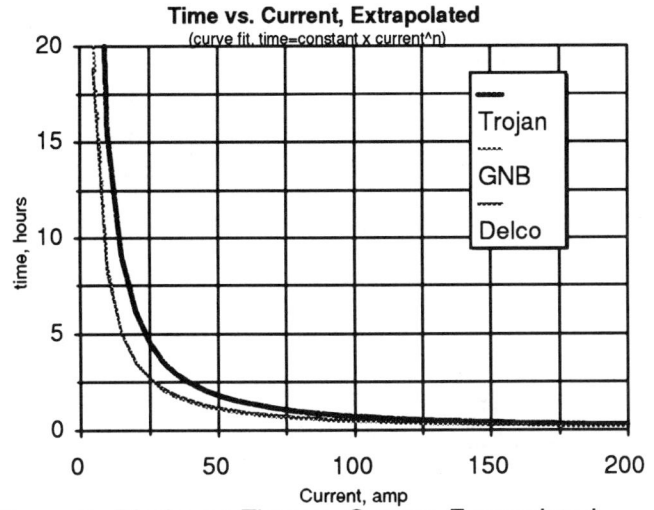

Figure 6. Discharge Time vs. Current, Extrapolated

With the data extrapolated, all three batteries can be compared to the requirements. That is, knowing the current draw of 51.6 amps, associated with the speed of 48km/h (30 mph), the time can be found and converted into an estimate of ZEV range. In addition, since battery cyclic life will be based on 80% depth of discharge, ZEV range for this condition can also be determined. This is shown in Table 3.

Table 3. Estimated ZEV range @ 48 kph (30 mph)

	100% DoD km (mi)	80% DoD
Trojan 5SH(P)	84.5 (52.5)	67.6 (42.0)
GNB AP115	50.7 (31.5)	40.5 (25.2)
Delco 105	54.9 (34.1)	43.9 (27.3)

This analysis indicates that all three batteries are acceptable. However, this analysis, aside from the obvious lack of data, has some significant concerns including: statistical reliability, the actual system the batteries are being used within, and temperature. The statistical battery data presented is typically nominal data. On average a given battery should be able to provide this performance. Since the battery stack is a series configuration to achieve 120 volts -- the weakest battery determines the stack capacity. With ten batteries, the battery stack, on average, will not have this capacity.

The 1993 **AMPhibian**'s controller, Curtis 1221B, has a built in low voltage protection system. That is, the controller will reduce the output of the motor to allow the battery low voltage to recover while still allowing the driver to maneuver the vehicle. This is a very practical feature. However, insufficient data has been collected to determine if the controller will even permit maintaining of ZEV 48 kph (30 mph) required speed for levels of depth of discharge less than the 80% DoD level. In short, even though the battery may be capable of delivering the current needed to achieve the desired minimum range, the vehicle system (e.g., controller) may dictate a larger battery.

By far the most significant concern is temperature. The above analysis is based upon assumption that the data provided was determined at room temperature conditions -- normal practice. However, it is well known that ambient-temperature, lead-acid batteries exhibit temperature-dependent capacity. In short, the lower the temperature, the lower the capacity. By using a linear approximation of this lead-acid behavior[1] and making the gross assumption that all lead-acid batteries behave the same, an estimate of ZEV range as a function temperature can be made. This can be seen in Figure 7.

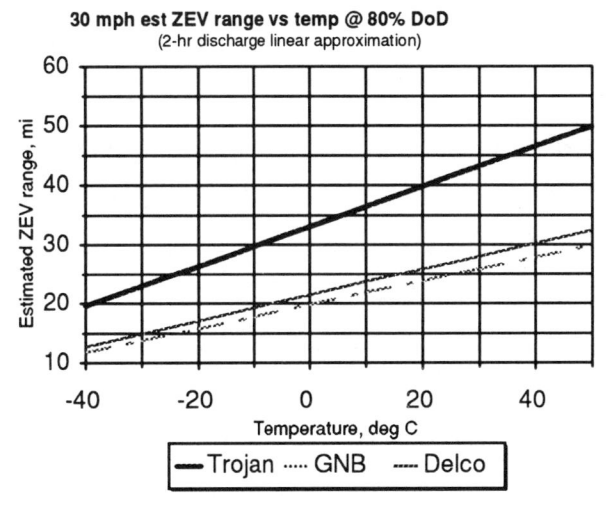

Figure 7. Ambient Temperature Effect on Estimated ZEV Range

Temperature effect has also been tabulated, in Table 4.

Table 4. Minimum estimated temperature to achieve 25 mile ZEV range @ 48 kph (30 mph)

	100% DoD (deg C (deg F))	80% DoD (deg C (deg F))
Trojan 5SH(P)	-38.8 (-37.9)	-24.0 (-11.14)
GNB AP115	0.905 (33.6)	25.7 (78.3)
Delco 105	-6.76 (19.82)	16.13 (61.0)

For the 1994 Challenge, scheduled for June, in lower Michigan, temperatures should be high enough, although very marginally for the GNB battery, for all three batteries to provide the 1994 ZEV range. However, only the Trojan 5SH(P) provides the desired range for the vast majority of the continental United States throughout the year -- both summer and winter. This practical consideration dictates the continued use of the Trojan 5SH(P). Additionally, APU output concerns also reinforce this decision. Considering all driving speeds, the estimated 1994 **AMPhibian** ZEV driving ranges are shown in the following figures:

---

[1]Keller, A. Scott, and Gerald Whitehead, "Thermal Characteristics of Electric Vehicle Batteries," SAE Technical Paper Series 911916

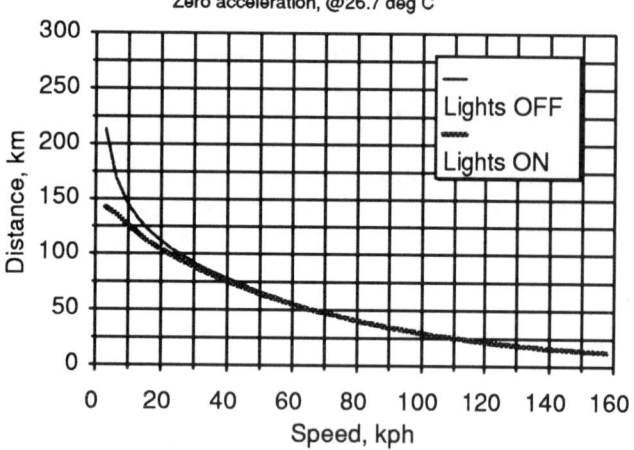

Figure 8. '94 **AMPhibian** ZEV Range (km) at 80% DoD

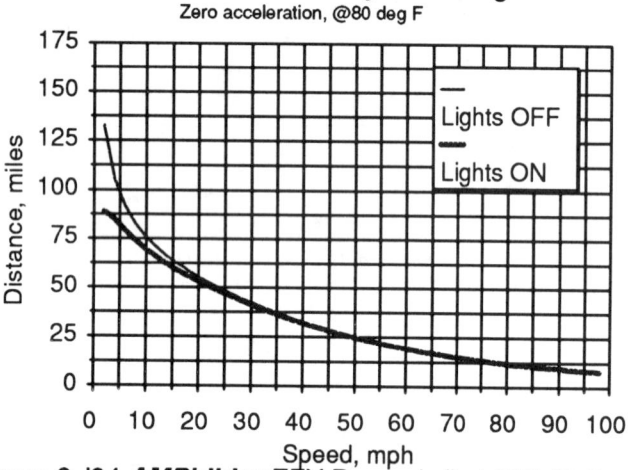

Figure 9. '94 **AMPhibian** ZEV Range (mi) at 80% DoD

It is interesting to note that lighting loads do not significantly alter typical city or highway speed ranges due to the relatively small contribution to total power loads. However, it should also be noted that heating, ventilation, and air conditioning (HVAC) loads are not included.

1994 AUXILIARY POWER UNIT (APU) - The number one objective was to minimize or eliminate the stalling condition and assess the performance. Consultation with the manufacturer about the stalling resulted in the incorporation of a easy fix to the simple carburetor (e.g., no accelerator pump, automatic choke) to the carburetor Main Air Jet. The Main Air Jet allows air to enter the main fuel pickup tube bleed holes and mixes with the fuel entering the venturi throat. The simple fix was a reduction in the Main Air Jet orifice diameter from 2.5 mm to 1.3 mm. This fix tremendously improved the engine's sudden load application problem. With this improved configuration, **AMPhibian**'s electrical economy with the APU running, was determined to be 37 km/kW-hr at 64 km/h (40 mph) and 13.8 km/kW-hr at 80 km/h (50 mph). This is based upon extremely limited data. Though limited, it does allow the updating of APU On predictions. The resulting prediction can be seen superimposed upon the ZEV estimates in the following figures.

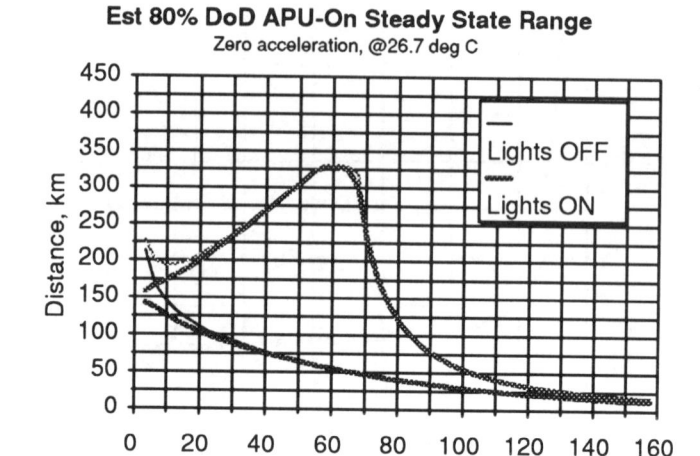

Figure 10. APU-On and ZEV Range (km) at 80% DoD

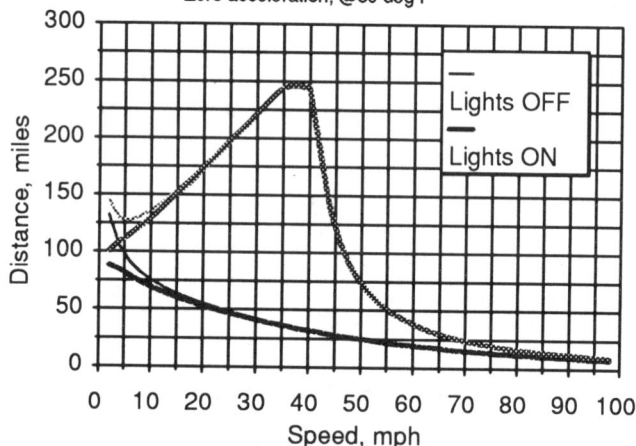

Figure 11. APU-On and ZEV Range (mi) at 80% DoD

Aside from the unusual curves, as compared to typical EV and ICE range predictions, the curves provide significant insight into the HEV capabilities and behavior. The obvious insight is that the modified 1993 configuration does not meet the 1994 Challenge requirement for 300 mile range @ 72 km/h (45 mph) in HEV operation. But more importantly the figure highlights the key vehicle performance drivers. That is, this data indicates the APU, as configured, can only provide approximately 9.1 kW or approximately 62.8 amps. Where the peak estimated range is driven by this APU output, to the left of the peak, the curve is driven by fuel capacity, the stock fuel tank presently, and to the right of the peak the range is driven by the battery capacity. That is, to the right of the peak the vehicle will still have fuel in the tank but is "out of battery." Additionally, to the left of the peak, the estimate is pessimistic in estimating potential range since this represents the APU-On continuously even if the battery is "full." Conceptually, the HEV range can be extended by managing (turning on and off) the APU so that it is not running when at/near full battery, thereby conserving fuel. This data and analysis also confirms the alternator supplier's estimate that engines, in this category, optimistically provide, "at best," about 90% of rated power. This means that for the **AMPhibian** to meet the 1994 Challenge requirements, the rated engine performance needs to be 16.8 kW (22.5 hp), as a minimum.

$$(95 amps)(120 volts) = 11.4 kW @ 45 mph$$

$$\frac{11.4 kW @ 45 mph}{9.1 kW actual / 13.4 kW rated} = 16.8 kW needed$$

The alternative to increasing the APU output is to increase the capacity of the battery. This is possible but unrealistic due to cost (e.g., Ni-Cad battery) or weight (i.e., larger Pb-Acid) considerations.

Modification of the existing 1993 engine to increase performance is possible. This is estimated to possibly increase the output by a maximum of two (2) hp by enlarging the venturi diameter from 21 to 23 mm, advancing the ignition timing up to 6 degrees, and milling the heads. Associated with this performance increase are the negative impact on kickback and knocking at 2000 rpm, according to the manufacturer. Even with two more horsepower, this is short of what is required. Hence, alternative engines were and are being investigated.

In general, the alternatives had to produce more power and had to be equal to or smaller in physical size than the existing engine. This lead the team to consider gas turbines or 2-stroke engines. A gas turbine powered APU, similar to Volvo's Environmental Concept Car (ECC) and the "clean" 0.8 liter, two-stroke, Orbital Industries, gasoline engine were investigated. Both where eliminated for the key reason of lack of near-term availability of these power plants for use in **AMPhibian** by June 1994. Had they been available, the team would still have eliminated them for budget considerations (i.e., unless they were reduced in cost or donated). This forced the team to canvass any source for conventional 4-stroke engines. There were several engines showing promise, but were eliminated on physical size limitations. For example, the Onan Performer 24 (P224) gasoline engine has a maximum rating of 17.9 kW (24 bhp) @ 3600 rpm. However, this engine would not fit within the existing "under-the-hood" compartment confines without significant design changes. The team identified four candidate alternate engines. The Onan Performer P220V, air-cooled horizontally opposed gasoline engine, the Honda GX620, air-cooled V-twin gasoline engine, and the Kawasaki FD620D; both carburetor and electronic fuel injection, liquid-cooled V-twin gasoline engines. The first three engines have ratings of 14.9 kW (20 bhp) @ 3600 rpm, while the Kawasaki FD620D electronic fuel injected version is rated at 16.4 kW (22 bhp) @ 3600 rpm. All engines appeared as though they would "fit under the hood."

Although all are undersized in power, any of the engines appeared to be *good* modular upgrades with the *best* upgrade being the Kawasaki EFI FD620D based upon what is currently, readily available. The Kawasaki versions also offered the potential for conventional water, waste heat, cabin heating in extreme cold environments.

The selection of the *best* alternate engine upgrade for the APU was greatly simplified by the generous donation of an engine by Kawasaki.

The EFI engine was integrated as a modular upgrade (i.e., very similar to the Vanguard installation). The fuel system includes the stock Escort fuel tank, the Escort fuel sending unit with the Kawasaki pump replacing the Escort fuel pump, the existing Escort fuel lines, the Escort fuel filter, and flexible tubing to mate with the engine fuel feed and return lines. The engine air intake uses the existing Escort air cleaner (cost) and fuel vapor canister along with the crank case vent and vapor recovery canister lines connected to be burned. The **AMPhibian** exhaust consisted of a custom made, stainless-steel exhaust manifold, UEC 3-way catalytic converter, an electric air pump (i.e., a "small" 12v, 200 psi portable tire inflation pump), ceramic exhaust header wrap, and the existing Escort tail pipe & muffler. The stainless-steel header and UEC converter were wrapped in the ceramic header wrap in an effort to aid quick light-off of the catalyst as compared to relatively expensive electric resistively heated catalysts. The cooling system was modified by removal of the belt driven cooling fan and replacing it with an electrically driven, 12 volt, thermostatically controlled, fan. It should be noted that the EFI components and configuration was left "as-is."

APU-On performance predictions based upon the new engine indicate successful compliance with the 1994 HEV Challenge requirements. With the estimated APU peak output of at least 93 amps and an additional 10.2 l (2.7 gal) of usable fuel the predicted range performance can be seen in the Figures 12 and 13.

Est 80% DoD APU-On Steady State Range
Zero acceleration, @26.7 deg C

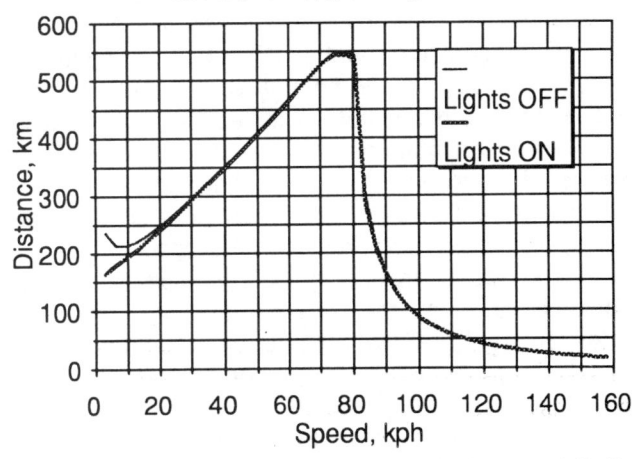

Figure 12. APU-On (Kawasaki) Range (km) at 80% DoD

Est 80% DoD APU-On Steady State Range
Zero acceleration, @80 deg F

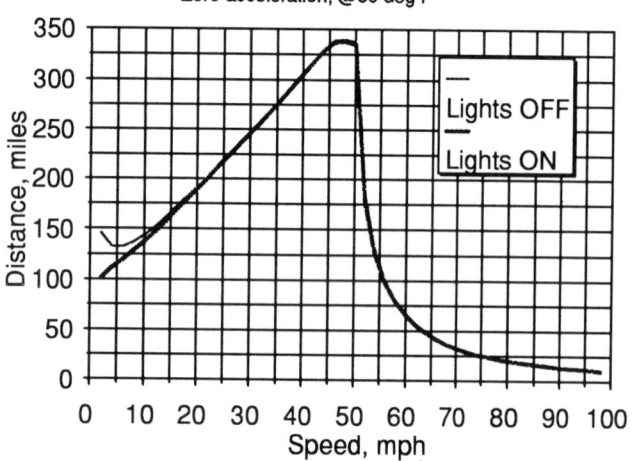

Figure 13. APU-On (Kawasaki) Range (mi) at 80% DoD

The additional fuel can be explained by the utilization of the stock Escort feed line incorporating the Kawasaki fuel pump as compared to the Vanguard utilizing the stock Escort fuel sending unit return line as a feed line.

1994 Motor Selection -- The 1993 *AMPhibian* exhibited marginal acceleration characteristics. Since the vehicle is a series configuration, the motor, transaxle and wheels were considered for changes to increase torque delivered to the ground. The easiest candidate was reduce the tire/rim combination diameter. This option was eliminated since the current tires are near their DOT ratings and smaller tires would only aggravate this condition. The next option was to change the transaxle to the Escort GT transaxle thereby providing higher gear ratios. The significant cost for limited gear ratio changes also eliminated this option. The best option occurred when the *AMPhibian* team was given the opportunity to buy and incorporate GE's new shunt motor and control system. This system was inexpensive, and could easily replace the existing series motor since the motor housing is identical.

This new system is comprised of a motor, controller, accelerator box, and digital dash display. Additionally, the built-in contactor and ignition key interface relay allows the system to be controlled with the existing 12 volt system, and are both included in the control box which is an environmentally sealed enclosure with an external heat sink. The low voltage control circuits, which include inputs for safety interlocks such as start, travel, and clutch switches, are accessed through one 21 pin commercial grade sealed connector. Connections for the dash display, accelerator potentiometer, and variable regen potentiometer are also in this connector. The control system uses a microprocessor to control IGBT power transistors which have a lower resistance in the on state than commonly used MOSFETs. The system has the ability to do both internal and external diagnostics and interface with safety inputs. Should an error condition exist, an error code number is displayed on the digital dash display, alerting the operator to the fault and leading the technician to the source of the problem. The dash display also displays battery state of charge and a run time hour meter. The microprocessor controlled system allows many of the system operating parameters to be adjusted and set for optimum user needs. The two most significant being regen current and field weakening. Field weakening allows the motor to spin at higher rpm and be more efficient during low load situations such as constant speed cruising. Regen allows the motor to act as a generator during deceleration and put energy back into the batteries. Regen can be set at a specified current level or the level can be controlled via a voltage input such as a potentiometer or voltage transducer.

HEV Strategy Mode - Given the nature of a hybrid electric vehicle, it is essential that there be some means of using of the onboard power supplies (battery pack and APU) to achieve the greatest range. It is unrealistic for the driver of the vehicle to be expected to drive the car and determine the most efficient use of available energy sources. The task at hand is to design a "turn-key," "hands-off" controller for the *AMPhibian* that will regulate the use of the APU based on different factors of the vehicle's performance.

The microprocessor to be used is the Motorola M68HC11. Also on the associated evaluation board are a real-time clock, all of the required support equipment (except 5 volt power supply), and access to all of the input/output ports. The micro-controller has the following specifications:[2]

    12 Kbytes ROM
    512 Bytes EEPROM
    512 Bytes RAM
    16-Bit Timer System
    8-Bit Pulse Accumulator
    Real Time Interrupt Circuit
    Computer Operating Properly (COP)
        Watchdog System
    8-Channel 8-Bit Analog-to-Digital Converter
    38 General-Purpose Input/Output Pins

Contained in the 12 Kbytes of Read Only Memory (ROM) are the monitor program that makes it possible to communicate with a personal computer (PC) and a number of utility routines. The 512 Bytes of Electronically Erasable, Programmable Read Only Memory (EEPROM) and 512 Bytes of Random Access Memory (RAM) are to hold the control code to be used in the *AMPhibian*. The EEPROM, once written, can only be changed using a PC, and any program stored there remains if power is lost. RAM is volatile memory, and will be erased if all power is lost to the chip, but there is a battery backup to protect against unforeseen

---

[2] Motorola MC68HC11E9 Technical Data (Motorola Inc., 1991) 1-1.

there is a battery backup to protect against unforeseen power losses to the Central Processing Unit (CPU). The 16-Bit Timer System simply counts the pulses of the CPU's internal clock. The default speed for the CPU clock is 8 MHz, so the real-time clock on the evaluation board is to used for timing purposes instead of counting the number of times that the CPU timer system overflows. The 8-Bit Pulse Accumulator is an event counter. Once activated, it increments its own counter each time a 5 volt signal is detected on its input pin. It will not be used in this version of the **AMPhibian** strategy control. The Real Time Interrupt Circuit can be used to interrupt a running program at specific times, based on input from the real time clock. The COP Watchdog System will interrupt a running program should a fault be detected in the CPU, and the Analog-to-Digital (A/D) Converter is to be used extensively for gathering data for the HEV strategy control. It will read four of the eight channels at one time, and it does so by comparing an input signal to the 5 volt reference high (V_{RH}). A number between 0 and 256, corresponding to the input level is then stored in one of the result registers. The General Purpose Input/Output (I/O) Pins are used to drive relays or other devices to implement the control code stored in memory.

The controller will be tasked primarily with maintaining the battery stack voltage at a usable level. The two causes for a low voltage on the battery stack are depleted batteries and a large power demand from the propulsion motor. Stop-and-go city driving will cause the voltage on the stack to drop, but the APU should be started only when the voltage is low for an extended period of time, due to prolonged high-speed driving or low battery charge. The basic strategy can be seen in the following table.

Table 5. Basic Strategy for Main HEV Program

Condition	Causes	Action
-Stack voltage low for a given period of time	-Highway speeds -Battery charge low -Long incline	-Start APU
-Stack voltage high for a given period of time	-Batteries charged -Return to normal driving	-Kill APU
-HEV "plugged-in"	-Charging batteries	-Disable APU -Control exhaust fan
-Battery compartment too hot	-Hot day -Discharging	-Exhaust fan to high speed
-Hood raised	-APU service	-Disable APU
-Oil pressure low	-Malfunction with APU	-Shut down and disable APU

The following flow charts show the implementation of the basic strategy for the 1993 APU engine:

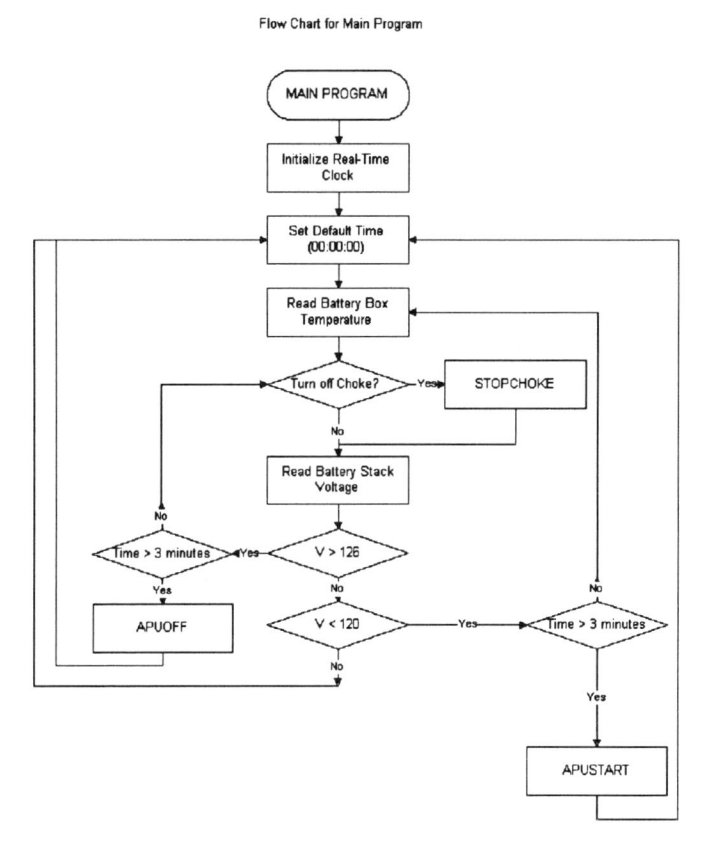

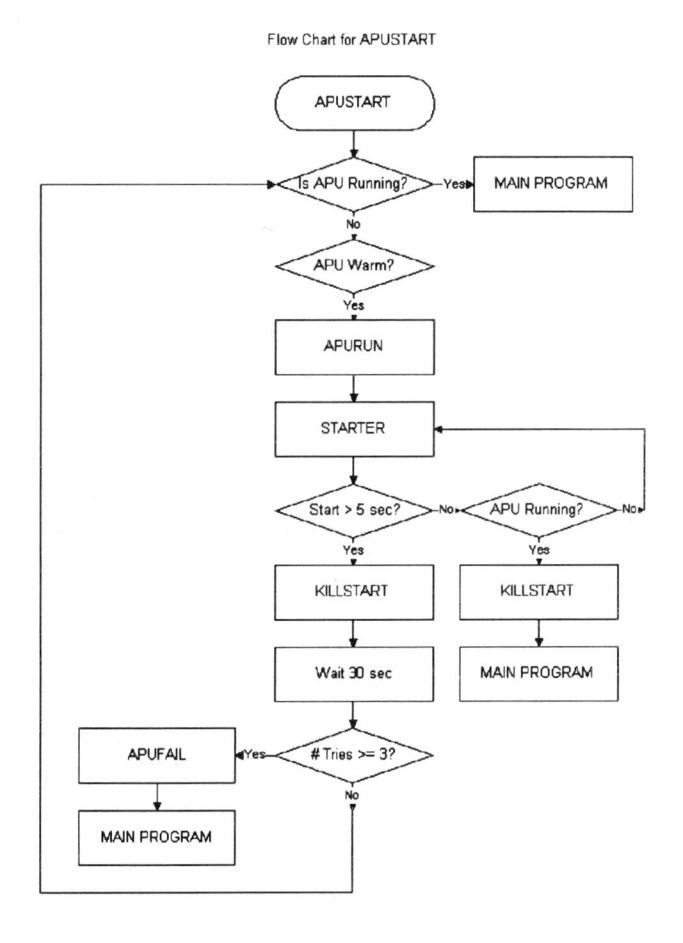

Flow Chart for Main Program

Flow Chart for APUSTART

Flow Chart for ADINIT

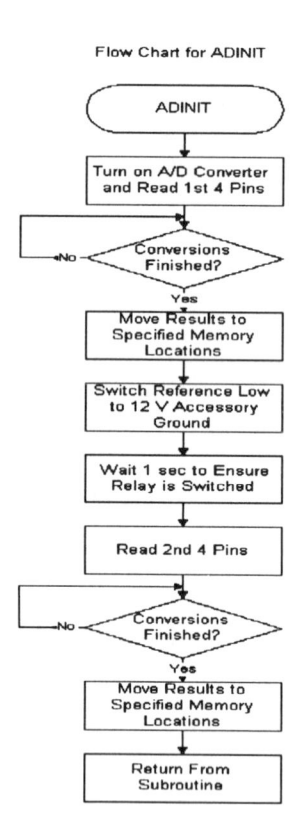

```
         ADINIT
            │
            ▼
  ┌──────────────────┐
  │ Turn on A/D Converter │
  │  and Read 1st 4 Pins │
  └──────────────────┘
            │
            ▼
      ╱Conversions╲  No
      ╲ Finished? ╱─────┐
            │ Yes       │
            ▼           │
  ┌──────────────────┐  │
  │  Move Results to │  │
  │ Specified Memory │  │
  │    Locations     │  │
  └──────────────────┘  │
            │◄──────────┘
            ▼
  ┌──────────────────┐
  │ Switch Reference Low │
  │  to 12 V Accessory   │
  │       Ground         │
  └──────────────────┘
            │
            ▼
  ┌──────────────────┐
  │ Wait 1 sec to Ensure │
  │   Relay is Switched  │
  └──────────────────┘
            │
            ▼
  ┌──────────────────┐
  │  Read 2nd 4 Pins │
  └──────────────────┘
            │
            ▼
      ╱Conversions╲  No
      ╲ Finished? ╱─────┐
            │ Yes       │
            ▼           │
  ┌──────────────────┐  │
  │  Move Results to │  │
  │ Specified Memory │  │
  │    Locations     │  │
  └──────────────────┘  │
            │◄──────────┘
            ▼
  ┌──────────────────┐
  │   Return From    │
  │    Subroutine    │
  └──────────────────┘
```

HVAC - If HEV's are to be truly practical vehicles, heating, ventilation, and air conditioning (HVAC) should be considered. If HVAC is considered for even mild temperature extremes, tremendous range impacts are incurred. For the extremes in temperature, loads can exceed 10 kW, with the present energy storage capacity, it is extremely challenging to provide climate control and maintain desired driving ranges, especially ZEV. Since the Challenge is scheduled to take place in June, in Michigan, there is minimal need for any heating or air conditioning. Additionally, since no Challenge points appear to be directly dependent upon a functioning HVAC, this key, practical HEV consideration is given a low priority for incorporation given budget constraints. However, due to the importance, the *AMPhibian* team did do a thermal survey in both EV and APU-On modes. The survey confirmed the lack of significant waste heat in EV mode as a possible cabin heating source. Additionally, only in APU-On mode did waste heat become significant. The team concluded that for EV mode the conceptual solution was to place/generate the heat (i.e., resistive) in only those areas requiring it vice cabin air heating. This included the resistively heated seats, floor areas where feet are normally placed, possible radiant heaters for the face/head areas, restively heated windows (i.e., similar to the rear window defoggers but placed on the front windshield) and heat pumps for warmer climes for some cabin air heating. For cooling, the use of heat pumps was considered the most likely solution along with conventional window tinting to reduce solar gain. For both extremes, thermally efficient windows and windscreens.

The *AMPhibian* has incorporated a sunroof to facilitate air flow. However, the biggest benefit of the sunroof is not thermal but photovoltaic. That is, the sunroof incorporates a 12V panel that should offset the 12 volt accessory battery stand loses.

BATTERY CHARGER - In order to promote a practical design, the 1993 *AMPhibian* had an on-battery charger. The advantages of the charger included:
  inexpensive, <$400
  light weight, approx. 4.5 lb
  multi-voltage, 120 to 250 Vac/50 or 60 Hz input
The disadvantages include:
  no isolation transformer
  poor reliability
  manufacturer is out-of-business
The key disadvantage is poor reliability. As evidenced by the charger being "broke." The team is attempting to fix the charger, but is having limited success since no technical data can be obtained from the manufacturer. It should be noted to mitigate the lack of an isolation transformer, the team purchased an off-board isolation transformer rated at 250 VAC 50 A. This transformer was relatively inexpensive (approx. $500) but weighs more than 200 lb. It was assumed that this was a viable solution since the EV/HEV infrastructure is not developed. That is, it was assumed consumers would have isolated power sources (e.g., garage), in the future, and would not have to carry this large mass aboard the vehicle.

The *AMPhibian* team investigated several off-board chargers. The team found that ferro-resonant float chargers used in the power utility, communications or industrial type load where suitable for EV/HEVs and were readily available and very efficient. Efficiencies range from a low of 85% to typically 90% with regulation at ± 0.25%. Although the regulation for Pb-Acid appears to be better then required, these chargers would seem to be very useful for Ni-Cad or other batteries. The significant draw back was cost and weight, typically in the mid $2000's and greater than 300 lb. Since this is a major area of concern, the team opted to purchased one of these expensive chargers. Additionally the team bought a back-up charger from a well known manufacturer of ferro-resonant chargers, for electric golf carts, for under $1000.

Additionally, a student has almost completed an on-board SCR charger for emergency use (i.e., away from "home"). The design attempts to correct those features deemed unreliable in the original 1993 charger. Its' incorporation as a practical EV/HEV backup will depend upon successful bench and trail testing.

WIRING - A partial *AMPhibian* schematic is shown in Figure 14.

  FUSES - The maximum DC current estimated to be drawn from the main batteries is about 350 A. Thus, a 400 A time delayed fuse is located on the positive lead from the battery before the circuit breaker. Since the steady state current draw will be about 175 A. The maximum amperage will be approached when accelerating quickly. Thus, this high current will only be drawn for a short period of time. Additionally, another fuse is planned to be placed within the battery stack to mitigate the distinct possibility of shorting the battery with dropped tools onto the stack, as experienced at the 1993 HEV Challenge.

In addition, since the alternator is the single most expensive component added to the conversion, each of the three phases are fused with a 60 amp fuse to the bridge rectifier. Between the rectifier and 120 volt buss is a 125 amp fuse.

WIRING - The battery and all high voltage areas are completely isolated from the chassis. This should prevent any shorts to the chassis. Cable no smaller than 1/0 should be able to handle estimated loads. To minimize losses, **AMPhibian** is wired with 2/0 cable between battery an motor controller. 3/0 cable between the motor controller and motor.

SHUNTS - Current shunts are installed in series with the three main components of the series hybrid configuration -- 1) the main battery stack, 2) motor controller and 3) APU output. These shunts are added to facilitate data collection and energy management (i.e., amp-hr meter). These are bare metal shunts and are rated at 500 Amps with corresponding 50 mV drop. Heavy rubber insulation has been used to cover these shunts to minimize safety concerns.

INSULATION BREAKDOWN VOLTAGE - The maximum voltage applied to the buss is expected to be the charging voltage of 144 Vdc. Accounting for a factor of safety, the insulation should be able to withstand 200 Vrms. Since the industry standard is 600 V, the breakdown voltage should not be a problem in any size wire used.

STRUCTURAL MODIFICATIONS - The only structural modifications to the **AMPhibian** is in the area of the suspension. The stock springs were replaced with custom made springs with approximately double the spring constant of the original springs. The springs were modified to compensate for the increased weight of the conversion vehicle. Non structural modifications include putting a hole in the roof for a sunroof, three 10.16 cm (4 in) ventilation holes were made in the floor of the car for battery box air circulation. These holes are sealed off from the passenger compartment.

SAFETY - **AMPhibian** has incorporated several safety features beyond those required by the Challenge. The main battery stack can be fully isolated as required, by a single throw, double pole circuit breaker. Additionally, the main battery stack is interruptable by an emergency disconnect switch and an ignition-keyed main contactor. The emergency disconnect switch is redundant to the keyed main contactor. It is anticipated that mass produced EVs and HEVs would solely rely on the main contactor and the conventional key switch. In the event of a collision, **AMPhibian** has addressed the potential of having emergency personnel (e.g., firemen) exposed to main battery stack voltage by incorporating the Escort's Fuel Pump Shut-off Switch to not only shut of the APU's electric fuel pump but also disconnecting the main contactor.

An area of concern is the battery compartment ventilation may need future modification. That is, the initial design placed the two air feed ducts just forward of the rear axle and the single exhaust duct just forward of the rear bumper with all three pointing down. The configuration was assumed to be fail safe in that it was thought that if the blower failed, the forward ducts would experience a higher pressure then the rear duct, hence causing induced flow through the battery box as the vehicle drove the down the road. However, a 2-D flow analysis at 80 km/h (50 mph) indicates the pressure is less for the feed tubes as compared to the rear duct. This may or may not be a problem but the analysis indicates assumptions simply need to be verified.

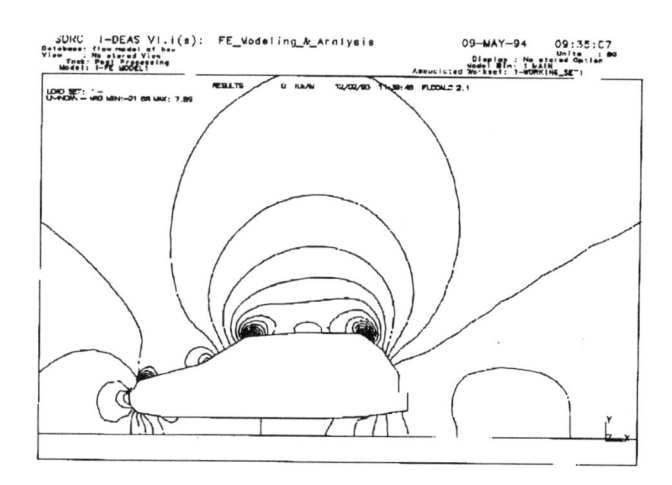

Figure 15. 2-D Flow Analysis Results@80 km/h (50 mph)

The main battery stack tie down approach was simplified while significantly improving safety by the incorporation of non-conducting webbing as compared to the aluminum grid structure used in the 1993 **AMPhibian**. Two inch polypropylene webbing was selected for load carrying capacity, ease of handling, and acid resistance (i.e., electrolyte spillage). The battery tie down straps were secured to itself by polyester (acid resistance) hook and loop (Velcro).

CONCLUSION - For the 1993 HEV Challenge, the United States Naval Academy considered safety, time, and resources and came to the conclusion that the conversion, series approach was the best option for success. Similarly for the 1994 Challenge, USNA considered the same constraints and opted for the same approach with modular upgrades with limited budget resources. The current Trojan 5SH(P) battery stack will again be used for the 1994 Challenge. The estimated ZEV range for the 1994 Challenge should easily be met by the **AMPhibian** in summer or winter conditions. The 1994 **AMPhibian** has been improved by modular increments. This included a higher output water-cooled engine and a higher output shunt motor/control system. 1994 **AMPhibian** performance predictions indicate compliance of the APU-On Range performance of 300 miles at 45 mph. The incorporation of a new HEV Strategy mode system has been developed. Improvements in safety have been accomplished through the incorporation of the Escort's fuel pump collision switch to the main contactor and APU fuel pump and battery tie down approach. Finally, the evolving United States Naval Academy HEV, **AMPhibian**, configurations for the Hybrid Electric Vehicle Challenge can be found in Table 6.

Table 6. **AMPhibian** Specifications

HEV type	Series, range-extender
Conversion Vehicle	
Make:	Ford Motor Company
Year:	1992
Model:	Escort
Body:	LX Wagon
Type:	Passenger
Exterior color:	Ultra Blue Clearcoat Metallic
Interior color:	Titanium
VIN:	1FAPP15JNW191198
1992 Escort retail cost	$10,000
Estimated conversion retail cost including Escort retail cost:	$24,000 to $28,000
Wheel base	250 cm (98.4 in)
Length	435 cm (171.3 in)
Height	136.1 cm (53.6 in)
Width	169.4 cm (66.7 in)
Front tread	143.5 cm (56.5 in)
Rear tread	143.5 cm (56.5 in)
Transaxle	5-speed manual overdrive
Type:	Type A: Standard MTX
Control:	Floor Shift
1st	3.416
2nd	1.842
3rd	1.290
4th	0.972
5th	0.731
Reverse	3.214
Final Gear Ratio	3.619
Steering	Manual rack-and pinion 23.7:1 overall ratio
Front suspension	Strut-type independent suspension with upper strut-mounted coil springs and stabilizer bar
Rear suspension	Strut-type independent twin trapezoidal links with upper strut-mounted coil springs and stabilizer bar
Brakes	Four-wheel hydraulic, power assisted front disc/rear drum
Passenger capacity	Two (2) 80 kg (177 lb) passengers
Minimum Cargo Capacity	20 kg (44.1 lb)
Minimum Carrying Capacity (passenger and cargo):	180 kg (397 lb)
Weight	
'92 Escort LXW GVWR:	1572 kg (3466 lb)
'92 Escort Front GAWR:	847 kg (1869 lb)
'92 Escort Rear GAWR:	759 kg (1675 lb)
'92 Escort Curb Weight (as delivered)	1126 kg (2483 lb)
Minimum Curb Weight less conversion components	894 kg (1972 lb)
1993 HEV Challenge Constraint GVWR + 10%	1729 kg (3812 lb)
1993 AMPhibian Curb Weight	1530 kg (3372 lb)
1993 AMPhibian GVW	1710 kg (3769 lb)
1993 AMPhibian Front GAW	717 kg (1580 lb)
Left Front	345 kg (760 lb)
Right Front	372 kg (820 lb)
1993 AMPhibian Rear GAW	813 kg (1792 lb)
Left Rear	423 kg (932 lb)
Right Rear	390 kg (860 lb)
1994 HEV Challenge Constraint GVWR + 15%	1808 kg (3986 lb)

1994 AMPhibian Curb Weight	1582 kg (3487 lb) [NTE 1620/1648 kg (3589/3632 lb)]
1994 AMPhibian GVW	1762 kg (3884 lb)
1994 AMPhibian Front GAW	831 kg (1832 lb)
Left Front	415 kg (914 lb)
Right Front	416 kg (918 lb)
1994 AMPhibian Rear GAW	931 kg (2052 lb)
Left Rear	467 kg (1030 lb)
Right Rear	464 kg (1022 lb)
1993 Motor:	
Make:	General Electric, Electric Vehicle Systems
Type:	DC Series-wound
Model:	5BT1346B50
Enclosure:	Blower Ventilated
Power:	15.59 kW (20.9 hp) @4707 rpm, 90V, 184A
Maximum Speed:	6500 RPM
Mounting:	Flange
Serial Number:	WG-9-345-WG
Rotation:	Clockwise
Mounting:	AU1840
Dimensions (L x D):	45.1 cm x 22.9 cm (17.75 in x 9 in)
Weight:	77.1 kg (170 lb)
1994 Motor:	
Make:	General Electric, Electric Vehicle Systems
Type:	DC Shunt Traction Motor
Model:	5BT134B104
Enclosure:	Blower Ventilated
Power:	20.1 kW (27 hp) @5000 rpm, 96V, 250A
Duty Cycle:	One hour @ 140 C
Maximum Speed:	6500 RPM
Mounting:	Flange
Serial Number:	
Rotation:	Clockwise
Reference Outline:	36B550574AH (SK93C50130)
Indicators:	One N.O. Thermostat
Insulation Class:	H (Total Temp 180 C)
Dimensions (L x D):	45.1 cm x 22.9 cm (17.75 in x 9 in)
Weight:	77.1 kg (170 lb)
Motor Ventilation Blower	
Make:	Attwood Corporation
Type:	In-line, ignition protected, axial-flow, 3 inch hose
Capacity	145 CFM
Current draw	3.2 amps @ 12volts DC
1993 Motor Controller	
Make:	Curtis Instruments, Incorporated
Model:	1221B-1074
Serial Number:	121295
Type:	MOSFET
Operating Frequency:	15 kHz
Standby Current:	< 20 mA
Standard Throttle Input:	0 - 500 ohms, ±10%
KSI Input Level:	from 8 V to 1.5 x max battery voltage
Input Voltage:	72-120 Vdc
Overall Dimensions:	27.7 cm L x 18 cm W x 8.0 cm H (10.9 in L x 7.1 in W x 3.15 in H)
Weight:	4.90 kg (10.8 lb)
Maximum Rated Current:	400 Amps
1994 Motor Controller:	
Make:	General Electric, Electric Vehicle Systems
Model:	
Serial Number:	
Type:	IGBT
Operating Frequency:	4 kHz
Accelerator Input:	5k - 0 ohms, 4.5 Vdc - 0.5 Vdc
Thermal Protection:	70 C
Ambient Temperature:	-40 C - +50 C
Modulation:	PWM
Logic Card:	Microprocessor
Repair:	Field
Creep Speed Adjustment:	2 - 15% Time On
Current Limit Adjustment:	Max minus 200A
Controlled Acceleration Adjustment:	27 - 68 sec
Minimum Field Current Adjustment:	0 - 31A
Maximum Field Current Adjustment:	0 - 31A
Internal Resistance Compensation:	11.44 Vdc - 0.74 Vdc
Rate of Field Current:	0.09 - 22.9 sec
Adjustment Method:	Softset
On-Board Coil Suppressors:	Yes
PMT:	Yes
SRO:	Yes
Reversed Battery Protection:	Line Control
Field Weakening:	Variable
Regenerative Braking:	Standard
Maximum REGEN Current:	250 Amps
Diagnostics:	Standard
Hour meter:	Standard
Stored Status Codes:	16 Codes and BDI
Battery Indication:	Standard
Input Voltage:	96-144 Vdc
Overall Dimensions:	cm L x cm W x cm H (in L x in W x in H)
Weight:	kg (40 lb)
Motor Current Limit:	500 Amps

Main Battery, Individual:	
Make:	Trojan Battery Company
Model:	5SH(P)
Type:	12 VDC, Conventional Lead-Acid (flooded)
Individual Overall Dimensions:	34.4 cm L x 17.15 cm W x 28.9 cm H
	(13 9/16 in L x 6 3/4 in W x 11 3/8 in H)
Individual Weight:	39 kg (86 lb)
Cycle Life @ 80% DoD	560
Percentage of Charge	Open circuit voltage
100%	12.60 V
75%	12.36 V
50%	12.18 V
25%	11.94 V
Capacity	165 amp-hr @ 20 hrs
Main Battery, Stack:	10 arranged in series
nominal voltage	120 VDC
Percentage of Charge	Open circuit voltage
100%	126.0 V
75%	123.6 V
50%	121.8 V
25%	119.4 V
Battery Enclosure Blower	
Make:	Attwood Corporation
Type:	In-line, ignition protected, 4 inch hose
Capacity	225 CFM
Current draw	4 amps @ 12volts DC
Circuit Breaker, Main Battery Stack	
Make:	Heinemann Electric Company
Model:	Series GJ
	GJ2 -- B3 - DU0250 - 01C
Delay:	Type 1
No. of Poles:	2
Current Rating:	250 A
Max Voltage:	240 V, 50/60 Hz, 125/250 Vdc
Interrupting Capacity:	18,000A @ 240 V, 50/60 Hz
	10,000A @ 125/250 Vdc
	14,000A @ 160 Vdc
	25,000A @ 65 Vdc
Agency Recognition or Approval:	240 V, 50/60 Hz and 125/250 Vdc: UL listed
Emergency Battery Disconnect Switch	
Make:	Curtis-Albright
Model:	SW203
Configuration	Single Pole with magnetic blowouts
Max thermal current rating (100%)	250A
Breaking current @ time constant 15 ms	1500A @ 120 Vdc
Typical voltage drop across terminals @ 100 A	8-12 mV
Weight:	0.8 kg (1.76 lb)
1993 Single Pole on/off Contactor (normally open)	
Make:	Curtis-Albright
Model:	SW20012DCCW(13220640)
Rating:	350A @ 50% duty, 250A continuous duty
12 Volt Accessory System	
DC/DC converter:	
Make:	Solar Car Corporation
Model:	Com-00100
Input Voltage:	90-130 Volts
Output Voltage	
0-85% Load:	14 V DC
Full Load:	12 V DC
Output Current:	30 Amps
Overall Dimensions:	17.2 cm L x 18.4 cm W x 10.2 cm H
	(6.75 in L x 7.25 in W x 4.0 in H)
Weight:	1.81 kg (4.0 lb)
12 Volt Battery	
Make:	GNB Incorporated
Model:	Pulsar 5P Racing Battery
Type:	12 Volt, Lead-Acid
Cranking Amps at 0 C (32 F)	220 Amps
Reserve Capacity:	23 min
Overall Dimensions:	25.6 cm L x 4.78 cm W x 21.6 cm H
	(10.09 in L x 1.88 in W x 8.5 in H)
Weight:	4.58 kg (10.1 lb)
1994 12 Volt Solar Panel (Sunroof)	
Make:	
Sunroof:	Webasto Sunroofs, Incorporated
Photo voltaic cells:	Solarex Corporation
Type:	
Serial Number:	ACADEMY01
Ref. Paddle Number:	278 103.9
Module Temperature:	23.8 C
Reference Voltage:	15 V
Cell Efficiency:	10.8 %
Irr. Level:	1 sun
Date:	7/21/93
Corrected to 25 C	
Voc:	20.7 V
Isc:	0.81 A
Pmax:	12.6 W
V @ Pmax:	17.1 V
I @ Pmax:	0.74 A
Efficiency:	10.8 %
Fill Factor:	75.0 %
I @ Reference V:	0.77 A

1993 APU Engine:	
Make:	Briggs and Stratton, Vanguard Series
Model:	350447
Type:	0075-01
Code:	92102611
Cycle:	4 stroke
Cylinders:	V-Twin, OHV
Bore:	72 mm (2.83 in.)
Stroke:	70 mm (2.76 in.)
Displacement:	570 cc (34.7 cu. in.)
Power:	13.4 kW/12.8 net kW (18 bhp/ 17.2 net bhp)
Breather System:	Closed, recirculating
Cooling System:	Pressure air-cooled, radial flow blower
Starter:	1) Positive engagement solenoid type starter with 12 volt negative ground. 2) recoil pull cord
1994 APU Engine:	
Make:	Kawasaki
Model:	FD620D
Type:	
Code:	
Cycle:	4 stroke
Cylinders:	90 deg V-Twin, OHV
Bore:	76 mm
Stroke:	68 mm
Displacement:	617 ml (37.7 cu. in.)
Power:	16.4 kW @ 3600 rpm (22 hp @ 3600 rpm) [SAE J1349]
Breather System:	Closed, recirculating
Cooling System:	Low pressure, recirculating water/glycol cooling
Starter:	Shift type electric starter
Dimensions (H x W x L)	624 mm x 448 mm x 556 mm
	(24.6 in x 17.6 in x 21. 9 in)
Dry weight:	41.5 kg (91.5 lb)
APU Exhaust Catalytic Converter	
Make:	United Emission Catalyst
Type:	3-way
APU Alternator (with fused, full wave bridge rectifier added):	
Make:	Fisher Technology, Inc.
Type:	permanent-magnet (Neodymium-Iron-Boron)
Phases:	3-phase (WYE connected)
Resistance:	0.06 Ohms (phase-phase)
Power:	13.5 kW, 150 Vpeak
Maximum Speed:	3600 rpm
Overall Dimensions:	
Diameter:	27.6 cm (10.87 in)
Length:	17.27 cm (6.8 in) includes a 3.81 cm (1.50 in) through shaft
Weight:	9.5 kg (21 lb)
On-Board Battery Charger	
1993 AMPhibian	
Make:	manufacturer is out-of-business
Type:	Transformerless, Silicon-controlled rectifier
Components/features:	dual SCR/Diode full wave bridge, quad op-amp gate drive, opto-coupled triac triggered, output current & voltage limiting
Input Voltage:	120-240 VAC (max 50 amp)
Input Frequency:	50-60 Hz
Output Voltage:	6-270 VDC
Output Current:	200 mA to 20 amp (pulsed)
Overall Dimensions:	20.5 cm L x 13.5 cm W x 12.0 cm H
	(8.07 in L x 5.31 in W x 4.72 in H)
Weight:	2.49 kg (5.5 lb)
1994 AMPhibian	(not incorporated)
Make:	student made
Type:	Transformerless, Silicon-controlled rectifier
Overall Dimensions:	24.1 cm L x 14.61 cm W x 8.26 cm H
	(9.5 in L x 5.75 in W x 3.25 in H)
Weight:	kg (lb)
1994 Off-Board Battery Chargers	
Primary Charger	
Make:	C & D Charter Power Systems
Model:	ARE130CE35F
Type:	Ferro-resonant
Input Voltage:	208/240/480 VAC
Input Current:	29/ 25/ 12.5 Amps
Input Frequency:	60 Hz ± 5%, single-phase
Float Range:	120-142
Equalize:	FL-150
Output Current:	35 Amps
Weight (shipping):	209 kg (460 lb)
Back-up Charger	
Make:	Lester Electrical of Nebraska, Inc.
Model:	07740-00
Type:	Ferro-resonant
Input Voltage:	115/230 ± 10% VAC
Input Frequency:	60 Hz
Output Voltage:	120 volts DC
Output Current:	25 Amps

Tires:	
Make:	Goodyear Tire & Rubber Company
Model:	Invicta GL BSL RP TL SI
Type:	Radial (low rolling resistance)
Size:	P175/65R14 81T
DOT Quality Grades	
Treadwear:	280
Traction:	A
Temperature:	B
'93 AMPhibian Tire Pressures	
Left Front	276 kPa (40 psi)
Right Front	324 kPa (47 psi)
Left Rear	345 kPa (50 psi)
Right Rear	290 kPa (42 psi)
'94 AMPhibian Tire Pressures	
Left Front	345 kPa (50 psi)
Right Front	345 kPa (50 psi)
Left Rear	415 kPa (60 psi)
Right Rear	415 kPa (60 psi)
rev/mile	908
Rolling Radius	11.11 in
Fuel capacity (stock fuel tank)	(45 l (11.9 gal)
'93 AMPhibian	approx. 38 l (10 gal) usable
'94 AMPhibian	48 l (12.7 gal)
1993 Gasoline mileage	38 mpg (modified FUDS "C")
1994 Gasoline mileage	22 mpg

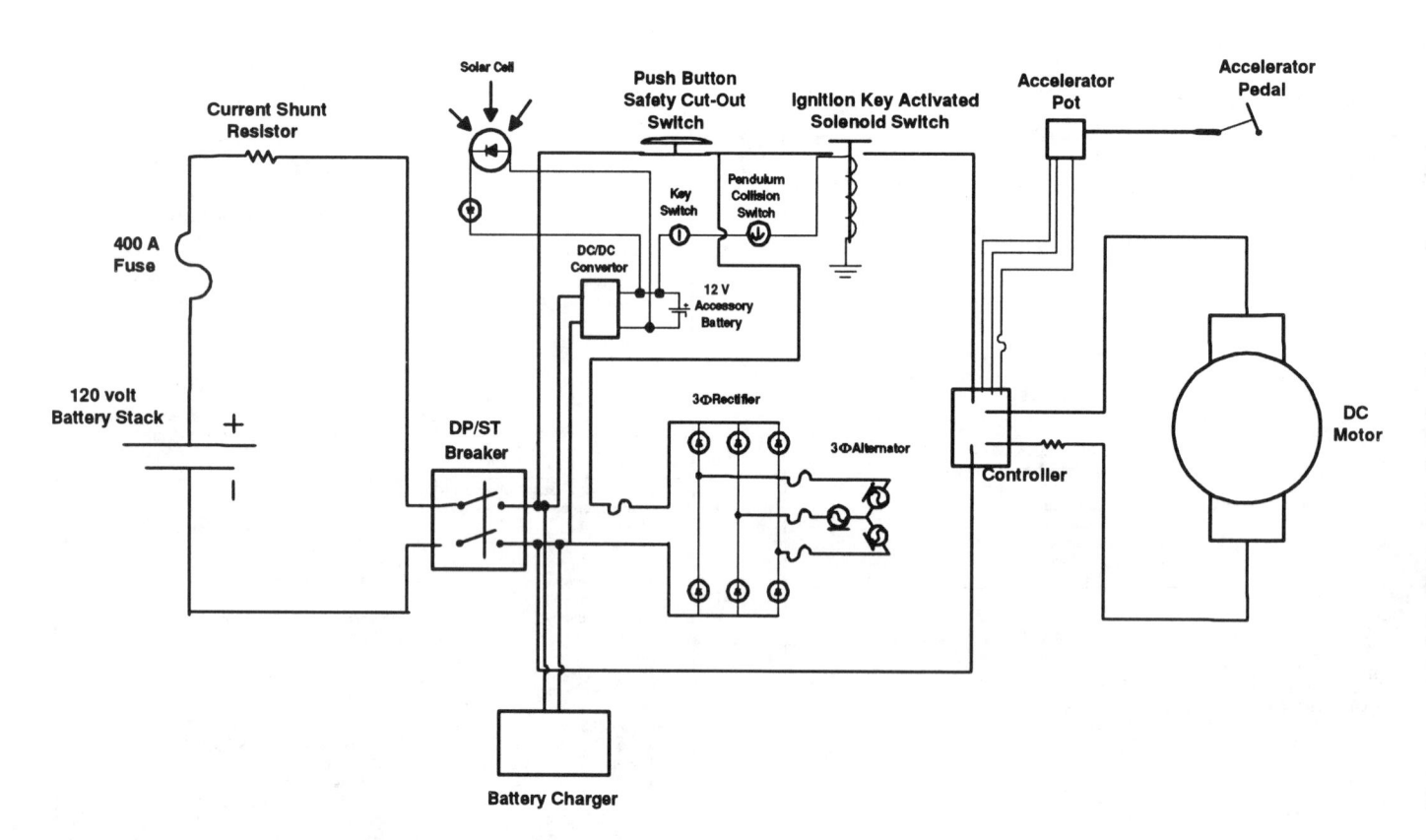

Figure 14. AMPhibian Electrical Schematic

Schematic of AMPhibian Microprocessor Controller

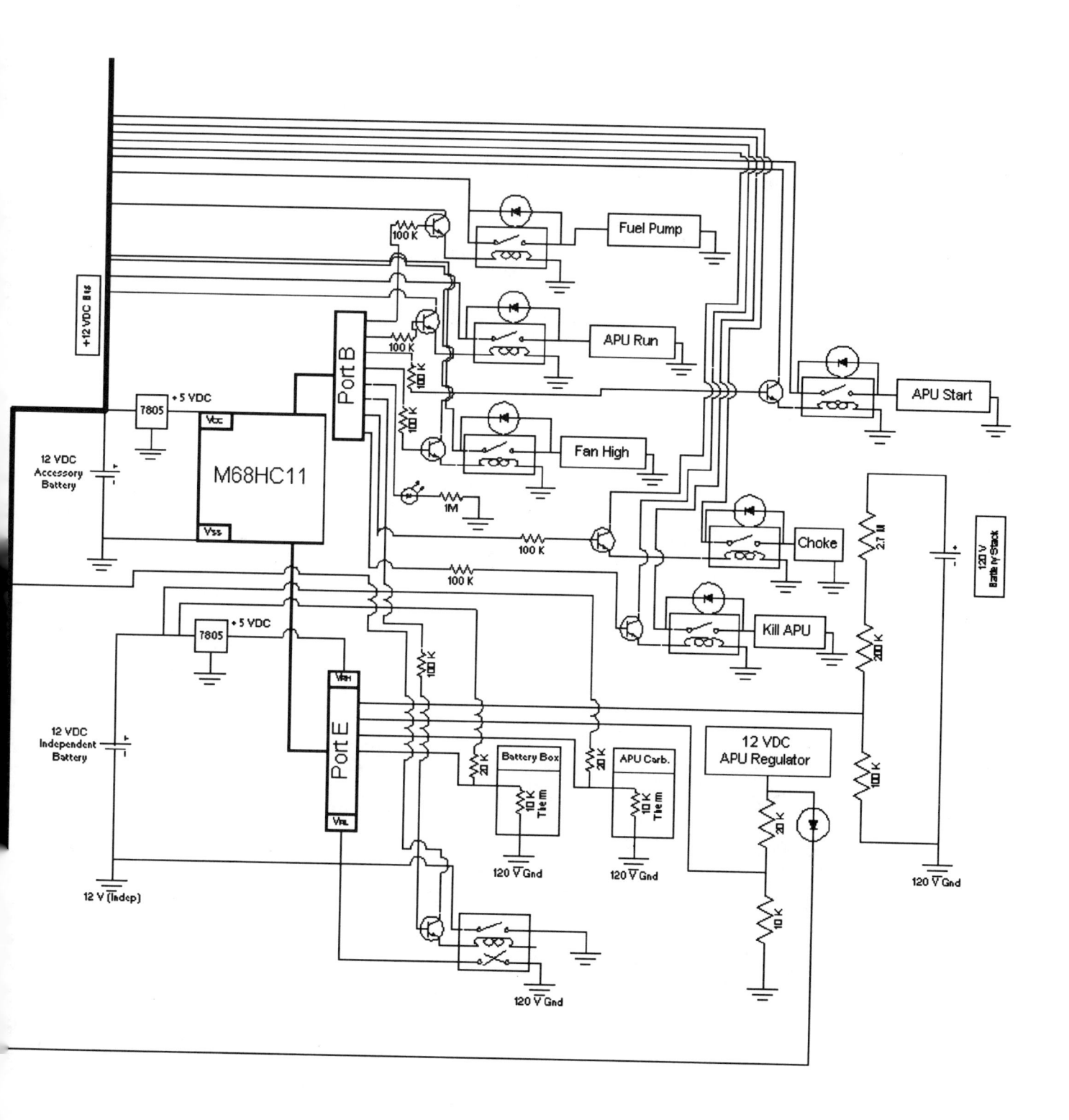

1994 Hybrid Electric Vehicle Challenge
Ground-Up Class

Technical Report
for the Joint Team of
Washington State University
and the
University of Idaho

Kirk Lupkes

ABSTRACT

A joint design team formed by students of Washington State University and the University of Idaho has designed and manufactured a passenger vehicle to compete in the 1994 Hybrid Electric Vehicle Challenge. The vehicle features an electric propulsion system supplemented by an internal combustion engine.

The WSU/UI HEV offers the potential of combining many of the benefits of electric vehicles with the performance capabilities of production automobiles. Additionally, in an effort to produce a truly marketable vehicle, safety features such as anti-lock brakes, traction control and a steel passenger safety cage have been incorporated into the design.

INTRODUCTION

The purpose of this paper is to give a technical description of the Washington State University/ University of Idaho hybrid electric vehicle. This vehicle will compete in the Ground-Up class of the 1994 Hybrid Electric Vehicle Challenge. The WSU/UI team will be competing for the second time in this competition, having achieved a seventh place overall finish in the field of twelve last year.

The 1994 WSU/UI HEV is a two passenger design based upon a steel space frame and roll cage. The heart of the vehicle is a 100 kW AC induction motor that drives the front wheels through a single speed gear reduction. Twenty-eight sealed lead-acid batteries contained within a thermally managed compartment provide electrical energy storage.

A 14.9 kW (20 hp) internal combustion (IC) engine and alternator comprise the Alternate Power Unit (APU) that makes this vehicle a hybrid. The IC engine is controlled automatically, operating to produce electric power only when the batteries have discharged 20% of their power. The APU is designed to provide the vehicle with enough power to keep it at a steady 50 miles per hour without further depleting the batteries.

The vehicle features a gasoline fired air heater that maintains the battery compartment at an elevated temperature for greater efficiency. It may also be used to heat the passenger compartment when desired.

Additional design features include a four wheel anti-lock brake system, adjustable regenerative braking, traction control, and a compact disc stereo system.

The performance design goals set for the vehicle included the following: a top speed of 75 miles per hour; an electric-only range of 80 miles; an HEV range of 500 miles at 50 mph; and a 0 to 50 miles per hour acceleration time of 7 seconds.

DRIVETRAIN

AC-100 DRIVE SYSTEM - A power electronics unit (PEU) and alternating current induction motor were acquired as a complete system from AC Propulsion, Inc. This state-of-the-art system has many outstanding features. The PEU houses a power inverter, auxiliary 12 VDC power supply, battery recharge circuitry, regenerative braking electronics, traction control electronics and all required cooling fans.

Power Electronics Unit - The PEU power inverter converts the direct current signal of the batteries into an alternating current signal for the induction motor. Vehicle accessories may also be operated by the DC power supply of the PEU, which provides up to 100 A of current (at 13.8 VDC). The PEU auxiliary 12 VDC system is electrically grounded to the vehicle frame, and all external control and data signals are chassis referenced for safety.

The internal recharge circuitry of the PEU will accept up to 20 kW from any 50 to 60 Hz power source between 100 and 250 VAC. This highly flexible charging system allows drive system components to double as battery charger components, eliminating the need for external chargers. In addition, the system maintains unity power factor across the entire range of power sources it can utilize.

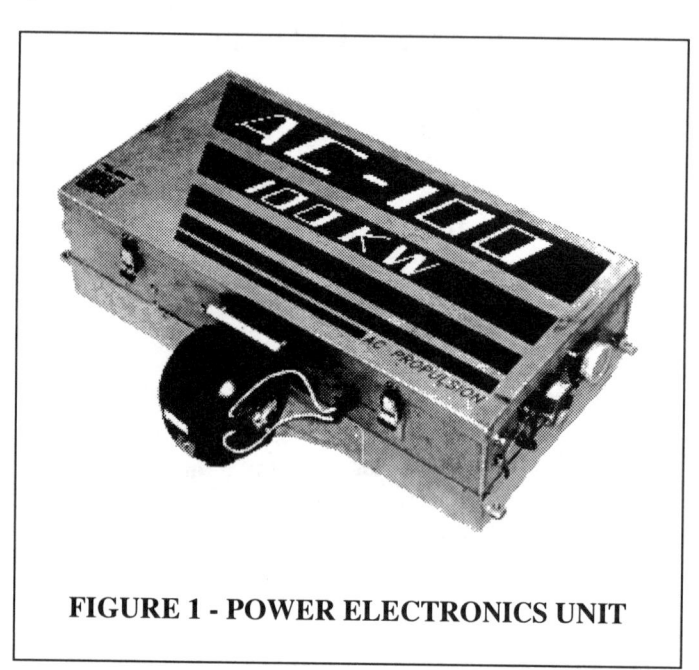

FIGURE 1 - POWER ELECTRONICS UNIT

The regenerative braking and traction control capabilities of the PEU are outstanding. The driver may adjust the magnitude of the regenerative braking, by means of a simple slide adjustment in the passenger compartment, from 0 to a maximum of 115.3 N-m (85 ft-lb) of torque. The regenerative braking is activated by the release of the accelerator pedal. The adjustability of this regenerative braking is an important feature, as many drivers would initially be unfamiliar with this type of braking system.

The traction control system monitors wheel speed through the same magnetic pick-up signals that the hydraulic anti-lock brake system uses. The PEU processes the wheel speed information and increases or decreases power to the induction motor as necessary to help maintain vehicle control.

The PEU and induction motor are forced-air cooled, each continuously operating its own centrifugal fan. The advantages of air cooling these components over liquid cooling include savings in weight, increased reliability and ease of maintenance.

<u>Induction Motor</u> - The 100 kW AC induction motor provides high energy density, efficiency and an element of safety inherent in the design. The lack of permanent magnets allows the PEU to control magnetic flux in the motor. This feature helps to maintain efficient operation across the wide range of loads encountered by a motor in a vehicle application. In addition, the rotational speed of the motor is not constrained by voltage; thus full power can be delivered across the entire range of rpm, reducing the need for a multi-ratio mechanical transmission. It produces a maximum shaft power of 89.5 kW (120 hp) between 6,500 and 10,000 rpm.

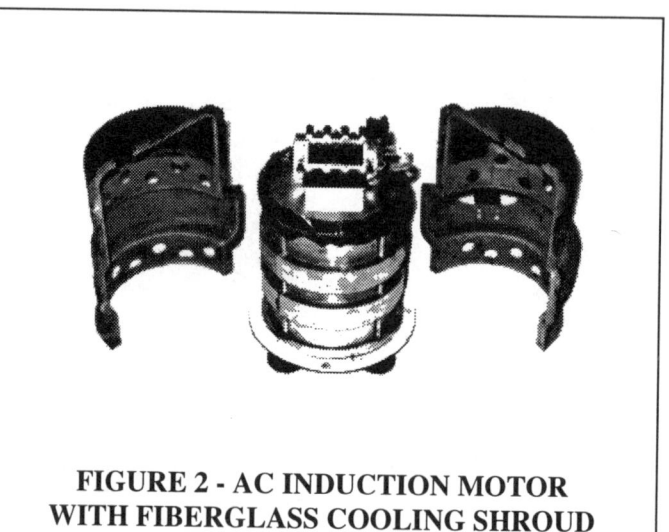

FIGURE 2 - AC INDUCTION MOTOR WITH FIBERGLASS COOLING SHROUD

The lack of permanent magnets also lowers material costs and increases disassembly safety -- the motor emf goes to zero when disconnected from the PEU. Induction type motors are durable and require little, if any, maintenance.

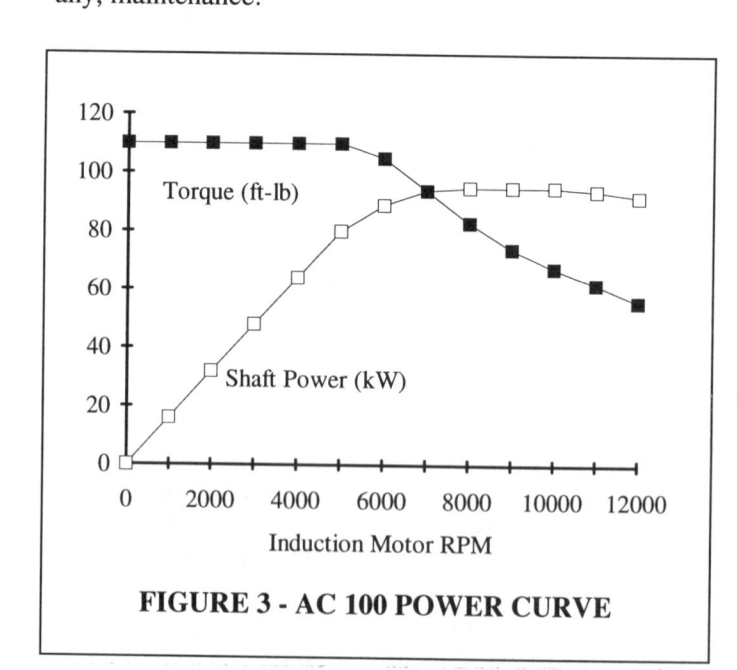

FIGURE 3 - AC 100 POWER CURVE

Both the PEU and induction motor are constructed from widely available components using standard manufacturing techniques. This system could easily be produced in large numbers should the demand exist.

GEAR REDUCTION - The nearly constant 149.2 N-m (110 ft-lbs) of torque delivered by the induction motor from 0 to 5000 rpm allows the use of a single speed transaxle. However, the complex design and manufacturing challenges associated with gear reduction units prompted the modification of a production transaxle to meet the requirements of the HEV.

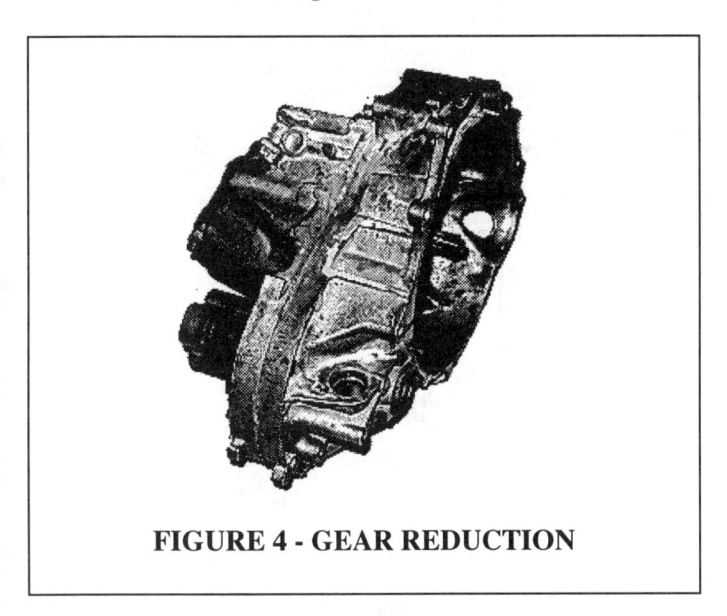

FIGURE 4 - GEAR REDUCTION

The optimum gear ratio to provide a balance of acceleration and a reasonable top speed was determined. A 1985 Chevrolet Citation front wheel drive transaxle, with a first gear ratio of 11.73:1, was selected. This ratio is close to the calculated optimum value and the drive axles used were also manufactured by Chevrolet. The transaxle was modified to deliver better performance and reduce its weight. The first gear sprocket was permanently engaged by welding it in place, and all other gear sprockets and shifting mechanisms were completely removed. The removal of these components greatly increased the lubricating oil capacity of the transaxle. The increased oil volume helps to dissipate the significant heat generated by the high rotational speeds the induction motor is capable of.

Both the aluminum transaxle housing and induction motor are bolted to a 1.59 cm (0.625 in) thick aluminum plate. This plate rigidly holds the components in alignment. Torque is transmitted from the induction motor to the input shaft of the transaxle via a Honda CRX clutch backing, a phenolic plate, and the Chevrolet Citation clutch backing. The CRX clutch backing fits over the male spline of the induction motor shaft, and the Citation clutch backing mounts to the transaxle input shaft. The phenolic plate is bolted between these. This plate isolates the induction motor's rotor from electric ground. The bolts connecting the clutch backings and phenolic plate have been designed to fail in shear to prevent the transmission of potentially damaging torque to the transaxle input shaft and gear reduction.

ALTERNATE POWER UNIT - The primary components of the APU are a 14.9 kW (20 hp) Kohler Command internal combustion engine and 15 kW alternator. These components and a 56.8 l (15 gal) fuel cell are mounted in the rear of the vehicle. This position was selected to simplify exhaust pipe routing and to separate the system from electric drivetrain components in the front of the vehicle.

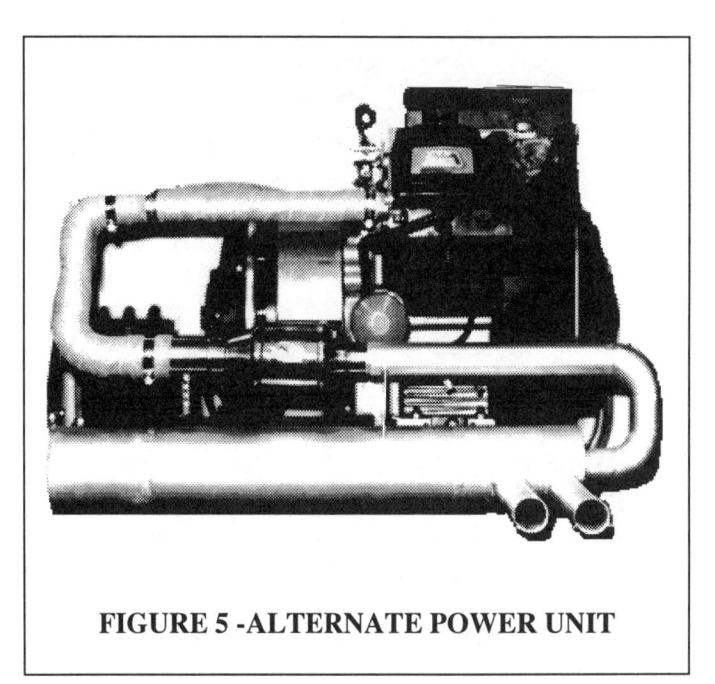

FIGURE 5 - ALTERNATE POWER UNIT

<u>Internal Combustion Engine</u> - The unleaded gasoline IC engine is a twin cylinder, four-cycle design that produces its maximum power at 3600 rpm. The engine operates continuously at this speed when the APU system is engaged, thus improving efficiency and reducing emissions. The engine's overhead valves and hydraulic lifters make it an excellent choice for the HEV, where low maintenance and reliability are important considerations. Another significant feature is the 207 kPa (30 psi) oil pump and full flow oil filter. The total engine displacement is 624 cc (38 cu. in.) with a compression ratio of 8.5:1. The oil capacity is 1.9 L (2 U.S. quarts) of 10w40 motor oil. An engine that uses unleaded gasoline was selected as this is by far the most widely available of the hydrocarbon fuels, and therefore the most convenient for consumers.

An electronic automatic control unit starts the engine when the batteries have reached a state of 80% charge. This control unit is capable of repeated cranking attempts on a thirty second cycle in the event that the engine fails to start on the first attempt. In addition, it will shut the engine down in response to engine overheat and low oil pressure. A manual override

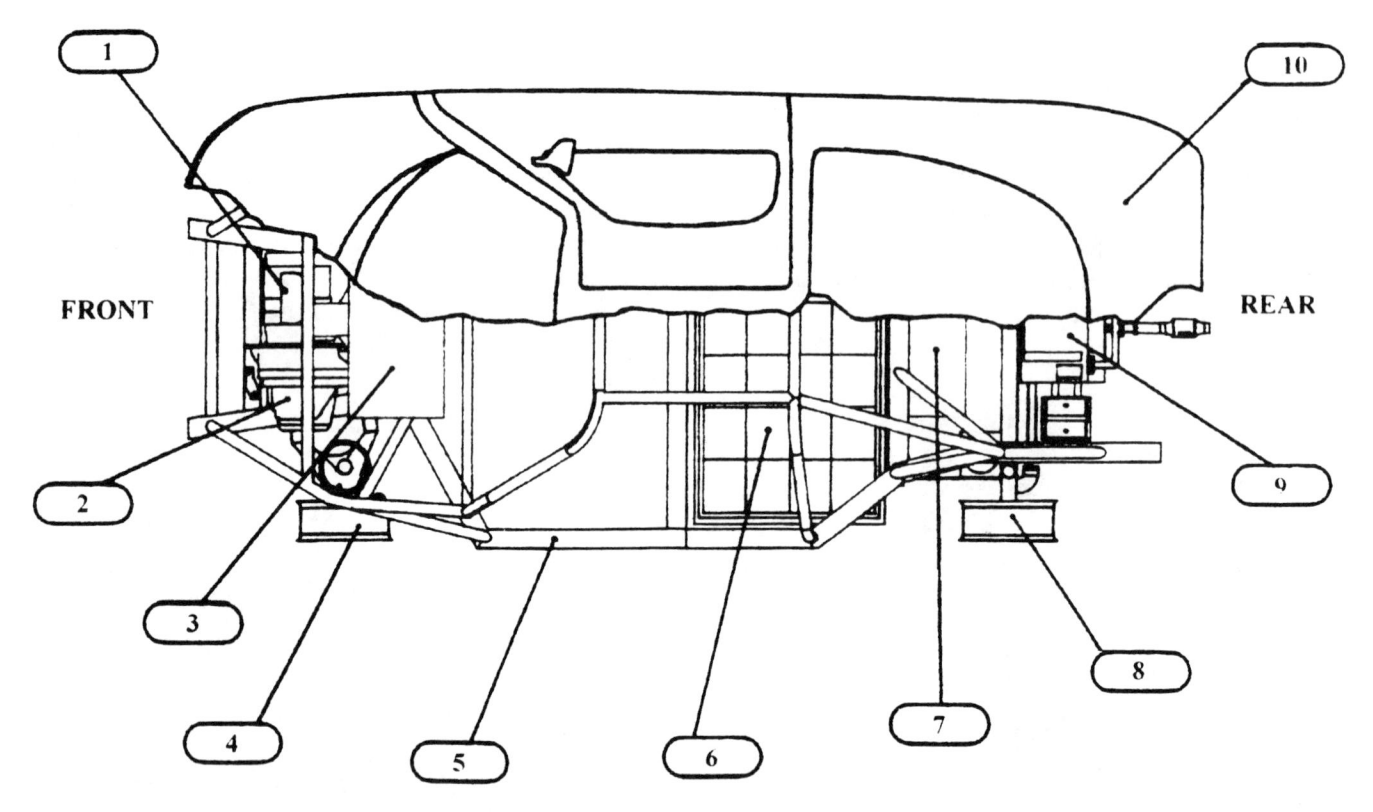

1. AC INDUCTION MOTOR
2. GEAR REDUCTION
3. POWER ELECTRONICS UNIT
4. FRONT RIM
5. MAIN FRAME RAIL
6. BATTERY COMPARTMENT
7. AIR HEATER
8. REAR RIM
9. ALTERNATE POWER UNIT
10. BODY (APPROXIMATE)

FIGURE 6 - COMPONENT LAYOUT

circuit is also employed to allow the driver to start and stop the engine as desired. This feature gives the vehicle the ability to charge the battery pack, regardless of its charge state, when an external electric power source is unavailable. Electric power for the automatic control unit, electric engine starter and APU cooling fans is supplied by a 12 V lead-acid battery mounted next to the engine in the rear of the vehicle. The battery charge is maintained by the 12 VDC PEU auxiliary power supply.

The APU engine is air cooled by a thermostatically controlled electric fan. The stock shaft-driven fan was removed to allow the engine and exhaust system to reach their most efficient operating temperature more quickly upon engine start. At 3600 rpm, the electric fan must provide 22.7 m^3/min (800 cfm) of cooling air in the enclosed environment of the vehicle shell to maintain a safe operating temperature. Ducting directs air from the exterior of the vehicle to the fan to meet this requirement. Passages through the rear of the vehicle shell also provide an exit for heated air.

As supplied by the manufacturer, the engine meets 1994 CARB emissions standards for its power range and also meets industrial noise regulations. The emissions

Component	Dimensions cm (in)	Weight kg (lb)
PEU	76.6 x 36.6 x 20.8 (30.2 x 15.2 x 8.2)	31.8 (70)
Induction Motor	38.1 x 30.5⌀ (15.0 x 12.0⌀)	41 (90)
APU Engine	43.2 x 42.2 x 46.2 (17 x 16.6 x 18.2)	11.6 (25.5)
APU Alternator	10.2 x 21.3⌀ (4.0 x 8.4⌀)	11.6 25.5
Air Heater	37.0 x 15.2 x 11.4 (14.6 x 6.0 x 4.5)	3.5 (7.7)
Battery Compartment	157.5 x 76.2 x 20.3 (162 x 30.0 x 8.0)	1984 (900)

FIGURE 7 - COMPONENT SUMMARY

are further improved by the use of catalytic purifier. By incorporating more platinum and palladium catalyst material into its honeycomb structure, the purifier is able to operate at the low exhaust temperatures of smaller displacement engines. Additionally, the purifier is positioned close to the exhaust manifold to achieve the highest possible operating temperature.

A venturi, supplemented by a small air pump, draws atmospheric air into the exhaust stream before it enters the catalytic purifier, providing additional oxygen for the chemical reactions. The purifier reduces carbon monoxide levels by approximately 95% when compared to untreated exhaust. Hydrocarbon and nitrous oxide levels are also significantly reduced by the purifier.

The noise level of the APU is held below 80 dB (A scale) at a distance of 15 meters from the vehicle by both the catalytic purifier and an exhaust muffler. This muffler is located after the catalytic purifier along the exhaust pipe route. The muffler incorporates both sound absorbing media and a baffle design to reduce the noise level across a broad range of frequencies. Other than the 2.54 cm (1.0 in) diameter exhaust manifold, all exhaust piping has a 4.45 cm (1.75 in) diameter.

A foam filled, 56.8 L (15 gallon) fuel cell contains the unleaded gasoline for the IC engine and the air heater (see ENERGY STORAGE for more information concerning the air heater). Located in the rear of the vehicle, the fuel cell is of one piece, seamless polyethylene construction and has a ball check valve to prevent fuel spillage in the event of vehicle roll-over. The filler hose connects to a small external door at the rear right side of the vehicle. Flexible rubber fuel lines connect the cell to a regulated electric pump mounted on the engine. Additional rubber fuel lines and a separate electric pump transmit fuel to the air heater.

Evaporative emissions from the fuel cell are captured in a charcoal canister. This canister is evacuated by a positive crankcase ventilation (PCV) circuit upon engine start. The PCV circuit also returns crankcase emissions to the induction tract for combustion.

Alternator - A 15 kW Fisher Electric Motor Technology alternator is driven by the IC engine to produce electric power. The small size, low weight and direct shaft mounting were factors that prompted the use of this alternator. The direct output shaft mounting eliminates the need for any pulleys or drivebelts, reducing weight and maintenance.

Like the IC engine, the alternator is air cooled. An additional thermostatically controlled fan that operates independently of the engine cooling fan is mounted on the alternator shell.

Neodymium magnets and a proprietary design enable the alternator to provide up to 15 kW of power. At 3600 rpm the alternator produces 400 volts (0-peak). The output signal is WYE connected 3 phase, and the total resistance of the alternator coil is 0.2 ohms.

The entire APU system, including the engine, alternator, fuel cell, and exhaust system, is incorporated into a mild steel (AISI 1020) sub-frame assembly that can be completely removed from the vehicle. This assembly will be mounted with rubber bushings to the main frame rails to minimize the transmission of engine vibration. A self-contained, removable assembly not only facilitates maintenance of the system, but also allows the HEV to accept other APU systems that incorporate different engines or fuels.

ENERGY STORAGE

Electrical energy is stored by twenty-eight 68.3 kg (31 lb) sealed lead-acid batteries wired in series. The batteries are arranged in a single layer and enclosed in a removable compartment located directly behind the driver and passenger seats. A single battery compartment was selected for safety, ease of maintenance, and thermal management considerations. It is well protected by the primary frame rails and roll cage hoops. As a single unit, the entire compartment can be lowered out of the vehicle, allowing the batteries to be serviced or replaced as needed without having to individually remove them from locations throughout the vehicle. Additionally, in an effort to improve efficiency, the interior of the battery compartment is heated to a temperature of 110°F during vehicle operation. A single, well insulated compartment requires less energy to maintain at this temperature than multiple compartments.

THERMAL MANAGEMENT - The air circulated through the interior of the battery compartment is heated to 110°F. By maintaining this temperature, both the efficiency and life span of the batteries is improved based on their ability to store energy. The electrolyte diffuses through the plates more readily at elevated temperatures and improves energy performance. At approximately 140°F, however, the expander in the negative plates chemically reacts and damages the battery. Therefore, a compartment temperature of 110°F was selected to give a reasonable increase in efficiency and life without harming the negative plates.

An unleaded gasoline fired air heater, typical of those used to heat the passenger compartment of vehicles with air-cooled engines, maintains the battery compartment at the elevated temperature. This axial flow heater draws in outside air, heats it, and directs it into the battery compartment air circulation ducts. An automatic monitoring system with temperature sensors inside the compartment controls the heater.

The actual exit temperature of the air from the heater is approximately 200°F. To avoid exposing the batteries to this high temperature, several intake ducts are located along the upper lid of the compartment to quickly and evenly distribute the heated air. The heater is capable of a 110 kg/h hot air throughput while delivering a heating capacity of 1800 Watts. At this operating rate, 0.24 l/h of unleaded gasoline is consumed. A 12 VDC power source, provided by the PEU auxiliary DC supply, is also required for the operation of the heater circulation fan and spark plug.

COMPARTMENT VENTILATION - To maintain a continuous flow of air through the compartment, a 0.425 m^3/min (15 cfm) brushless DC fan expels air from the rear of the battery compartment where the internal circulation ducts terminate. This fan operates whenever the batteries are being charged or the vehicle is being driven to prevent the build up of dangerous levels of hydrogen gas, a by-product of the chemical reactions in lead-acid batteries. The continuous evacuation of air from the compartment necessitates repeated operation of the air heater.

The initial heating of the battery compartment to 110°F on a 60°F day takes about ten minutes. To maintain the temperature, the heater must operate for an additional ten minutes every hour. At this rate, the heater uses less than one liter of unleaded gasoline during a five hour drive. In addition, the thorough combustion of the gasoline in an oxygen rich environment reduces the emission of harmful pollutants from the air heater to a level below that of the APU system.

MATERIALS AND MANUFACTURE - The battery compartment is constructed from high density foam and plywood core material sandwiched between layers of fiberglass. The plywood forms the majority of the structural base of the compartment, with foam being utilized to save weight in areas that did not require high strength. An internal grid of core material surrounds each battery, with channels left through the grid to allow for circulation of heated air. The outer wall structures and internal grid are coated in fiberglass, giving an extremely strong, rigid structure with excellent thermal and electrical insulating properties.

VEHICLE WIRING

The high voltages and currents associated with the electric drive system required careful selection of wire gauges, insulation and shielding materials. National Electric Code (NEC) standards are met by all conductors in the WSU/UI HEV. These standards were used to determine the insulation required to give a breakdown voltage of twice the maximum expected voltage in all conductors. All high voltage electrical cables have THWN type insulation. This insulation type is moisture and petroleum resistant, and flame-retardant. The insulation is protected by molded rubber grommets where conductors pass through openings in panels and bulkheads.

Figure 8 summarizes the maximum voltage and current encountered by the critical signal paths and the corresponding wire gauge and fuses selected for each. The primary electrical system components and associated signal paths and fuse locations are shown in Figure 9 on the next page. The figure shows fuse locations as an indication of electrical position only.

Conductor Route	Peak Voltage	Current (125% of max.)	Wire Gauge	Fuse Rating
Generator - Batteries	400 VDC (peak)	47 Amps	6 AWG	60 Amp (3 fuses)
Batteries - PEU	12 VDC • 28 = 336 VDC	100 A* • 1.25 = 125 A	2 AWG	400 Amp
PEU - AC Motor	Battery Output Limited	400 A*	2 AWG	none
Charging Port - PEU	220 VAC • √2 V = 311 VDC	20 kW / 311V • 1.25 = 80.4 A	6 AWG	125 Amp

(* indicates value supplied by AC Propulsion, Inc.)

FIGURE 8 - CRITICAL SIGNAL PATHS

FIGURE 9 - WIRING SCHEMATIC

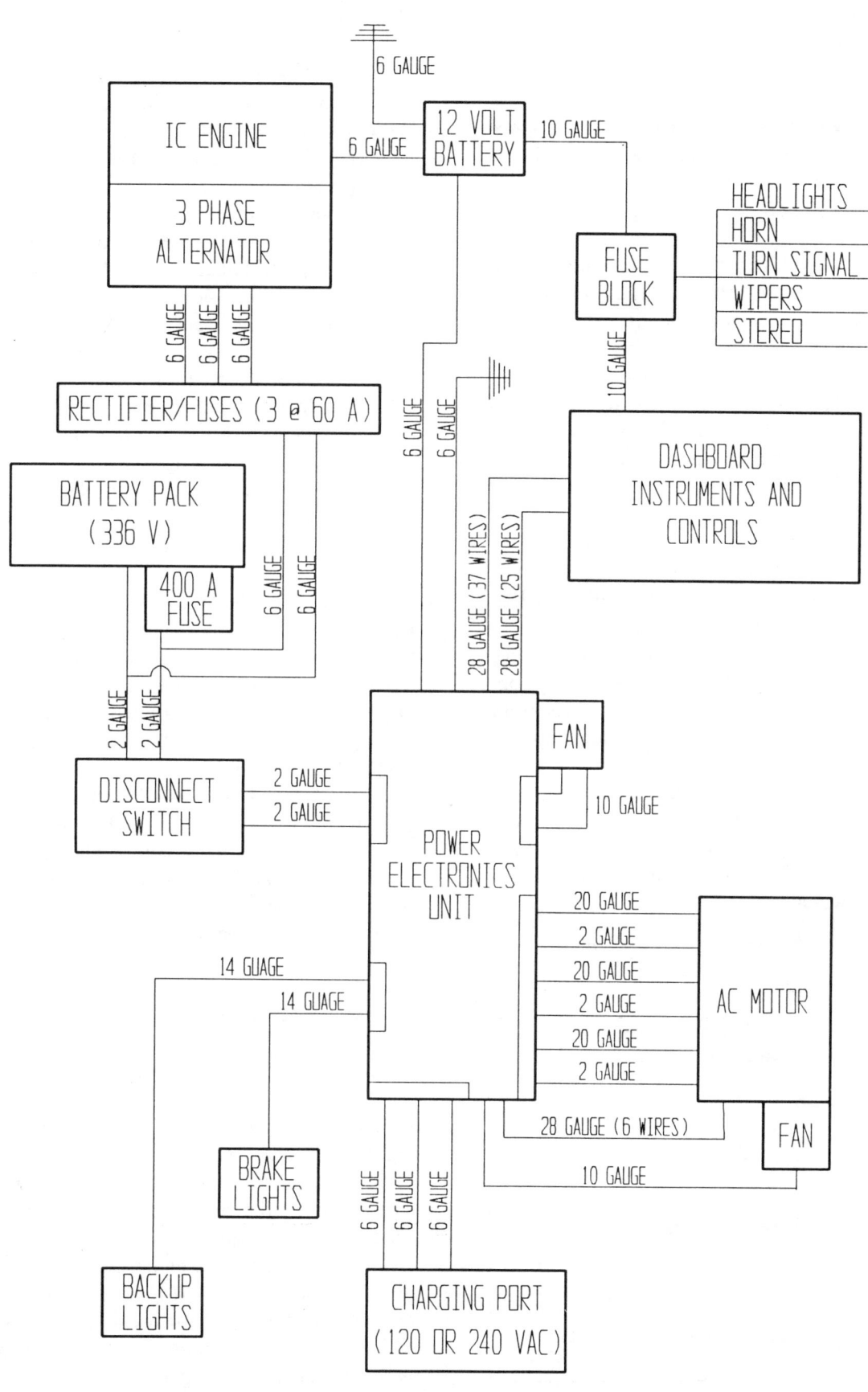

Fuses for the PEU-AC motor and charging port-PEU signal paths are located within the PEU casing. The PEU-batteries signal path fuse is located at the rear of the battery compartment. This fuse is contained within an insulated, sealed housing. The three 60 A fuses of the alternator-batteries signal path are also enclosed in an insulated housing located in the rear APU compartment.

All high voltage/high current conductors are shielded by both a thin layer of aluminum and also a braided steel casing. This combination helps prevent the emission of electromagnetic radiation across a broad range of frequencies.

Two high voltage/high current conductors pass through the passenger compartment. These conductors pass between the seats, connecting the PEU and battery compartment. They are securely fastened to the floorpan, and a heat formable thermoplastic center console encloses the conductor route for safety. All other high voltage/high current conductors are completely separated from the passenger and storage compartments of the vehicle by insulated firewalls and bulkheads.

The batteries are wired in series in the battery compartment with 2 AWG wires. All battery terminals, including those of the single 12 V battery of the APU, are covered with THHN terminal guards.

The vehicle has a manually operated battery disconnect switch that can be locked in the off position. The switch completely disconnects the batteries from the electrical systems of the vehicle, thus allowing for safe maintenance.

A three pole, molded case switch rated at 250 VDC was wired in series to be able to safely disconnect the 336 VDC battery pack. The switch is located behind the seats in the passenger compartment and is clearly labeled "Battery Switch".

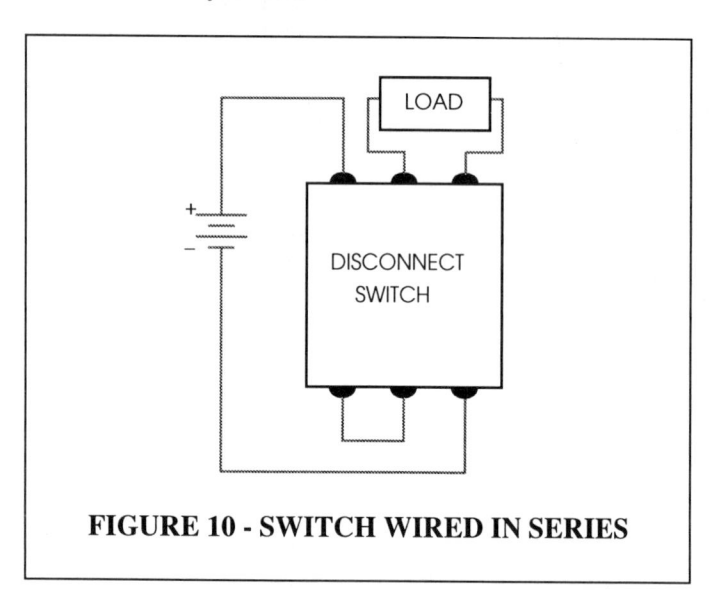

FIGURE 10 - SWITCH WIRED IN SERIES

OPERATION MODES

The vehicle has three modes of operation: standard hybrid, zero emission (ZEV), and APU engaged. The driver may select any operation mode at any time by means of a simple three position key switch.

STANDARD HYBRID - In this operation mode, the vehicle will operate on electric power only until the batteries have reached an 80% charge state. The APU system will then automatically start to maintain the battery charge level and permit operation of the vehicle without further depleting the batteries.

To control vehicle speed, an accelerator pedal in the form of a potentiometer is used to send a 0 to 5 volt signal to the PEU. In response to this input, the PEU varies the AC signal sent to the induction motor from 0 to 120 Hz, thus changing the rotational speed of the motor. The driver controls vehicle direction by means of a simple three button controller connected to the PEU. Forward, reverse or neutral may be selected only when the vehicle is at rest.

The APU is controlled by an electronic system that senses the charge state of the batteries. Engine start and stop is completely automatic, requiring no action by the driver.

ZEV MODE - In ZEV mode, the APU is not allowed to function at any time. In this purely electric operation mode, all systems other than the APU function as they would in the standard hybrid mode. This feature is included in the vehicle to allow the driver to prevent the IC engine from starting or to shut it down should it already be operating.

APU ENGAGED - The APU may be operated at any time by the driver by selecting the APU engaged mode. This feature allows the driver to charge the batteries at any time they fall below a 100% charge state; thus the vehicle is capable of recharging its battery pack even when an external electric power source is unavailable.

FRAME

The primary criteria in designing the vehicle frame was deflection under the static and dynamic loads encountered under normal driving conditions. Both rail and space frame designs were considered, and it was decided that a combination of the two types would best support the estimated loads of the vehicle with a minimum of deflection.

The main beams that comprise the rail frame part of the design provide support for the concentrated loads of various drivetrain and suspension components. The stiffness of the space frame, on the other hand, provides resistance to torsion and other out of plane forces that

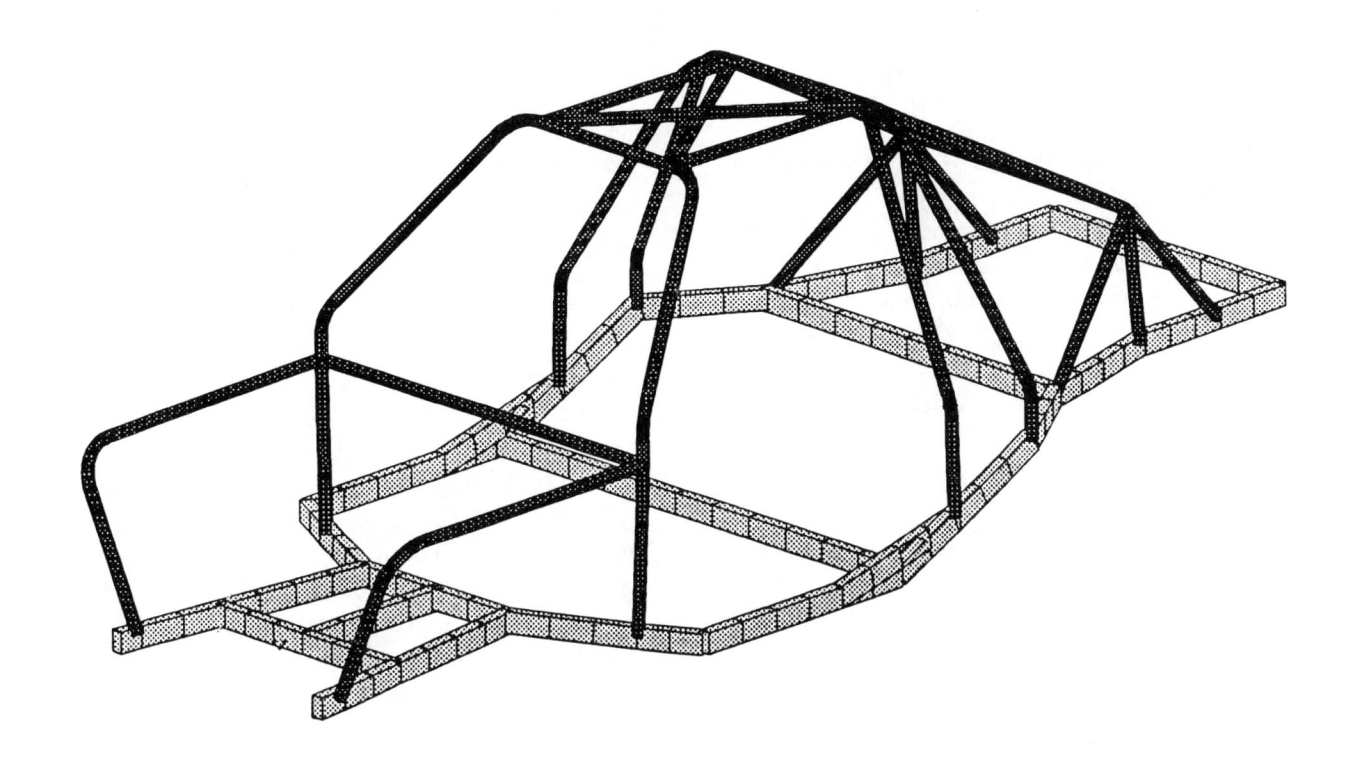

FIGURE 11 - FRAME AND ROLL CAGE

act on the vehicle. In addition, the roll cage requirements set by the organizers of the 1994 HEV Challenge were also incorporated into the space frame portion of the design.

The AISI 1026 steel box channel selected to function as the main frame rails measures 3.81 cm x 6.35 cm (1.5 in x 2.5 in) with a wall thickness of 0.391 cm (0.154 in). The AISI 1026 pipe members that form the roll cage are all seamless, with a 3.8 cm (1.5 in) outside diameter and a wall thickness of 0.31 cm (0.12 in). The estimated weight of the frame was calculated to be 134.7 kg (297 lb).

FRAME ANALYSIS - The frame design was optimized for stress, weight and deflection through the use of finite element analysis software. To simulate the affects of a dynamic loading situation, a static load equal to twice the estimated weight of the vehicle was applied to a computer model of the frame. This 32,484 N (7,300 lb) load was distributed throughout the frame to simulate the major components of the vehicle such as the battery compartment and drivetrain components. Reaction forces were applied to the model at points where suspension components would be attached. The analysis revealed that the pipe members of the space frame were subjected to a maximum bending stress of 177 MPa (25,700 psi). The maximum axial stress experienced by the pipe was 14.7 MPa (2,100 psi). The main frame rails were subjected to a maximum bending stress of 105 MPa (15,200 psi). The axial load on the main rails was zero in this idealized analysis. Under these loading conditions, the maximum deflection of any frame member was 0.686 cm (0.27 in.). This deflection occurred in members comprising the top of the roll cage over the rear of the passenger compartment.

Another important consideration in the design of the frame was accommodating the suspension components. The vertical loads transmitted by the coil springs and shock absorbers in both the front and rear of the vehicle are distributed to the main frame rails through the space frame. The pipe members of the space frame transmit these loads through tension. The lower A-arms of the front suspension are mounted directly to the main rails. The control arms of the rear axle pivot about brackets mounted to the main rails as well.

The frame was sanded and painted to help prevent corrosion. Also, the aluminum housing of the gear reduction is separated from the frame by non-metallic vibration mounts.

MATERIALS AND MANUFACTURE - Low carbon mild steel (AISI 1026) was selected as the frame material for a number of reasons. This material is strong, with a yield strength of 331 MPa (48,000 psi). The good ductility and weldability makes this steel an excellent choice from a manufacturing standpoint. Comparable steel is commonly used in many current production vehicles, and all frame members are commercially available in the geometry and dimensions specified.

SUSPENSION

The vehicle will incorporate front independent suspension and a trailing arm type rear axle. MacPherson struts are utilized in the front, while separate coil springs and shock absorbers suspend the rear. Suspension components from a front wheel drive 1984 Chevrolet Celebrity wagon were used to save design and manufacturing time. The Celebrity has a GVWR over 1814 kg (4000 lb), which exceeded the highest estimates of the HEV weight with passengers. The front lower A-arms, front axles, disk brake rotors and brake calipers of the Celebrity have been incorporated into the HEV design. The rear axle and control arms have also been used.

All necessary mounting brackets and tabs needed for the suspension components were manufactured from 0.476 cm (0.188 in) hot rolled AISI 1020 steel plate.

FRONT SUSPENSION - The front lower A-arms pivot about tabs welded to the main frame rails of the HEV. Care was taken to orient these control arms exactly as they were in the Celebrity. Both the springs and shock absorbers of the original MacPherson struts have been replaced with components more suitable for the 1633 kg (3600 lb) HEV.

The struts were positioned so that the axis defined by the top of the strut and the ball joint passes through the centerline of the tire where it meets the road. This design helps to prevent torque steer during acceleration. Ideally, all vertical loads encountered by the suspension will be transmitted directly to the struts and not the lower control arms.

In addition, the front sway bar from the Celebrity has also been used to prevent excessive roll-over while cornering. The sway bar is held in place by two brackets welded to the main frame rail that passes in front of the passenger compartment.

REAR SUSPENSION - The rear suspension consists of a dead axle with control arms, a panhard rod, coil springs and shock absorbers. As a trailing arm type suspension, two control arms pivot about brackets located in front of the axle. These arms allow the axle to translate vertically. The panhard rod locates the axle in the proper lateral position while allowing vertical translation as well.

The rear trailing arm brackets also carry part of the battery compartment loads. These loads result from the normal cornering and linear accelerations of the vehicle.

Twin coil springs and shock absorbers suspend the rear axle. The springs are seated in brackets welded directly to the main frame rails. The shock absorbers mount to brackets welded to pipe members comprising the rear of the roll cage.

HYDRAULIC ABS BRAKE SYSTEM - A four wheel anti-lock disc breaking system from a 1989 Mercury Cougar was adapted to the HEV. Magnetic pickup collars from the Cougar were attached to Chevrolet Celebrity disc brake rotors. The master cylinder and electronic processor from the Cougar were also utilized. By monitoring the rotational speed of the wheels, the processor is able to regulate braking and prevent wheel lock-up.

Due to space limitations forward of the passenger compartment, the Mercury Cougar master cylinder is located in the rear of the vehicle. It is operated remotely by an additional smaller master mounted to the steel firewall in front of the passenger compartment.

Standard automotive hydraulic brake lines and connectors are used throughout the braking system. Also included in the vehicle design is a manual, hand-operated, cable actuated parking brake.

STEERING - A manual Chevrolet Citation rack-and-pinion steering system was adapted for use in the HEV. A manual system was selected for several reasons. Compared to a hydraulically assisted steering system, it is lower in weight, requires less space and does not need an external power source. All three were important considerations in designing the HEV.

WHEELS AND TIRES - In order for the traction control system to operate properly, a 61 cm (24 in) diameter tire was needed. Goodyear Invicta GS 195/60 R15 tires were selected to meet this size requirement. The tires were a good compromise between traction and low rolling resistance, and the HEV weight is well within their maximum load rating.

Fifteen inch diameter aluminum wheels were selected for their low weight and functional design that helps draw air out of the wheel well to aid in brake cooling.

MATERIALS AND MANUFACTURE - The front A-arms and rear axle are formed from stamped mild steel sheet. This manufacturing technique is inexpensive, fast, and produces strong, lightweight components. The front sway bar and rear panhard rod are extruded round steel that was shaped to form the necessary geometry. All other suspension components, including the springs and shock absorbers, are manufactured by standard fabrication techniques.

All brake system and steering system components are commercially available and do not require specialized manufacturing techniques. The wheels and tires are widely available, off-the-shelf components as well.

BODY

A one-piece, foam-core fiberglass sandwich was constructed to function as the vehicle body. This

FIGURE 12 - ARTIST'S CONCEPTION OF BODY

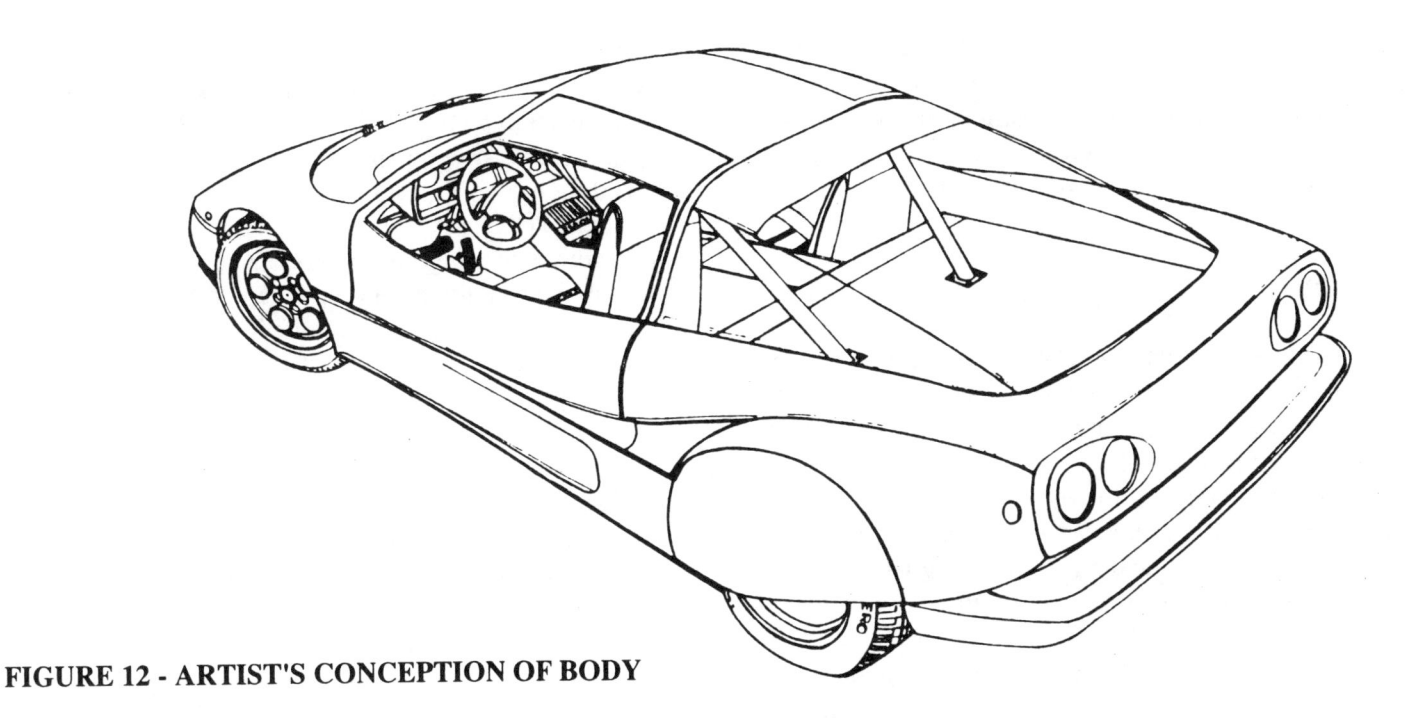

material combination was selected for its high strength-to-weight ratio and formability. Shaping the vehicle in foam allowed a smooth integration of the body with the selected windshield and exterior lighting assemblies.

BODY DESIGN - In designing the body, priority was given to reducing frontal area and producing an overall aerodynamic shape. Frontal area was reduced by tapering the passenger compartment inward above the passenger's shoulders and limiting the height of the vehicle to 132 cm (52 in). The steeply raked windshield and rounded profile of the body also help reduce the overall drag of the vehicle. The roof line behind the passenger compartment slopes away at only a 20° angle from horizontal to reduce separation of the airflow from the surface of the body.

The body mounts to the frame and roll cage at eight different locations. At each mounting point, a 16 gauge mild steel sheet with two protruding 0.95 cm (0.375 in) bolts has been fiberglassed onto the inner surface of the body. A steel strap passes over the pipe member of the roll cage at each mounting point. A rubber vibration mount separates the pipe member from the body, and the two protruding bolts hold the strap in place.

Bumpers constructed from ASTM A36 steel were also incorporated into the body. Both front and rear bumpers attach directly to the main frame rails and utilize shock absorbing mounts.

WINDOWS AND EXTERIOR LIGHTING - The vehicle features a 1994 Mazda MX-3 laminated glass windshield. This windshield was selected as it had the general dimensions needed and the lamination makes it safer in the event of an accident. It was narrowed slightly at the top to integrate smoothly with the HEV shell. Side and rear windows are formed from custom shaped Plexiglas.

Integrated headlight/turn signal and taillight/turn signal assemblies from Mazda vehicles are also included in the shell. These assemblies provide all required passenger vehicle forward, rear and side exterior lighting.

MATERIALS AND MANUFACTURE - A wood jig composed of cross sections of the body at 30 cm increments was first constructed. Foam material was then attached to this jig and formed by hand to give the material a smooth, continuous surface. The windshield, headlights and taillights were also integrated into the foam shape. Two layers of fiberglass embedded in epoxy resin were applied to the entire exterior surface of the foam layer. The epoxy resin was selected for its excellent strength-to-weight ratio and low emissions of harmful fumes when being applied. The fibers of each layer were positioned at 45° angles to the center line of the vehicle and perpendicular to each other to improve the rigidity of the shell. After the outer layer had hardened, the wood jig was cut away from the inside surface of the foam. The inner side was then coated with a single fiberglass layer.

Passenger compartment doors and all other needed access doors were cut in the shell after the inner fiberglass layer had hardened. These removed sections were then utilized in the drape forming process used to manufacture the Plexiglas rear and side windows. Body filler was applied to the entire exterior fiberglass surface and then sanded to achieve a smooth finish. Finally, the body was painted.

A fiberglass body of this type could be produced quickly in large numbers by creating a full scale female die. The fiberglass and epoxy could then be sprayed into the die and removed upon hardening.

INTERIOR

The interior of the vehicle accommodates two adult passengers and provides ample storage space for cargo. A cab-forward interior was designed to make vehicle operation comfortable for drivers in the fifth to ninety-fifth percentile range for body size. Bucket seats from a 1990 Mazda MX-6 were selected for their wide range of adjustability, low seat height and integrated headrests that reduce the potential for injury in a rear end collision. These seats were also easily modified to accept a five-point seat belt harness as required by the HEV Challenge Rules. A tilt steering column is included in the design for driving comfort and to facilitate entrance and exit from the vehicle.

The visual field of the vehicle occupants was an important consideration in the design. The effort to utilize a cab-forward design required a large, steeply raked windshield. The windshield size was selected to ensure that vehicle occupants are able to see both a point on the ground 9.2 m (30 ft) ahead of the vehicle, as well as a point 7.3 m (24 ft) high and 27.5 m (90 ft) ahead. The rear window was positioned to give the driver a clear view of a point 0.61 m (2 ft) high and 15.25 m (50 ft) behind the vehicle. A center mirror, and right and left side-view mirrors are also incorporated into the design. Side windows extend from the front windshield to the extreme rear to provide a clear view of any areas behind and to the sides of the vehicle not visible with the mirrors.

INSTRUMENTATION - A clustered instrument console has been placed in clear view of the driver. The instruments visible through the upper arc of the steering wheel include a speedometer, kWh meter, odometer, and induction motor and Power Electronics Unit temperature gauges. APU system fuel level and temperature gauges are located to the left of the center cluster. A single LED status light will also indicate other potential problems with the APU system. This LED will signal the driver to check the more detailed status LED's that are part of the APU control system located in the rear of the vehicle. To the right of the center cluster, discharge/charge rates for the battery pack and APU system are monitored by separate ammeters. Additional controls and instrumentation located in a center console include the APU controller switch (off/on/auto), vehicle direction controls (forward/reverse/neutral) and regenerative braking controller. The center console will also house heater and fan controls and the compact disc player/stereo receiver. Finally, for driver and passenger convenience, a two cup drink holder is also included in the design.

HEATING AND VENTILATION - Warm air for window defrosting and passenger compartment heating is supplied by the air heater mounted in the rear of the vehicle. A forced air ventilation system draws the heated air down a tunnel passing along the floor on the passenger side of the vehicle. This ventilation system may also be used to draw in fresh air through intakes located at the base of the windshield.

MATERIALS AND MANUFACTURE - The dash is constructed from a foam and fiberglass sandwich similar to that used for the body shell. Again, this material was selected for its excellent strength to weight ratio and its ability to be easily formed into complex shapes. All instrument consoles are formed from machinable phenolic plastic and are removable from the dash.

A mild steel sheet metal floor panel is welded to the main frame rails and completely separates the passenger compartment from the road surface. This panel and the seats are supported by two steel cross members spanning the entire distance between the main frame rails. A sound insulating material and fire-resistant carpet covers the interior surface of the panel. All other interior surfaces are covered in fire-resistant fabric or vinyl.

CONCLUSION

The Washington State University/University of Idaho hybrid electric vehicle is a safe, economical solution to the challenges facing the automobile industry. For an estimated 95% of its operation time, the WSU/UI HEV will be capable of operating as a purely electric vehicle. With the APU, it also has the range capabilities that today's drivers take for granted.

The vehicle is well suited to the period of transition that both manufacturers and consumers will be facing as emissions legislation comes into effect. Its hybrid design gives electric vehicles reasonable performance and also utilizes the distribution system of unleaded gasoline that is already in place.

Electric vehicles are the future of personal transportation. However, until energy storage technology improves to an acceptable level, the hybrid electric vehicle will remain the best alternative to current internal combustion powered vehicles.

ACKNOWLEDGMENTS

The author would like to thank all the WSU/UI HEV team members, particularly the following team leaders who were such a tremendous help in preparing this report:

Ron Dennis	Jeff Smutny
Steve Harold	Mike Harold
Mark Weller	Alan Chapman
Craig Muth	Dennis Niehenke
Dave Paul	Greg Meboe
Jason Dundas	Cyril Stockman

REFERENCES

"A High Power, Sealed Lead Acid Battery for Electric Vehicles," D.B. Edwards and B. Carter. *Journal of Engineers for Industry* (ASME), Vol. 112, August 1990.

"Concepts of the New Hybrid Electric Vehicle Frame," J. Colgan, T.W. Fong, R. Haggart, B. Hammond and S. Scyphers. Dept. of Mechanical Engineering, University of Idaho, Fall 1993.

"HEV Gear Reduction Final Report," J. Adams, C. Hopper, D. Krumpelman and D. Miller. Dept. of Mechanical Engineering, University of Idaho, Fall 1993.

"HEV II Suspension Design: Where the Rubber Meets the Road," James Eidman, Kevin Howerton, Mike Porter, Eric Thompson, Kelly Willett. Dept. of Mechanical Engineering, University of Idaho, Fall 1993.

"Thermal Systems Management," D. Baker, D. Clark and E. Pek. Dept. of Mechanical Engineering, University of Idaho, Fall 1993.

"1994 HEV Design: Shell and Interior," P. Andrew, G. Davis, J. Montez, M. Richman and M. Sonius. Dept. of Mechanical Engineering, University of Idaho, Fall 1993.

Washington University's Parallel Hybrid Electric Vehicle

Michael Vorhies
Malcolm Early
Adrian Martin
Thomas Birchard
Tim Orcutt

ABSTRACT

Vehicle emission concerns and alternatives for commuter transportation have created new interest in electric vehicles. The lack of range of the typical electric vehicle (EV) is overcome by hybrid electric vehicles (HEV). A 13.4kW Briggs and Stratton Vanguard gasoline engine is placed in a parallel configuration with an Advanced DC series wound electric motor yielding 14.9kW continuous and 59.7kW peak. The electric motor is controlled by a Curtis PMC controller. The vehicle meets the EV and HEV range requirements (483km), emission goals, and modern vehicle performance expectations of the 1994 SAE Hybrid Electric Vehicle Challenge.

INTRODUCTION

This is the second year for the Hybrid Electric Vehicle Team from Washington University in St. Louis. The HEV Team converted a 1992 Ford Escort station wagon into a parallel configuration HEV and have named their entry *HEVolution*. HEVolution is designed as an urban commuter vehicle utilizing the parallel configuration to extend the vehicle range. City driving is performed completely in EV mode while highway driving is done in HEV mode.

POWER TRAIN CONFIGURATION

The power train configuration for HEVolution is a parallel configuration using an electric motor (EM) for primary propulsion, while using an internal combustion engine (ICE) to increase vehicle range (See Figure 1). All acceleration is done in EV mode using the EM only. The electric motor is a high torque, low RPM Advance DC series wound 120V motor controlled by a Curtis PMC 120V 400 amp motor controller. The motor and controller combination provide enough torque and power to accelerate HEVolution from 0 - 85 kph in 20.2 seconds. Power is supplied by 22-12 volt lead acid batteries arranged in two parallel sets of eleven batteries in series to produce a pack voltage of 132V. HEVolution uses Optima Model 800 batteries. The Optima batteries are marine application batteries which carry a 56 amp-hr rating and can store 7.4kW-hr, enough for EV mode.

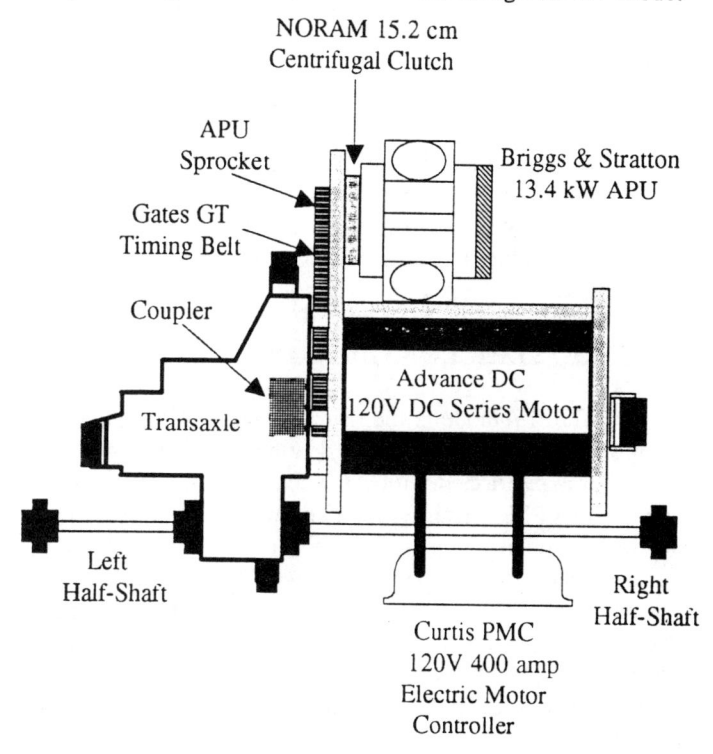

Figure 1: HEVolution's Power-Train Configuration.

EV mode is not efficient at speeds above 48.3km/hr. Therefore, the HEV configuration uses an ICE to directly drive HEVolution. The ICE, also called an auxiliary power unit (APU), is a Briggs and Stratton 13.4kW Vanguard V-twin two cylinder gasoline engine. The ICE is coupled to the same shaft as the EM via a Gates GT timing belt for maximum energy transfer and is engaged by a Noram centrifugal clutch at 1800 RPM. The belt sprocket ratio between the EM and the ICE is 1:1 (See Figure 2). The only modifications to the stock Briggs and Stratton ICE is the addition of a rotary solenoid to the carburetor butterfly valve. The rotary solenoid permits a microprocessor to control the ICE.

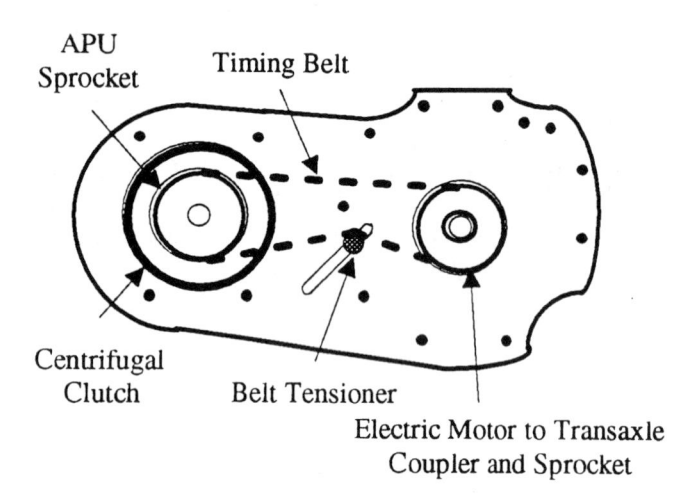

Figure 2: Transaxle Face Plate. In EV mode, the centrifugal clutch freewheels and the EM directly drives the transaxle. In HEV mode, the EM freewheels while the ICE engages the centrifugal clutch and drives the transaxle through the belt.

Both the ICE and the EM are housed in a mounting system designed to utilize the existing engine mounts in the Escort built out of 6061 Aluminum. The face plate in Figure 2, and spacers are used to connect the EM to the transaxle. The EM serves as a rigid member and connects to the transverse engine mount seen in Figure 1. The coupler-belt sprocket is made of 1018 Steel. This coupler exceeds the minimum space requirements of the stock Escort clutch forcing its elimination. The ICE is directly mounted to the face plate.

ELECTRIC VEHICLE MODE CONTROL STRATEGY

The EV mode is the primary mode of operation for driving. Acceleration and urban driving (typified by stop and go traffic) is performed with the electric motor. HEVolution uses a Curtis PMC electric motor controller and an Advanced DC Series wound motor as the electric drive train. To send the accelerator pedal response to the PMC controller, the existing Ford throttle body has been modified to accept a potentiometer. As the angle of the accelerator pedal increases, so does the resistance in the potentiometer. When the pedal has no deflection, the throttle body closes a switch turning off the PMC controller.

As HEVolution is accelerating, it is driven like any other car. No modifications have been made to the steering and the braking inputs from the stock Ford Escort. When the driver wants to shift, he uses a speed shifting technique where, before gear disengagement, the accelerator pedal is released. The transaxle is re-engaged once the shaft speed of the electric motor slows down to the proper engagement speed of the new gear.

HEV MODE CONTROL STRATEGY

HEVolution uses the HEV mode as an "overdrive-cruise control" feature. Once the APU is engaged the vehicle speed will remain constant, acceleration is achieved only through the electric motor. The key to HEVolution's parallel control strategy is matching the APU crankshaft speed to the existing electric motor speed (See Figure 3). To engage HEV mode the driver toggles a switch and the microprocessor initiates a sequence of commands.

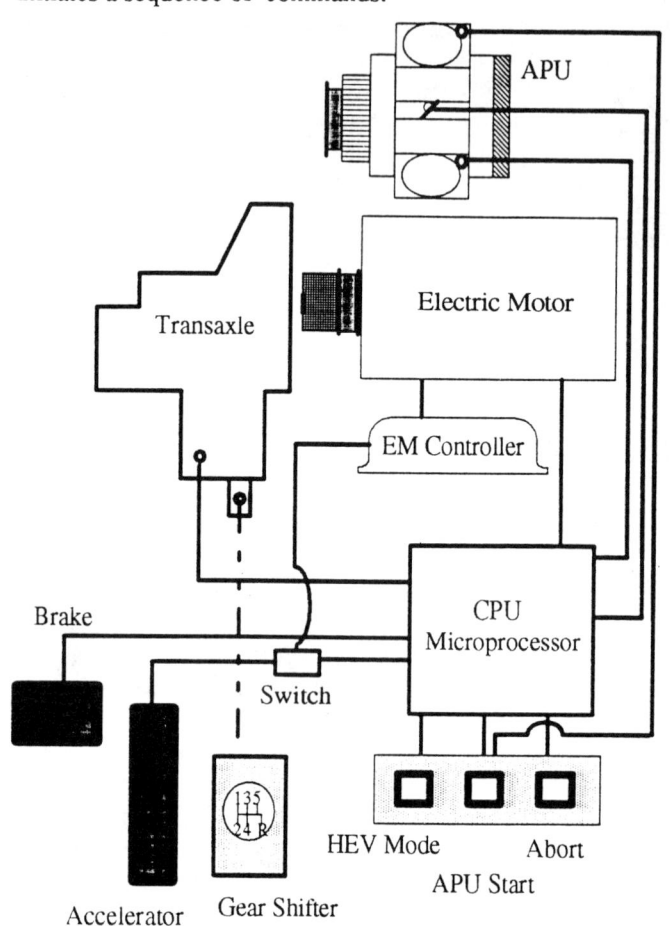

Figure 3: HEVolution's Control System Diagram for EV and HEV Mode.

A panel adjacent to the dash holds three switches that activate the following; "HEV mode," "APU Start," and "Abort." The panel is positioned to permit easy driver access. Also mounted on the dash board is an engagement light indicator.

Once the driver has turned on the HEV mode, the microprocessor measures the electric motor shaft speed and the vehicle speed. The microprocessor then calculates which gear the transaxle is in and what the required APU shaft speed should be to engage the transaxle in this gear. It also sets the carburetor butterfly valve to the start position. The microprocessor then lights the dash mounted LED that prompts the driver to start the APU.

The driver presses the "APU Start" button until the APU fires up. The microprocessor compares a vacuum switch signal input from the APU to the position of the "APU Start" button. If the APU is running and the driver has released the "APU Start" button, the microprocessor sets the APU at the idle speed.

To disengage the transaxle and the electric motor, the accelerator must be released (see speed shifting sequence discussion under Electric Vehicle Mode Control Strategy). In HEV mode, the initial release of the accelerator signals the microprocessor to keep the controller in the off position. The microprocessor now relates the accelerator movement to the rotation of the butterfly valve on the APU. The microprocessor compares the APU shaft speed and the necessary engagement speed. Since the centrifugal clutch sprocket and the electric motor sprocket have a 1:1 ratio, the microprocessor is simultaneously sensing the APU and the electric motor shaft speeds. Once the driver has accelerated the APU to the same shaft speed of the transaxle, the microprocessor lights a dash indicator. This prompts the driver to engage the transaxle in the same gear.

To disengage HEV mode, the driver can step on the brake, press the "Abort" button, or release the accelerator pedal. Any of these commands for disengagement turns off the APU and turns on the PMC controller. The next time the accelerator is depressed it will control the electric motor.

EMISSIONS CONTROL STRATEGY

Emissions control strategy consists of two major points 1.) Catalysis of harmful emissions 2.) Optimization of air/fuel ratio.

A three-way platinum/rhodium catalyst is utilized for the control of CO, HC, and NOx emissions. The three-way catalyst is unique in its ability to allow both oxidation and reduction reactions to occur on the same substrate. The catalyst dimensions have been optimized for the given flow rate of exhaust of about 1200 l/min. After consultation with Johnson and Matthey, a cross-sectional area of 72.6 cm^2 and a length of 21.6 cm was selected. These dimensions are the best compromise among the following set of goals: 1.) minimization of significant thermal gradients 2.) provision for sufficient residence time 3.) minimization of back pressure.

It is necessary to maintain an air/fuel ratio of 1.07:1 based on a mass ratio such that the average cylinder temperature increases by approximately 10°C operating at the chosen air/fuel ratio as opposed to operating at a 1:1 ratio. The Carnot equation shows an increase of approximately 9% in efficiency for lean operation as opposed to stoiciometric. Operating at this higher temperature results in lower exhaust emissions at the cost of a lower power efficiency, since volumetric efficiency is reduced during higher operating temperatures. A temperature profile of the catalyst has been estimated based on thermocouple measurements.

Since the engine will not be running at only one particular load in this configuration, it is a relatively difficult task to provide the chosen air/fuel ratio. Inherent in the emissions dilemma is the compromise between carbon oxides and nitrogen oxides, since NOx is favored at higher temperatures and carbon oxides are favored at lower temperatures.

A fuel injection system was designed for the engine, however the complexity of this system was deemed excessive. A simple constant velocity carburetor was sufficient for the fuel metering needs, and more compatible with this particular air intake system. HEVolution utilizes a RAM-AIR intake system with a diffuser nozzle that is more compatible with this carburetion system.

The struggle between emissions and overall power efficiency has become an engineering trade-off. The primary goal is to propel the vehicle using a small engine. The philosophy being, why try hybrid if you're going to lug around a big engine? The smaller engine means poorer emissions performance but smaller volumetric output.

FUEL EFFICIENCY CONSIDERATIONS

HEVolution uses gasoline instead of the other available fuels for several reasons. The use of a carburetor essentially nullifies any advantages of using either ethanol or methanol, considering that the carburetor is designed for gasoline. A 12 gallon fuel cell provides sufficient capacity to meet the minimum range requirement of 483km. Efforts were made to reduce overall vehicle weight to increase both fuel and electric efficiency. Also, Goodyear Tire Company has suggested that HEVolution use Invicta GL tires with a tire pressure at 344kPa.

ELECTRIC EFFICIENCY CONSIDERATIONS

Several modifications were implemented on HEVolution to increase the electric efficiency (See Figure 4). A forced convection heat sink was added to the electric motor controller. This change was motivated by the thermal cutout of our electric motor controller. When the controller temperature becomes excessive, the controller automatically reduces the current output. This feature extends controller longevity, but decreases vehicle performance. The heat sink was added to help cool the controller. Forced convection was added due to excessive engine compartment temperatures during ICE operation. Two small muffin fans were added to increase the heat transfer by forcing air across the heat sink.

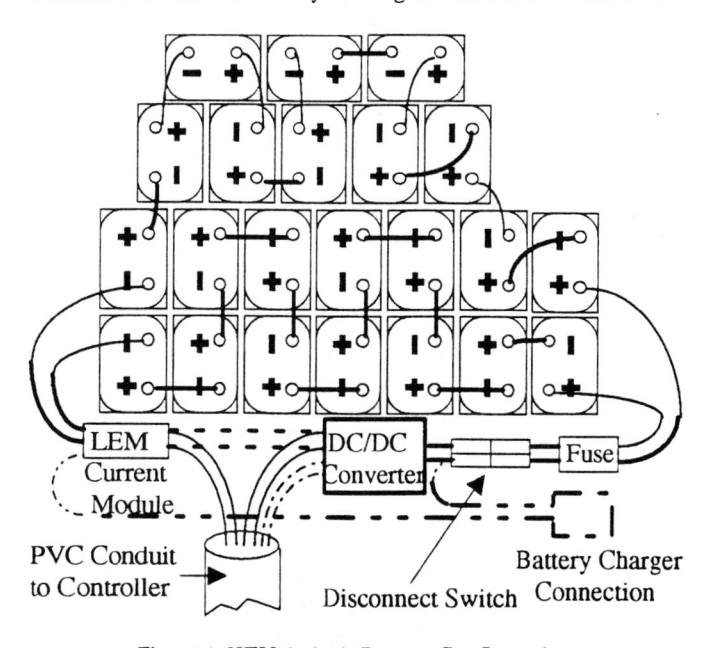

Figure 4: HEVolution's Battery Configuration.

The number of batteries was also increased to improve the electric efficiency. Considering the characteristics of the batteries, the electric motor, and the electric motor controller led to a change from 20-12 volt batteries, to 22-12 volt batteries. Lead acid batteries have the characteristic that the Amp-Hour capacity is a function of the discharge rate. As the discharge rate increases, the Amp-Hour capacity decreases. It is therefore desirable to keep the discharge rate as low as possible. The electric motor controller is rated at 400 Amps maximum and the electric motor is rated at 144 Volts maximum. The number of batteries in the battery pack was analyzed for both situations using the power required to propel the vehicle at 88.5 kph.

The following calculations demonstrate that by adding two batteries, (and the corresponding 36 kg) the current per battery has decreased by 21% and the power loss in the batteries has been reduced by 17.4%.

$P = VI$

$P = 14.4$ kW

$V_{20} = 120$ V nominal

$I_{20} = \dfrac{14.4 \text{ kW}}{120 \text{ V}} = 120$ Amps

Current per battery $= \dfrac{120 \text{ Amps}}{20 \text{ Batteries}} = 6$ Amps/Battery

$V_{22} = 132$ V nominal

$I_{22} = \dfrac{14.4 \text{ kW}}{132 \text{ V}} = 109.1$ Amps

Current per battery $= \dfrac{109.1 \text{ Amps}}{22 \text{ Batteries}} = 4.96$ Amps/Battery

21% reduction in current per battery

Power Loss :

$P_{Loss} = I^2 R$

$\dfrac{P_{Loss22}}{P_{Loss20}} = \dfrac{I_{22}^2}{I_{20}^2} = \dfrac{109.1^2}{120^2} = 0.826$

The only drawback to this design is the additional weight which is an acceptable penalty. In practical application, these results will not be as significant because the vehicle is not always driven at a constant speed.

VEHICLE STRUCTURE AND MODIFICATIONS

HEVolution's major components are outlined in Figure 5. The back seat was replaced with a sealed and ventilated battery bank. Five fans draw air thorough the floor and out the rear side panel. Inside the battery bank is a thermal switch which activates four of the fans if the inside temperature reaches 30° C. One fan runs continually to vent any hydrogen gas build up created by recharging or discharging the batteries.

The battery bank is enclosed with aluminum sheet metal and sits on a 1018 Steel tube rack. The fuel tank is mounted outside the vehicle directly below the battery pack. The fuel tank is a standard racing cell with a puncture proof bladder which holds 45.4L of gasoline. The fuel tank is connected to the ICE via stainless steel tubing and Swagelock fittings.

A result of removing the stock Ford 1.9L four cylinder engine, two additional modifications were made to replace the lost alternator and power brakes. A DC to DC converter was designed and built to step down the battery bank voltage from 132V to 12V. The 12V leads are used to recharge the auxiliary battery in the engine compartment. A vacuum pump was added to create a vacuum in the power brake boost allowing the power brakes to be used rather than just manual brakes.

As a safety precaution, a 9.1kg ABC dry chemical fire protection system (FPS) was installed. The FPS is a modified version of what is used on coal mining equipment. An ABC dry chemical was chosen because it's more environmentally friendly when discharged in comparison to Halon systems which damage the Ozone layer. Four nozzles are placed through out HEVolution in critical fire/explosion areas; two nozzles are in the battery bank, one nozzle points at the ICE, and the other nozzle is aimed at the EM controller. In addition, a roll hoop was installed to protect the driver and passenger during a roll-over crash, made of 440mm mild steel tubing.

SUSPENSION

The vehicle suspension had to be modified to accommodate the significant additional weight from the 22 lead acid batteries. Static and dynamic loads on the suspension system differ from the stock vehicle in magnitude and location. Static considerations involve the ride height, center of gravity, roll center, and roll moment. Dynamic considerations include the damping effects, cornering ability, and avoidance of an oversteer condition.

Total weight of HEVolution is about 10% greater than the stock Escort, with a weight bias of about 40% front and 60% rear. Weight distribution of the stock Escort is approximately 60% rear, 40% front. Front axle weight of HEVolution maintains a 1:1 ratio with the stock Escort, while the rear axle is about twice that of the stock model.

The installed struts are GAB high performance struts with adjustable damping rates. The front struts have damping rates ranging from just below the stock damping rates to 15% over the stock rates. The rear struts have damping rates ranging from just below stock values up to 150% over stock values. The struts have modified spring perches to accommodate custom made springs as well as adjustable collars for variable ride height.

The replacement springs were custom made with a 63.5mm inner diameter and will work with the GAB struts.

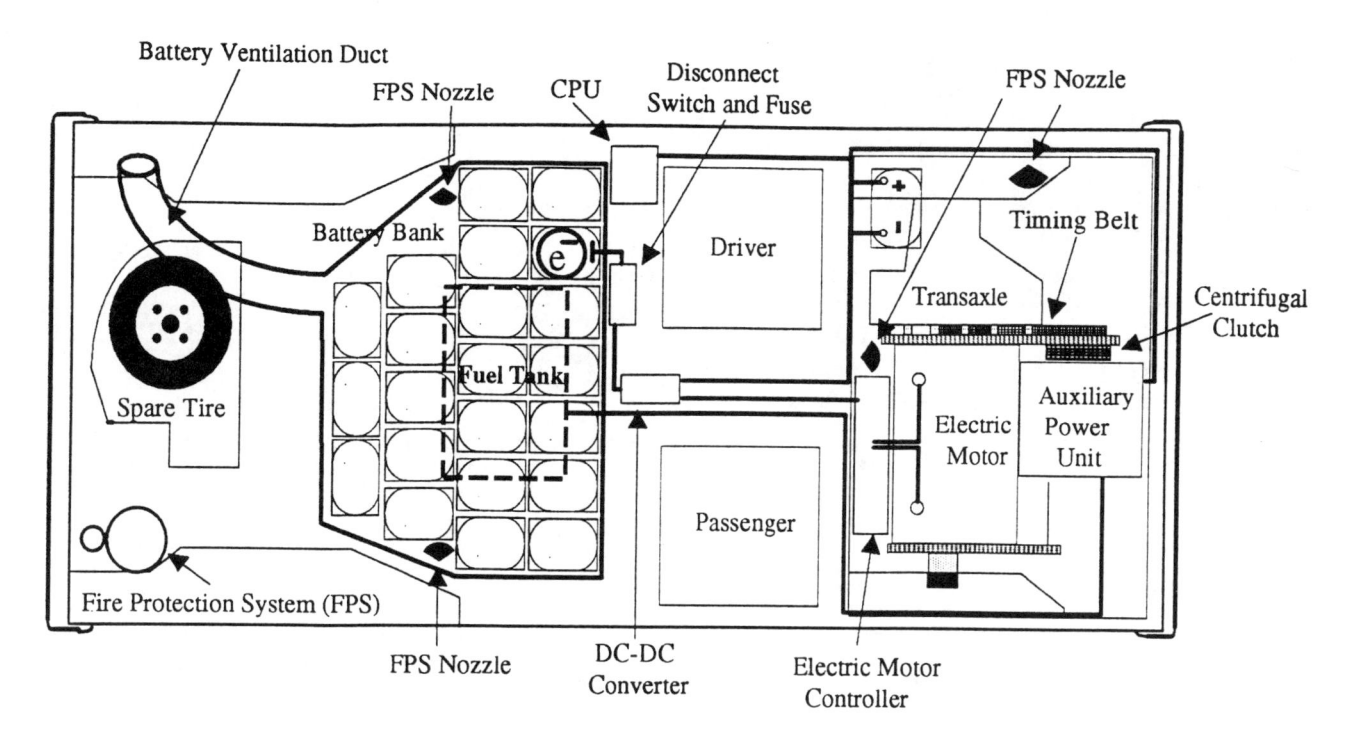

Figure 5: HEVolution's Overall Vehicle Schematic.

Front springs have a progressive rate based on the weight distribution of the vehicle. High performance anti-sway bars were fabricated by Addco Industries Incorporated. The bars have adjustable end links to provide roll stiffness. A larger anti-sway bar was installed in front to help prevent an oversteer condition.

MANUFACTURABILITY

Manufacturability was a prime consideration during the design of HEVolution. Stock parts or "off the shelf" components were used where ever possible to simplify and expedite vehicle manufacturability. The few custom pieces in HEVolution could be easily manufactured if the vehicle were in mass production.

The major components are the battery storage shell, the APU and electric motor mount, and the sprocket on the transaxle. There are also several lesser ancillary items; wiring, conduit from the battery pack to the electric motor, the exhaust system, and several common sensors.

CONCLUSION

This year's parallel configuration was chosen over a series configuration due in part to funding limitations. This is a more difficult configuration to control yet it yields a higher overall energy efficiency.

The control system to move from EV to HEV is cumbersome and in need of refinement. HEVolution's control strategy required some means to match APU shaft speed and transaxle speed. Failure to closely match both shaft speeds would exceed the tensile strength of the drive belt. This could be overcome by reinstalling the stock clutch between the electric motor and the transaxle. This would disengage the electric motor and reduce the shock the drive belt experiences.

The 1994 WU-HEV team effort was beset with severe financial limitations. The team pulled together through this adversity and worked toward the common goal of a successful showing in Detroit in June.

Outside the design lab, HEVolution participated in the Arizona Public Service Electric 500 posting the fastest one mile oval lap speed of 110.3 kph. HEVolution has also stopped at several local middle schools to talk to the students about science and engineering and the environmental effects of transportation.

This parallel configuration can be a viable design for HEVs. This is admittedly not a "power" vehicle at highway cruising speeds, but it does meet its intended goals as a commuter vehicle.

The 1994 Weber State University Hybrid Electric Vehicle

William B. Nelson, Student, and David A. Erb, Faculty Advisor
Weber State University

ABSTRACT

Engineering technology students at Weber State University designed and built a hybrid electric conversion of a Ford Escort station wagon. The design is a parallel configuration utilizing the stock engine and transmission. An electric motor is coupled to the rear of the transmission main shaft. The design is simple, reliable, and performs well in all modes of operation.

Optimizing energy efficiency was an important consideration. Testing has shown that the vehicle's electric efficiency is approximately double its gasoline efficiency. Therefore, design efforts were directed toward taking full advantage of electric operation. Nickel cadmium batteries provide the high energy capacity needed.

INTRODUCTION

Recent emissions regulations have increased public interest in hybrid and electric vehicles. The California Air Resources Board (CARB) has scheduled a progressive phase-in of four vehicle classes, with increasingly restrictive emissions levels: Transitional Low Emission Vehicles (TLEV), Low Emission Vehicles (LEV), Ultra Low Emission Vehicles (ULEV), and Zero Emission Vehicles (ZEV). Beginning in 1998, two percent of all cars sold in California must be ZEV. [1] Many engineering professionals believe that part of the solution to the emissions problem will be the hybrid electric vehicle (HEV). An HEV is equipped with an electric motor as its primary source of propulsion and an internal combustion engine to provide extended range. Although an HEV is not a ZEV, it could be considered a ULEV with a ZEV mode of operation.

During the 1991-92 and 1992-93 academic years, a team of engineering technology students at Weber State University (WSU) designed an HEV conversion for a 1992 Ford Escort station wagon. The original design was based on the following criteria:

1. simplicity,
2. manufacturability,
3. ease of conversion, and
4. low cost.

The HEV conversion proved very successful in the 1993 HEV Challenge, scoring an overall second place in the competition. Among other accomplishments were:

    First Place Cost Assessment Event
    Most Manufacturable Vehicle Award
    Most Environmentally Friendly Vehicle Award
    First Place Emissions Event
    SAE Design Excellence in Engineering Safety
      Award
    Second Place Electric Efficiency Event
    Second Place Overall Range Event

A new team of engineering technology students has recently completed the second phase of this HEV conversion. Building on the success of the 1993 WSU HEV team, the 1994 team set out to win the 1994 HEV Challenge. The team determined that improvement of the vehicle's overall energy efficiency was the best way to achieve their goal.

The purpose of this report is to outline the design of the WSU HEV conversion, including all aspects of design that were considered and the basis upon which those not used were rejected.

FUEL AND ELECTRIC ENERGY EFFICIENCY

Data obtained on the WSU HEV in the 1993 HEV Challenge revealed an interesting comparison of energy efficiency. Figure 1 shows this comparison graphically. Under highway driving conditions, the efficiency in gasoline mode was measured at 18.37 km/L. Efficiency measurements taken for highway driving in ZEV mode yielded 4.65 km/kWh. Using the energy content of gasoline (9.03 kWh/L), the electric energy efficiency can be converted to an equivalent kilometers per liter value for the purpose of comparing electric efficiency with gasoline efficiency. The conversion is as follows:

$$(4.65 \frac{km}{kWh}) \times (9.03 \frac{kWh}{L}) = 41.99 \frac{km}{L} \quad (1)$$

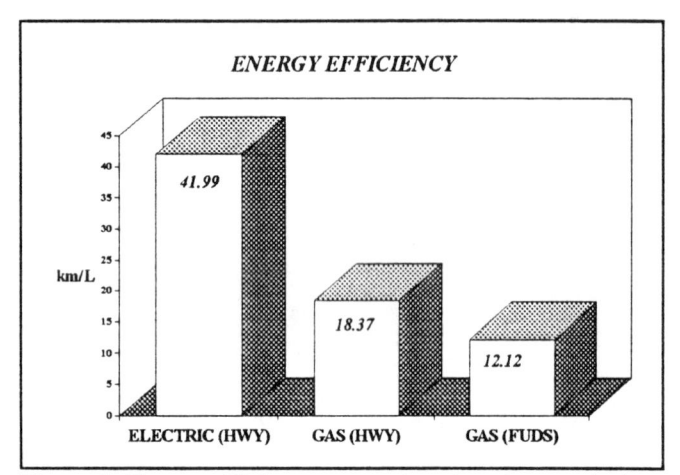

Figure 1 Energy efficiency comparison based on equivalent kilometers per liter.

The data used for this comparison were generated during several of the 1993 contest events. These events, while not fully controlled, did involve very similar operation. However, for full confidence in the comparison, results from a more controlled test are desirable. Fortunately, such results exist. The Idaho National Engineering Laboratory (INEL) dyno-tested a converted Chrysler minivan, using the Federal Urban Driving Schedule (FUDS). This 2318-kg vehicle achieved electric energy efficiency of 3.38 kWh/km, equivalent to 30.5 km/L of gasoline. [2] The INEL minivan's efficiency can be used as a conservative proxy for the electric efficiency achievable by the 1572 kg WSU HEV in a FUDS test. Since the WSU HEV achieved 12.12 km/L on gasoline in a controlled FUDS test at the 1993 HEV Challenge, the assumption of significantly enhanced vehicle energy efficiency appears well-founded.

The WSU team is well aware of the difficulty of arranging suitable dynamometer time for controlled testing of all contest vehicles. However, if this resource can be found, FUDS testing in ZEV mode would significantly enhance the usefulness of data generated in future HEV Challenges.

In light of the above, HEV design should aim for achievement of maximum ZEV range. Range extension should then be provided by the cleanest, most efficient method possible, up to whatever level is needed for overall vehicle utility. This approach will tend to maximize energy efficiency while minimizing global emissions.

It is expected that the gasoline and electric efficiencies of the WSU HEV will decrease slightly in the 1994 HEV Challenge, due to a five percent increase in overall vehicle weight. However, a large increase in the vehicle's overall energy efficiency is expected because the electric efficiency is much better than the gasoline efficiency. With increased energy capacity of the energy storage system (see the BATTERIES section under POWERTRAIN CONFIGURATION), the vehicle's electric range is expected to increase by about 32 km. The additional electric range will allow the vehicle to take more advantage of the preferred energy efficiency of ZEV mode.

POWERTRAIN CONFIGURATION

After a thorough review of the 1994 HEV Challenge rules, the primary design criteria for the 1994 WSU HEV power train became:

1. simplicity,
2. manufacturability,
3. ease of conversion, and
4. optimum energy efficiency.

The resulting design is a parallel configuration that meets all of the above criteria (see figure 2).

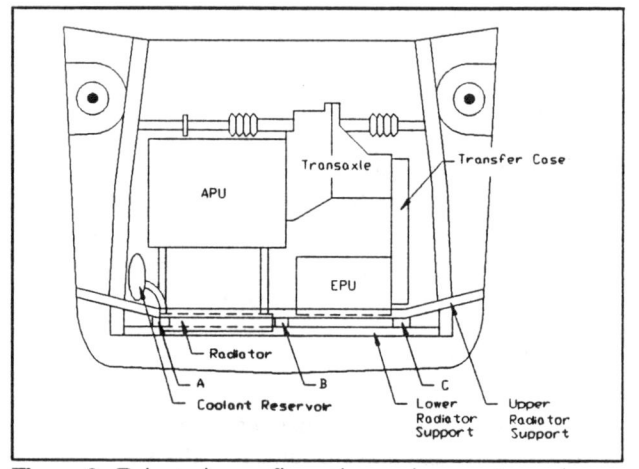

Figure 2 Drivetrain configuration and support members A, B, and C.

ELECTRIC POWER UNIT - The electric power unit (EPU) for the 1993 Weber State University HEV was selected based on an acceleration goal of 0 to 72 km/hr in 15 seconds and a cruising speed goal of 64 km/hr for at least 64 km. An Advanced DC Motors model 000-256-4001 electric motor and Curtis PMC model 1221B controller were selected for their ability to operate within these parameters. [3] The option of changing to an AC motor was considered early in the development of the 1994 WSU HEV. This option was dropped for the following reasons:

1. efficiency loss in converting from DC to AC, which partially offsets any gain in motor efficiency;
2. added weight and high cost of DC-AC inverter; and
3. satisfactory performance of the Advanced DC model 000-256-4001 in the 1993 HEV Challenge.

In the 1993 HEV Challenge, the WSU HEV placed second in the electric energy efficiency and range events with 4.65 km/kWh and 62.4 km, but was unacceptably slow in ZEV acceleration. [4] System voltage in the 1993 vehicle was 72V, which is the low end of the voltage range specified for the Advanced DC Motors model 000-256-4001. [5] Figure 3 shows that motor performance can be dramatically improved with increased voltage.

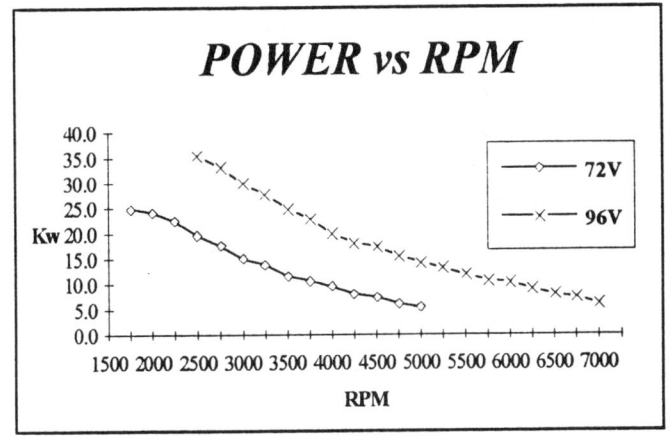

Figure 3 Advanced DC Motors model 000-256-4001 power curves.

Because the Advanced DC motor was proven reliable in the 1993 HEV Challenge and performance could be improved with increased system voltage, it was decided that resources would be best directed toward improving the vehicle's energy storage system.

REGENERATIVE BRAKING - Efforts toward optimizing the vehicle's energy efficiency led to an investigation of the feasibility of a regenerative braking system. The engineer who designed the motor was contacted. It was his recommendation that the motor not be used directly for regenerative braking. To realize the highest efficiency of this particular electric motor, the brushes are placed at a 9 to 12 degree angle (see figure 4). At this angle, a normal brush temperature of 130°C is realized. When the motor is used as a regenerative source, brush temperatures exceed 220°C, causing the brushes to burn out prematurely. [6]

A regenerative braking system supplied by Solar Car Inc. was also considered. This system consists of a 4.5 kW alternator and clutch assembly attached to a tail shaft on the rear of the motor. [7] The Advanced DC Motors model 000-256-4001 is available with an optional two inch tail shaft to mount the assembly (see figure 4).

An analysis was done to evaluate this option. The analysis was meant to answer two questions:

1. How much deceleration can 4.5 kW of regenerative braking supply?
2. How much electric energy can be recovered?

The analysis showed that 0.1135 G is the largest deceleration that 4.5 kW will achieve. Figure 5 shows the G levels for speeds from 0 to 105 km/hr when a constant regenerative braking power of 4.5 kW is applied. A total vehicle mass (including occupants and cargo) of 1808 kg was assumed for these calculations. At these low deceleration rates, it would be necessary to assist the regenerative braking with the service brakes in order to stop the vehicle over a reasonable distance.

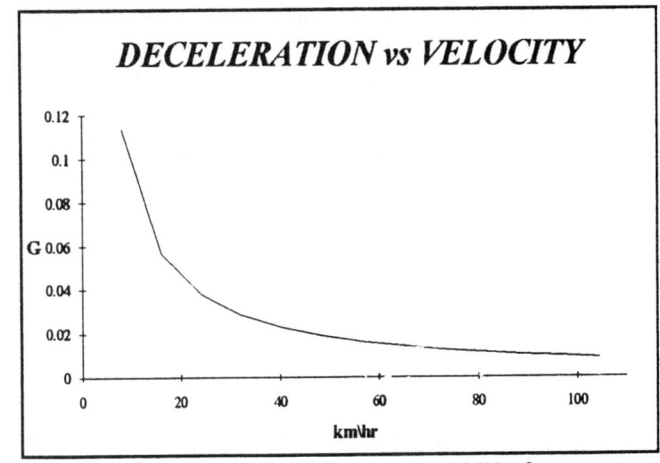

Figure 5 Deceleration vs Velocity at 4.5 kW of regenerative braking power.

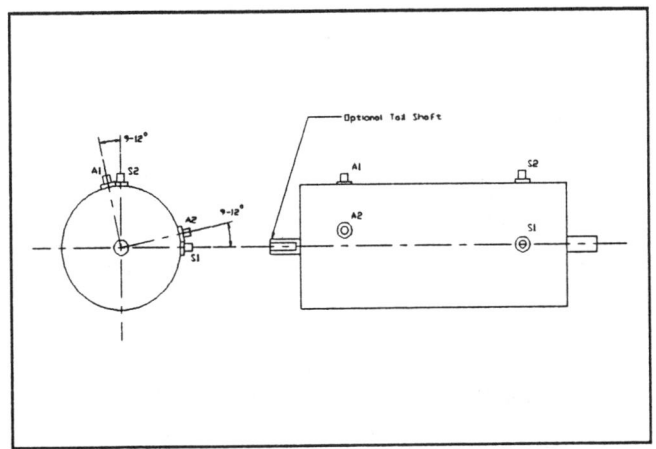

Figure 4 The motor cannot be used for regenerative braking because of the 9-12° brush angle.

The Road Rally event of the HEV Challenge was simulated to estimate the amount of energy that could be recovered. The following assumptions were made:

1. 200 stops from 56 km/hr,
2. 75 m stopping distance, and
3. a suitable RPM would be maintained through down shifting to maintain a constant 4.5 kW from the alternator.

It was calculated that an average braking power of 30 kW would be required to stop the vehicle from 56 km/hr in 7.4 seconds over a distance of 75 m. Of the 30 kW, 4.5 kW would be charging the batteries. If the batteries accept 80 percent, the energy recovered would be

$$\frac{4.5\,kW}{30\,kW} \times 80\% = 12\%. \qquad (2)$$

Total energy recovered is 0.0074 kWh/stop, and 200 stops produce about 1.5 kWh. With an electric energy efficiency of 4.65 km/kWh, the vehicle's range would increase by about 6.5 km.

This regenerative braking package would add about 18 kg to the weight of the vehicle. In order to add this regenerative system, the battery pack would have to be reduced by one module to stay within the weight limit set forth in the 1994 HEV Challenge rules. Each of the chosen batteries (see BATTERIES section), has a mass of 23.2 kg and contributes 1.2 kWh. With an electric efficiency of 4.65 km/kWh, one battery contributes 5.6 km to the vehicle's range.

Although the range gained by either option is about the same, the battery has other benefits. There would be no need to replace the electric motor with one equipped with the optional tail shaft needed to mount the regenerative system (see figure 4). Also, reducing the number of batteries reduces the system voltage which reduces the power output of the electric motor (see figure 3). Based on these observations, it was decided that the benefits of keeping the battery outweighed those of the 4.5 kW regenerative braking system and the option of regenerative braking was dropped.

BATTERIES - Battery selection was based on the following criteria:

1. a battery pack weight limit of 416 kg,
2. increasing system voltage from 72 to at least 96 volts, and
3. maximizing energy capacity.

Batteries considered for use in the 1994 WSU HEV include nickel metal hydride (NiMH), sodium sulfur (NaS), zinc air (Zn air), zinc iron (ZnFe), nickel cadmium (NiCd), and lead acid (Pb acid). All of these except NiCd and Pb acid were eliminated early because of availability and price. Further, only one NiCd pack could be found at a price within the project budget. Table I summarizes the Pb acid and NiCd batteries considered. The projected range was based on the vehicle's electric energy efficiency in the 1993 HEV Challenge.

Because of the vehicle weight limit specified by the contest rules, 416 kg was set as the maximum for battery weight. To improve the acceleration of the vehicle in electric mode, the system voltage needed to be increased from the 72V system of the 1993 vehicle. It was determined that the electric efficiency of the vehicle was much better than the gasoline efficiency (see FUEL AND ELECTRIC ENERGY EFFICIENCY section). For this reason, it was important to obtain the highest possible energy capacity. The Nickel Cadmium batteries provided the best mix of overall battery pack weight, system voltage, and energy capacity. The even power distribution characteristics of the nickel cadmium batteries also make them well-suited for use in a hybrid electric vehicle (see figure 6).

Table I Batteries considered for the 1994 WSU HEV.

	Pb Acid SCS-200	Pb Acid 27 TM	Pb Acid 24 TM	NiCd STM5-200
Energy Density	35 Wh/kg	35 Wh/kg	35 Wh/kg	52 Wh/kg
Unit Mass	28.1 kg	23.6 kg	20.4 kg	23.2 kg
Total Mass	393.7 kg	377.4 kg	408.2 kg	371.2 kg
Voltage	84 V	96 V	120 V	96 V
Energy Capacity	13.9 kWh	13.2 kWh	14.3 kWh	19.3 kWh
Projected Range	64 km	61 km	66 km	90 km

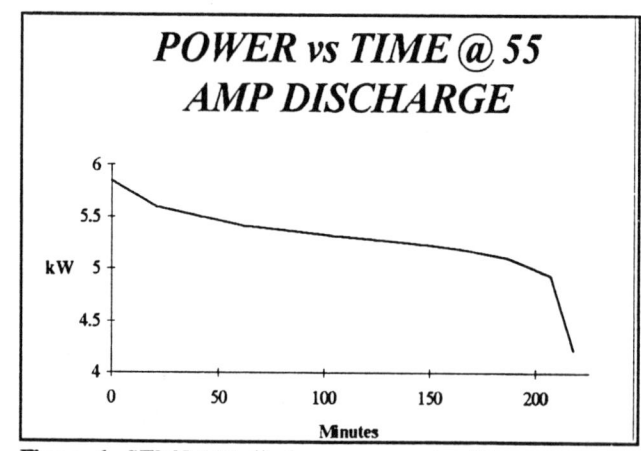

Figure 6 STM5-200 discharge curve at a 55 amp discharge rate.

Like many battery materials, cadmium is toxic. In order to maximize the environmental benefit of the HEV, worn-out batteries are sent back to the manufacturer for recycling.

AUXILIARY POWER UNIT - Several gasoline powered internal combustion engines were considered for use as the auxiliary power unit (APU) for this HEV.

1992 Original Equipment Escort Engine - This engine was chosen as the APU for the following reasons:

1. the original packaging could be used,
2. the stock emissions systems could be used,
3. there would be no loss of performance in gasoline mode, and
4. a significant cost savings.

This engine produces 65 kW @4400 rpm and 146 N-m torque @ 3800 rpm. [8]

Kawasaki 454 - This was a 4-stroke, 2-cylinder, liquid cooled motorcycle engine. There were many advantages to using this engine. In fact, early in the development of the vehicle, it seemed to be one of the best options. The Kawasaki was very light (only about 34 kg) and had more power than the 3-cylinder automobile engines considered. This engine would produce 37.3 kW @ 9500 rpm and 41.2 N-m torque @ 8000 rpm. The use of this motor was rejected because its crank rotation was opposite of that required of the transmission. To make this engine work, major precision machining of transmission components would have been necessary. Emissions control hardware would also have to be designed, built, and tested. The team simply lacked the resources, most notably time, required to make this option viable.

Other Engines Considered - Also considered were three and four cylinder Geo Sprint engines and a three cylinder Daihatsu engine. These engines were rejected because of the high cost of adapting the necessary computer controls.

TRANSMISSION MODIFICATIONS - The APU transfers power to the front wheels through the original Escort transaxle. A simple transfer case links the electric motor to the same transaxle. The overdrive gears and cover were removed from the transaxle tailshaft and replaced with a 9.5 mm (3/8-inch) aluminum plate to locate and support the electric motor as shown in figure 7.

The transfer case contains a Morse type HV chain drive. A 25.4 mm (one-inch) wide chain with 9.5 mm (3/8-inch) pitch connects two 88.9 mm (3.5-inch) pitch diameter sprockets. This system is designed to handle 71.6 kW, well above the expected service of about 45 kW. The sprocket on the electric motor is a press fit with a 6.35 mm (1/4-inch) key. The sprocket on the transaxle (in place of the fifth gear) is splined to match the existing shaft. Splines on the sprocket were matched to the shaft by electron discharge machining. The case cover was made with 20 gauge sheet metal and welded to a 6.35 mm (1/4-inch) thick by 25.4 mm (one-inch) wide steel flange. [3]

The chain manufacturer specified that the sprockets had to be aligned at a distance of 329.29 mm +1.397/-0 (12.946 in. +0.055/-0) and had to lie within the same plane within +0.559/-0 mm (+0.022/-0 in). [9] The transfer case and electric motor mounts were designed so that both mounted directly to the transaxle and nothing else. This allowed assembly as a unit before installation in the car. Using a coordinate measuring machine, they were assembled and adjusted until specifications were met. All bolts used either nylon lock nuts or liquid lock-tight.

CONTROL STRATEGY

One objective of control strategy was to retain stock control characteristics in all modes of operation. Except for the mode selector switch, power switch and clutch lockout, driving the WSU HEV is very similar to driving a stock Escort.

DRIVING MODES - The WSU HEV has three driving modes. They are: zero emissions vehicle (ZEV), gasoline, and hybrid electric vehicle (HEV). When in ZEV mode, the vehicle is powered by the EPU. In gasoline mode, the vehicle is powered by the APU. In HEV mode, the EPU and the APU are combined.

CHANGING MODES - To operate in ZEV mode, the link between the gasoline engine and the transmission is

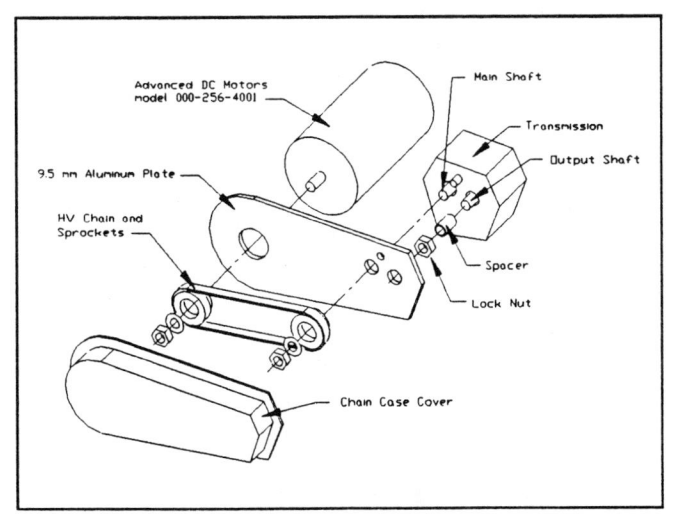

Figure 7 The overdrive gears and cover were removed and replaced with the electric drive components.

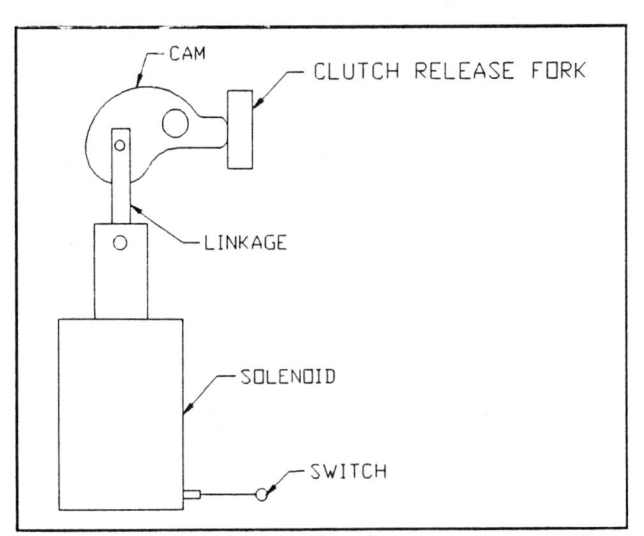

Figure 8 A clutch lockout is used to disconnect the link between the APU and the transmission.

disengaged with the clutch lockout shown in figure 8. The driver uses the foot pedal to release the clutch as in a normal gear change. A toggle switch mounted on the dashboard energizes a solenoid which moves a cam that holds the clutch release fork in the disengaged position. The EPU then drives the main shaft independent of the internal combustion engine.

With the clutch locked out, the mode selector switch in the EPU position, and the main power switch on, the WSU HEV is ready to drive in ZEV mode. In ZEV mode, shifting is accomplished without a clutch by modulating the throttle to match gear speeds. Because of the relatively low inertia of the motor, the synchronizers match gear speeds quickly. Shifting is very smooth and, with a little practice, is easier than shifting with the clutch.

Operation in gasoline mode is almost identical to driving the stock Escort. With the clutch lockout disengaged, the main power switch on, and the mode selector in the APU position, the WSU HEV can be started and driven in stock fashion.

To operate in HEV mode, the mode selector switch is set to the HEV position and the APU is started. With both power units operational, a single gear is selected for a straight line acceleration run. Because of the drastic difference in torque characteristics between the two power units, the team expects some difficulty shifting gears in this mode. This mode was included only to improve performance in the acceleration event of the 1994 HEV Challenge and to test the possibilities of dual power unit operation in future versions of this parallel hybrid vehicle.

ELECTRICAL COMPONENTS OF CONTROL - All power lines are of sufficient size to carry the needed current and are protected by the proper size fuses. All modes will be controlled through a 12V source provided to a main contractor and relays. The 12V source is a regular lead acid automotive battery mounted under the hood as originally equipped. The main contractor and relays are enabled or disabled according to the chosen mode of operation. The controlled items are:

1. the main contractor, which enables the EPU to be energized,
2. the DC/DC relay, which enables the DC/DC converter to charge the 12V lead acid battery from the main battery pack while in ZEV mode,
3. the cooling pump relay, which enables the motor controller to be cooled,
4. the alternator relay, which enables the alternator to charge the 12V lead acid battery while in gasoline mode,
5. the ignition switch relay, which enables the APU to be started,
6. the clutch lockout relay, which enables the lockout solenoid to be energized, and
7. the battery cooling relay, which enables the battery cooling fans to operate.

Table II State of main contactor and relays in each mode of operation.

	ZEV	Gas	HEV
Main Contactor	CLOSED	OPEN	CLOSED
DC/DC Relay	CLOSED	OPEN	OPEN
Cooling Pump Relay	CLOSED	OPEN	CLOSED
Alternator Relay	OPEN	CLOSED	CLOSED
Ignition Relay	OPEN	CLOSED	CLOSED
Clutch Lockout Relay	CLOSED	OPEN	OPEN
Battery Cooling Relay	CLOSED	OPEN	CLOSED

Table II shows the states of the main contractor and each relay for each mode of operation. A wiring diagram is shown in the Appendix identifying high voltage connections to the drive system, low voltage connections for accessories, and all system safety devices.

EMISSIONS CONTROL

The WSU HEV is powered in gasoline mode by the stock 1.9L Escort engine, equipped with all stock emissions controls. The 1993 Escort is very similar to the 1992 model on which the WSU HEV is based, and complies with CARB TLEV standards. A production version of the WSU HEV would be equipped with 1993 or later Escort emissions controls.

Leaving the stock emissions system intact took advantage of time and money already spent by Ford and proved to be a good solution in the 1993 contest. The WSU HEV won first place in the emissions event of the 1993 HEV Challenge. The 1994 vehicle uses the same emissions strategy.

SUSPENSION MODIFICATIONS

Increased weight required that the suspension be modified. Suspension modification objectives were:

1. stock ride height,
2. safe handling characteristics, and
3. simplicity.

New springs were specified to meet modification objectives. Spring constants were calculated for a 453 kg increase in rear axle weight and a 181 kg increase in front axle weight. Spring wire diameters were calculated based on the following:

1. increased axle weight,
2. stock coil end configuration, and
3. stock number of coils.

The new front and rear spring constants are 35.2 and 28.9 N/mm., respectively. Linear springs replaced the stock variable springs. The resulting shear stresses of 579 and 510 MPa are within the spring manufacturer's specification of 620 MPa. Corresponding spring free lengths are 289.6 and 381.0 mm. Steering and suspension geometry, mounting points, and linkages remain stock. [3]

Initial test drives revealed a need for additional damping. A set of off-the-shelf aftermarket struts was purchased and installed. These provide excellent damping. With the modified springs and struts, vehicle handling is very similar to stock, in spite of the large added mass.

BRAKE MODIFICATIONS

The absence of engine vacuum while operating in ZEV mode causes a serious problem with the vacuum assist braking system. Stopping the 1808 kg vehicle within reasonable distances without the vacuum assist would be difficult. Safety considerations also dictate a need for uniform operational feel in the various modes. To provide safe, consistent braking, an electric vacuum pump and reservoir were installed. A diagram of this system is shown in figure 9. The vacuum pump operates off the 12V lead acid battery whenever the vacuum sensor switch senses a low vacuum.

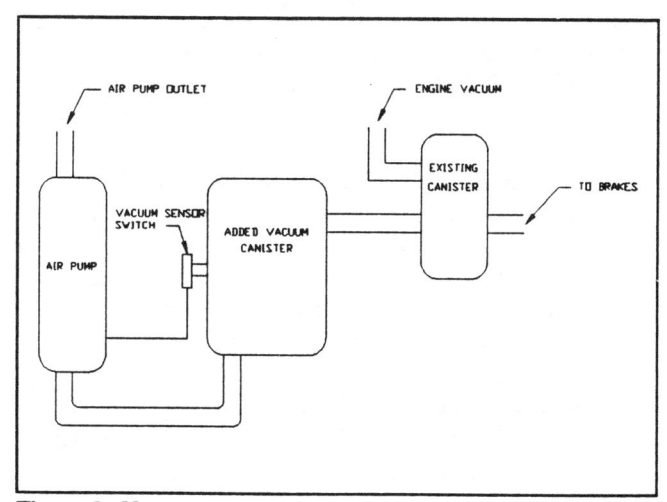

Figure 9 Vacuum assist for ZEV mode.

STRUCTURAL MODIFICATIONS

The center and side vertical members (members A, B, and C in figure 1) between the lower and upper radiator supports were removed to make room for the electric motor and radiator. Three brackets were made from 20 gauge steel to replace the members that were removed. The new brackets are of equal or greater size and strength than those removed. [3] There were no other structural modifications made to the vehicle.

BATTERY PLACEMENT AND ENCLOSURE

The battery enclosure was designed and positioned based on the following criteria:

1. proper cooling,
2. passenger safety, and
3. even front to rear weight bias.

The enclosure contains 16 nickel cadmium batteries positioned in four rows with four batteries in each row.

COOLING - In order for the chosen batteries to deliver peak performance, they must be properly cooled. The manufacturer required that the rows of batteries be at least 15 mm apart and that there be at least 12 mm between the batteries and the walls of the enclosure. Spacing was accomplished by building a gridwork of 12.7 mm (1/2 inch) square tubing (see figure 10). The bottom of the enclosure was lined with 1.588 mm. neoprene, wrapping it over the top of each 12.7 mm spacing tube. The tube and neoprene together provide 15.88 mm between rows of batteries and 14.29 mm between the batteries and the walls.

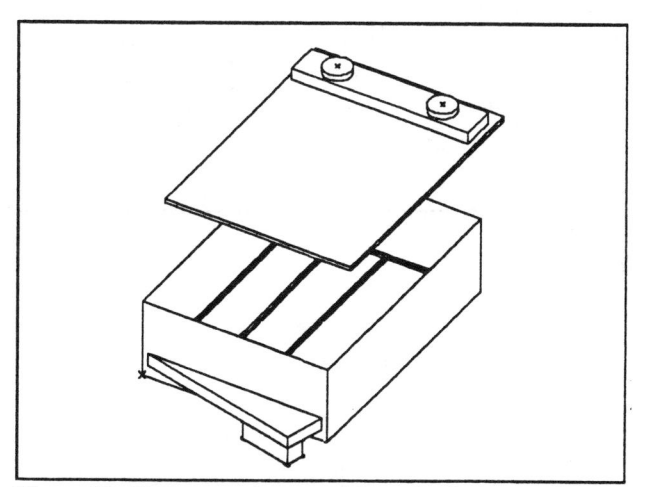

Figure 10 Battery enclosure with wedged plenum for even distribution of air flow.

In addition to spacing requirements, the manufacturer recommended 30 m^3/hr air flow per module. For 16 modules, this works out to a total of 480 m^3/hr. To account for pressure drops in the system, two brushless DC fans rated at 357 m^3/hr each were chosen, for a total rating of 714 m^3/hr. It was determined that this would more than account for the expected pressure drop. The fans are located at the exhaust ports in order to create a negative pressure in the enclosure. With a negative pressure, any microholes in the box or ducts will draw air into the system, minimizing the possibility of any hydrogen gas leakage inside the cabin.

To provide equal pressure for even distribution of air flow, a wedge shaped plenum was attached to the front of the enclosure as shown in figure 10. Fresh air is drawn from the floor on the driver side of the vehicle and flows into the large volume end of the wedged plenum.

PASSENGER SAFETY - To assure passenger safety, it was important that the batteries be properly ventilated to remove the hydrogen and oxygen gas produced during charging and discharging. The STM5-200 NiCd batteries come with a water filling system which also ventilates the H_2 and O_2 gases from the batteries. This system connects the fill plug/vents of eight batteries in a hydraulic series with 10 mm flexible tubing. When the batteries need topping-off, water is added to the system, filling cell by cell to a pre-determined level. When the eight battery hydraulic circuit is full, water exhausts at the downstream end, which vents to the outside of the vehicle through the rear bumper. The fill end of the circuit is equipped with a check valve to prevent gases from backflowing into the cabin.

The main component of this system is the water filling plug/vent. This plug/vent combination ensures the automatic regulation of electrolyte level when topping-off, and the exhaust of gases whenever gassing occurs.

The water fill/vent system is also equipped with a flame trap security system to prevent a flame from travelling into the fill/vent tube if a spark or flame should happen to occur near the exhaust outlet. This system consists of two deaerators in series at the end of each circuit. Gases cross a water column of at least 5 cm. The flame security system is shown in figure 11. [10]

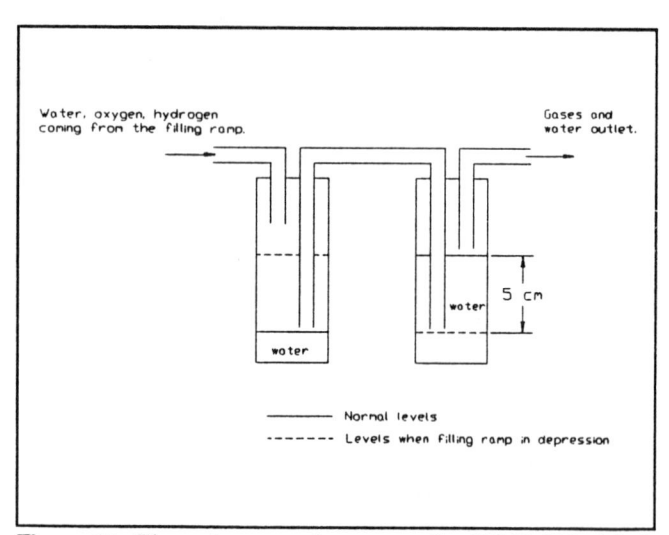

Figure 11 Flame trap security system for STM5-200 nickel cadmium batteries.

For passenger safety, the battery enclosure must contain the batteries in the event of an accident. The main concern was that the enclosure be sufficiently strong to keep the batteries out of the driver and passenger compartments. A finite element model was created to help analyze the strength of the container. A 40 G acceleration was used as a basis for the load used in the analysis. It has been established that this is the threshold of human tolerance. [11] The model showed a maximum deflection of 4.3 mm for a front/rear impact, and 9.22 mm for a side impact. Stress levels ranged from 0 to 446 MPa which is within the ultimate tensile strength limit for structural steel of 450 MPa. [12] The FEA distortion diagram for a front/rear collision is shown in figure 12.

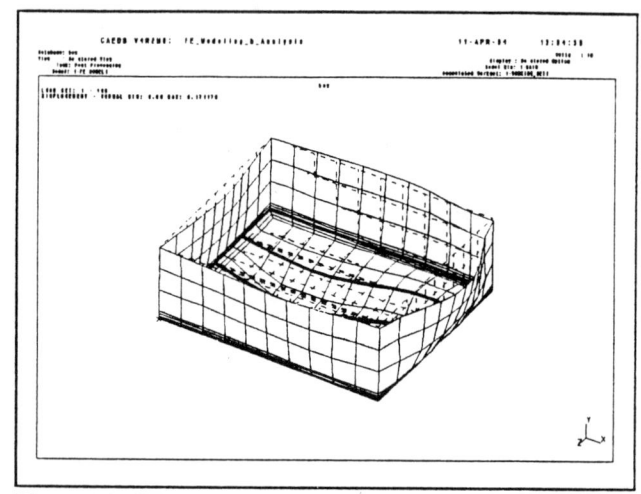

Figure 12 FEA model showing distortion of battery box in a 40G front/rear crash.

The batteries are strapped securely to the square tube grid so that the force in a collision would be applied to the grid structure rather than the sheet metal walls of the container. The use of the square tube grid not only simplified the design and manufacture of the box, but also provides the strength needed to restrain the batteries in the event of a collision.

WEIGHT DISTRIBUTION - Most of the mass added to the vehicle is due to the batteries and battery enclosure. The battery enclosure was placed as far forward as possible in the rear seating and cargo area. This positions more of the added mass towards the center of gravity for a lower polar moment of inertia, resulting in better maneuverability. For ideal handling, the front-to-rear weight bias should be approximately 50/50. [13] The position of the battery enclosure improves the original front/rear weight bias from 59% / 41% to about 51% / 49%.

COOLING

Other than the batteries, which have already been addressed, three main areas of the vehicle required cooling. They are:

1. gasoline engine,
2. electric motor, and
3. electrical components.

A 1979 Ford Fiesta radiator was installed to provide cooling for the gasoline APU. The new radiator was located forward of and below the stock location, to provide space for the electric motor. Radiator capacity was increased by replacing the stock core with a three row, high efficiency core. The filler, inlet, and outlet locations were altered to improve coolant flow. A 254 mm, 850 m^3/hr electric fan was mounted on the radiator to ensure proper air flow.

The electric motor is cooled by an internal fan which draws ambient air. Three separate air boxes are mounted to the EPU. Intake flow area is 64.5 square centimeters. Exhaust flow area totals 53.2 square centimeters. Air intake boxes are detachable for servicing the EPU and the foam air filter elements.

The electric motor controller also requires cooling. Initially no special effort was made to cool the controller. During the first stages of testing, it was found that the controller was getting quite hot, and was not operating efficiently. A heat sink and radiator system was designed to solve the problem. An electric pump moves liquid coolant through the heat sink to a dedicated radiator located in the engine compartment. After this cooling system was installed, the tested electric range of the vehicle nearly doubled. All other heat-producing components are cooled by natural convection. [3]

CHOICE OF MATERIALS

Materials used in manufactured parts were chosen based on the following:

1. performance,
2. manufacturability, and
3. cost effectiveness.

Three materials were used: steel, aluminum, and neoprene.

Twenty-two gauge sheet steel and 1.588 mm wall 12.7 mm square tubing was used in the battery box. A finite element model of the box was used to determine material thickness (see figure 12). The corrosion characteristics of aluminum preclude its use around batteries. Composites were also considered but ruled out because of equipment limitations. A polyurethane coating was used to protect the battery box from corrosion.

Twenty gauge sheet steel was used in the transfer case cover. This choice proved providential when the cover required modifications.

Other parts built with 20 gauge sheet steel include:

1. duct transitions,
2. electric motor restraining strap, and
3. radiator support brackets.

Twenty-eight gauge galvanized sheet steel was used for the cargo box, ballast boxes, and electric motor air ducts. Light weight and manufacturability were the main reasons for this selection.

The roll bar was made of 4130 seamless tubing. Strength, weight, and Sports Car Club of America (SCCA) safety guidelines were primary reasons for this choice.

Aluminum was chosen for the transfer case plate since weight considerations ruled out the use of steel.

Neoprene was used extensively in the battery enclosure, due to its corrosion resistance, elasticity, and electrical insulating characteristics. [3]

VEHICLE MANUFACTURABILITY

Design centered around in-house fabrication capabilities wherever possible. The main manufacturability criterion was simplicity. To facilitate design and construction, conversion was separated into three main areas:

1. transfer case,
2. battery pack, and
3. wiring-electrical.

By keeping the manufacturing processes simple, fabrication was faster, easier, and less expensive. This allowed earlier completion of the vehicle, extending the test time under actual operating conditions. Simple processes also allow quicker recovery when components fail during testing. [3]

ERGONOMICS

Very few ergonomic changes were made in converting the Escort. Stock items that remain unchanged include the front seats, carpeting, side panels, dash board, and steering wheel. The rear seat was removed to provide room for the battery box. The vehicle also has the original heater/defroster which functions normally in gasoline mode but lacks heat in ZEV mode. Air conditioning is not provided. The stock clutch, brake, and throttle pedals; gear shift; and parking brake lever are used for their original functions.

Additional levers, switches, and gauges were selected and mounted for function, safety, ease of use, and appearance. These additions include a fire extinguisher activation handle, located in the coin tray between the seats. This location provides easy access and decreases the chance of accidental discharge. Emergency stop buttons are located just inside the front corner of the side windows, accessible to the driver, passenger, and course workers. They disconnect all electric power when depressed. An electric motor overheat light is mounted on the dashboard to the left of the steering wheel, visible to the driver. Voltage and amperage gauges are mounted below the climate control panel. [3] The clutch lockout toggle switch and the mode selector switch are mounted on the dashboard near the shift lever.

SUMMARY

Over the past two years, the WSU HEV teams have designed and produced a simple, manufacturable, and energy efficient HEV conversion. Numerous aspects of the design required extensive research, analysis, and collaboration. The most significant design choices were:

1. use of the existing Escort engine and transmission,
2. attaching the EPU to the rear of the transmission main shaft, and
3. the use of nickel cadmium batteries to obtain an increased energy capacity.

The WSU HEV team feels that the current design is the best combination of all available design options that achieve the design criteria set forth in this report.

ACKNOWLEDGMENTS

The authors would like to thank the other members of the 1994 WSU HEV team (Steve Barker, Charles Beauregard, John Long, Mike McKay, Jeremy Palmer, Steve Shoop, Brad Stephens, and Kerry Walker) for their assistance in gathering the information included in this report and for their many hours of hard work and dedication to this project. The student team would like to thank the faculty advisors (Dave Erb, Kerry Tobin, and Kermit Reister) for all their help and support, and the 1993 HEV team for their hard work and dedication to the project last year. Special thanks are in order for Kory Yelderman and Steve Mott of the 1993 team, whose availability and willingness to answer questions have gone far beyond the call of duty.

REFERENCES

1. M.J. Riezenman, "Electric vehicles: why now?", IEEE Spectrum. November 1992; pp. 18-21.

2. A.F. Burke and R.D. MacDowall, "Track and Dynamometer Testing of the Eaton DSEP Minivan and Comparisons with Other Electric Minivans", SAE Technical Paper Series, 910243; 1991.

3. D.A. Erb, L.C. Garcia, and R.H. Leland, "Weber State University - WILDCAT", SAE Technical Paper Collection SP-980; 1994.

4. "1993 Ford HEV Challenge Awards/Results"; June 1993.

5. Advanced DC Motors Inc., Model 000-256-4001 Motor Technical Specifications.

6. Gary Dierof, design engineer, Advanced DC Motors Inc., telephone consultation; October 1993.

7. Solar Car Corporation, "Electric Vehicles and Components" catalog no. 1093; 1993.

8. Motor Vehicle Manufacturers Association of the United States, Inc., "Manufacturers Motor Vehicle Specifications 1992", Manufacturer - Ford Motor Company, Vehicle Line - Escort; Issued August 15, 1990, Revised November 15, 1991.

9. Morse Industrial, Emerson Power Transmission Corp., catalog PT-88; 1988.

10. Saft Nife, "Technical Handbook - Nickel-Cadmium Air Cooled Block Unit Type"; January 1994.

11. L.M. Patrick, "Human Tolerance to Impact - Basis for Safety Design", SAE Technical Paper Series, 650171; 1966.

12. A. Higdon et al., Mechanics of Materials (New York: John Wiley and Sons, 1985) pg. 664.

13. T.W. Birch, Automotive Suspension and Steering Systems (Fort Worth: Saunders College Publishing, 1993) pp. 551-561.

APPENDIX

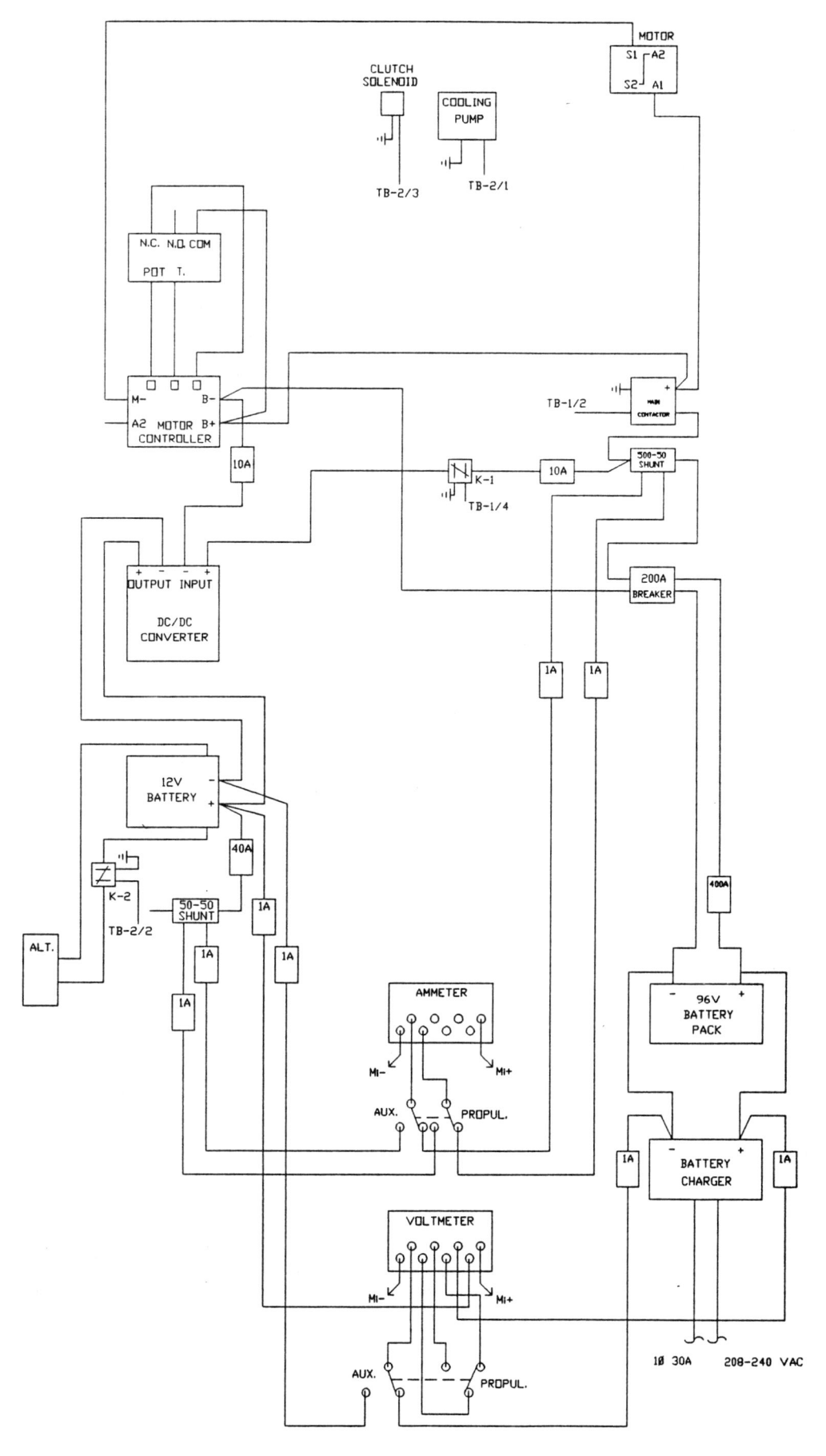

Figure 13 Wiring Diagram

West Virginia University
1994 HEV Challenge Technical Report

The Design and Development of an Efficient Series Hybrid Electric Vehicle at West Virginia University

Wayne Taylor and Chris Atkinson (Faculty Advisor)

ABSTRACT

This paper presents the design and specifications of West Virginia University's (WVU) entry in the 1994 Hybrid Electric Vehicle (HEV) Challenge and details some of the results from the competition. Several major criteria were considered in designing the vehicle and making upgrades and modifications to the 1993 design. Emphasis was placed upon the use of readily procurable components of proven reliability. The vehicle is easily manufacturable in the short term and has a practical, low cost design. Environmental issues played an important role in the design, namely in the choice of batteries, the design of the auxiliary power unit, and the development of the overall vehicle control strategy in all modes of operation.

INTRODUCTION

The Ford Motor Company developed the HEV Challenge to ignite student interest in current environmental issues pertaining to air pollution. Today's automobiles have created a significant air pollution problem in metropolitan areas. Colleges and Universities chosen to compete in the competition were asked to develop an HEV that meets or exceeds the requirements set out in the challenge. Teams had the option of converting a 1992 Ford Escort Station Wagon or building a vehicle from the ground up.

WVU chose to upgrade and refine its design of the 1992 Ford Escort Station Wagon converted for the 1993 competition. The 1993 WVU design earned the Most Efficient Overall Vehicle Award in that year's competition.

The vehicle had an electric motor drive system as a main power source and incorporated an alternator driven by an internal combustion (IC) engine as a supplemental energy source (see Figure 1). This vehicle was capable of operating in zero emissions mode (ZEV), which utilized only battery power, for operation within city limits. Outside city limits, the vehicle used the hybrid mode (HEV) to extend its range.

POWER TRAIN CONFIGURATION

Due to its inherently efficient design, the conceptual design for 1994 changed very little from 1993, although many refinements and improvements were implemented. This design utilized a series hybrid configuration. A diagram detailing all power connections of the vehicle is shown in Figure 2. While a true series hybrid has the auxiliary power unit (APU), batteries, and the motor connected as a "linear" string of components, this hybrid series configuration had the APU directly connected through a controller to both the batteries and the electric motor. This differed from the typical series as well as from a parallel configuration where the IC engine and motor are separately coupled to the transmission, or the wheels.

The disadvantages of a series configuration are considered to be a relatively poor efficiency due to the multiple mechanical-to-electrical-to-mechanical energy conversion procedure and a poor power-to-weight ratio. However, an advantage is the ability to attain maximum efficiency for the IC engine resulting in low emissions. The series configuration also minimizes the number of mechanical connections and hence mechanical losses in the overall drive system.

The IC engine chosen was modified to produce a constant 18 kW output at 3200 rpm. A belt drive was used to couple the engine to a 3 phase alternator, which was then connected to a rectifier to convert the alternating current to direct current. A computer kept the APU output power constant by controlling the power flow to or from the batteries in HEV mode.

BATTERY SELECTION - The main factors taken into consideration in the selection of batteries were: weight, cost, energy storage, environmental impact, safety, and recyclability. The decision was made to replace the Optima 800, the batteries used for the 1993 competition. These were originally chosen for their recyclability, energy storage, and cost. However, these spiral wound gel lead-acid batteries were found to suffer from polarization at the electrodes under conditions of high charge and discharge experienced during typical operation such as that found during the Federal Urban Driving Schedule (FUDS). Thus the decision was made to use a flooded lead-acid battery since a high current charge and the "sloshing" of the electrolyte due to vehicle movement would mitigate electrode polarization. Due to cost vs. energy storage considerations, GNB Stowaway lead-acid batteries were used. These batteries had a high energy storage density which lowered the battery weight contribution, and a quicker recharge rate which increased charging efficiency.

WVU's battery configuration was changed from the 1993 configuration of three parallel strings of eight batteries to a configuration of one series string of ten batteries, weighing a total of 272 kg. This configuration supplied a nominal 120 V to the system.

MOTOR SELECTION - It was estimated that the vehicle would require approximately 5-8 kW of power to operate at a steady 45 mph, depending upon aerodynamic factors, small variations in road grade, rolling resistance and internal mechanical losses. Up to 40 kW would be required to achieve acceptable acceleration for short periods of time. This necessitated an electric traction motor with high transient peak power compared to continuous or long term maximum power. These considerations, along with the general philosophy of producing a highly reliable, readily manufacturable and low cost vehicle, prompted the specification of a series-wound DC motor of 14 kW continuous and 51 kW peak power rating at 120 V DC and approximately 40 kW at 96 V DC. This motor, supplied by Advanced DC, had the advantage of low cost and high reliability at the expense of a slight weight and efficiency penalty compared to an AC induction motor. This motor had a relatively similar speed range (0-3600 rpm) as the original Ford Escort engine (0-5500 rpm), which made the retention of the existing transmission both feasible and practical.

ALTERNATOR SELECTION - The alternator selected for the vehicle was a custom-made Permanent Magnet Synchronous (brushless) Alternator from Fisher Electric Motor Technology, Inc. The alternator was constructed of a rare earth type permanent magnet with a high coercive force which allowed for a high power-to-weight ratio. The alternator had a maximum power output of 18 kW at 3200 rpm, with an efficiency of approximately 91-95%. The alternator was equipped with 106 and 145 volt taps and had a mass of 16.3 kg. The voltage regulation for the alternator, (determined by taking the difference between the no-load voltage and the full load voltage and dividing it by the no-load voltage), was given as 27.5%.

ENGINE SELECTION - Four main criteria were considered while selecting the IC engine for the APU. The team wanted an engine that was efficient, produced low emissions, had a good power to weight ratio and a reasonable cost. After discovering that major improvements could be made to increase the efficiency and performance of the engine used in the 1993 design, it was decided to utilize the same engine for the 1994 design after a full tear down, inspection, and rebuild.

The engine was the Kawasaki FD620D 4-cycle light industrial engine, derived from the V-twin motorcycle engines. This was chosen because of its reliability, ease of modification, and familiarity to the team. Being of all aluminum construction, the engine offered a very good power-to-weight ratio. Following mechanical problems experienced with the methanol fueled engine in 1993, the decision was made to operate the 1994 engine on reformulated gasoline (RFG) at the stock compression ratio of 9.5:1.

TRANSMISSION ADAPTER PLATE - The decision made by last year's team to use the existing 5 speed Ford manual transmission remained unchanged for the 1994 team. The operating range of the transmission and the DC motor were relatively close and worked well. In order to join the two components, an adapter plate was fabricated to make the connection. This was achieved by taking two 25 mm thick aluminum plates and designing one to fit on the transmission and the other the electric motor. After the mating surfaces of the two plates were milled for good contact with each other, they were mounted to their respective components. The electric motor and transmission were then meshed and bolted together.

VEHICLE CONTROLLERS AND CONTROL STRATEGY

CONTROLLER SETUP - Two controllers were used in the WVU HEV (in addition to the engine controller), the Auxiliary Power Controller (APC) and a Motor Controller. The APC controlled power distribution from the APU to the batteries or the motor controller, and the motor controller was used to deliver power to the DC motor.

The final controller setup was a load leveling configuration and allowed for charging of the batteries in hybrid mode if the power required by the motor was less than the power produced by the APU. Extensive testing was conducted to examine this load leveling effect.

AUXILIARY POWER CONTROLLER - The APC (designed and constructed in house at WVU by members of the HEV team) was the heart of the Hybrid Electric Vehicle. It maximized the APU electrical power delivered to the system (batteries or motor controller), while operating the APU at a constant speed. This permitted the IC engine to be optimized for best performance for the lowest possible emissions.

The APU power delivered to the system was governed through the use of pulse modulated control (PMC). Here the DC power supply from the rectifier on the APU was converted into a pulse train by chopping the DC signal. The magnitude of the power delivered to the system was then dependent on the pulse width of the pulse train. As the pulse width increased, the average power delivered to the system increased. The primary purpose of the PMC was to isolate the APU from changes in the load (the power demanded by the DC motor), which change with driving conditions. The APC, constructed with MOSFET power electronics, was rated at 400 amps at 200 volts, which was roughly sufficient for the full power requirements of the DC motor.

The APC was controlled by an on-board computer that monitored system parameters and maintained the APU at a constant speed by manipulating both the APU throttle and the PMC.

ENGINE CONTROLLER - The IC engine used in 1993 was supplied with a carburetor, but was converted to ported fuel injection to improve combustion and therefore reduce emissions. The 1994 engine was fitted with throttle-body fuel injection and used inputs from the manifold air pressure (MAP) sensor and coolant temperature (CLT) sensor to determine the rate of fueling. Additionally, an exhaust gas oxygen (EGO) sensor was fitted and an interface to the stock electronic control unit constructed to ensure that the fueling was held within a very narrow range around stoichiometric for optimum operation of the three-way catalytic converter.

VEHICLE CONTROL STRATEGY

The vehicle operated in three modes, electric only, hybrid mode and APU activated mode. In electric only, or ZEV mode, the power demanded by the driver was measured by the position of the (standard) throttle pedal, and the power transmitted to the electric traction motor was regulated by the Motor Controller. In hybrid or HEV mode, once the batteries dropped below a 70% depth of discharge (DoD), as measured by both a proprietary state of charge meter (Curtis) and by measuring the net energy flow out of the battery pack, the APU was switched on automatically. The engine controller had a starter circuit that started the engine. In the event that the engine failed to start or stalled, the start circuit reset and attempted to restart the engine. However, if three consecutive start attempts failed, an engine fault light alerted the driver to the possibility of mechanical problems and any further attempts to start were prevented.

In HEV mode, the APC automatically directed power to the electric drive motor as demanded, or directed power to recharge the batteries in the event that road conditions, vehicle speed and driver demand did not match or exceed the power available from the APU. Once the batteries recharged to approximately 85% of full charge (as measured by no-load battery voltage and integrated power flow into the batteries), the APU switched off automatically. In this fashion, the APU cut in and out during HEV operation, but typically stayed on for the order of 30 minutes to 1 hour, depending on power demands. In APU mode, the APU started automatically but the driver was warned if the batteries were at an elevated state of charge, in which case they could not operate efficiently as a load absorbing device.

The scheduling strategy, the automatic APU starting, and the overall vehicle control functions were performed by a 80486-based single board computer (CPU) mounted in the rear of the vehicle. The CPU also performed many vehicle and subsystem monitoring functions, including battery state of charge, instantaneous power flow into or out of the batteries, and various engine parameters.

VEHICLE STRUCTURE MODIFICATIONS

Major modifications made to the Escort structure included the construction of a battery box (see Figure 3) and the addition of a roll bar. Minor structures added or changed consisted of a structure to contain and insulate major underhood electrical components, the PMC enclosure, the modification of the upper radiator mount, the transmission adapter plate described above, the electric traction motor mount, and the alternator mount and belt drive system.

BATTERY BOX - A section of the floor behind the front seats was removed to accommodate a heavy duty steel battery box. The existing cross-member and uni-body frame rails remained intact and unmodified, while the floor material was replaced by the battery box. Two hangers, made of 3.2 x 25.4 x 50.6 mm steel C-channels, were attached under the battery box, which ran from the rear cross member to the vehicle floor. The floor of the battery box was constructed of 3.2 mm steel and the walls made from 16 gauge steel, considerably thicker than the original sheet metal removed from the vehicle. The bottom of the box could hold fifteen batteries, and an additional nine could be accommodated on a second level using a 14 gauge aluminum bracketing system. The batteries were secured with nylon straps. The lid, used to contain liquids and fumes in case of an accident, was made from 16 gauge aluminum, and was sealed with neoprene rubber. Additional safety measures included in the box were: blow-out discs to vent the box in the event of an extreme pressure build up, rubber sheets to electrically insulate the box, and a forced air circulation system to minimize hazards during charging.

ROLL BAR - Occupant roll-over protection was achieved with an SCCA approved roll bar, manufactured by Autopower Industries. It consisted of a single hoop with aft supports and cross braces. The hoop mounted directly to the chassis rails behind the front seats, while the rear supports dropped down from the top of the hoop at an angle of 55 degrees from vertical and mounted to the rear strut towers. Cross braces designed specifically for the application were placed to avoid interference with the battery box and to provide mounting points for the five point harnesses. The design exceeded specifications set out in the HEV regulations.

Finite element modeling indicated that the addition of the battery box and roll bar had significantly increased structural rigidity and strength of the HEV over the original Escort structure.

PMC ENCLOSURE - The PMC's and other electrical equipment were located in a purpose-built enclosure under the hood of the vehicle. The structure was built considering safety, aesthetics, and access to its contents. The PMC enclosure base was constructed using a Lexan sheet to provide mounting points and electrical insulation to the PMCs, contactor, and other electrical components. The remainder of the enclosure was made of light gauge aluminum and served to prevent accidental human contact with the components and provided air ducts for cooling.

UPPER RADIATOR MOUNT - The upper radiator mount was made removable to provide easy access to the APU and engine bay. The support was then attached to the sides of the engine bay with steel brackets. The center support member was remade from aluminum. The new piece performed the duty of the original and provided additional mounting points for the new radiator.

EMISSIONS CONTROL STRATEGIES

EXHAUST SYSTEM - The exhaust system was designed to adhere to the environmental aspect of the competition regulations. The system was completely leak free and heat shielded for passenger safety. The exhaust system was rerouted to accommodate the addition of a battery box, fuel tank, and the reorientation of the IC engine. A 38 mm diameter alloy pipe was used to allow for easy flexibility and all joints were welded to insure full containment of exhaust gas.

CATALYST - A close coupled 3-way catalyst was used to reduce exhaust emissions. The catalyst was located within 300 mm of the exhaust manifold to take advantage of reaction inducing heat from the engine. To further increase the efficiency, an insulating fiberglass blanket was wrapped around the catalyst to ensure fast light-off on startup.

EVAPORATIVE EMISSIONS - The total emissions system in the production 1992 Escort consisted of two separate systems: a rearward system which tapped vapor not yet in the fuel lines, and a forward system which controlled vapors in the engine compartment.

The rearward system consisted of piping that ran from the fuel tank to a vapor collection tube located next to the fuel filter. Vapor left the fuel tank through its own separate line and was piped into the vapor tube where it condensed back to liquid, returning through a second line to the fuel tank.

The forward system incorporated a carbon canister and piping that was positioned between the fuel filter and the intake manifold. A solenoid purge valve in the line directed the flow of the vapor from the incoming fuel line to the carbon canister, where the vapor was filtered. When the car was started, a second solenoid opened on the outlet side of the carbon canister, allowing the vaporized fuel to be sucked into the intake manifold. This volatilized fuel was then burned.

The standard evaporative emissions system was retained intact for the 1994 converted vehicle, changed only by the addition of the solenoid purge valve. The vehicle thus had a fully sealed fuel system with an evaporative emissions system.

FUEL SYSTEM

The RFG fuel system consisted of eight components essential to the engine's operation. The components of the system were the fuel tank, fuel pump, fuel filter, stainless steel fuel rail, fuel injector, intake manifold, pressure regulator, and Teflon tubing fuel lines with stainless steel braiding.

FUEL CHOICE - The team chose to use RFG due to its designation as an alternative fuel and in anticipation of its future widespread use. In addition, RFG does not require engine modifications for normal operation.

FUEL TANK - The tank had to be large enough to carry enough fuel to meet the HEV Competition criteria. The HEV criteria required that the vehicle travel approximately four hours in hybrid mode. The amount of fuel required to travel this time with a calculated maximum injector flow rate of 1.55 g/s was 23.32 kg of fuel. The amount of fuel corresponding to this was approximately 29.76 liters.

The fuel tank used in the 1993 design was reused in the 1994 design since it still met all requirements and was reliable and already in place. The fuel tank used was the Aerotec Labs (ATL) Sports Cell. This tank has a volume of 45.5 liters, and dimensions of 514 x 448 x 240 mm. Structural modifications to the vehicle remained the same for the 1994 design and the fuel cell remained in what was the spare tire well of the original vehicle.

FUEL INJECTOR - The Kawasaki engine employed a single fuel injector in a throttle body. The injector, manufactured by Nippondenso, was adequate to deliver the 1.55 g/s fuel flowrate required to produce 15 kW of mechanical power based on a mean engine thermal efficiency of 22%.

FUEL COMPONENTS - The fuel pump was relocated to the top of the fuel cell and inside the fuel cell cover. This change of location from underneath the floor panel below the driver's seat to its present position was for safety and practical reasons. The inertial fuel pump shut-off switch was also re-installed for safety reasons.

FUEL RAIL - The fuel rail acted as a fuel reservoir for the injector. The fuel rail was 152 mm long with a 16 mm inner diameter. The injector mount was machined out of the same 16 mm stainless steel and welded to the custom intake manifold that was fabricated for the project in the 1993 design. The fuel lines were made of Teflon, and had a stainless steel shield on the outside. The lines were replaced and rerouted through the center tunnel of the vehicle's structure to decrease the chances of damage from obstacles in the road and other possible hazards. The lines running from the fuel cell were changed from an inner diameter of 11 mm to 9.5 mm, and the lines running from the fuel rail to the injectors were altered from an inner diameter of 9.5 mm to 12.7 mm.

COOLING SYSTEM - The cooling system for the IC engine consisted of the original radiator purchased as part of the unmodified 620 cc Kawasaki 620D engine, and a small DC electric fan. The fan was rated to cool up to a 2 liter engine and had a 262 mm diameter. The fan was purchased from Dunham Bush of Ontario, Canada.

SUSPENSION

The suspension system on the original vehicle was a McPherson coil spring over strut arrangement and was used on all four corners of the vehicle. Since changes to the suspension had to be documented for safety requirements, design and analysis efforts focused on redesigning the coil over strut system to support the added weight.

The original Escort model had a curb weight of 1094 kg with a weight distribution of 59/41 front to rear. The coil over strut arrangement had a spring constant of 14.9 kN/m, with a spring length of 373 mm. The conversions to HEV made for the 1993 design caused the weight of the vehicle to rise to 1730 kg, with a 50/50 weight distribution. Refinements for the 1994 design brought the weight to approximately 1600 kg with the same weight distribution. Changes made to the chassis of the car lowered the lowest point on the vehicle by an additional 51 mm.

These modifications caused the ride height of the rear of the vehicle to drop considerably to approximately 76 mm. To remedy this problem, a redesign of the rear springs was undertaken.

A simple change of the spring constant to compensate for the additional weight would have made the constant too high, creating an unacceptable ride. The solution was to increase both the spring constant and the physical dimensions of the spring. An aftermarket spring available for the Escort with a spring constant of 20.6 kN/m allowed for the additional weight, but drastically reduced the ride quality. A spring of the standard length was found to be insufficient to correct the ride height deficiency.

A spring was then designed using the 20.6 kN/m spring as a model. The result was a spring that had the same number of turns, wire thickness, and outer diameter as the original, but with a different pitch. By adding 10 mm to the pitch, the height capabilities of the spring were enhanced, which allowed height rules to be satisfied as well as provided an acceptable ride. The new spring had a spring constant of 28.9 kN/m with an overall length of 432 mm, and passed all applicable buckling tests adequately.

The loading on the front suspension was not changed radically. Therefore, there were no modifications made to this system.

BRAKES - The increase of the vehicle's curb weight from the unmodified 1090 kg to approximately 1600 kg prompted the decision to upgrade the performance of the brake system. The rear drum brakes were removed and replaced with disk brakes and the front rotors were exchanged for larger size rotors from a 1992 Ford Escort GT. Components purchased to improve the braking performance and resist locking included spindles, calipers, rotors, brake pads, and the proportioning valve from an Escort GT.

The rear braking system consisted of new brake lines, spindles, rotors, and calipers with a modification of the existing parking brake. The rotor outer working diameter was 251 mm and the inner working diameter was 184 mm. The addition of a flexible brake line to conform to the changes was necessary.

The front braking system changes included the replacement of spindles, rotors, and calipers. The outer working diameter of the vented cast iron rotors was increased from 235 mm to 251 mm with the inner working diameter remaining constant at 144 mm. The vehicle stopping distances were radically reduced by these upgrades.

TIRES & RIMS - The upgrade of the brakes necessitated the use of larger rims. To maintain the proper ride height and rolling resistance, lower profile tires were also necessary. The 1993 design used the stock 13 inch rims and Goodyear 175/70R13 tires. The upgrade was made to aluminum BBS 6.5 x 14 inch rims and Goodyear Invicta GL 175/65R14 tires. These changes increased the ride height by only 3.8 mm, an acceptable tolerance.

VEHICLE MANUFACTURABILITY

One of WVU's original goals for the 1993 vehicle was to design a vehicle that could go into production within a short period of time. This goal was maintained for the 1994 vehicle by using the 1993 design as a basis upon which upgrades and improvements could be made. The main components of the car removed during the 1993 modification were the engine, the rear seat, and the floor under the rear seat. These components were replaced by a DC motor, smaller IC engine, alternator (replacing the engine), and a battery box (replacing the rear seat and floor). The braking system was upgraded to that of an Escort GT. The rear suspension system was upgraded but the original coil over strut system was retained. The transmission and steering were kept largely intact and used in the 1994 upgrade.

Another major point, and largely the design philosophy of the project, was to design the vehicle for short term production using relatively low cost components that are proven reliable and easily obtainable. The major portion of the original conversion cost was due to a few components, including the IC engine, the DC motor, the power flow controllers, the alternator, and the battery pack. These components were available at a relatively fixed price with the exception of the battery pack.

In the case of the batteries, the price varied depending on the type of battery chosen. There were many factors involved in choosing the batteries, but the one factor that swayed the overall decision to use lead acid batteries was their recyclability. Since protecting the environment was the overall issue in the competition, it was beneficial to use a battery that could be readily recycled.

INSTRUMENTATION

Many instrumentation changes were made to the Escort when it was converted to a HEV for the 1993 competition. Since these changes proved to offer good monitoring of the systems, only minor modifications were made to incorporate these existing changes into the 1994 design. Instrument changes to the Escort to monitor the vehicle in all modes of operation were performed.

The dash instrument cluster was upgraded to the Escort GT specification and other gauges were added to the center console to monitor power flow and other electrical parameters. Additional parameters were also monitored by the CPU. The speedometer retained the original connection to the transmission. The fuel gauge displayed the RFG fuel level. The analog temperature gauge monitored the IC engine coolant temperature, and the tachometer read the electric motor speed using a diode-photo transistor setup.

Several other gauges were displayed on the middle console such as a voltmeter which measured the voltage across the battery pack. An ammeter displayed current going into the batteries over time, and a state of charge meter displayed power integrated over time entering or leaving the batteries.

Other system temperatures monitored by the CPU included three thermocouples on the batteries, one on each of the PMC and motor controller, one on the alternator, and another on the IC engine. By utilizing and expanding the preexisting instrumentation, the driver and passenger were given a complete overview of the status of the HEV subsystems.

CONCLUSION

The design philosophy behind the 1994 WVU HEV was to utilize relatively low cost, easily procurable components of proven reliability. The main goal of the 1994 WVU HEV team was to optimize the 1993 design by making any changes and refinements necessary, and spending more time for testing. The relocation of several major systems components greatly improved packaging and enhanced under-hood aesthetics. This vehicle also showcased the use of advanced MOSFET power electronics for vehicle control, and had a progressive computer control strategy.

Once the results of the 1994 competition were released, several lessons were learned concerning the performance of the design. During the Federal Test Procedure (FTP) emissions testing at Ford Motor Co.'s Certification Laboratories in June, 1994, WVU's Escort achieved the results shown in Table 1. The HEV emissions test resulted in a small net positive state of charge increase,

Table 1: Emissions Standards and Results

Emissions g/mile	U.S. Tier I Standards	WVU Results
HC	0.41	0.046
NMHC	-	0.043
CO	3.4	0.305
NO_X	1.0	2.476

as well as an indicated 44 mpg fuel consumption. This consumption over the 17.75 km of the FUDS test indicated an apparent overall thermal/mechanical efficiency of 22%. In other words, of the energy provided by the combustion of the fuel, 22% was recovered at the wheels of the vehicle.

The close coupled catalytic converter configuration significantly reduced levels of hydrocarbons and carbon monoxide but levels of oxides of nitrogen exceeded federal standards, mainly due to the heat retentive insulation used to wrap the converter. Apart from the problems associated with the overheating of the catalyst, the engine mounts failed repeatedly throughout the competition, resulting in the premature withdrawal of the vehicle from the Range Event. Future engineering plans involve the design and construction of a significantly more robust integrated engine, alternator and traction motor mount. Furthermore, despite apprehensions about the series configuration, this implementation proved to be fairly efficient.

WVU's Ford Escort Station Wagon met all specifications and requirements for the 1994 HEV Challenge. Being a converted vehicle (as opposed to a ground up vehicle), the car can be easily manufactured, and was just as structurally sound and reliable as the original Escort. The series hybrid arrangement of this vehicle was designed specifically to produce minimum engine emissions during HEV operation, when it uses an alternative fuel (RFG). Because the vehicle produced low emissions and used recyclable batteries, it was environmentally safe, a major requirement that will pave the way for the acceptance of future hybrid electric vehicles.

ACKNOWLEDGMENTS

The West Virginia University Hybrid Electric Vehicle Team would like to thank our chief sponsor, the Monongahela Power Company, a division of the Allegheny Power System, for their continued support throughout the 1993 and 1994 competitions. We are also grateful to the Kawasaki Motors Corporation for the donation of an engine, and to BBS of America for the donation of wheel rims.

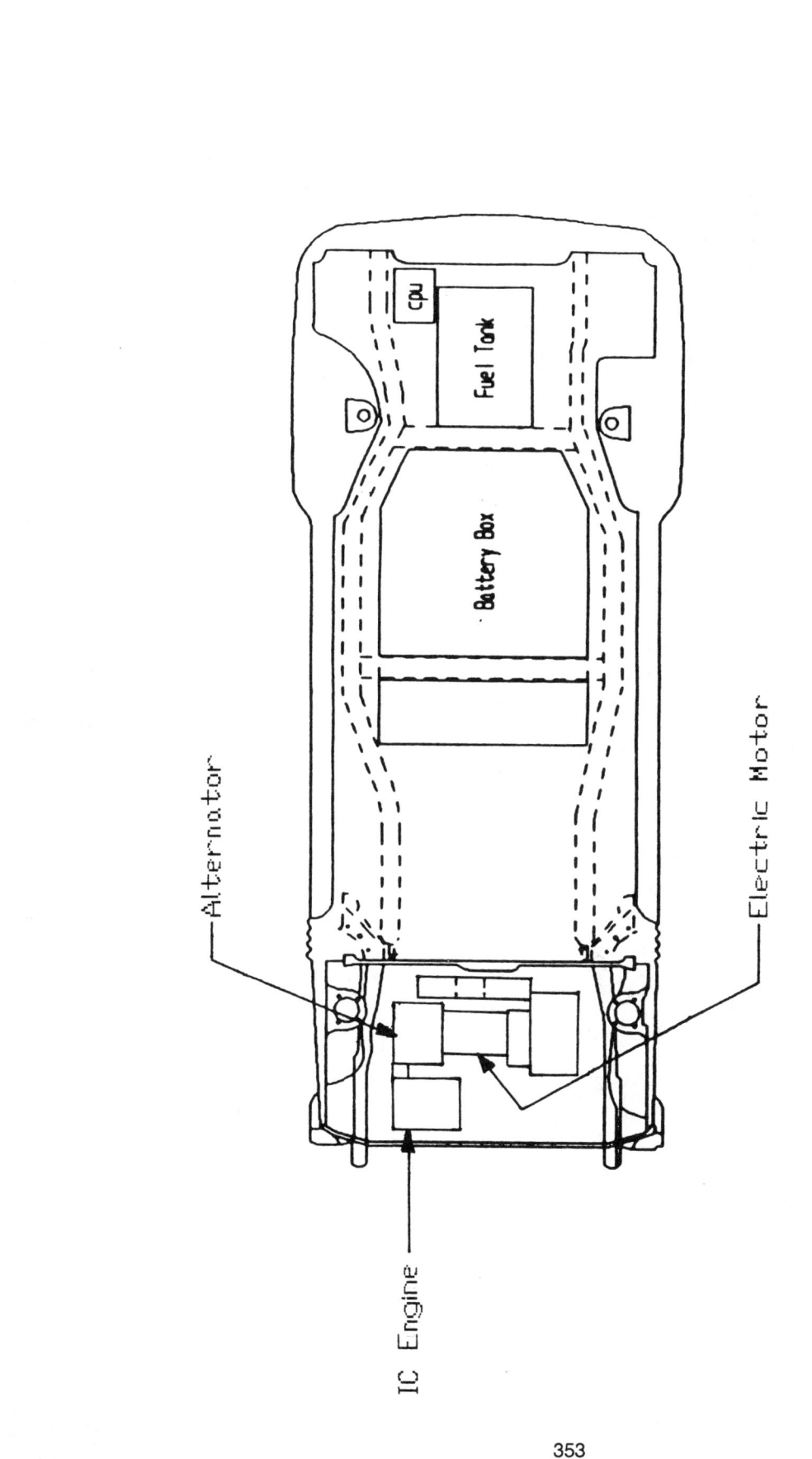

FIGURE 1: CONVERTED FORD ESCORT WAGON — TEAM #28
WEST VIRGINIA UNIVERSITY
ELECTRICAL & COMPUTER AND MECHANICAL & AEROSPACE ENGINEERING

FIGURE 2: POWER TRAIN CONFIGURATION — WEST VIRGINIA UNIVERSITY TEAM #28 — ELECTRICAL & COMPUTER AND MECHANICAL & AEROSPACE ENGINEERING

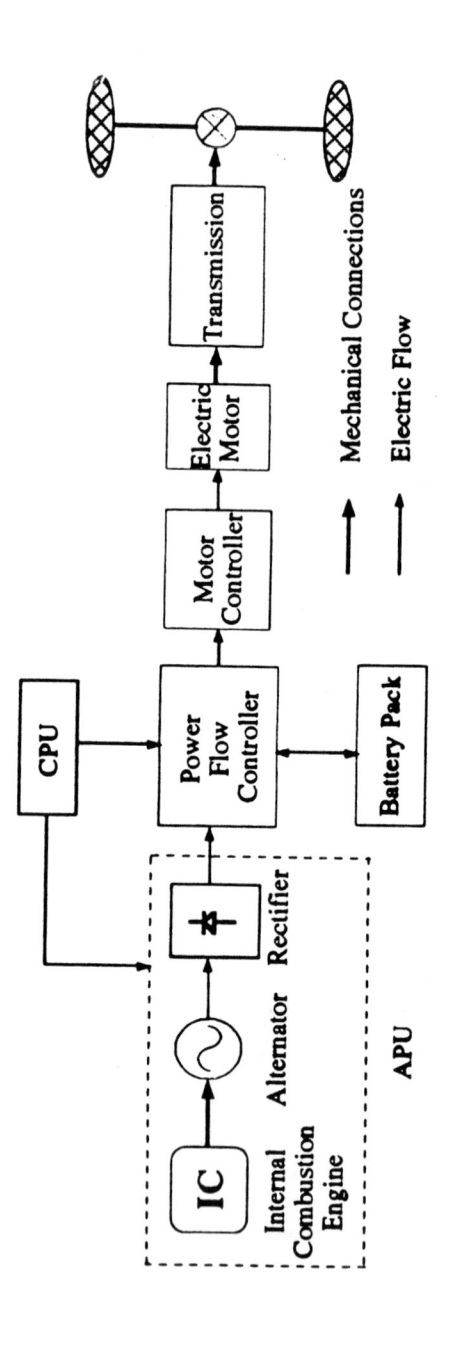

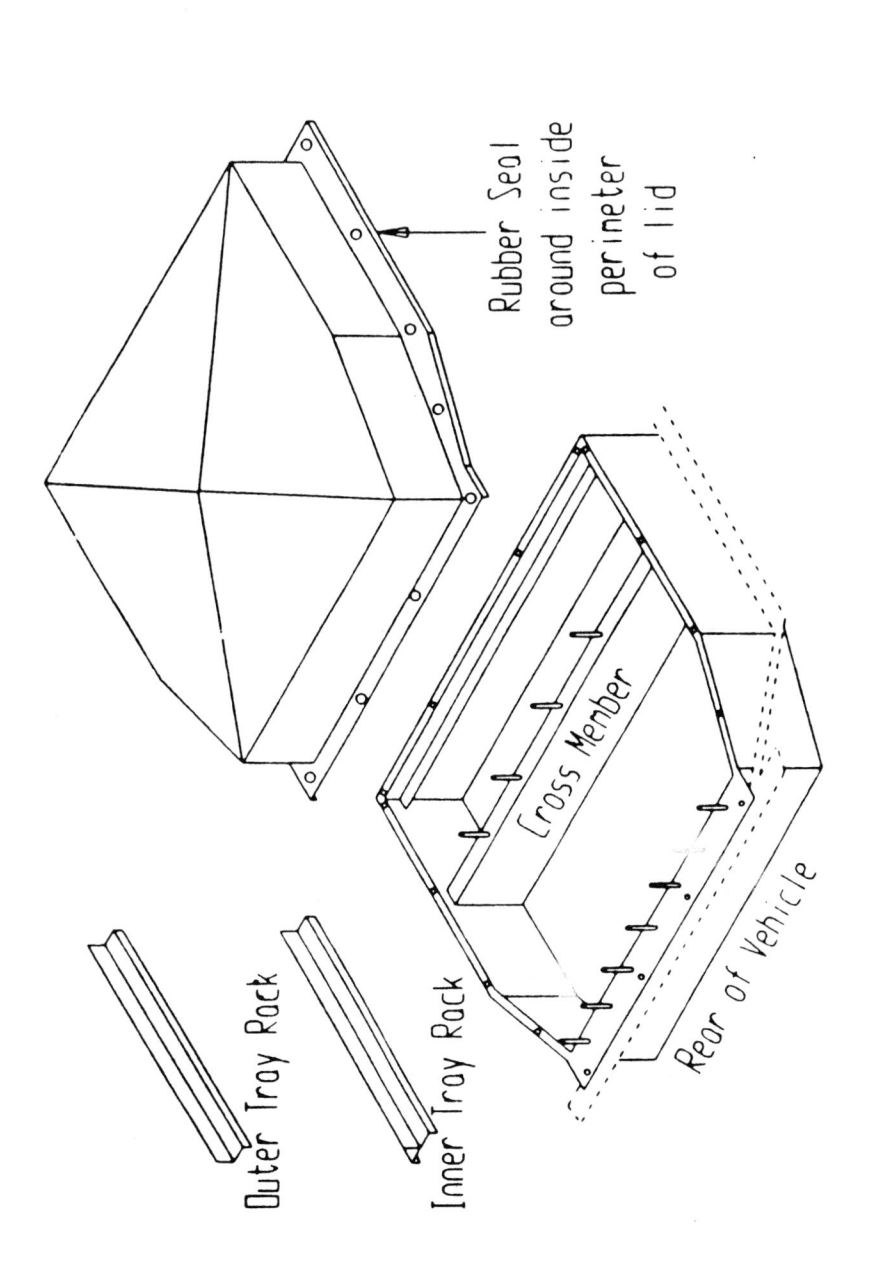

FIGURE 3: BATTERY BOX — TEAM #28

The University of Wisconsin - Madison Paradigm Hybrid Electric Vehicle

Patrick Barber, David Bell, Richard Bonomo, Scott Costello, Barton Heldke, Nathan Hendon, Clark Hochgraf, Robert Rossi, and C. Thomas Wiesen

University of Wisconsin - Madison

ABSTRACT

The University of Wisconsin's entry into the 1994 Hybrid Electric Vehicle Challenge is a stock-production, 5-speed Ford Escort Wagon modified to operate as a hybrid electric vehicle. The Escort's original powertrain has been removed and replaced with a series-configuration hybrid electric drivetrain consisting of a conventional air cooled 50 HP 220V 3-phase AC induction motor and a 20 HP 4-stroke, 2-cylinder, electronic fuel-injected engine. This paper discusses the vehicle's powertrain configuration, control strategy, design modifications, efficiency, materials selection, and vehicle manufacturability.

INTRODUCTION

Personal high speed transportation has grown deep roots in society. Although today's pure gasoline or diesel engine vehicles suffer from excessive tailpipe emissions, they dominate the ever growing vehicle market. As communities become aware of the direct health impacts of tailpipe pollutants, legislation is being created to force the development of zero emission vehicles. The problems with present day zero emission vehicles include inferior driving range and refueling inconvenience. The Hybrid Electric Vehicle (HEV) offers the same range as a conventional vehicle with the zero-emissions of a pure electric vehicle.

In its simplest form, the Hybrid Electric Vehicle reduces the total emissions per mile by allowing for a number of zero emission miles to be driven. Beyond this, the dual power sources in a HEV can be strategically controlled to further minimize emissions when the engine is operating. This control strategy allows for the use of a engine of lower HP than that of a conventional car. The HEV project team at the University of Wisconsin, Madison (Team Paradigm) designed and built a series HEV, named "The Paradigm", to demonstrate the benefits and the implications of a hybrid drivetrain in a conventional vehicle.

Team Paradigm recognizes that in the process of building a product, the driving force behind the design is the needs and wants of the customer. Team Paradigm defined the customers as the HEV Challenge organizers and the typical automobile consumer. The HEV Challenge organizers defined their requirements by way of the official HEV rules. Of 1,000 points available in the challenge, 450 points are strongly related to emissions and economy, 300 points for performance, and the remaining 250 points all allocated for the engineering design of the vehicle. These rules were analyzed using a weighted multi-attribute utility model to define the design criteria and provide a guide for choosing component and configuration options. An initial model was used to assist with the 1993 vehicle design. A new model, based on the 1994 HEV Challenge rules, was used to determine what changes were necessary.

Along with the multi-attribute utility model, some component and configuration decisions were made based on the results of computer simulations. Various simulated vehicles with different components and configurations were pitted against each other in computer simulations of the challenge's dynamic events. The results were in the form of challenge points and were used as a guide for selecting components and system configurations.

POWERTRAIN CONFIGURATION

HYBRID ELECTRIC DRIVETRAIN CONFIGURATION OPTIONS - Hybrid electric vehicle (HEV) drivetrain configurations may be generally classified by how power is transmitted to the drive wheels and how the power is divided between the two power sources.

A HEV may be classified as either "parallel" or "series" depending on how power is transmitted to the drive wheels. In a "series" drivetrain, an electric motor is directly connected to the transmission. The alternate power unit (APU), usually in the form of an internal combustion engine, drives a generator, which in turn charges the vehicle's battery or supplies energy directly to the electric drive system. In this configuration the engine is not directly connected to the transmission. A

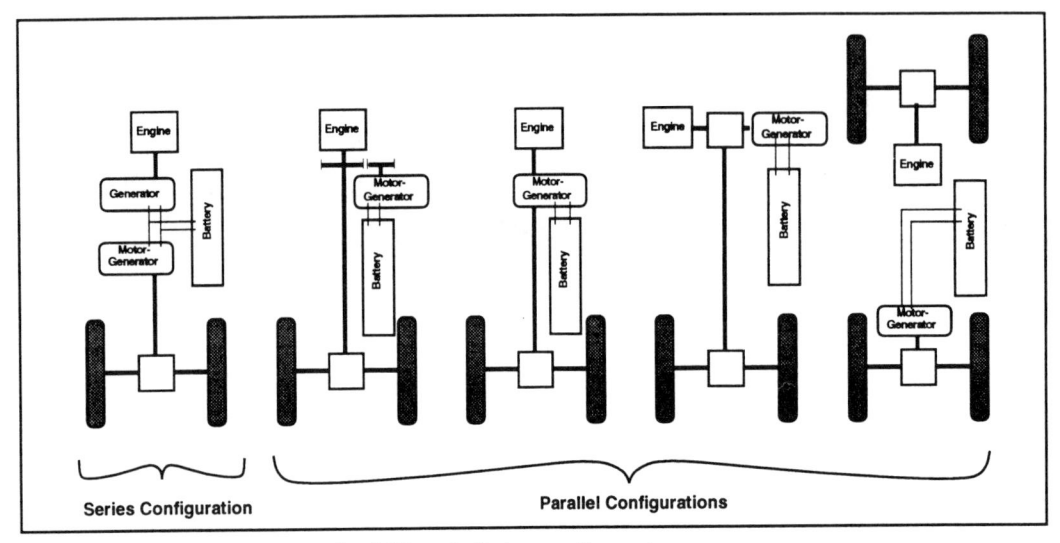

Figure 1. Various types of Hybrid Electric Drive configurations.

"parallel" drivetrain consists of both an electric motor and an APU each having a direct mechanical link to the vehicle's drivetrain. A "parallel" drivetrain may be integrated to supply power to one pair of wheels or may have electric drive on one pair of wheels and APU drive on the other pair.

A HEV is a dual power source vehicle somewhere between a conventional internal combustion engine and a battery powered pure electric vehicle. There are three major classifications of HEVs with respect to the power division between the two power sources. A "power assist" HEV uses an engine that is smaller than a conventional automobile's because it provides only enough power for sustained steady state operation. A small battery serves to provide the extra power needed during acceleration and for traveling short distances. A "range extender" has a large battery for a significant electric range and a small APU that serves to recharge the battery, but may not have enough power for sustained high speed operation on APU power alone. A "dual mode" HEV can operate like a "power assist" having enough APU power for sustained high speed operation, as well as a large battery for sustained pure electric range.

TEAM PARADIGM'S DRIVETRAIN SELECTION - The configuration selected for the Paradigm was a series dual mode. The configurations described above were evaluated based on the configuration's estimated ability to contribute points to the team's score in each event of the HEV Challenge, consumer and environmental practicality issues.

Estimated point contributions were arrived at by assessing the impact of the relevant factors (e.g. total acceleration, lateral stability, etc.) for each event on the possible score. Each factor was broken down into measurable attributes (peak engine power, drive motor torque, electrical efficiency, etc.), and each attribute was subjectively assigned a relative weight based on how the attribute contributed points to each event. The weight was then multiplied by the points for the event to arrive at a point score for each event. The points were summed up across all of the events to arrive at a total score for each attribute. The attributes with high point totals were given the highest consideration when comparing configuration and component alternatives.

Well-designed parallel configuration drivetrains offer improved APU system efficiency over their series counterparts because they lack extra losses present in the electric conversion process. However, the high complexity involved in maintaining a constant engine power output in a parallel configuration made it less desirable than a series design. Maintaining constant engine power output provides improved APU efficiency and lower emissions.

Consumer and environmental practicality also influenced the configuration decision. Parallel systems are additive when using both power sources. This means that when using energy from just the APU or just the battery, power available to the driver drops considerably. A series configurations performance is reduced only slightly enhanced when the APU is turned off as the APU only serves to minimize voltage droop under heavy load. There will be a slightly larger performance loss when the battery has a very low charge. A potential scenarios for future diving would be that drivers would have to shift from Hybrid mode to pure electric mode when entering high pollution areas. A parallel configured vehicle would have a significant performance drop when entering these areas where as shifting operating modes in a series vehicle would appear virtually seamless to the driver in terms of performance.

ELECTRIC DRIVETRAIN OPERATING VOLTAGES - Team Paradigm's choice of a three-phase 220 VAC electric drive system was driven by the need to maximize the efficiency and reduce the weight of the electric portion of the powertrain. A 220 VAC system requires 330 VDC from the battery which was a difficult requirement to satisfy. The high voltage provides better performance and efficiency than the 100 VDC motors used in most electric vehicles on the road today. There is a tension in EV design between the drive system designers who are attracted to the lower operating currents and the higher motor speed of high-voltage systems, and the battery designers who are attracted to the lower cell number requirements of low-voltage systems. In the case of Paradigm, the battery voltage specification was driven entirely by the requirements and preferences of the drive system designers.

POWERTRAIN BATTERY SYSTEM - As with most electric vehicle projects, the main obstacles constraining Team Paradigm's efforts to build a high-performance electric vehicle were the weight, volume, and storage capacity limitations common to almost all commercially-available battery systems. Two batteries were chosen for use in the 1993 Paradigm HEV. The first was desirable but immature molten-electrolyte lithium technology. The second battery, a less intimidating but more fully-developed starved-electrolyte lead/acid system served as a backup option to cover the high risk involved in obtaining the lithium battery. The range of practical battery choices available to us was significantly reduced by our selection of a 220VAC motor, and thus the requirement of a 330 VDC battery bus voltage; most batteries designed for use in electric vehicles assume a 100 VDC platform, and are intractably large for use in a high-voltage system. Technical and logistic constraints forced us to use the "second," i.e. lead/acid battery in the 1993 Challenge. Budgetary considerations have caused us to continue to use this battery in the 1994 Challenge.

Lead/Acid Battery System - Despite their low energy density, the lead/acid battery is the most widely used battery in electric vehicles today, because it is the most widely available and least expensive to obtain. Since they are in practice modular, and available in a wide range of voltages and sizes, lead/acid batteries can be connected in series or parallel. This provides a wide range of voltage and capacity combinations, while still displaying versatility for placement into a vehicle. It was this versatility and low cost that persuaded the Paradigm team to select a lead/acid system as the "contingency" battery for the 1993 HEV Challenge, and to continue to use it in the 1994 Challenge.

Team Paradigm's backup battery consists of 31 12-Volt UPS 12-95 modules donated by Johnson Controls, Inc. of Milwaukee, WI. The total battery weight is approximately 920 pounds. The battery has an open circuit voltage of 394 volts DC when fully charged and "relaxed" (2.12 volts nominal per cell) and a rated energy storage capacity of 9.8 kWh at a three-hour discharge rate. Johnson Control's in-lab testing results indicate that a UPS 12-95 module is capable of 300 cycles to 80% depth of discharge at a five-hour discharge rate.

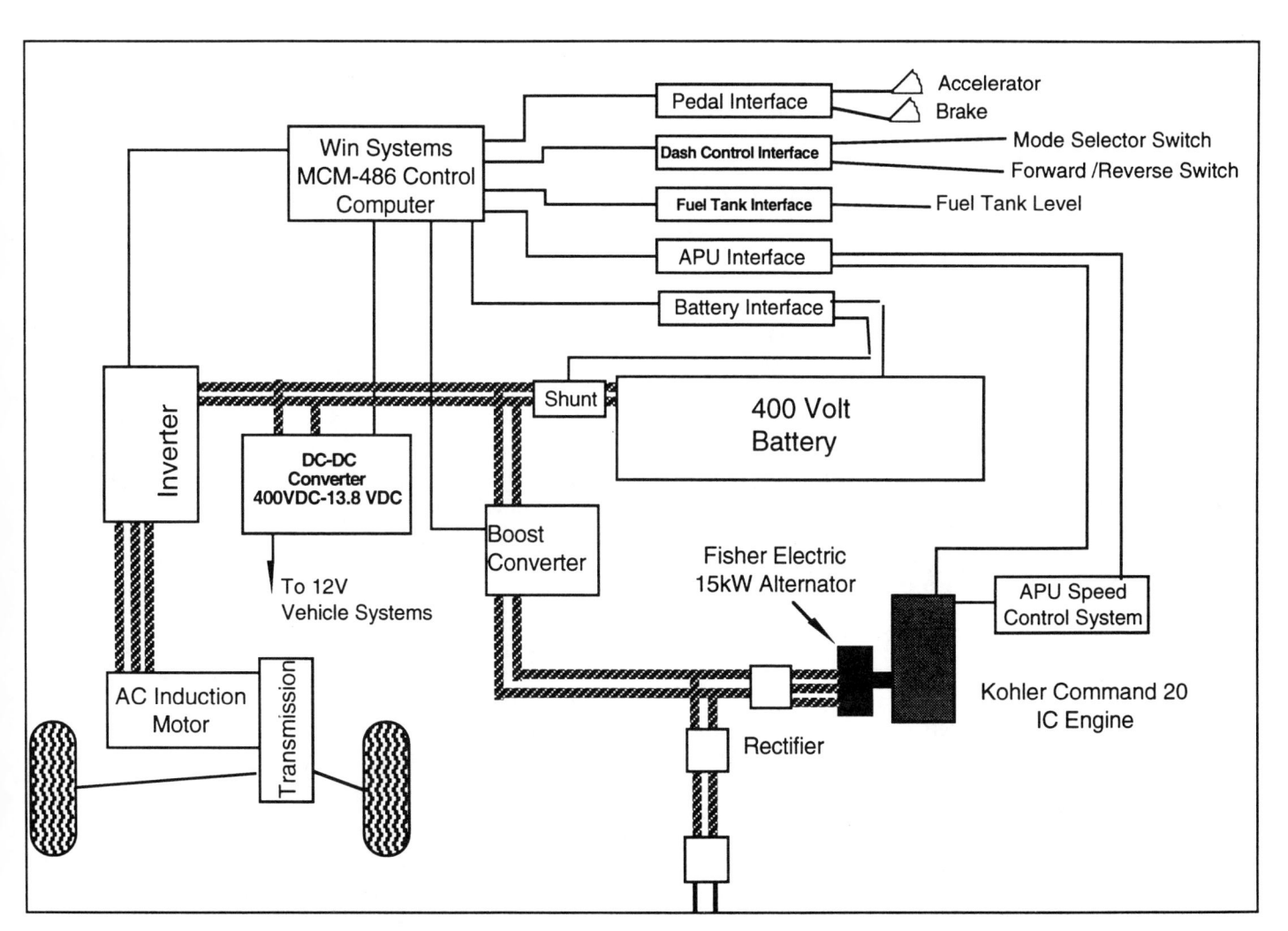

Figure 2. The Team Paradigm series drivetrain schematic.

The 31 modules are series connected, and a Gould-Shawmut 500 A DC-rated fuse is included in the module string to be the "last resort" protection against short circuits.

THE ENGINE - On any vehicle, the average power demand (or energy consumption) of the drivetrain is considerably lower than the peak power demand. Peak power is required only at maximum acceleration. At cruising speeds the drag force is mainly due to aerodynamic drag. The average power consumption of a moving vehicle may be approximated by:

$$P = v \left(C_d \cdot A_f \cdot \frac{1}{2} \rho v^2 + W \cdot C_r \right)$$

where v = vehicle velocity (ft/sec)
C_d = vehicle body drag coefficient
A_f = projected frontal area of the vehicle
r = the density of air
W = the vehicle weight
C_r = coefficient of rolling resistance

Only 14 HP is required to maintain a 50 MPH cruise speed. Although far greater power is required to provide adequate acceleration, the *average* energy drain on the powertrain at cruising speed will generally be less than 16 HP.

The engine in Paradigm's series-configuration drivetrain is operated at a power level just above that required to maintain a steady state cruising speed. The vehicle draws the power needed for acceleration from the main battery even when the engine is running and set up to charge the battery. On a time-averaged basis, the engine power can charge the battery through the alternator, if desired.

The Paradigm H*EV*'s engine is a Kohler Command 20, manufactured by the Kohler Corporation of Kohler, WI. This overhead-valve, 624cc V-twin engine is typically used in garden tractors and engine generator sets. It was selected over comparable engines from Briggs & Stratton, Kawasaki, and others on the basis of the following:

- the engine's overhead valve, compact V-twin design
- the availability of technical support from Kohler
- the availability of financial support from Kohler
- Kohler's Wisconsin-based location and workforce
- Kohler's UW alumni contacts

The Command 20 has been modified by the installation of a sequential-port electronic fuel injection system. The fuel injection system is a modified version of the 1300 series EFI system manufactured by Pacer Industries of St. Louis, MO. This software-controlled system allows maximization or engine performance and efficiency through changing fuel injection and spark advance maps. Multiple fuel types may be used with this system.

THE ALTERNATOR - A 15 kW, 310 V three-phase AC alternator, manufactured by Fisher Electric Motor Technology Inc. of St. Petersburg, FL, is attached to the output shaft of the engine to convert its mechanical output into electrical energy. The alternator has a permanent magnet design and provides a peak efficiency of 94%.

The alternator mounts directly to the engine; the stator mounts to the engine case, and the rotor is mounted directly to the output shaft of the engine. Since under-hood space is at a premium, rare earth (neodymium) magnets are used to reduce the alternator size and weight. The entire engine-alternator unit is located under the hood of the Paradigm HEV, in front of the electric motor and transmission. This location allows for effective cooling of both the alternator and the engine.

THE POWER ELECTRONICS - Power electronics are used extensively in a HEV to control tractive effort, on-board generation of electricity and battery charging. The Paradigm HEV utilizes two major power electronic control devices: a torque-controlled AC motor drive and a boost converter. Smooth control of the vehicle acceleration and regenerative braking is achieved through the use of a power electronic motor drive. De-coupling of the traction load from the engine /alternator load is accomplished by the Boost converter. By utilizing the ability of these converters to rapidly control power flow, it is possible to increase efficiency and reduce engine emissions.

Paradigm's motor drive system, a torque-controlled inverter, converts the battery system's DC power into the AC power required to run the motor. This drive uses a flux vector torque control scheme, which avoids the jerking phenomenon known as "cogging ." Team Paradigm currently has three drives available for development and experimentation: A modified Indramat RAC 2.2 industrial spindle drive, a Magnetek VCD 703 industrial spindle drive, and an Indramat MVD 1.1 vehicular drive. All interface in a similar manner to our on-board controls.

For smooth natural operation under high accelerations, the precise control of torque is useful. With a torque controlled drive, regenerative braking can be readily implemented up to the torque limits of the drive. High quality torque control is important at low speeds to insure safe operation while parking and maneuvering the vehicle.

The boost converter is a current-regulating power supply, designed and built by the Paradigm team. This device is used to de-couple the alternator load from the transient powertrain demands of the vehicle. De-coupling allows the engine to operate at a continuous load level reducing engine emissions. Fluctuations in battery voltage during regeneration and hard accelerations will not affect the engine load.

The boost converter may also serve as an on board intelligent charger when the car is parked. It can be set up to accept a DC input from the charge port and inject a controlled charging current into the battery.

THE ELECTRIC MOTOR - When used with a one-gear transmission, the motor must be able to work over a wide speed range. In choosing between AC or DC motors, field weakening operation became an important issue. The permanent-magnet DC motor can not be easily field weakened and is also structurally speed limited to around 9000 RPM. The AC machine on the other hand has good controllability, a wide speed range and good efficiency at high speeds. A 2 pole AC induction motor design was chosen over a 4 pole design because it has higher efficiency and wider speed range in the field weakening region. Paradigm's 220 VAC three-phase induction motor is a stock 25 HP AC 3-phase industrial motor manufactured by Lincoln electric, which weighs about 200 lbs. Though rated for only 25 HP, the motor has routinely delivered in excess of 100 HP for short intervals without any signs of damage.

THE TRANSMISSION - The Paradigm HEV utilizes the original Escort transaxle, but it has been rotated 90° such that the bell housing is now located above the driveshafts. This rotation provides more room in the engine compartment for other components.

The first two speeds of the original 5-speed transmission are used. First gear provides high acceleration and performance, with a top speed of 65 MPH. Second gear provides high powertrain efficiency and a higher top speed than available in first gear.

Due to the transmission's new orientation and high speed, an oil distribution system has been added. An external electric gear pump supplies oil to the high speed input shaft, thus supplying adequate gear lubrication.

CONTROLLER STRATEGY

THE NEED FOR A CONTROLLER - A standard automobile utilizes two torque control interfaces: an accelerator and brake pedal. The Paradigm team wanted the same devices to provide torque control in our HEV, such that Paradigm HEV would have the driving feel of a "regular" car. These specifications required special control electronics, due to the intricacies involved in achieving variable torque control of a three-phase induction motor.

WHAT IS USED FOR THE CONTROLLER - The Paradigm HEV uses two powerful computers to control its drivetrain and several other critical systems. The first is a microprocessor integrated into the control electronics of the electric motor drive(s). It provides precision torque control for acceleration and regenerative braking. The second computer is an on-board 50 MHz Intel 80486-based MCM-486DX computer manufactured by WinSystems of Arlington, TX. The MCM-486 is used to monitor on-board systems, inform the driver of the vehicle's status, and provide top-level control of several systems. It oversees and preconditions the driver's brake and accelerator input to the motor drive, and is also the brains behind the engine's on-off strategy, the battery charging strategy, and the battery thermal control system.

THE CONTROL STRATEGY - Paradigm's overall vehicle control system was designed with efficiency, low emissions, practicality, and safety as primary objectives. It was desired that Paradigm HEV drive consistently and predictably, while the control system worked to provide high energy efficiency and minimize engine emissions.

BATTERY CHARGING - Using the boost converter as a charging unit would allow implementation of computer-based current-controlled battery charging. The MCM-486 has been equipped with voltage and current sensors to monitor the battery's condition during routine operation. These same sensors would also allow the

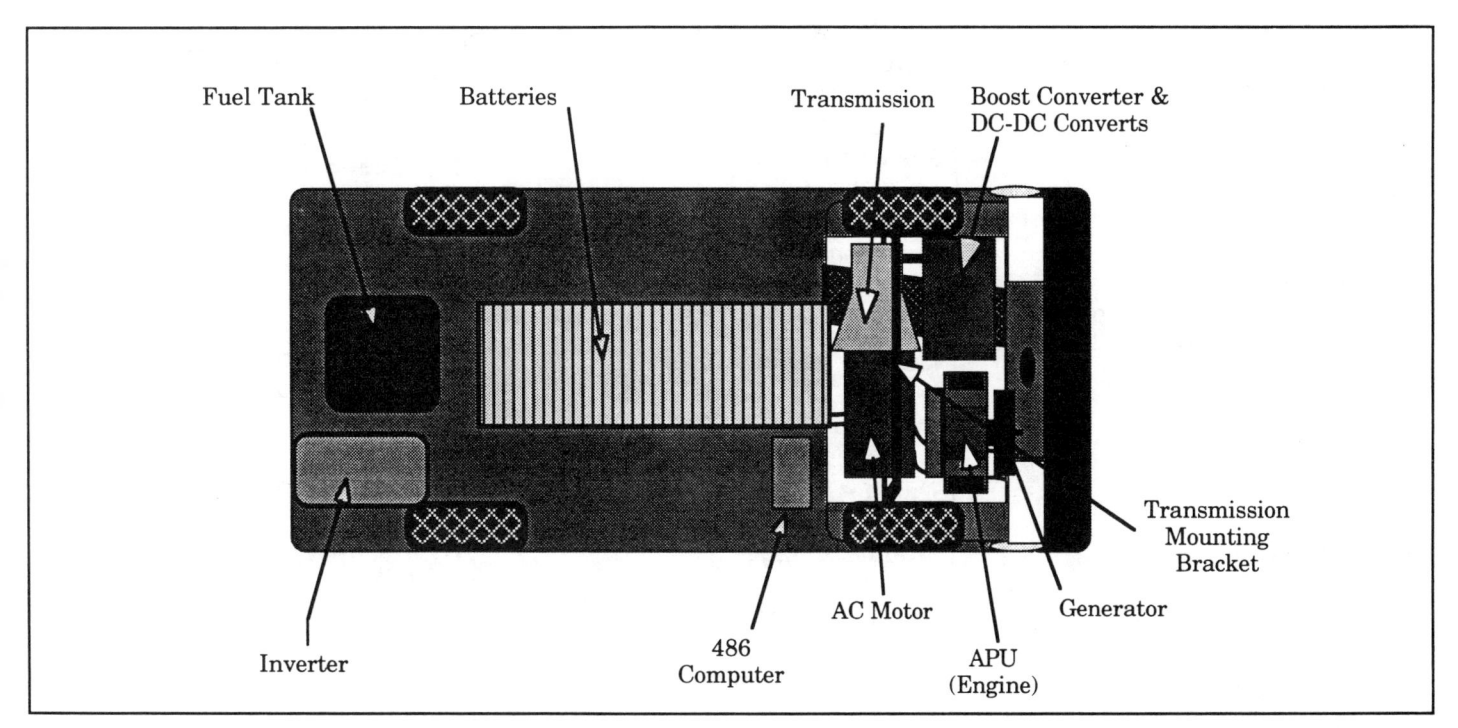

Figure 3. Component layout in the Paradigm HEV Ford Escort Wagon.

Paradigm HEV at a Glance

GENERAL
Frame Type
  1992 Ford Escort Wagon LX
Weight
  3650 Lb.
Tires
  Goodyear Invecta P185 70R13
Chasis Material/Construction
  Steel/Unit construction
Fuel Capacity
  12 gallons
Wheelbase
  98.4"
Track
  56.5"

HEV Stategy
  Series configuration, single drive motor, power from battery or APU, APU - engine / generator set
Braking System
  Electric regeneration braking with hydraulic assist

APU
Engine
Model / Manufacturer
  Kohler Command 20, Kohler Corp., Kohler, WI
Engine Type
  624CC, V-Twin, overhead valve w/ hydaulic lifters
Fuel Type
  Reformulated gasoline
Fuel Delivery
  Sequencial port electronic fuel injection
Power Output
  15 kW (20 hp) @ 3600 rpm
Fuel Consumption
  N/A
Cooling
  Forced air
Weight
  92 Lb.

Generator
Model / Manufacturer
  Fisher Electric Motor Technology, St. Petersburg, FL
Construction
  Permanent magnet AC alternator
  Neodymium permanent magnet
Output Voltage
  310 VAC at 3000 RPM
Power Rating
  15 kW
Weight
  26 Lb.

DRIVELINE
Electric Motor
Model / Manufacturer
  Lincoln Electric
Motor Type
  3-phase, AC induction, 2 pole, low leakage inductance for wide speed range
Speed Rating
  Maximum speed: 7000 rpm
  Base speed: 3650 rpm
Voltage
  220 VAC, 3-phase
Power Rating
  Continuous: 18.6 kW (25 hp)
  Peak: 37.3 kW (50 hp)
Torque Rating
  102 ft.-lb.
Cooling
  Air cooled

Transaxle
Model / Manufacturer
  1992 Ford Escort Transmission
  Manufactured by Mazda
Type
  5-speed manual, modified to accommodate electric motor
Final drive ratio
  3.62:1
Lubrication
  External dry sump with oil cooling

BATTERY
Model / Manufacturer
  UPS 1295, Johnson Controls
Construction
  Lead acid series modules
Number of Modules
  31
Module Voltage
  12
Energy Capacity
  6 kWh
Open Circuit Voltage
  398 VDC

Weight
  970 Lb.
Cycle Endurance
  1000 cycles @ 80% DOD

POWER ELECTRONICS
Model / Manufacturer
  Rexroth Indramat
Construction
  Bipolar transistor
Control Scheme
  Vector flux torque control
Maximum Current Rating
  172 Amperes
Switching Frequency
  1.8 kHz
Power Rating
  55.6 kW
Output Voltage
  220 VAC, 3-phase
Weight
  100 Lb.

CONTROL COMPUTER
Model / Manufacturer
  MCM-486, Win-Systems
Operating System
  QNX 2.4, Quantum Software
Items Controlled
  APU on/off stategy, pedal-inverter interface, driver feedback systems

PERFORMANCE
0-40 MPH
  Less than 8 seconds
Top Speed
  70 MPH
Braking Stategy
  Series regeneration with hydaulic assist. First 2 inches of pedal travel to control regeneration, remaining pedal travel allows for hydaulic braking system (unmodified stock Escort braking system)
Zero Emission (ZEV) Range
  30 miles
Hybrid (HEV) Range
  N/A
Fuel Economy
  29 MPG

MCM-486 to be used as an intelligent charging unit, capable of following almost any charging strategy.

ENGINE SPEED, TORQUE, AND ON-OFF CONTROL - Simple control schemes are used to control the engine's torque, speed, and ignition. Paradigm has three user-selectable operating modes:

- Zero Emission mode (ZEV)
  electric only
- Hybrid Electric mode (HEV)
  electric + engine as needed
- Engine On   (APU ON)
  electric + engine on full-time

In HEV mode, the MCM-486 computer monitors the battery's current and voltage output, and starts the engine when the battery's state of charge drops below a specified level. It turns the engine off again if the battery state approaches full charge. When the engine is running, the MCM-486 analyzes the vehicle's powertrain output and selects an engine power level appropriate to the current driving conditions. The optimum engine torque and speed appropriate to this power level are set by sending a current command to the boost converter and a speed command to a Kohler speed control system on the engine.

In APU ON mode, the engine remains on regardless of battery state of charge. This mode is required for emissions testing for the HEV Challenge. In contrast, ZEV mode keeps the engine shut down regardless of battery state of charge.

USER CONTROL INTERFACES - Dash-mounted switches allow Paradigm's driver to "shift" the vehicle between forward, reverse, and neutral and to select the vehicle's operating strategy (ZEV, HEV, or APU ON). Both of these switches are monitored and acted on by the MCM-486. A large panic button on the dash may be used to simultaneously disconnect the main battery and shut down the APU in an emergency.

ACCELERATOR AND BRAKE INTERFACE - The Escort's original accelerator and brake pedals have been modified to send torque commands to the drive electronics via the MCM-486 with the addition of a linear potentiometer to each pedal. The free travel of the brake pedal was increased, such that the first two inches of brake pedal travel provide pure regenerative braking. After the first two inches, the hydraulic and regenerative brakes work in parallel, and full hydraulic braking power is still available. Routing these signals through the 486 allows for adjustment of the pedal interface gains, as well as implementation of the logic required for reverse "gear", safety overrides, and reverse torque cutoff near zero speed.

DATA LOGGING - An extensive logging system is included with the control software. The purpose of this logging system is to monitor system operation and allow evaluation of performance with the goal of using this information to assist in improving the design. This logging system has provided important data, including battery voltage, current, and vehicle speed. Among the uses being made of the data is an attempt to model the battery performance to provide operation points for efficient use of the APU.

EMISSION CONTROL STRATEGY

TAILPIPE EMISSIONS - The engine's non-transient operation allows for reduced emissions because the fuel injection system may be run under closed loop air-fuel mixture control. Pollutant concentrations are directly related to the air-fuel ratio. Therefore, maintaining a nearly stoichiometric air-fuel ratio minimizes pollutant concentrations. (Ref. 1, Bosch)

A standard 3-way catalyst was chosen for the exhaust system. The catalyst is designed for conventional automobiles running unleaded gasoline.

Implementation of other pollution control strategies, in particular exhaust gas recirculation (EGR), was not feasible due to development time constraints and the number of modifications needed to adapt such systems to work with the Kohler Command 20.

EVAPORATIVE EMISSIONS - Evaporative emission control is used on this vehicle. Positive crankcase ventilation is being used to lower these evaporative emissions. The original activated charcoal canister from the Escort can lower evaporative emissions from the fuel tank.

CHOICE OF FUEL - Reformulated gasoline was chosen as the fuel source primarily for its high energy characteristics. Reformulated gasoline and M85 generally provide similar CO, NOx, and THC emissions, but reformulated gasoline provides approximately 75% more energy per gallon. While the Paradigm HEV fuel injection system is capable of burning alternate fuels, the use of alternative fuels is not practical until the APU efficiency can be enhanced to obtain a 300-400 mile range between refueling without significantly increasing the size of the fuel tank. Research into increasing APU efficiency is on going.

FUEL AND ELECTRICAL POWER CONSUMPTION

Range was determined by dynamometer and on-road measurements. In ZEV mode, the approximate range of the car at 40-50 MPH is 25-30 miles. As of "press time" the only fuel consumption data available were for the carbureted version of Paradigm's APU

VEHICLE STRUCTURE AND DESIGN MODIFICATIONS

BATTERY LOCATION - As with most electric cars, the hybrid electric vehicle has a large battery weight burden. Not only are these batteries extremely heavy, they also take up a considerable amount of space. These batteries must be placed in the vehicle along with the electric motor, IC engine, and many other systems and controls. The batteries must be safely placed in the vehicle, while still providing room for the driver, passengers, and cargo. All the space found in the vehicle is valuable.

There are several different configurations which could have been chosen to hold a large battery pack. Batteries could be placed in the rear cargo area or back seat area of the Ford Escort Wagon with little or nothing done to the structure of the vehicle. However, the vehicle handling would be compromised due to the extra weight in the rear of the vehicle.

The original front-rear weight bias of the Ford Escort is 59% front, 41% rear. This is advantageous for a front-wheel drive vehicle, such as the Escort, because weight in the front provides traction on the front driving wheels. It gives the steering a good "feel" and also provides for better braking due to the greater force present on the front wheels.

Placing the batteries in the back seat and rear cargo areas places a large load in the rear of the vehicle. This reduces the front weight bias and could even cause a rear weight bias. For a front-wheel drive car, this would create reduced understeer. Reduced understeer could cause steering to be unresponsive and handling to be different than standard passenger cars. This could cause a potential danger if the driver is unaware or is unable to compensate for the vehicle's tendency to oversteer.

The Paradigm Hybrid Electric Vehicle has its batteries running along the centerline of the car, spanning from the front drive suspension subassembly to the rear axle suspension subassembly. This keeps the front-rear weight bias as close to the original weight bias as possible as well as keeping the center of gravity as low as possible. This also leaves the rear seat and cargo area free from batteries, providing plenty of space for equipment and extra passengers. Figures 3 and 4 illustrate the position of the batteries in the vehicle.

THE BATTERY CASE - The function of the battery case is to provide adequate protection for the batteries, using the least amount of internal space in the car. Secondary functions of the battery case include providing added structural rigidity to the car, protecting the passenger compartment from the high voltage of the batteries, and adequately supporting the battery weight.

A rectangular section of the floor of the car was removed, equidistant from the centerline of the vehicle. This hole, measuring 15 by 80 inches, spans from the fire wall to the rear axle. A large, 16 gage sheet metal cover, 18 inches in height, was placed around the area removed. This cover is welded to the floor and reinforced with sheet metal ribs on the underside. This provides a cover for the battery tray and replaces the floor that has been removed, adding stiffness to the overall cross-sectional shape of the chassis.

The battery tray itself is constructed from 6061-T6 aluminum angle. This material provides high strength at reduced weight over a comparable tray made of structural steel angle. There are two layers of batteries, with 20 batteries in the bottom tray and 11 on the top tray. These batteries are protected from the elements by a large, removable polypropylene cover. The polypropylene provides high toughness and has a high heat distortion temperature, so it will not deform under discharge temperatures. The batteries are secured with flat braided nylon straps, which hold the batteries securely in place in case of a roll-over.

To give the battery tray adequate strength, as well as provide for torsion rigidity, A36 structural steel was selected for the main supports. It provides good toughness, is inexpensive, is easily welded and is readily available. The aluminum battery tray is bolted to these main structural steel members. These cross members on the main structural beams are bolted in four positions to the underside of the vehicle. The front structural member bolts through a main structural sub-frame box section and through the floor. Reinforcing steel sheets provide extra strength on the top of the floor where the battery case is bolted. The rear structural member bolts to an extension of the structural box section just in front of the rear trailing arm mounts.

GAS TANK - The fuel container used in the Paradigm HEV is a racing-type fuel cell manufactured by Aero Tec Laboratories of Ramsey, NJ. The cell consists of a molded, seamless, plastic bladder placed in a custom-built steel container located in the rear of the vehicle. This puncture-resistant fuel cell is also equipped with a foam baffling and a filler safety check valve.

The fuel cell is located in the former spare tire well. The well was modified to allow for the cell's installation. A high impact polystyrene bulkhead isolates the tank from the passenger compartment (a contest requirement).

The close proximity of the fuel tank to the rear of the vehicle has been of significant concern to the team. To alleviate the hazard posed by a rear end collision, a crush zone consisting of an aluminum honeycomb matrix has been added between the rear bumper and the fuel

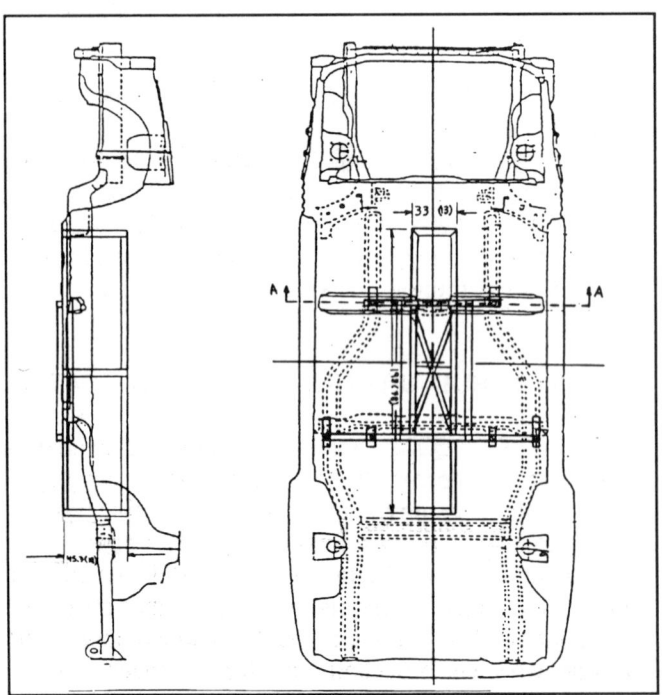

Figure 4. Battery location Schematic

cell.

THE ROLL BAR - Under normal circumstances, the Escort pillar system can support the weight of its structure if its cargo contents are not securely fastened to the chassis. With the additional weight of the batteries secured to this vehicle, the roll bar provides structural support in case of roll over, protecting the passengers from failure of the existing pillar system.

The roll bar also provides extra torsional stiffness to the vehicle chassis. The chassis stiffness is reduced due to the installation of the batteries in the floor of the vehicle. The frame bolts to the chassis in four positions. The rear mounts are located on the rear strut towers. The front mounts are on the floor just behind the front seats. This front mount was placed as close as possible to the floor frame rail without compromising the strength of the frame rail.

POWER TRAIN STRUCTURAL CHANGES - The original stock engine and its major subsystems (i.e. ignition electronics, fuel delivery system, cooling system and computer control) were removed from the car leaving only the original transaxle. The clutch pedal was also removed from the car. The coupling between the motor and transmission was designed to be permanent. Gears for forward and reverse are not needed because a switch for these modes reverses the flow of electricity in the motor reversing its rotational direction.

POWER SWITCH - The current from the battery flows to the drive motor through a standard General Electric industrial 250 amp dc-rated circuit breaker found in the engine compartment. This switch is equipped with an electric actuator to allow for automatic switching of the breaker, and a shunt trip to allow for manual (or automatic, if desired) direct tripping of the breaker in an emergency. Normally open contactors around the breaker allow the driver to temporarily bypass the breaker via a resistor set to pre-charge the capacitors in the high-voltage circuit before turning the breaker on. The breaker's thermal trip is set for 250 A; the magnetic (instantaneous) trip level is adjustable, and set to 1000 A. The breaker trip characteristics are such that the breaker should trip on its "thermal" limit before the (very expensive) fuse in the module string which makes up the main battery opens.

SEAT BELTS - As stipulated in the HEV Challenge rules, a 5 point racing harness was installed for the driver and front passenger. The two rear seat belts were kept as originally manufactured.

SUSPENSION MODIFICATIONS

Due to added weight of the batteries and power electronics, etc., the Paradigm HEV curb weight exceeds 3200 lb. After adding passengers and cargo, the gross vehicle weight is in excess of 3600 lb. This is 150 lb. greater than the rated GVW of 3466 lb.

The greatest problem with this added weight is its vehicle stability. The Paradigm HEV approach to this problem was to add stiffer springs in the front and rear. This added suspension stiffness has been shown to help the Paradigm HEV in handling. The car won its class in a SCCA autocross event during the fall of 1993 competing against several gasoline powered vehicles. On a very tight turning course, the car has shown exceptional handling capabilities for a station wagon.

CHOICE OF MATERIALS

The materials chosen for the *Paradigm HEV* were based on cost and availability. One of the intentions of the HEV challenge is to produce a practical vehicle. This practical approach limited the use of exotic materials such as structural composites in the construction of the vehicle.

The following is a list of basic materials used in the construction of the vehicle:
- Steel
    - Battery cover
    - Fuel Tank cover
    - Driveline mounting
    - Seat frames
- 6061 T6 Aluminum
    - Battery tray
- 3003 Aluminum
    - Engine intake manifold components
- High density Polyethylene tubing
    - Fuel line
- Aluminum Honeycomb
    - Crush Zone
- Polypropylene
    - Battery Terminal covers
    - Water proof cases for
        - Boost converter
        - Main battery switch
- Copper
    - Battery bus bars

MANUFACTURABILITY

One major guideline used in selecting components was that the components we use should have a demonstrated history of volume manufacturing. The three phase AC inverter using IGBT's or BJT's is already a well established industrial product. With regards to manufacturability it is much simpler to construct than an inverter which uses massively paralleled MOSFET's, each of which must be individually matched.

The AC induction machine has a long history of volume manufacturing and reliable operation in harsh environments. DC machines that we considered would use permanent magnets which cost around $80 per pound and are difficult to reliably mount into the large rotor size required for a high horsepower machine.

The chassis modifications needed to produce The Paradigm were not made using mass production techniques due to cost and complexity reasons, but were designed so that they could be reproduced using conventional industrial processes. The following areas would need to be added or modified for its assembly: battery box, engine compartment, fuel tank, structural modifications, inverter mounting, electrical display, and added control buttons.

Other factors contributes to the manufacturability of the Paradigm including the standardization of fasteners and the modularity of components. Since the original Escort uses primarily metric fasteners, every attempt was made to use common metric size fasteners. Wiring for electronic controls of the engine, electric AC drive, and other electrical devices incorporated standard connections that allow the component to be easily installed or removed from the vehicle, reducing assembly time. It also allows for ease of maintenance.

BATTERY BOX - For future vehicle development, the main battery box would most likely be make either from stamped sheet steel or of a recyclable plastic composite. Due to limited manufacturing capabilities here at the University, the battery box was produced by spot welding steel sheet metal. The box was welded directly to the center forward part of the car's frame. The bottom side of the box is modified with steel tubing and is designed to hold the weight of the batteries. The bottom mounts directly to the car's body.

Construction of the battery case would best be accomplished in two pieces, the main battery box and the bottom battery sled. The main battery box would best be produced through stamping operation in the same manner that the bottom of the car's chassis was produced. This would be more economical and time efficient than by welding the sides together. This main battery case would be directly welded to the car's frame when the body is being welded together. For appearance and safety, the battery case would be covered with either a plastic vinyl sheet or carpeting similar to the car's interior.

The bottom of the battery case would be constructed with a large sheet of steel with a tubular steel structure to mount and support the weight of the batteries. This would bolt directly into the car's frame on its underside.

ENGINE COMPARTMENT - The only major component found in the engine compartment of the Paradigm HEV that would also be located in a standard Escort engine compartment is the transaxle. The mass produced vehicle would require the addition of an APU, generator, electric motor and their respective mounts.

The electric motor is located between the engine compartment bulkhead and the APU unit. The motor and transaxle unit would be located using three mounting points. One mount, between the motor and frame, is attached directly to the car's frame and would be installed during the main body welding of the car. A second mounting location is at the bottom of the transaxle on a longitudinal bracket between the lower radiator support and the frame cross member. This mount would be bolted to the vehicle just after the final welding processes. The motor would be attached to the transaxle before installation, and these two pieces would be added before the addition of the APU. The third mounting point for the motor and transaxle combination would be on a cross bar connected between the strut rails. This member is affected mainly by motor and braking torque reactions and would be added just after the installation of the motor and transaxle.

Mounts for the APU would consist of aluminum box tubing running longitudinally between the lower radiator support and the frame cross member. Aluminum angle attaches the motor to the primary mounting member. Rubber isolators are used to reduce vibrations transmitted to the vehicle. The mounts would best be added just after the car's body is welded together.

The APU rests on the mounts in an upright position. This allows easy placement of the engine from the top of the engine compartment. Gas lines enter as in the original vehicle design. The alternator, attached directly to the APU, is located about where the original engine's radiator was. The electrical output wires feed through the underside of the car to the inverter located in the trunk.

FUEL TANK - The fuel tank would best fit between the original spare tire storage compartment and battery box under the car. This would require some design modifications to the existing tire compartment and rear cargo area, but the basic manufacturing operations would be the same.

STRUCTURAL ADDITION - Addition of a roll bar was a contest safety requirement. If it is determined that the existing pillar structure can support the weight of the car with battery in event of a roll over, than the addition of a roll bar would not be needed. A mass production vehicle specifically designed for electric vehicles would most likely include a strengthened pillar structure. However, if a roll bar is needed, it could be assembled by bending and adding mounts to a large steel tube mounted directly to the car's frame. Assembly of the roll bar would be performed during the main welding of the car body before addition of the roof.

ELECTRONICS DISPLAYS - Because of the nature of this prototype car, their were numerous sensors required the monitor different systems condition. An LED screen and 486 computer was required to monitor these systems.

The screen for these system was added the dash just right of the console and the 486 unit was placed in the glove box. In a final production version of the car, the LED would be placed in the original console and the 486 placed under the passenger's seat. Placing of the LED screen would be feasible because a majority of the original monitoring sensors would not be needed for the new car production and hence the screen size would be minimal. Their are two reasons the original computer was placed in the glove box, these are due to the size of the test system and the need to have good accessibility to the unit for repair and modifications. The size of the unit could be reduced significantly in future models allowing for its placement under the passenger's seat. The unit would be installed at the same time the wire harness is added to the car.

INVERTER - The inverter was placed in the rear of the car due to its size and the desire to simplify weatherproofing of the unit. This allows for good maintenance for servicing and in installing in the car during manufacturing. The main unit was made from a modified unit bought off the shelf.

ERGONOMICS

DASH DISPLAY - The dash display will not change in location or design but an 8 in. X 5.5 in. back lit black and white LED will be added on the dash between the driver and passenger. The flow of system information has been designed so that each of the running parts of the software package can communicate with the driver or passenger via the display. Operating parameters needed in testing the car are also able to be displayed there.

DASH CONTROLS - Most of the existing controls in the car were left alone except for the addition of a mode selector button, shifting, main power shut off (battery) and fire suppression system.

The mode selection button consists of a simple rotary dial switch with four possible choices. The position of the wording is placed directly above the need position for the switch. No pictorial icons were used in order to allow the driver a clear understanding of the exact function of its position. In future modifications, these wordings need to be back lit for easy viewing at night. The location is placed on the right side of the steering wheel which allows for easy viewing by the driver and for its operation. The switch is designed so that the driver only has to rotate its position and not have to perform any other operation when changing modes.

Shifting of the car is accomplished by an electrical actuator attached at the transmission to the shift lever. A forward-neutral-reverse switch is mounted on the dash to allow easy visibility and driver accessibility. The 486 computer acts as a watchdog and locks out a direction reversal when the car is moving.

The main fire suppression system control is located on top of the battery box between the driver and passenger, 5 inches from the dash. In this location it can be operated by either the driver or front seat passenger. The system is manually operated to ensure that it works even when the car is not in operation or in the event of a power failure. The handle is red, T-shaped pull knob and is clearly marked *fire suppression system*.

CONTROLS DISPLAY - Due to the number of different subsystems needed to be monitored by the driver and passenger during the contest, a backlit LCD screen was added. The system is designed to display information quickly and accurately. The screen allows the system designer the flexibility to add or remove needed information as required by the driver for a particular event. The screen also enables viewing at night.

WARNINGS - Warnings of problem systems will be done two ways, audio and visual. Their will be three different levels for audio signals: one, two and three beeps. The number and duration of beeps increase as the severity of the problem increases. The one beep will be for situations in which a reading is about to enter a critical region. Once the reading enters a critical region, two beeps lasting one second will occur. For systems about to fail, one long beep lasting 1 second out of every 3 sec. will be initiated. Visual signal will be displayed on the LED screen and with red lights on the original dash. The display reading will blink with the lettering alternating black and white. Addition red lights will be installed in the original dash to warn the driver of critical systems.

HEATING AND VENTILATING SYSTEM - The original heating element was removed from the car. The car lacked air conditioning when manufactured and a system is not expected to be installed. The original ductwork was removed and new ducting was added to allow for passenger compartment ventilation.

SEATS - The original Escort seats were modified for use in the Paradigm HEV. The seat frames were cut to accommodate the battery box. They were then welded back together and covered with cloth tailored to fit the modified seats. The front seats were cut down 2.5 in. in their breadth because of the battery box width and were remounted. This provides a problem because the driver is no longer center with the steering wheel, however it is barely noticeable after a short time driving the vehicle. The floor pedals have been adjusted to remain centered with the driver. Longitudinal movement and tilt control of the front seats were maintained for the front passenger and driver. Either the car's width will have to be increased or the battery box width decreased for improved comfort in a future production vehicle. The rear seats had a 14.5 in. X 8 in. section removed from the center of the seat.

ACCELERATOR & BRAKE PEDAL - The accelerator & brake pedals have been moved left of their original position because the seats were moved to accommodate the battery box. This keep the alignment of the driver's right leg straight with the pedal. The size and movement of the pedals are the same as the original pedals.

PARKING BRAKE - The location of the parking brake has been moved to the floor behind the driver's feet. The location still allows the driver to apply good upward pressure to activate the system. A disadvantage of this position is that is does not allow for good driver visibility during operations of the car.

OTHER ISSUES

Light - The interior and exterior lighting systems have not been modified.
Noise - The car is expected to operate at levels less than the 80 dB limit required for the HEV Challenge.

REFERENCES

1. Bosch, Robert, <u>Automotive Electric / Electronic Systems,</u> Robert Bosch

HEV Paper Design

Advanced Hybrid Electrical Vehicle Design
California State University, Chico
Mechanical Engineering and Manufacturing Department

K. Earnest	J. Lee
T. Bingaman	H. Li
D. Boyer	J. Poon
M. Charonnat	B. Proshold
B. Elmuhtadi	C. Wilcox
R. Gage	
K. Golemo	C. Allen (Advisor)

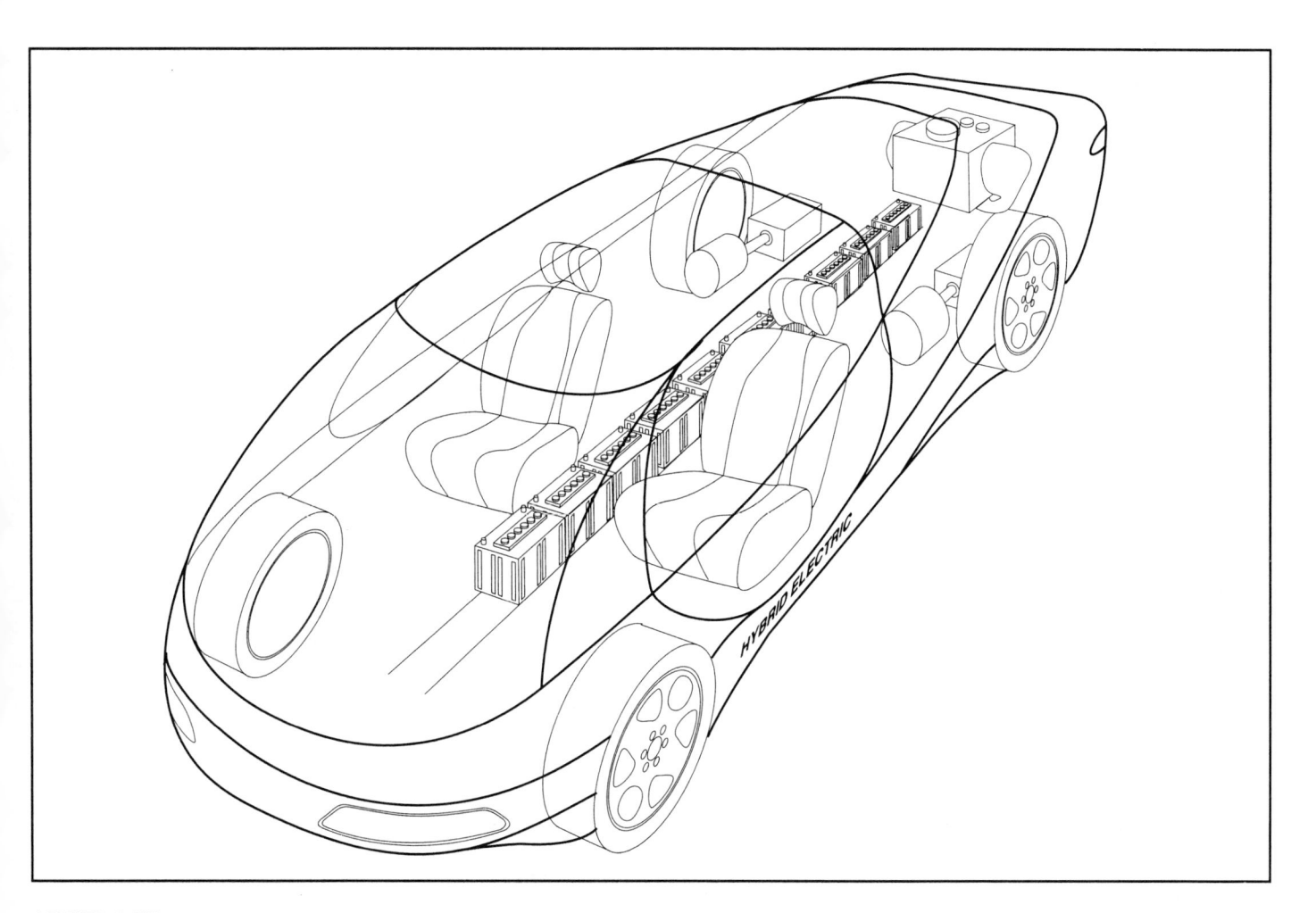

ABSTRACT

A hybrid electric vehicle was designed to compete in the 1992 DOE Hybrid Electric Vehicle Design Contest. Special effort was given to minimizing the Coefficient of Drag (Cd) to lower power requirements. Several models were wind tunnel tested and the target Cd of 0.15 was eventually achieved. A 24 kW (32 hp) single shaft gas turbine using ethanol as a fuel, directly coupled to a generator supplies the auxiliary power. The 245 kg (540 lb) battery pack is made up of ten 12 volt 46 amp hour lead-acid batteries. Power is delivered through two high efficiency, rear mounted 18.6 kW (25 hp) brushless DC motors with 8:1 gear reductions. Vehicle curb weight is approximately 680 kg (1500 lb).

The gas turbine exhaust emissions at 24.6 m/s (55 mph) are Nitrous Oxide 0.0038 g/mile, Carbon Monoxide 0.0038 g/mile, and Hydro Carbons 0.02 g/mile. These are values beyond those required for ULEV vehicles. The turbine requires no oil, since the use of hydrodynamic air bearings totally eliminates the need for conventional lubrication. The elimination of oil saves on maintenance costs and pollutants.

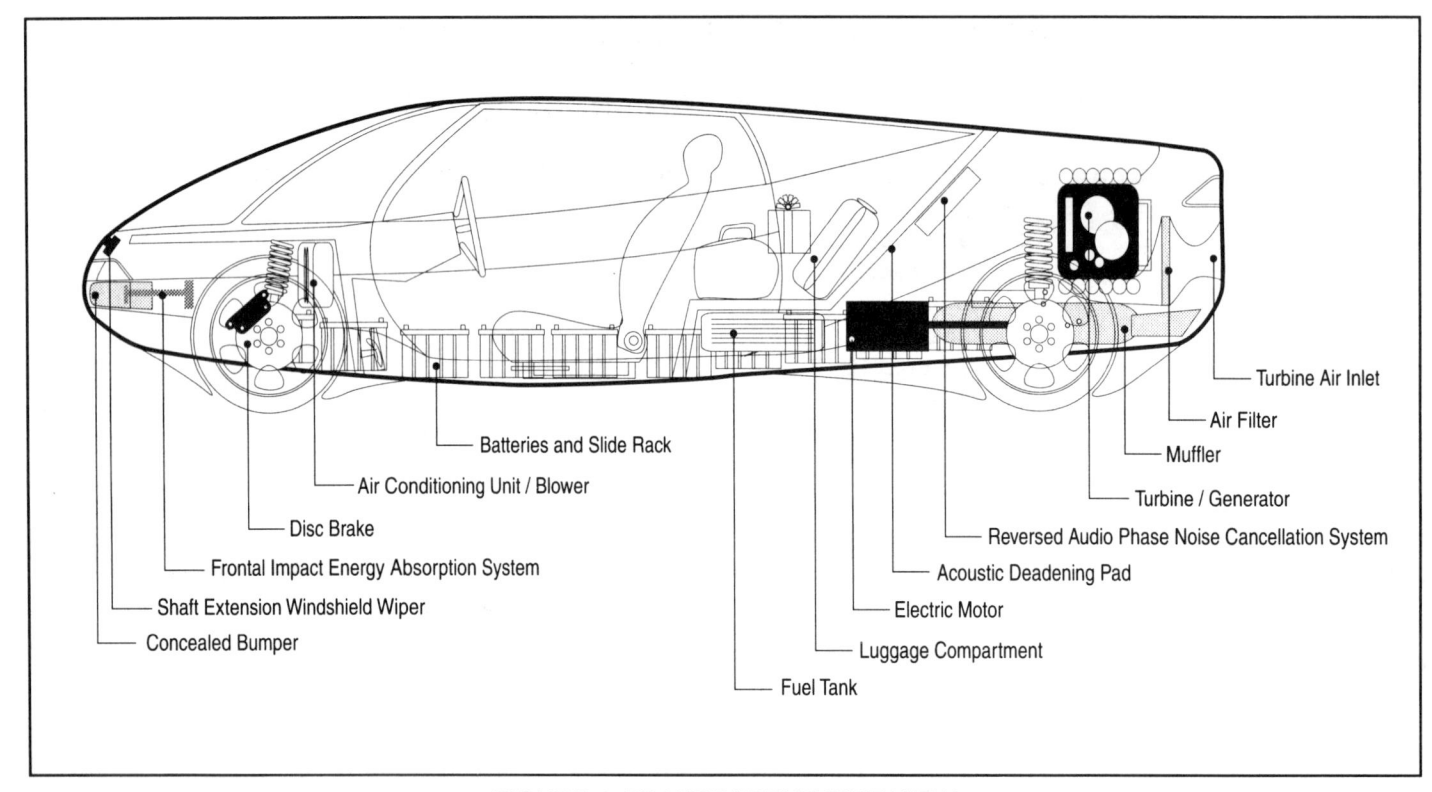

FIGURE 1: TRANSPARENT SIDE VIEW

INTRODUCTION

The 1992 DOE Hybrid Electric Vehicle Paper Design Contest was undertaken as a class project by the Mechanical Engineering Department's Systems Design class at California State University at Chico.

The class undertook this project with the premise:
1) The contest criteria and guidelines will be met with innovative application of practical, available or soon to be available (1993) components.
2) Everything will be as environmentally clean as possible.
3) Maintenance free and low maintenance designs will be given greater priority.

AERODYNAMICS

In order to meet the range requirements, while in battery only mode, it is necessary to keep vehicle drag low - thus minimizing the weight of the battery pack, which is the single heaviest component in the vehicle.

Aerodynamic drag has more influence over the final weight of the vehicle than all other factors combined. This is due to the fact the power requirement increases as the cube of the vehicle velocity. Therefore, any small increase in the aerodynamic drag can have a major influence upon power requirements.

The vehicle is designed to travel 50 miles on a battery charge, therefore power requirements directly affect the battery size and weight.

Aerodynamic drag depends upon the vehicle frontal area and the coefficient of Drag (Cd). Studying various configurations, it was found that a realistic minimum frontal area of 1.49 square meters (16 square feet) was acceptable.

The drag coefficient is dependent on:
1) flow separation.
2) vehicle shape.
3) surface discontinuities.
4) extent of laminar boundary layer.

To minimize these factors the following measures were taken.
1) The body will taper back from the largest cross-sectional area at angles less than 5° to 7° to decrease the strength of the adverse pressure gradient.
2) A teardrop shape was chosen because it has the best aerodynamic three dimensional shape for unrestricted airflow.
3) Separation tends to occur at discontinuities or minor surface irregularities. These will be minimized by keeping exterior hardware to a minimum, making each protuberance as smooth as possible, and placing the exterior hardware as close together as possible so that any separation produced by one will flow over the other, minimizing the effect of the second.
4) A turbulent boundary layer delays separation when compared to a laminar boundary layer. With the high Reynolds numbers encountered, and the free stream turbulence from other vehicles, turbulent boundary layers will exist; thus the HEV is designed for this condition.

The drag coefficients of numerous low drag ground vehicles were examined, and a target value of 0.15 was chosen as a practical balance between the state-of-the-art, and the truly exotic. For comparison, the GM Impact electric car has a value of about 0.19, and the theoretical minimum for a flat ground vehicle is about 0.07.

The teardrop body shape achieves low drag numbers, unfortunately it also has its own drag problems that need to be dealt with: The ground beneath the vehicle tends to stagnate the airflow underneath the car and increase drag and lift. This can be overcome by ensuring the bottom surface is very smooth. A smooth "inlet" to the bottom surface exists so that airflow is unrestricted beneath the car, and the rear wheels are fully faired inside the vehicle.

The ground clearance and underside shape are designed to improve the flow under the vehicle. Additionally, sharp front and rear glass angles are eliminated to further reduce drag. The wheels are narrow, and the front wells are contoured to improve airflow. When not in use, the windshield wiper is retracted inside the vehicle so as not to trip the boundary layer of air in the critical nose section. All lights fit flush with the body panels.

Present state law does not allow it, but the side mirrors could be replaced with a "heads-up" video display of the sides and rear of the vehicle. This would eliminate a significant parasitic drag, and provide the driver with constant surveillance of the surroundings. Also eliminated is the diverting of attention away from the primary driving function. The general profile of the car will have an airfoil-like front end. This creates a favorable pressure gradient and helps the boundary layer overcome the increase in cross-sectional area. Once maximum thickness occurs, the cross sectional area will be decreased gradually to delay separation.

The rear of the car will not be a point or a thin trailing edge. Instead, this area is allocated for the engine (figure 1). This design produces beneficial car space; it might even decrease the area of separation, because the angle at which the surface is decreasing need not be as great.

The bottom surface near the front of the car will allow air to flow underneath the car. This will decrease lift and drag by producing non-turbulent air underneath the car. Also, a negative angle of attack will be designed for the overall car shape to reduce the lift on the vehicle.

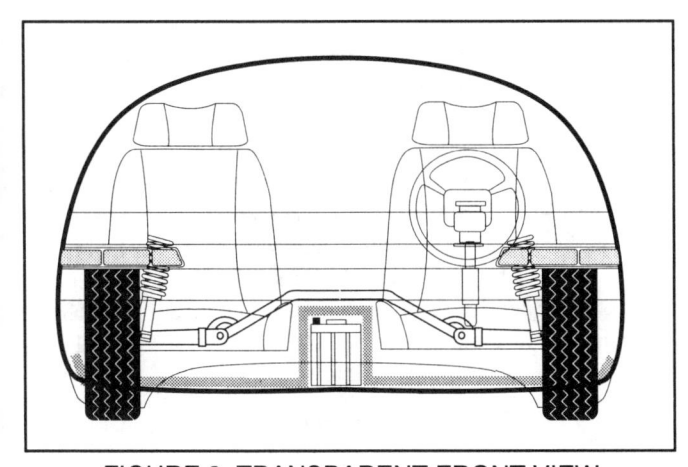

FIGURE 2: TRANSPARENT FRONT VIEW

The car's side view (figure 1) shows the negative angle of attack, an airfoil-like front end, and the "clean inlet" to the bottom surface. The top view (figure 8) shows a larger front wheel base as compared to the rear.

The placement of the turbine in the rear produces two benefits. First, the internal drag will be small because it is not disrupting flow near the front of the car. Secondly, the suction may be used to keep the boundary layer attached to the rear surface where separation would generally occur.

Extensive wind-tunnel testing and evolutionary modeling of the vehicle shape, eventually produced a wind tunnel confirmed Cd of 0.15 (this model included a left side mirror).

BATTERY SELECTION

The selection of an appropriate battery pack for the HEV was a vital portion of the design. The batteries are the largest and heaviest single component in the entire vehicle, thus requiring careful consideration and research.

In order to determine battery type and size, two sets of limitations were applied; first, the restrictions outlined in the contest rules, and second, self imposed restrictions. The contest rules limited the field to "near-term" batteries thereby excluding Silver-Zinc, Silver Cadmium, Silver-Iron, and non-terrestrial batteries. Additionally, the batteries selected must have an industry accepted life of at least 300 recharge cycles (80% Depth of Discharge). The self imposed restrictions limited the field of possible batteries to low-cost, high-power-density batteries that could be easily purchased and installed.

With these limitations and design concepts in mind, Nickel-Cadmium batteries were considered, but rejected due to low depth of discharge, coupled with high cost. It became clear that the most logical type of batteries to use were sealed Lead-Acid. Once the type of battery was determined, the selection from available models was made based on minimizing size, weight, and number of cells required to provide the necessary power, as well as being able to document the 300 cycles requirement.

Vehicle performance requirements dictated the battery pack must be able to provide 5,000 W for a one hour period (5 kWh). Also the electric motors required a minimum of 72 and a maximum of 120 volts.

The batteries selected are Deka Solar series #8G24. These are 12 volt sealed lead-acid, with an approximate life of 350 cycles. The three hour discharge rating for current is 19 amps. This design has ten batteries connected in series to produce a total of 120 volts and 45 amps (one hour rating). This gives a total output of 5,400 watts for one hour. Each battery weighs 24.5 kg (54 lb) and is 0.27 x 0.17 x 0.25 meters (10.875" x 6.75" x 9.875") giving a total pack weight of 245.5 kg (540 lbs), and a volume of 0.12 cubic meters (4.2 cubic feet). The one hour power density is 22.0 Wh/kg (48.5 Wh/lb).

NOTE: Significantly lighter batteries were considered, but only verbal statements as to approximate cycle life could be obtained, therefore they were not used.

The battery compartment is ventilated with a 12 volt continuous duty brushless DC motor powered fan that supplies 10 CFM. The fan operates whenever the batteries are charging or discharging.

This battery pack provides an inexpensive, reliable source of power to the HEV while both in hybrid and electric-only modes of operation.

DRIVE MOTOR SELECTION

The electric motor is the heart of an electric automobile. It must be chosen with regard to the type of automobile and intended driving uses.

The motors chosen for this design are ultra-high efficiency brushless DC motors made by UNIQ Motors. They have the advantage of:

1) Highest power densities available, approaching 3.3 kW/kg (2 hp/lb).
2) No maintenance brushless DC type.

3) Ultra high efficiency design, virtually flat torque curve.
4) Efficient heat transfer and thermal protection using forced air cooling.
5) Sealed precision bearings.
6) High pole count - low cogging effect.

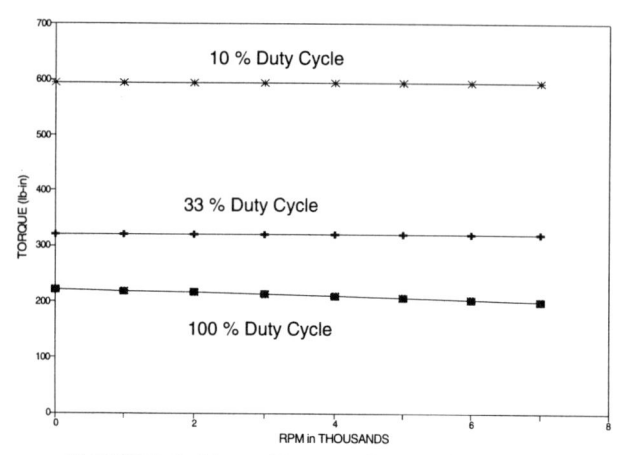

FIGURE 3: Motor Torque-Speed Curves

These DC motors have a flat torque curve - virtually the same starting, operating, and peak torque (figure 3). This eliminates the need for a transmission. Since a manual transmission may have a power loss of from 5 to 10 percent, eliminating it is a advantage.

These motors are high speed motors which supply maximum power output at 785 radians/s (7500 rpm). To increase generator efficiency and limit vehicle speed an 8:1 reduction gear is used on each motor. The continuous operational speed is limited to 30 m/s (67 mph).

BATTERY CHARGING SYSTEM

The on-board battery charging system can utilize either 120 or 240 VAC. Permanently wired into the battery pack, it features a light-weight transformerless, full sine wave (i.e. low AC harmonics) system. The system is micro-processor controlled, and the battery's charge and temperature are monitored.

A "two-step" charging system is used to optimize the charging rate. A high current (25-30 Amps) is applied until the microprocessor senses the gassing point, then a constant voltage is applied keeping the voltage at or just below the gassing point. The current will fall off as the cells become "fully charged". Once fully charged, the system will automatically switch off.

The charger will have a controlled maximum output of five kilowatts. Complete charging of the battery pack will be processor controlled, allowing rates of maximum battery charge acceptance. Thus charging of the battery pack is well under 8 hours whether the supply is 120 or 240 VAC.

There are several safety features in this system. The vehicle cannot be started as long as the power cord is plugged into the vehicle. Two battery isolation switches, one on the dashboard, the other next to the power cord receptacle must BOTH be in the on position. Either switch placed in the OFF position allows the immediate and complete electrical isolation of the batteries. The charger itself has an internal cooling fan to prevent overheating. The battery compartment fan is continuously operating during charging.

CHASSIS

The chassis is a modified-ladder aluminum alloy frame integrated with a high-strength structural-fiber inner structure surrounded with a semi-monocoque skin. This extremely strong and rigid structure protects passengers, batteries, and fuel tank while providing excellent stability for the drive train.

OCCUPANT SAFETY-For occupant protection, the front of the HEV has 31 (0.0254 m diameter, 0.305 m long) aluminum tubes. This bundle of tubes is designed to collapse and fully absorb a 8.9 m/s (20 mph) frontal impact with an immovable object (figure 4).

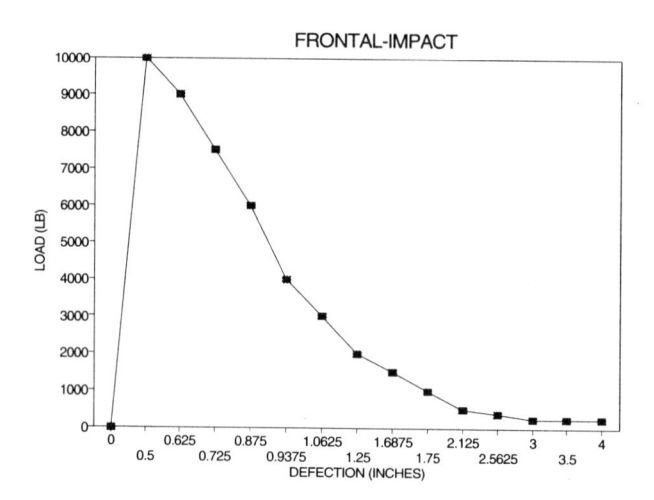

FIGURE 4: FRONTAL IMPACT SIMULATION

Driver and passenger are each further protected from a frontal impact by air bags in addition to the standard three point restraints. These BOSCH systems utilize an accelerometer and use digital signal micro-processor technology to provide light-weight, reliable crash protection.

Two roll bars, one just forward of the seats, the other slightly behind the seats, form a safety envelope for the occupants in the event of a roll-over. Square carbon steel tubes (44.45 mm, 1.75 in) within the gull wing doors interlock with the two roll bars to provide added safe-guards in case of side impact. Deflection testing of these beams established that they would deflect 0.13 m (5 inches) when impacted with a 782 kg (1720 lb) point load traveling at 2.7 m/s (6 mph) (figure 5).

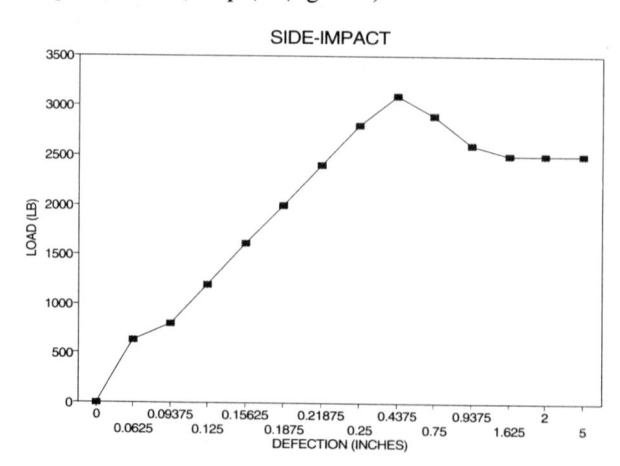

FIGURE 5: SIDE IMPACT SIMULATION

The roll bars are 0.0762 (3 in) alloy steel tubes. The safety advantages of these large diameter tubes, offset any weight penalty.

Roll-over was modeled on COSMOS/M, a finite-element package. A 10000 lb point load was applied 45° to the horizontal plane. This simulated a 6 g load on the corner of the vehicle roof - the worst case. Results were, negligible deflection vertically and 0.44 m (17.5 in) horizontally (figure 6). This design means the occupants receive protection from side, Roll-over, and frontal impact.

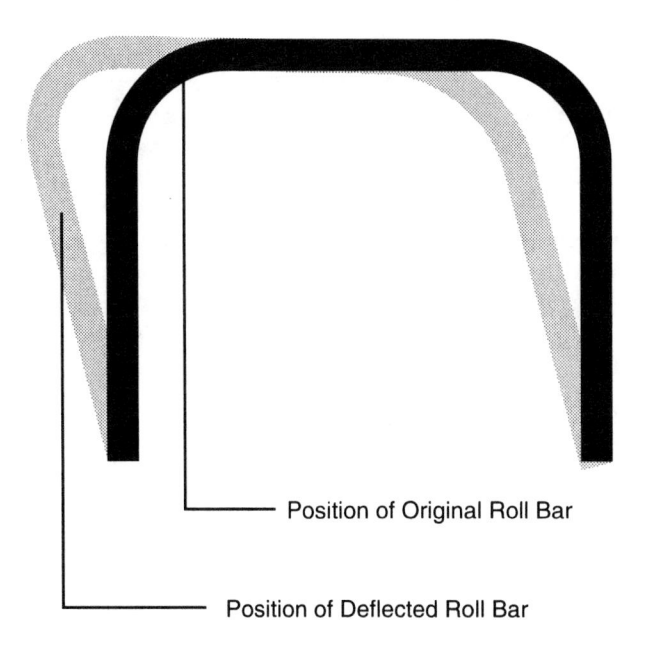

FIGURE 6: Roll-OVER SIMULATION

SUSPENSION-The hybrid electric vehicle features 4 wheel fully independent suspension providing excellent ride and handling characteristics. The rear suspension uses semi-trailing arms with coil spring/shocks (the drive motors are mounted on the same axis as the trailing arms to prevent binding). The front suspension utilizes damper struts with coil springs. As a result, the vehicle has a low center of gravity, large trunk capacity, stability, and control in any driving or road condition.

STEERING/BRAKES-Steering is a rack and pinion system. Braking is by front discs and rear drums. The disc brakes are vented for optimal cooling. The drum brakes provide efficient, economical braking and have a secondary cable actuating system to also provide a parking brake.

AUXILIARY POWER UNIT

Selection of the heat engine (APU) was based on power to weight ratio, ease of meeting emission standards, simplicity of design, cost, and ease of maintenance. Based on the criteria of cost, fuel cells were quickly dropped from consideration.

The system configuration became important in the determination of what type of APU would meet our criteria. The big question: Does the APU transfer mechanical power directly to the wheels, or drive the vehicle with electrical power?

Several configurations were examined, and on the basis of weight, size, emissions, and simplicity of design; the decision was made to use electric motors to directly drive the HEV. This configuration has the advantages of:

1) Lower weight of the vehicle through the elimination of a mechanical linkage between the heat engine and wheels.
2) Allows the engine to run at some optimum steady state speed, thus a higher efficiency and reduced emissions.
3) Flexible charging system could easily be employed that allows the APU, as well as external sources to recharge the on-board batteries.

APU SELECTION-A small single shaft gas turbine was chosen as the best solution for this configuration. The characteristics that favored gas turbines:

1) Engine Weight and Size
2) Lower Operating Pressures
3) Lower Fuel Consumption
4) Lower Emissions)
5) Quieter Operation
6) Wide Range of Possible Fuels
7) Mechanical Simplicity
8) Mechanical Balance
9) Low Maintenance
10) Cost

TURBINE SELECTION-The gas turbine chosen for this design is a NoMac gas turbine. Designed specifically for hybrid electric vehicles, and is now in final development. It has a single stage centrifugal compressor with single stage radial axial inflow. Compression ratio is 3:1. This turbine is small, compact, relatively dust resistant, and easy to manufacture.

ENGINE WEIGHT, SIZE, AND OPERATING PRESSURES-The gas turbine and generator set weigh 1.5 kg per kilowatt (2.5 lbs per hp). An equivalent piston engine without generator would range from 1.8 to 2.7 kg per kilowatt (3.0 to 4.5 lbs per hp).

The compactness of the gas turbine is because it is built around a single rotating shaft. It is air cooled further reducing the overall weight of the system.

LOWER FUEL CONSUMPTION-The gas turbine is designed to operate at steady state, hence higher efficiency, and lower fuel consumption than a piston engine operating at varying speeds.

The reheating in the gas turbine significantly reduces fuel consumption and lowers emissions due to more complete combustion. The use of ethanol significantly reduces the emission of nitrous oxides.

MORE SILENT OPERATION AND DIFFERENT FUEL TYPES-Quiet operation of the turbine exhaust is due to its continuous pressure output. Vibrational noise of piston engines is significantly greater than that of a dynamically balanced turbine engine for the same output. This turbine can easily run on a variety of fuels, whereas a piston engine is more specifically designed for a single fuel type due to the timing and carburetion needed for good combustion.

MAINTENANCE REQUIREMENTS-Because the unburned fuel and combustion by-products contaminate the oil in a piston engine, the recommendation is to change oil every three thousand miles. The bearings in the NoMac turbine never come in contact with the combustion process, as they do in the piston engine. The NoMac turbine uses hydrodynamic air bearings therefore requiring no oil at all.

COSTS-Gas turbines have fewer parts than a piston engine therefore they cost less to manufacture. The NoMac turbine has lower: maintenance, fuel consumption, and emissions making it a cost effective engine. As a prediction biomass fuels will soon be produced more economically, making the turbine even more cost effective.

TURBINE EFFICIENCY-The life of a turbine is very dependent upon thermal stresses, with the lower operating temperatures of ethanol, life expectancy should rise. NoMac predicts reliability similar to that of a household refrigerator.

Small gas turbines have been characterized by having low thermal efficiency, resulting in high fuel consumption. The NoMac uses its exhaust through a proprietary, true counter flow circumferential recuperator to heat the compressed air before combustion to increase performance and fuel efficiency. The recuperator is built around the turbine unit and is 85 percent effective. This design means few losses due to piping and ducting and can absorb even the smallest heat losses in the turbine unit. The recuperator allows this small turbogenerator to have an overall thermal efficiency of 30 percent and an ethanol fuel consumption of 0.405 kg/kWh (0.66 lbs/hph) (about 90 mpg at 55 mph in turbine only mode).

PHYSICAL SPECIFICATIONS-This turbine comes with a generator mounted to its single shaft to generate electricity. The shaft rotates at 10,000 radians/sec (96,000 rpm), this high speed reduces the size of the generator. The complete turbine unit with generator and recuperator, weighs about 36 kg (80 lbs). The size of the unit is extremely compact 0.66 m (26 in) length, 0.39 m (15.4 in) width, and 0.30 m (11.8) height.

FUEL SELECTION-The fuel types specified for the contest were ethanol and methanol, both which are extremely clean burning fuels. They have about 2/3 and 1/2 respectively, the energy per unit volume of that of gasoline. Both fuels are potential aids in the reduction of emissions, and could be used in reciprocating piston engines and small gas turbines. A goal of the design was to reduce the weight of the vehicle, based on the amount of fuel needed to meet the 300 mile test, ethanol was chosen as the fuel based upon energy density. This choice also required the least amount of modifications to existing fuel delivery systems; due to the corrosiveness of methanol.

EMISSIONS-Emissions from the vehicle have been predicted using a model of natural gas burning in a turbine. This assumes that ethanol could be used as efficiently as natural gas. With ethanol as the fuel, combustion temperatures will be below the formation temperature of nitrous oxide (NOx), this should result in very low NOx emissions. Since the turbine has an air to fuel ratio of 150:1, carbon monoxide levels are also low. The projected numbers from NoMac are Nitrous Oxide 0.046 g/kWh, Carbon Monoxide 0.030 g/kWh, and Hydro Carbons 0.24 g/kWh. This means at 55 mph the emissions are Nitrous Oxide 0.0038 g/mile, Carbon Monoxide 0.0025 g/mile, and Hydro Carbons 0.02 g/mile.

GOALS MET-The heat engine choice achieves the goals. Extremely low weight and size, low emissions, nearly maintenance free, and more than adequate power. The turbogenerator could recharge the battery system with less emissions than that of a fossil fuel power plant, therefore it could be classified as a zero emissions vehicle. The fuel consumption is low, the vehicles mileage should be in the range of 38 km per liter (90 mpg), thus only requiring a minimal amount of fuel, saving weight. This turbogenerator is an excellent choice as an APU for a Hybrid Electric Vehicle. It allows for a car that runs smoother, has lower maintenance, and does not ask the customer to sacrifice performance in order to help combat air pollution.

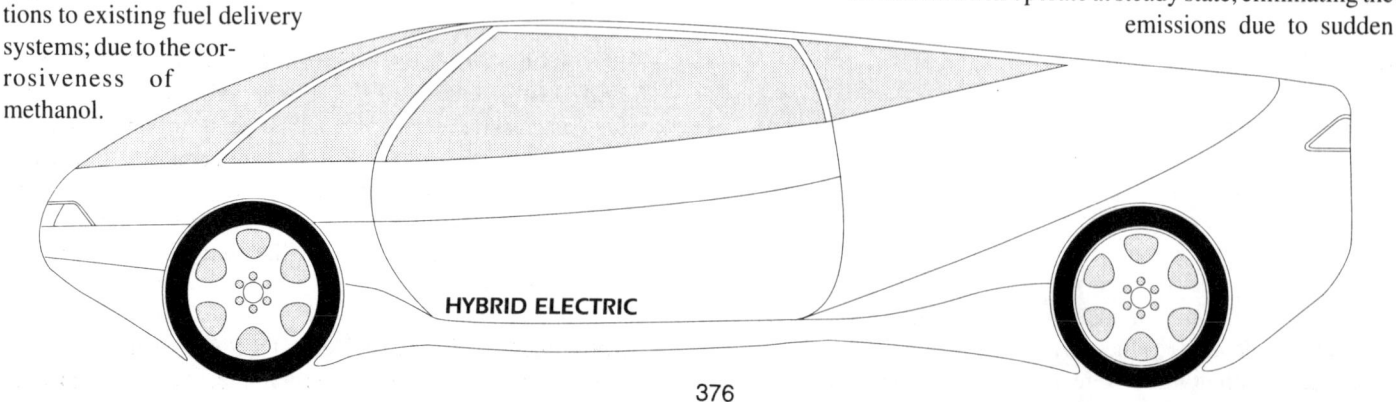

HYBRID/EMISSION CONTROL SYSTEM

To control a vehicle with two power sources, a specific control strategy has to be adopted. One of the main objectives of this vehicle is to reduce pollution, therefore special care was given to the emission control system.

The control system is divided into two parts :
1) Emission Control System
2) Propulsion Mode Control System

Each part will be discussed in detail and the control strategy behind it, will be explained.

EMISSION CONTROL SYSTEM-Automobiles pollute the air by the evaporation of fuel from the fuel tank, combustion by-products from the engine and exhaust gasses. The most damaging pollutants are Hydrocarbons, Carbon Monoxide and Oxides of Nitrogen. This HEV will have zero emissions while operating in the electric mode only. The use of a turbine as an alternative power unit reduces the emissions compared to a piston engine.

The turbine will operate at steady state, eliminating the emissions due to sudden

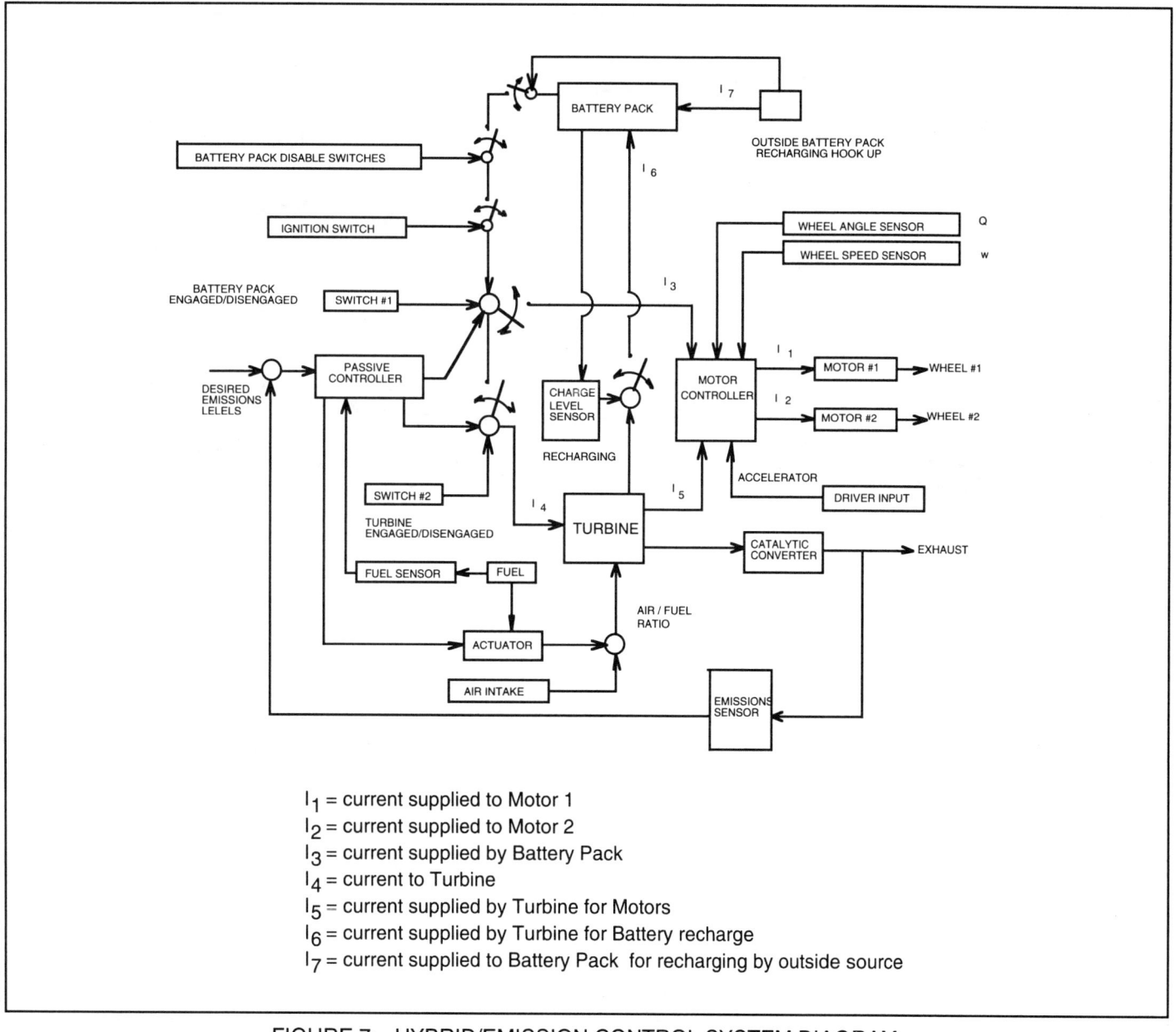

FIGURE 7: HYBRID/EMISSION CONTROL SYSTEM DIAGRAM

I_1 = current supplied to Motor 1
I_2 = current supplied to Motor 2
I_3 = current supplied by Battery Pack
I_4 = current to Turbine
I_5 = current supplied by Turbine for Motors
I_6 = current supplied by Turbine for Battery recharge
I_7 = current supplied to Battery Pack for recharging by outside source

changes in engine RPM.

The emission control system is supported with an on-board diagnostic system that continuously monitors tail-pipe emissions. If the emission level is higher than the preset standard, a controller will adjust the air to fuel combustion ratio in the turbine. This will bring the pollutant level back into specification. If the standards are not attainable, the controller will stop the turbine and implement battery only operation. This passive strategy cannot be changed by the occupants. The turbine must then be serviced by an authorized diagnostician before it can be restarted with emission levels within acceptable standards.

A fuel tank control system is used to reduce evaporative emissions in the HEV. The fuel tank vapor control system allows 10-12% expansion of the total tank volume by locating the filler neck so that an airspace is created at the top of tank, when it is filled. While sealing the fuel tank and allowing the 10-12% expansion the tank vapor control system reduces the amount of fuel evaporation from the tank. A nozzle is used to click on and off when the tank is being filled with fuel.

The tail-pipe emissions are reduced using a Honeycomb catalytic converter. The catalytic converter consists of a ceramic core that is coated with a microscopically thin layer of the catalytic agent (Platinum). The honeycomb configuration of the core gives a huge surface area so pollutants are forced to come in contact with the catalytic agent. The catalytic converter will oxidize hydrocarbons and carbon monoxide and reduce them to harmless water vapor and carbon dioxide. These measures ensure the emission control system is always operating at maximum efficiency.

SWITCH MODE CONTROL SYSTEM-The vehicle operational mode will be determined by the operator or the passive control strategy. The passive control strategy is dominant in the case of conflicting signals.

Through a three way switch the rider will have a choice of operational mode. This can be overridden by the passive control system which monitors the emission level of the vehicle using the on-board emission diagnostics and control.

If the emission levels are not within the specified limits, the controller will first try to adjust the turbine combustion process

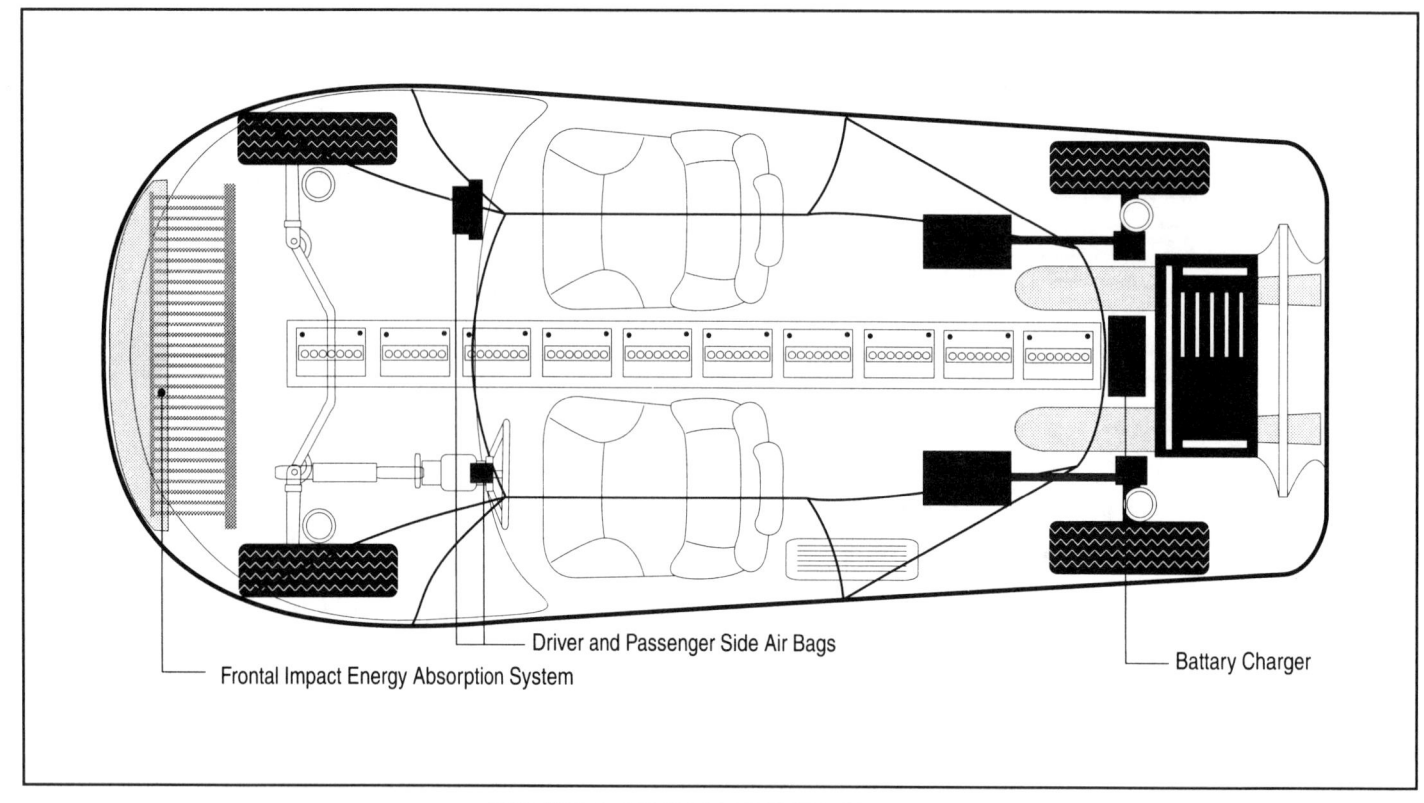

FIGURE 8: TRANSPARENT TOP VIEW

to meet the standards. If the standards are not met after maximum adjustment, the controller will shut off the turbine and rely on the batteries as the sole power source until examined and certified (by a diagnostician) within standards at which point, the driver will again have control over the operational mode.

When the switch is in the turbine mode - such as when problems prevent battery use - the drive motors use the power from the generator.

Battery mode is when the drive motors are only powered by the batteries - perhaps because of local restrictions on internal combustion engines.

In the normal or hybrid mode, power will be supplied from both the generator and the batteries. The control system will optimize the battery charging and discharging based on depth of discharge history and present vehicle power requirements. By using depth of discharge history as a charging criteria, the expected life of the battery pack should be extended because of minimizing "quick charging".

MOTOR SPEED CONTROL-Motor speed control is with a solid-state chopper circuit. Motor current is limited by altering the time duration of voltage at the motor. This pulsing technique is achieved with silicon-controlled rectifiers (SCRs). The SCR can best be thought of as an electronic switch that turns the voltage to the motor on and off at a rate that is determined by the position of the accelerator pedal. When the accelerator pedal is initially activated, a few pulses of voltage and current per second are applied to the motor. As the accelerator pedal is further depressed, the pulse duration and width increases at a fairly linear rate. By varying the number of power pulses per second and the duration or width of each individual pulse, the chopper circuit effectively limits the current applied to the motor.

ELECTRONIC DIFFERENTIAL-A motor at each rear wheel eliminates the differential, but it is still necessary that motor speeds be controlled separately to compensate for speed differences when turning. This will done by monitoring the reduction gear housing torque on each wheel and controlling each motors speed to equate the torques.

MISCELLANEOUS

This vehicle has several features that save power when in battery only mode. Neither the heater nor the air conditioner can be operated if the doors are open. It is recommended that the vehicle be initially cooled or heated in other than battery only mode. This will greatly extend the battery range. The air horn has a reserve tank that can only be charged when the turbine is running or by external power.

The high pitched whine from a turbine can be irritating, this has been overcome with a Reversed Audio Phase Noise Cancellation System (RAPNCS). This system monitors the noise in the engine compartment and generates a anti-wave (sound waves that are 180° out of phase) to cancel out these waves. The resulting sounds are of a lower amplitude and frequency.

The air conditioner is a compact Non-CFC unit that uses about 1.5 kW (2 hp).

Higher voltage lighting (24 or 48 volt) was considered to lower the wiring weight due to lower amperage hence smaller diameter wire. This idea was discarded due to limited availability of higher voltage lighting (aircraft applications mostly).

Tires are Goodyear low rolling resistance P165/65R-14, that have a coefficient of rolling resistance of about 0.006. Tire rolling resistance increases dramatically at speeds above about 75 mph due to increased hysteresis. This is another reason for limiting vehicle speed (to significantly limit battery drain).

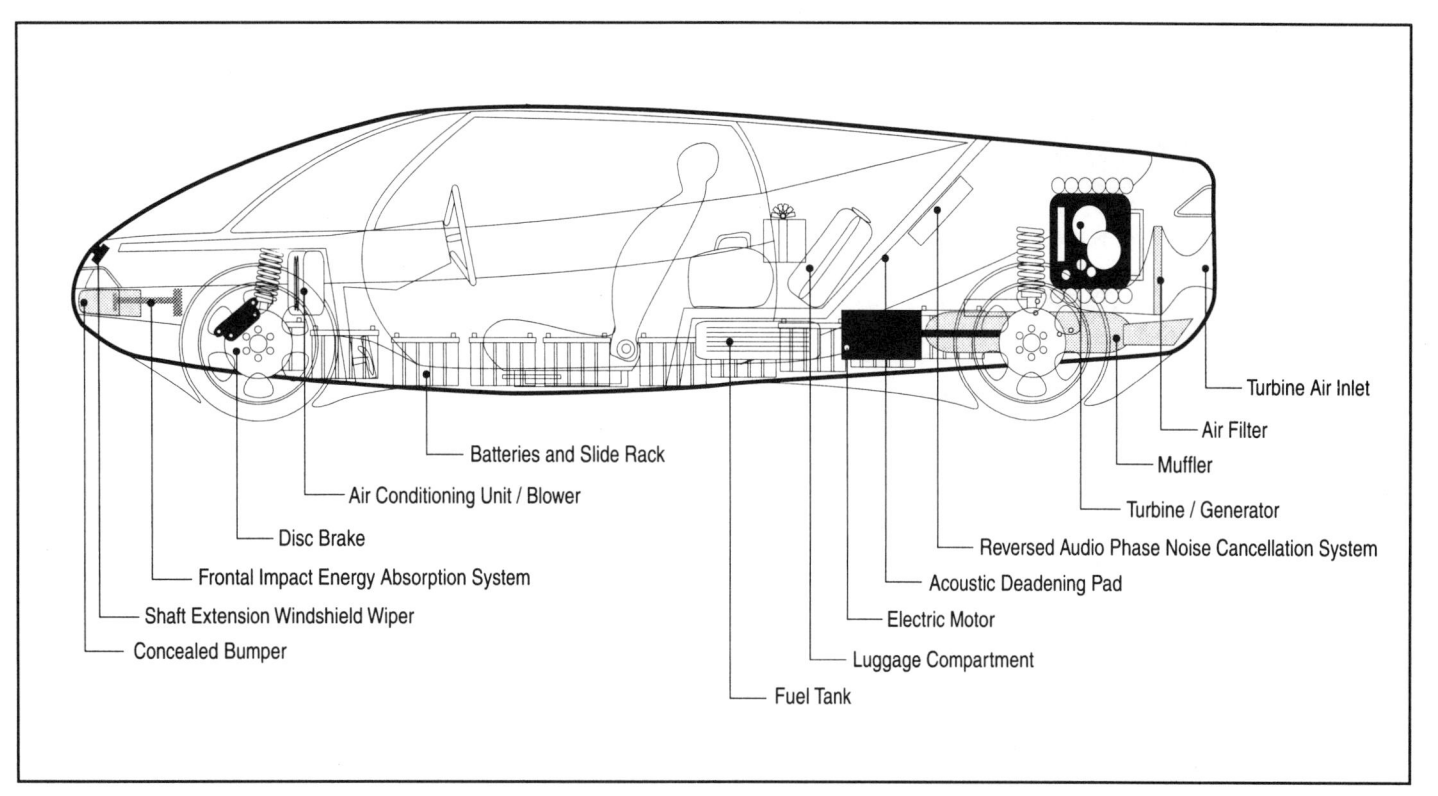

FIGURE 9: TRANSPARENT SIDE VIEW

SPECIFICATIONS

GENERAL

Rear-Engine, Rear Wheel Drive, 2-Door, 2 Passenger, Hybrid Electric

MAJOR EQUIPMENT

Air Conditioning, Dual Air Bags, AM/FM-Stereo/Cassette

ENGINE

NoMac Turbogenerator, Single stage, Centrifugal Compressor, Single Stage, Radial Inflow Turbine, Permanent Magnet Generator, 96,000 RPM, 24 kW Continuous at 1 atmosphere and 35°C (95°F), 30 kW Peak, 30% Efficiency from Fuel-in-to-Electricity-Out

MOTORS

(2) High Speed Brushless DC 18.6 kW (25 hp) Motors,

DRIVETRAIN

Direct Drive with 8:1 Gear Reduction, Traction-Control System

STEERING

Rack and Pinion
2.8 Turns Lock to Lock
Turning Radius 9.8 m (32 ft)

FUEL

Ethanol, Methanol, Kerosene, Jet Fuel, Diesel, Propane, Natural Gas, Unleaded gasoline, Butane

MEASUREMENTS

Wheelbase	2.54 m	(100 in)
Track Front	1.52 m	(60 in)
Track Rear	1.27 m	(50 in)
Curb Weight	674 kg	(1485 lb)
Weight Distribution	F/R	55/45
Ground Clearance	0.018 m	(7 in)
Coefficient of Drag		0.15
Fuel Capacity	19 l	(5 gal)
Cargo Capacity	0.5 m^3	(18 ft^3)

SUSPENSION

Independent Front with Damper Struts, Coil Springs, Independent Rear with Semi-Trailing Arms

BATTERIES

(10) Deka Solar, Sealed Lead-Acid
12 Volt, 45 Amp Hour
24.5 kg (54 lb) each
Recharging Time 240 VAC   3.5 hrs
Recharging Time 120 VAC   8.0 hrs

FRAME/CHASSIS

Frame and Chassis Type: Aluminum Alloy Modified Ladder Type with High Strength Structural Fiber Sub-Frame and Semi-Monocoque Outer Skin

WHEELS and TIRES

Goodyear Low Rolling Resistance P165/65 R-14, Cast Aluminum Wheels

PERFORMANCE

hybrid   0 to 60 mph   13 sec
battery   0 to 60 mph   13 sec
turbine   0 to 60 mph   13 sec
Hybrid Range   813 km   (505 mi)
Battery Range   89 km   (55 mi)
Turbine Range   724 km   (450 mi)

REFERENCES

Abbott, I.H. and von Doenhoff A.E., *Theory of Wing Sections*, McGraw-Hill, NY, 1949

Blevis, Deborah L., *The New Oil Crisis and Fuel Economy Technology*, Quorum Books, NY, 1988

Deka Solar, "Manufacturers Data Sheets", distributed by Stored Energy Company, Spring House, PA, 1992

Judge, W. Arthur, *Small Gas Turbines*, The Macmillian Company, NY, 1960

Mackay. Robin, "Gas Turbine Generator Sets for Hybrid Vehicles", Journal Article No. 920441, Society of Automotive Engineers, Warrendale, PA, 1992

NoMac Energy Systems, "Manufacturers Data Sheets", NoMac Energy Systems, Rolling Hills Estates, CA., 1992

Odgus, J., *Gas Turbine Fuels and Their Influence on Combustion*, Turnbridge Wells, Kent & Cambridge, MA, 1986

Paul, J.K., *Methanol Technology and Application in Motor Fuels,* Noyes Data Corp., NJ, 1978

Robert Bosch Corporation, "Manufacturers Data Sheets", Robert Bosch Corporation, Broadview, IL, 1992

Setright, L.J.K., *Some Unusual Engines*, Mechanical Engineering Pub. Inc., London, England, 1975

Society of Automotive Engineers, *Engines, Fuels, Lubricants, Emissions, and Noise.* Vol 3, Report No. SAE J1297, Warrendale, PA, 1990, Jun 1989

Unique Mobility, Inc., *DR Series Motor Data Sheet*, Rev. A, UNIQUE MOBILITY, INC., Englewood, CO., Nov, 1991

Deka SOLAR

The Deka Solar series of valve-regulated, gelled-electrolyte batteries is designed to offer reliable, maintenance-free power, for renewable energy applications where frequent deep cycles are required and minimum maintenance is desirable.

Applications
Water pumping • Residential • Communications
Cathodic protection • Remote monitoring • Refrigeration
Lighting • Aids to navigation • Wind generation

Specifications
- Voltage .... 12 volts nominal
- Plate alloy .... Lead calcium
- Element, post ...... Threaded stud or "flag" terminal, forged bushing
- Container/cover ... Polypropylene
- Charge voltage ... Cycle 2.30 to 2.35; Float 2.25 to 2.30
- Specific gravity ... 1.280
- Electrolyte .... Sulfuric acid thixotropic gel
- Vent .... Self sealing (2PSI operation)

Cycling Ability
Number of cycles vs. depth of discharge at +20°C (77°F) discharge with 20 hour rate

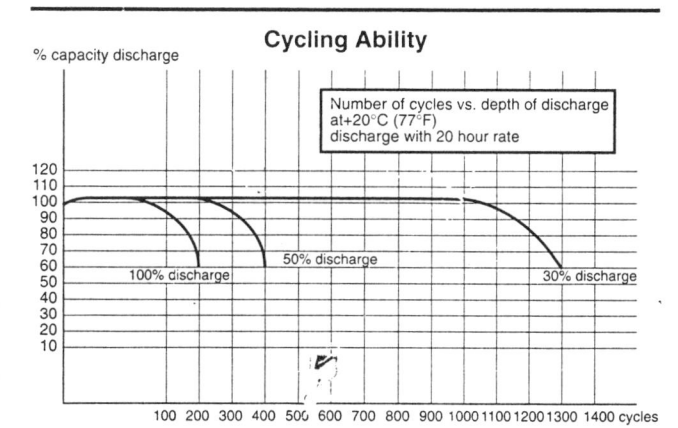

The solar battery excels in cycling applications.

CAPACITY WITHDRAWN	CYCLES
100%	200
50%	400
10%	2,000
1%	20,000

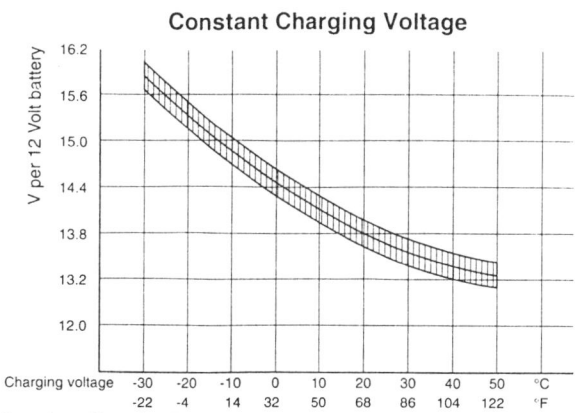

Constant Charging Voltage
Shown is the constant charging voltage in relation to the ambient temperature. The bandwidth shows a tolerance of ± 30mV/cell. This constant voltage is suitable for continuous charging and cyclic operation. In a parallel stand-by mode it always keeps the battery in a fully charged state; in a cyclic mode, it provides for rapid recharging and high cyclic performance.

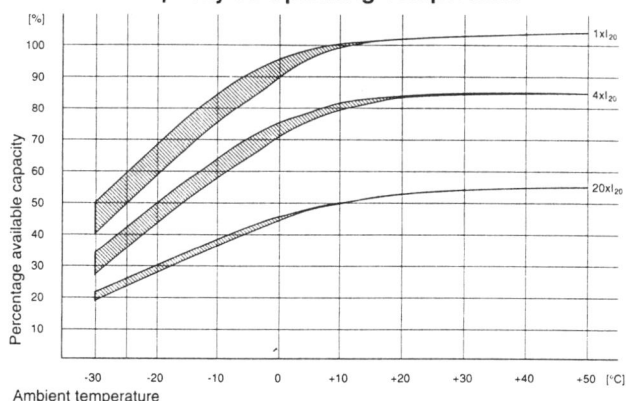

Capacity vs Operating Temperatures
Shown are the changes in capacity for wider ambient temperature range, giving the available capacity, as a percentage of the rated capacity, at different ambient temperatures, for 3 different load examples, with uninterrupted discharge to the appropriate discharge cut-off voltage. The values for the upper edge of the curve were obtained from charging at an ambient temperature of +20°C with a voltage limit of 2.3 V/cell. For the lower edge, charging was carried out at the specified ambient temperature. The curves show the behavior of the battery after a number of cycles.

Terminal Information
T874 | T881 | T872 | SAE

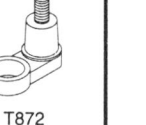

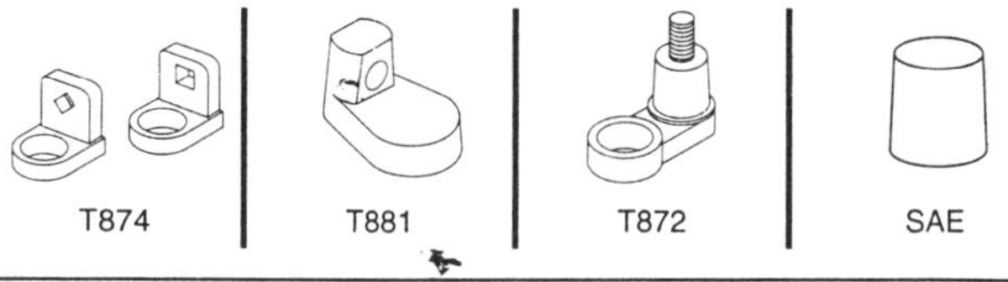

Discharge Amps per 12 volt unit to 1.75VPC at 80°F

Type No.	Foot Notes	Terminal	5 Min	10 Min	15 Min	20 Min	30 Min	60 Min	90 Min	3 Hr	6 Hr	8 Hr	20 Hr	24 Hr	48 Hr	100 Hr	Approx. Wt. Lbs. (Kgs.)	L	W	H	Pallet Qty.
8GU1	H	T874	90	70	50	42	32	20	14	8	4.34	3.37	1.55	1.31	0.698	0.364	24(10.9)	7³/₄(197)	5¹/₈(130)	7¹/₄(184)	120
8G22NF	FH	T881	135	95	80	70	50	30	23	12.34	6.84	5.31	2.4	2.04	1.09	0.565	39(17.7)	9³/₈(238)	5¹/₂(140)	9¹/₄(235)	72
8G24	CEH	T872	215	160	120	100	74	45	33	19	10	7.81	3.5	2.94	1.55	0.8	54(24.5)	10⁷/₈(276)	6³/₄(171)	9⁷/₈(251)	44
8G27	CEH	T872	255	195	150	125	95	56	41.5	24	12.75	9.81	4.3	3.6	1.94	0.98	64(28.9)	12³/₄(324)	6³/₄(171)	9⁷/₈(251)	36
8G4D	CH	SAE	485	375	290	245	185	114	83	47	26.7	20.5	9	7.71	4.14	2.1	135(61.2)	20³/₄(527)	8¹/₂(216)	10(254)	16
8G8D	CH	SAE	600	470	350	305	235	140	102	59	33	26	11.25	9.58	5.2	2.65	168(76.1)	20³/₄(527)	11(279)	10(254)	12
8G30H	CH	T872	295	215	165	138	105	62	46	26	14	10.67	4.75	4.0	2.16	1.09	73.5(33.37)	13⁹/₁₆(345)	6¹³/₁₆(173)	10³/₁₆(258)	30

ALL RATINGS ARE AFTER 15 CYCLES AND CONFORM TO B.C.I. SPECIFICATIONS.

IMPORTANT CHARGING INSTRUCTIONS:
WARRANTY VOID IF OPENED OR IMPROPERLY CHARGED.
Constant under or overcharging will damage any battery and shorten its life! Use a good constant potential, voltage-regulated charger. Solar panel should have an in-line voltage regulator. **Charge to at least 13.8 volts but no more than 14.1 volts at 68°F (20°C).** Do not charge in a sealed container.

Footnotes:
- C - Includes handles
- E - Combination terminals, offset with 5/16" stainless stud and wing nuts
- F - Combination terminals, offset post with horizontal hole, 5/16" bolt and hex nut
- H - Polypropylene container and cover

Distributed by: SEC

Bill Brooks
Stored Energy System
Spring House, PA
(215) 654-9334

© EPM 1992 Printed in U.S.A.

The Concept of a Sporty HEV for Commuting and Leisure

Petros Frantzeskakis
Dennis Dionatos
Bruno Forcione
Stas Katsiaras
Harry Kekedjian
George Metrakos
Achilles Nikopoulos
Chris Psinas
A.J. Whitman
Concordia University

ABSTRACT

The demand for economical, efficient and environmentally clean high performance vehicles has reached an all time high in today's automotive society. A vehicle design using advanced materials to reduce weight is proposed in this report. A parallel hybrid vehicle concept was found to benefit from the best of all possible drive configurations including an electric and alternative fuel drive. It was established that lead-acid batteries best suit the needs of the proposed design, due to their low cost and satisfactory performance. The procedure for modifying a 1.9 ℓ engine to run on methanol (M85) fuel was outlined and the performance was projected. The control strategy with circuit diagrams for the different modes of operation were outlined. A mathematical model for the different drive configurations was developed followed by a simulation and optimization. Imposed performance requirements for acceleration, gradability and range were achieved in all modes of operation. Also being discussed is the feasibility and the environmental impacts of such a vehicle.

INTRODUCTION

The proposed hybrid electric vehicle (HEV) concept benefits from the best of all possible drive configurations including electric, alternative power unit (APU) and hybrid modes. An electric configuration is advantageous because of its potential emission-free driving. In dual-mode operation the electric motors provide high torque at low speeds, while the internal combustion engine (ICE) delivers high torque at higher speeds. [1]

The primary purpose of designing an HEV is the need to be less dependent on fossil fuels. Thus in a mode where repetitive stopping and accelerating is required, such as city traffic, an electric motor is a perfect alternate power source to consider. The main advantage of using this design is that air pollution in the urban area is decreased (ideally eliminated once electric vehicle use becomes widespread).

Electric vehicle technology is the most important automotive accomplishment which will enable zero-emission vehicles (ZEVs) to exist. The governments' willingness to pursue such technologies will be a relief to the urban community allowing for cities to "breathe" again. California has already passed legislation mandating that 2% of an automaker's sales in 1998 to be of ZEVs and 10% in 2003. Electric vehicles will meet these requirements being the most viable choice in terms of available technologies. [2]

In this design paper an attempt will be made to convey our interpretation of how a Hybrid Electric Vehicle should be conceived and developed. Vehicle safety is an essential aspect in the design and an environmental assessment will be a top priority before completing this vehicle. Also an investigation of the resources for proper disposal and recycling of used batteries will be effected. Development of a viable electric vehicle is the first target. The following goals were also considered to be very important:

• lowered cost of conversion, allowing effort to be placed on durability, driveability, and safety, and
• lowered cost of production,

permitting the implementation of HEVs alongside

gasoline vehicles. Figure 1 is the proposed layout if our ideas will lead us to produce such a vehicle in the future.

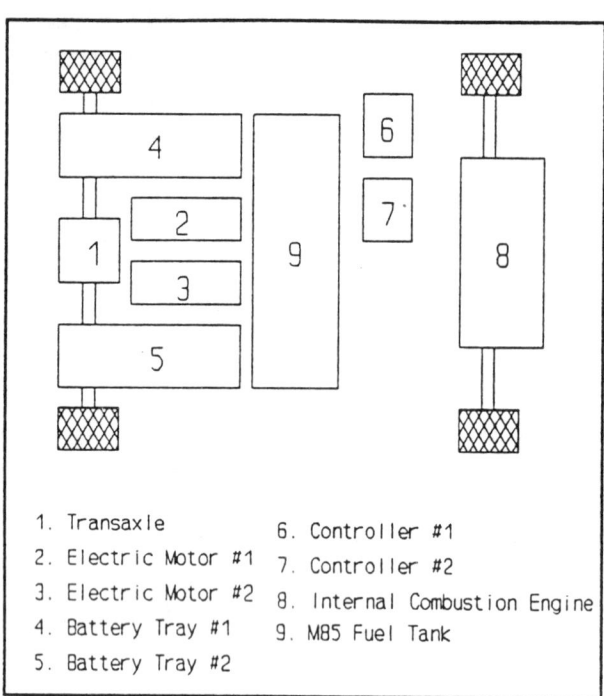

Figure 1 Layout

1. Transaxle
2. Electric Motor #1
3. Electric Motor #2
4. Battery Tray #1
5. Battery Tray #2
6. Controller #1
7. Controller #2
8. Internal Combustion Engine
9. M85 Fuel Tank

VEHICLE DESIGN

To comply with strict environmental and fuel efficiency requirements, a low vehicle weight should be a design target without contributing significantly to the overall cost. At the same time a lightweight small exterior body must be designed without sacrificing interior styling, comfort and the general aesthetic appeal. An emphasis will be placed on safety and reliability, while minimizing the amount of major modifications that will be required for the mass-production of such a vehicle. An aluminum space-frame sheathed in reinforced plastic composite panels is the best approach to realise these goals. [3] [4]

A multi-cell interlocked body of extruded aluminum members will be used in a slip-fit arrangement which are then adhesively bonded and riveted together. [5] These methods of joining components will withstand high static and fatigue strengths dealing with both shear and peel loads (figure 2).

Having a higher strength-to-weight ratio as compared to steel, the aluminum body (taking into account the larger volume) will produce a weight savings of at least 25%. [6] The frame will be fitted

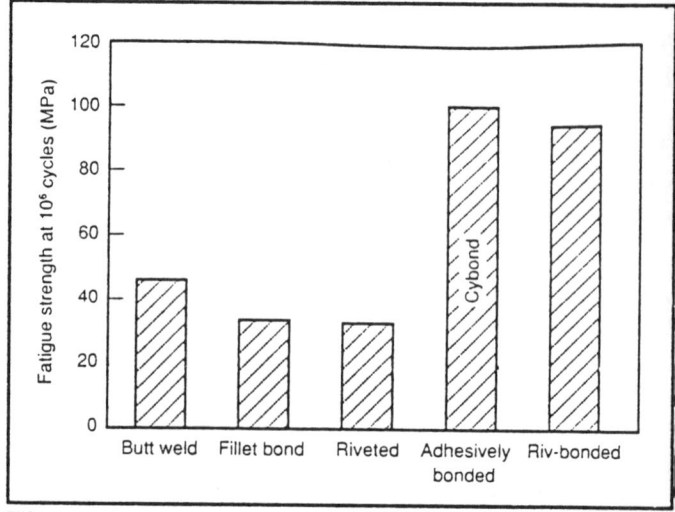

Figure 2 Joining method fatigue strengths for space frame 6xxx alloys

with an existing vehicle exterior comprised of few protrusions for improved aerodynamic efficiency. Ideally, the door and mid-section components will be reaction injection moulded (RIM) reinforced panels resulting in a 20% weight savings (as compared to steel) while still providing a stiff and a safe overall structure. Plastic composites are also corrosion resistant thus eliminating the need for weld sealers, pre-coated metals and E-coating. [7]

SUSPENSION - One of the goals in this design is to create responsive and spirited handling by incorporating a simple but effective suspension technology. Both the front and rear suspensions will be composed of glass-fibre composite transverse leaf springs leading to a savings of 20 kg; replacing the functions of the coil spring, lower suspension arm and some of the stabilizer bar's functions. The leaf spring will eliminate the conventional rear strut towers and consequently increase interior storage space.

COMPONENT PLACEMENT - A vehicle based on our proposed design with an anticipated curb weight in the range of 900-1000 kg is illustrated in figure 3. The internal combustion engine driving the front wheels will be placed in the front compartment whereas the electric motors will be coupled to a central transmission and differential driving the rear wheels. Enclosed within the

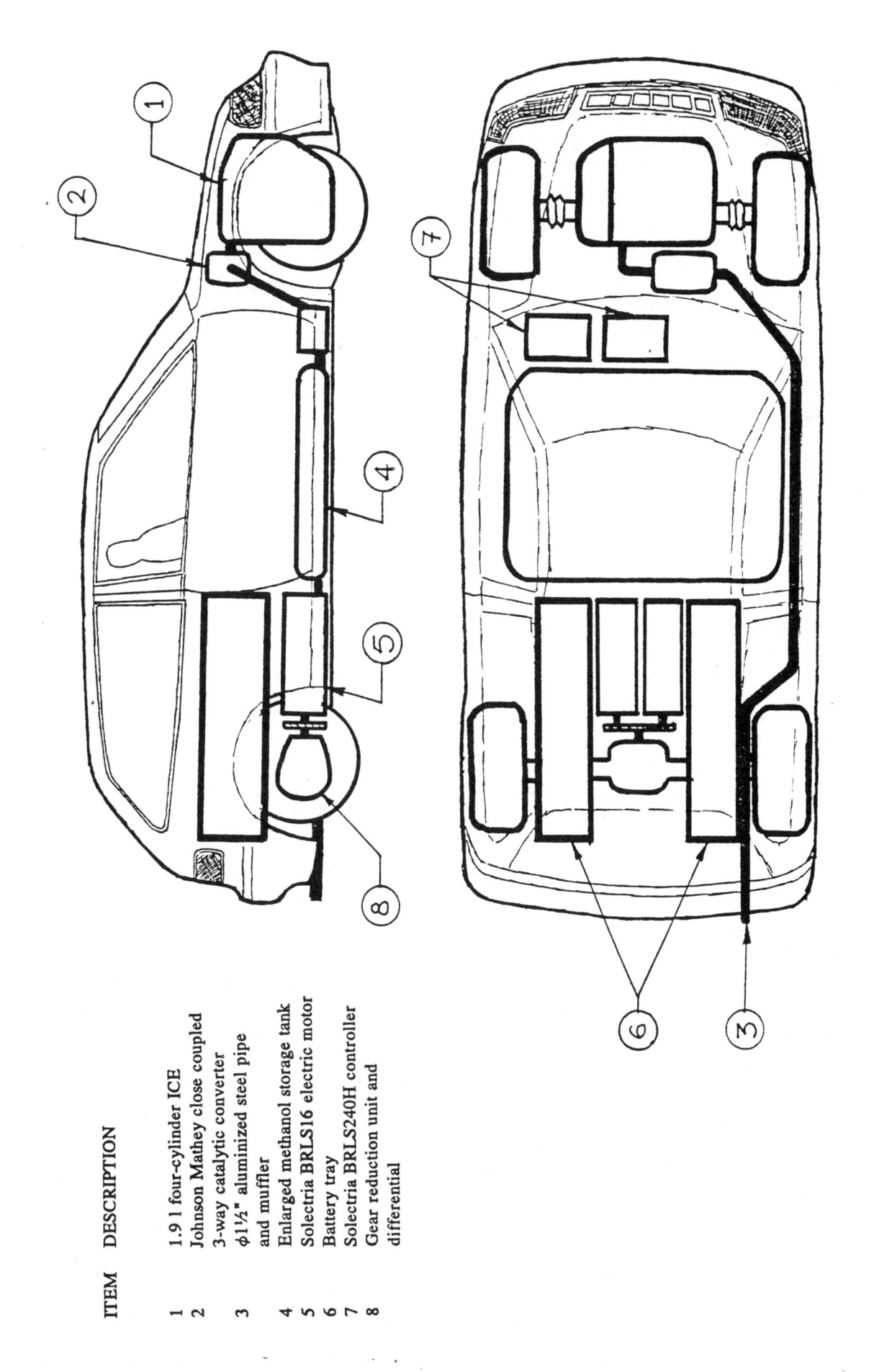

ITEM	DESCRIPTION
1	1.9 l four-cylinder ICE
2	Johnson Mathey close coupled 3-way catalytic converter
3	ϕ1½" aluminized steel pipe and muffler
4	Enlarged methanol storage tank
5	Solectria BRLS16 electric motor
6	Battery tray
7	Solectria BRLS240H controller
8	Gear reduction unit and differential

Figure 3 Component Placement

interior, the shielded battery packs will be securely attached to the structure in the longitudinal direction on either side of the differential. Although isolated from the passenger compartment, the battery packs' crashworthiness is crucial. By means of a stronger member, collision impact forces will be softened and battery tray motion will be delayed long enough to allow for the compression of the vehicle's extremities.

COMFORT - Occupant comfort is a critical aspect in an automotive design. An interior composed of a cockpit-like control cluster will allow easy entry and exit to and from the vehicle accommodating a minimum of two people. Since convenience options and accessories greatly influence the comfort and pleasure levels in the driving experience, a centralised practical energy-efficient lighting system, a radio system and HVAC unit will be included.

CLIMATE CONTROL- To control a more efficient automobile one has to re-examine many systems, such as climate control. It is desired to modify the air conditioning system to ensure that we have a capable, productive and practical vehicle in today's energy conscious world. Automotive air conditioner's are presently using the CFC-12 refrigerant (Freon), which is not environmentally friendly. Slowly being phased out, Freon will be replaced by the hydrofluorocarbon HFC-134a. A conventional cooling system will be incorporated in this vehicle with HFC-134a. It will be required though, to increase the new refrigerant's cooling performance in order that the new system is as energy efficient as with Freon. Some sections of the system will be replaced with a new compressor and a multi-flow heat exchanger to increase the environmentally friendly performance of the hydrofluorocarbon HFC-134a. Conventional serpentine condensers can be replaced with more efficient multi-flow condensers. A comparison of three condensers heat soaked for two hours until the interior temperature reaches 60 °C is represented in figure 4. [8]

Another possible modification will be the inclusion of a solar-panel operated, waterproof automotive ceiling fan. This will aid to further decrease the compartment temperature.

In order to keep passengers cool in the summer, conserving the interior's look and also minimizing

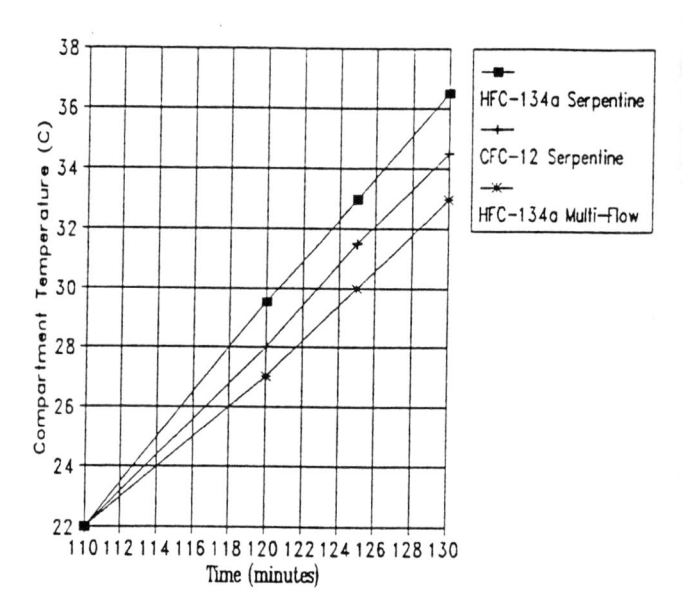

Figure 4 Condenser Analysis

the air conditioner's cooling load, the following products will also be used:

- solar control glass - lowering UV transmissions
- solar coated windshields - reducing solar heat gain and UV.

Two heating units will be incorporated in this design. In APU mode the vehicle's heating will operate similar to that of a conventional system. In the EV mode an electric heater will provide the majority of the heating. In HEV mode the exhaust gases will be used as a heat exchanger and surplus heat will be provided by the electric heater. The HVAC system will be powered from an interfaced DC-DC converter tapped from the main battery supply at 12V.

BRAKES- Safety will be assured with a four disc vented brake setup. Using an ITT Teves compact single unit brake assembly (2 kg lighter than conventional system) along with composite aluminum rotors and lightweight low drag calipers, a reliable, low brake drag system will be produced. [9]

OVERALL - Aluminum/magnesium alloyed flush-designed rims with low resistant tires will compliment the overall performance criteria. This vehicle will give the buyer/user the benefits of current technology and versatility as well as a taste of future innovations.

DRIVETRAIN DEVELOPMENT

ELECTRIC DRIVE SELECTION- The electric drive for Concordia's hybrid vehicle will propel the rear wheels through a custom built transmission using two electric motors. Choosing an electric motor was the first step in the design layout. Electric motor technology for EVs can be grouped into five classes:

- Synchronous-type motor
- Induction motor
- Series wound DC motor
- Shunt wound DC motor
- Brushless DC motor.

Synchronous and induction motors both require AC voltage and have some inherent advantages to their design. These motors are generally smaller and have greater efficiency than DC motors. AC motors are widely used in industry and therefore are mass produced and typically cheaper than comparative DC motors. A major disadvantage is that AC motors require AC current which must be converted from the DC battery pack aboard the vehicle. Most AC motors are made to run on higher industrial voltages and therefore, special costly and heavy electronic inverters as well as a high voltage battery pack or an additional stepping up of voltage, is required. In our view, these added weight and space requirements far outweigh any benefits derived from using AC motors. [10]

Series wound motors have their field coil connected in series with the armature. The current in the armature is the same current that flows in the field windings. The torque in the armature increases faster than the current applied. This is necessary since it will allow for reasonable acceleration torque required to move the vehicle from rest. [11]

Shunt wound motor have their field windings connected in parallel with the armature. High resistance in the field coils allow the field to absorb approximately 5% of the current through the armature. The torque produced is directly proportional to the current. Any increase or decrease on the load in the armature has no effect on the speed at which the motor turns. The stall torque is relatively low and therefore this motor is not commonly used as a vehicle drive motor. [12]

Compound-wound motors use both a series and a shunt field coil, and therefore have many of the advantages of both motors described above. Different configurations and proportions often provide for very good characteristics for an EV motor. [13]

Brushless DC motors behave similar to brush-type DC motors, however the operation of the commutator and brushes is replaced by semiconductors and position sensors. This allows for greater efficiency and speed. [14]

Permanent magnet motors behave very much like compound-wound motors but they usually have a higher stall torque and are very efficient. Permanent magnets provide for the magnetic field and therefore a field coil is not required. In addition, the resistive losses incurred by the coils is eliminated. [15]

Taking into consideration the power, weight and efficiency requirements of the various motors available, the brushless DC permanent magnet (PM) motor offered us the best solution. It is the brushless DC permanent magnet (PM) motor which offers us the best solution for power, weight and efficiency considerations. The Solectria BRLS16 PM motors have been selected to drive the vehicle's rear wheels. These motors with the appropriate controller supplied by Solectria provide for, tractive power as well as for regenerative braking, converting kinetic energy during braking into electric energy for recharging the battery pack. [16]

Solectria's motor was chosen over the next most suitable motor made by Unique Mobility due to the significantly higher costs of the latter.

ELECTRIC DRIVE CONFIGURATION- The electric drive configuration (Figure 5) will use two Solectria motors in order to produce enough stall torque to accelerate the vehicle at an acceptable rate. Power will be transmitted from the motors to the transmission via standard 630 type motorcycle chains and sprockets. Each motor is connected independently to its own sprocket which is then connected to the transmission.

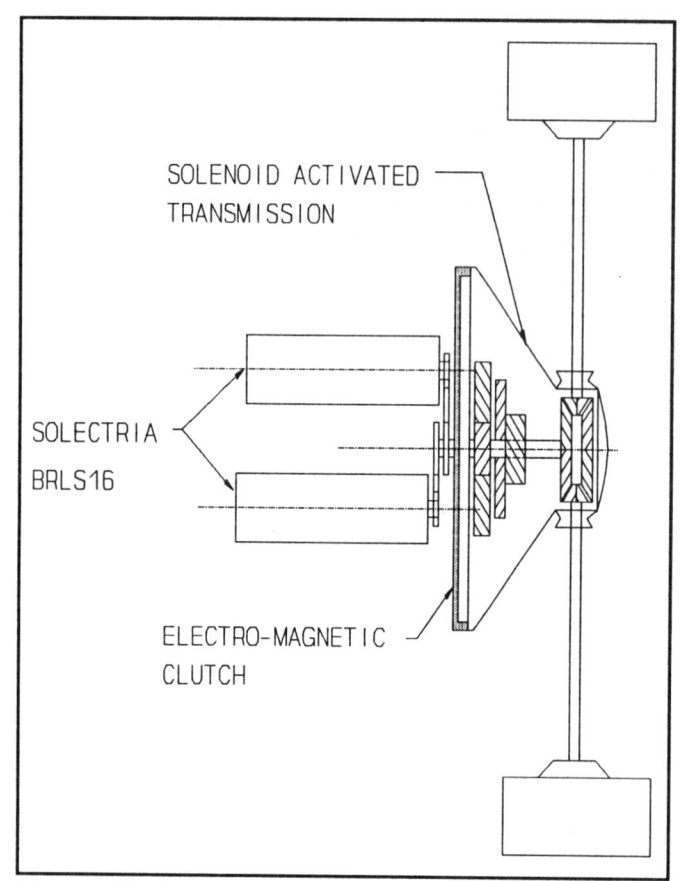

Figure 5 Electric Drive Configuration

In order to accommodate the specific needs, in terms of gear reduction, for the final drive from the motor output to the wheel, the electric drive system transmission will be custom designed. The transmission gear ratios will be selected with the help of a computer based simulation.

This two speed gear box will have a low gear that is used for acceleration and a cruising gear for higher speeds. Gear changes will be engaged by the driver with electronic controls on the steering wheel.

When the driver has selected a gear, a signal is sent to an electro-magnetic clutch. This clutch consists of a drum and a rotor assembly that is filled with ferrite powder. [17] The driver's signal will send current to a coil which magnetizes the powder and allows the rotor and drum to turn together. The drum is connected to the sprockets from the PM motors and the rotor is connected to the geartrain.

Disengaging the clutch sends a signal to a set of solenoids that operate a selection lever in the gear box which will mesh the appropriate driver selected gears. Power is returned to the clutch at this point to magnetize the ferrite powder and couple the motors and transmission. Further description of how this system is controlled, can be seen in the controls section.

The transmission housing will also function as the main support for the two motors. This design will eliminate excess bracket weight and will allow the entire drive system to perform as a single compact unit.

The output from the two speed rear transmission will be connected to the rear differential. Half-shafts, connected at both ends of the rear differential using constant velocity joints, will drive the rear wheels. This design allows the use of the transverse composite leaf.

Vehicle operation, in the electric mode, will consist of two electric motors operating synchronously using two controllers. These two controllers will receive a signal from one potentiometer which will control the PM motors' operation. In hybrid mode the electric motors will provide additional low-end torque. Along with the ICE, high load demands such as hill climbing and during hard acceleration will be accommodated. The ICE has a specific operating range where it is most efficient, therefore, it could be used over a long period of time, especially under highway cruising conditions where the speed is relatively constant. The electric motors would be used in the city and as an additional power unit for either rear wheel or four wheel drive operation.

PARALLEL DRIVE CONFIGURATION - Whereas series configuration uses a fuel driven generator to produce electricity for the electric motors, parallel drive configuration is established by directly coupling the ICE with the transaxle. Parallel configuration was chosen over series configuration because of overall efficiency considerations. A series configuration requires an alternator to convert mechanical energy to electrical energy and then the electrical energy is converted back to mechanical energy through the controller and the electric motor. This configuration introduces two additional efficiencies as compared to the one drivetrain efficiency of parallel configuration. The advantage of the series configuration is that the ICE is kept at a constant speed and therefore the exhaust emissions are easier to control. This benefit is overshadowed by the lower overall operating efficiency. This can be summarised in the following calculations:

Series:

Electro-mechanical efficiency is obtained for a Fisher Electric Motor Technology Inc. alternator (0.92). [18] Combined electric motor and controller efficiency is for the Solectria Corp. BRLS16 motor and BRLS240H controller (0.92). [19]

$$\eta_{overall} = \eta_{elec-mech} * \eta_{elecdrive} * \eta_{drivetrain}$$

$$\eta_{overall} = 0.92 * 0.92 * 0.90 = 0.76$$

Parallel:

$$\eta_{overall} = \eta_{drivetrain} \approx 0.90$$

The parallel configuration requires that the ICE is run at variable speed and variable load. These changes lead to transient effects which requires a closed loop fuel deliver strategy in order to achieve stoichiometry and minimum exhaust emissions.

BATTERIES

BATTERY OVERVIEW - Selecting batteries for electric vehicle (EV) or hybrid electric vehicle (HEV) applications is a lengthy process of negotiating trade-offs between battery characteristics in order to reach the best available compromise. Battery cell and module characteristics must be compatible with the vehicle requirements for a particular battery pack to be used. Since space and voltage requirements vary significantly for various vehicle configurations, no single cell or module can be expected to be suitable for all vehicles.

HEV battery systems must be able to:

• withstand continuous charging and discharging on a daily basis,
• function under extreme working conditions such as high and low ambient temperatures,
• withstand vehicle vibrations and
• allow for fast recharging.

Despite a considerable research effort into the development of advanced batteries for electric vehicles, the only practical battery systems commercially available are still lead-acid and nickel-cadmium (nicad). A direct comparison (Figure 6) between similar lead-acid and nicad systems reveals that lead acid provides a comparable performance at a fraction of the nicad system's cost. [20]

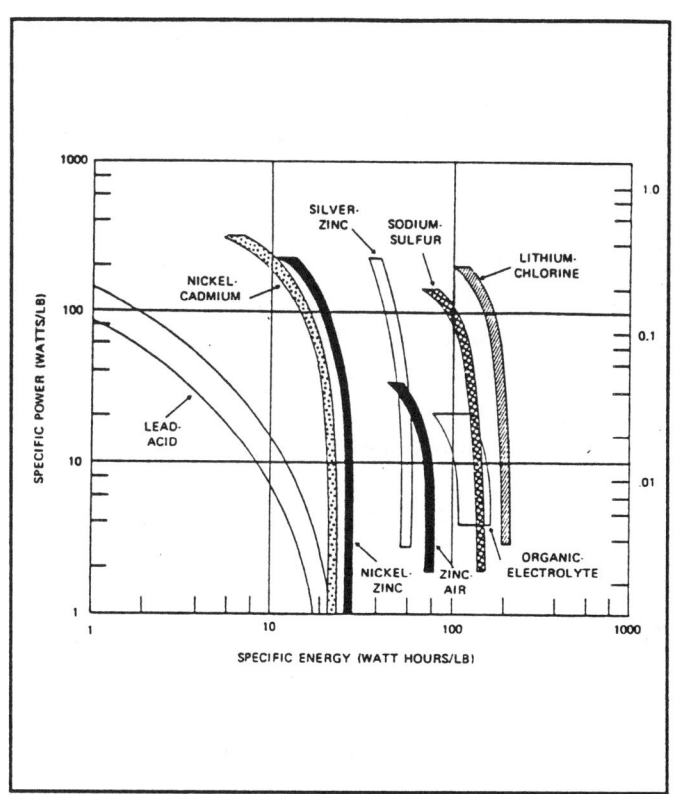

Figure 6 Battery Characteristics

Another disadvantage of alkaline batteries is the higher cost of nickel, and nicad batteries would need approximately 70% more cells in order to produce the same battery voltage as lead-acid batteries. In general, nicad systems are better cold weather batteries and are able to endure more abusive working conditions such as deep discharging and fast charging. However, even with the possible addition of a thermal management system for the colder climates of the northern regions, the cost and module weight of lead-acid make them more attractive than the nicad batteries. It is the simplicity in design of the lead-acid module that allows for low cost production on a local basis. This, in turn has resulted in the development of a solid industrial network of production, distribution and recycling in Canada and the United States. [21]

HEV BATTERY REQUIREMENTS - The preferred lead-acid battery technology, for EV and HEV use, are of the deep cycle sealed lead-acid

(SLA) type. SLA batteries are relatively maintenance free. They eliminate the need for periodic watering because of built-in recombination properties and do not spill acid even when overturned. If extensive overcharging occurs, the battery casings are designed to operate as a "check-valve" allowing excess pressure to escape, while preventing atmospheric particles from penetrating inside. Adversely, SLA batteries do not allow for a fill up of water after venting, resulting in a premature battery failure due to dehydration. Premature failure due to gassing is not a major concern since gassing can be significantly reduced with proper recharging techniques, the recombinant properties incorporated in the SLA design, and the use of a gelled electrolyte or absorbed electrolyte design (in certain models).

BATTERY SELECTION - The power and energy density requirements of a vehicle can be estimated through vehicle simulations (see performance section).

The Delco, Stowaway and Optima modules possess roughly the same capacity, however the Optima has the highest energy density.

Power requirements will finally depend on the target use of the vehicle, the overall efficiency desired, the volume and weight constraints, and the availability of batteries in the range of power and energy required. This project's goal is to design and build a vehicle at the lowest cost with high efficiency and excellent performance characteristics. The vehicle is to be built and modified with commercially available, off the shelf accessories that would be within the reach of the ordinary consumer. These are the main reasons for the use of lead-acid batteries to drive the Concordia HEV (CHEV).

The CHEV is to be driven using twin DC motors on a 144 V bus with a maximum battery weight restriction of 320 kg. The weight restriction translates into a module weight of no more than 27 kg, and does not allow for a second pack of fourteen batteries to be connected in parallel for increased capacity. Batteries have been first analyzed based on their capacity in the required weight range. There are several types of 12 V modules commercially available, including the spiral-wound Optima 800, the tubular plate Chloride 6EF78, the flat plate Delco Voyageur and Stowaway ST-154. Their characteristics are summarised in Table 1.

Table 1 Battery Data

PARAMETERS	DELCO VOYAGER	STOWAWAY ST-154	OPTIMA 800	CHLORIDE 6 EF 78
MODULE VOLTAGE	12	12	12	12
APPROX. WEIGHT (kg)	24	27	18	26
BCI GROUP SIZE	27FMF	27	--	--
CAPACITY (Ah) (C/8 RATE)	105	105	105	85
ENERGY DENSITY (Wh/kg)(C/8 RATE)	52.5	46.67	70.0	39.23
EV CYCLES (80% DOD)	300	350	50	350
UNIT COST (U.S.$)	85.00	90.00	75.00	114.00

For a 12 module battery pack, the total weight of each analyzed battery is:

- Delco   : 288 kg,
- Stowaway : 324 kg,
- Optima   : 216 kg and
- Chloride   : 312 kg.

A preliminary calculation relieves that the CHEV would require a continuous draw current of 45 amps to maintain a constant velocity of 72 km/h (45 mph) with a bus voltage of 144 V. Figure 7 depicts typical discharge curves that show the Optima lasting longer than the Delco Voyager.

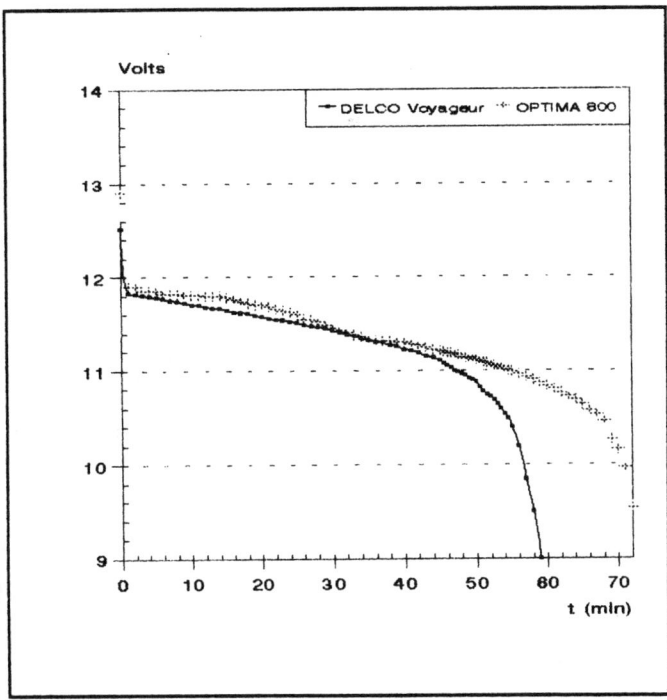

Figure 7 Discharge curves with 45A continuous draw

Although a continuous draw of 45 amps is not realistic in terms of a true driving cycle, it offers a good indication of the duration of the charge available from the batteries when subjected to a fast discharge. The continuous discharge method for battery testing does not take into account other factors such variations in wind resistance, acceleration from rest, temperature and other factors which greatly affect battery performance and capacity. However, under heavy stop and go routines, regeneration can increase capacity by as much as 20% depending on the amount of stopping that takes place.

There are other methods of decreasing the amperage required to propel the vehicle. Increasing bus voltage would reduce the amperage drawn by the motors and result in longer discharge times as demonstrated in Figure 8. A smaller current discharge considerably affects the discharge times. The Optima 800 could last up to 1.35 hours with an amperage drain of 40 amps.

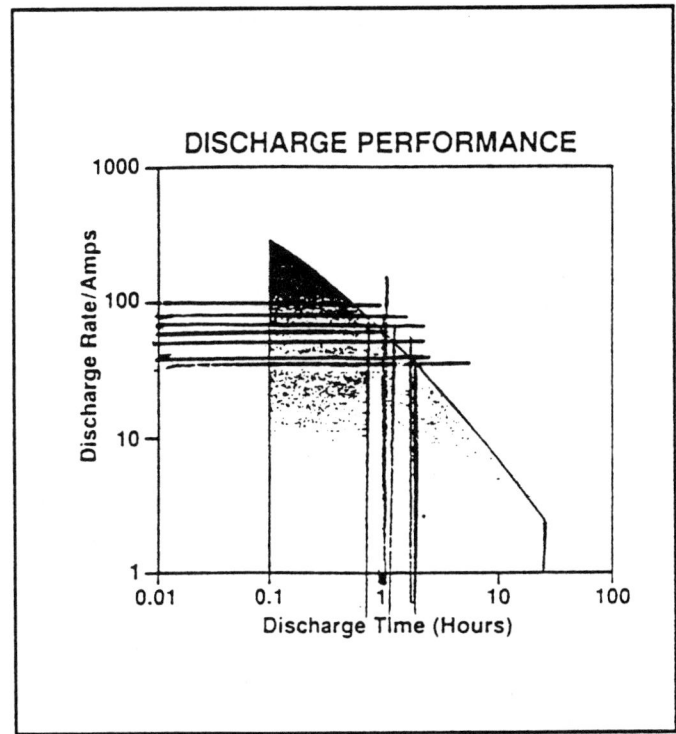

Figure 8 Optima 800 discharge performance

The limitations as to how high the bus voltage can be increased depends on the safety considerations, the maximum permissable weight and the electric motor controller characteristic.
The CHEV team decided that a 144 V and even 156 V bus system would better use the limited capacity of the single series string of lead- acid batteries while still maintaining a reasonably safe operating voltage.

Selection of the SLA battery system will be made based on the batteries that best balance performance, cycle life, capacity and weight. The relatively low SLA battery cost allows for actual testing of the vehicle with different battery packs. Concordia's final decision will be based on these testing results.

CHARGING- The recommended charging practice used for charging sealed lead-acid batteries.

is by "tapering" the input current. Once the current "tapers" to zero the charger automatically shuts off. Two-step "taper" chargers are not recommended for sealed batteries since the second "taper" tends to overgas the batteries. Charging of sealed batteries can never be totally uniform with a taper charger, since the current will taper to the batteries that are the least discharged. It is for this reason that open batteries are often "boiled" in order to bring up the capacity of the pack. This is not possible with sealed batteries since there is no way to replenish diminished water levels. The total possible capacity of the pack is therefore never realized and the batteries are being cycled at a lower than full capacity. [22]

The nominal voltage of the CHEV is a 144V bus powering the electric drive system. The charger that will be used will be a 220V dc, 30A taper charger. Most manufacturers recommend that the batteries be discharged to a depth of 80%. Although it is possible to resuscitate batteries that are completely discharged, such a practice reduces their cycle life damaging them with a possible risk of reversing their polarity. Fast charging is possible to a certain extent, however, it must be kept in mind that recharging is the reversal of the chemical process, therefore it should not be excessive.

METHANOL (M85) FUEL SYSTEM

Methanol has been proposed as an alternative fuel for reducing harmful automotive emissions in areas of poor air quality. Concordia SAE students' experience with methanol has resulted in the decision to use an M85 1.9ℓ engine. The basic requirement for the conversion of a vehicle to a methanol-based fuel system is to ensure that the fuel will not corrode any of the storage and delivery components. Methanol is highly corrosive to aluminium and rubber, and therefore will damage the standard gasoline fuel supply and delivery system.

To provide corrosion protection, certain fuel supply components must be methanol compatible. The aluminum fuel tank shall be made of stainless steel, fitted with a high-flow, methanol compatible fuel pump, and a methanol resistant fuel level sensor. Teflon fuel lines, a methanol compatible fuel rail and fuel pressure regulator will be used.

The stock 1.9ℓ fuel injectors must be replaced with higher-flow methanol-compatible injectors.

Complete optimization of this conversion requires software modifications, to the standard 1.9ℓ calibrations, in the EEC (Engine Electronic Controller). If this information is unobtainable an electronic controller will be built and calibrated, but the efficiency and reliability will be less than by just using the standard EEC and consequently modifying the present tables.

ENGINE CONTROL MODIFICATIONS- Electronic spark timing will be used to fully optimize the combustion efficiency of the engine and decrease HC and NO_x emissions. Due to the higher-flow injectors required, the injector on-time schedule will vary from the standard gasoline calibrations. Exhaust gas recirculation will be used to lower combustion chamber temperatures during high-load conditions. The air to fuel ratio (A/F) will be modified, from the stock gasoline calibration, enabling the engine to operate in a closed loop environment using methanol as the fuel.

Calculation of the A/F ratio for M85 is required for the re-calibration of the stock engine. M85 contains 85% methanol and 15% gasoline by volume.

The stoichiometric equation for M85 can now be written as:

$$0.85 \, CH_3OH + 0.15 \, C_8H_{16} + 3.075 \, (O_2 + 3.76 \, N_2)$$

$$\rightarrow 2.05 \, CO_2 + 2.9 \, H_2O + 3.075 \, (3.76 \, N_2)$$

The A/F ratio for M85 is calculated as follows:

$$r = \frac{mass_{air}}{mass_{fuel}} = \frac{(3.075 \times 32) + (3.075 \times 3.76 \times 28)}{0.85(12+16+4) + 0.15((8 \times 12)+16)} = 9.6$$

POWERPLANT MODIFICATIONS- To increase the mean effective pressure (MEP) of the engine two options were considered. The first option is to decrease the volume of the combustion chamber by either milling the cylinder heads or changing the pistons. The second is to increase the intake air pressure entering into the cylinders by supercharging or turbocharging. The idea to increase the engine's compression ratio was rejected because it involved the risk of an improper engine reassembly and was deemed too simple in terms of educational value. Hence it was decided to supercharge or turbocharge the engine.

Mechanical Supercharging will increase the air density where the supercharger is directly engine driven. The advantages of this system is that:

- the supercharger responds immediately to engine load changes,
- it is relatively simple and is installed on the cold side of the engine and
- the hot engine exhaust gases are not involved in the process.

Supercharging's main disadvantage is that it is being driven by the engine, causing an increase in fuel consumption. It was decided that the installation of an exhaust gas turbocharger (EGT) would be the best option (Figure 9). [23]

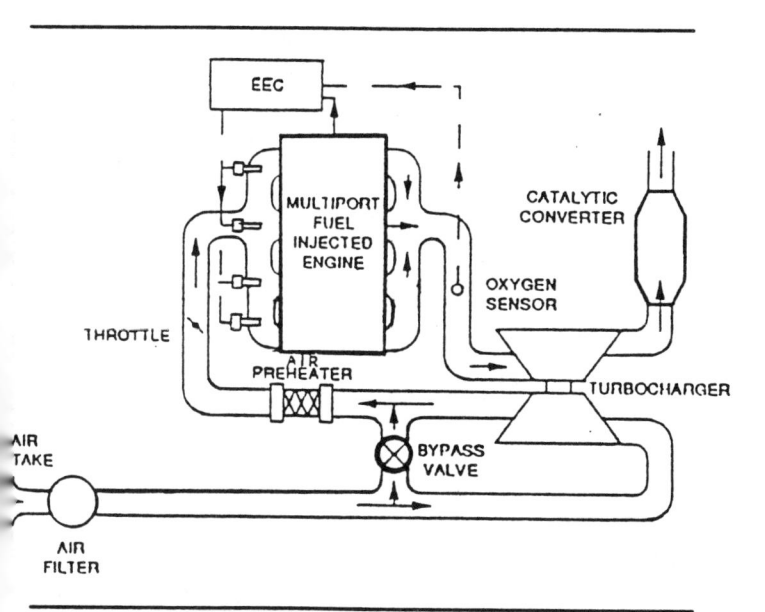

Figure 9. Exhaust Gas Turbocharger

An EGT will amply increase engine power, reduce fuel consumption as compared to a normally aspirated engine of on equal power rating, and also offer an improvement in the exhaust gas emissions. [24] Disadvantages of EG turbocharging are:

- the installation of high temperature resistant materials,
- poor response time to engine load changes and
- a decrease in torque at low engine speeds

affecting the car's driveability. The anticipated performance of the turbocharged engine as compared to the naturally aspirated engine, is shown in Figure 10.

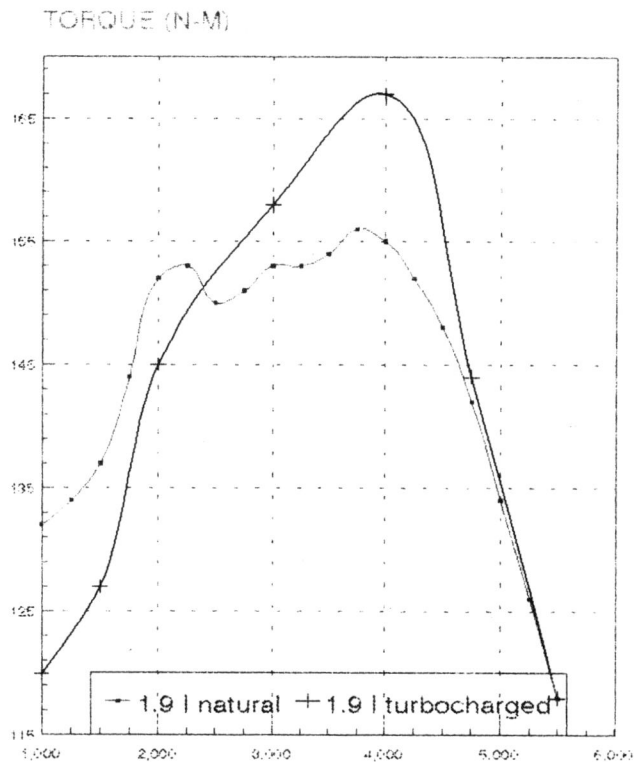

Figure 10 Engine performance

IMPROVED STARTING- Methanol poses some difficulties in cold weather starting because of its high heat of vaporization and low vapour pressure. Given our previous experience with cold starting, during the Methanol Marathon, an air preheater will be installed in the intake air manifold. The heater used is manufactured by NGK Spark Plugs Ltd., and is powered using a 12 volt supply. The air preheater is regulated by a coolant temperature sensor. When the coolant reaches a threshold temperature the sensor sends a signal to the controller which switches off the relay. The controller which is connected through the ignition switch of the vehicle, will operate once the key is turned on and the cooling water temperature is within the required range. Figure 11 shows the circuit diagram for the intake air heater unit. [25] The heater is only on for a few seconds, therefore the energy drain on the battery is low.

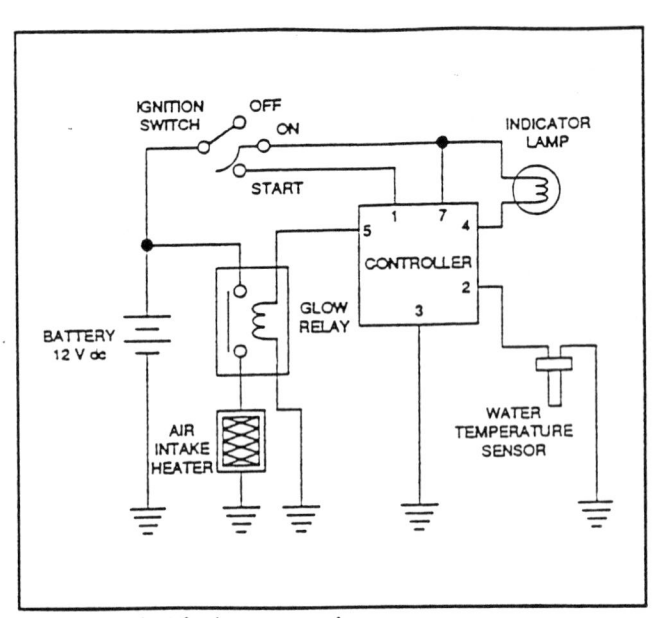

Figure 11 Air heater unit

CONTROLS

A vehicle of such nature requires a clever control scheme. Three modes of operation under consideration are the internal combustion engine (APU) front wheel drive, the brushless DC rear wheel electric drive (EV) and the combined four wheel drive (HEV). The main challenge emerges when trying to synchronize the two drive systems for a four wheel drive operation.

APU MODE- The 1.9ℓ ICE will be controlled by the EEC. The EPROM chip containing the fuel delivery and spark advance tables will be re-calibrated for methanol using an EPROM burner. In this manner, the performance of the engine can be maximized. Aside from this modification, the operation of the front wheel drive system will not differ from that of a typical system with all the electrical components running on a conventional 12V battery.

EV MODE- The electric drive unit in the rear will require its own control strategy. A schematic of the system can be seen in Figure 12. Two Solectria controllers will be powered by a separate 144V battery pack and each will drive a Solectria BRLS16 motor. This battery pack and accessories will be completely isolated electrically from the electrical system in the front and from the frame of the vehicle for safety precautions. The controllers are turned on when the key switch is in position and the electric drive switch inside the vehicle is selected.

HEV MODE- A microcontroller will be introduced to the system in order to synchronize and ensure smooth operation between the front and rear drive systems. The microcontroller reads signals from the front and rear wheel tachometers as well as from the accelerator pedal potentiometer and then sends the control signal to the Solectria controllers. Speed control is achieved by varying the armature voltage by means of chopper or pulse width modulation control. The signal coming from the brake pedal potentiometer is connected directly to the Solectria controller for controlled regenerative braking.

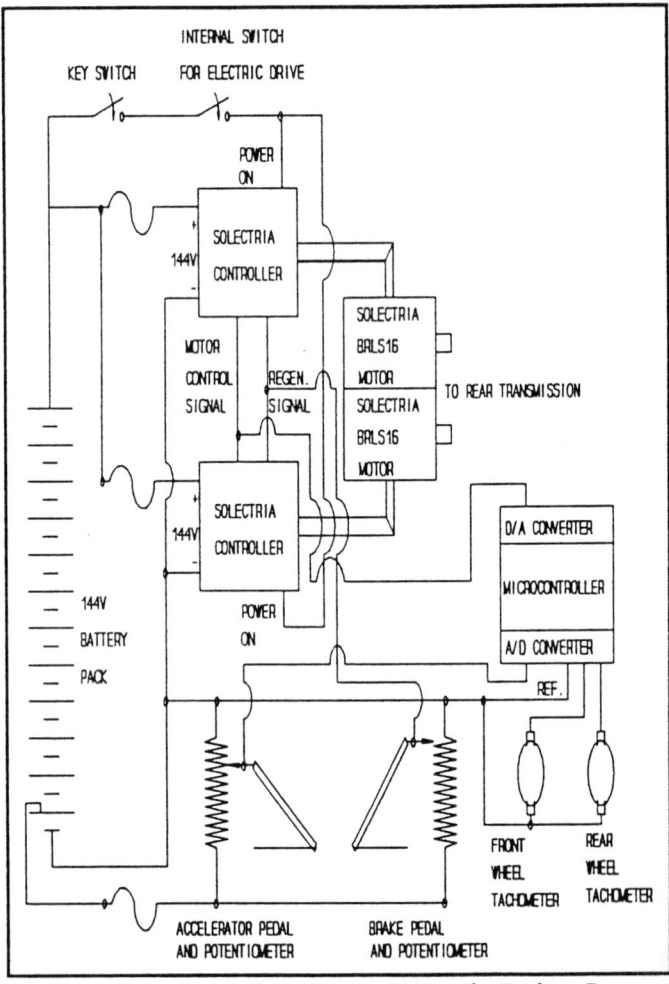

Figure 12 Schematic of Rear Electric Drive System

The two electric motors will be coupled to the two speed solenoid activated transmission through an electro-mechanical clutch. A schematic of this system can be seen in Figure 13. Gear shifts can be controlled electronically with shift switches on the steering wheel. Each gear shift switch disengages the clutch and activates a solenoid to engage the desired gear.

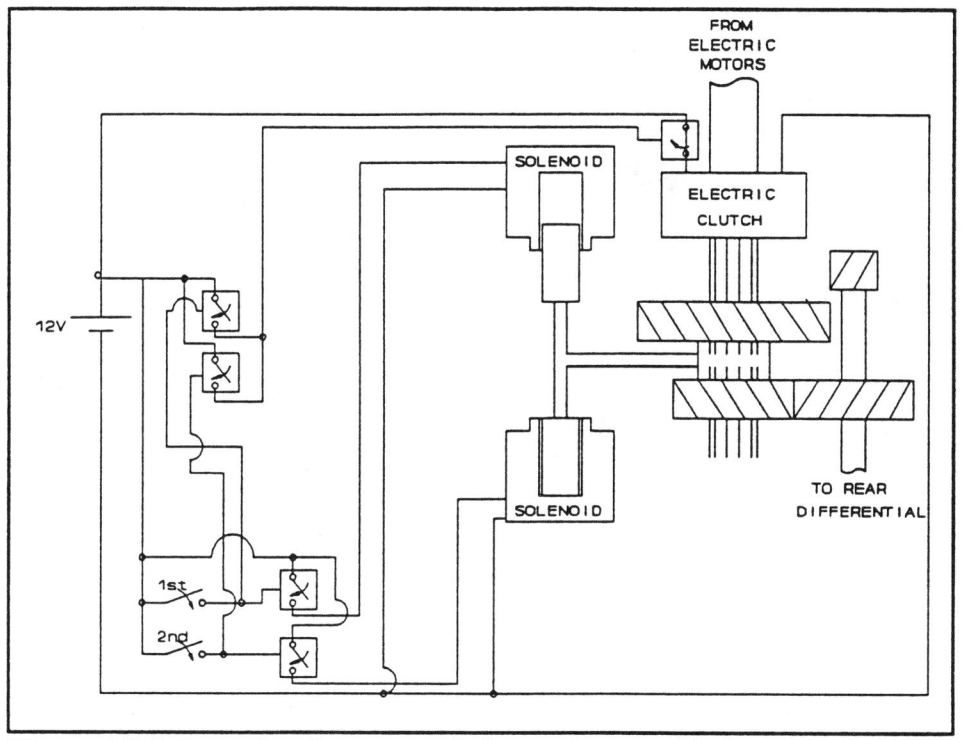

Figure 13 Schematic of two speed electronic transmission

Additional features such as a delay between the disengagement of the clutch and the activation of the solenoid may be required, thus the schematic is only a preliminary design. The final design will be accomplished through rigorous testing of the unit before implementation.

MATHEMATICAL MODEL

A mathematical model is essential in order to analyze the performance characteristics of the HEV. The primary concerns are acceleration, capabilities to negotiate grades and range at various speeds. The two determinant factors in the HEV's potential performance lies in tractive effort and resistive forces. The tractive effort is the net force developed at the wheels which can be used to accelerated the vehicle.

Tractive effort has two limiting factors:

- the characteristics of the drive and transmission units and
- the coefficient of road adhesion and the normal load on the driven axis.

With reference to Figure 14 the maximum tractive effort limited by road adhesion can be calculated using the following equations. [26]

Front wheel drive:

$$W_f = \frac{l_2}{L}W - \frac{h}{L}\left(Ra + \frac{aW}{g} + Rd \pm W\sin\theta_s\right)$$

$$F_{maxf} = \mu W_f$$

Rear wheel drive:

$$W_r = \frac{l_1}{L}W + \frac{h}{L}\left(Ra + \frac{aW}{g} + Rd \pm W\sin\theta_s\right)$$

$$F_{maxr} = \mu W_r$$

The main resistance forces encountered by vehicles:
- aerodynamic resistance force,
- rolling resistance force and
- climbing resistance force

will be analyzed as per the following procedures.

Close attention will be paid to the coefficient of

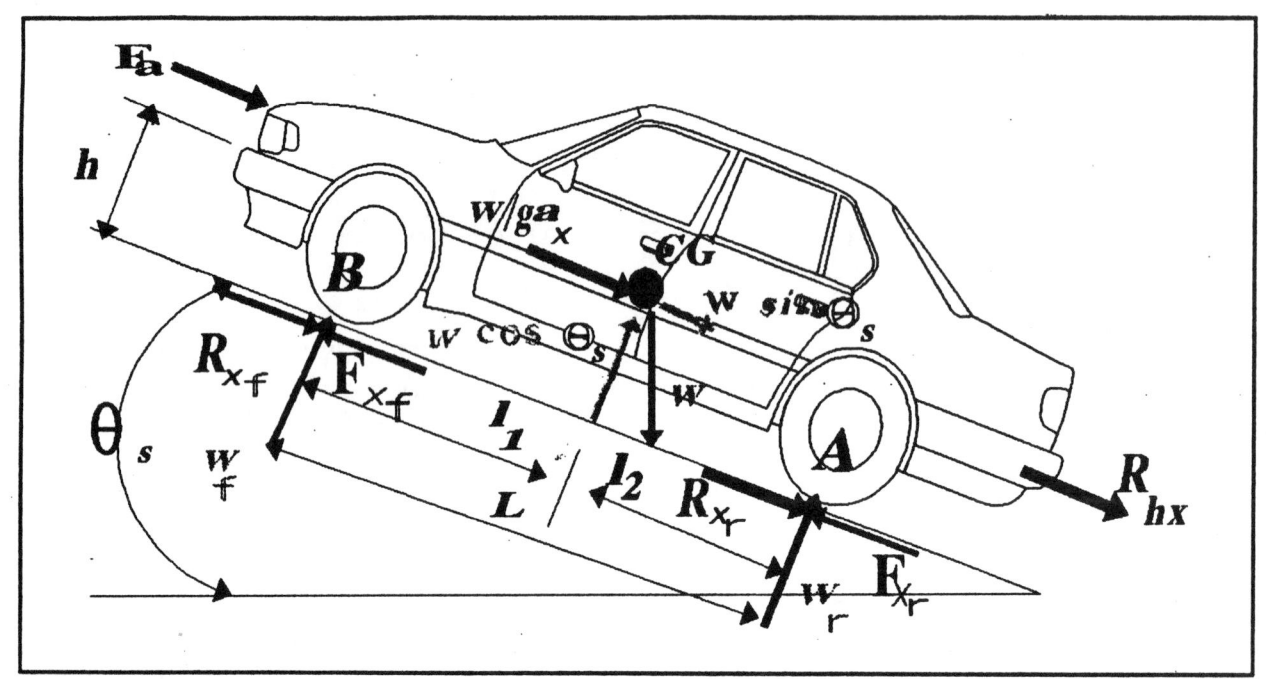

Figure 14 Tractive effort

drag and frontal area of the HEV, since these two parameters have significant effects on the vehicle's performance even at moderate speeds of 60 mph. Proper design analysis of these two design parameters will result in better fuel economy in APU mode and lower energy demands on the battery packs in EV mode.

The general trend by which the frontal area and coefficient of drag should be varied is seen by the aerodynamic resistance, F_a, equation:

$$F_a = \frac{\rho}{2} C_d A_f (V_{HEV} \pm V_{wind})^2$$

The rolling resistance, F_R, can be calculated by the following equations,

$$F_R = \mu_r (mg)$$

$$\mu_r = F_{crr} + \left(0.0015 * \frac{V_{HEV}}{27.77}\right)$$

Where the coefficient of rolling resistance (μ_R) is a function of tire pressure, contact surface, tire radius and vehicle speed (V_{HEV} in m/s).

Climbing resistance is the force created by the grade change of the road (F_{CR}), calculated using the following equation:

$$F_{cr} = \cos(\theta_s) mg$$

The total resistive force is calculated by taking the sum of the climbing, aerodynamic, and rolling resistances.

$$F_{resistive} = F_{cr} + F_r + F_a$$

The tractive effort is calculated by taking the difference between the force developed at the wheels, by the drive unit, and that of the total resistive forces.

$$F_{tractive} = F_{dw} - F_{resistive}$$

The acceleration of the vehicle can be then be calculated as follows:

$$acceleration = \frac{F_{tractive}}{m_{eff}}$$

$$m_{eff} = mass_{HEV} + \frac{J_{eff}}{r^2}$$

where, J_{eff} is the effective inertia and m_{eff} is the effective mass and r is the effective radius of the wheel.

In APU the following effective mass (m_{eff}) can simplified by multiplying the vehicle mass by the

objective functions are placed on the vertical axis having one scale. The design parameters are drawn on four horizontal scales and placed one below the other. Figure 16 demonstrates the graphical optimization process which was used for EV mode. This method was also used to optimize the APU and HEV modes.

This figure, clearly indicates the direction each design parameter should be varied in order to obtain the optimal objective functions. As a next step, different weights were assigned to the objective functions and calculations were made to optimize the design parameters to achieve the required trade-offs. The main concern in EV mode is the acceleration, due to the imposed requirement of the competition to accelerate from 0 to 95.6 km/h (60 mph) within 15 seconds. Therefore, the majority of emphasis is placed on this objective function. The optimal design parameters obtained from Figure 15 are as follows:

- vehicle weight: 950 kg
- 1st gear ratio: 9.2
- selected EV power: 71 hp
- tire roll. resistance: 0.005.

A realistic approach was taken when the optimal design parameters were selected. As an example, the design parameter of weight was selected to be 950 kg in order to be within the projected weight required to build the HEV. The proposed weight is as follows:

- 160 kg driver and passenger
- 250 kg battery packs
- 110 kg EV drive-train
- 200 kg APU drive-train
- 230 kg chassis and body

The two Solectria Electric motors would require a nominal bus voltage of 156 volts in order to obtain the 71.5 hp. The BRLS240H controls can be tuned to run at this voltage, however the manufacturer prefers to have the controller operating at 144 volts. Through testing the appropriate modification will be made to the controller, ensuring that there is no danger in running at a 156 nominal voltage.

The optimal 1st gear ratio is 9.2, any increase or decrease of this proposed value will led to a poorer acceleration. Through this optimization, it was discovered that in order to obtain the required acceleration, a two gear transmission unit had to be incorporated in the electrical drive-train. This clearly shows the importance of the optimization process. The second gear ratio was selected to be 4.8, enabling the vehicle to achieve a maximum speed of 106 km/h (66 mph).

A coefficient of rolling resistance of 0.005, for a dry paved surface will be obtained with the use of Goodyear's Invicta GLR P175/70 R13 82S which are low rolling resistance, high pressure tires.

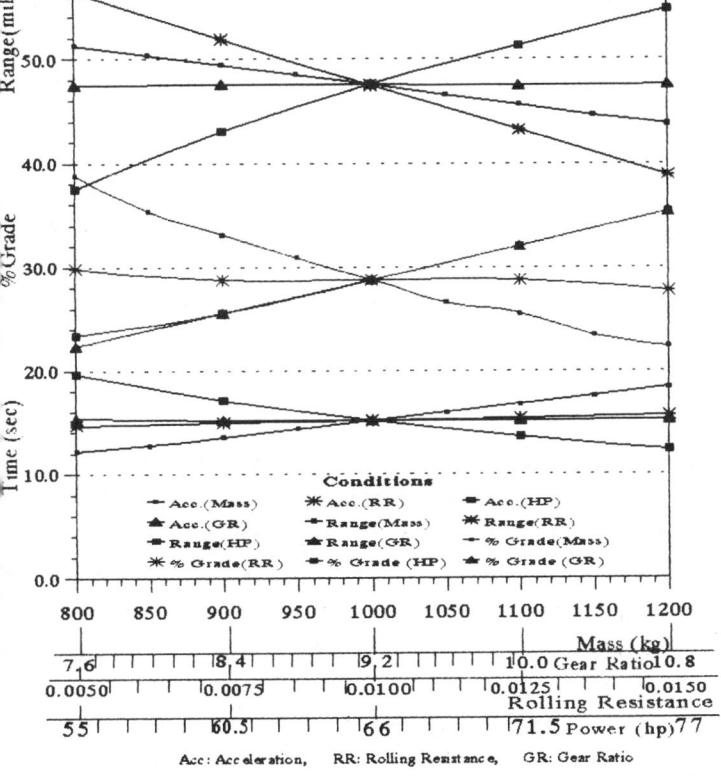

Figure 16 Graphical EV Optimisation

PERFORMANCE

The optimal design parameters were selected and thereafter introduced to the Simulation program. The simulation revealed that the acceleration requirements, 0 to 96.5 km/h (60mph) within 15 seconds, for the three modes of operation were met. Figure 17 displays the performance curve of the three modes. The HEV mode is shown to have the best acceleration. This is mainly due to the fact that the HEV combines the high stall torque of the electric motors with the inherent high-end torque of the ICE. The performance in this mode of operation makes the vehicle competitive with many high performance vehicles of today.

Table 2 Summary of performance results

	DRIVING MODE		
	APU	EV	HEV
Acceleration 0-96.5 kmh (sec)	9.9	12.5	6.4
Gradability (% Grade)	30.9	34.5	54.6
Range (km)	482	96.8	–

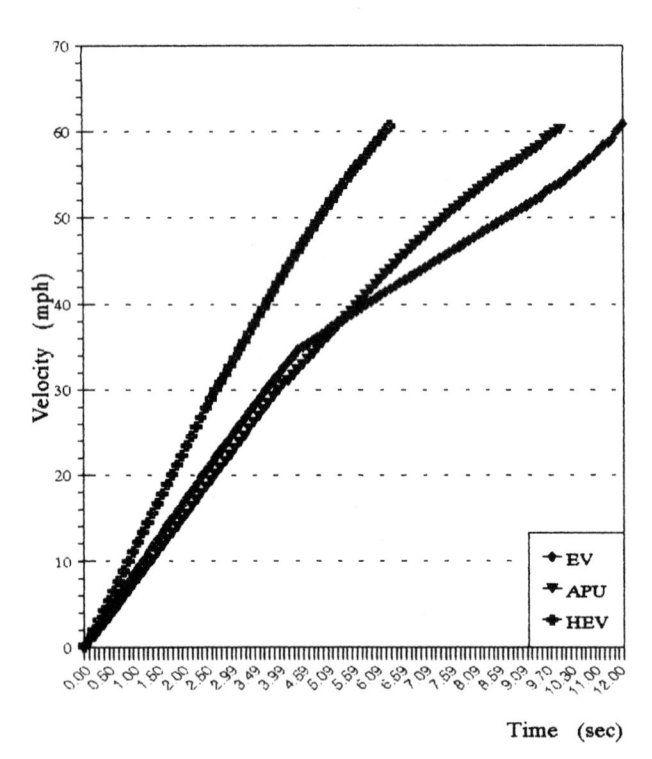

Figure 17 Performance curves for three modes of operation

The gradability achieved in each mode of operation was above the imposed 15% grade. Gradability was analyzed from rest, and had as a limiting value, a set minimum performance of 0 to 24 km/h (15mph) within 15 seconds.

The projected range in EV mode was 96.8 km (60.2 miles) at a constant speed of 72.5 km/h on a levelled surface. Taking into account the volume ratios between M85 and gasoline for equal energy output, it was estimated that a highway fuel efficiency of 11.3 ℓ/100 km (25 mpg) could be achieved. The calculated range in APU mode is 482 km (300 miles) using a 55 ℓ (12 gal) fuel tank.

ENVIRONMENTAL ASSESSMENT

An increase in public awareness regarding environmental issues and rising oil prices has brought about the need for cleaner-burning and energy efficient alternatives to conventional internal combustion engine vehicles (ICEVs). The main objective of such vehicles is to minimize the consumption of fuel and the production of exhaust emissions.

Presently, HEVs are among the most promising in a group of alternative automotive propulsion systems whose design and operation requires currently available and relatively affordable near-term technologies. Another advantage is their ability to combine the best features of ICEVs (versatility, high performance and range levels) with those of EVs (no tailpipe emissions and quiet operation).

VEHICLE TAILPIPE EMISSIONS - HEV emissions and fuel consumption levels can vary greatly and are difficult to quantify, mainly do to different patterns and the exhaust emission technology currently available. The dual-energy supply of the hybrid permits the substitution of electric energy for some of the petroleum fuel which would otherwise be consumed by a conventional vehicle. Therefore, although some fuel is consumed by the APU and emissions are produced, their levels are greatly reduced.

The previously-discussed parallel configuration will permit a substantial reduction of fuel consumption of the HEV as compared to series hybrids by eliminating the additional electro-

following factor γ_m (which is a function of the overall drive ratio ξ_0):

$$m_{eff} = mass_{HEV} \, \gamma_m$$

$$\gamma_m = 1.04 + 0.0025 \, \xi_o^2$$

The time required to accelerate to a certain speed can be calculated by:

$$t = m_{eff} \int_{v1}^{v2} \frac{dv}{F_{tractive}}$$

This can easily be integrated by the use of a digital computer.

SIMULATION

Using the governing mathematical equations a simulation was developed in FORTRAN. When the program was completed, actual testing was performed in order to determine the validity of the simulation. A stock 1992 Ford Escort Wagon was loaded to 1550 kg and acceleration tests were performed. The time required to accelerate from (0 to 45 mph) was obtain for several trials. A 10 second nominal value was obtained. The design parameters of the Ford Escort were used as inputs for the simulation. Results of the simulation revealed that the vehicle would accelerate from 0 to 72 km/h in 10.2 second. This ensured that the simulation was performing in a similar manner to the actual system objectives, hence validating the model. A second simulation was developed using a TUTSIM simulation package. Using TUTSIM the open loop system transfer function of the electrical mode of operation was simulated. The results of the TUTSIM simulation were closely related to those obtained from tests using the FORTRAN simulation. In Appendix A the TUTSIM simulation block diagram developed for the EV mode is illustrated. Figure 15 outlines the torque vs. speed curve of the Solectria motor. This was used along with Figure 10 to perform the simulations for the three modes of operation.

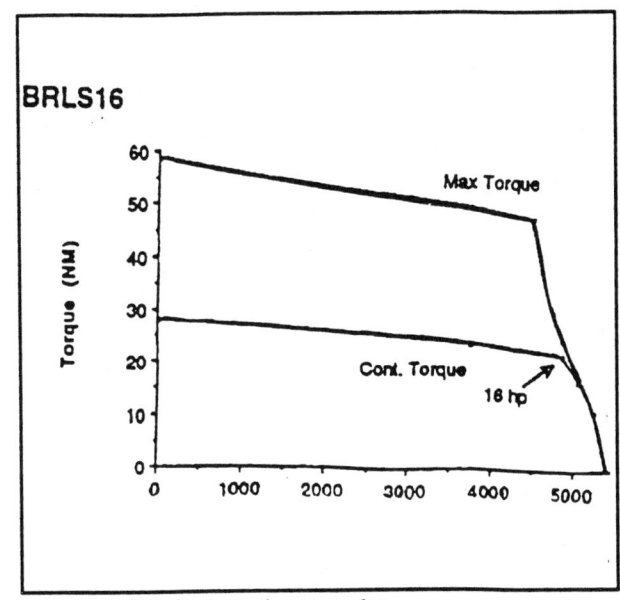

Figure 15 Solectria speed-torque curve

OPTIMIZATION PROCESS

The optimization process involved varying several design parameters to see an impact on the vehicle's performance. These design parameters were as follows:

- vehicle weight
- 1st gear ratio
- power of the drive unit
(2-Solectria BRLS16)
- tire rolling resistance

The coefficient of drag and frontal area are important design parameters, however they will not be varied in this optimization process. By varying the design parameters within some determined limits, the impact on some chosen system objectives can be investigated. This is achieved by a graphical method where some nominal values of the design parameters are assigned. The system objectives are:

- time required to accelerate from 0 to 96.5 km/h (60 mph),
- range of the vehicle at 72 km/h (45mph) cruising speed and
- maximum obtainable grade.

Each objective function is obtained by varying one design parameter from left to right, while holding the remaining nominal values constant. To make the graphical representation compact the

mechanical efficiency.

In the ZEV operating mode, hybrids are "clean" vehicles since they do not emit any tailpipe emissions, making them very attractive for city use. In the simultaneous APU and HEV mode, the range of loads and speeds at which the APU is required to operate smoothly and responsively is restricted. The engine is therefore run at a relatively fixed operating point in its torque-speed map so as to achieve maximum efficiency and thus yielding particularly low emissions. This driving condition where speed and load do not vary considerably is most desirable for the control system to maintain stoichiometric Air/Fuel ratio and consequently an optimum operating condition in terms of emissions. The HC, CO, and NO_X emissions are anticipated to be well below the FTP standards. [27]

ENERGY GENERATION CONSIDERATIONS - In a more global perspective, NO_X, SO_X and PM emission levels are very sensitive to the type of fuel and technology used by power plants to generate the electricity as well as to the emission control systems.

In order to maximize the environmental benefits of HEVs and EVs, electricity generation should be derived from relatively clean power sources such as hydro-electric power, wind power and plants that burn cleaner fuels such as Natural Gas. According to some studies, the future and partial substitution of conventional vehicles by HEVs and EVs will permit a dramatic reduction of CO and HC and to a lesser extent NO_X in such polluted urban centres as the L.A. basin. [28]

BATTERY CONSIDERATIONS - Due to the extensive use of batteries and the growing concerns for the safe disposal of the spent batteries, near future key developments are expected to focus on such issues as the toxic nature of battery discharge products and possible ways to recycle battery materials.

The hydrogen gas produced in the HEV's sealed battery trays will be safely disposed of by properly venting it out of the vehicle and into the atmosphere. Numerous recuperation facilities now exist for the recycling of certain battery elements. For example, lead used in the storage battery industry is derived from lead obtained from reworking the scrap from worn out batteries. Other elements found in lead-acid batteries such as antimony, arsenic and tin can be extracted with techniques such as reverberatory refining. [29]

CONCLUSIONS AND RECOMMENDATIONS

In this report, we have accomplished our preliminary goals for designing a practical Hybrid Electric Vehicle that would also be attractive to the average consumer. As demonstrated in this report, the design of such a vehicle is not a trivial task, especially when it comes to meeting the performance demands of today's vehicle standards.

The design incorporates several advanced techniques in order to maintain a low overall vehicle weight while maximizing interior storage space. This facilitates the task of placing the batteries without significantly compromising interior space, comfort and styling.

Development of a suitable drivetrain involved rigorous analysis of various electric drive systems as well as making the best selection between parallel and series configurations. The Solectria BRLS16 electric drive system was chosen based on overall cost, efficiency and power. A parallel configuration using two separate drive systems was seen to be the most practical in terms of overall efficiency. This provides an emission-free electric drive unit primarily for city driving where emissions from ICE's are relatively high and an alternative power unit running on methanol for high speed extended highway driving where emissions from ICE's are relatively low. In addition to this, the proposed hybrid mode converts the vehicle into an attractive high performance 4 wheel drive vehicle.

Recognizing that a hybrid electric vehicle already makes note of the fact that today's battery technology still imposes a strict constraint on the range of an electric vehicle, the most logical battery choice would come from the lead-acid family. Two batteries were selected to have the highest energy density in this class of batteries and upon discharge tests the advanced spiral wound technology of the OPTIMA 800 series proved to last significantly longer at a reduction in weight and comparable costs. This was important in keeping both the weight and cost of the vehicle low.

Taking advantage of methanol's lower emissions would involve replacing all components coming into contact with it to avoid any corrosion as well as to modify the fuel delivery characteristics of the fuel injection system. To make full use of methanol's ability to operate at higher compression without producing knock, exhaust gas turbocharging was found to be the most suitable method in fulfilling this option resulting in a higher performance curve and thermal efficiency.

A control strategy was developed that would not compromise the integrity of each drive system's autonomy. This is best accomplished by allowing the conventional electronic engine control of the ICE to perform its usual tasks after being re-calibrated for methanol and introducing a microcontroller to the electric drive system to assure proper synchronization in 4 wheel drive operation. The design of an electronically controlled transmission was also laid out to be used for improved acceleration and climbing ability.

The use of a valid mathematical model in conjunction with a computer simulation provides an overview of what is involved and where complications may arise. Together, they form a useful tool for projecting the performance of a particular system as well as the impact of varying certain design parameters. Although the mathematical model has its limitations in terms of accuracy, it is a helpful instrument for making comparisons and can guide the designer toward developing an optimal design. The computer simulation was used to optimize vehicle parameters in order to minimize the number of manufacturing iterations that would be required to reach a final design. A final simulation using optimized parameters revealed that the proposed vehicle design would meet the imposed vehicle performance requirements. In fact, further assessment of the vehicle performance reveals that this vehicle would be very competitive with today's hybrids in acceleration, range and gradability in each operating mode.

Finally, the environmental benefits of the proposed vehicle would include exhaust emissions well below the FTP standards. Although the idea of an all purpose, emission free vehicle being around the corner, would be naive, the hybrid electrical vehicle concept paves the way for such an eventual goal, while significantly reducing the environmental hazards of today's vehicles.

REFERENCES

1. News Trends, "Scanning the Fields for Ideas", Machine Design,
2. "Propulsion Technology: An Overview", Automotive Engineering, July, 1992, pp. 29-33
3. E.P. Patrick and M.L. Sharp, "Joining Aluminum Body Structure", Automotive Engineering, vol. 100, No. 5, pp. 31-33
4. S. Ashley, "The Shape of Cars to Come", Mechanical Engineering, May 1991, pp 37-42
5. E.P. Patrick, pp. 31-33
6. S. Ashley, pp 37-42
7. E.P. Patrick, pp. 31-33
8. J.F. Lindsay and M.H. Rashid, Electromechanics & Electrical Machinery, Montreal, Prentice Hall, 1989, pp.115-23
9. Ted Lucas and Fred Reiss, How To Convert To An Electric Car, Crown Publishers Inc., NY, New York, 1975
10. Ted Lucas and Fred Reiss, How To Convert To An Electric Car, Crown Publishers Inc., NY, New York, 1975
11. Ted Lucas, pg. 30-50
12. Ted Lucas, pg. 30-50
13. Ted Lucas, pg. 30-50
14. Yasuhika Dote and sacan Kinoshita, Brushless Servomotors Fundamentals and Applications, Clarendon Press, Oxford, 1990, pp. 1-12
15. Ted Lucas, pg. 30-50
16. Ted Lucas, pg. 30-50
17. Nicholas Bissoon-Dath, "Technical Highlight", Car And Driver, Vol. 34, No. 9, March, 1989, ppg. 99
18. Fisher Electric Motor Technology Inc., Pamphlet
19. Solectria Electronics Division, Pamphlet
20. Karl V. Kordesch, Batteries, Marcel Dekker Inc., New York, 1977, pg. 6
21. Robert Bosch, Automotive Handbook, Robert Bosch GmbH, 2nd Ed., 1986, pg. 427
22. M. Allen, "Battery Chargers", Popular Mechanics, Vol. 168, No. 9, Sept., 1991, pp. 30-31
23. M. Smith, C. Lisio, and T. Krepec, Alcohol-Clean Automotive Fuelconclusion Drawn From "Methanol Marathon competition", Proceedings form CSME Forum SCGM 1992, Concordia University, pp. 389-393

24. Robert Bosch, Automotive Handbook, Robert Bosch GmbH, 2nd Ed., 1986, pp. 302-303

25. M. Smith, pp. 389-393

26. Thomas D. Gillespie, Fundamentals of Vehicles Dynamics, Society of Automotive Engineers, USA, 1992, pp. 11-14

27. A. F. Burke, "Hybrid/Electric Vehicle Design Options and Evaluations", Electric & Hybrid Vehicle Technology SP-915, Society of Automotive Engineers Inc., Warrendale, PA, 1992, pg. 62

28. Q. Wand, M.A. DeLuchi, D. Sperling, "Emission Impacts of Electric Vehicles", Journal of Air and Waste Management Association, Vol. 40, Sept, 1990, pp. 1275-84

29. C.D.S. Tuck, Modern Battery Technology, Ellis Horwood Ltd., West Sussex, England, pg. 186

APPENDIX A

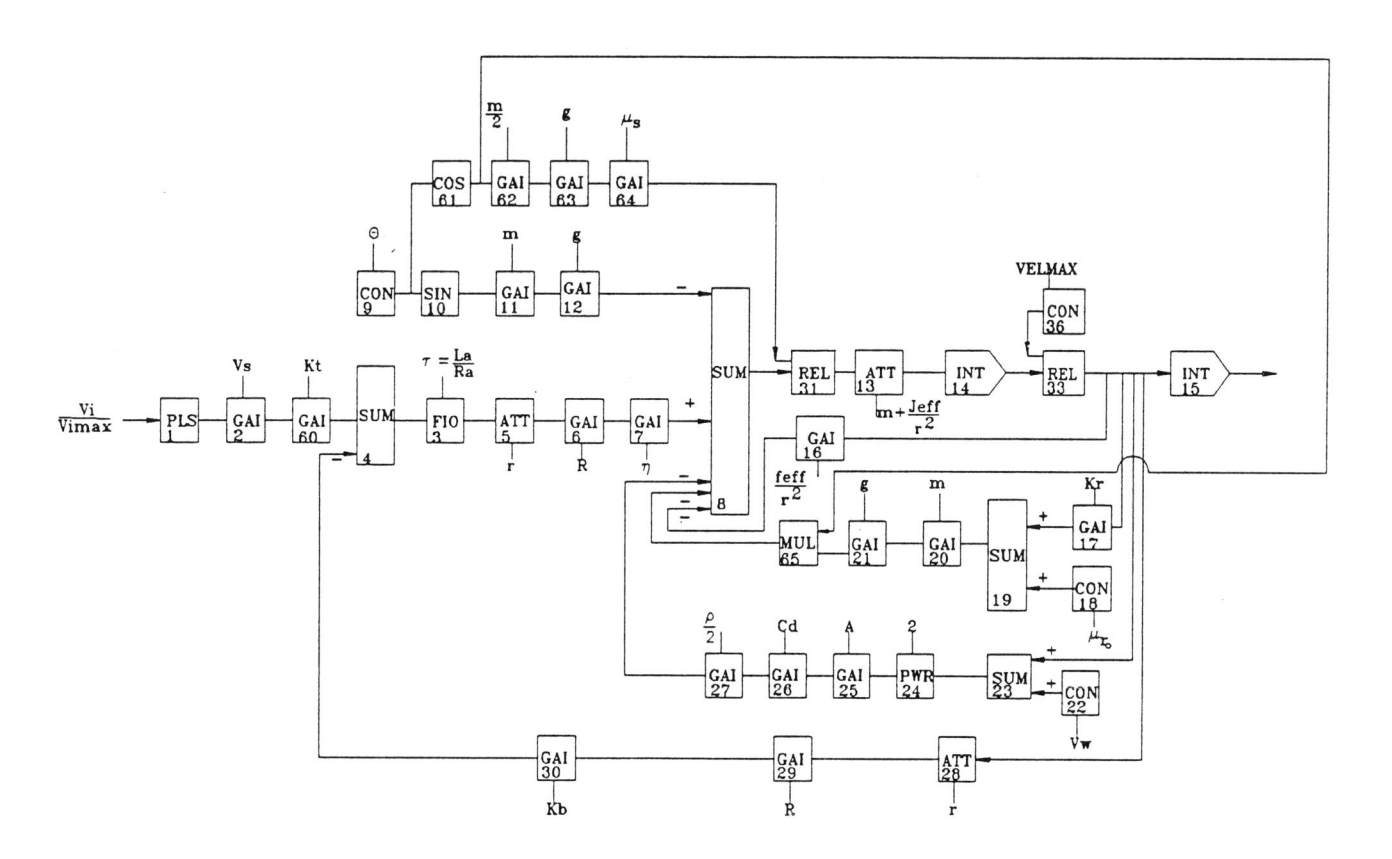

TUTSIM BLOCK DIAGRAM

Vi	Voltage from throttle position sensor
Vimax	Maximum voltage from throttle position sensor
Va	Armature voltage to DC motor
Vs	Maximum armature voltage (supply voltage)
Kb	Back emf constant
Kt	Torque constant
Ra	Armature resistance
La	Armature inductance
Tm	Torque developed by DC motor
R	Complete gear ratio through geartrain
r	Wheel radius
Fdw	Force developed at wheel
m	Mass of vehicle
g	Acceleration of gravity
Θ	Angle of incline of vehicle
Finc	Incline force
feff	Effective friction coefficient on all bearings = $n_1 f_1 + n_2 f_2 + \ldots$
Ff	Friction force from bearings
Jeff	Effective Inertia due to gears and wheels = $n_1^2 J_1 + n_2^2 J_2 + \ldots$
Kr	Increase in rolling resistance coefficient vs. vehicle speed
μ_r	Coefficient of rolling resistance
μ_{ro}	Coefficient of rolling resistance
Vw	Face wind
ρ	Air density
Cd	Drag coefficient
Fa	Aerodynamic Forces (Drag)
A	Frontal Area of Vehicle
Kv	Velocity coefficient
$\ddot{x}$	Vehicle acceleration
$\dot{x}$	Vehicle velocity
x	Vehicle position

Lehigh University Hybrid Electric Vehicle

Senior Students of Mechanical Engineering
Engineering Department, Lehigh University

Douglas Patridge
Robert Bittler
Walter Fader
Scott Fredericks
Mark Garber
Peter Goodale
Kevin John
Kamal Patel

William Rickelman
George Saghiyyah
Peter Shuleski
Thomas Schmidlin
G. C. Seibert
Robert Shankland
James Zaiser

ABSTRACT

The goal of this paper was to create a feasible design for a hybrid electric vehicle, one that will meet with wide acceptance among both manufacturers and consumers.

The goals we set for the optimum design of a hybrid-electric vehicle include a lightweight frame, good acceleration and braking characteristics, excellent safety conditions, and the use of as many existing systems as possible. By incorporating slightly modified existing systems into the design, we can expect a high acceptance of the vehicle among manufacturers. Many factories are already able to produce the various components that make up the car, and with only minor adjustments, other plants may easily convert to the construction of hybrid electric vehicles.

Having achieved all of our goals, we feel that we have a working solution to the design of a hybrid electric vehicle, and that this design may be easily implemented at many manufacturing plants.

In part prompted by the California Emissions Law there is renewed interest in electric automobiles. These vehicles offer the promise of a pollution free, low noise, and perhaps low cost, means of common transportation. Several companies have introduced or plan to introduce in the near future, some type of electric car.

The first source of an electric car dates from the 1920's when, then as now, battery weight compared to energy storage capacity and recharge capability has restricted the use of electric cars to relatively short range, specialized purposes for example golf carts, delivery trucks, and factory delivery systems. Existing and near term vehicles can at best provide transportation ranges of 150 miles with recharge times of from 15 minutes to eight hours.

Until battery technology progresses to the point where performance payload and convenience is comparable to the current standard for automobiles, it is doubtful that electric cars will receive consumer acceptance as a primary means of transportation.

As an interim measure, the hybrid elctric vehicle has been suggested to bridge the gap between standard internal combustion automobiles and purely electric vehicles. Hybrid vehicles combine the major components of each of these two automobiles into a more environmentally sound form of transportation that will hopefully meet with wide consumer acceptance.

PERFORMANCE SPECIFICATIONS (Walt Fader)

In order to size the batteries, heat engine and electric motor, an estimate was made at the drive wheels (assuming rear wheel drive) of peak and average power and torque, and total required energy consistent with the Argonne National Laboratory specification of a 15 second acceleration time to achieve 60 mph, and 300 mile range with a sustained velocity of 55 mph using no more than 7.1 gallons gasoline or equivalent M85 or E100 fuel.

To determine these parameters, the EPA highway fuel economy driving schedule was fed into a computer model. The test is performed on a chassis dynamometer which simulates both road load power due to aerodynamic drag and rolling resistance, and inertial effects. The chassis dynamometer utilizes a power absorption unit to apply a load to the drive rolls for simulating road load power. Flywheels or other means are used to simulate inertial effects. In order to evaluate the capability of the vehicle design to meet the 300 mile range specification, a analysis was done to determine the total energy required to negotiate the 10.2 mile drving cycle. A second by second tabulation of the results is given in the results section, in which the driving cycle was repeated for the total 300 mile range. Average velocity for the cycle was 48.3 mph, giving a driving time of 6.2 hours. The result was a total required energy of 56 kW-hr at the wheels. Using an assumption of 50 percent mechanical and electrical efficiency for the engine, alternator, battery, and motor assembly gives approximately a 110 kW-hr total delivered by the electrical system to the motor. At the wheels, the vehicle produces an average power of 12.1 hp (9 KW), and with a tentatively assumed tire diameter of 20 inches an average torque of 78 ft-lb (106 Nm).

As a means of verifying the results of the driving schedule calculations, energy output for a conventional engine powered automobile was calculated. For an automobile which gets 35 mpg, using gasoline with a heating value of 19,000

BTU/lb with an overall efficiency of 22 percent, a total of 67 kW-hr energy output is needed for a 300 mile trip. This value corresponds fairly well to that which was calculated from the EPA driving schedule. The driving schedule does not include sufficient acceleration to test the limitations of torque and power produced by the motor, so a maximum acceleration schedule must be developed.

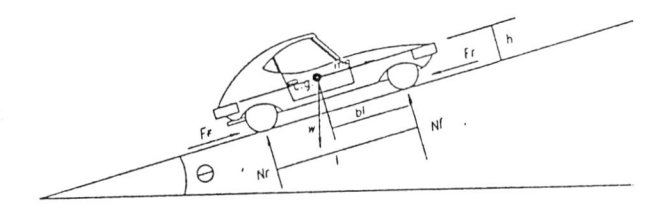

Figure 1. Free Body Diagram

Acceleration of the vehicle is limited by the amount of friction that can be developed by the rear wheels (assuming rear wheel drive, $F_f = \mu N_R$), without spinning. By manipulating equations of forces acting on the automobile, the following expression for maximum acceleration with no velocity which eliminates both the friction and normal forces may be obtained:

$\sum M_F : N_R l - W \cos\theta\, bl - F_d y_d = mah$

(where y_d is the vertical component of the center of drag)

$\sum F : F_f - F_r - F_d - W \sin\theta = ma$

$$a = \frac{b*\mu*\cos(\theta) - \sin(\theta) - \frac{F_r}{W} - \frac{F_d}{W}(1 - \frac{y_d l}{\mu})}{1 - \frac{\mu*h}{l}}$$

Specifications require that the vehicle have the capability of accelerating from rest to 60 mph in 15 seconds or less. With a constant acceleration, a value of 0.182g is necessary, however, this scenario is not desirable. Instead, the acceleration should decrease with time so the acceleration at

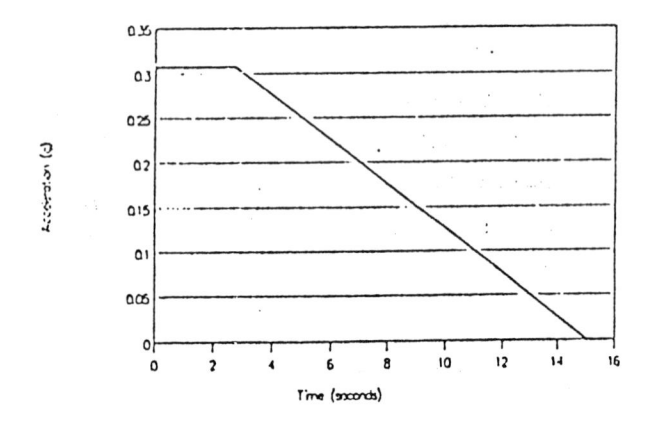

Figure 2. Acceleration History

60 mph is zero. Although the maximum speed will be only 60 mph, power consumption will be minimized. Also, the required peak sustained speed of 55 mph is fulfilled. A starting acceleration of 0.364g results from a linear variation. This value exceeds the allowable value of 0.310g without wheel spin, though a higher value may be possible for higher friction coefficient. In other words, a non-linear history of acceleration must be developed. For the assumed acceleration schedule, a peak power of 56 hp (42 kW) and a peak torque of 590 lb-ft (800 N-m) are developed.

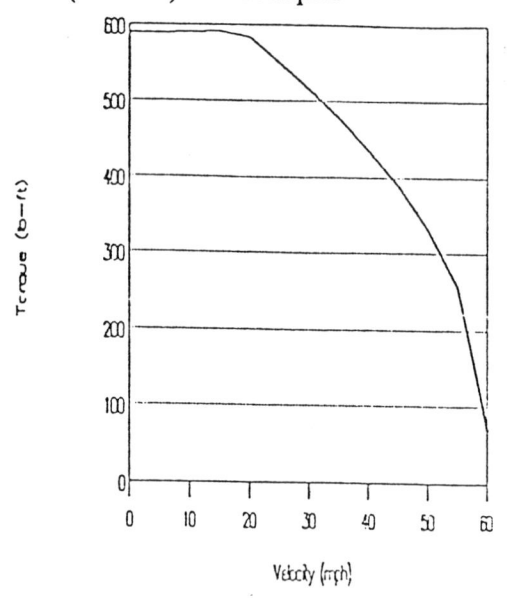

Figure 3. Required Torque Output, Initial Figures

Another stipulation placed on the Hybrid Electric Vehicle involves the ability to climb an incline of up to 15 percent grade (8.53 degrees). A hill introduces another term into the equation of motion; $F_g = mg\sin\theta$. On a grade of 15 percent, maximum acceleration without wheel spin will be 0.173g, as obtained from the equation of motion. Power of the motors available for traction force is reduced on the hill, consequently, the maximum speed of the vehicle will be somewhat less than 55 mph.

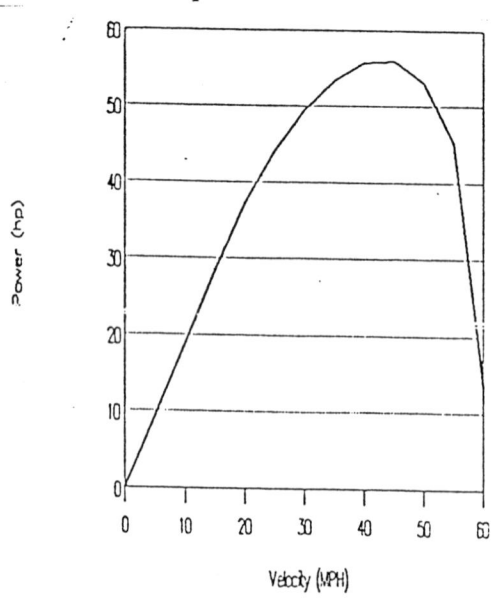

Figure 4. Required Power Output, Initial Figures

RESULTS-Parameters for an Assumed Acceleration Schedule to Achieve 60 mph in 15 seconds.

Grade (%)	0	0	0	0	0
Time, t, (sec)	0	2.2	4.6	7.6	15.0
Acceleration, a, (g)	.308	.308	.262	.185	0
Velocity, v, mph	0	15	30	45	60
Inertia, ma, lbs	678	678	576	407	0
Rolling Res, F_r, lbs	29	29	29	29	29
Drag, F_d, lbs	0	3	14	30	54
Traction, F_f, lbs	706	708	618	466	84
Power, P, HP	0	28	49	56	13
Power, P, KW	0	21	37	42	11
Torque, T, lb-ft	588	591	515	388	70
Torque, T, N-m	798	802	698	527	95

Summing the forces acting on the vehicle in the direction parallel to the slope result in the equation of motion,

$$F_f - F_r - F_d - F_g = ma$$

where,

F_f is the friction (i.e. traction) force developed between the road and the tires

$F_g = mg \sin\theta$, is the component of the vehicle weight parallel to the slope.

$P = F_f v$ is the total mechanical power required at any speed.

Power (in hp) required to overcome the rolling and drag resistance may be expressed as,

$$P_r = (C_R mg + 0.0025 C_d A v^2) v / 375,$$

where C_R is the coefficient of rolling resistance and has a value of approximately 0.013, A is the frontal vehicle area in ft^2, and v is the vehicle velocity in mph.

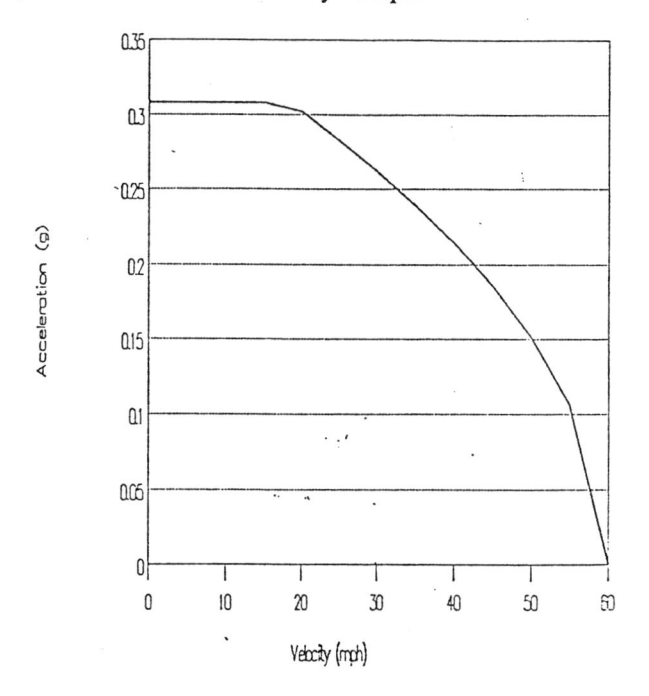

Figure 5. Acceleration Schedule

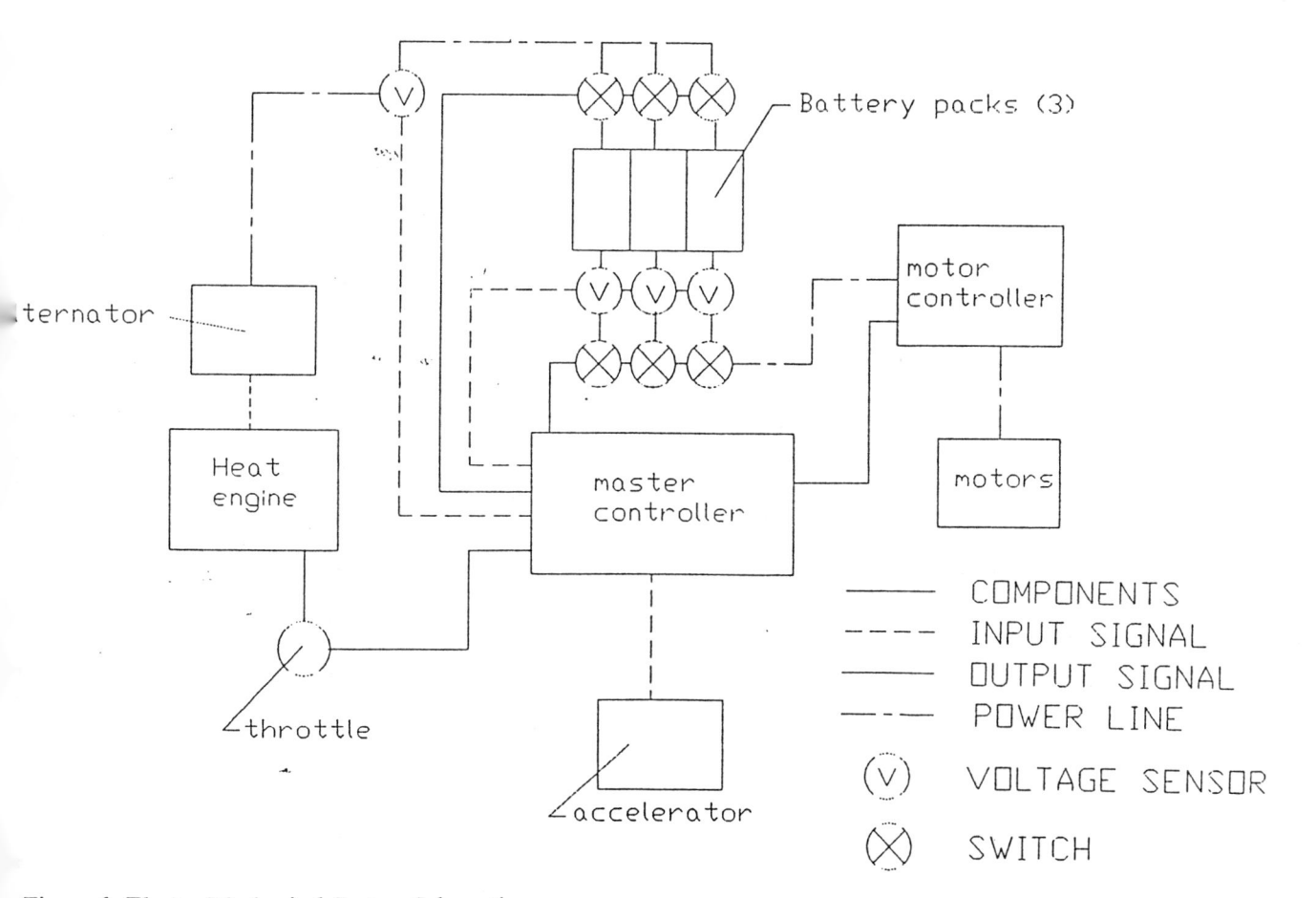

Figure 6. Electro-Mechanical System Schematic

ASSUMPTIONS- A vehicle weight of 2200 pounds.

The center of gravity located in the midpoint of the wheel base and low to the ground.

A coefficient of friction between the wheels and the road of $\mu = 0.65$.

No slope ($\theta = 0$).

Frontal area $A = 1.8$ m^2.

Air density $\rho = 1.225$ kg/m^3.

Drag coefficient $C_d = 0.31$.

SUMMARY OF SPECIFICATIONS-
- average power = 12.1 hp (9 kW)
- average torque = 78 ft-lb
- peak power = 56 hp (42 kW)
- peak torque = 590 ft-lb
- energy output for the 300 mile trip = 56 kWH

MASTER CONTROLLER (Kevin John)

Figure 6 (previous page) shows the basic configuration of the electro-mechanical system, in which the master controller is responsible for overall operation. It's functions are:

1. Receives commands from the driver controlled "acceleration pedal" and sends a signal to the motor controller.
2. Monitors battery depletion and switches battery packs as needed.
3. Determines when the heat engine is to be run and whether at peak power or peak efficiency.
4. Determines which battery should be charged.

The responsibilities of the master controller can be broken-up into two semi-independent systems; the battery recharge/switching system and the speed control system.

The recharge/switching system is primarily concerned with maintaining the power required by the motors. This power may be variable which is determined by the speed control system. Hence the input of power requirement for the recharge/switching system is given by the speed control system. This is where the two systems are interdependent.

The strategy of the recharge/switching system is a feed back loop. The master controller will monitor the charge in each of the three battery packs. The pack with the greatest charge will be used to drive the motors. Once this pack is depleted to a given minimum charge the demand is switched to the battery with the greatest charge at that time. There is now a need to recharge a battery. At this point the IC engine is started delivering mechanical energy to the alternator, which is converted to electrical energy. The master controller then sets the switch to charge the depleted battery to it's fullest capacity. Once this charging is completed the master controller then determines which, if any, of the batteries need to be charged. If there is a depleted battery the master controller then sets the switch to charge that battery. If there is no need for charging the IC engine is set to idle if the battery under demand is under a certain charge range and turned off if the battery charge is above this value. A flow chart of the system is given in Figure 7.

The speed control system is less complicated. An input is received from a device that converts a mechanical signal(delta X) to electrical. This signal is read by the master controller and is interpreted as a power requirement. The master controller then determines from the motor characteristic curve, and the speed feed back, a suitable torque and speed. This torque and speed is relayed to the motor controller. A new voltage and frequency is produced by the vecotr motor controller to either change the speed and or torque depending upon the demand on the vehicle. The feed back for this system is the driver, who determines if the vehicle needs more torque to go up a hill at the same speed or to increase speed on a level surface.

The controller design calls for a custom controller. This controller will be programmable so that the battery/recharge algorithm can be programmed along with the speed control system. The speed control will be programmed with data tables for the motor torque vs. speed(RPM) curves at various frequencies. When a new power demand is given to the controller it will interpolate a torque and speed from these tables, and send it to the motor controller. This technology is currently available through Allen-Bradley corporation, by using an Eprom for the various motor characteristics.

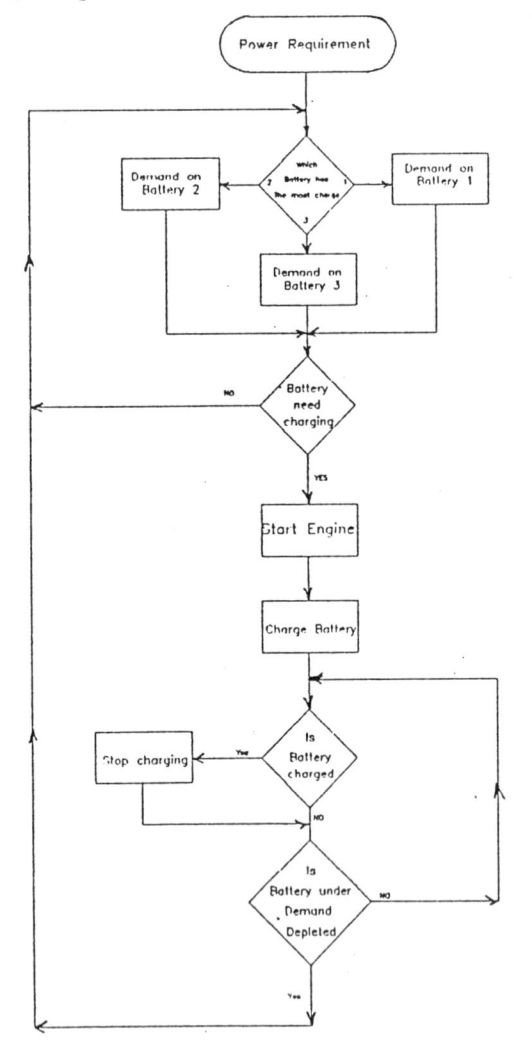

Figure 7. Flowchart of Battery Recharge/Switching System

WIRING SYSTEM (Jim Zaiser)

Vehicle wiring must conform to the DOE/ANL specifications which require that "ampacity must exceed the highest operating current by at least 25% with a breakdown voltage of at least two times the peak ground referenced voltage", and "routing so that there be no inadvertant physical contact with high voltage equipment or components." In addition, "all high power connections shall be made through connectors rated at greater than 125% of the peak current through the connector."

WIRING CONFIGURATION AND ROUTING-Wiring routing and location was done using AutoCAD. The strategy used in developing the system was to generate routing diagrams on two layers of components. The layout is shown in Appendix A, Figures A1 and A2.

WIRING SPREADSHEET-In Appendix A, Tables A1, A2, and A3, all of the wires in the car, including code numbers so that they can be identified, peak power, voltage and ampacity including the required specifications overloads, the total lengths, necessary insulation, and projected cost.

ENGINE (Robert Shankland & Scott Fredricks)

ENGINE-The heat engine in a series HEV is to provide power to the generator. Since a series HEV will not be used to power the car mechanically as in a parallel drive system, the series system has been chosen so that the additional weight of the batteries will be offset by a savings in weight from the omission of the dual drive system. The weight saved from omitting the linkage from the heat engine to the wheels is estimated to be 150 lbs.

The IC engine's fuel injection system, upon receiving a signal from the master controller will operate in one of four modes, depending on the required control strategy; off, idle, peak efficiency and peak power. A feedback loop between the engine and the controller will maintain engine RPM in one of the four modes. The engine will be off for short trips or until the first of the three battery packs is depleted. The engine will be idling, providing just a trickle source of power to the generator, when the engine is already on and charging is not required. The engine will run at its peak efficiency setting for the majority of the time and at this setting must be able to produce sufficient power to the generator so that a single pack of 13 KWHr capacity can be charged in 1 hour. The peak power setting will only be used when the driver puts excess demand on the vehicle and the batteries are being discharged faster than they can be charged at the peak efficiency setting. In order to fully realize the benefits of the hybrid vehicle, the engine should not have to operate at the peak power setting for long periods of time. Because the engine essentially operates at one speed, efficiency and power can be optimized and emissions can be minimized without worrying about the engine having to perform well over a broad range of RPM.

Possible choices for heat engines include, 2-cycle and 4-cycle spark ignition engines and compression ignition engines. Because of the specification restriction in M85 or E100 fuels, the compression ignition engines were eliminated from consideration. The 2-cycle engines were also eliminated because they require oil additives and emit unacceptable levels of emissions. The engine selected is a four-stroke SI engine, running on 4 cylinders and will be liquid cooled. The weight of the engine is 196 pounds and will displace 67 cubic inches.

FUEL-Methanol is the fuel we have choosen for our application. Some reasons for this choice is that it can be produced inexpensively from coal, wood, or biomass garbage and that the emission of NO_x and CO are reduced. The formula for methanol is $CH_4O(l)$ and it has a heating value of 8600 Btu/lb (20.0 MJ/kg) which is about half that for gasoline. The blend we will be using is not pure methanol but it will consist of 85% methanol and 15% hydrocarbons (or gasoline), which increases the heating value to 9800 Btu/lbm. The advantages of mixing methanol and gasolines is that it will increase the heating value, decrease the cold-start limit to 0 degrees F, and also produce a visible flame when ignited.[*]

ALTERATIONS-For the engine stated above there only need be a few minor alterations, this is mainly due to the highly corrosive nature of methanol. The fuel tank must be made of stainless steel and the float level in the tank needs a protective circuit. The fuel lines need to upgraded from rubber to Teflon tubing covered with a braided stainless steel mesh around it. Methanol tolerant fuel pumps as well as injectors need to be installed (Figure 8).

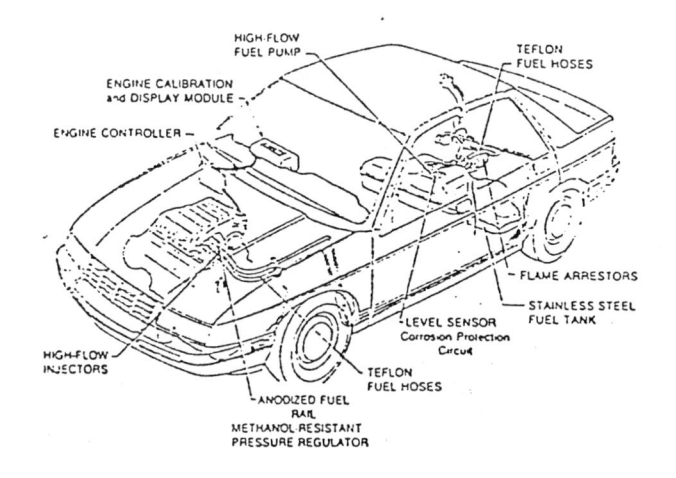

Figure 8. Engine Alterations

The flame arrestor is needed because of the high combustibility of methanol and air over a broad range of mixtures. This device is to be bolted to an engine's cylinder head, inside the inlet elbow leading to the cylinder. One flame arrestor is place in the inlet to each cylinder. The purpose is to prevent the fuel in the intake to backfire and cause serious damage. There is currently one on the market designed and manufactured by Woods Energy Products, Inc., which has been tested into stationary natural gas burning engines and has showed that its restriction of the flow rate is minimal.[**]

ELECTRONIC CONTROL UNIT-The system will be controlled by an electronic control unit(ECU) which will in turn be controlled by the master controller. The ECU will monitor the temperature of the cylinder, air intake, and monitor the exhaust from the cylinder so as to regulate the air and fuel to best fit the needs of the operator. The ECU needs to be calibrated for M85 running, things such as spark advance, equivalence ratio, and valve timing have to be changed. The sensor will tell the ECU whether to make adjustments when engine running at either idle, peak power, or peak efficiency.

---

[*]   Road & Track, "Alternative Fuels...", Nov 1989, v.41.
[**] Design News, "Flame Arrestor...", 3 Nov 1988.

EXHAUST-

Table 1
Effects of Changes in Methanol/Gasoline Mixture on Emissions

Exhaust, g/mi	M0	M85 % Change from M0
NMHC	.24	-42.3
OMHCE	.29	-37.0
NMOG	.24	14.4
CO	2.8	-31.3
NO_x	.40	23.0
Exhaust Toxics, mg/mi		
Benzene	15.0	-84.0
Formaldehyde	2.8	436.0

NMHC: non-methane HC
OMHCE: organic material hydrocarbon equivalent
NMOG: mon-methane organic gases

The above data was determined from 19 early prototype flexible/variable fueled vehicles, representing seven technology types.[*] Therefore, the numbers above are not a good representation because on a few engines there could have been very large or small emissions percentage which would affect the data adversely.

In the design of an emission system the fuel/air ratio has a major affect on the choice for a catalyst. Another factor that influences the choice of this catalyst system is the ability for it to minimize the formaldehyde emissions. Advances in electrically heated catalysts help the catalytic coverter to reach a temperature range where the formaldehyde is minimimum at a faster rate. Due the emissions of large amounts of formaldehyde a special catalytic converter will need to be installed. There is such a converter on the market and it is produced by Engelhard Corp. It is supposed to reduce the emissions from 15 mg/mile to 1 mg/mile which is well below California Standards.[***]

MOTOR DRIVE SYSTEM (Bill Rickelman)

AC vs. DC MOTORS-While in the past dc machines were the most widely used motors for various applications, most experts now seem to agree that ac induction motors are preferable for electric vehicle applications. According to J. Herbert Johnson, senior advanced development design engineer, A.O. Smith Electrical Products Co., the advantages of the ac induction motor include, possessing more power in less space, no need for failure prone brushes, smaller sizes and a flatter efficiency vs. speed curve.[**] Since the primary electrical power source is batteries which provide dc current it would seem that a dc motor might still be the better choice, since an ac motor would require an inverter. The inverter also serves as the motor controller. A dc motor also requires a motor controller for speed variations; and the dc controllers are simpler and more efficient than ac controllers. In the overall drive system, the ac drive is more efficient, and cheaper, than the dc drive system.

While the complexity of the ac controller is undesirable, every area which gives greater efficiency is important. The ac drive system has been implemented in many electric vehicles (such as GM's Impact).

MOTOR CONTROLS-There has been much improvement in the area of ac drives. Much of this revolves around the development of semiconductors. Peter Cleaveland writes,"The continuing improvement in the performance of power semiconductors, coupled with their declining prices, is making variable speed ac drives more and more attractive in the larger sizes".[*] The three different types of ac drives are:
- Pulse-width-modulated (PWM)
- Variable voltage input (VVI)
- Current source input (CSI)

The PWM is the most complex, and also the most efficient. Some advantages of the PWM over the VVI and CSI, include: high dynamic response for high performance use, very wide speed range, smooth motor control to and through zero speed, displacement power factor of 96%, regardless of speed and virtual elimination of line notching.[**] The newer PWM drives are more efficient than the older VVI and CSI drives, because their power components have lower losses. The best PWM drives use a bipolar switching method rather than the older unipolar switching method. The bipolar switching method results in better approximation of a fundamental sine wave and makes more effective use of the ac motor.[***]

Improvements on the dc drives is in the digital dc drives which is an upgrade from the analog dc drives. The digital drive keeps the motor speed sampled repetitively at short-term intervals, rather than continuously as in the analog drives. The digital methods can not, however, eliminate the brushwear and the high speed commutation limits of the motor, though brushless dc motors are available.

SPECIFICATIONS-Listed below and taken from the performance goals of the vehicle accelerating from 0-60 mph in 15 seconds, are the peak power and torque requirements at each of two rear wheels with a tire diameter of 20 inches, which are to be driven by electric motors. The required torque and horsepower for a sustained speed of 55 mph is also given. The motors must be capable of operating anywhere within these limits continuously over the three hundred mile range for about 9 hours.
- Peak Power, 20 hp
- Peak Torque, 407 ft*lbs
- Power @ 55mph, 7 hp
- Torque @ 55mph, 22 ft*lbs

These are specifications for a motor at the axle of each of two rear wheels with a tire radius of 20 inches.

Characteristics to be considered in addition to basic

---

[*] Automotive News, "Methanol/Gasoline...", Apr 1992.
[**] Business Week, "Giving a Clean...", 19 Jun 1989.
[***] IEEE Spectrum, "Pursuing Efficiency", Nov 1992.

[*] I&CS, "Variable Speed...", Jan 1990.
[**] I&CS, "Selecting the Right...", Apr 1989.
[***] I&CS, "Selecting and Applying...", Jul 1989.

performance are, size and weight, reliability of the motor, durability, maintenance and efficiency.

PROPOSED SYSTEM

Step-up Transformer-The first component in the drive system will be a step-up transformer. The motor control drive chosen for the vehicle needs an input voltage of 325 VDC to produce the 240 VAC needed to run the ac motor. The high voltage will be achieved through the step-up transformer that will increase the expected voltage of 66 VDC from the battery packs or alternator to the necessary voltage.

Motor Control Drive-The next component will be the actual motor control. As discussed earlier in this section a PWM control is the best choice in both performance and efficiency. In addition to this a vector control would help even further in the motor performance as discussed in the next paragraph. AC vector control from Thor Technology can be combined with a standard induction motor to provide both a transistorized PWM inverter with a microprocessor based vector control employing proprietary current regulator technology. A schematic of the vector control operation along with a description of the process can be seen in Appendix B. The Thor Series 7000 can operate from speeds of zero to 14,000 rpm with full rated torque and incorporates many drive protection/fault trips such as overcurrent, over/undervoltage and over temperature. The size of the control drive is 1.32 ft^3 and it has a 95% efficiency rating.

Motor-Through the current belief by experts in the field of electric vehicles that an ac motor is better suited than a dc motor it was decided to use an ac induction motor for the vehicle. An ideal stock motor was found at C-Tac Inc. in Cleveland, Ohio. They have a 20 hp ac inverter duty motor with constant-torque and 10:1 speed range, 6-60 Hz. If a PWM controller is used with vector control, the constant-torque operation is rated down through stall. The 60 Hz speed is 1800 rpm which would propel the vehicle at the required maximum speed of 60 mph. Each motor has a volume of 1.74 ft^3 and a weight of 211 lbs. Some additional features include a motor blower cooling independent of shaft speed, two motor overload thermostats (a warning at 130 degrees C and motor shutdown at 155 degrees C) and a totally enclosed construction providing maximum durability and reliability. The efficiency rating on the motor is 86.5%. The base cost for one motor is $1700.

BATTERIES (George Saghiyyah)

In designing a hybrid electric vehicle an in depth analysis and selection of the battery system is essential. This analysis will consist of collection of data about different battery types and systems. Analysis of this data and comparison of the different batteries and battery systems will provide the basis for the selection of the optimum battery for the design.

The batteries must conform to the Argonne National Laboratory specifications, sections 14.0 and 15.0. The performance requirements of 38.8 kWhrs total energy storage capacity and 30 kW peak power also must be satisfied.

The battery types studied include: lead/acid, nickel/cadmium, nickel/metal hydride, nickel/iron, zinc/air, zinc/bromine, and sodium/sulfur. Very little information was found on the lithium/iron monosulfide battery, so it was not considered. There is a table in Appendix C summarizing the results of the research into the batteries.

The lead/acid and nickel/cadmium (NiCad) batteries are by far the most widely used in todays electric and hybrid electric vehicles. This is simply due to the fact that these batteries are available at the present time. They, however, do not represent the latest in battery technology. The lead/acid batteries are inexpensive when compared with other types, but using lead/acid batteries in the car would require between 2150-2600 lbs (depending on who they are purchased from) of just batteries to fulfill the energy storage requirement. For the weight consideration alone, the lead/acid battery was dropped from the list of possibilities. The weight for NiCad batteries would be about 1500 lbs. Cadmium, however, is an environmentally hazardous material to dispose of, thus the NiCad battery was also eliminated from the list, since the purpose of the car is to be environmentally friendly.

The nickel/metal hydride and nickel/iron batteries were compared with each other. From information obtained from Eagle Picher Industries Inc. concerning both of these battery types, the nickel/iron battery was deemed the better choice for this application. This was due to its slightly higher specific energy and significantly longer life cycle.

Eagle Picher Industries Inc. makes one module type of the nickel/iron battery type, the NIF-200. For the three packs needed for the vehicle, eleven NIF-200 modules connected in series are required per pack, or thirty-three modules in total. The total weight for the batteries would be 1850 lbs. The module operates at 6 Volts, and a nominal capacity of 200 Ah. The price for a module is $2000, but for such an order, the price would be $1500 per module, or a total of $49,500 for the three packs. The cost, combined with the weight and the volume of all these modules is quite substantial, thus eliminating the nickel/iron battery from the list.

The zinc/air and zinc/bromine are interesting. The two of them are still in the development (advanced) phase, thus it may not be possible to purchase either of them. They exhibit medium to high specific energy, are potentially low cost systems, and operate at or close to room temperature. The zinc/air battery has a low specific power and a low life cycle due to dendrite growth during the charge/discharge processes. Bromine is very difficult to seal due to its very reactive nature and at the same time it is hazardous to inhale.

The sodium/sulfur battery is the only remaining candidate. This battery was chosen, to some extent, by the process of elimination. Sodium/sulfur batteries were only found to be manufactured in the United Kingdom and in Germany, creating the problem of obtaining specific information. From what was studied, however, the sodium/sulfur seemed to be the optimum battery. It exhibits the highest specific energy of all considered battery types, has adequate specific power, an adequate life cycle, and a potential low cost. The cells of this battery type also hold their charge, or will not self-discharge. The battery, however, must be kept at about 660°F, which would require insualtion, and it contains molten sodium and sulfur which are very dangerous materials.

Sodium/sulfur batteries manufactured by Silent Power Ltd. will comprise the battery system for this vehicle. Calculations at this stage (without actual manufacturers information) indicate that a total of 18 modules of the type PB-MK3 will be required, or six per battery pack connected in series and supplying 47.4 Volts. The batteries would have a total weight of about 1160 lbs. This is an estimate due to the lack of more accurate information regarding this battery type.

ALTERNATOR (Mark Garber)

In the HEV, the alternator will supply electrical power for battery recharging and, depending upon demand, will also provide power directly to the motor controller.

TOTAL POWER DEMAND-For charging of the batteries alone, Figure 9 below shows the required battery power demand as a function of charge time for battery capacities of 13, 26, and 39 kWHR. Nominal peak efficiency power output of the heat engine is 20 HP (15 kW) at 2500 RPM (262 rad/s). This corresponds to a torque of 42 ft-lb (57 N-m). Alternator efficiencies can be as low as 75%*, which gives a nominal alternator output power of 15 HP (11.25 kW). The power necessary to run the loads from Table 1 is about .5 kW. Therefore the batteries will have an allowable 10.75 kW supplied to them. From Figure 9, the corresponding charge times at nominal alternator output are 1.21 HR for one battery, 2.42 HR for two batteries, and 3.63 HR for all three batteries.

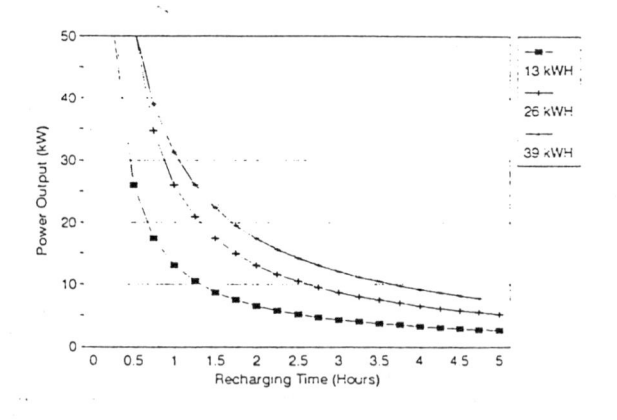

Figure 9. Alternator Power Output vs. Recharging time of Batteries

Summarized below in Table 2 is the typical electrical load for an automobile*. Heated backlight (rear defroster) constitutes about 33% of the total electrical load. Comfort control, internal entertainment, lighting, minimum demand, demand due to wet weather, and average of intermittent loads constitutes the other 67%. Since the specifications for the HEV do not require a heated backlight, the total current demand on the alternator for these loads is only 41.4 Amps instead of 61.4 Amps. This table is based on a typical automobile alternator of 12 volts. Therefore these loads would only require a peak power demand of .497 kW.**

In the case where the alternator will also have to provide power directly to the motor controller and recharge the batteries, the charge times will be different. In this case, the alternator will have to provide 5.22 kW to the motor controller for nominal motor usage. This means that the alternator will only be supplying 5.53 kW for battery recharging. From Figure 9, the corresponding charge times are 2.35 HR per battery. Because the motor will be running at a much higher voltage than the alternator, a step-up transformer will be necessary to make the components compatible.

Table 2. Typical Electrical Load for a 2-1 Engine Passenger Car

Description	Current (Amps)
Comfort control	7.0
Internal entertainment	1.2
Lighting	15.0
Minimum demand	6.5
Demand due to wet weather	7.5
Heated backlight	20.0
Average of intermittent loads	4.2

ALTERNATOR SPECIFICATIONS-The nominal angular velocity of the engine is 2500 RPM. This corresponds to a peak efficiency input torque of 42 ft-lb, and a peak efficiency input power of 20 HP (15 kW) to the alternator. For peak performance by the engine, the alternator must be able to accept a peak power supply of 40 HP (30 kW).

If an efficiency of 75% is assumed, then the nominal power output of the alternator is 15 HP (11.25 kW) with a peak power output of 30 HP (22.5 kW). If a 12 volt alternator is used (typical car alternator), then a nominal current of 938 Amps, and a peak current of 1875 Amps will have to be generated. Since this is an unrealistic amperage for the batteries to accept, the voltage must be increased to reduce the current generated.

The torque on the alternator depends upon the torque on the engine, the speed of the engine, and the speed of the alternator. The ratio of engine RPM to alternator RPM is equal to the ratio of alternator torque to engine torque. The angular velocity of the alternator is equal to the RPM of the engine multiplied by the gearing ratio (radius of crank pulley to radius of alternator pulley).

ALTERNATOR SELECTION-Based upon the specifications of the alternator and the research done, it has been determined that a 48 Volt, 15 HP (11.25 kW) alternator will be adequate for the HEV. A 48 Volt alternator will reduce the nominal current being generated to 234.4 Amps. United States Energy Corp. can supply such an alternator with a 75% peak efficiency at 2500-4000 RPM. The alternator would be belt driven, include a regulator to convert AC to DC, weigh approximately 50 lbs, and cost about $3000.

FRAME (Peter Goodale)

The frame and faring are to be designed to meet the specifications on length, width, rollover, and crash protection as well as cargo area.

The initial form of the body and frame was sketched from a combination of some prototype electric cars including the Blitz* and Honda CRX**. These sketches were then digitized and entered into SDRC's (Structural Dynamics

---

* IEEETOIA, "An Experimental Study...", May/Jun 1992.
** International Harvester, "Electrical Alternator", 1972.

---

* Autoweek, Luca Ciferri, 7 Sep 1992.
** Road & Track, Kim Reynolds, Oct 1992.

Research Corporation) MECAD software package called IDEAS (Integrated Design Engineering Analysis Software). The frame was created using box beams as members. The faring was created in seven skin sections that best fit the frame while maintaining an aerodunamic shape.

To create the faring, splines were created and meshed to form skin groups. Each skin group was then made into a skin and given a thickness of 0.039 inches. The skin groups used in the faring are given in Figure 10.

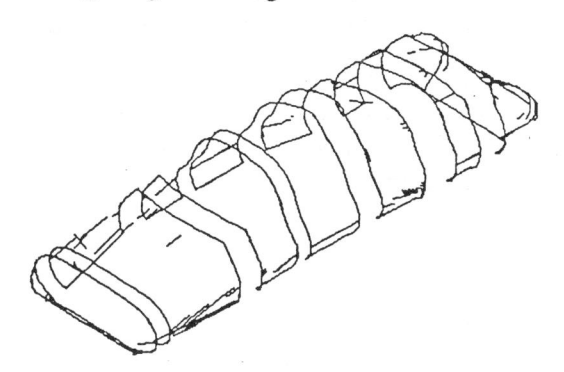

Figure 10. Exploded View of Faring Skins

Solid primitives of blocks and cylinders were used to represent other components with significant weights, such as the engine, batteries and motors. These components were then placed in the frame to determine the spacial requirements and overall mass properties according to specifications supplied by each respective group. An undercarriage was also created and attached to the frame. The simplified assembled model of the car with all of the components is shown in Figure 11. A shaded isometric view of the car with 75 percent translucency applied to the faring is shown in Appendix D Figure D1. A preliminary dimensioned drawing of the frame is shown in Appendix D Figure D2.

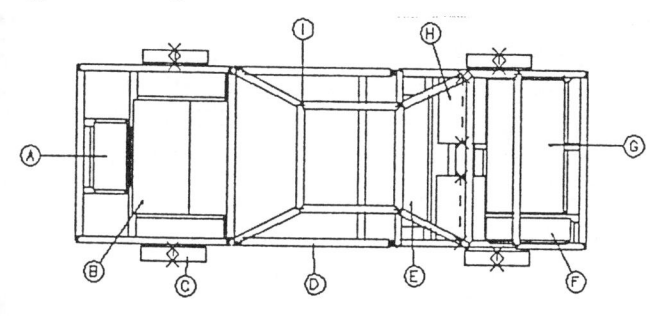

Figure 11. Assembled Model of HEV
Key: A-Master Control   F-Gas Tank
      B-Engine   G-Batteries
      C-Wheel   H-Motor
      D-Door   I-Frame
      E-Storage

A finite element analysis is being done on the frame using beam elements to determine it's structural integrity under different loading conditions. An initial linear static analysis was done to determine the deformation of the frame under the weights of its components including a driver and passenger (200 pounds each). The maximum deflection using Aluminum 2014-T6 was .0353 mm. A finite element model of the frame with its nodal constraints and forces due to static weight loads is shown in Figure 12. A plot of the deformed shape superimposed on the original frame is shown Figure 13. The corresponding deflections and stresses are shown in the graphs in Appendix D Figures D3 and D4.

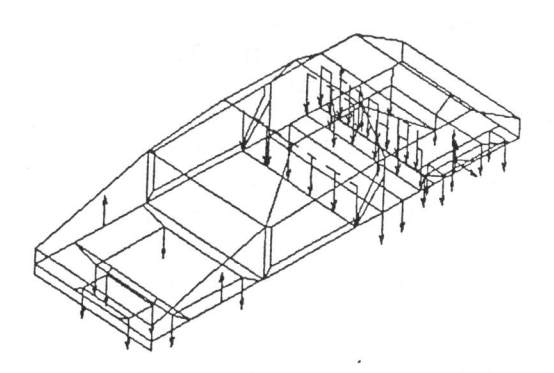

Figure 12. Structural Loads and Restraints

A refined model and different modes of analysis such as buckling and plastic analysis are necessary to analyze for impact and rollover as per government specifications. A drag coefficient for the car is to be determined which is expected to vary between .19 and .4.[*]

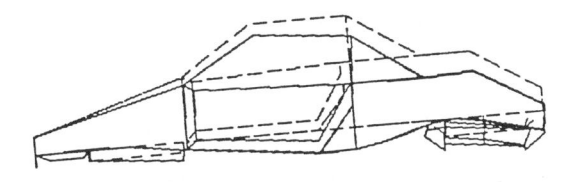

Figure 13. Deformation Due to Component Weights, Scale *5000

SUSPENSION SYSTEMS (Rob Bittler)

The main goal of any type of suspension system is to absorb road irregularities and to counter side loads that occur during cornering. Most car suspension systems consist of linking members, springs, shock absorbers, wheels, and tires that come together to effectively achieve this goal. This is done while always keeping in mind the need for compatibility with the steering and braking systems. The early models of suspension systems exlusively used non-independent systems. The advantages of non-independent systems include being very sturdy, inexpensive, and simple to install. The drawback are that they are very heavy and bulky. Also, if one of the wheels is displaced when in use, there is an undersireable reaction of the other wheel. In many cases, this could lead to a problem known as wheel shimmy. The introduction of independent front suspension (IFS) was a turning point in better ride and comfort. The elimination of the front axle running across the frame made it possible to locate the engine further to the front and lower in the car, leading to a more comfortable ride. The height of the center of gravity of the sprung mass was lowered, creating better directional control of the vehicle. IFS

---
[*] Autoweek, Luca Ciferri, 7 Sep 1992.

also reduced the vibration problems in the front suspension, along with completely eliminating the problem of wheel shimmy. By reducing the unsprung mass by over 50%, IFS allows more space available for a softer spring. Although IFS systems are more complicated and expensive than non-independent systems, they are much lighter and offer superior performance. The three most used IFS systems are the Twin A-arm (also call double wishbone), transverse leaf springs, and MacPherson struts, and all of these have been considered as the possible design for the HEV.

Passive systems use only the linkage geometry to achieve the goals of ride and handling. The state-of-the-art active systems use sensors, valves, and electric motors or hydraulic controls to adjust the spring and damping rates under differing driving conditions. Although these systems achieve the best performance, they are much more complex, heavier, and more expensive than passive systems. Another disadvantage of active systems is that they need energy to operate, which would deprive the HEV of much need battery power.

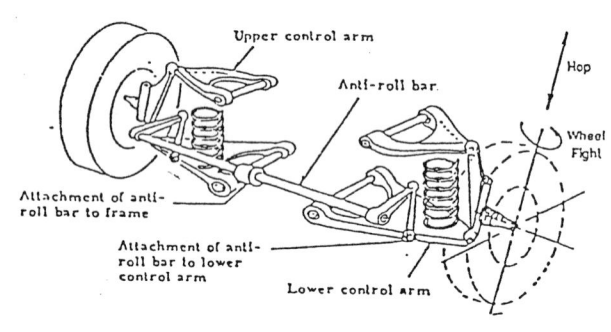

Figure 14. Twin-A Arm Suspension

CHOICE OF SUSPENSION SYSTEM FOR HEV-The choice of which suspension system best suits the HEV is made clearer when using a design evaluation matrix. The weight of the system, performance of the system, and cost of the system were the three criteria used in the selection process. The weighted value is used to show the importance of each criteria to the HEV. The weight of the suspension system (weighted importance of .45) is very important because every additional weight attached to the HEV will necessitate a higher power output from the batteries to achieve a specified velocity. The less the suspension weighs, the better. The performance (weighted importance of .45) of the systems is also a very crucial criteria. The system needs to be able to adequately handle the unique weight distribution of the HEV, and provide a comfortable ride to the passenger. The cost of the system (weighted importance of .10) is not an utmost concern at this point, but nonetheless should not be completely overlooked. Five different systems are rated to determine which one is the best design for the HEV. A front suspension system with a Twin A-arm configuration, shown in Figure 14, was clearly the best choice when rating these criteria.

There are not nearly as many choices when considering the rear suspension. The most common system used with the Twin A-arm system is the live rear solid axle suspension, where the two back wheels are rigidly attached to the axle. This is a classical combination that is strongly related to the "magic numbers" in chassis design that will be discussed in the analysis of the system. For this reason, the solid axle will be used in the the rear suspension for the HEV. The solid axle rear suspension can be seen in Figure 15.

ANALYSIS OF SUSPENSION SYSTEM-Many hybrid electric vehicles, such as the Honda CRX, have experienced some problems in the handling of the car.[*] This is due to the fact that the suspension system has not been altered to adapt for the unique weight distribution of the HEV. (This weight distribution includes batteries, alternators, control box, motors, and an IC engine.) Developers have long used a set of "magic numbers" in optimizing many of the variables in suspension systems. These numbers have been based on previous experience using a trial and error basis. On of these "magic numbers" presently being used is a ratio comparing the rear suspension spring constant to the front suspension spring constant to be at or around 1.21. In order to determine if this value is useful in the HEV, it was necessary to analyze the suspension system for this HEV.

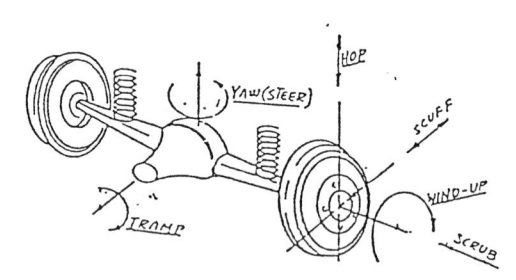

Figure 15. Solid Axle Suspension

By creating a computer system dynamic model of the HEV with its unique weight distribution, it was possible to analyze the mode shapes and a response for a given road input. The main criteria used to determine a "favorable response" is from the fact that bounce motion is preferred to pitch motion in the automobile. Also looked at is the settling time of the response. Using a five inch bump as the input and a time delay of 0.1 seconds from when the front wheel goes over the bump to when the back wheel does (this corresponds to a traveling speed of 57 MPH), the spring ratio of 1.21 was analyzed. Figure 16 shows that the nodal displacements are out of phase much of the time, which would lead to an undesireable pitch motion.

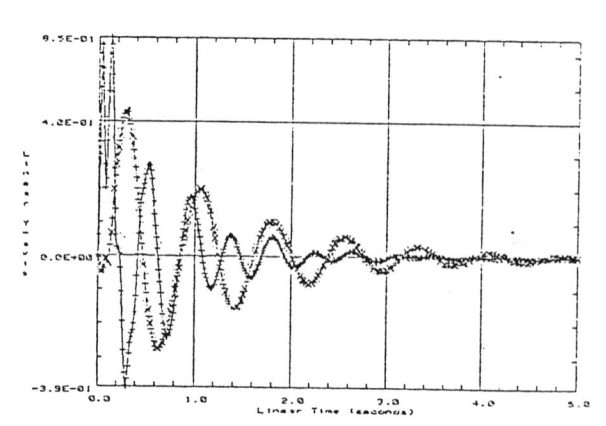

Figure 16. Time Response for Nodal Displacements

After playing with the spring ratio, the optimal design was

---
[*] Fundamentals of Machine Component Design, 1991.

determined to be a rear spring with a spring constant of 2.25 times that of the front spring (see Figure 17). As evident from the figure, the displacement become in phase at 1 second, which will give a favorable response with bounce, not pitch motion. These response characteristics repeated for various road inputs ranging from potholes to ramps. Therefore, it was concluded that the optimal suspension system would have a back/front spring ratio of 2.25. Then using a damping ratio of .4 (this was chosen because it is a "magic number" that provides the best handling of the car), it was possible to determine the optimal constants for this HEV by having a settling time of under 3 seconds. The front suspension should have spring constants of 105 lb/in, and damping constants near 5 lb/in/sec for each wheel. The solid axle rear suspension should have a spring constant of 236 lb/in, and a damping constant around 20 lb/in/sec.

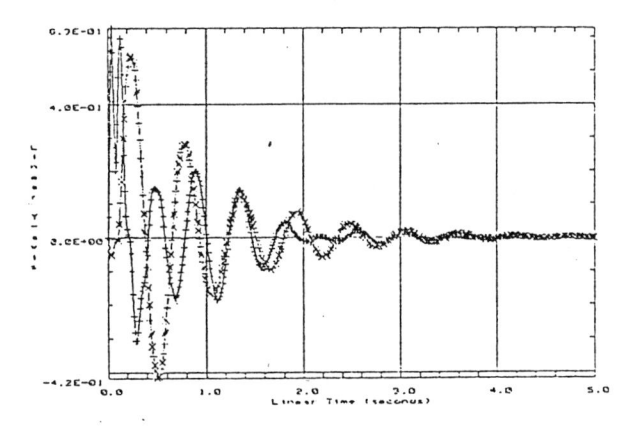

Figure 17. Time Response for Nodal Displacements, Optimized

STEERING (G.C. Seibert)

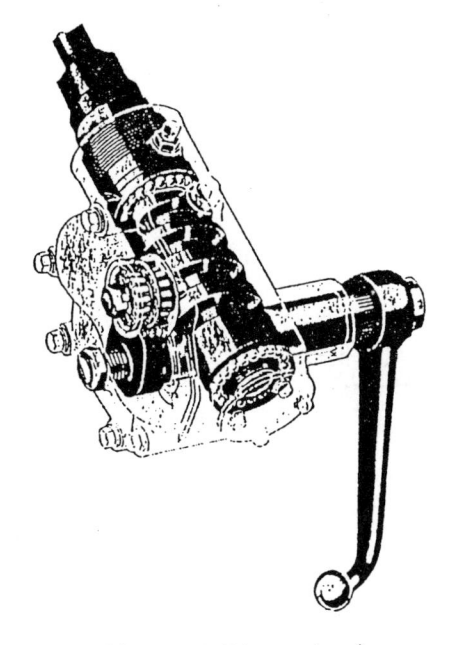

Figure 18. Pitman Arm*

In the interest of conserving power and simplifying the design, a conventional non-power assisted steering system has been chosen. Considerations for the system are as follows:

TYPES OF STEERING MECHANISMS-There are two proven steering mechanisms that will be explored here. The first mechanism employs a worm gear which is attached to the end of the steering shaft. The rotation of the worm gear in turn rotates the second gear which is attached to a Pitman arm. As the Pitman arm is rotated, it causes the translation of the tie rod which in turn rotates the steering arms attached to each wheel hub.

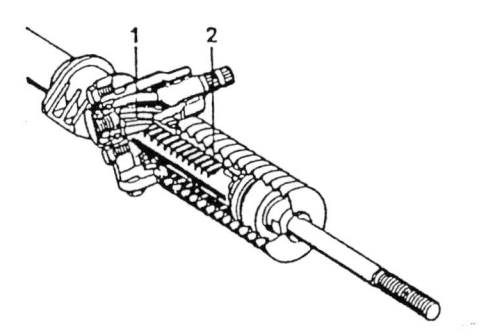

Figure 19. Rack and Pinion Steering*
Key: 1-Pinion 2-Rack

The second steering mechanism is a rack and pinion system. This system is used in most automobiles currently in production. A small gear, or pinion, is attached to the end of the steering shaft. This gear, as it is rotated, causes the translational motion of the rack. As the rack is translated, it rotates the steering arms as in the previous design, thereby steering the automobile.

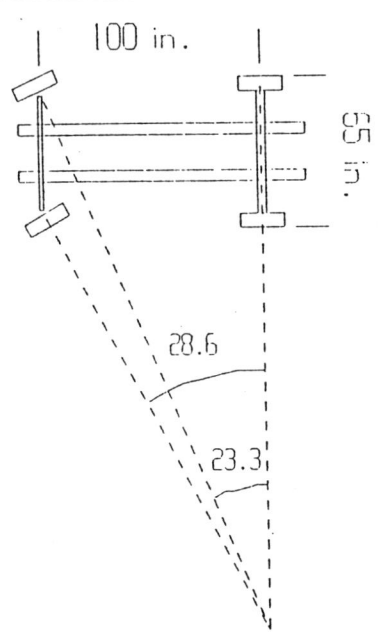

Figure 20. Toe Out on Minimum Turn Radius

TURN RADIUS-The turn radius for the vehicle has been set at twenty feet (well within the maximum turn radius of 39.4 feet*), which would allow a complete U-turn in a street forty feet wide. This requires a total angular rotation of the front wheels of 28.6° from the straight ahead. An angular

---

* Fundamentals of Automobile Chassis....

* Automotive Handbook, 2nd Edition.

of this size requires that, for a twenty inch tire, a minimum clearance of five inches be provided from the inside wall of the tire to the wheel well.

SYSTEM CHARACTERISTICS-The essential features of the steering system are: 1. Steering Arm, 2. Drag Link, 3. Idler Arm, 4. Tie Rod (Rack), 5. Steering Wheel, 6. Steering Shaft, 7. Steering Box (Pinion), 8. Pitman Arm. Automobiles generally require two to three steering wheel rotations to turn the road wheels through their maximum angle. Because this is to be a non-power assisted system, three steering wheel rotations will be specified to reduce the necessary applied forces. Steering ratio is defined as the ratio of steering wheel rotation to the actual wheel rotation. For this system, with three steering wheel rotations needed for the lock-to-lock rotation of the front wheels, the steering ratio is 18.88.

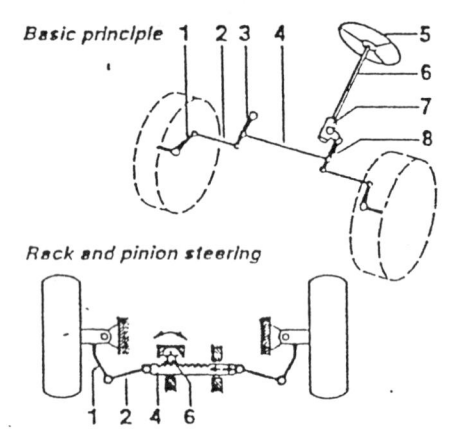

Figure 21. Steering Assembly*
Key: 1-Steering Arm 5-Steering Wheel
2-Drag Link 6-Steering Shaft
3-Idler Arm 7-Steering Box
4-Tie Rod 8-Pitman Arm

The steering arm is to be 6 inches long. It will nominally be at an angle of 21.9° from a line parallel to the center-line of the car when the wheels are facing straight ahead. This will require the rack to travel 20.94 inches when the wheels are rotated from extreme-left to extreme-right, and vice-versa. For a rack movement of 20.94 inches, and a steering wheel rotation of three, the required pitch radius of the pinion is 0.286 inches.

The weight distribution of this automobile is approximately 25% Front to 75% Rear. This means that, of the 2500 lbs total weight, 625 lbs of it will be focussed on the front wheels. The following equation gives the approximate value of the static torque required to turn a non-rolling tire.

$$T_s = \frac{\mu W^{\frac{3}{2}}}{3 P^{\frac{1}{2}}}$$

---

*Automotive Handbook, 2nd Edition.

Where: μ = Coefficient of Friction
W = Load
P = Tire Pressure

For this case, μ can be taken as 1.0 for tire interaction with dry concrete or tarmac. The load on each front wheel is 312.5 lbs, and the tire pressure is assumed to be 35 psi. From these values, it can be calculated that the torque required to turn the wheels when the car is at rest is T_s = 311.26 lb-in. In turn, the required force to translate the rack is approximately 104 lbs. This would require a torque of 29.72 lb-in. to be applied by the 0.286 inch radius pinion. For a 14 in diameter steering wheel, the driver would need to apply 4.25 pounds of force to turn the wheels when the car is a rest.

In order to meet the auto-centering stipulation, the suspension system would have to allow for a caster angle of +3°. This would tilt the top of the kingpin axis towards the back of the car, thus the wheels are actually being pulled (not pushed), and would tend to achieve a straight-ahead position.

SAFETY-The steering column and shaft should be hinged at some point to a structural member on the frame of the car. This will allow the steering column to pivot upwards in the case of a front-end collision, thereby reducing the risk of impaling the driver.

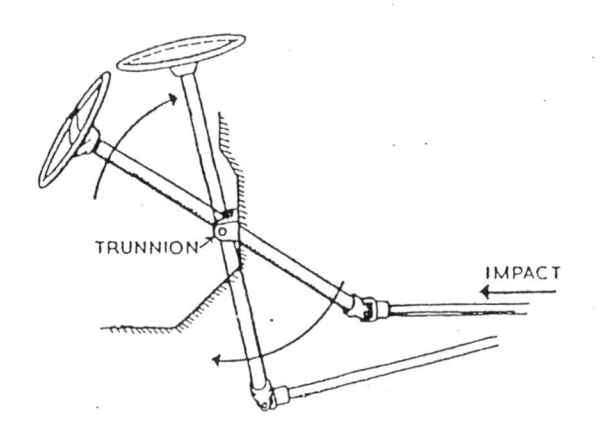

Figure 22. Steering Column Attached to Frame*

CONNECTION OF STEERING SHAFT TO STEERING MECHANISM-The length of the steering column in a conventional automobile is about 48 inches and makes an angle of about 30° with the horizontal. Because of this angle, it is necessary to employ universal joints to connect the steering shaft to the actual steering mechanism.

BRAKES (Peter Shuleski)

Specifications from the Department of Energy and Argonne National Laboratories give the limitations and requirements for the braking systems for the Hybrid Electric Vehicle. Each of these is addressed in the following paragraphs. The emphasis is on safety, simplicity and minimum weight.

Although the car is unconventional, braking need not be

Man and Motor Cars. any different than that of current

systems. This approach offers such advantages as it will be familiar to existing repair mechanics, will not need new technology and will provide an established reliability. Current power assisted brakes will not be considered in order to save the additional weight and energy such a system would require.

BRAKING SYSTEMS-As required by the specifications, two independent braking systems will be incorporated into the design of the car. With disk brakes in the front and drum brakes in the rear with an independent hydraulic system controlling each.

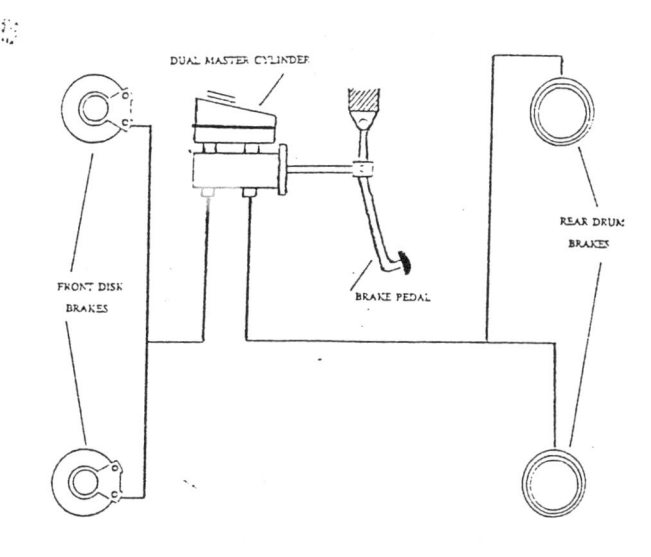

Figure 23. Layout of Brake System

The disk arrangement is favored for the front of the car because of its ease of maintenance, linear stopping power and greater heat dissipation than the drum type.* On the other hand a rear drum brake is favorable because of its ability to be cable operated, for emergency and parking brake use, its long service life and moderate heat dissipation characteristics. Both system appear to weigh about the same with only a slight weight savings in the disk brake system.

For this application the mounting of the brake components will be much the same as that of a modern vehicle. The pedal and master cylinder will be located on the fire wall at the front of the vehicle while the actual brakes, rotor, drums etc., will be located in the standard location at the end of the suspension where the tire mounts to the vehicle.

Both systems will be designed to stop the car independently. As stated earlier, the rear drum brakes will incorporate a lever operated cable as a parking brake/emergency brake for parking on grades or if both front and rear hydraulic systems fail.

BRAKING CHARACTERISTICS-Some of the variables that will affect the brake performance are:
- Overall car maximum mass, 1200 kg.
- Weight distribution on each wheel, 75% rear, 25% front.
- Range of friction coefficients between brake pad and rotor/drum, 0.3 for wet conditions up to 0.9 for dry cases.
- Road surface to wheel friction coefficients, range from 0.5-0.7 for rain slick roads and up to 0.9 for rough dry surfaces.
- Braking system response, time for complete system to

---
* Fundamentals of Machine Component Design, 1991.

pressurize and the pad to come into contact with the rotor/drum.
- Heat buildup is system, heat generated after repeated cycling of braking system may alter characteristics of braking.
- Braking Factor as affected by the rubbing velocity of rotor.
- Diameter of brake rotor or drum will be 0.254 m.
- Average pedal force applied at brake pedal is 400 N.
- Car wheel base is 2.4 m.
- Center of gravity is 1.8 m back of front tires and 0.504 m from the ground.

All of these variables will need to be accounted for in order to have the car follow the prescribed stopping distance curve, shown below.

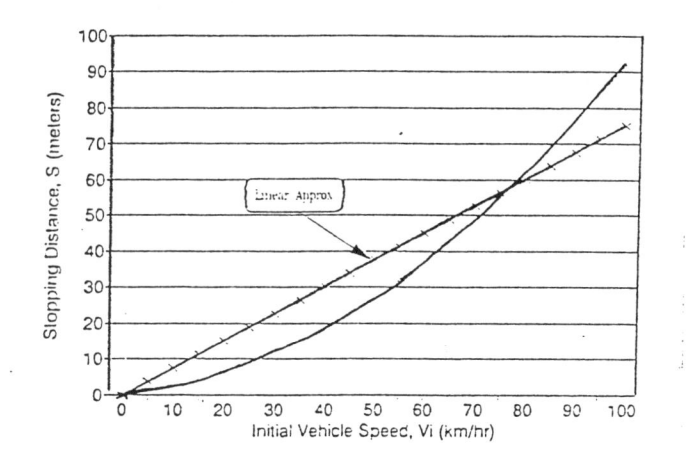

Figure 24. Brake Effectiveness Test
Brake Systems, Australia (ADR 35/00)

By separating the given curve down into small increments of constant deceleration it was noted that the average deceleration never exceeds 5 m/sec^2. This deceleration was then plotted versus the time needed for the distance specified. A generous constant time of 2.5 seconds was added to the deceleration time to account for driver response as well as system response time. This plot is shown below as, Average Deceleration During Braking Time.

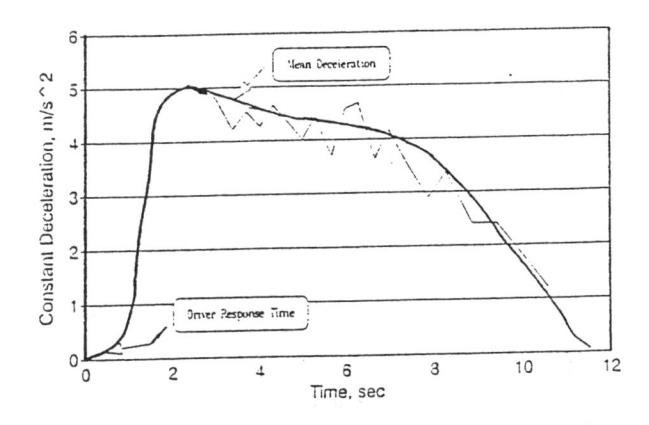

Figure 25. Average Deceleration During Braking Time

A dynamic analysis of the braking system shows that during a deceleration equal to 9.8 m/sec^2 approximately 21%

of the vehicles weight is shifted from the rear of the car to the front. Therefore by noting the static and decelerated weight distribution of the car on each axle a graph of the maximum deceleration allowable before skid can be plotted versus the percent braking supplied by that axle. This maximum is also dependent upon the coefficient of friction between the road surface and the rotating tire, these lines are also included in the graphs. A graph for each axle is included in the following text.

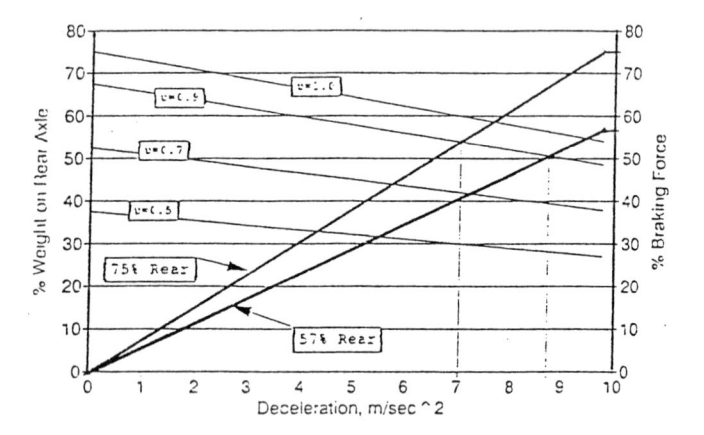

Figure 26. Weight on Rear Axle and Braking Forces

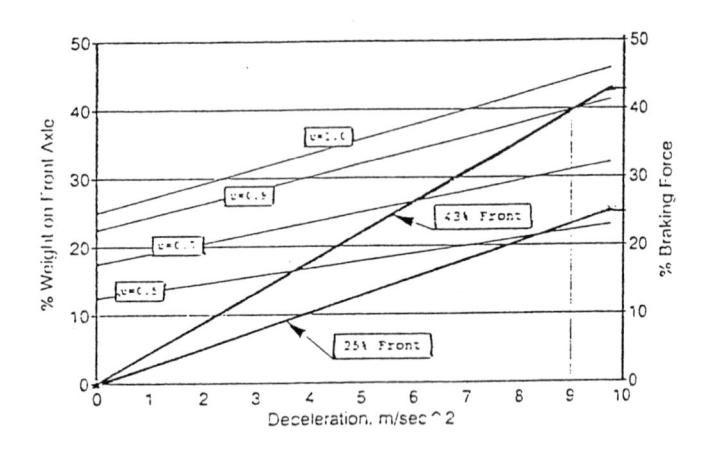

Figure 27. Weight on Front Axle and Braking Forces

The initial assumption was made that the proportion of braking force supplied by each axle was equal to the static weight on that axle, 75% rear and 25% front. This proved to be a problem in that the maximum deceleration before skid of each axle were no where near the same value. Ideally the maximum deceleration of each axle should be arranged so that the rear tires would begin to skid just before the front tires. This will allow for more steering control by the driver in the event of a panic skid stop. This requirement is met when the proportion of braking by the front and rear is adjusted to 57% rear and 43% front. This will provide a maximum deceleration of 9 m/sec^2 for the front tires and a slightly lower value for the rear of 8.125 m/sec^2.

SELECTION OF BRAKE COMPONENTS[**]-From interpretation of the braking characteristics it was determined that a deceleration of slightly less than 9 m/sec^2 is capable before sliding on dry ground with the following component specifications:
- Pedal force = 400 N
- Pedal ratio = 6:1
- Master cylinder bore = 5 cm
- Rear wheel cylinder bore = 2.00 cm
- Front caliper piston diameter = 4.162 cm

A maximum of 9 m/sec^2 was selected because as stated in the requirements, each system must be capable of stopping the vehicle within the required distance. This deceleration combined with the less significant rolling and wind resistance will achieve this goal.

REGENERATIVE BRAKING-No specific requirements on motor selection have been made which accommodate regenerative braking. At this time there is no intention of incorporating this type of braking system into the vehicle. Although it will be considered when a second iteration of the overall design takes place. However, considerable energy savings could be realized, decreasing required battery capacity if some of the energy used in friction braking can be recovered.

ALTERNATE ENERGY SYSTEMS (KAMAL PATEL)

Energy, for the LUHEV is to be provided by chemical batteries. Two future alternatives to current chemical batteries are the mechanical battery and fuel cells. Both, the mechanical battery and fuel cells are in developmental stages and these power sources will not considered for the present LUHEV proposal. Also, it is understood that mechanical batteries are specifically excluded by the specifications, and the use of fuel cells would require a major change in the configuration of the overall vehicle design. The material presented here is intended to give an overview of possible future alternatives to conventional chemical batteries.

MECHANICAL FLYWHEEL BATTERY-Development of a flywheel surge power unit has been conducted by R.C. Flanagan and C. Aleong at the University of Ottawa, Canada and Unique Mobility Inc. has been manufacturing prototypes in Englewood, Colorado. The essential parts of the system consists of a rotor, a motor/alternator, bearings, a vacuum system, and power electronics.

Electrical power, supplied by an external source, is applied to the motor which drives a flywheel. The high speed rotation of the composite flywheel wheel will result in stored mechanical energy. To recover this energy, the flywheel engages the motor which now acts as an alternator and provides electrical output power to a load. The whole system is enclosed in a sealed vacuum container due to the high tip

---

[*] Automotive Chassis, "Braking Dynamics...", 1991.
[**] Bosch Automotive Handbook, 1991.

speeds of the flywheel."

Development has evaluated advantageous disc geometries, production and manufacturing costs, residual ring and hub stresses, and materials. A high power density permanent magnet brushless motor is employed with the required power electronics. Rotor temperature caused by bearing friction, rather than drag, will determine the upper pressure level acceptable in the system. Magnetic bearings and ceramic bearings are currently being evaluated."

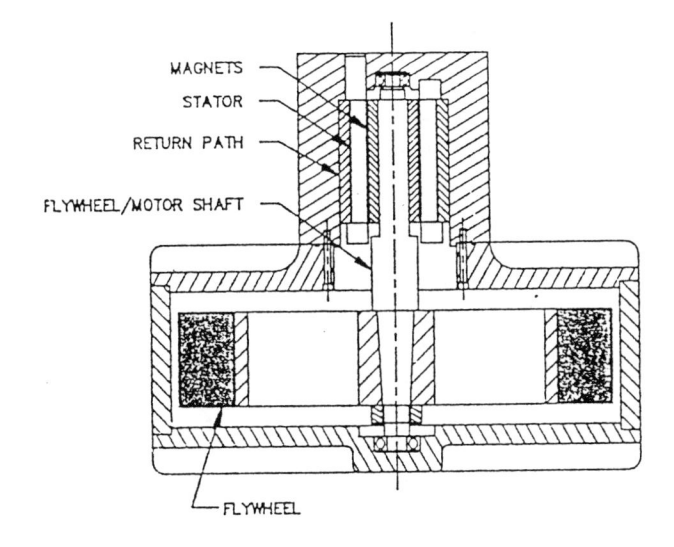

Figure 28. Diagram of a Flywheel*

Twenty four rotor prototypes have been evaluated. Results show the possibility of obtaining specific energy densities of 60 Wh/kg. Decision analysis of the prototypes revealed S2-Glass/Exas-Carbon Thick Ring Disc as being the top rated rotor. For this specific rotor, design speed is 30,000 rpm with a fatigue life of 100,000 cycles. It occupies a swept volume of 789 cubic inches with an outside diameter of 16.940 in.*

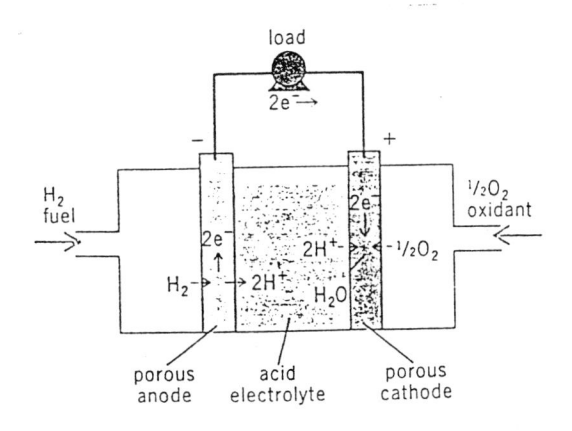

Figure 29. Diagram of a Fuel Cell

FUEL CELL OPERATION-"Attempts to solve the transportation-related chemical noise pollution have until now largely centered around battery powered electric vehicles. These attempts have been rather unsuccessful due to the limited energy which can be stored in a reasonably sized battery which in turn severly limits vehicle performance and range. In addition, battery recharge is time consuming. In contrast, the fuel cell, using methanol as a fuel, will be capable of a full day's operation on a reasonably sized fuel tank with refueling accomplished much like present internal combustion vehicles"."

A fuel cell is a device that converts chemical energy of reactants (a fuel and an oxidant) directly into low voltage DC electricity. A fuel cell effects this conversion through electrochemical reactions. Fuel cells are converters only, and will operate as long as they are fed with a suitable fuel and oxidant and the reaction of the products are removed. It consists of a fuel electrode (anode) and an oxidant electrode (cathode), separated by an ion-conducting electrolyte. The electrodes are connected electrically through a load by a metallic circuit. Electric current is transported by flow of electrons, whereas in the electrolyte it is transported by the flow of ions. By electrically connecting a multiplicity of unit cells, it is possible to form a fuel cell battery of any desired voltage and current output.**

The efficiency of a fuel cell differs from that of a heat engine. One it does not decrease sharply as the size of the unit is decreased, and secondly, the fuel efficiency does not appreciably change if the fuel cell operates at part load. Fuel cells have at least twice the efficiency of internal combustion engines and at least one and a half times the efficiency of diesel engines. The energy density of a fuel cell is greater than that of a battery. An increase in fuel cell energy density can be achieved merely by increasing the size of the fuel tank. "Taking into account efficiency of use, methanol in a tank for use in a fuel cell represents 1900 Wh/kg; whereas gasoline for an IC engine corresponds to 900 Wh.kg. In contrast, lead-acid batteries are less than 40 Wh/kg, advanced batteries with the most optimistic specifications only being 200 Wh/kg." A fuel cell can be recharged much faster than a battery since it is only necessary to refill the fuel tank, much like an IC engine. Another consideration is that the extensive network of gasoline stations may be converted to handle fuels such as methanol.**

An electric car powered by a direct methanol-air fuel cell is attractive. One of the products of the oxidation reaction of methanol at an anode is carbon dioxide. With use of an alkaline electrolyte, the formation of carbonates is a serious problem. Taking this into account, a practical fuel cell must operate with an acidic or solid electrolyte. "To date, the most practical catalyst for both the methanol-oxidation reaction and the oxygen reduction reaction are all Pt-based materials." R. Manoharan and J.B. Goodenough of the Center of Materials Science and Engineering present an alternative for each reaction. A search for new platinum-free catalytic material for both methanol anode and oxygen reactions of a direct methanol-air acid fuel cell has been made. A 50-50 ordered alloy NiTi has been found to be stable in acid solution and to be active electrocatalyst for the methanol oxidation reaction.***

"Fuel cells using non petroleum fuels such as methanol can potentially provide the advantages of an ICE in a power

---

*   25th IECEC, "Design of a Flywheel...", 1990.
*   34th International Power...", "Methanol Fuel...", 1990.
** Fuel Cells Handbook, 1989.
*** 26th IECEC, "Electrodes Without...", 1991.

system, but with clean and quiet operation. Because a fuel cell will deliver power as long as a fuel and air is supplied, the range is limited only by amount of fuel carried and can directly replace the ICE." A study sponsored by the Department of Transportation was conducted to determine the feasibility of a fuel cell/battery hybrid power source. Based on this study, the Departments of Energy and Transportation have initiated a fuel cell/battery powered bus system program.[*]

In 1986, Los Alamos National Laboratory conducted an extensive study for the Department of Transportation on the feasibility of powering standard size and small buses with fuel cells operating on methanol fuel. The study concluded that a hybrid fuel cell/battery powered bus can be built with available technology to meet typical inner city service requirements. "Energy Research Corporation (ERC) and its team partners, Bus Manufacturing USA, Inc. (BMI) and the Los Alamos National Laboratory (LANL) were selected to develop the urban mass transit bus design utilizing air-cooled phosphoric acid fuel cell (PAFC)/Battery hybrid system." They have concluded that the PAFC/battery hybrid power source for a city bus is competitive with conventional diesel engines on the basis of cost, with the real benefits from reduction in noise, pollutants, and use of nonpetroleum fuel.[**]

Lehigh University's Hybrid Electric Vehicle Design meets with all specifications set forth by the Society of Automotive Engineers (SAE), Department of Energy (DOE), and Argonne National Laboratories (ANL) as detailed in the DOE/ANL Hybrid Electric Vehicle Paper Design Contest. In addition, it will theoretically meet the emission standards to be classifed as an Ultra Low Emission Vehicle (ULEV).

A further study is currently being done to insure that the design meets with all Department of Motor Vehicles safety standards. Since our vehicle is similar in design to current vehicles however, we are confident that these standards will be met. Constuction of a hybrid electric vehicle is scheduled to begin in January of 1993 following this design. The project will be led by a small group of students who actually participated in the original design.

Finally, we would like to give special thanks to Dr. Oszoy and Dr. Sarubbi for their help and guidance with this project. We would also like to thank the many faculty members and students who aided this design with objective insights and constructive criticism. Lastly, we would like to thank the numerous companies which provided us with information on component designs and specifications.

---

[*]   23rd IECEC, "Integration of a Fuel...", 1988.
[**] 34th International Power..., "Methanol Fuel...", 1990.

APPENDIX A

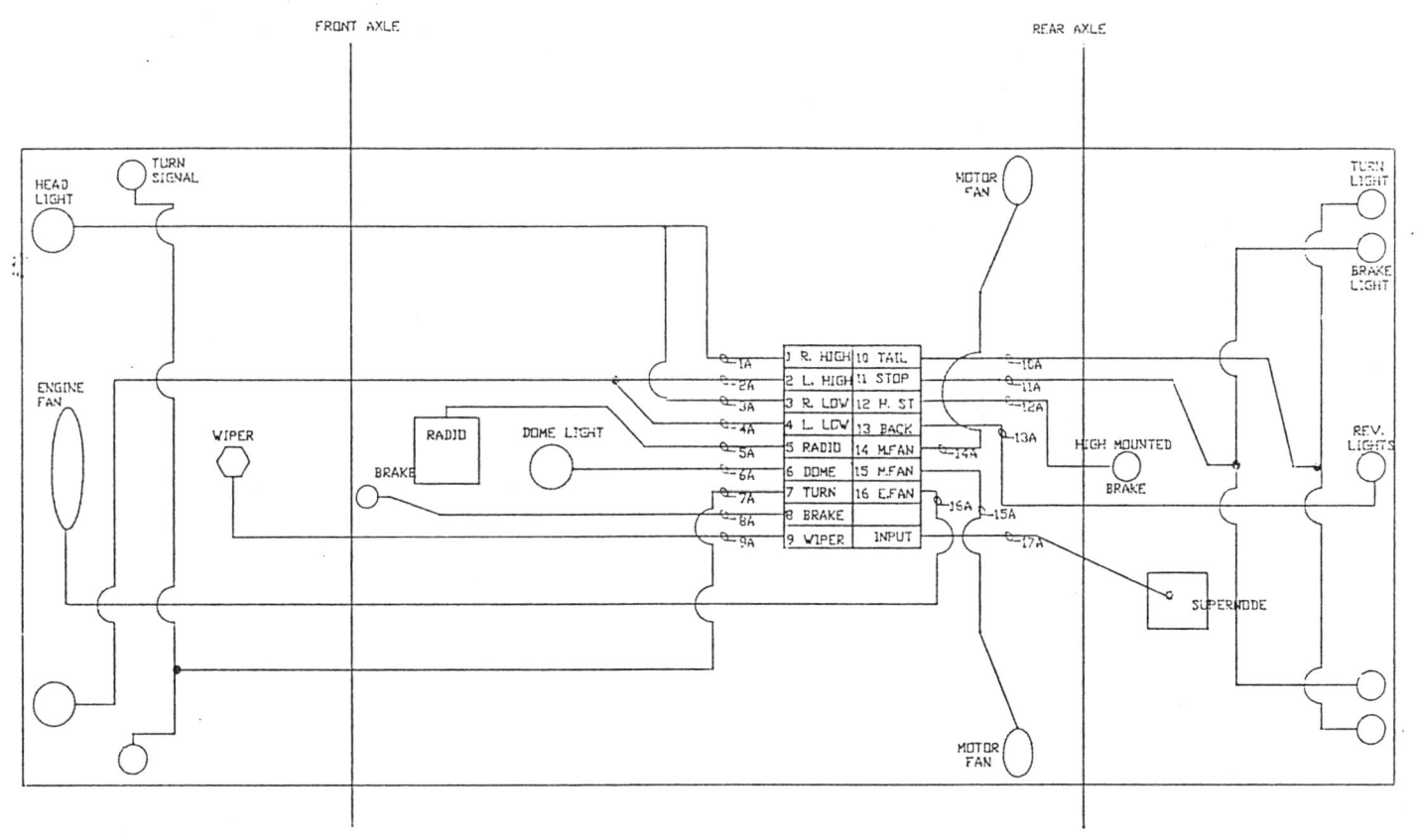

Figure A1. Accessory Wiring Diagram

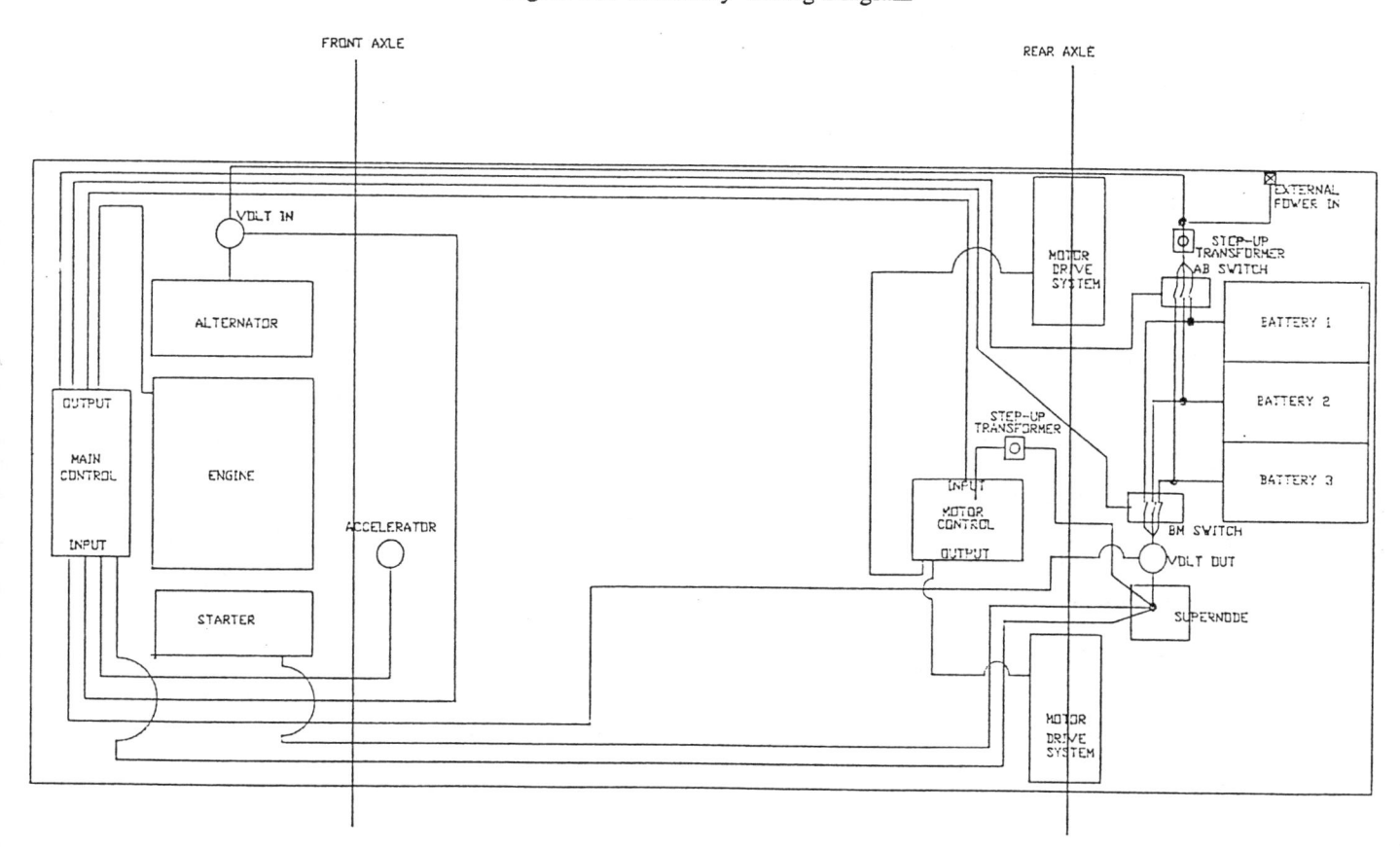

Figure A2. Main and Controller Wiring Diagram

Table A1

MAIN COMPONENTS WIRING

Number	Name	Length (in)	Volts	Current (amp)	Power 125% (amp)	Insulation
1M	Alternator to AB Switch	150			0	
2M	External Power to AB Switch	30			0	
3M	Switch to Node Battery 1	24	66		0	
4M	Switch to Node Battery 2	36	66		0	
5M	Switch to Node Battery 3	48	66		0	
6M	Node Battery 1 to Battery 1	12	66		0	
7M	Node Battery 2 to Battery 2	12	66		0	
8M	Node Battery 3 to Battery 3	12	66		0	
9M	Node Battery 1 to BM Switch	48	66		0	
10M	Node Battery 2 to BM Switch	36	66		0	
11M	Node Battery 3 to BM Switch	24	66		0	
12M	BM Switch to Supernode	25	66		0	
13M	Supernode to Left Motor	40			0	
14M	Supernode to Right Motor	45			0	
15M	Supernode to Engine Starter	150			0	
16M	Supernode to Main Controller	160	12	10	120	

Table A2

CONTROLLER WIRING

Number	Name	Length (in)	Volts	Current (amp)	Power 125% (amp)	Insulation
1C	Voltmeter Reading Out Of Batteries	160				
2C	Voltmeter Reading Into Batteries	40				
3C	Accelerator to Main	30				
4C	Main Controller to AB Switch	150				
5C	Main Controller to BM Switch	150				
6C	Main Controller to Motor Controller	125				
7C	Main Controller to Engine Throttle	30				
8C	same as #6C					
9C	Motor Controller to Right Motor	30				
10C	Motor Controller to Left Motor	30				

Table A3

ACCESSORY COMPONENTS WIRING

Number	Name	Length (in)	Volts	Current (amp)	Power 125% (amp)	Insulation
1A	Fusebox To Right High Headlight	130	12	10	120	
2A	Fusebox To Left High Headlight	130	12	10	120	
3A	Fusebox To Right Low Headlight	130	12	10	120	
4A	Fusebox To Left Low Headlight	130	12	10	120	
5A	Fusebox To Radio	100	12	10	120	
6A	Fusebox To Dome Light	85	12	15	180	
7A	Fusebox To Turn Light	130	12	5	60	
8A	Brake Sensor to Fusebox	100	12	5	60	
9A	Fusebox To Windshield Wiper	110	12	15	180	
10A	Fusebox To Rear Turnlights	70	12	15	180	
11A	Fusebox To Brake Lights	70	12	5	60	
12A	Fusebox To High Brake Light	50	12	5	60	
13A	Fusebox To Backup Light	70	12	5	60	
14A	Fusebox To Motor Fan Right	40	12	15	180	
15A	Fusebox To Motor Fan Left	40	12	15	180	
16A	Fusebox To Engine Fan	140	12	15	180	
17A	Supernode to Fusebox	50	66	30	1980	

APPENDIX B

HOW THE VECTOR CONTROL WORKS

- A field current is induced into the rotor bars by the armature flux.
- An armature current is set by the outer loop speed regulator to produce the required torque.
- The field current is positioned 90° from the armature current.
- The vector sum of the field current and armature current is calculated to produce a reference stator current.

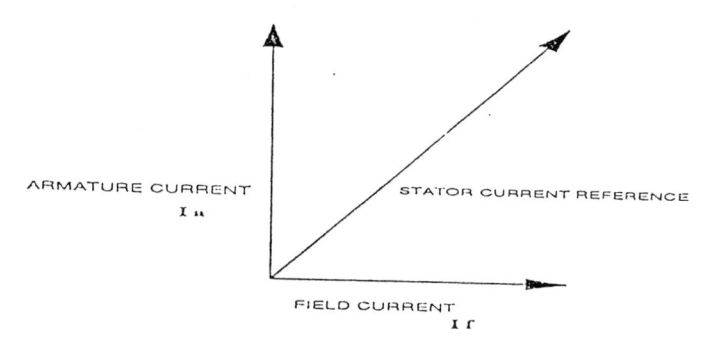

Figure B1. Sample Diagram of Current Geometry

- The reference stator current is converted to three stator phase currents separated by 120°.
- The encoder is used to provide closed loop velocity, position and torque control.
- Rotor position, as monitored by an encoder, is then used to commutate the stator phase currents as the motor rotates.
- The encoder position signal is also used to provide a feedback signal to the outer loop speed regulator.

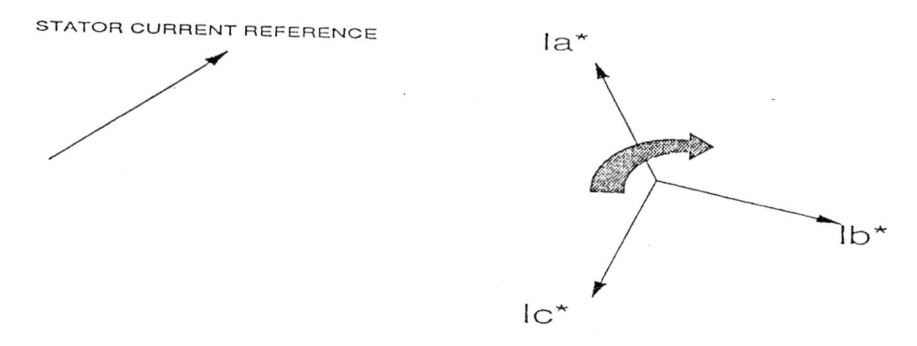

Figure B2. Diagram of Stator Current as Phase Currents

ω_{ref} = speed reference
ω = actual speed
θ = rotor position
$\Delta\theta$ = transfer function which converts rotor position to speed

I_a^* = A phase stator reference current
I_a = A phase stator current
$I_{a/f}$ = armature/field current

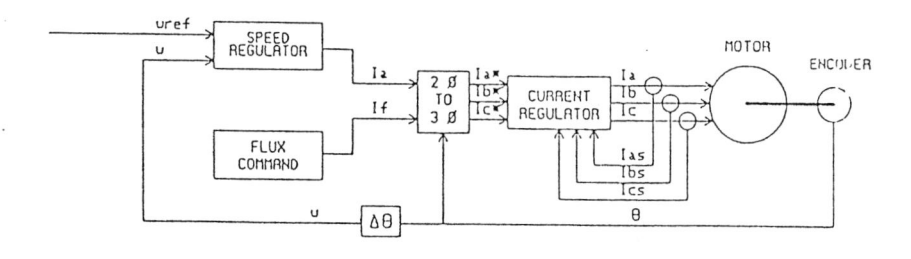

Figure B3. Vector Control Diagram

APPENDIX C

Table C1. Characteristics of Various Types of Batteries

BATTERY	MODEL	WEIGHT	ENERGY	POWER	CYCLE
Nickel/Iron	NIF-200	55	23.2	45	918
Nickel/Metal Hydride	RMH-320	15	26.7	40	800
Zinc/Air	na	na	~45.4	~5.5	na
Zinc/Bromine	ZBB-5/48	178.2	35.9	18.2	334
Sodium/Sulfur	B-11	557	36.8	69.1	592
	PB-MK3	64.2	35.9	40.9	795

APPENDIX D

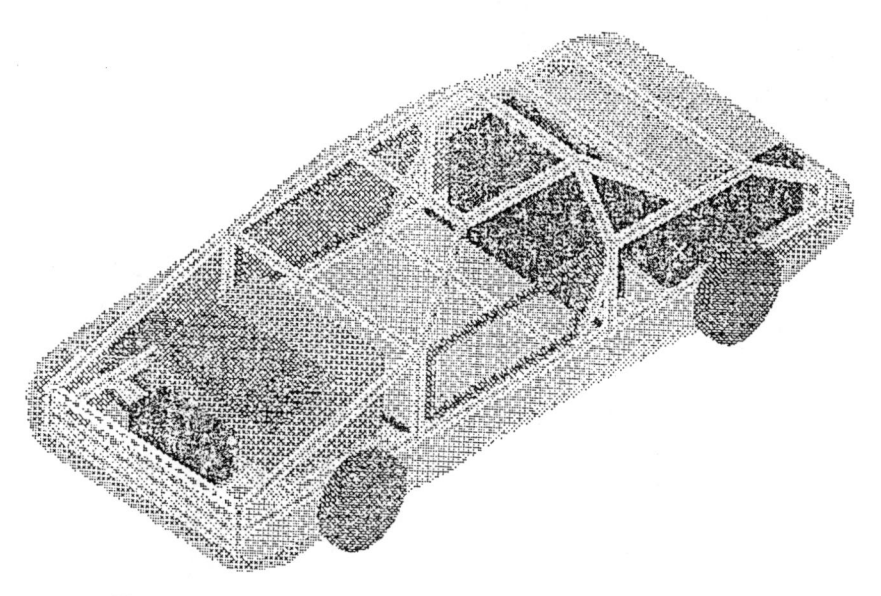

Figure D1. Isometric View of Assembled Car and Components

Dimensioned Frame

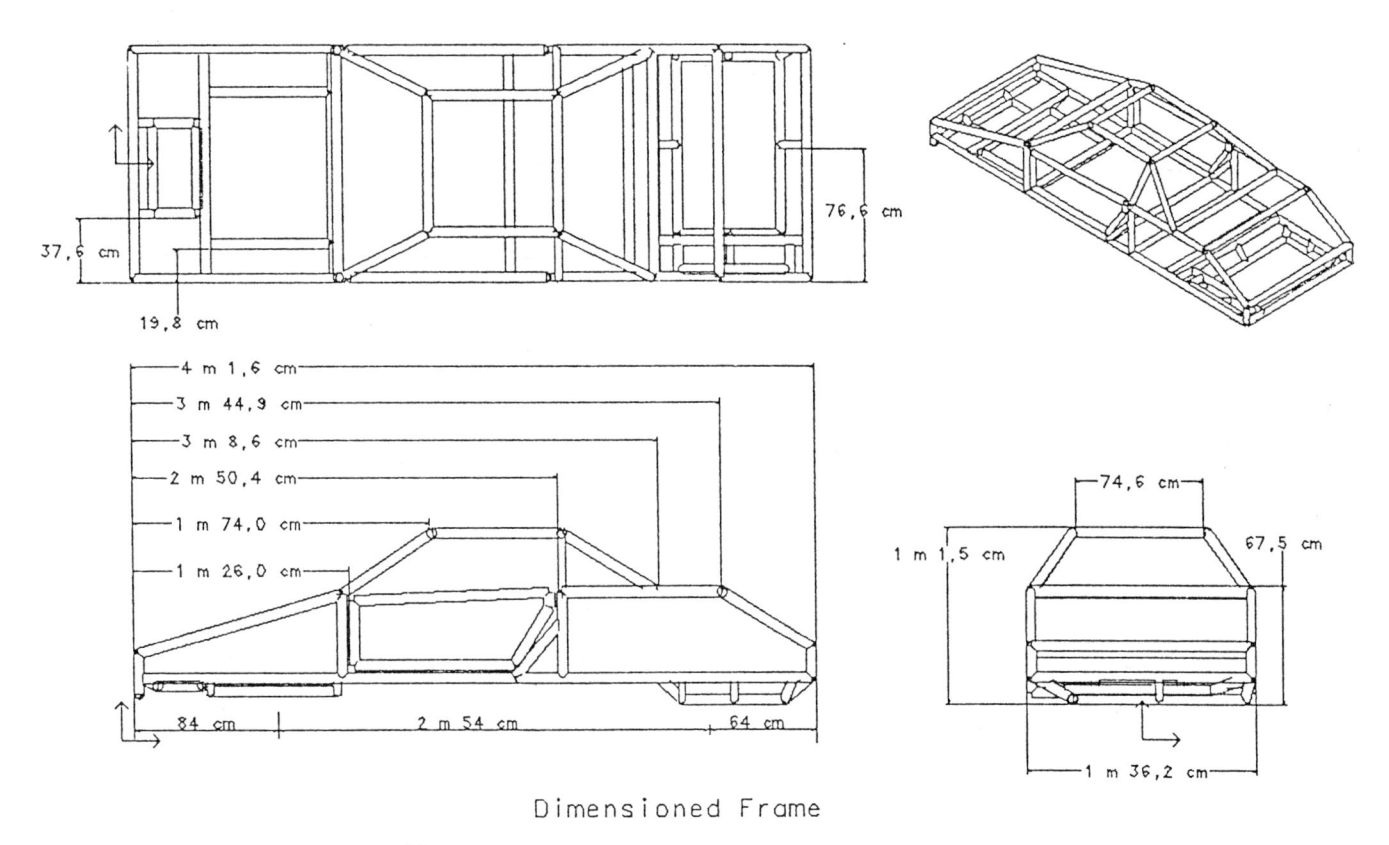

Figure D2. Preliminary Drawing of Dimensioned Frame

425

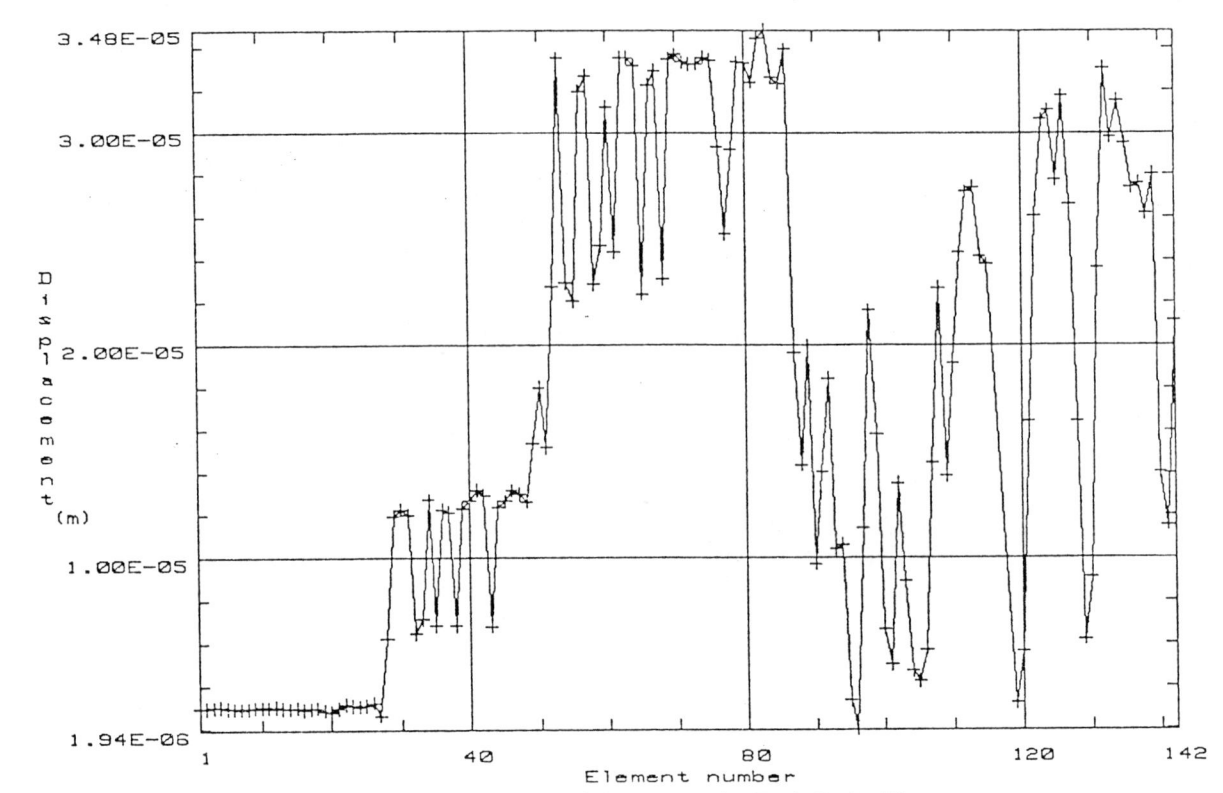

Figure D3. Maximum Diplacements for Each Beam Element
Plot of Function(s) Current Magnitude Displacement

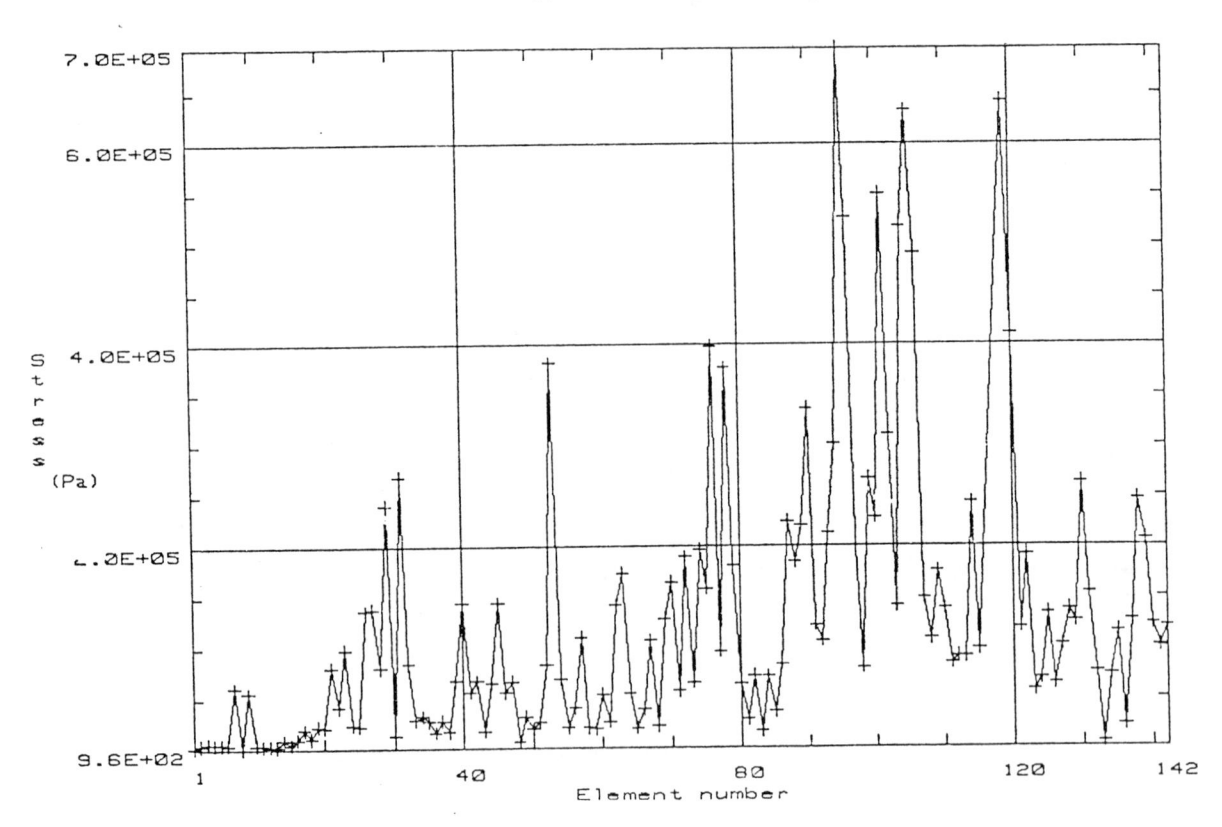

Figure D4. Maximum Stress in Each Element
Plot of Function(s) Current Von Mises Stress Top Surface

REFERENCES

23rd Intersociety Energy Conversion Engineering Conference, 1988, Vol II, "Integration of a Fuel Cell/Battery Power Source in a Small Transit Bus System", Pandit G. Patil-U.S. Department of Energy, Clinton C. Christianson-Argonne National Laboratory, and Samuel Romane-Georgetown University.

25th Intersociety Energy Conversion Engineering Conference, 1990, Vol IV, "Design of a Flywheel Surge Power Unit for Electric Drive Vehicles", R.C. Flanagan and C. Aleong-Mech Engr, at University of Ottawa, Ottawa Canada, and W.M. Anderson and J. Olberman-Unique Mobility Inc., Englewood, Colorado.

25th Intersociety Energy Conversion Engineering Conference, 1990, Vol IV, "Evaluation of a Flywheel Hybrid Electric Vehicle Drive", R.C. Flanagan and M. Keating-Mech Engr, at University of Ottawa, Ottawa, Canada.

26th Intersociety Energy Conversion Engineering Conference, 1991, Vol III, "Electrodes Without Platinum for a Methanol-Air Acid Fuel Cell at Ambient Temperatures", R. Manohorn and J.B. Goodenough-Center for Materials Science and Engineering, ETC.

34th International Power Source Symposium, 1990, "Methanol Fuel Cell Power Source For City Bus", C.V. Chi, D.R. Glenn, S.G. Abena-Energy Research Corporation.

Automotive Chassis, Vol AE-3, "Braking Dynamics, Brake Requirements & Design", Chrysler Institute of Engineering, 1971.

The Automotive Dictionary, Jennings, William Dogan Annual Publications Associates, New York, 1969.

Automotive Handbook, 2nd Edition, Bosch, Society of Automotive Engineers, 1991.

Automotive News, "Methanol/Gasoline Blends and Emissions", Apr 1992.

Battery Reference Book, Crompton, 1990.

Business Week, "Giving a Clean Fuel the Old College Try", 19 Jun 1989 n3111.

Car Magazine, "Pole Vault", Oct 1992.

Design News, "Flame Arrestor Protects Engine From Random Misfires", 3 Nov 1988.

Eagle Picher Industries Inc., Information on Nickel/Iron and Nickel/Metal Hydride Battery Systems.

Electric Power and Light, "Battery Advances Will Boost Electric-Vehicle Sales", Jul 1992.

Fuel Cells Handbook, A.J. Appleby-Texas A&M University and F.R. Foulkes-University of Toronto, van Nostrand Reinhold, New York, 1989.

Fundamentals of Automobile Chassis and Power Transmission, Kuhn and Plumridge, American Technical Society, Chicago.

Fundamentals of Machine Component Design, Juvinall and Marshek, John Wiley & Sons, New York, 1991.

I&CS, "Selecting and Applying Motor Controls", Murphy, Jul 1989.

I&CS, "Selecting the Right AC Drive", Hudak, Apr 1989.

I&CS, "Variable Speed Drives: What's New?", Cleaveland, Jan 1990.

IEEE Spectrum, "The Great Battery Barrier", Nov 1992.

IEEE Spectrum, "Pursuing Efficiency", Nov 1992.

IEEE Transactions on Industry Applications, "An Experimental Study of Alternator Performance Using Two-Drive Ratios and a Novel Method of Speed Boundary Operation", Vol 28, No 3, May/Jun 1992.

International Harvester Truck Service Manual for Scout II, "Electrical Alternator", 1972.

Man and Motot Cars, Black, W.W. Norton & Company Inc., New York.

Mechanical Engineering, "Electrical Vehicles: Getting the Lead Out", Dec 1992.

Road & Track, "Alternative Fuels: Fill 'er Up, But Whatever With?", Vol 41, Nov 1989.

Society of Automotive Engineers, Meeting in Dearborn Michigan, Oct 1992. Notes taken by Douglas Patridge.

Vehicle Body Engineering, Pawlowski, Business Books Ltd, London, 1969.

Whitford, Jim; Program Manager, Nickel/Iron Division, Eagle Picher Industries. Notes taken during several conversations between Mr. Whitford and George Saghiyyah.

Texas A&M Hybrid Electric Vehicle

Murali Arikara
Blake Dickinson
Jim Lee
Steven Moore

Mike Pollard
Doanh Tran
Lance Wright

ABSTRACT

Electrically driven vehicles (EDV) can be an effective replacement for the internal combustion engine (ICE) if the propulsion system is transparent to the driver. A successful EDV needs the same acceleration, range and rapid refueling capabilities of a present day automobile. A hybrid electric vehicle can meet these goals and still provide significant reductions in tailpipe emissions over ICE vehicles.

INTRODUCTION

Currently the air quality of most cities falls below the Environmental Protection Agency's (EPA) minimal standards. According to the EPA, the largest single contributor to urban air pollution is transportation. A solution to the problem of air quality which is currently being pursued is the reduction of vehicle emissions. California and other states mandate the increased use of vehicles which have very low emission, as well as vehicles which produce no emissions, also known as Zero-emission vehicles (ZEV).

The California Air Resources Board (CARB) has mandated that 52% of the vehicles sold by each manufacturer in California be either Low (48%), Ultralow (2%) or Zero (0%) emission vehicles by the year 1998. These percentages are to increase until the year 2003 when 100% of the vehicles sold will fit into one of these three categories. Vehicles are categorized by the amount of Nonmethane Organic Gases (NMOG), Nitrogen Oxides (NOx) and Carbon Monoxide (CO) they produce. A Low-emission vehicle can produce 0.047 g/km of NMOG, 0.12 g/km NOx, and 2.11 g/km CO, while the Ultralow-emission vehicles can only produce 0.025 g/km, 0.12 g/km, and 1.06 g/km of NMOG, NOx, and CO respectively. The Zero-emission vehicle can produce no emissions of any type (1).

Several solutions have been proposed to meet the CARB mandates. Possible solutions include increasing the efficiency of Internal Combustion Engines (ICE), Electric Vehicles (EV) powered by batteries or fuel cells, and vehicles propelled by electric/ICE and Battery/fuel cell hybrid configurations. Each of these solutions have merits and problems associated with them.

The primary merit of improving the efficiency of vehicles powered by ICEs is that a new refueling network would not be needed. Improvements would result from using lighter materials, better aerodynamics, and improvements in engine efficiency. A problem with this solution is its dependence on oil. Transportation accounts for a majority of the oil consumed in the United States, over half of which comes from foreign countries.

Electric vehicles are the only vehicles currently capable of meeting the ZEV criteria. Batteries and fuel cells, operating on pure hydrogen, are power sources which are capable of zero-emission operation. Attributes of electric vehicles include high efficiency, simplicity, and quietness. Problems associated with battery powered EVs is that they have only a limited range and require long periods of time to recharge the battery. Fuel cell powered EVs have difficulty storing enough hydrogen to provide sufficient range. This problem can be overcome by using liquid fuels which are hydrogen rich and separating the hydrogen out. The result is a vehicle which has a slow response, due to the hydrogen separation process, and the ZEV capability is lost.

Hybrid vehicles try to maximize the benefits of the preceding solutions while minimizing the weaknesses. An electric/ICE hybrid can increase the range of an electric vehicle and use an alternative fuels to reduce the dependence on oil. A problem associated with this type of hybrid, however, is that the ICE will still produce significant amounts of pollutants when operated.

Battery/fuel cell hybrid systems offer all the advantages of EVs powered solely by each system without the problems. Using a reformer, a fuel cell can use any fuel which is hydrogen rich and provide desirable range and refueling capabilities. The fuel cell provides the base power required to propel the vehicle and the battery is used when extra power is required for hill climbing and acceleration. The problem of sluggishness experienced by liquid fueled fuel cells is eliminated by the battery which is capable of rapid power changes.

Fuel cells which are fueled with reformed liquid fuels will not meet the ZEV criteria. However, fuel cells are highly efficient, and the high efficiency will minimize the amount of pollutants produced by the vehicle. Battery/fuel cell powered vehicles can easily meet the Ultralow-emission vehicle criteria.

DESIGN CRITERIA

The vehicle design process took into account a number of criteria to be met. Foremost, the parameters presented in the Paper Design Contest Announcement were considered. These design criteria were scrutinized and the following list was constructed:

The hybrid vehicle must:
* Have a maximum of two power sources.
* Have the ability to operate in a zero emissions mode at a flip of a switch.
* Have a total range that exceeds 480 km with over 80 of these kilometers in a zero emissions mode.
* Have a maximum allowable fuel capacity of 45.8 liters of M85 or 40.1 liters of E100.
* Be capable of standard highway cruising speeds.
* Be a practical commuter vehicle for at least two occupants and some cargo.
* Display usable acceleration for strenuous city driving and demonstrate the ability to climb a 15% grade from rest.

In addition, the design team included specific concerns which were felt to be important criteria. Foremost of these concerns is that the vehicle must have an expansive safety system which is not limited to cutoff switches and fuses. Secondly, the vehicle must be easy to operate. The operator should not have to be knowledgeable in engineering to drive the car. There should be little difference in user requirements than what would be expected from a gasoline ICE car.

VEHICLE PERFORMANCE

The HEV should exhibit a total range of 480 km with 80 km of this range in electric only mode. Acceleration from 0 to 96 kph in 15 seconds is required in each operational mode: electric-only, hybrid, and APU-only. To achieve these specifications the vehicle can carry a maximum of 45.8 liters of M85 or 40.1 liters of E100. Power for the vehicle can come from one APU, either fuel cell or heat engine, and up to two battery packs.

The Texas A&M HEV is designed to meet or exceed all required design specifications. For the design to be successful, the HEV has to be energy efficient. Energy efficiency requires low mass, low rolling resistance, and low aerodynamic drag. The Texas A&M HEV uses the latest in plastics, composites and light metals to reduce the empty vehicle weight. The APU, batteries and motor/controller were chosen to reduce weight. The rolling resistance was kept to a minimum by using high pressure tires with a low rolling resistance coefficient. The aerodynamic drag was lowered by reducing the vehicle's profile area and thus the coefficient of drag. All body panels on the car were made to fit flush and the air intake for the radiator and condenser is small. The bottom of the car is smooth since there is no need for exhaust piping.

The power as a function of speed needs to be determined to find the range and acceleration capabilities of the vehicle. The vehicle specifications used to calculate the power requirements are the following:

Vehicle Curb Weight:	1300 kg
Payload (2 Passengers + Luggage):	200 kg
Frontal Area:	1.7 m^2
Rolling Resistance Coefficient:	0.006
Drag Coefficient:	0.32

Power required at the wheels to overcome the driving resistance is calculated as follows:

$$P_W = P_{Ro} + P_L + P_{St}$$

The power required to overcome rolling resistance is determined by:

$$P_{Ro} = f*m*g*v$$

The aerodynamic drag is calculated by:

$$P_L = 0.0386*\rho*C_d*A*v^3$$

The climbing resistance is calculated using:

$$P_{St} = m*g*\sin\alpha*v$$

Power requirements due to rolling resistance, aerodynamic drag and climbing resistance as a function of speed are shown in Figure 1.

Figure 1
HEV Power as a Function of Speed

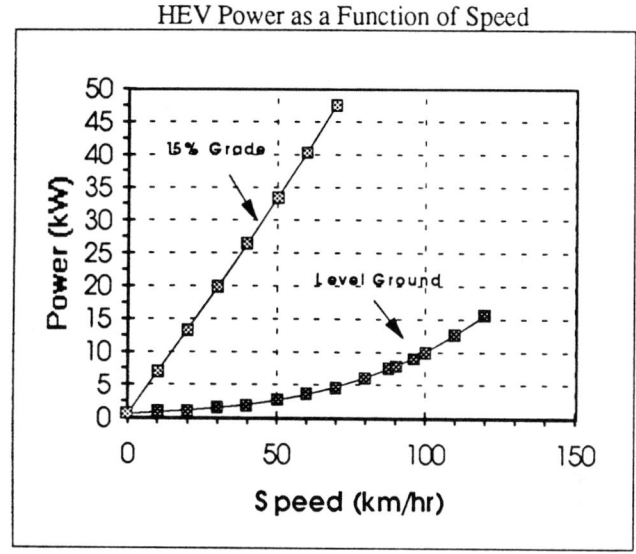

Figure 2
HEV Acceleration as a Function of Speed

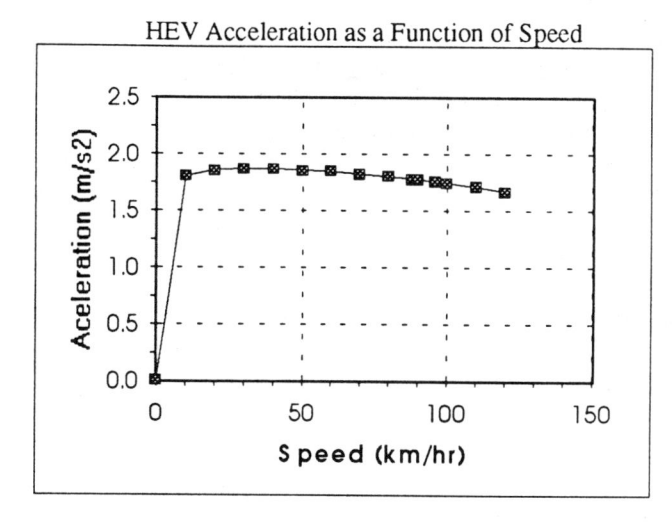

Figure 1 shows that at 88 kph the HEV requires 7.4 kW at the wheels. To climb a 15% grade at 60 kph the HEV requires 40 kW at the wheels. To calculate the power needed from the HEV's power system the efficiencies of the components are needed. Typical efficiencies and accessory loads follow:

Transmission Efficiency:	90 - 95%
Motor Efficiency:	90 - 95%
Controller Efficiency:	94 - 98%
Accessory Loads:	500 W

The power system would need to provide 9.5 kW at 88 kph on level ground.

The power available to drive the wheels is a function of motor torque and vehicle speed. The characteristics of an electric motor is that it has constant torque to a certain base speed and constant power there after. The wheel power available is calculated by:

$$P = (M*u*\eta*v)/r$$

The remaining power $P-P_w$ is available to accelerate the vehicle. The acceleration of the vehicle is determined by:

$$a = (P-P_w)/(v*m)$$

The acceleration capability of the vehicle is shown in Figure 2.

This acceleration profile results in a 0-60 mph time of slightly less than 15 seconds. Given the electric motor limitation of 12,500 rpm, the 0.562 m diameter P-165/65 R 14 tires, and the imposed top speed of 75 mph, drive train characteristics can be determined. From these parameters, the overall gear ratio (11:1), peak motor torque (80 Nm), and peak motor power (80 kW) are necessary for the vehicle to meet the acceleration criteria. The HEV is required to meet the acceleration criteria in all three modes of operation. Sizing the battery and APU to meet the acceleration power requirement will be presented in following sections.

The range of the HEV will be determined on the Federal Test Procedure Highway Cycle (FTPHC). A simulation of the Texas A&M HEV on the FTPHC was performed using the SIMPLEV computer program written by G.H. Cole (2). Using the vehicle parameters described previously, the simulation was run to determine the propulsion system power requirements over the highway cycle. Figures 3 and 4 show the speed and propulsion system power requirements of the HEV, including a 500 W auxiliary load average (stereo, lights, cooling, etc.).

Figure 3
Required Speed on FTHPC

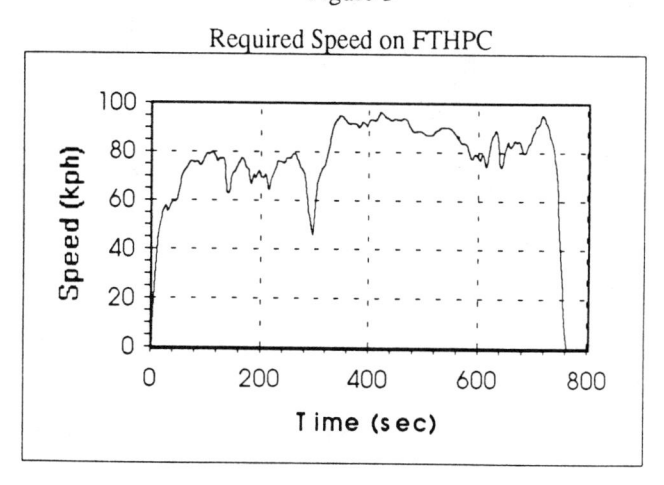

Figure 4
HEV Required Power on FTHPC

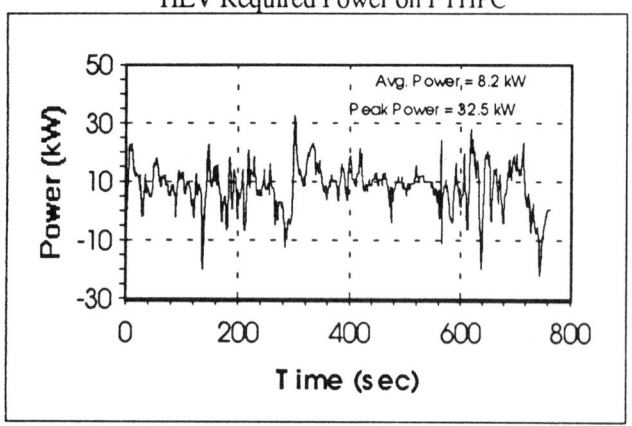

The power load averages 8.2 kW and peaks at 32.5 kW for the average test speed of 77.6 kph. These are the powers required from the power system. The motor requires a peak power of approximately 30 kW and continuous rating of 8.0 kW. The range of the HEV in electric only and APU mode will be determined in their respective sections.

ELECTRIC DRIVE TRAIN

The electric drive train selection is based on efficiency, size, cost, ruggedness and control strategy. The efficiency can be broken into two parts: peak efficiency and part load efficiency. Peak motor efficiency is important because it determines the cruising speed. Part load efficiency is important because an EV also needs to operate at high efficiency in the city. Generally, AC motors have higher power and volume densities compared to similarly sized DC motors. The cost of the motor is determined by the initial cost and its power capabilities ($/kW). The controller electronics cost must be included. Ruggedness is required in any equipment used on automotive applications. The shock, vibration, and corrosion that automotive components must withstand is severe. An ideal electric motor would be one that required no maintenance. Since there is no direct transfer of electric power to rotating parts in the AC motors, little maintenance is required. In comparison, DC motors have brushes and commutators, which are subject to wear and tear upon usage. Modern DC motor controls are much smaller and simpler than AC motor controls. Future AC motors will have reduced size of controls. An advantage of AC controls is that they can be used to charge the battery as well as control the motor. In a DC setup, a separate charging system has to be used with the battery. An electric vehicle's drive train can be greatly simplified with the use of high speed motors. A high speed motor eliminates the use of a transmission. The maximum speed of the DC motor is limited by the length of the arc produced at the junction of the commutator and brush. However, the speed of an AC motor is dependent on the frequency of operation and number of poles.

Table 1 Motor Comparison

Performance Criteria	DC Brush Type	Brushless DC (PM)	Switched Reluctance	Induction
Peak Eff(%)	85-89	95-97	<90	94-95
Eff : 10% Load	80-87	73-82	?	93-94
Max. RPM	4,000-6,000	4,000-10,000	>15,000	9,000-15,000
Cost/Peak Shaft kW ($/kW)	20-30	5-20	?	2.75-5.0
Relative Cost of Electronics	1.00	3.7-6.0	4-10	2.5-3.0
Ruggedness	Good	Fair	Excellent	Excellent

Considering all the above factors, the AC motor was chosen for this application.

SELECTION OF AN AC MOTOR - Two AC Induction Motors will be used in the design vehicle. The selection of two motors is concluded based upon the calculations for the amount of power necessary to propel the vehicle from 0 to 60 mph in 15 seconds. Calculations show that the car requires 80 kW in order to accomplish this requirement. For normal cruising speeds, 20 kW is needed. The specifications for a single motor follow:

- Nominal Power   20 kW
- Max. Power   45 kW
- Nominal Torque   20 Nm
- Max. Torque   45 Nm
- Nominal Speed   4,000 rpm
- Maximum Speed   15,000 rpm
- Weight   40 kg
- Peak Efficiency   95%

The controller designed for the AC motor had the following specifications:

- Max. Current   500A
- System Voltage   120 - 300V
- Peak Efficiency   98%

AC MOTOR SETUP - One AC motor will be attached to the left front half axle and the second to the right front half axle. The motors will be connected via tooth belts. A speed reduction of 11 to 1 is produced by the tooth belt pulley system. Since the motors are attached independently to each half axle, they act as a differential for turning.

BATTERY

The battery selection process involved finding a battery that would provide the power density for acceleration and energy density for range necessary to meet the performance goals. The battery would have to be relatively inexpensive in initial and operation costs. Operational costs includes the life of the battery and its energy efficiency. Furthermore, an acceptable battery needs to be environmentally friendly. An environmentally friendly battery is non-toxic, safe, and easily recycled. Finally, batteries designed for an electric vehicle should be modular and rugged.

A good way to determine the battery energy capacity is to discharge the battery at constant power. A Ragone plot is constructed by plotting specific energy (Wh/kg) as a function of specific power (W/kg). A Ragone plot of several batteries chemistries tested at Texas A&M University is shown in Figure 5.

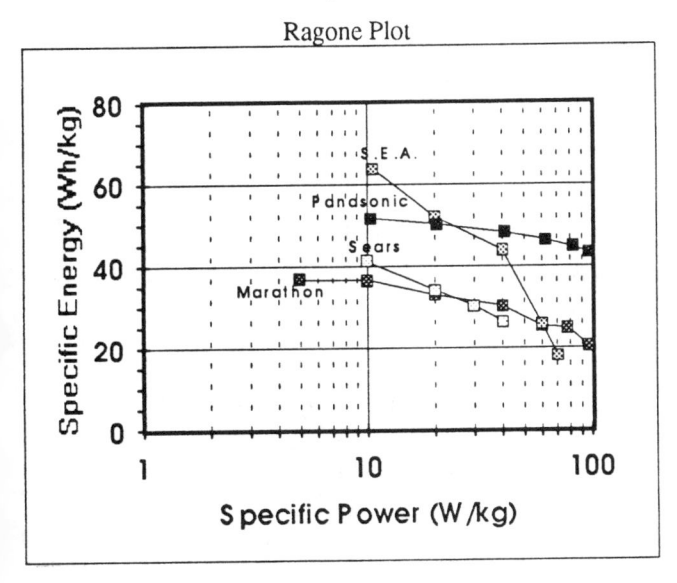

Figure 5
Ragone Plot

As previously calculated, the propulsion system needs to provide 9.5 kW to propel the vehicle at a constant 88 kph (55 mph). It has been decided that the battery pack should weigh no more than 200 kg in order to minimize the vehicle weight. The power density the battery would have to provide in battery-only mode would then be 47.5 W/kg. At this power density, the Panasonic NMH (Nickel Metal Hydride) would provide the best energy density (almost 47 Wh/kg), the S.E.A. Zn/Br (Zinc Bromine) would provide 40 Wh/kg, a Ni/Cd (Nickel Cadmium) battery manufactured by Marathon would have a capacity of 28 Wh/kg, and the Sears deep cycle marine lead acid battery would provide approximately 25 Wh/kg. The following table describes the range attainable from each battery technology based on a 200 kg pack.

Table 2 Comparison of Battery Ranges at 88 kph

Battery Chemistry	Specific Energy (Wh/kg)	Energy Capacity (kWh)	Discharge Time (hr)	Range (km)
NMH	47	9.4	0.99	87.1
Zn/Br	40	8.0	0.84	74.1
Ni/Cd	29	5.8	0.61	53.7
Pb-Acid	25	5.0	0.53	46.3

The HEV must have a 50 mile (80 km) range based on the Federal Test Procedure Highway Cycle (FTPHC). It has been shown that the range of a particular vehicle for various driving cycles can be determined by knowing the average and peak power of each cycle (3). The average specific power required of the battery can be calculated from the average power and battery mass. The energy capacity of the battery at this rate can be determined from the Ragone plot. Furthermore, the battery's discharge time and vehicle range can be found. Note must be taken in that the vehicle will fail the driving schedule earlier than predicted by the Ragone plot because of peak power requirements placed on the battery. At this point, the battery would still have energy left in it but could not be extracted at the desired rate. The point at which the battery fails the driving cycle is related to its available peak power. A plot of specific peak power as a function of depth of discharge (DOD) from Texas A&M is shown in Figure 6.

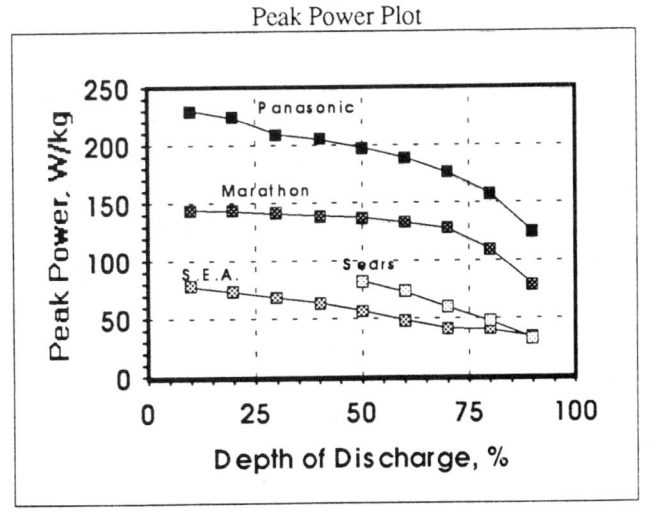

Figure 6
Peak Power Plot

From the peak power required for a driving cycle, the specific power required of the battery can be determined. Applying this requirement to the peak power plot, the DOD at which the battery will fail the driving schedule is determined. The fraction of the DOD multiplied by the battery discharge time derived from the Ragone plot will give the resultant discharge time. With a Ragone and peak power plot, the range of any EV on any driving cycle can be accurately approximated.

The Texas A&M HEV was found to demand an average of 8.2 kW and a peak of 32.5 kW from the propulsion system on the FTPHC. The specific average and peak power requirements work out to be 41 W/kg and 162 W/kg respectively. Table 3 describes the simulated capabilities of the Texas A&M HEV with several battery types using the technique described above.

Table 3 Comparison of Battery Ranges on FTHPC

Battery Chemistry	Specific Energy (Wh/kg)	Energy Capacity (kWh)	Usable DOD (%)	Discharge Time (hr)	Range (km)
NMH	48	9.6	75	0.88	68.1
Zn/Br	44	8.8	0	0	0
Ni/Cd	30	6.0	60	0.44	34.1
Pb-Acid	27	5.4	0	0	0

Table 3 shows than none of the batteries listed can meet the minimum range requirements. This is due to the peak power requirements placed on the batteries. If the peak power requirement was removed and only average power was required, both the NMH and Zn/Br batteries could propel the HEV for 80 km on the FTHPC. An ultracapacitor bank was designed to smooth out the power fluctuations of the cycle. The ultracapacitors will be described in detail in the ultracapacitor section.

Realistically, the battery has to be inexpensive. The total battery cost is based on the initial cost of the battery, its energy efficiency and its life of usable capacity. The battery must have a good calendar life as well as a good cycle life. A good calendar life means the battery can be idled or shelved without permanently damaging it. Higher cycle life results in a lower overall cost of the battery. Also, the battery must be energy efficient and have a low self-discharge. HEV batteries ought to be modular in that they should be flexible in their construction so that they can be placed anywhere in the vehicle. A HEV battery should have good durability. A durable battery is reliable, rugged and maintenance free. A durable battery can be over-charged, over-discharged and requires very little maintenance. It also has to be chemically-, electrically-, and fire-safe. The battery should be designed to be easily recycled or reusable. Another important criterion for batteries is that their SOC can be easily and accurately determined.

The following tables list some of the non-performance criterion important for EV batteries.

Table 4 Battery Criteria

Battery Type	Energy Eff. (%)	Cycle Life	24 hr Self Discharge (%)	Cost (US$/kWh)	Availability	Recharge Time (hrs)
NMH	65 - 75	500 - 2000	10 - 25	200 - 1500	Mid - Term	2 - 6
Zn/Br	70 - 75	200 - 2000	20	150 - 300	Near - Term	4 - 6
Ni/Cd	65 - 70	500 - 2000	5 - 10	200 - 1500	Present	2 - 6
Pb-Acid	75 - 85	50 - 1000	5	50 - 200	Present	6 - 12

Table 5 Battery Criteria

Battery Type	Recycleability	Safety	Durability	Modular	SOC Determinability
NMH	Moderate	Excellent	Moderate	No	Hard
Zn/Br	Strong	Moderate	Strong	Yes	Easy
Ni/Cd	Weak	Moderate	Moderate	No	Hard
Pb-Acid	Strong	Moderate	Weak	No	Hard

To attain the performance goals required of the HEV, only the NMH and Zn/Br battery packs can be used without increasing the battery weight. In terms of non-performance criterion, the two batteries have similar characteristics. The Zn/Br has the advantage of a much lower cost, near term availability and modularity. Based on these advantages, a 200 kg Zn/Br battery in parallel with ultracapacitors will be used to propel the Texas A&M HEV in battery-only and hybrid modes. The 200 kg Zn/Br battery will have a capacity of 10 kWh at 240 volts.

ALTERNATIVE POWER UNIT

The APU can be any type of heat engine or fuel cell. It must provide the range and rapid refueling requirements of an HEV. It must propel the vehicle from 0-60 mph in under 15 seconds. The APU must be able to operate on either methanol (M85) or ethanol (E100).

The choice for the APU will be made on efficiency, environmental friendliness, durability, and compatibility with the rest of the vehicle. The ICE efficiency is limited by the Carnot cycle and has a theoretical maximum efficiency of 40% to 50%. In practice an optimized ICE will have a range of thermal efficiencies between 20% - 30%. In a dynamic situation like the HEV would experience on the FTHPC in APU only mode, the optimized ICE would provide a practical efficiency of 25%. The efficiency of an advanced mechanical drive train (transmission, differential, etc.) is around 80%.

Table 6 ICE Drive Train Efficiency

Component	Efficiency (%)
Engine	25
Transmission	80
Overall	20

A fuel cell does not have a Carnot cycle efficiency limitation. The theoretical maximum efficiency is between 80% and 90% based on Gibbs free energy. The fuel cell can be designed to have a practical efficiency of between 10% and 70% based on the power density needed. It has been shown in lab tests that the difference in efficiency of a fuel cell running dynamically or steady-state is only a few percent (4). The motor, controller and differential of an advanced EV can transmit between 80%-90% of the power to the wheels.

Table 7 Fuel Cell Drive Train Efficiency

Component	Efficiency (%)
Fuel Cell	50
Motors, etc.	80
Overall	40

An environmentally friendly APU should have low emissions, create minimal noise pollution, and produce little waste over its operating life. Due to combustion inefficiencies and high temperatures, ICE operation will emit hydrocarbons HC, CO and NO_X. In addition, the combustion process will produce CO_2, which is not a regulated pollutant, but a potential greenhouse gas. In contrast, a fuel cell, shown in figure 7, directly converts the chemical energy of H_2 to electricity.

Figure 7

Fuel Cell Schematic

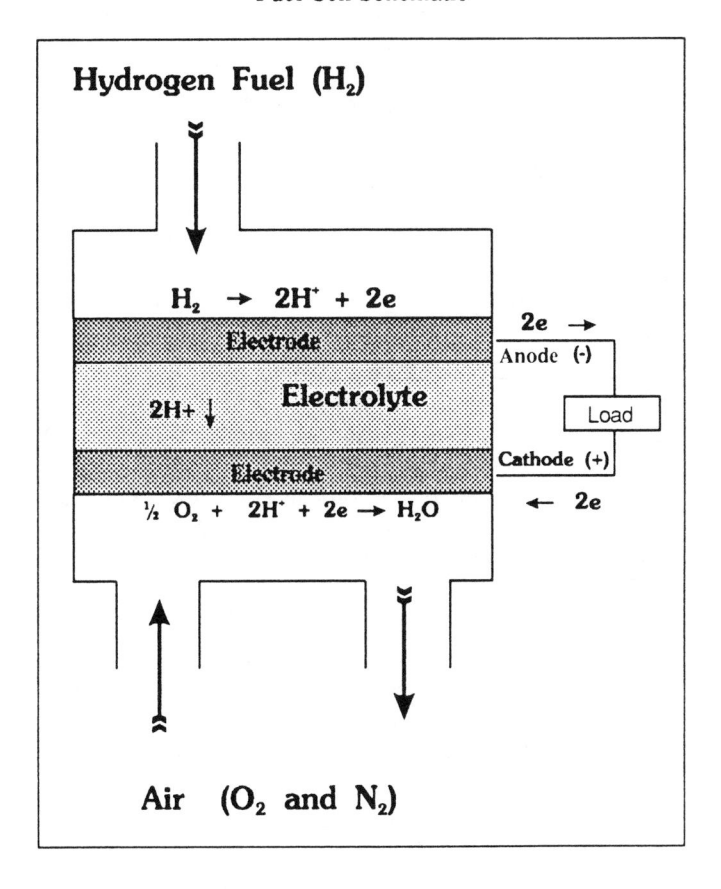

Hydrogen from the reformer and oxygen from air are continuously fed into the fuel cell separated by an ion conducting electrolyte. The H_2 is split into two electrons and a hydrogen ion by a catalyst reduction reaction at the anode. The ion then passes through the electrolyte to the oxygen side. The electrons cannot pass through the electrolyte and are forced to take an external electrical circuit that leads to the oxygen side. The electrons can provide useful work as they pass through the external circuit. When the electrons reach the oxygen side they combine with the hydrogen ion and oxygen creating water. A fuel cell emits no pollution when it runs on hydrogen. However, the rules of the competition require the use of methanol or ethanol. At this time fuel cells cannot run directly on these fuels but methanol can be reformed into H_2 and some trace pollutants. Figure 8 shows a comparison of emission from an ICE on gasoline and fuel cell running on reformulated methanol. NO_X formation in the low temperature fuel cells considered for transportation is not a problem.

Figure 8
ICE and Fuel Cell Emissions

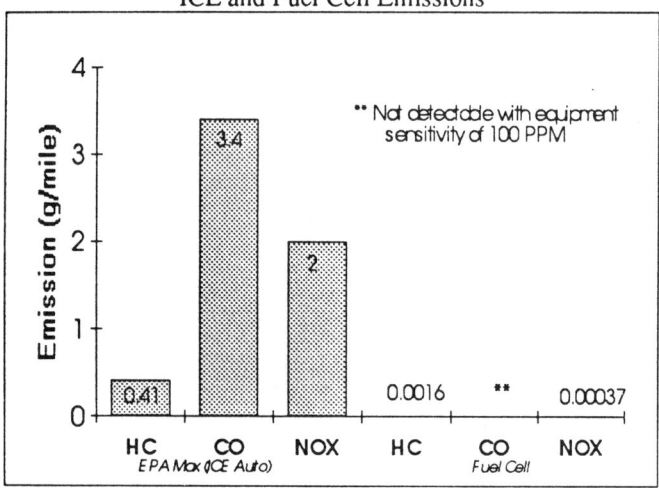

The ICE requires the use of a muffler to reduce noise pollution, while a fuel cell is silent during operation. The ICE needs regular maintenance over its operating life. The maintenance includes oil, filter and spark plug changes. If not disposed of properly the oil can harm the environment. The filters and spark plugs are not recyclable at this time. The fuel cell requires little or no maintenance and produces no waste.

The durability of ICEs has been thoroughly proved over the years. The fuel cell has run for thousands of hours in the lab with little or no performance degradation. Both the fuel cell and ICE can supply instantaneous power, but the fuel cell is more compatible with an electric drive. Based on the superior fuel efficiency and near zero pollutant emissions, the fuel cell was chosen to be used as the APU.

Two fuel cells types are currently considered for use in transportation, the phosphoric acid and proton exchange membrane (PEM). The phosphoric acid fuel cell has an operating temperature around 200°C and requires a heater to start. Power densities are around 50 W/kg at 50% efficiency. This fuel cell works very well with reformulated methanol. The PEM fuel cell has an operating temperature of between 50°C and 80°C and is self-starting. Power densities are over 100 w/kg at 50% efficiency. The PEM fuel cell needs a CO_2 scrubber to operate on reformulated methanol. The PEMs self-starting capabilities and its higher power densities make it an obvious choice for use in the HEV.

A 20 kW fuel cell will be used to provide the average power necessary for FTPHC. The fuel cell is capable of much higher powers, but at lower efficiencies. A 200 kg fuel cell will be used to provide the average power at 50% efficiency. For a 240 volt system, the fuel cell stack will consist of 400 cells with 120 cm^2 per cell, based on 0.65 volts at 750 ma/cm^2 for a single cell (5). The fuel cell stack will be air cooled using part of the air supplying the oxygen needs of the cathode. Part of the water produced by the fuel cell will have to be recalculated to keep the electrodes moist.

The reformer will provide 5.1 kg of H_2 from 12.1 gallons of methanol. One kilogram of H_2 releases 33 kWh of energy based on lower heating value. A PEM fuel cell at 50% efficiency can deliver 84.2 kWh to the drive system. The Texas A&M HEV requires an average of 8.2 kW from the propulsion system on the FTHPC. The drive time would be 10.2 hrs at an average speed of 77.6 kph. The HEV would exhibit a range of 800 km (500 miles) on the FTHPC.

ULTRACAPACITOR

The vehicle power system must absorb the enormous electrical demands of the electric motors during acceleration. The electric motors can exhibit power draws exceeding 75 kW in extreme conditions. Power for acceleration can be drawn from the fuel cell and battery during hybrid operation. For strenuous transients, power can be drawn from an ultracapacitor bank. The ultracapacitor bank is connected to the main system voltage node located before the motor controller unit. The ultracapacitor bank is charged continuously whenever the power demands do not exceed the maximum output of 40 kW of the battery and fuel cell. When power demands exceed the limit of the power sources, the ultracapacitor bank can supply additional power until the demand ceases or the ultracapacitors discharge. At that time power levels drop down to the maximum output limit of the fuel cell and battery. In such conditions, available power is limited to about 40 kW and the driver is notified to ease his demands on the vehicle. During fuel cell only operation, the vehicle relies heavily on stored charge in the ultracapacitor bank to provide the power for acceleration. The fuel cell is responsible to recharge the ultracapacitors when power demands drop. Note that the nominal cruising power demand is only about 9.5 kW, giving the ultracapacitor bank plenty of excess power from the fuel cell to recharge. In addition, the ultracapacitor bank receives charge during regenerative braking.

The ultracapacitor bank will compose of a multitude of near-term available carbon-based non-aqueous electrolyte ultracapacitors. The bank will run at 240 volts and consist of five elements of sixty-eight ultracapacitors per element. With each ultracapacitor weighing 300 g, the total weight will be 102 kg. Each kg of ultracapacitors will provide an energy density of 5 Wh/kg and a power density of 750 W/kg. Using a bank of 100 kg of ultracapacitors, an available power of 75 kW can be achieved. Combined with the approximate outputs of the battery and fuel cell, the ultracapacitor bank can provide the necessary power and energy to accelerate the vehicle 0 - 60 mph in 15 seconds. Acceleration will begin to cease as available power and energy is consumed from the ultracapacitor bank and then levels off at 40 kW.

Figure 9

Schematic of Hybrid Control System

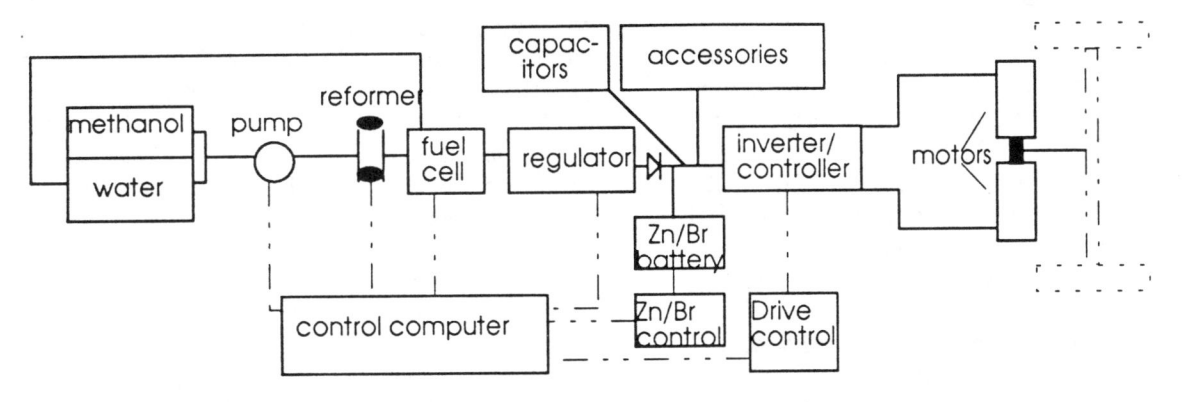

The ultracapacitor bank can be mounted in an electrically insulating container structurally capable of supporting the required 100 kg. The container would have a volume of approximately 100 liters and could be shaped in a variety of configurations. A circuit breaker and fuse could be mounted on the container for safety precautions. A discharging resistor switch would also be provided to guard against the danger of a massive accidental discharge. This precaution would be installed inside the container to prevent the danger of an exposed discharge resistor.

REFORMER

The reformer is the device which takes in methanol and water and produces a hydrogen-rich gas. The reaction requires several catalysts on a catalyst bed and is governed by the following equation:

$$CH_3OH + H_2O + Heat = CO_2 + 3H_2$$

The required water is obtained by condensing vapor present in the fuel cell exhaust gasses and the heat is obtained from gasses exhausted from the fuel cell which contains up to 20% hydrogen.

Methanol and water enter the reformer and are vaporized and mixed. After mixing they are exposed to the catalyst beds and the reformation reaction takes place. The resulting hydrogen rich gas then enters the fuel cell where electricity is produced. Once the exhaust gasses leave the fuel cell they enter the reformer where the remaining hydrogen is burned. This produces the required heat for continued reaction. Next these gasses are passed through a condensing heat exchanger which collects water and preheats the methanol and water entering the reformer reaction chamber.

The reformer consists of two parts. The first is the chamber where the reformation reaction takes place. The chamber is a cylinder which has a diameter of 0.3 meters and a height of 0.4 meters. The weight of the chamber is 80 kg. The second part of the reformer is the heat exchanger which preheats the methanol and water and condenses the exhaust gasses. It is a standard condensing counter flow heat exchanger which is about the size and weight of an automobile radiator (6).

HYBRID CONTROL SYSTEM

The battery/fuel cell hybrid control system would ensure optimum vehicle performance with little user requirements. The hybrid system controls the interaction between the vehicle power sources of a zinc/bromine battery and a fuel cell. Safety, reliability, and performance are the key components of the control system strategy. The system would have to be simple to operate to ensure easy and comfortable driving.

The schematic of the control system is given above as Figure 9. The control computer is responsible for both the fuel cell and battery operation. The fuel cell operation demands that the controller must operate the methanol pump, methanol reformer, fuel cell, and pulse width modulator regulator (PWM). The methanol pump delivers proper amounts of fuel to the reformer, which is connected to the fuel cell itself. The PWM regulator is responsible for delivering power from the fuel cell out into the system. These components must all be closely monitored by the controller for voltage, current, and temperature to ensure proper efficiency.

The battery component of the hybrid power source has not only the main control computer connected but its own dedicated battery control system. Battery operation relies on control of the Zn/Br electrolyte pumps and control of its operation valve.

Central Control Computer - The control computer, referred to as the controller, is the main control unit. The controller operation relies on various feedback sensors, however, the user has the ultimate control over the system in the form of an operation mode switch. The mode switch forces the controller to operate in one of three modes: the battery only mode for zero emissions, the fuel cell mode, and the optimized hybrid mode. The operation mode switch determines the duties of the controller. During the hybrid operation, the controller has the greatest range of influence over the vehicle systems because the fuel cell and Zn/Br battery must be properly integrated for optimized performance. The voltage output of the fuel cell PWM and the battery are constantly monitored along with the current demands of the system. The controller must then use the PWM to force the system voltage above or below the voltage of the battery. Using this strategy, one is able to configure the system to rely heavily on the fuel cell for cruising power and the battery for transient speeds. The controller will also communicate with the battery controller to determine charging demands of the battery.

The fuel cell and reformer require computer control for startup and shutdown. The reformer must have a specific amount of time to begin hydrogen production for the fuel cell during heat-up. During this startup time, car operation will rely on battery power or stored energy in the ultracapacitor banks. Once the vehicle ignition is turned off, the reformer must shut down. During the shut down period, hydrogen is still being produced for a specific amount of time. This hydrogen will be used by the fuel cell to charge the battery and ultracapacitor banks. Any excess power will have to be consumed by a power resistor.

The controller will have the ability to adjust to user demands of the operation mode switch. In the battery only mode, fuel cell operation is inhibited and shut down. The fuel cell only mode would cause the battery system to similarly shut down. Cutoff relays are provided for in the system to cut both the battery or the fuel cell off electrically.

The control computer is powered by a standard 12 volt system. The computer will incorporate a standard 12 volt battery and a DC to DC converter to draw power from the main system voltage. In addition, car accessories such as lights and windshield wipers will be powered off this system. The ignition switch relies on power from this system to begin vehicle operation. This system can be shut off using the emergency cutoff switches.

Temperature and voltage measurements are part of the safety features of the controller. If the temperature of any of the components becomes too high, a user warning light will come on. At dangerous temperatures or unusual voltages within the system, the controller will throw the main cutoff relay that will disconnect electrically the fuel cell and battery. A dash-mounted emergency cutoff switch is also provided to the user. A manual emergency cutoff switch is mounted on the exterior of the vehicle for additional safety precautions. Frame mounted accelerometers provide instantaneous cutoff signals in the event of a crash. If one of the drive motors fails, the CCC will issue a command to shut down the other motor to prevent the vehicle from running erratically.

Zn/Br Control - The Zn/Br battery requires a control system to constantly monitor battery conditions and make adjustments. The battery controller must determine system voltage compared to battery voltage. When the system voltage exceeds the battery voltage, the charging valve must be positioned for charging mode. During regenerative braking, the valve must also be positioned in the charging mode. The valve must be returned to normal operation mode position when current demands are being made on the battery. The Zn/Br Control (ZBC) is responsible for positioning the valve under varying conditions and regenerative braking. During hybrid vehicle operation, the fuel cell can provide charging power to the battery in some conditions. Charging will occur when the fuel cell voltage is higher than that of the battery voltage. This condition exists during steady vehicle cruising, braking and stopping. The fuel cell is also able to supply charging power when the vehicle is stopped.

The ZBC will constantly communicate with the central control computer (CCC). Operation of the electrolyte pumps and the battery cooling system will be determined from the operation status of the CCC and the outputs of the ZBC unit. A battery electrolyte leak sensor will cause the ZBC to call for a dashboard warning light and battery shutdown. The ZBC is also responsible for supplying vital battery information to the CCC such as charging status, voltage and temperatures.

Drive Control - The drive control unit controls the regenerative braking system. The drive control (DRC) receives signals from the CCC on the system voltage status and the motor controller status. The DRC transmits signals to the CCC on regenerative status so that the ZBC can configure the battery to the charging mode. Regenerative braking will be engaged as soon as the vehicle operator applies pressure to the brake pedal. Mechanical braking will engage with additional pressure, providing all the force at near-zero speeds. The DRC can be configured to engage regenerative braking whenever pressure is released off the accelerator pedal. This is a useful and efficient option in city driving conditions.

Inverter and Motor Controller - The inverter and motor controller unit has two principle functions. First, the system DC voltage must be converted to AC voltage for the drive motors. Next, the inverter and motor controller (IMC) must control the motor speed in response to input from the accelerator pedal and the CCC. Third, the inverter will serve as a battery charger for both 110 VAC and 220 VAC. Finally, the IMC is responsible for harnessing the regenerative braking power. Control for this function is provided by the CCC and the DRC. Under normal hybrid operation, manual battery charging will not be necessary because the fuel cell will provide the charging power.

The entire system is flexible in that one may decide to upgrade fuel cell capacity or battery capacity. The control system would need only slight adjustments in such upgrades. The system is user-friendly, having minimal user controls. The user would only have to contend with the mode switch, ignition switch, emergency shutoff switch, and dashboard warning lights to operate the control strategy. The control system maintains a constant temperature and voltage monitoring safety feature combined with emergency switches and dashboard warning lights. The CCC communicates at all times with the other units to provide an integrated system with maximum efficiency.

SAFETY

A major concern about the design of the Texas A&M HEV are the safety requirements. Safety is a major issue because human life is important. As a result, every effort has been made to ensure the safety of the driver and the passenger in the unfortunate event of an accident.

The Texas A&M HEV meets all federal crash test standards. Likewise, the HEV meets all federal braking requirements. This will be accomplished with anti-lock breaks and regenerative breaking. The HEV also possesses all the needed external lighting and mirrors for safe driving. In short, the car has all the safety requirements of a normal automobile. The real safety measures are those pertaining to the battery, methanol storage, reformer, ultracapacitors, and fuel cells.

The battery is a major threat to the occupants in an accident because of its mass. This potential threat has been eliminated by placing the zinc bromine electrolyte tanks on the underside of the car in between the axles as shown in Figure 10. The tanks will be housed in an aluminum cage in order to protect it from flying debris and other objects. An energy management system will be used so that the front end of the vehicle will not absorb the kinetic energy of the battery during the first 60 ms of an accident. With this constraint, the base vehicle (car without battery tray) and the battery will impact at different time, and as a result the instantaneous loading on the vehicle's chassis will be reduced. This will be accomplished with tray hangers that have shock absorbers attached to them (7). In the event that the tanks do dislodge from the frame in a severe accident, they should not pose a threat to the occupants because they should fall away or slide underneath the front seat area. The battery will also be electrically floated to reduce the risk of electrical shock. In effect, someone would have touch both a positive and a negative lead to get shocked. If the battery is not electrically floated, then contact with the chassis could result in a shock. Finally, the electrolyte tanks will have a leak detector. If a leak is detected, the driver will be notified by a warning light, and the battery pump will be shutdown.

The storage of methanol is just as important as the placement of the battery. The problem with methanol is that it is toxic. This hazard has been dealt with by placing the methanol/ CO_2 scrubber storage container in the rear of the car. The exact location will be discussed in the next paragraph. The lines that will transfer the methanol to the reformer will be rubber and have a stainless steel mesh covering in order to help prevent breakage during an accident. This measure should reduce the possibility that the occupants will get sprayed by the methanol during an accident. The storage tank will also have a leak detector. If a leak is detected, the driver will be notified with a warning light so that a corrective measure can be taken.

The fuel cells, methanol storage tank, battery pump, and ultracapacitors will be stored in the back seat area, and the reformer will be stored just behind them. All of the components will be bolted to the frame to prevent forward movement during an accident. In addition, a steel wall, something similar to a firewall in an automobile, will be installed behind the front seat compartment. The wall will protect the occupants in the unlikely event that one of the components comes free and slides forward. The height of the wall will be shorter than the front seat which means that it will not impede the drivers vision. In addition, a flat steel sheet will be placed perpendicular to the wall, and it will cover all of the components. This sheet will protect the occupants in case of a roll over, and it will prevent the occupants from being sprayed by methanol during a crash if the methanol should happen to leak. There will also be a Plexiglas wall behind the reformer so that all of the components will be completely enclosed. This enclosure will prevent the fuel cells, methanol storage tank, battery pump, reformer, and ultracapacitors from reaching the occupants. The components were positioned in the back seat area so that most of the car's weight would be positioned in between the front and rear axles. This was done to insure a low polar moment, and as a result, the car will not have a tendency to spin during lane changes. The components were also positioned so that a low center of gravity was produced. This is desirable because it will make the car have a low roll moment.

To further protect the occupants, temperature sensors will monitor the major components of the electric car. If any of the components becomes to hot, a warning light will appear warning the driver of the danger. If the temperature becomes too excessive and poses a threat to the driver, the controller will shutdown the vehicle. Also, if the driver feels that the car needs to be shutdown, then he can flip the electrical cutoff switch located on the dash. There will also be an emergency external electrical cutoff switch which will be located above the rear bumper. In addition, the vehicle will have an accelerometer and a roll-over sensor. In the event of a sudden deceleration or a roll-over all the components will be shutdown. These sensors will be installed so that if a crash occurs, the components will not keep operating and possibly cause harm to the driver. For example, if a crash occurs, the methanol pump would stop pumping which would greatly reduce the possibility that the occupants would come into contact with the methanol. Besides these sensors, all of the electrical equipment will have standard fuses so that they do not get damaged during a power surge.

The safety precautions taken stemmed from previous reports involving electric vehicles (8). Those cases were studied and their preventive measures were incorporated into the Texas A&M HEV. The precautions, along with the federal crash test standards, lets the driver worry about driving and not about his life expectancy.

Figure 10

Hybrid Electric Vehicle Schematic

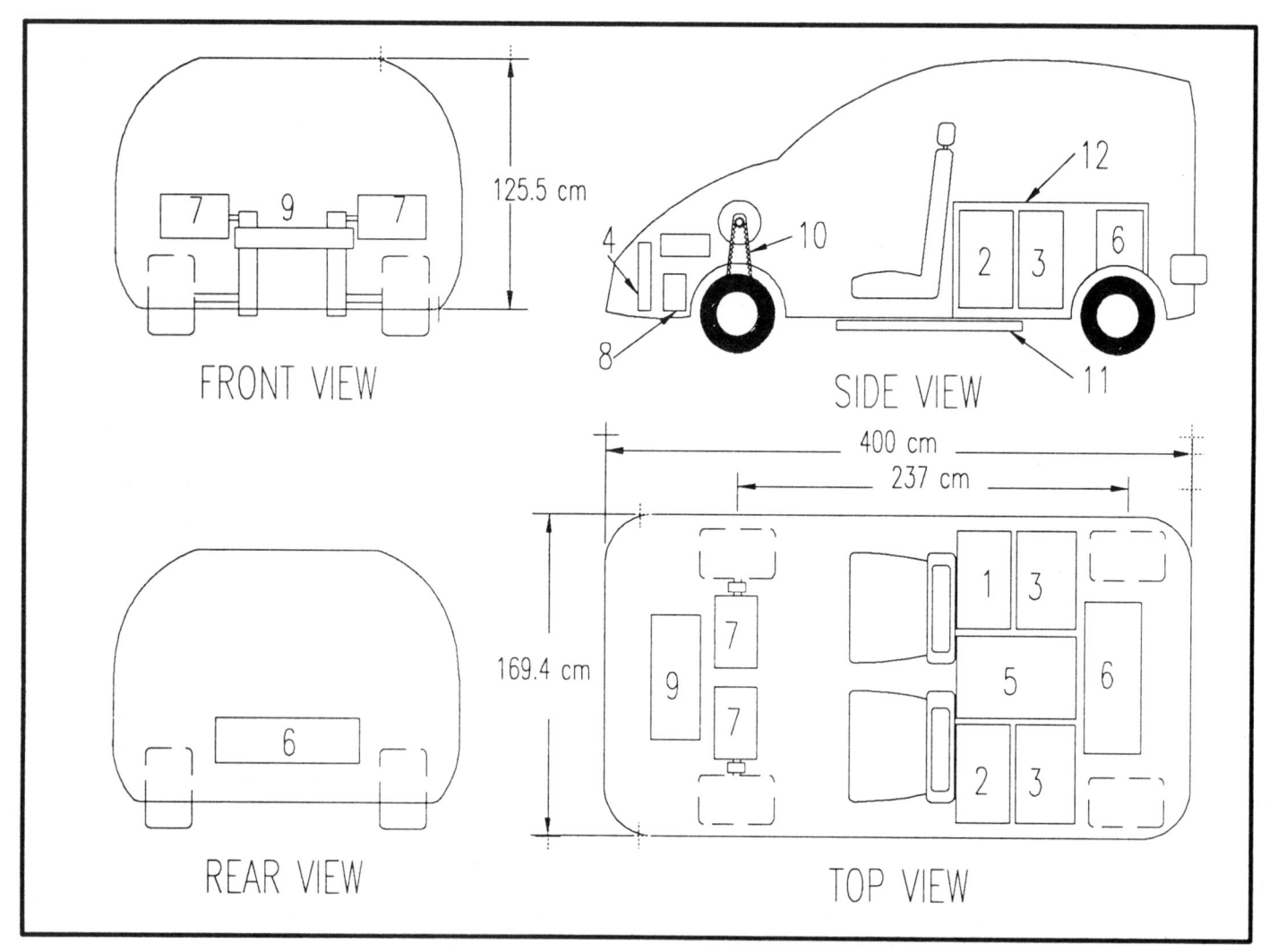

1. Battery Stack
2. Capacitors Contaiment Unit
3. Methanol Tank
4. Radiator
5. Fuel Cells Stacks
6. Reformer
7. A.C Induction Motor
8. Heat Pump
9. Motor Controller
10. Tooth Belt
11. Zinc bromine Tank
12. Steel Wall Protection Unit

ERGONOMICS AND STYLING

The design of this hybrid vehicle includes many of the practical and friendly features that make the Texas A&M HEV feel like a normal car. The driving controls such as the accelerator, brake pedal and steering wheel are in the same location as in a conventional automobile. The steering is tilt and is power assisted by an electric motor powering a hydraulic pump. For energy conservation, an accumulator is utilized so that the hydraulic pump runs only part of the time. The emergency brake is located on the far left side of the floorboard. With this out of the way, there is plenty of space between the two bucket seats for a compartment to store small items such as change and wallets, as well as two holders for drinks.

The dashboard is uncluttered and easy to read, with the speedometer taking the obvious center of attention. Because this vehicle does not have an IC engine and transmission, there is no need for oil, transmission, or coolant temperature and pressure gauges or a tachometer. This car does require, in addition to a methanol gauge, a state of charge meter and a water (for the reformer/fuel cells) gauge. Other instrumentation included is a high-beam indicator and specialized warning lights.

An HEV does not have to be totally utilitarian. For instance, this car has such niceties as an electric sunroof, electric side mirrors, lumbar-supported seats, power windows, cruise control, alarm system, and a stereo/CD player. Other non-necessary features include a cigarette lighter/socket, an ashtray, and an electric rear window defroster.

For passenger comfort, a computerized climate-control system is standard equipment. The driver is able to program the on-board computer to start the heating or cooling cycle at a certain time, so that the vehicle is at a comfortable temperature at the time he enters the vehicle. A remote activation switch is available that will allow the driver to activate the climate control from a considerable distance. For economy, the heating or cooling cycle will end after a certain, specified period of time unless the driver enters the car.

The climate control system includes a reversible high-efficiency, variable speed heat pump to cool in the summer. Some or all of the heat in winter, depending of driving conditions is provided by the heat pump. This heat pump is driven by a 1.1 kW electric motor. In the winter, the heat pump provides all of the transient heat until heat from the reformer (assuming it is turned on) is built up and can be vented. On very cold days, resistive heating must be used in place of or in conjunction with the heat pump.

More techniques for efficiently cooling the vehicle are employed than for its heating. Airco brand solar control film is used on the windows to provide 45% reflectance of incident radiation, yet still maintain a high transmissivity (70%) of visible light. This, combined with an insulated roof, provide a reduced steady-state load. If the vehicle is programmed or remote activated to cool, a computer first turns on a fan to pull in outside air until there is only a small thermal gradient. Then the A/C turns on, operating on recirculation mode until the driver arrives.

The exterior style of the car is attractive and aerodynamic. This car is a hatchback with smooth curves. The headlights are flush, and the side mirrors are low profile to reduce drag. Attractive alloy wheels and tinted glass add to the aesthetic value.

ENVIRONMENTAL BENEFITS

The environmental benefits of powering vehicles with battery/fuel cell hybrids are far-reaching. Obvious benefits take the form of reductions in air pollution and oil spills. More subtle benefits include reduced noise levels, reduced petroleum products in our lakes and rivers, and an improved economy.

A battery/fuel cell hybrid vehicle has two sources of emissions. The most direct type of emission is the exhaust from the fuel cell. The second source is the emissions created at a power station to generate the electricity which charges the batteries when the vehicle is not in use.

A comparison of the gasses exhausted from the fuel cell system to those exhausted from an ICE show that the fuel cell system produces substantially fewer pollutants. An ICE produces 0.25 g/km of unburned hydrocarbon, 1.2 g/km of NOx and 2.1 g/km of carbon monoxide. A Phosphoric Acid fuel cell/reformer, fueled by a methanol, only produced 0.001 g/km of unburned hydrocarbons, 0.00023 g/km of NOx and undetectable amounts of carbon dioxide. This represents a 99.4% reduction of unburned hydrocarbons, a 99.9% reduction in NOx, and a 100% reduction in carbon monoxide (9).

Comparing the power station emissions to those of an ICE again indicate substantial reductions in most pollutants. Unburned hydrocarbons would reduce from 0.25 g/km to 0.006 g/km, carbon monoxide would reduce from 2.1 g/km to 0.031 g/km, and NOx would reduce from 1.2 g/km to 0.75 g/km. The only pollutant which would increase is sulfur dioxide, which would increase from 0.25 g/km to 1.8 g/km. This is because many power stations in the United States burn high-sulfur coal to produce electricity (1).

The benefit of reduced oil spills will result from the reduced demand for petroleum products. As the number of battery/fuel cell powered hybrid vehicles increases the demand for gasoline will begin to decline. This will cause the amount of oil produced to decrease, as well as the amount shipped, reducing the chance that spills will occur during transport.

Another source of oil spills which will be eliminated are the spills produced by dumping used oil. Electric motors have very few moving parts and therefore do not require as much lubrication as ICEs. Since there will be no oil to change, there will not be no used oil to dispose of. Electric motors will not shed as much oil onto the road surface as ICEs, which will reduce the amount of oil washed into lakes and rivers each time it rains.

The lack of noise is another environmental benefit. The only two significant sources of noise in a hybrid vehicle are wind noise and the motor controller which produces a soft high pitched whine at low speeds. The noise reduction would make busy streets quieter and reduce the need for sound walls adjacent to freeways in residential areas.

Battery/fuel cell hybrid vehicles would improve the domestic economy by reducing the trade deficit. Since the fuel cell in a hybrid vehicles operates on methanol which can be produced from energy sources which are plentiful in the United States, billions of dollars which are being paid to other countries for oil will stay in the country. This money will help strengthen the economy and improve our quality of life.

HYBRID VEHICLE SPECIFICATIONS

DRIVE TRAIN
Type: Front Wheel Drive, one motor per front wheel
Final drive ratio: 11:1
Typical Efficiency: 90% - 95%

MOTOR
Type: High efficiency, twin AC Induction motors
Peak Power: 45 kW
Nominal Power: 20 kW
Torque: 40 Nm
Typical Efficiency: 90-95%

BODY/CHASSIS
Type: Unit Body
Suspension: 4-Wheel Double Wishbone
Steering type: Rack and Pinion
Tire size: R14 165/65
Wheels: 14" Alloy
Tires: Low rolling resistance, high pressure, all season tire

DIMENSIONS
Wheel base: 240 cm
Length: 400 cm
Height: 127 cm
Width: 170 cm
Track, front/rear: 147 cm/147 cm
Curb weight: 1300 kg
Gross weight: 1500 kg
Weight distribution
(w/driver, % f/r) 43/57
Drag coefficient.: 0.32

ELECTRONICS
Type: Dual MOSFET inverters

Maximum current: 400 amps
Maximum system voltage: 240 Volts
Frequency range: 0 - 500 Hz
Battery charger: Computer controlled, integral with dual inverter package

ELECTRICAL POWER SYSTEM

BATTERIES
Type: Zinc - Bromine
Voltage: 240 volts
Capacity: 50 ahr, 10 kWh
Battery pack weight: 200 kg

FUEL CELLS
Type: PEM
Nominal Power: 20 kW
System Weight: 200 kg

PERFORMANCE
Motor speed at 60 mph: 10,000 rpm
0 - 60 mph: 15 seconds in all three modes
55 mph cruising range: 800 km
Top speed: 120 kph

NOMENCLATURE

A   Frontal area (m^2)
APU   Alternative power unit
CARB   California air resources board
CCC   Central control computer
Cd   Coefficient of drag
CO   Carbon monoxide
CO_2 Carbon dioxide
DOD   Depth of discharge (%)
DRC   Drive control
EV   Electric vehicles
f   Coefficient of rolling resistance
FTPHC   Federal test procedure highway cycle
HC   Hydrogen -carbon chains
HEV   Hybrid electric vehicle
H_2 Diatomic hydrogen
H_2O Water
ICE   Internal combustion engines
IMC   Inverter motor controller
km   Kilometer
kph   Kilometers per hour
M   Motor torque (Nm)
m   Mass of vehicle (kg)
Ni/Cd   Nickel cadmium battery
NMH   Nickel metal hydride
NMOG   Non methane organic gases
NO_x Nitrogen monoxide
P   Power available at drive wheels (kW)
Pb-Acid   Lead acid battery
Pl   Aerodynamic drag power (kW)
Pro   Rolling resistance power (kW)

P_{st}	Climbing power (kW)
P_w	Power required at drive wheels (kW)
PEM	Proton exchange membrane
PWM	Pulse Width Modulator
r	Wheel radius (m)
u	Overall gear ratio
SOC	State of charge (%)
v	Velocity (m/s)
VAC	Voltage - alternating current
Zn/Br	Zinc bromine battery
ZBC	Zn/Br control
ZEV	Zero-emission vehicles
α	Climbing angle
η	Efficiency
ρ	Density of air (kg/m^3)

REFERENCES

1 Riezenman, Michael J., "Electric Vehicles", *IEEE Spectrum*, November 1992, pp. 18 - 21.

2 Cole, G. H., "SIMPLEV: A Simple Electric Vehicle Simulation Program", June 1991.

3 Hornstra, F., "A Methodology to Assess the Impact of Driving Schedules and Drive Train Characteristics on Electric Vehicle Range"

4 Swan, D. H. and Dickinson B. E., Paper to be presented to SAE, Center for Electrochemical Systems and Hydrogen Research, August 1992.

5 Ferreira, C., Center for Electrochemical Systems and Hydrogen Research, 1992.

6 Kevala, Russ, J., Daryl M. Marinetti, 1989, "Fuel Cell Power Plants for Public Transport Vehicles", *Society of Automotive Engineers Technical Paper No. 891658*.

7 Sandor Palvoelgyi and Peter K. Stangl. *Crashworthiness of the electric G-Van.* pg99-105.

8 Journal of Energy Resources Technology. June 1991, Vol. 113/103-104

9 Ramono, Samuel, L. Dean Price, 1990, "Installing a Fuel Cell in a Transit Bus", *Society of Automotive Engineers Technical Paper No. 900178*.

APPENDIX

Texas A&M Battery Data (Summer '92)

Chemical Couple	Ni-Cd	Ni-MH	Ni-MH	Pb-Acid	Pb-Acid	Pb-Acid	Pb-Acid	Zn-Br
Manufacture	Marathon	Ovonics	Panasonic	Chloride	Delco	Exide	Sears	S.E.A.
Model	44SP100	C-cell	HHR140A	3ET205	M27MF	GC-5	96522	ZBB-5/48
No. Cells	20	1	4	3	6	3	6	32
No. Modules	20	1	4	1	1	1	1	1
Weight (kg)								
As Tested	36	0.0832	0.126	32.8	25	30	26	85
Per Module	1.6	0.0832	0.0313	32.8	25	30	26	85
Volume (L)								
Actual	13.4	0.024	0.0352	12.054	10.47	11.21	10.62	95.25
Usable	14.7	0.0251	0.0364	13.2	12.23	12.63	11.68	113.58
Voltage	25	1.2	1.2	6	12	6	12	50
Charge Method	CI	CI	CI	CI/CI/CV	CI/CV/CI	CI/CV/CI	CI/CV/CI	CI
Overcharge (%)	23	17	15 - 17	15 - 20	5	15 - 20	15 - 20	10 - 15
C/3 Capacity								*C/5
Ahr	46.3	3.58	1.31	149.4	54.5	162.2	75.9	104.5*
Whr	1133	4.20	6.47	845	635	935	879	5115*
Energy Efficiency (%)	64.4	74.0	68.7	69.0	83.7	78.6	84.9	68.4
Coulomb Efficiency (%)	75.2	87.7	80.4	78.0	97.0	86.1	89.5	84.1
Voltage Efficiency (%)	85.7	84.4	85.4	88.5	86.3	91.3	94.9	81.4
Sp. Energy (Wh/kg)	35.4	49.5	51.4	25.8	25.4	31.2	33.8	60.1
Vol. Energy (Wh/L)	84.6	174.8	183.8	70.1	60.6	83.4	82.8	53.7
Peak Power, W/kg (at 50% D.O.D.)	>138	102	197	***	55.3	***	82.3	56

Ragone Plot
Texas A&M Data (Summer '92)

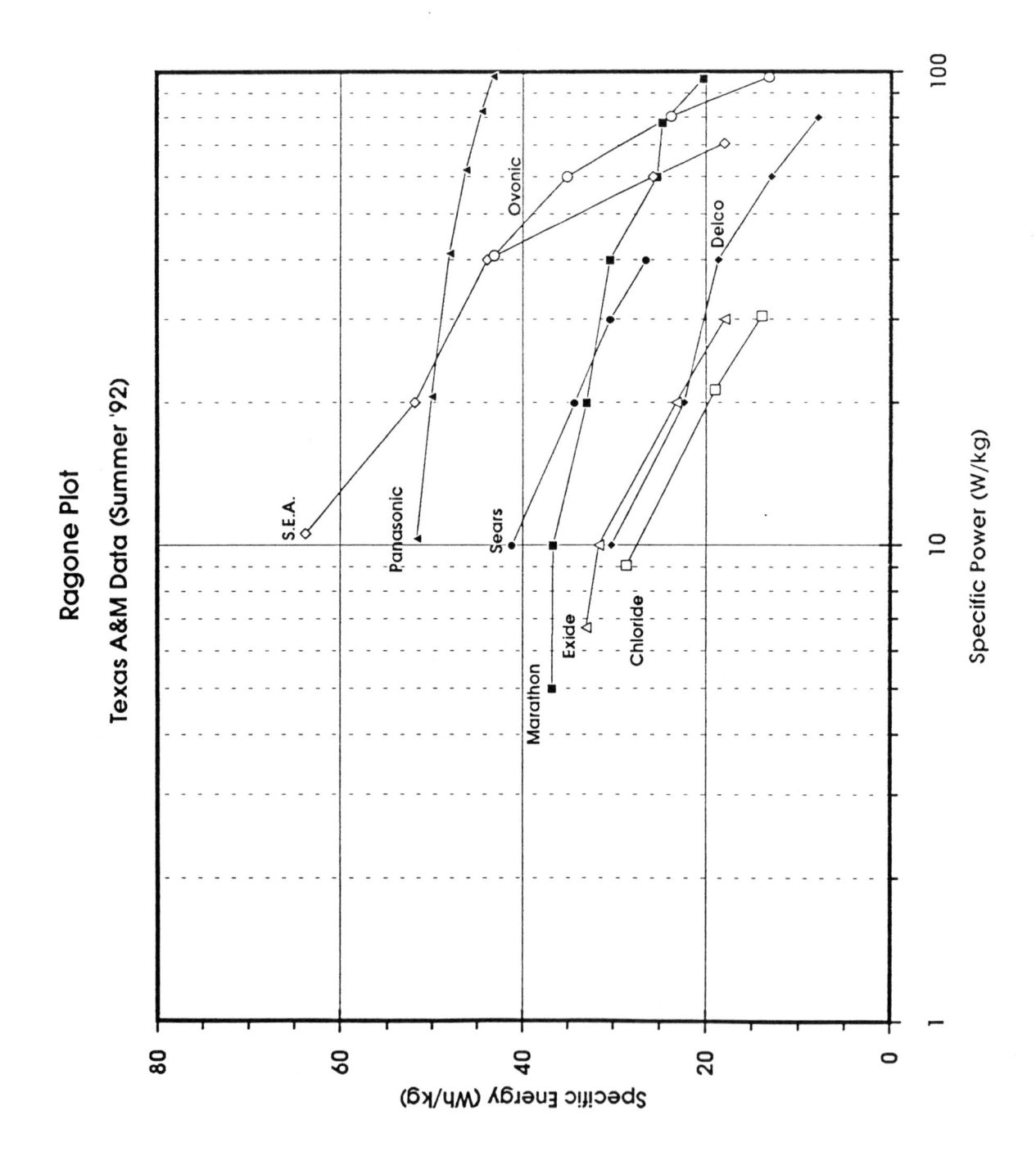

Available Peak Power as a Function of D.O.D.

Texas A&M Data (Summer '92)

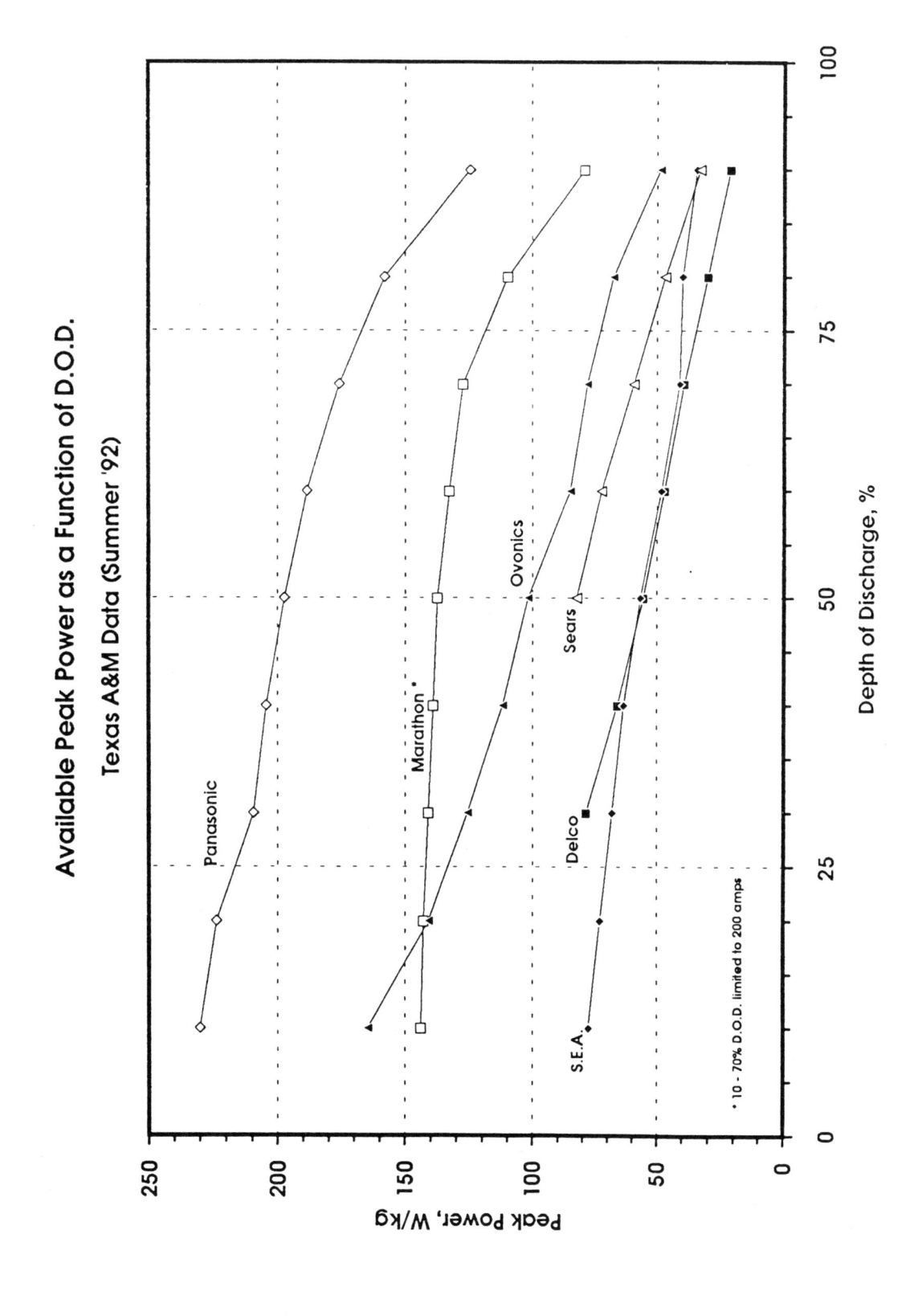

Self Discharge
Texas A&M Data (Summer '92)

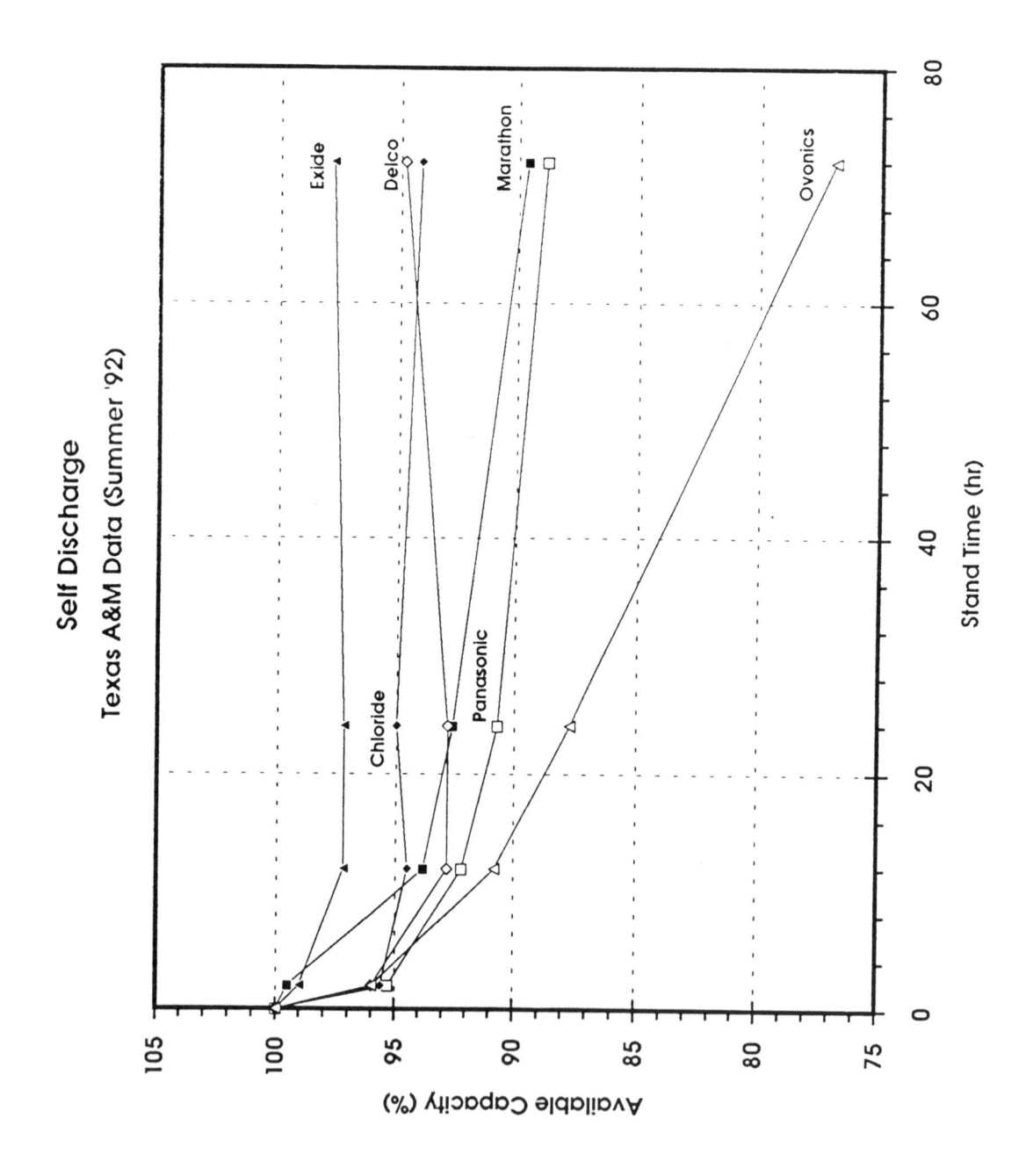

Average Cell Voltage as a Function of D.O.D.

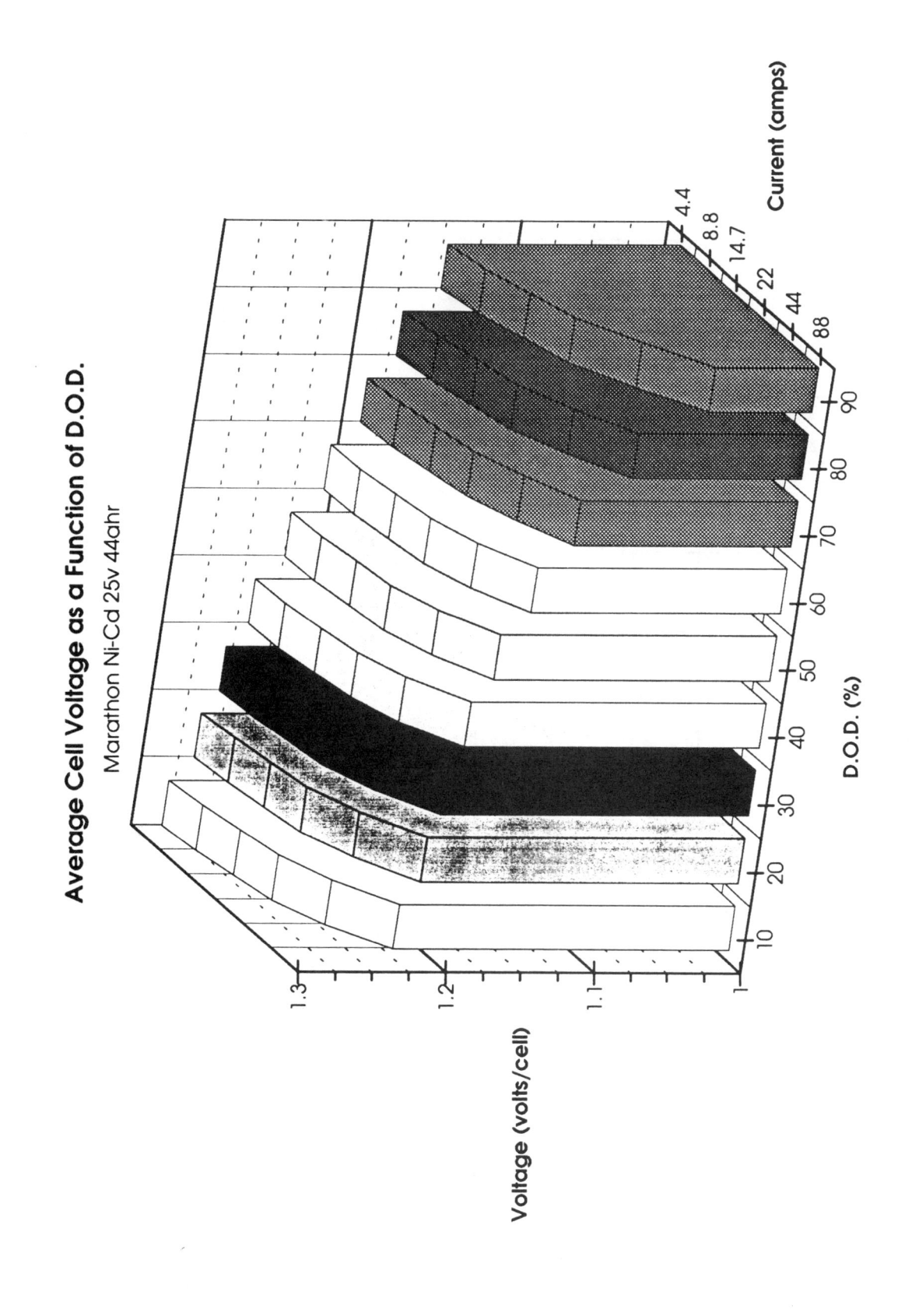

Marathon Ni-Cd 25v 44ahr

Texas A&M (Summer '92)
Comparison between Steady-State and Dynamic (Modified SFUDS) Cycles

Power System	Cycle Type	Avg. Power (watts)	Ahr	Whr	Wh/kg	Energy Eff.	Coulomb Eff.	Voltage Eff.
Phos-Acid Fuelcell	Steady-State	1007	20.99	1007	***	54.18	***	***
Phos-Acid Fuelcell	Dynamic	1008	22.05	1008	***	51.62	***	***
Ni-Cd Battery	Steady-State	320	45.9	1125	35.2	73.3	85.0	86.2
Ni-Cd Battery	Dynamic w/o/regen	320	32.9	800	25.0	50.1	60.6	82.7
Ni-Cd Battery	Dynamic w/regen	320	34.7	842	26.3	53.1	64.3	82.6
Pb-Acid Battery	Steady-State	250	61.0	721	28.8	81.8	93.6	87.5
Pb-Acid Battery	Dynamic w/o/regen	250	37.2	438	17.5	81.8	94.6	86.5
Pb-Acid Battery	Dynamic w/regen	250	35.6	412	16.5	80.7	95.2	84.8
Pb-Acid Battery	Steady-State	300	160	939	28.6	74.1	87.1	85.1
Pb-Acid Battery	Dynamic w/o/regen	300	105	605	18.4	71.6	80.1	89.4
Pb-Acid Battery	Dynamic w/regen	300	103.9	595	18.1	70.4	78.8	89.3

Advanced Parallel Hybrid Vehicle Design

Brady O'Hare and Norm Salmon

Western Washington University
Vehicle Research Institute

ABSTRACT

The parallel hybrid vehicle outlined in this paper was designed to achieve superior performance over conventional vehicles in the areas of emission control and overall power consumption. The safety, performance, range, drivability, and comfort of this vehicle will meet or exceed that of a desireable modern automobile. The environmental effects of manufacturing, the tail pipe emissions during the vehicles life, and the disposal of the vehicle were all considered with equal weight. This unique vehicle takes a fresh approach to hybrid vehicle technology with the introduction of the Dunstan Drive, a differential hydrostatic drive (CVT), and the dual tire - variable camber suspension system.

CHASSIS

The vehicle chassis design will be of similar construction to previous Viking experimental vehicles with features borrowed from the monocoque aluminum chassis of Viking VI.* The passenger compartment is designed to provide the highest torsional stiffness to ensure proper suspension performance. This central structure is designed to withstand deformation in a crash protecting the occupants. The forward and rear sections of the chassis are made of crushable aluminum honeycomb. The bumper and toe board bulkhead will act as reaction members for the honeycomb. These deformable regions are designed to even out the g-loading on the occupants during a 100 m/sec (approx.) head on crash. To achieve this, the deformable structures must crush at a controlled rate.

Picture 1. Viking 6 Before Crash Test (Front View)

The occupants will be protected by 16.5" (420 mm) of frontal crush, exceeding the requirement of the contest by 4.7" (120mm).

Based on the Viking VI crash reports, knee restraints will provide added protection to both occupants. Knee restraints will be made of a non-CFC polyurethane foam molded into a polyvinyl cover. The polyurethane crushes with little elastic

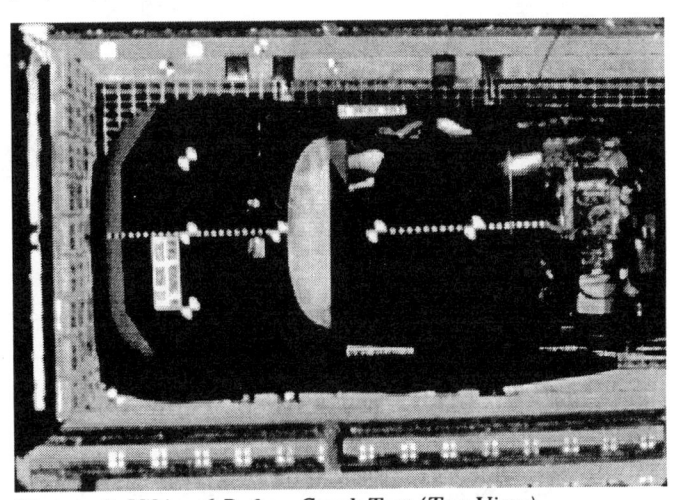

Picture 2. Viking 6 Before Crash Test (Top View)

deformation absorbing kinetic energy from the occupants.

Passive restraint air belts will be incorporated into the interior, originally designed by M. Fitzpatrick of Fitzpatrick Engineering and CalSpan. The belt uses an inbelt gas generator.* The belt is a standard three point design with a small inflator in the shoulder harness. The primary function of the air bladder is to prevent the occupants heads from rotating forward, greatly reducing the possibility of neck injuries. The gas generator in the belts is manufactured by Thiokol. It weighs 6 oz., fully loaded, and emits 12 grams of propellant per belt at 1600 degrees F.

Viking VI underwent a head on impact test, using an air belt system. The test was conducted by Dynamic Science on May 29, 1980. The head injury criteria numbers were 552 for the driver and 286 for the passenger. All of the measured injury parameters were well within the FM VSS 208 injury

---

*A light weight two-passenger automobile combining improved crashworthiness, good performance, and excellent fuel economy, Michael R. Seal and Michael Fitzpatrick

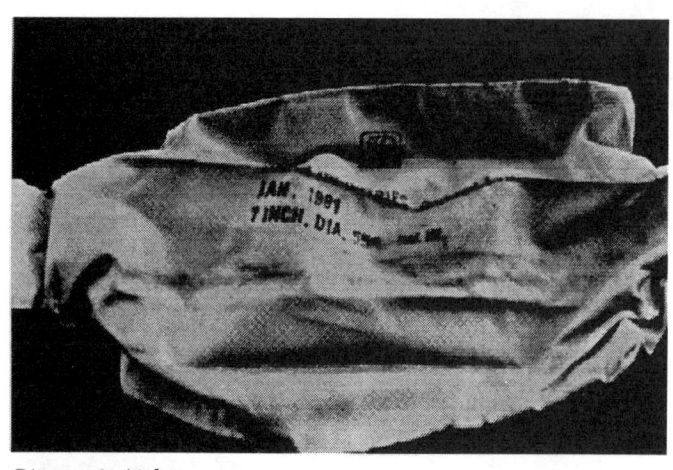

Picture 3. Airbag

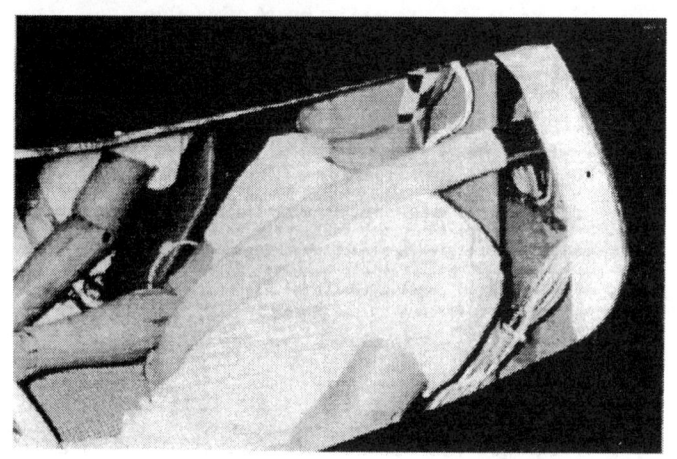

Picture 4. Airbelt on dummy

of causing the oncoming vehicle to crush. Computer codes will be run to verify our calculations before construction.

Adhesive bonding techniques will join the aluminum sheets together. Adhesive bonding will increase the strength and stiffness of the chassis by distributing the loads evenly. Rivets will be used during fixturing and will be left in place to prevent peel of the adhesive during a crash. Adhesive bonding has been used in the aircraft industry for many years and more recently in the automotive industry.*

The true cost of aluminum is up to three times that of steel. However by optimizing the properties of aluminum the chassis will be as stiff as a steel monocoque at one half the weight.* In addition, aluminum will not require coatings to prevent oxidation. The surface of the aluminum forms an oxidized layer that protects it from further erosion.** Not requiring additional surface treatment saves weight and reduces harmful chemical treatment. The aluminum chassis biggest advantage is the increased fuel economy through weight savings. The chassis possesses a high value and will be economically desirable to recycle.

Fitzpatrick Force Limiters are located on the upper attachment point of each seat belt. When the occupants move forward in a crash situation the seat belt attachment pulls a strip of mild steel through a set of rollers at 90° to the direction of the seat belt force. The deformation of the steel absorbs the energy of the occupants. Viking VI, a light weight two passenger automobile combining improved crashworthiness, good performance, and excellent fuel economy.***

FORWARD LIGHTING – Forward lighting in today's

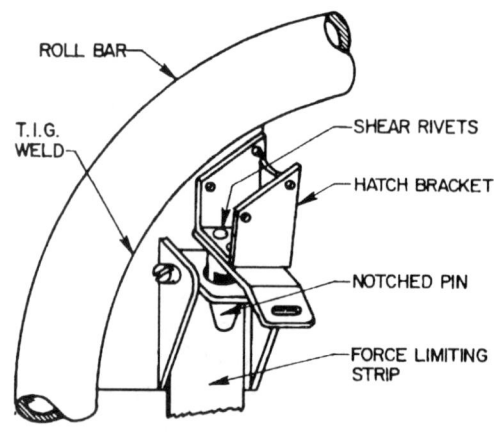

criteria. The crash velocity was 41.2 mph (66 kph), with a rebound velocity of 3 mph (5 kph). This test, although conducted in 1980, provides valuable proof that ultralightweight vehicles can be crashworthy. Viking VI weighed 1200 lbs (see Appendix D.)

The roll bar and front rollover hoop will be made of 7075-T6 Aluminum tubing (1.8" x .125"). The door side-guard beam will be 7075-T6 Aluminum (1" x .0625"). The monocoque surrounding the passengers will be made of 2024-T3 sheet .078" (2mm) thickness. The fore and aft crushable

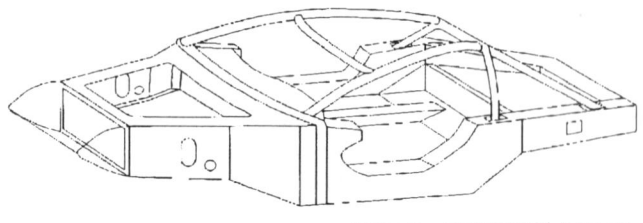

Figure 1. Chassic Drawing

sections will be .039" (1mm) thickness.

Localized reinforcement of the floor pan and side sills will be 7075 T-6 aluminum folded into hat sections of .06" (1.5mm) thickness. The side boxes will be 6 inches deep (152mm) filled with 100 psi (7Kg per cm²) aluminum honeycomb. This area is designed to minimize plastic deformation and evenly distribute the load of a side impact, with the goal

Figure 2: Fitzpatrick Force Limiter

cars is limited to conventional bulbs or sealed Halogen bulbs. However General Electric is working on a new system called High Intensity Discharge (HID), also called the Arc Discharge system. The General Electric automotive market development team claims "General Electric's first application of Arc Discharge lighting will target the automotive forward

---

*Road and Track "Ferrari 408" Dec. 1988 by Dennis Simanitis
**The Science of Engineering and Materials 2nd addition, Donald R. Askeland
***Michael R. Seal and Michael Fitzpatrick The Ninth International Technical Conference on Experimental Safety Vehicles

lighting market in the 1992 - 1994 time frame."* If this system does come about and meets D.O.T. approval, we will use it. If not, halogen bulbs will be used.

Lasting six times longer than conventional bulbs, the arc discharge system will outlive the average automobile. Four times more efficient than halogen bulbs the system can produce twice the light with one half the power consumption. Being only one half the size of halogen bulbs, High Intensity Discharge allows more freedom for styling. Smaller headlights will lower hood lines, improving aerodynamics.

Replacing the carbon filament in traditional sealed bulbs, Arc Discharge uses a controlled arc or spark between electrodes. From there the light is reflected off a parabola shaped mirror and out a lens. Poor vision conditions, such as rain or fog are penetrated better by means of ultraviolet light, which the (HID) will emit. Research has been done by both Saab and Volvo testing the ability of ultraviolet light to make certain paint and dyes glow. A practical application would be the illumination of distant road signs and markers.

The High Intensity Discharge system incorporates the use of ambient light sensors. Taking in light conditions ahead, the sensor sends a signal to a servo which varies the intensity and type of light.**

INERTIA SWITCH – An inertia switch disables the fuel pump after a collision. Normaly the switch is closed, and only opens with a sudden change in inertia , as in an auto accident . The switch is wired into the circuit of the fuel pump. Breaking the circuit of the fuel pump will stop fuel flow to the engine, reducing the risk of fire.

MERCURY SWITCH – Acting as a closed switch in the fuel pump circuit, the mercury switch will disenable the fuel pump in case of a roll over situation. The mercury in the switch is good conductor, which opens and closes the switch as the position changes.***

Picture 5. Interior dash mockup

*The GE Automotive market development team, Worldwide Automotive Lighting department
**Auto headlights for the '90s to use ARC Discharge system, Wyse Landau Public Relations Kelly Blazek
***Auto Electricity, Electronics, Computers, by James Duffy.

BODY

The body shape is a major factor in overall vehicle efficiency. Aerodynamic drag rises exponentially with speed and the power requirements rise proportionally. The coefficient of drag (Cd), a measurement of the goodness of shape, and the frontal area together are the variables that influence overall aerodynamic drag. These were both of major concern in the design of the body shape. To achieve a low Cd the vehicle must separate the air with minimum disturbance. It would be desirable to have laminar flow over the entire vehicle, but this does not work in practice. The best that can be achieved practically is to maintain non-stalled turbulent flow over the length of the vehicle. Preventing turbulent flow must be taken into account in the overall design. The "fast back" design is used to ensure attached flow clear to the rear of the vehicle. The rear track is narrower than the front allowing the body to taper toward the rear. The rear floor pan gently rises toward the rear bumper cover. The rear end tapering is done in order to reduce the trailing vortices.* The tapered sides also help to reduce drag in a side wind which effectively reduces the frontal area. The under body will use covers under the engine and battery tray that will fit flush with the rest of the smooth underpan. The side mirrors are placed on long stalks to ensure that the mirrors are out of the boundary layer of the body, if they were to remain in the boundary layer they could stall flow the entire length of the body. Occupants seats are also placed as close as practical to cut down on the frontal area of the cockpit. It is recognized completely that covered wheels will reduce drag, but they are omitted from this design for practical service and to maintain a practical turning radius.

Dupont's latest breakthrough has opened up a whole new use for lightweight polymers. The body will consist entirely of recyclable thermoplastic. Bexloy W, uses glass fiber reinforcement with a modified polyethylene terephthalate (PET) matrix, it provides parts with a painted appearance directly out of the mold. Bexloy K, also a glass reinforced PET, is engineered to withstand epoxy coat oven temperatures. Bexloy K will be used on panels that require a class A finish. (Dupont Customer Service)Finishing material will be a polyurethane with a water crosslinking agent, which helps to reduce the volatile organic chemical (VOC) levels during processing. Based on density PET is six times lighter than steel, enhancing both vehicle performance and economy. Plastic panels are produced at a fraction of the cost of steel panels due to the inexpensive tooling and lower processing energy. This makes styling changes much more economical due to the fact the tooling amortization time is shorter.

As the amount of plastic used in vehicles rises the need for practical plastic recycling is critical. PET is the most recycled plastic in the country today.** One of the keys to plastics recycling is design for disassembly (DfD). The BMW Z-1 is an example of a vehicle using DfD. The car's skin can be removed in 30 minutes and recycled at the local recycling center.***

* Automobile Aerodynamics Theory and Practice for Road and Track, Geoffrey Howard 1986 p. 83
**Industrial Plastics Theory and Application 2nd Addition Terry L. Richardson (Dupont customer service)
***S/EV 92 proceeding vol #1 Dr. David Stephensen, Sr Counsel, Brodeur and Partners

BATTERIES

Excellent power density is the most important characteristic of a hybrid battery. The battery must be able to discharge at a power level that provides the necessary performance for pure electric driving. This is the primary reason for selection the Saft P/N VP230 KHB nickel cadmium cell. Many batteries were considered, but no mass produced battery can provide the power density of a Ni/Cd. Each cell has a nominal voltage of 1.26 volts, 160 cells are put in series to reach the necessary buss voltage of 201.6 volts. The Unique Mobility motor operates at highest efficiency at 200 volts, the inductive charging system also requires a 200 volt minimum. Each cell is 1.060" x 3.180" x 8.20" and weighs 2.09 lbs. The total weight on the battery is 334.3 lbs. This battery provides not only for the proper bus voltage, but also provides enough to power the motor at its rated horsepower. Nickel cadmium batteries have a mostly flat discharge curve so voltage stays constant during discharge, not affecting the performance. The cells selected offer an energy density of 38.46 W-hrs/Kg and a power density of 191 W/Kg.

Recycling the batteries is an economic benefit for the manufacturers. Recycling helps stabilize the price and supply of cadmium. Cadmium is a by-product of zinc production, so it is not in danger of running out. Almost 50 percent of the cadmium produced is used in batteries. Nickel cadmium batteries have been recycled by the Saft Nife company for 20 years. All manufactures recycle spent batteries. It should not be a problem to recycle batteries since they will be serviced by specialist who will exchange the batteries. European recyclers have the ability to recycle 4.5 tons of batteries a year, only 3.8 tons are recycled each year.

Currently initial cost is one of the limiting factors for the Nickel Cadmium battery, because they are not produced in high volume, price is elevated. If production was similar to lead acid batteries price would drop to two times that of lead acid, however in the long run nickel cadmium batteries last much longer. Life expectancy is 2000 cycles or 8 years. The vehicle could travel 100,000 miles if the vehicle traveled 50 miles per cycle.*

Picture 6. Battery

---

*S/EV 92 proceedings Volume I State of the Art High Performance EV Battery Arne O. Nilsson

The battery pack will be stored in a fully contained case completely isolated from the occupants. The casing will be rotationally casted high density polyethylene (HDPE). HDPE provides excellent resistance to sodium hydroxide, the batteries' electrolyte. HDPE also offer excellent dielectric strength (18,000 - 20,000 V/mm), impact strength of 1.0 J/mm, and a slow burning rate of 25-26 mm/min.. The battery pack is located behind the occupants. Batteries will be serviced from under the vehicle in four manageable modules weighing 85 lbs. each including the case. Battery case ventilation is critical to prevent build up of hydrogen. The ventilation will include two Comair/Roton brushless DC fans (Flight 80 series) which are powered on 12V from a DC/DC converter. Each fan moves 33 cfm per minute and will operate whenever the vehicle is running or charging. The case will have one fan at the bottom which intakes air and one fan at the top that exhausts. The upper and lower fan operating in series will cause greater static pressure, causing more air to flow due to the high impedance of the battery case.* The fans are exhausted directly to a low pressure region on the body just behind the B pillar. The exhaust will double as a drain for electrolyte in the event of a rollover.

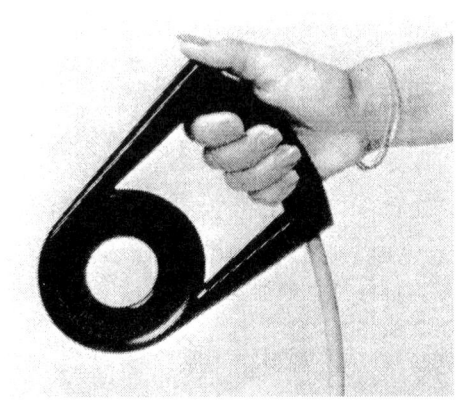

Picture 7. Inductive Paddle

Battery recharging is achieved using two systems, both of which come from Hughes Power Systems. The primary charging system is the Hughes Inductive Coupling System. The system includes a 6.6kW off-board 220 volt AC charger located in the home. An "inductive paddle", the primary coil, is inserted into a port in the front of the vehicle which contains the secondary coil. Power is transferred through this transformer via a magnetic field. This current regulated system is capable of accepting power levels from 1.5kW to 25kW. An on board charge controller monitors the battery functions, determined charge algorithm, and communicates to the off-board charger through the inductive coupling. This charger is capable of a complete recharge in less than two hours.

Inductive charging offers many advantages. It is lightweight in that most of the power electronics are located off-board. The single vehicle port can accept varying power levels, making it suitable for world wide acceptance and standardization. On-board chargers prove difficult to meet Underwriters Laboratory (U.L.) standards are also difficult to incorporate into a ground-fault connector. The inductive coupler can be used in any weather condition without the

---

*Comair/Rotron 1990 catalog

possibility of shock or arcing. The user friendly design and safety will add to setting the consumers mind at ease about recharging. Hughes has received a Gold Award for Design Exploration at the Industrial Design Excellence Awards for the device.

As an infrastructure for recharging does not exist,* on-board charging capability is essential. A 1.5kW 110 volt AC Hughes portable charger would be located in the trunk. This charger is capable of recharging in under eight hours, so it would only be used when the primary system was not available.

AIR CONDITIONING

Design considerations for an air conditioning system included minimizing cooling load and radiant heat gain, increased system efficiency, minimum weight, and incorporation of the newly accepted ozone friendly HFC-134a (Teteraflourethane).

The load requirements for the system are reduced with the aid of solar control glass from PPG industries. Solextria 7010 solar control glass will be used for all vehicle windows. This blue tinted glass absorbs 52% and reflects 6% of UV and IR radiation,** leaving 42% solar energy transferred.

Vehicle exterior and interior color would include only light shades to minimize vehicle heat absorption. The thermoplastic exterior also resists heating due to its increased R-value. Careful attention to insulating the cab as well as having a small cab volume will contribute to lowering the load.

A special air conditioning compressor, for use with HFC-134a, was chosen from Seiko Seiki of Japan. The SS-96A unit is all aluminum, weighing just 2.4kg (5.4lbs). The unit is capable of varying its displacement from 20-90cc/rev which allow it to operate consistently without cycling. Refrigerant flow is automatically adjusted to engine speed which reduces the demand from the engine at high and mid ranges. The compressor is connected with the input shaft of the transmission using a magnetic clutch. A multiple v-belt is used to allow for slippage during maximum acceleration.***

HFC-134a refrigerant requires larger condensing capacity than the traditional R-12 refrigerant. With the use of a parallel flow condenser the performance will meet or exceed the performance of R-12. The parallel flow condenser replaces the traditional serpentine or series condenser. The new condenser is half the weight, has 13% less frontal area, and allows for 20% less refrigerant to be used.****

BRAKING

Braking in a hybrid vehicle is accomplished with two separate systems, a traditional hydraulic system and regenerative braking, which share information from a common ECU and wheel speed sensors.

Reducing drag through the use of high pressure low rolling resistance tires is a serious safety compromise. Braking efficiency is directly related to the tire coefficient of friction and contact patch area. The dual tire and variable camber suspension systems take advantage of both low rolling resistance of a high pressure tire and the safety of a standard vehicle when both tires are deployed. The soft compound outer tire provides the necessary additional stopping power so that safety is not compromised. Two electrical brake switches are used. The first signals the brake light immediately after the brake pedal is touched. The second is tuned to come on later in the pedal stroke. When the pedal is depressed hard the second switch is tripped and opens the air solenoid, via the ECU, which interfaces all eight tires with the road. Initial test using a G-analyst show braking performance of the Viking 21 at -.6G using low quality bias-ply tires. This provides acceptable braking, but substantial improvement is expected with minor tuning.

Because of the large wheels used with the dual tire system 10.5 diameter discs can be used, which help to reduce pedal effort due to the longer effective lever arm. Reduced pedal effort is important due to the fact power brakes are omitted in the design. Discs are fabricated from an aluminum silicon carbide alloy from Duralcan USA. The material offers a density of 61.5% less than grey cast iron which allows unsprung weight to be substantially reduced. Prototype discs constructed at the VRI show a weight savings of over seventy five percent of iron discs. The material which is 20% silicon carbide by volume offers 74% greater thermal conductivity, 52% greater specific heat, and 20% greater yield strength than grey cast iron. Wear resistance is equal to or greater than grey cast iron due to the extremely hard silicon carbide which is a ceramic material. Special pads to prevent glazing are available from Allied Signal Friction Materials Group.

An opposed piston caliper is used on each disc. The calipers are all aluminum with a stainless steel piston. In order to reduce rolling resistance it is important for the pads not to drag against the disc. This is accomplished by modifying the bore of the caliper. The groove that retains the lathe cut piston seal is widened to twice it's original width. The seal is replaced with an additional round seal inserted directly in front of it in the bore. This allows for greater deformation of the lathe cut seal This additional deformation allows the seal to draw the piston back from the pad. This technology was proven on the Viking XX Solar race car, which crossed both the United States and Australia using a modified caliper like the one suggested.

The hydraulic system is split diagonally. The right front is connected with the left rear and the left front is connected with the right rear. Biasing is done using proportioning valves. The system is split this way to provide adequate braking if one system fails.

Regenerative braking is a feature designed into the Unique Mobility motor and controller. Regenerative Braking effectively turns the motor into a generator which charges the battery by converting the kinetic energy of the vehicle into electrical potential. This energy would be otherwise dissipated as heat through the brake discs. When the driver steps on the

---

*Solar and Electric Vehicles proceedings volume 1 October 1992 Inductive Charging Presented by Dick Bowman and Fred Silver Hughes Power Control Systems
**Automotive Engineering May 1990 Product Briefs p 96
***Seiko Seiki vane rotary compressor tech sheet 1-99110-03-cc/ga/f
****Popular Science July 1990 Ozone friendly cooling The Automotive Solution Brian Nadel p.60-63

brake pedal no hydraulic braking occurs until the regenerative braking is on fully. This would be used much like compression braking is used to slow a conventional car with a manual transmission. Regenerative braking power falls off with speed, so stepping deeper into the brakes actuates the hydraulic portion of the system in addition to the regenerative braking. There is some concern with the safety of maximizing regenerative braking. Under certain conditions it is possible to lock the wheels. The solution, however, is not to set a limit on the regenerative power, but to take advantage of recapturing this lost power. By resetting the potentiometer to 0 volts when the ECU senses a locked wheel, via the wheel speed sensors, the regenerative braking can be disabled. In such a panic situation the hydraulic ABS system can safely stop the vehicle on its own. Without a system like this, the driver would have to lift his/her foot to reset the pot. This is not a natural tendency for untrained drivers. With this system the driver can apply the brakes and the vehicle reacts.

INTERNAL COMBUSTION ENGINE

The engine chosen is a 750cc flat three cylinder four-stroke used in the BMW K75s motorcycle. Having a flat three offers a number of inherent advantages. Compared with upright engines the flat three has a very low center of gravity. The low profile makes for a low hood line which helps improve forward visibility and aerodynamics.

Picture 8. BMW Engine (front view)

The largest proportion of engine internal friction is created by the piston rings. Ring friction is directly proportional to the swept ring area. Area is reduced by approximately 11 percent over a four cylinder with the same bore and stroke ratio. The bore and stroke of the engine is 67mm x 70mm. The undersquare bore to stroke ratio is important in reducing crevice volume. This engine is a rare exception in a motorcycle market dominated by four cylinder four valve per cylinder engines with bore to stroke ratios drastically oversquare, making them most unsuitable for meeting strict emission requirements.

Ring crevice volume is reduced by as much as 28% over available four cylinder 750cc engines. Crevice volume is the area above the top compression ring and the piston top. This volume traps a portion of the air-fuel mixture during the compression stroke, the combustion flame cannot enter this area and the fuel goes unburned. Unburned fuel makes up a

Picture 9. BMW Engine (side view)

majority of HC emissions, with residual HC in the exhaust gasses making up the rest.* In addition to the geometrical reduction in crevice volume inherent in the three cylinder head land rings will be installed. Head land rings have an "L" cross section. The ring comes flush to the top of the piston further reducing the crevice

Combustion chamber surface area to volume ratio is better in a three cylinder and improved more due to the undersquare bore to stroke ratio. Less surface area reduces the thermal losses, raising the Brake Mean Effective Pressure (BMEP), which is directly related to engine efficiency. The combustion chamber would be modified to remove any squish area. Squish area helps to reduce detonation, but increases HC emissions. Due to the increased Octane rating of M-85 the squish area can be removed with out inducing detonation, thus improving HC emissions.

The engine has dual overhead cams. Cam timing can easily be changed with out making custom camshafts. Cam timing will be adjusted to automotive range having the intake open 15 degrees before TDC and closing 55 degrees after BDC. Exhaust opens at 55 degrees before BDC and closes 15 degrees after TDC. This is a baseline for dynamometer testing where optimum cam timing will be set using a vernier adjustment on the cams. This technique has been done successfully at the VRI on the Subaru Legacy DOHC four valve prototype and an experimental five valve engine designed for Chrysler. Cam profiles will be reground to provide a balance between maximum power and emissions.

The valve train friction is substantially less than other .75 liter engines with 16 valves. The BMW three cylinder has two valves per cylinder, and a total of six valves to actuate. Four valve per cylinder engines have the advantage of high swirl and improved breathing at low rpm, but in a hybrid the electric motor offers it's maximum torque at stall making up any loss over a four valve. The engine, due to the Dunstan Drive (CVT), operates in a narrow range for maximum power and doesn't need to be adapted to transient conditions like a vehicle with a manual or standard automatic transmission. The two valve engine incorporates less parts, creating less friction

*Fuel Economy in Road Vehicles Powered by Spark Ignition Engines, Ch by JT Kummer, edited by John C. Hillard and George Springer 1984 pg.61.

and cost less to build and maintain.

Because the engine's performance depends on the transmission and not entirely on the transient reactivity of the engine and it's controls, features like active manifolds are not necessary. The engine power can be tuned narrowly. This is done with resonant pressure wave super charging. Volumetric efficiency of up to 120 percent can be achieved during resonance. The tuned length is determined by the following equation:

$$L = \frac{tVs}{N}$$

Where: L = resonant length of intake runner
t = 90 degrees intake and 85 degrees exhaust
This is determined by the mean average time constant in crank angle degrees for the pulse to travel the length of the runner and back.
Vs = Intake = 1100 ft/sec the speed of sound in air
Exhaust = 1800 ft/sec the speed of sound in exhaust gases
N = Engine RPM

Both intake and exhaust are tuned to 6600 rpm. The intake length is 14.17 inches and the exhaust is 24.55 inches.*

The engine fires every 240 degrees giving even power pulses, thus there is no torsional roughness. However, the three cylinder creates a primary rocking couple. This is removed using a balance shaft in the BMW. The balance shaft smooths the engine, but does create a slight increase in added bearing friction. Consideration will be given to remove the balance shaft to reduce friction and weight. It may be possible to isolate engine vibration without a balance shaft using liquid filled tuned engine mounts. The liquid, a rheologic fluid, varies it's viscosity with electrical current. The engine mounts would be put into a closed loop circuit with a piezo crystal which senses engine vibration, thus dampening the vibration precisely.

Engine cooling will also be modified to insure proper cooling over the whole engine. This will be accomplished with the use of an external water manifold. The manifold will have replaceable jets which can be tuned on a dynamometer using a thermocouple to read the temperature on each cylinder.

The ignition will use individual coils on each spark plug. Direct distributorless ignition allows for precise control of spark advance with extremely high output voltage (80kV), to help cold start of the M-85 fuel. Spark advance will be controlled by the fuel management computer supplied by Accel. M-85 has very high octane so compression can be raised to 11 to 1 with out inducing detonation The spark event can also come earlier. There are limits, pre-ignition is a particular problem with M-85. Individual detonation sensors will be used on each cylinder to sense detonation and pre-ignition. Spark will be cut back to the cylinder with the problem, not affecting the others. Preignition is the ignition of the air-fuel mixture prior to the spark event. It is caused by surface ignition usually on an overheated plug, carbon deposit, or hot spots in the combustion chamber. M-85 requires a very low percentage of its total lower calorific value (13%) to dissociate through formaldehyde to carbon monoxide and hydrogen. The broken down hydrogen combusts with relatively low surface ignition temperature. Platinum tipped spark plugs are very prone to surface ignition due to their catalytic reaction and must be avoided in M-85 fueled vehicles. Tests show that a plug with a nickel alloy ground and center electrode performs best. A longer insulator (2.5mm) also offers a lower rise is surface ignition temperature with increased spark advance.* Lowering the plug heat two ranges will also help reduce surface ignition temperature.** Combustion chamber modifications will be made to remove any sources of possible pre-ignition. Sharp edges and corners will be eased. Spark plug threads and the internal threads of the head will be matched so that none are exposed inside the chamber. Polishing of the combustion chamber will also help to reduce pre-ignition.

SUSPENSION SYSTEM

The suspension system will be the same suspension as the Viking 21, a high performance parallel hybrid, designed by Dr. Michael Seal. The variable camber suspension and dual tire system have multiple patents pending. The system has proven road worthy under the harshest conditions. Viking 21 charged to victory through mud, ice and snow at the inaugural running of the Pikes Peak Hill climb for solar, electric, and hybrid vehicles. The car averaged 40.8 mph over the 5.5 mile course hitting a top speed of 49 mph up the 10 percent grade of Pikes Peak.

Picture 10. Viking 21 victory at Pikes Peak

The dual tire assembly provides for extremely low rolling resistance with the safety of a standard passenger car tire. Although the tires used look small they are capable of supporting a 2320 lb. vehicle. The inside tire is a hard compound DOT-approved motorcycle tire which provides a coefficient of

---

*The Sports Car, Collin Campbell, pg. 54, 1969.

*Pre-Ignition Phenomena of Methanol Fuel (M-85) by the Post Ignition Technique Toshiyuki Suga, Shinichi Kitajima, and Isao Fuji Honda R and D Co.,Ltd. SAE paper 892061 presented at the International Fuels and Lubricants Meeting and Expo. Baltimore, Maryland, September, 1989.
**Further Development of the Methanol-Fueled Escort Roberta J. Nichols Ford Motor Company SAE paper 830900

rolling resistance of approximately .008 at the rated pressure of 42 psi. The outer tire is a softer compound DOT approved racing radial. The inside tires are in contact with the road most of the time. The outer tires are used during braking, cornering, and maximum acceleration.

The outer tires come in contact with the road in two ways. The first is through the natural change in geometry of the suspension during cornering. The suspension is a parallel wishbone configuration. During a corner the parallel links cause the wheels to move into positive camber. Statically the

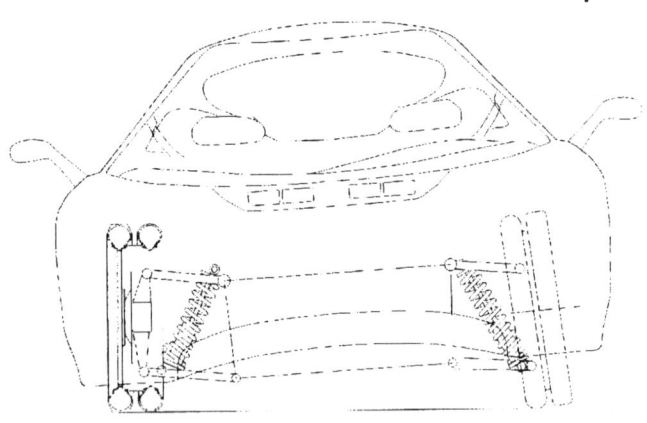

Figure 3. Suspension System showing geometry's natural lean

wheel is negative. When in a corner, the outer loaded wheel is move positive relative to the static position, thus the loaded wheel is at zero camber relative to the road. A total of six tires are in contact with the road during hard cornering. After the Viking 21 had run up Pikes Peak there were marks left in the mud on the corners where six tread marks were clearly visible.

The second way the outer tires make contact is by the actuation of a pneumatic cylinder on the upper control arm of the suspension. This device is positively controlled using an air valve that is actuated by microprocessor control. When the brakes or the vehicle is in a maximum acceleration mode the main ECU sends a signal to the air valve to open. Air is supplied by a small high pressure air compressor (200 psi) and is stored is a pressure reservoir to reduce cycling of the compressor.

The dual tire system offers an additional level of safety having two tires gives the vehicle the ability to run on a flat tire. Viking 21 tested the run-flat ability unintentionally finishing the last part of the Pikes Peak run with the right rear inside tire flat. The problem was not even discovered until the car had driven back down the mountain. There is no need for a spare tire.

The suspension also offers extremely low unsprung weight. The aluminum hub carriers weigh only 3.5 lbs. Brake discs will weigh only 3.25 lbs using aluminum matrix composite material. The wheel itself is a breakthrough in weight

Picture 11. Dual tires in negative camber

savings weighing only 15.4 lbs including the tires. The wheel is carbon fiber and epoxy and weighs 4.75 lbs, this is built with prototype tooling and weight could be reduced further. Though light it is extremely strong and able to take abuse as proved on Pikes Peak. Total unsprung weight is approximately 32 lbs per corner including the unsprung portion of the drive axle.

Picture 12. Suspension

DUNSTAN DRIVE

The Dunstan Drive is an infinitely variable differential hydrostatic transmission. This device is a new concept in automotive transmissions and has U.S. and foreign patents pending. The transmission is the design of Design Engineer Phil Dunstan who spent 33 years at the Boeing Company working on projects from the lunar rover to hard mobil launchers.

Picture 13. Dunstan Drive

The differential hydrostatic transmission uses a variable swash plate pump coupled to a fixed swash plate motor. Unlike a conventional hydrostatic there is almost no relative motion between the hydrostatic pump and motor at 1 to 1, effectively locking up at high speeds where it is nearly 99 percent efficient. The extremely high pressure and excellent sealing rings inherent in the Dunstan patent system allow for high efficiency at maximum reduction. The transmission is capable of transmitting the full torque of both the I.C. engine and Electric motor combined. The unit chosen will accept up to 220 ft/lbs of torque and can operate up to 5000 RPM. The unit was designed for testing in experimental vehicles at the VRI. Many of the more complex parts were machined in the Western Washington University Machine Metal Lab on the CNC milling machine by Eric Friesen, student.

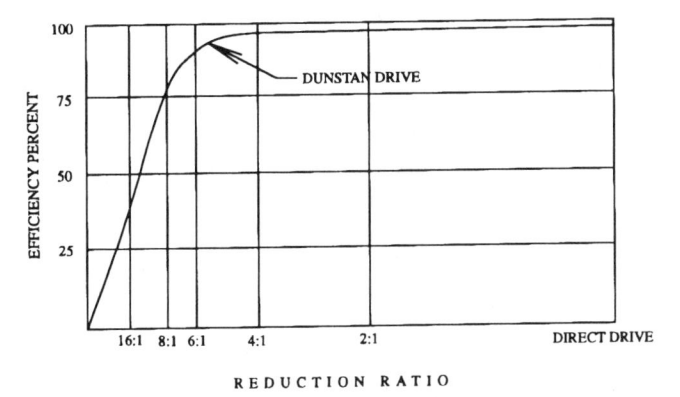

*Figure 4. Dunstan Efficiency Graph
Information provided by Phil Dunstan*

Fuel Economy improvements in the Dunstan Drive are not due to reduced friction. The mechanical efficiency is slightly less than a manual transmission. The transmission makes it's improvement by precisely matching the speed ratio to the required road load horsepower. The result is a 10 to 15 percent increase in fuel economy. Brake specific fuel consumption is made nearly linear over the entire speed range.

Current transmissions either automatic or manual have fixed gear ratios, unlike the Dunstan. This makes it impossible to match their engine to the vehicle load requirement across the entire speed range. Because of the mismatch, the engine operates at less than optimum efficiency impairing fuel economy.*

The Dunstan Drive also offers decreased acceleration times. Because it is a CVT, there is no shift interval. The transmission speeds with the motor until the motor reaches peak horsepower. The final drive ratio limits the motor from revving to its peak initially. The transmission control electronics hold the engine at maximum horsepower, until the transmission reaches 1 to 1 at top speed during a max accel cycle. At steady state high cruise the transmission drops engine RPM to maximize economy.

The Dunstan Drive is an excellent candidate for a hybrid vehicle. It's excellent efficiency combined with it's seamless speed transition make it the perfect candidate. With a driver transparent control system an automatically controlled transmission is mandatory. The Dunstan Drive can operate like no other transmission to optimize ratios for the electric motor, I.C. engine, and the pair operating in tandem. Software can also take into account for maximum acceleration, cruise, and even can be utilized to affect problem areas of emission controls. (See appendix A for a detail mechanical description and cut away)

*p 20 David Cole, Fuel Economy in Road Vehicles Powered by Spark Ignition Engines Edited by John C. Hillard and George S. Springer 1984

ALTERNATIVE FUELS AND EMISSION CONTROLS

M-85 is considered to be both practical and safe blend for automotive use as it improves cold start and, burns with a visible flame. The risk of ignition in the fuel tank is greatly reduced. Air saturated with neat methanol vapor contains approximately 13 percent alcohol which is in the explosive range. Air saturated with gasoline is too rich to ignite.* The M-85 mixture reduces the probability of ignition with the addition of gasoline. In order to further reduce the risk of ignition in the fuel tank a spark arrestor is added between the filler and the tank. The spark arrestor is a fine wire mesh that would dissipate the heat of a spark faster than it could ignite the air fuel mixture in the tank.

M-85 offers a number of advantages as a fuel, but there are number of disadvantages that must be overcome. M-85 offers greater power potential than gasoline. Although gasoline has more energy per pound than M-85, M-85 has a far higher latent heat of vaporization and it is an oxygenated fuel. Having oxygen in its molecular structure is one reason that it has a power advantage over gasoline. With the stoichiometric ratio occurring at 7.62 to 1 vs. 14.6 to 1 in gasoline there is a greater volume of fuel flowing into the combustion chamber at any given time. More fuel means less air, but between having oxygen in it's structure and the natural intercooling effect, M-85 comes out with greater power. With gasoline a 25 degree C drop in charge temperature occurs with a rich mixture. Neat alcohol can drop the intake charge by 50 degrees C, this can improve volumetric efficiency by about 10 percent.** The low latent heat of vaporization is a blessing to power but a curse to cold start. The vapor pressure of M-85 will vary depending on if finished gasolines or pure hydrocarbons are used. Pure hydrocarbons or simple hydrocarbon mixtures can be used to raise the vapor pressure to facilitate cold start, however there is a penalty of increased evaporative emission.*** A carbon canister will be used to deal with evaporative emissions. The fuel system is sealed to prevent evaporative emissions. The fuel condensed in the carbon canister is released to the engine manifold via a purge valve when the vehicle is running.

Fleet vehicles running on neat methanol have exhibited 10 percent lower brake specific fuel consumption than gasoline at their power peak on an energy equivalent basis.****

Fuel injectors, fuel lines, fuel sender unit, and high pressure fuel pump would all be methanol specific. Engine oil used will be formulated to combat the harmful effects of methanol.

Methanol is highly toxic so warning labels would be placed on all fuel related components and the filler cap.

*An Overview of the technical Implications of Methanol and Ethanol as Highway Motor Vehicle Fuels Frank Black U.S. Environmental Protection Agency SAE Paper 912413
**The Sports Car, p 24-25, Collin Campbell, 1970
***Vapor Pressure Characteristics of M-85 Methanol Fuels Robert L. Furey and Kevin L. Perry GM Research Labs. SAE Paper 912415
****Further Development of the Methanol-Fueled Escort SAE Paper 830900 Presented to the Second International Conference on Automotive Engineering,Tokyo, Japan, Nov. 1983

A methanol vehicle will have to meet the same emission standard as gasoline powered vehicles for hydrocarbons (HC), carbon monoxide (CO), nitrogen oxides (NOx), and also eventually aldehyde and ketone emissions. Starting in 1993 methanol fueled vehicles will be required to meet HCHO standards of 15 mg/mile for formaldehyde.

Formaldehyde emissions are mostly a result of unburned fuel self igniting in the combustion chamber during the exhaust stroke and a small portion oxidizes in the exhaust manifold.* The greatest formaldehyde emissions take place between 250 and 275 degrees C in the catalysts,** so it is no coincidence that HC emissions are also greatest during this region before catalyst light off. About 75 percent of engine HC emissions occur with in the first 200 seconds of the FTP-75 (bag 1) emission test. In order to prevent HC and formaldehyde it is important light off the catalyst as fast as possible. Quick heating of the exhaust system is also important because the fuel management system operates in closed loop until the oxygen sensor has reached over 260 degrees C where closed loop operation takes over to precisely control the air fuel mixture. An electrically heated oxygen sensor will be used to cut the time necessary.

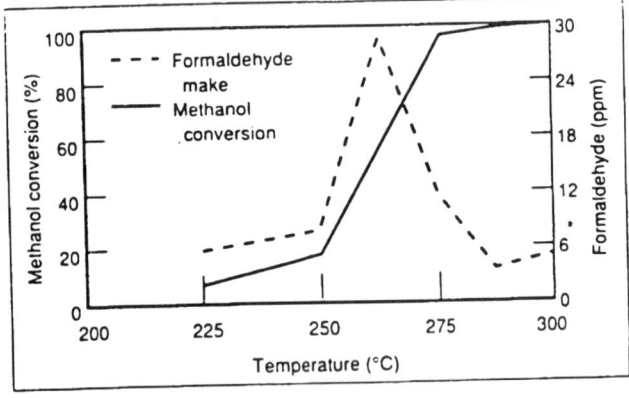

Figure 5. Formaldehyde Graph
Reprinted with permission of Johnson Matthey Catalytic Systems Division, North American Technical Center, Dr. Barry Cooper, from "The Future of Catalytic Systems"

Fast light off will be accomplished using a close coupled catalyst. The catalyst will be located at the collector of the resonant tuned exhaust manifold. Close coupled catalyst offers better performance on HC and formaldehyde than under-floor catalyst. During cold start test at Allied Signal Catalyst Group show that close coupled catalyst have 10 percent of the methanol and formaldehyde emissions of a vehicle with under-floor catalyst. Emissions during hot start were 20 percent of the under-floor catalyst vehicle. The catalyst will be a Pd three-way catalyst, which offers better conversion performance due to higher operating temperature than Pt/Rh catalyst.* The three-way catalyst accomplishes both oxidation and reduction. To further aid the light off time a set of reed valves will let heated air on to the catalyst substrate. The air will be heated with a highly efficient positive temperature coefficient ceramic heater located remotely. This heater will also send heat to the intake manifold and can provide a back up source of heat to the passenger compartment.

Primary vehicle heating is achieved using latent heat storage, this is also a primary emission control for cold start. Barium hydroxide, a salt with a low melting point (78 degrees C), is encased in flat metal fins inside an insulating vacuum, the unit is called a heat battery. The salt has a latent heat capacity of 89 W-hr/Kg which is nearly three times the energy density of the Ni-Cad batteries used. The salt can stay liquid for up to three days. The energy to liquify the salt comes from waste engine heat from the cooling system.**

LATENT HEAT

The latent heat battery is in series with the engine and the heater core. This closed loop system will operate any time the engine is running using an electric coolant pump. A by-pass will be used any time the heater will be used alone.** This system will provide excellent heating while the vehicle is operating as an electric only and will not affect the vehicle range. Using a electric 2500 watt electric heater will use about

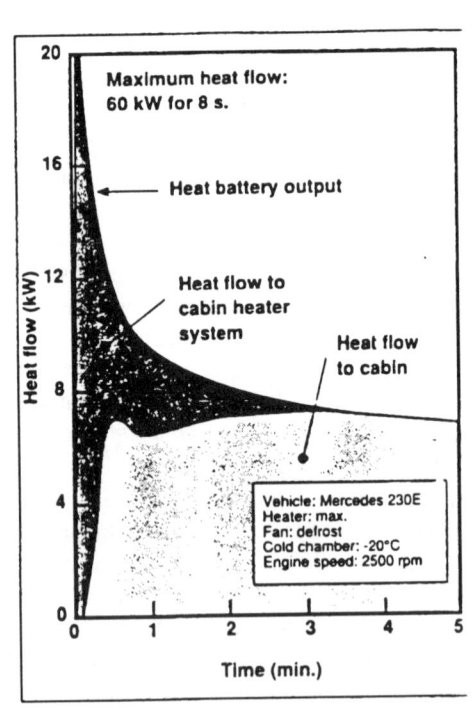

Figure 6. Latent Heat Graph
Reprinted with permission from SAE Paper No. 910305
©1991 Society of Automotive Engineers, Inc.

---

*Comparison of Unburned Fuel and Aldehyde Emissions from a Methanol-Fueled Stratified Charge Engine N. Scull C.Kim and D.E. Foster SAE Paper 861543 presented at the international Fuels and Lubricants Meeting and Expo. Phil. PA, Oct. 1986

**The Future of Catalytic Systems Automotive Engineering, April 1992 Dr. Barry Cooper

*Automotive Catalyst Strategies for Future Emission Systems W.Burton Williamson, Jerry C. Summers, and John A. Scaparo, American Chemical Society Symp. Ser. 495 1992 p 35-39.

**Latent Heat Storage, Automotive Engineering, Feb. 1992 pp.58-61

the same power it takes to drive the vehicle at 40 mph. The heat release from the heat battery is instant; 50-100 kW can be released during the first 10 seconds of operation. This instant heat will provide for excellent defrosting. In order to offset the low heat rejection of the I.C.E. and the fact that the I.C.E. may go long periods without running, there will be an electrical heating coil to the heat battery. This will allow the vehicle to "charge" the heat battery during wall charging. This is a more efficient way to heat the vehicle, because the heat battery offers better energy density than the electrical battery. The vehicle will still have an electrical heater, as mentioned before, in the case that the heat battery's capacity was used up.

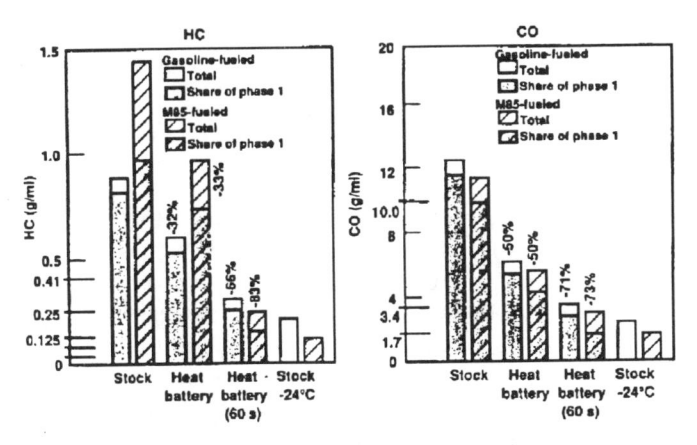

Figure 7. Latent Heat Storage
Reprinted with permission from SAE Paper No. 910305
©1991 Society of Automotive Engineers, Inc.

The stored heat improves cold start and when used in combination with the close coupled catalyst and heated oxygen sensors will provide substantial reductions in HC, CO, and formaldehyde. The heat battery has undergone test by the EPA. Test show a drop of 40 percent in HC emissions and 50 percent reduction in CO in M-85 vehicles. (see figure 7) The test vehicles show that LEV standards for HC and CO at standard temperatures can be met at -7 degrees C using the heat battery.

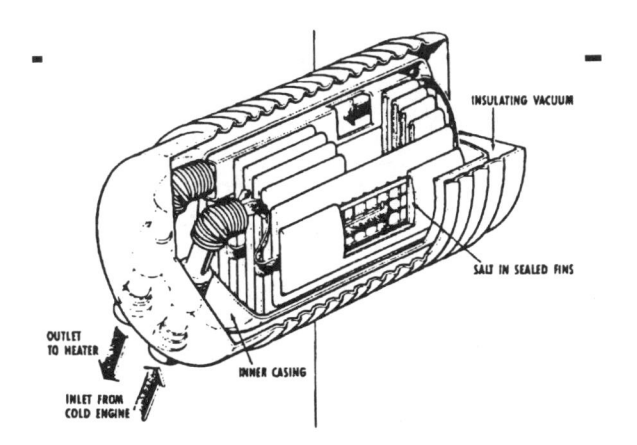

Figure 8. Latent Heat Battery
Reprinted with permission from SAE Paper No. 910305
©1991 Society of Automotive Engineers, Inc.

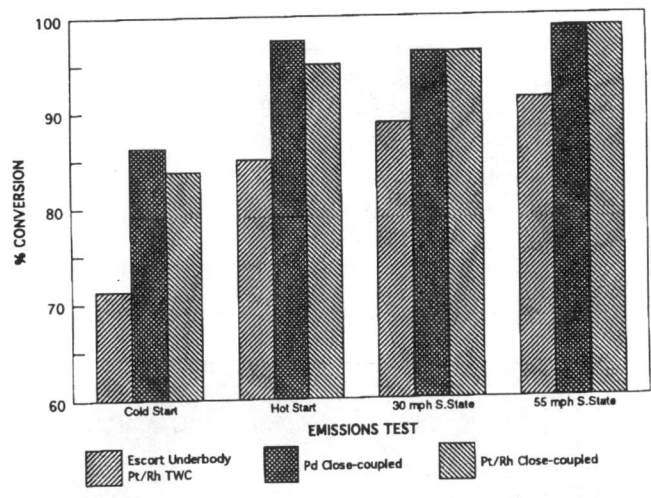

Figure 9. Emission Test Bar Graph
Reprinted with permission from Catalytic Control of Air Pollution, Mobile and Stationary Sources, Silver, R.G., Sawyer, J.E., Summers, J.C., eds. pp 26-41; Copyright© 1992 by the American Chemical Society.

Air fuel calibration will initially be set at stoichiometric, because lean calibration tends to lead to poor NOx conversion. NOx is a function of combustion chamber temperature. NOx will be controlled with the use of exhaust gas recirculation (EGR). EGR is a source of an effectively inert gas used to cool combustion. Some natural EGR takes place during the exhaust intake balance point, however this is highly unregulated. A digitally controlled EGR valve will be developed to precisely control NOx. The valve will be controlled by the fuel management system which will have the optimum setting to control NOx for every speed and load which will be mapped out on the chassis dynamometer.

ELECTRIC MOTOR

Unique Mobility's SR180L D.C. motor is used. This rare-earth permanent-magnet brushless motor uses a thin laminated core stator. This construction permits high speed operation with low losses. The motor is much lighter and efficient than conventional motors because of this design. The motor offers better than one horsepower per pound during intermittent operation (1-3 min.). The permanent magnets are neodymium iron boron "Super magnets". This permanent magnet offers the highest flux strength to weight available. The motor uses 18 poles to provide for the highest energy density and efficiency. The high pole count also reduces the amount of magnetizing steel, which reduces the heating losses. The motor operates between 88 and 92 percent during all modes of operation. The motor weighs 52 lbs.

D.C. motors require commutation. This is accomplished by the motor controller electronically switching the polarity of the stator using pulse width modulation. A Hall effect sensor is used to provide armature position information to the motor controller. D.C. current is transformed to three-phase using six half-bridges in the controller. Three main conductors and a logic interface cable connect the motor to the controller.

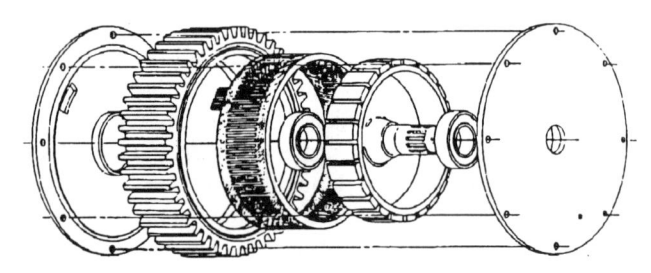

Figure 10. Breakdown of Unique Mobility SR180L/CR20-300

Picture 14. Unique Mobility SR180L/CR20-300 (52 lbs.)

The controller provides functions specifically designed for electric vehicle operation. The motor controller provides forward and reverse, pulse by pulse current limiting, and regenerative braking. The regenerative braking circuitry boost the voltage up so that the battery will accept the charge. Regenerative braking is effective to 0 RPM, but the power falls off with speed. The controller can be preset with a regeneration limit, but this will be controlled in closed loop with the ABS braking system mentioned earlier. Regeneration is limited when the battery is at a full charge. A tach signal and thermocouple are also provided.

Heat build up is a limiting factor on how much and how long the motor produces peak power, because of the low thermal mass of the motor. The motor and controller are water cooled which help to increase the duration of intermittent peak power and increase reliability. The water cooling system will be in parallel with the engine. During normal operation the systems will operate separately with their own radiator. When additional cooling is required water from the engine will be used to cool the electrical components. The electric motor maximum winding temperature is 300 degree F. The electric motor cooling will open into the main I.C.E. cooling when the temperature reaches 220 degrees F, which is the operating temperature of the I.C.E.. During off peak horsepower conditions the heat rejection from the electric motor is very low and the electrical component cooling system will be sized accordingly. The coolant will be circulate in the I.C.E. with the stock water pump and with a small electric pump in the electrical component cooling system.

DRIVE SYSTEM – The drive system is a compact front drive unit. This allows for ample trunk space in the rear and also situates the electric motor so that it is able to achieve maximum regenerative braking.

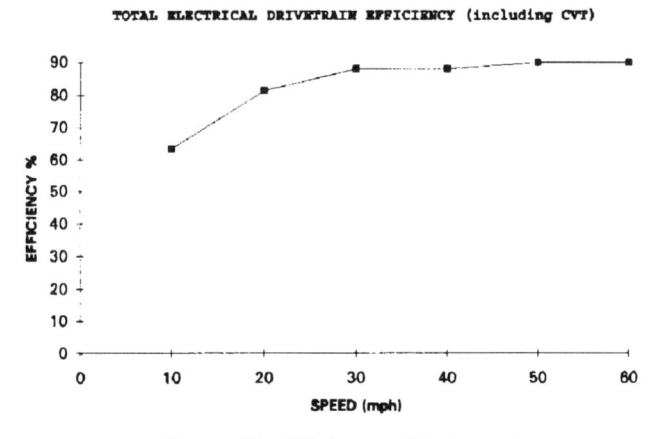

Figure 11. Efficiency of Drivetrain

The electric motor and I.C.E. drive into a transfer case. The transfer case will be made of a cast aluminum shell and use a Morse H-V silent link chain. The H-V chain is highly efficient (99.7%) and is designed to take the high speed high horsepower requirements. The chain is lubricated using oil pressure created in the Dunstan Drive, the two share a common oil pan which uses common engine oil. Each shaft will enter the case through an oil seal the shaft will be supported on both sides of the case with a bearing. The sprockets provide an opportunity to match the speed ratios of the engine, electric motor and the transmission. Both the engine and electric drive to the transmission at a 1.2 to 1 ratio. The transfer case has two openings to allow the drive axle and the air conditioning compressor shafts to pass through. The axle shaft opening is a large round opening in the center of the case to allow the axle to move through the steering and suspension travel without fouling the case. The air conditioning compressor shaft passes though the case to a pulley. The pulley is driven by the input of the transmission at a ratio of 1.4 to 1. A belt is used to allow slippage during maximum acceleration.

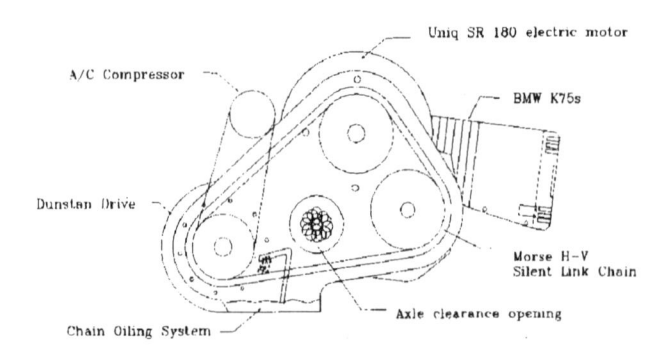

Figure 12. Transfer Case

The I.C. engine drives into the transfer case via the stock dry clutch. A short drive shaft is used from the clutch to the transfer case. The clutch is actuated using a pneumatic cylinder controlled by a four way air valve from Versa valves. The air valve is controlled by the ECU. It is an on/off operation with little slippage. The usual mode of starting the engine will be engaging the clutch while the vehicle is in motion, the engine will start using the electric drive motor when the vehicle is at rest. The 12 volt starting system, charging system, and magneto are removed as they are not needed. The electric motor works as the engine's flywheel. When the clutch is decoupled the motor stops it instantly because no longer has a flywheel attached. During deceleration the clutch is decoupled to prevent compression braking. The fuel injectors are shut off and the ignition is disabled to improve emissions during deceleration.

The electric motor is mounted directly to the transfer case. The motor turns whenever the vehicle is in motion. Because it is a permanent magnet motor it will create a back EMF if the motor is not in use. The motor controller must always be powered or this generated current will damage the controller. It is for this same reason that the vehicle must never be allowed to roll when the transmission is not in neutral. If the controller is unpowered the generated current has no path and will destroy the controller. This current will be used to charge the battery during I.C.E. only operation.

The Dunstan Drive is bolted to the transfer case on the input end and bolted to the differential case on the output end. The output shaft drives a helical gear which drives the differential at a 4.4 to 1 ratio. The differential is located in the center of the vehicle. The axle shafts are equal length so torque steer is reduced. The axles are parallel to the road under static conditions, this reduces drag caused by operating the constant velocity joints at an angle.

VEHICLE PERFORMANCE – Full Throttle acceleration from 0-60 mph was determined by the following:

$$t = \frac{V}{A} \text{ (sec)}$$

Instantaneous acceleration was plotted using the equation:

$$a = \frac{P \times g}{M1}$$

Where P = surplus tractive effort =
wheel thrust - (rolling drag + aero drag)

Wheel thrust is limited by an overall gear ratio of 18.48 to 1. As the CVT is shifted from 3.5 to 1, to 1 to 1 torque drops and horsepower rises to maximum at @30 mph and remains constant thereafter. Wheel thrust exceeds the tractive limit up to @45mph in hybrid mode during maximum acceleration. Traction control is provided by regulating the power into the electric motor. Wheel speed is sensed using the same sensors used with the ABS braking system.

Rolling Resistance

$$\text{Tire drag} = F = \frac{Crr \times \text{Weight of Vehicle}}{100}$$

$$F = \frac{.703 \times 1759 \text{lbs}}{100} = 12.654 \text{ lbs}$$

Power = F x speed

Rolling drag is plotted in Figure 13 as horsepower.*
Aerodynamic drag in horsepower is found by the equation:

$$D = \frac{Cd \, A \, V^3}{146600} \text{ (Horsepower) **}$$

Cd = .26 estimated using 1/10 scale model

A = 18.4 sq. ft. determined by frontal scan

Fgure 13. Road Horsepower

Picture 15. One-tenth scale model in WWU wind tunnel (front view)

Picture 16. One-tenth scale model in WWU wind tunnel (side view)

*GM Sunraycer Case History, The Wheels, Tires, and Brakes Chester R. Kyle 11/2/88 Lecture
**The Sports Car, Collin Campbell, 1970

A clarification of the formula for instantaneous acceleration is:

$M1 \text{ (lb)} = M(1+\alpha+\beta) \times N/V \text{ (RPM/MPH)}$

Where:
M1 is the true effective mass of the vehicle for different speeds.
(α) is a correction factor for the polar moment of inertia of the rotating parts of the engine and or electric motor.
(β) is a correction factor for the polar moment of inertia of the road wheels

M is assumed to be 1759 lbs. + 150 lbs. for driver based on the addition of all known component weights plus chassis weight base on experimental vehicles built at Western Washington Universities Vehicle Research Institute (Viking VI 1200 lbs, Viking VII 1250 lbs, Viking VII 1450 lbs, and Viking 21 mark I mule chassis (steel tube frame) @ 1800 lbs.)

Maximum Range on M-85 at 55 mph is determined by the following:

$$MPG = \frac{\text{Total Fuel weight}}{\text{BSFC} \times \text{Road HP}} \times \frac{\text{Speed}}{\text{total gallons of fuel}}$$

Where:
Total fuel weight is 79.16 lbs (35.9 kg) based on the .79 and Gasoline .75 .*
BSFC (Brake Specific Fuel Consumption) is estimated to be 1.08 lbs/ hp-hr at part throttle.
Road load HP at 55 mph = 7.28 hp
Total gallons of M-85 = 12.1 gallons
Estimated Range on M-85 = 553.75 miles
Estimated M-85 fuel economy = 45.76 MPG

Vehicle Range on Electric only is estimated by the following equation:

$$\text{Range} = \frac{\text{kW / hrs in battery}}{\text{Road HP} \times .746} \times \text{vehicle speed}$$
$$\text{system eff.}$$

Where:
kW / hrs in battery 160 Saft Ni-Cad Cells P/N VP230KHB with an energy density of 38.46 W-hrs/Kg and 36.54 W-hrs/cell = 5.846 kW/hrs figured on a 5 hour rate.
Road HP = see Figure 13.
System eff = product of drive train efficiencies electric motor and controller: see Figure 11.
Transmission : see Figure 4.
Transfer case : 99 percent

0-60 estimated times (see Figures 16-18):
Electric only .... 10.9 sec
IC only .... 14.93 sec
Hybrid .... 7.75 sec

---

*An Overview of the Technical Implications of Methanol and Ethanol as Highway Motor Vehicle, Frank Black, SAE paper 912413

CONTROL SYSTEM

Vehicle controls are all operated through one single programmable computer. This computer comes from Vesta and is programmable with a lap-top computer using the Forth programming language. The Forth language is similar to machine language in that it offers the fast processing necessary for real time control. Software has been developed and tested on the Viking 21 drive system with success.

Picture 17. Computer Control System

The computer control system consists of a main ECU, driver LCD readout, and a relay board. The system is modular and can be expanded to cover additional systems from as simple as turn signals to the complete drive system coordination.

The control of the start up procedure is done by the ECU. When the driver inserts to key the 200 volt relay is switched on. The ECU is powered up by a DC to DC converter from Vicor. DC to DC converters are used to power up all the 12 and 24 volt accessories in the vehicle. The ECU actuates the soft-start relay which sends power through a resistor before going to the controller, which prevents the full buss voltage from instantly entering the capacitive circuits in the Unique Mobility controller. The LCD prompts the driver to select forward or reverse mode. When the direction is selected the motor controller is enabled and the logic circuit is switched on. The vehicle is ready to go at this point. The start up procedure takes approximately 5 seconds.

Picture 18. VICOR DC to DC Converter

All driver input to the system is received through the accelerator and brake pedal. The accelerator communicates need for power from the driver through a linear potentiometer

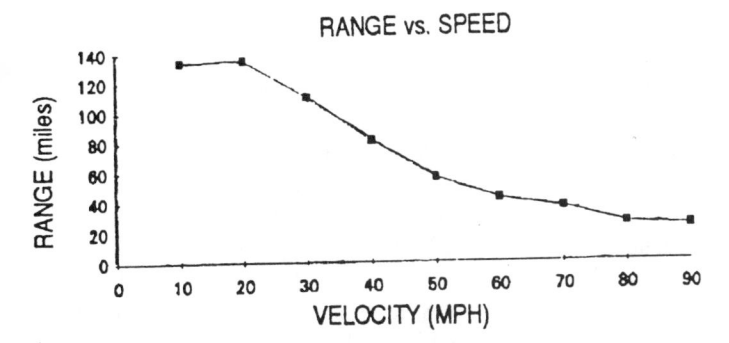

Figure 14. Range On Electric vs. Speed Graph

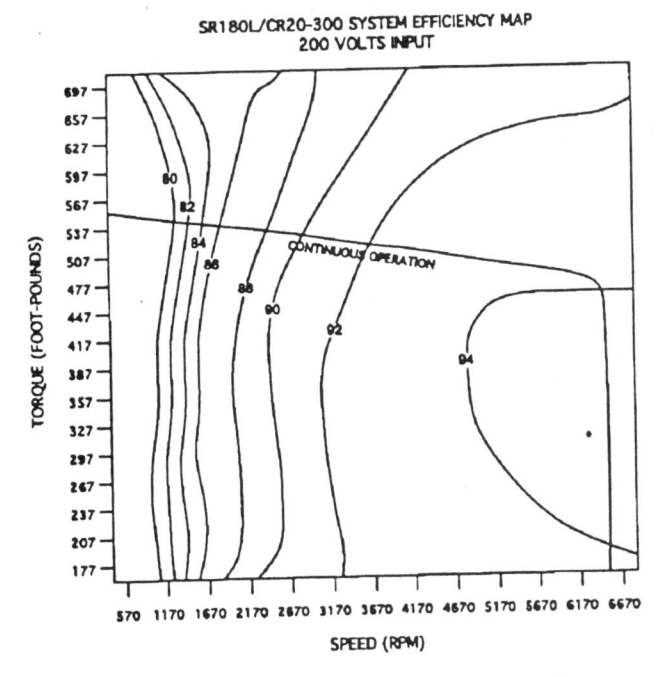

Figure 15. Unique Motor Efficiency Graph

and a microswitch that activates when the accelerator is pushed to the floor. The electric motor controller gets speed information directly from the ECU. The I.C. engine throttle is controlled with a stepper motor. The stepper motor has dual safety springs and the current to the stepper motor is cut directly via the emergency battery pull off, located inside the cockpit and out side the body. This "fly-by-wire" system gives the computer complete control of the drive system output.

When pulling away from rest the vehicle will use electric drive only, except when the accelerator is floored. When the accelerator microswitch is depressed the vehicle delivers the maximum power of both power units. This is analogous to the downshift of an automatic transmission. The vehicle will provide brisk performance on electric only, but the additional performance of the I.C. engine will provide power for passing and accident avoidance.

The vehicle provides adequate range on electric only to make most short trips. Even at speeds of 50 mph the range is

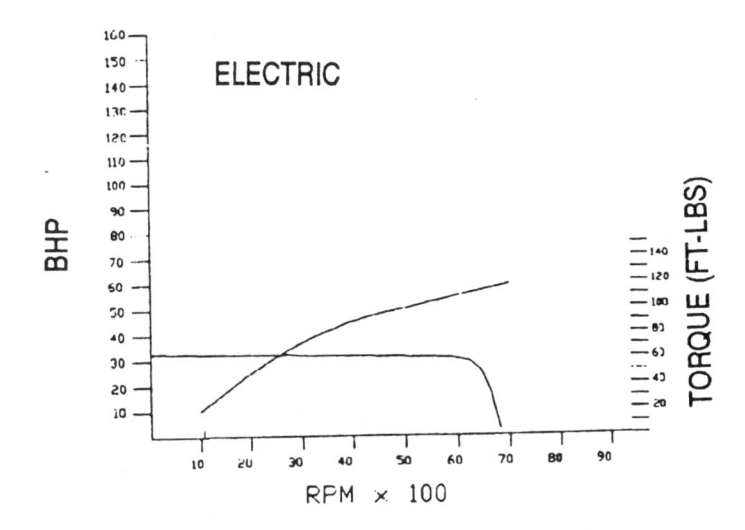

Figure 16. Electric Power and Torque

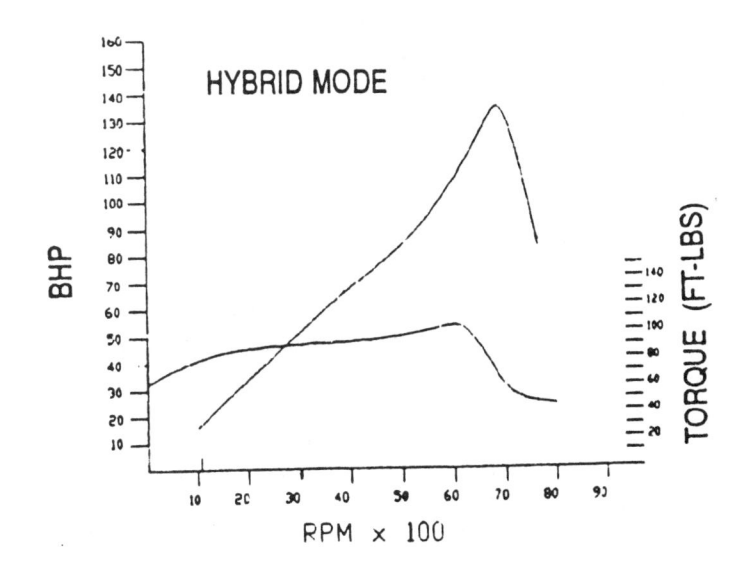

Figure 17. Hybrid Power and Torque

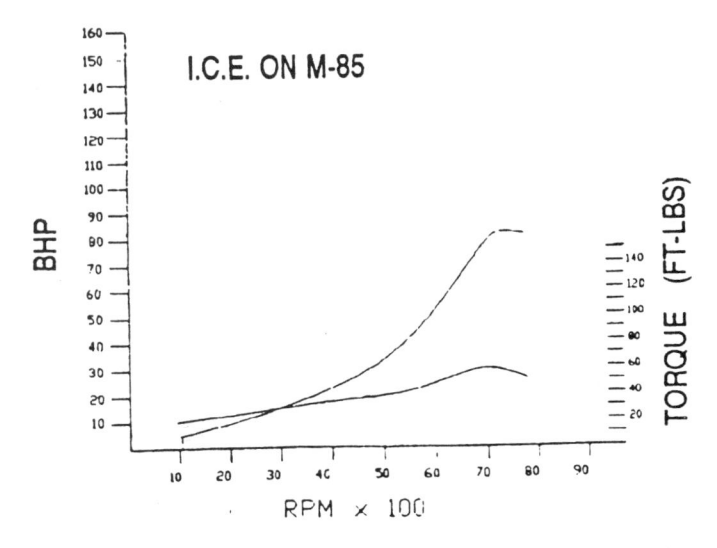

Figure 18. Internal Combustion Power and Torque

56 miles and approximately 100 miles at city speeds. Driving as a pure electric offers the lowest possible emissions in critical areas. The vehicle will never be driven to it's maximum range on pure electric drive, as this implies running the battery flat, which is detrimental to the battery life. The depth of discharge (DOD) has a direct relation on the life of the battery. In order to achieve extended battery life, the battery will never be drawn below 80 percent DOD in any situation. The range as an electric allows the vehicle to be fully functional when operating within a zero emission zone without depending on the I.C.E. .

A navigation computer will be interfaced with the ECU to provide information about zero emission vehicle zones or ZEV zones. Global positioning satellites can locate the vehicle accurately within 50 feet. The navigation software will be updated as legislation dictates new ZEV zones. The navigation computer will also help drivers find their destination accurately, saving power and reducing emissions.

Outside of ZEV zones the vehicle will operate as a hybrid. Hybrid operation is largely dependent on the battery state of charge. The state of charge is determined by a Brusa amp-hour meter, which measures the current flow in and out of the battery. This information will be stored in the RAM memory of the ECU. The amp-hour reading will be crossreferenced with the battery voltage to obtain an accurate estimate of the state of charge. When the battery is fully charged there is a slight rise in the nominal voltage. The amp-hour counter is reset at this point. Stored battery information will be available to service personnel for review.

The vehicle will operate as a pure electric at all times except maximum acceleration during the top 20 percent of charge. This allows short trips to be made with virtually no emissions. This is especially beneficial to HC and formaldehyde emissions due to not having to go through cold start. Recharging efficiency drops as the battery reaches capacity, so it is not efficient to keep the battery topped off constantly. After the battery has dropped below the 20 percent mark the vehicle will operate as a pure electric, pure I.C. , or in a recharging mode. At low speeds the vehicle will operate as a pure electric until the battery reaches 80% DOD. Ni-cad batteries can be consistently drawn down to 80% DOD without adversely affecting their service life. At this discharge level the vehicle would start the I.C.E. and continue to it's destination. Since most low speed trips are in the city and the distances are usually short the I.C.E. would be rarely used to complete a trip. Aerodynamic drag starts to rise significantly at 30 mph for this vehicle, but total system efficiency rises partially balancing the increased drag. At speeds above 40 mph range drops off such that is not practical to operate at a sustained duration. The ECU will switch over to I.C.E. operation at speeds above 40 mph. This operation will not take place instantly to insure smooth operation a digital timer will signal the ECU to switch after 10 seconds, this allows for electric operation in stop and go traffic without having the engine operate for short bursts. Short operating times for the I.C.E. are very poor for emissions, because the catalyst is not lit off and the engine operates in a open loop mode. The I.C.E. operation gives the vehicle the ability to operate at high speeds for long duration. Highway operation is where the I.C.E. is most efficient and caused the least problems with emissions. When operating between cities it will be desired to operated as ZEV in both the city of origin and the city of destination. During highway operation the vehicle will recharge the battery using the regenerative capability of the electric motor. The computer will send a voltage between 0 and -10 volts to the controller, where -10 volts is maximum regenative. The regeneration rate will be adjusted depending of the level of discharge in the battery. The limiting factor will be the load induced on the I.C.E.. Loading the engine allows the engine to operate at greater efficiency due to lower throttling losses, but NOx is a problem under high engine load. Dynamometer testing will be done to determine the maximum regeneration without compromising NOx emissions.

ENVIRONMENTAL BENEFITS

Fifty percent of the petroleum used in the United States today is imported, adding to the trade deficit. Sixty four percent of that was used in transportation.* Even with improvements in fuel efficiency of petroleum our petroleum dependence will increase. Use of non-petroleum base fuels, such as natural gas and alcohols is one partial solution. Yet another partial solution is electric vehicles. A hybrid vehicle will incorporate both of these solutions.

The vehicle described in this paper will be able to operate as pure electric in sensitive intercity areas shifting the emissions to power plants. Power plants are not generally in urban areas and many power plants such as hydroelectric or nuclear plants do not contribute to air pollution. Shifting the problem does not completely solve the emission problem, but it makes it more manageable. Power plants can be held accountable for their emission much easier than in individual vehicles. Power plants are that use combustion operate using continuous external combustion, unlike batch cycle engines, they operate at a much lower average temperature. NOx pollution is function of high temperature combustion and is not a problem in external combustion. Utility generators are also very low on carbon monoxide and hydrocarbons. Nearly one third of today's power plants do not create emissions.

- 55% Coal
- 10% Natural Gas
- 4%   Oil
- 10% Hydroelectric
- 21% Nuclear**

When this vehicle does use its I.C.E it operates in situations that are best suited for the lowest emissions. The I.C.E. does not have to run at idle, deceleration, or in many instances where the engine cycles through the complete rev range. The Dunstan Drive gives the vehicle an added advantage over conventional I.C.E. vehicles. It allows the I.C.E. to be optimized for any speed or load which improve both emissions and fuel consumption. Because of the engines small size it also uses less oil and coolant, therefore there is less consumption and less hazardous waste.

The design takes in to account for the vehicle total cost on the environment. Environmental costs, including the effects of manufacturing, vehicle recycling, and environmental issues were major factors in material selection. Another major concern was design for disassembly (DfD).

This vehicle will have the ability to positively change the environment without compromising the consumer's needs for safety, performance, range, drivability, and comfort.

---

*Hybrid Vehicle Program Plan Nov 1992, Office of Transportation Technologies Conservation and Renewable Energy, U.S. Dept of Energy
**S/EV 91 Proceedings, October 26-27

APPENDIX A • DUNSTAN DRIVE
Infinitely Variable Differential Hydrostatic Transmission
Reprinted with permission of Phil Dunstan

Construction

The main drive body (1) is constructed of eleven separate aluminum sections which are bolted together (see sec. B-B). The main drive body contains eight pistons (2) connected to a fixed angle swashplate (3) which is fixed to output shaft (4). The main drive body, also, contain 8 pistons (5) connected to variable swashplate (6) which is connected to case (7) by trunnion pins (8). The variable swashplate pivots on pins (8) to vary the swashplate angle which varies the hydraulic displacement of the unit. The main drive body, also, contains the cylinder barrels (9), piston seals (10), fluid reservoirs (11) & (12), fluid flow passages (13), valve ports (14), and bearing mounting (15) for the output shaft (4).

The fixed displacement unit and the variable displacement unit are interconnected through common reservoirs (11) & (12) and both function as either pump or motor depending upon the operational mode.

Operation

The drive operates in four different modes.
1. Neutral
2. Direct drive
3. Speed reduction/torque multiplication
4. Reverse

Neutral

In the neutral the variable angle swashplate is adjusted to equal the angle of the fixed displacement unit. In this position the fluid displacement of both units are equal.
As input shaft (16) is rotated by the motor, the main drive body which is fixed to it, also, rotates. The variable unit which has its swashplate fixed to the case, functions as a pump. Pumping fluid from reservoir (11) to reservoir (12), the pressure developed in reservoir (12) drives the fixed displacement unit as a motor which discharges fluid back to reservoir (11). Since the fluid displacement of the fixed and variable units are the same, output shaft (4) is driven at the same speed with respect to the main drive unit as the speed of the input shaft with respect to the case. The valving is such that the output shaft is rotating in a reverse direction, resulting in the output shaft having zero speed with respect to the case.

Direct drive

In this mode the variable swashplate is adjusted to a position perpendicular to the shaft which maintains the pistons in a fixed axial position, and there is no displacement of fluid as the main drive body is rotated. Resistance torque at the output shaft results in the fixed displacement unit functioning as a pump, pumping fluid from reservoir (11) to reservoir (12). Since return of the fluid to reservoir (11) is blocked by the variable unit, the fixed unit is hydraulically locked and all energy is transmitted mechanically to the output shaft.

Reduced speed/increased torque

In this mode the variable swashplate can be at any desired position between neutral and direct drive.

The following explains one half speed as an example. The angle of the variable swashplate is adjusted such that the fluid displacement per revolution is one half that of the fixed displacement unit. This allows the variable unit, which is functioning as a motor to discharge to reservoir (11) one half the fluid the fixed displacement unit is capable of pumping. Therefore, the fixed displacement unit output shaft is hydraulically forced to make one half revolution for every full revolution of the input shaft. The fluid under pressure which is being displaced from reservoir (12) to reservoir (11) through variable displacement unit acting as a motor produces a torque between the main drive unit and the case (which is the interconnections of the swashplate) which is equal to and added to the input torque resulting in twice the input torque being available at the output shaft.

Reverse

In reverse mode the variable swashplate is adjusted to an angle such that the displacement of the variable unit is greater than the displacement of the fixed unit.

This results in the output shaft being driven in a reverse direction with respect to the main drive body at a speed faster than the speed of the input shaft with respect to the case, resulting in the net speed of the output shaft with respect to the case, being in a reverse direction.

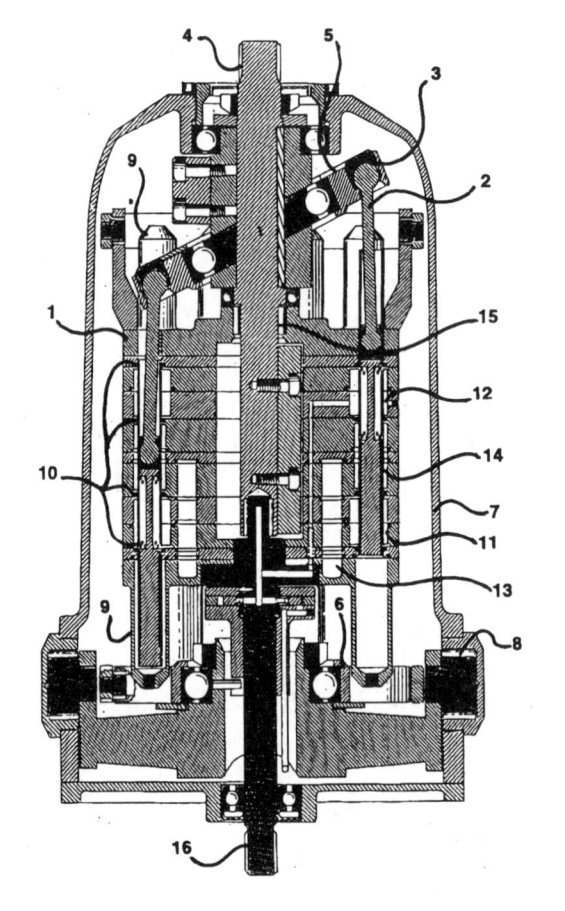

APPENDIX B • 0-60 CALCULATIONS

0-60 ELECTRIC

TORQUE (ft.lb)	MPH	MOTOR RPM	OVERALL RATIO	TORQUE Ft lb	THRUST (lb)	DRAG (lb)	P	N/V	MASS FACTOR	M_1	$a = P \times g M_1$
65	10	2753	18.48	1201	1281	-13.85	1256	274	2.15	4104	1.484
65	20	5507	18.48	1201	1281	-17.84	1263	274	2.15	4104	1.489
58	30	6600	14.77	857	913.7	-23.7	890	220	1.7	3245	1.670
58	40	6600	11.07	642	684	-32.3	652	165	1.3	2482	1.745
58	50	6600	8.80	513	546	-43.4	503	132	1.2	2291	2.088
58	60	6600	7.38	428	456	-56.8	400	110	1.1	2100	2.410

10.886 sec

0-60 INTERNAL COMBUSTION ENGINE

TORQUE (ft.lb)	MPH	MOTOR RPM	OVERALL RATIO	TORQUE Ft lb	THRUST (lb)	DRAG (lb)	P	N/V	MASS FACTOR	M_1	$a = P \times g M_1$
21	10	2753	18.48	388	414	-13.85	386	274	2.20	4200	4.98
41	20	5507	18.48	758	809	-17.84	791	274	2.20	4200	2.43
65	30	6600	14.77	960	1024	-23.7	1000	220	1.8	3341	1.53
65	40	6600	11.07	720	768	-32.3	735	165	1.4	2672	1.7
65	50	6600	8.80	575	614	-43.4	570	132	1.3	2482	1.99
65	60	6600	7.38	480	511	-56.8	455	110	1.2	2291	2.306

14.93 sec

0-60 HYBRID

TORQUE (ft.lb)	MPH	MOTOR RPM	OVERALL RATIO	TORQUE Ft lb	THRUST (lb)	DRAG (lb)	P	N/V	MASS FACTOR	M_1	$a = P \times g M_1$
86	10	2753	18.48	1589	1695	-13.85	1681	274	2.20	4200	1.6
106	20	5507	18.48	1958	2089	-17.84	2071	274	2.20	4200	1.6
223	30	6600	14.77	1817	1938	-23.7	1914	220	1.8	3341	1.29
223	40	6600	11.07	3262	1452	-32.3	1420	165	1.4	2672	1.02
223	50	6600	8.80	1082	1162	-43.4	1119	132	1.3	2482	1.09
223	60	6600	7.38	908	968	-56.8	911	110	1.2	2291	1.15

7.73 sec

APPENDIX C • SPECIFICATIONS I

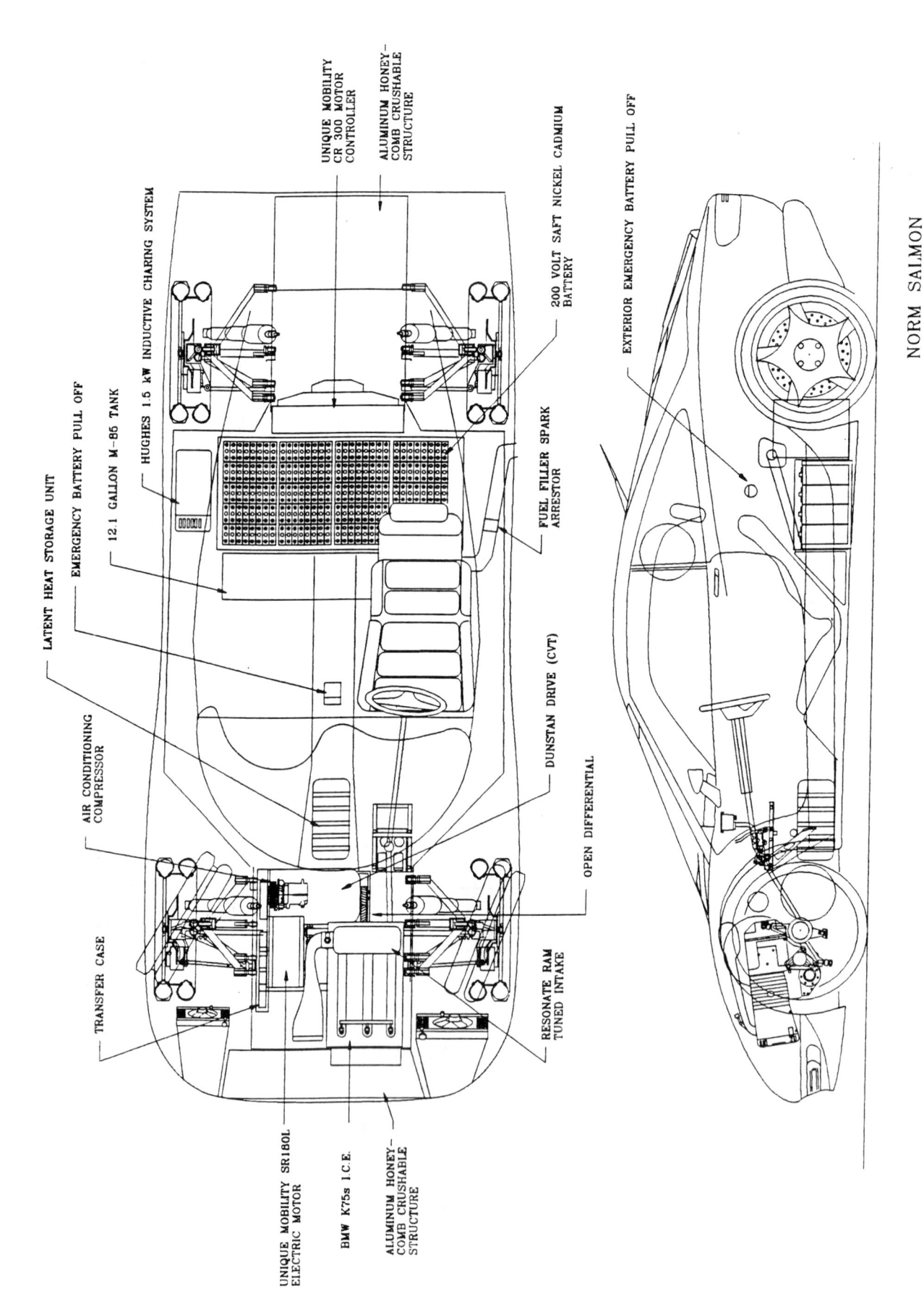

APPENDIX D • SPECIFICATIONS II

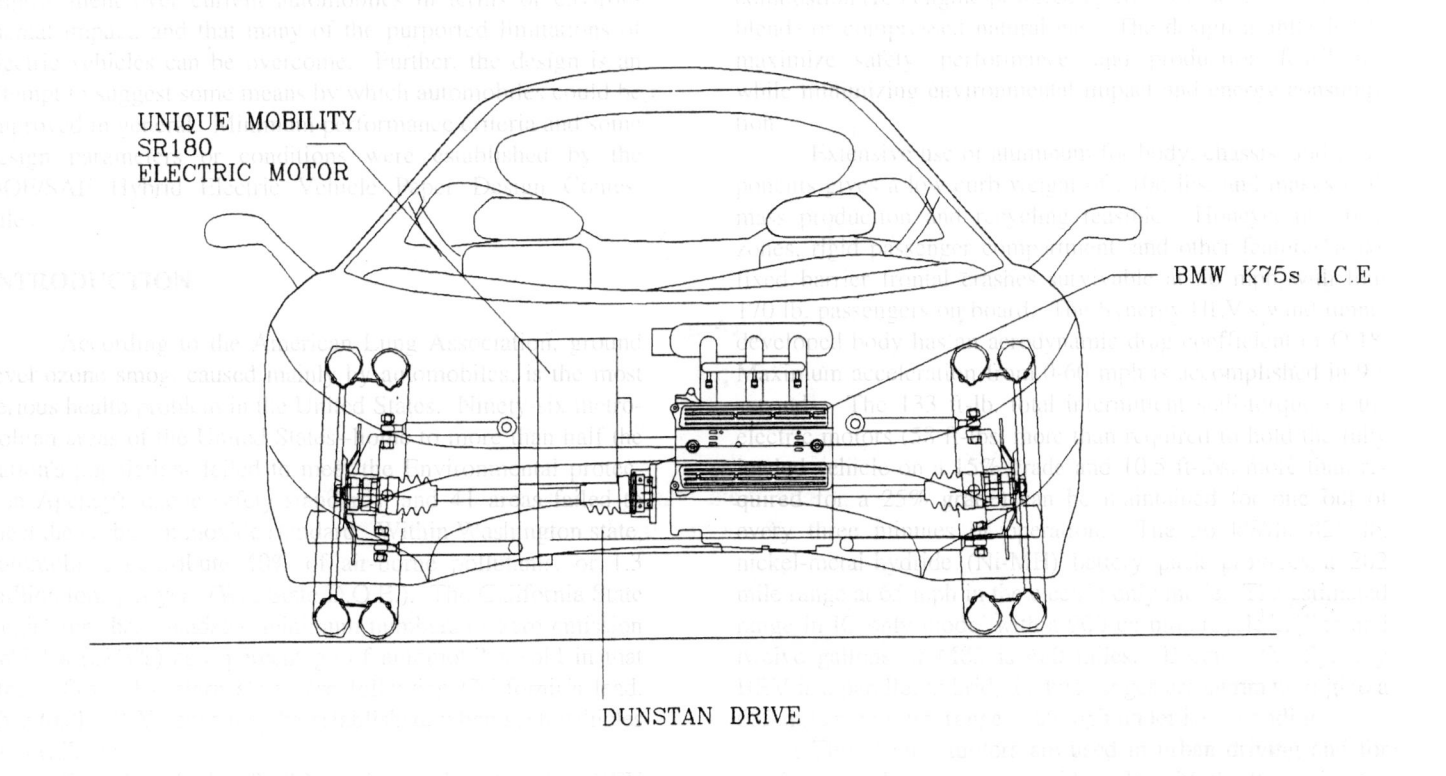

UNIQUE MOBILITY
SR180
ELECTRIC MOTOR

BMW K75s I.C.E

DUNSTAN DRIVE

APPENDIX E • SPECIFICATIONS III

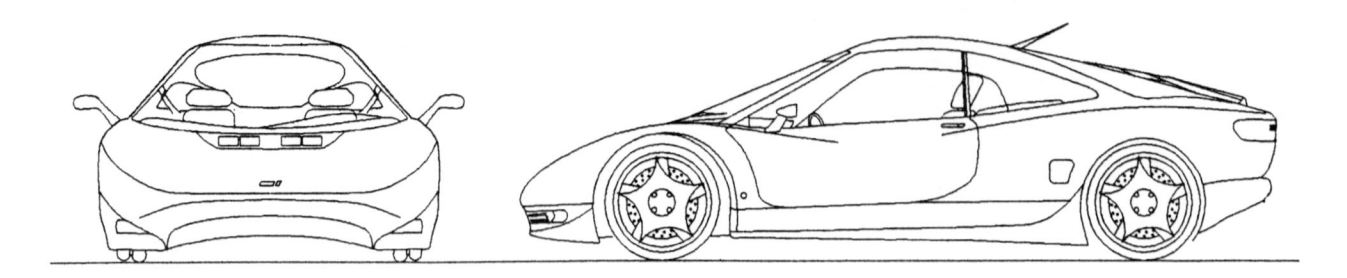

ENGINE:
    BMW K750:   Transverse, all aluminum, 3 cylinder, dual over head cam, 2 valve per cylinder.
    CAPACITY: 750cc, bore and stroke 67mm X 70mm
    FUEL & IGNITION: 3 accel EFI systems, 1 oxygen sensors per cylinder., individual detonation sensors, multi-point sequential timed injection, direct high energy ignition (80 Kw)
    MOTOR: UNIQUE MOBILITY INC. SR 180L /CR20-300 at 200V watercooled, 3 phase brushless DC.
    MAX HP: @61 at 7000rmp (1 min overload)
    MAX TORQUE: 47 ft/lb. at stall
    MAX TORQUE: 67 FT/LB. at intermittent stall
    WEIGHT: 52 lb.
    MOTOR CONTROLLER: UNIQUE MOBILITY CR300 watercooled
    WEIGHT: 48lb.

TRANSMISSION: DUNSTAN DRIVE differential hydrostatic

SUSPENSION: DUAL TIRE SYSTEM parallel link A arms with Bima hydraulic cylinders, carbon fiber leaf springs.

STEERING: RACK & PINION

BRAKES: BMW ABS aluminum metal matrix composite disks.
    UNIQUE MOBILITY regenerative braking.

WHEELS & TIRES: CARBON FIBER DUAL TIRE
    INSIDE: Yokohama WT 995 2.5x17in.
    OUTSIDE: Michelin Comp K 2.5x17in.

DIMENSIONS:
    OVERALL LENGTH   145 in.
    WHEEL BASE   92 in.
    HEIGHT   46 in.
    WIDTH   63.2 in.
    F TRACK   51.5 in.
    R TRACK   48.6 in
    DRIVER EYE LEVEL   34.4in (87.3cm)
    CURB WEIGHT   1759 lb.
    PERFORMANCE: 0-60 (ELEC) - 10.886 sec
        0-60 (IC) - 14.93 sec.
        0-60 (HYBRID) - 7.75 sec.

FUEL: M-85

BATTERIES: SAFT NICKEL-CADMIUM

RANGE: In Hybrid mode – 553.75 miles at 55 mph
    In electric mode – 110 miles at 30 mph
    or 45 miles at 55 mph

APPENDIX F • ADDITIONAL PICTURES

Picture E-1. Side view after crash test

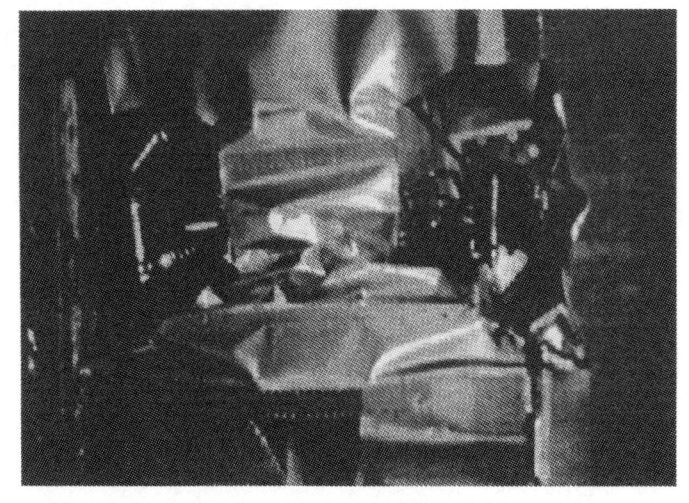

Picture E-2. Front view after crash test

Picture E-3. Model 3/4 side view

Picture E-4. Model 3/4 front view

Picture E-5. Model front top view

Picture E-6. Model side view

APPENDIX G • ACKNOWLEDGEMENTS

Co-authors, Norm Salmon and Brady O'Hare, wish to acknowledge the following people for their help and support while writing "Advanced Parallel Hybrid Vehicle Design."

- Dr. Michael Seal
- Eileen Seal
- Nate Rodriguez
- Jack Bobbin
- John Arbak
- Brian Van Kleeck
- Jeff Van Kleeck
- Bill Roe
- Gavin Cambell
- Hilary Noe

The Synergy HEV:
Design for a Hybrid Electric Vehicle

Timothy Moore
Vehicle Research Institute,
Western Washington University,
Bellingham, WA

December 1992

ABSTRACT

This paper describes the design for a hybrid electric vehicle (HEV) with the intent to show that HEV's would be an improvement over current automobiles in terms of environmental impact, and that many of the purported limitations of electric vehicles can be overcome. Further, the design is an attempt to suggest some means by which automobiles could be improved in general. Minimum performance criteria and some design parameters or conditions were established by the DOE/SAE Hybrid Electric Vehicle Paper Design Contest rules.

INTRODUCTION

According to the American Lung Association, ground level ozone smog, caused mainly by automobiles, is the most serious health problem in the United States. Ninety-six metropolitan areas of the United States--home to more than half the nation's population--failed to meet the Environmental protection Agency's ozone safety standards, and 41 areas failed to meet the carbon monoxide standard. Within Washington state, automobiles contribute 40% of air-borne pollutants, or 1.3 million tons per year (WA State D.O.E.). The California State Legislature has mandated minimum numbers of zero emission vehicles (ZEV's) as a percentage of automobiles sold in that state. Several eastern states are following California's lead. Eventually ZEV zones may be established, when such vehicles are available.

Electric vehicles (EV's) are best suited to fulfill ZEV requirements, as EV technology and required infrastructure are currently the most developed alternative. Because of their high efficiency, EV's also have the potential to dramatically reduce energy consumption and redistribute automotive energy demands to a diverse array of sources, including renewable ones. The most evident drawbacks of EV's have been limited range and performance, long charging times between uses, and insufficient safety features.

The Synergy HEV design overcomes these limitations by combining the fast fuel-up characteristics of an internal combustion engine with high-efficiency low-drag EV design and impact-absorbing light-weight materials.

SUMMARY

The Synergy HEV is a four passenger, parallel-hybrid electric vehicle utilizing two electric motors and one internal combustion (IC) engine powered by M85 (a methanol-gasoline blend) or compressed natural gas. The design is intended to maximize safety, performance, and production feasibility, while minimizing environmental impact and energy consumption.

Extensive use of aluminum for body, chassis, and components gives a low curb weight of 2106 lbs. and makes both mass production and recycling feasible. Honeycomb crush zones, rigid passenger compartment, and other features make fixed barrier frontal crashes survivable at 56 mph with four 170 lb. passengers on board. The Synergy HEV's wind tunnel developed body has an aerodynamic drag coefficient of 0.18. Maximum acceleration from 0-60 mph is accomplished in 9.3 seconds. The 133 ft-lb. total intermittent stall-torque of the electric motors (58 ft-lbs. more than required to hold the fully loaded vehicle on a 15% grade and 10.5 ft-lbs. more than required for a 25% grade) can be maintained for one out of every three minutes of operation. The 26 kW/h, 825 lb. nickel-metal-hydride (Ni-MH) battery pack provides a 202 mile range at 65 mph in the electric only mode. The estimated range in IC only mode, with a 900 cc motorcycle engine and twelve gallons of M85 is 480 miles. Because the Synergy HEV is a parallel hybrid, the two ranges are summed to give a 682 mile maximum range at 65 mph under ideal conditions.

The electric motors are used in urban driving and for maximum performance in hybrid mode, with the IC engine for higher speeds (above 40 mph), acceleration, hill climbing, and extended travel. This parallel hybrid drive system is optimized for the characteristics of both the electric and IC power plants; the electric motors maintain near peak efficiency in urban conditions, while the IC engine utilizes high energy density fuels for increased loads and range at freeway speeds. There are none of the inefficiencies associated with converting and storing the output of the IC engine, as in a series hybrid, since it is used to drive the rear wheels directly. An on-board geographic positioning system (GPS) maintains the electric only mode while the vehicle is in established zero-emission-vehicle (ZEV) zones, making the Synergy well suited to strict emissions requirements.

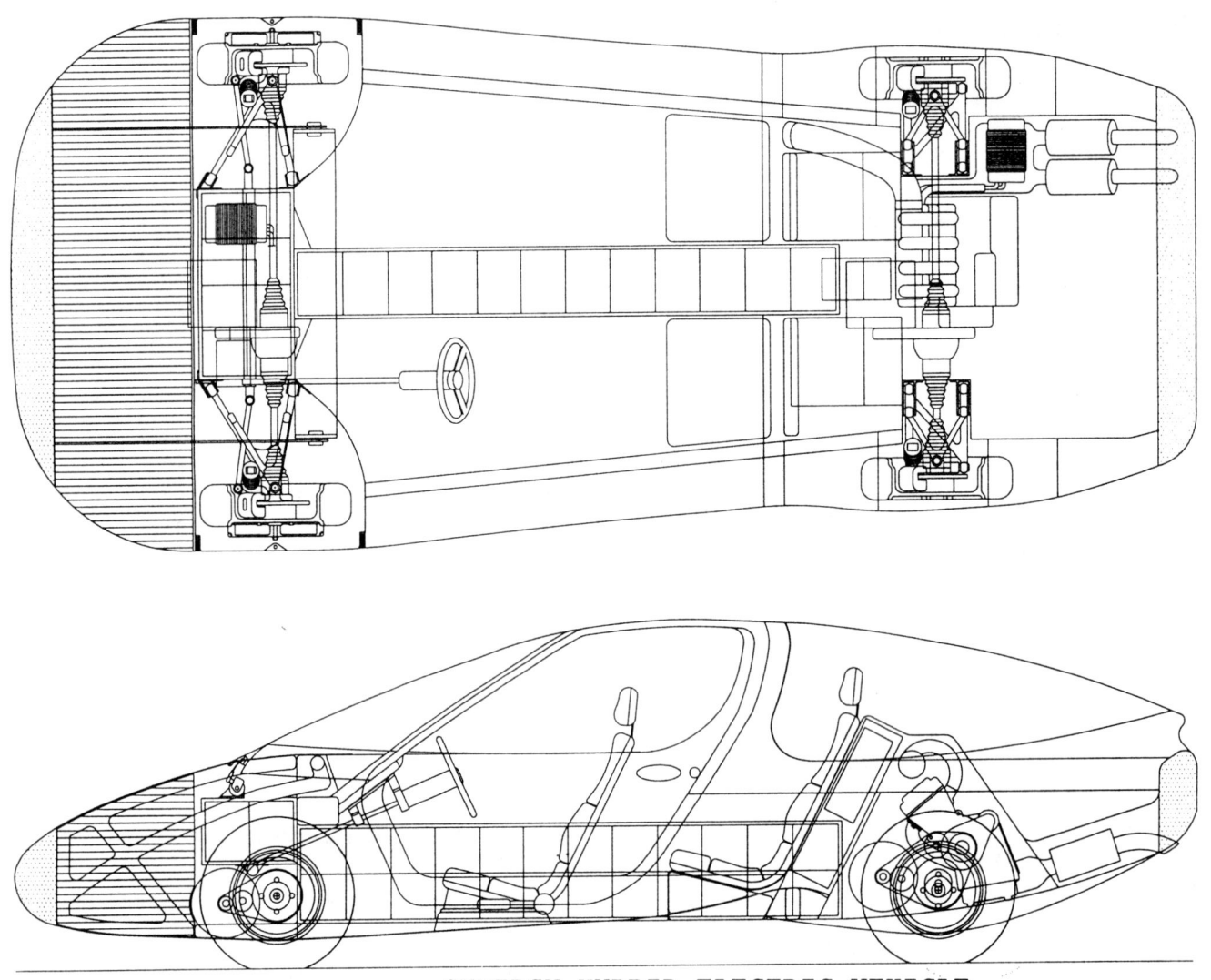

SYNERGY HYBRID ELECTRIC VEHICLE

COMPONENT LAYOUT

Electric motors drive both front and rear axle differentials through constant-mesh helical gears. Driving the front and rear axles through differentials eliminates the potential for torque-steer from a motor failure associated with dual motor configurations, which provide differential action by independently driving opposing wheels at one end of the vehicle. An additional benefit is the availability of full-time four wheel drive for foul weather and road conditions. Controllers for each of the electric motors are located immediately behind the rear seat, where they would not likely be damaged in the event of a minor collision. The battery packs for the electric drive system are located in the center tunnel and above the front electric motor. The heat exchanger for motor and controller cooling is located in a NACA duct, which rises under the passenger side rear seat. A latent-heat battery (LHB) for defrost assist is located forward of the glove compartment (in series before the electric drive heat exchanger).

The IC engine drives the rear axle differential through a constant-mesh helical gear with an overrunning clutch on the extended engine crank shaft. One reason for locating the IC engine behind the rear axle is that the space provided for it can be easily converted to cargo space if the vehicle is also to be produced as a pure electric. The heat exchanger (radiator) for the IC engine is located in a NACA duct which rises under the driver side rear seat. The IC engine warm-up assist LHB is located above the catalytic converter along side the engine (in series before the ICE heat exchanger). Two, six gallon fuel tanks straddle the cooling-system NACA ducts on either side of the center tunnel under the rear seats.

The multi-function electronic control unit (ECU) is located in a well protected area between the rear seat backs and close to the motor controllers, IC engine, and rear latent heat battery which interact with it. The windshield wiper system is housed below the hood line under a hinged aerodynamic cover flap with the blades resting on the extended lower edge of the windshield. The full-sized spare tire lies flat over the jack and lug wrench in a recess under the cargo space.

477

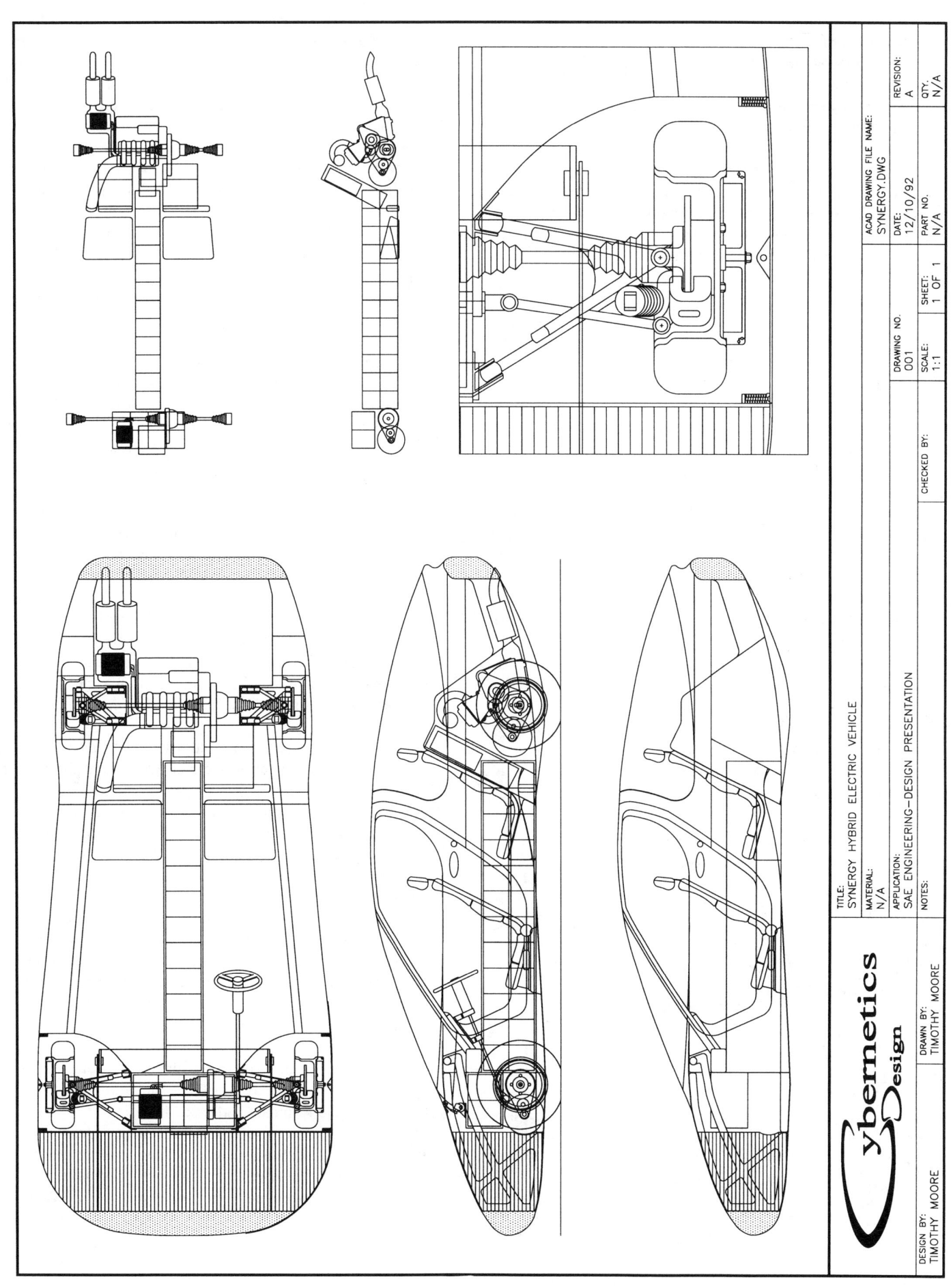

POWER PLANTS

Electric Motors:

Unique Mobility SR180L, liquid cooled, brushless, DC, motors with laminated stators were chosen for their relatively high power to weight ratio and high efficiency. The combined intermittent (1 min. per 3 min.) peak power of the electric drive system is 136 bhp @ 6,500 rpm with a system voltage of 200 volts. The motor controllers limit power at excessive motor temperatures, control a cooling pump for both motors and controllers, and provide regenerative braking in addition to controlling motor speed. Performance specifications supplied by the manufacturer are as follows:

Rated Power:
Continuous @ 6,500 rpm .... 35.0 kW
Continuous @ 6,500 rpm .... 47.0 bhp

Peak Torque:
Continuous @ 6,500 rpm .... 38.0 ft-lb
Continuous Stall .... 46.6 ft-lb
Intermittent Stall (1 min / 3 min) .... 66.6 ft-lb
Intermittent @ 6,000 rpm (1 min / 3 min).. 54.0 ft-lb

Maximum No-Load Speed @ 200 Volts .... 6,850 rpm
Motor EMF Constant .... 29.0 keV / krpm
Motor Torque Constant .... 2.45 Kt in-lb / Amp
Maximum Winding Temperature .... 300 deg F
Winding DC Resistance l-l @ 25 deg F .... 0.028 ohm
Winding Inductance @ 25 deg C .... 40 µH
Number of Poles 18
Rotor Inertia .... 0.200 in-lb / sec^2
Motor Weight (liquid cooled) .... 54.0 lbs

Controller Maximum V input .... 200 Vdc
Controller Minimum V input .... 30 Vdc
Controller Minimum No-Load Current........ 0.150 A
Controller Input Capacitance .... 19,040 µF
Controller Weight (liquid cooled) .... 48 lbs

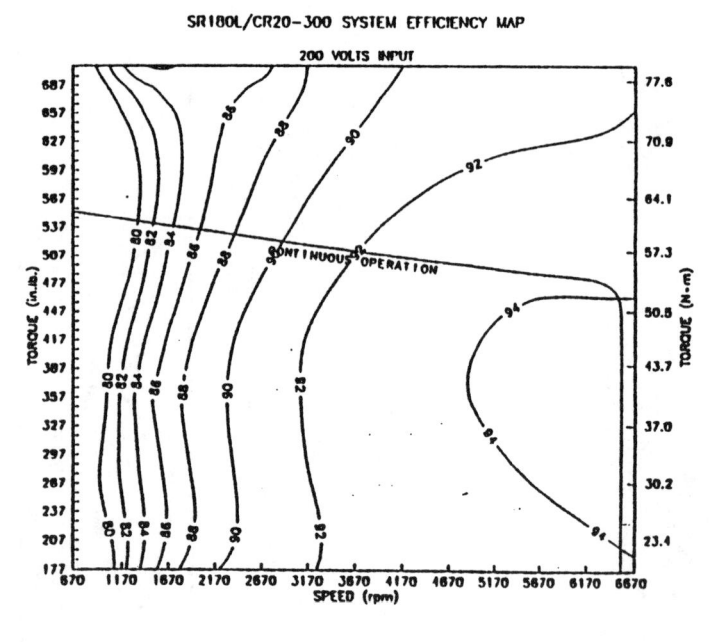

IC Engine:

The IC powerplant is a 900 cc, in-line, four cylinder, water cooled, DOHC, four valve per cylinder, Honda motorcycle engine, chosen for its high power to weight ratio. Rated power is 120 bhp @ 10,500 rpm, with a peak torque of 64 ft. lbs. @ 8250 rpm. The stock compression ratio is 11.0:1 with a bore and stroke of 70.0 and 58.0 mm, respectively. The complete engine and transmission unit weighs 147 lbs.

For this design the stock engine would be modified to shift the torque and horsepower peaks to a lower rpm range, reduce emissions, prepare the engine for methanol (M85 as specified by the DOE/SAE paper design competition rules) or alternatively natural gas combustion, improve the life expectancy of the engine, and remove the excess weight and bulk of the transmission. The following modifications would be made to accomplish these improvements:

- Regrind the camshaft to shorten valve duration, improve low rpm torque, lower the rpm at which torque and horsepower peaks occur, and cut output power by 20-40%, improving the engine's life expectancy. The anticipated range of best performance is 3,000 to 6,500 rpm.
- Raise the compression ratio to 14:1 (as permitted by M85), improving combustion efficiency.
- Install Haltech programmable fuel injection system to optimize fuel/air ratio over the entire rpm range of the engine.
- Install alcohol resistant injectors, stainless steel fuel lines, flash-chromed fuel pump, and molded plastic fuel tanks would eliminate the corrosive effects of M85. In ambient temperatures less than 40 degrees F, a tank warming jacket preheats the fuel (when vehicle is operating) to maintain M85 vapor pressure minimum of 0.1 in. bar.
- Provide for shut off of the fuel injection to the number two and three cylinders by the ECU for the extended range mode.
- Fabricate an intake manifold with dual tuned-length runners for resonant-ram effects at 3,960 rpm (50 mph) and 4,752 rpm (60 mph).
- Fabricate an exhaust manifold with tuned-length runners for a resonant-ram effect at 3,960 rpm (50 mph) and a close-coupled catalyst built into the collector.
- Install an exhaust gas recirculation system (EGR), three way catalysts with an oxygen sensor, and an electric airpump, each controlled by the electronic control unit (ECU), including feedback to the fuel injection system.
- Remove the secondary drive shaft, clutch, and transmission components, and cut off and cap the transmission portion of the case, dropping the weight from 147 lbs. to approximately 93 lbs.

The IC engine has a barium hydroxide heat storage system to speed its warm-up time, and runs on two of four cylinders in the automatic extended range mode. Using the IC engine only at freeway speeds will maintain improved levels of volumetric and thermal efficiency, eliminate engine idling, and reduce exhaust emissions. The GPS will prevent the IC engine from being operated in established ZEV zones.

GEAR RATIOS

A drive ratio of 5.29:1 for the electric motors was chosen based on the maximum useful motor speed of 6,512 rpm, maximum desirable vehicle speed of 82 mph, maximum anticipated hill grade of 25%, and maximum gross vehicle weight of 3,027 lbs (including four 170 lb passengers and 241 lbs. of cargo).

A fixed direct drive ratio of 5.28:1 for the IC engine was based on the estimated range of best performance after modification (3,000-6,500 rpm), the 40 mph hybrid mode minimum speed, the electronically limited maximum speed of 82 mph (to protect electric motors), and a typical cruising speed of 65 mph.

HILL CLIMBING ABILITY

**Torque Required to Hold Vehicle
(15% grade, with four 170 lb passengers):**

$$Tq = \frac{Wt * Sn * Rr}{De * Dr}$$

Where:
Tq = Torque required to hold vehicle (ft-lbs) = 74.7
Wt = Curb weight + four 170 lb passengers (lbs) = 2,786
De = Driveline efficiency = 0.97
Dr = Drive ratio = 5.29
Sn = Sine max. grade angle = 0.148
Rr = Drive tire radius (ft) = 0.93

Peak continuous torque available from electric motors is 93.2 ft-lbs total (18.5 ft-lbs. in excess of that required to hold the vehicle on a 15% grade). Peak intermittent torque at stall (1 min. per 3 min.) is 133 ft-lbs. This is 58 ft-lbs. more than required to hold the fully loaded vehicle on a 15% grade and 10.5 ft-lbs. more than required for a 25% grade.

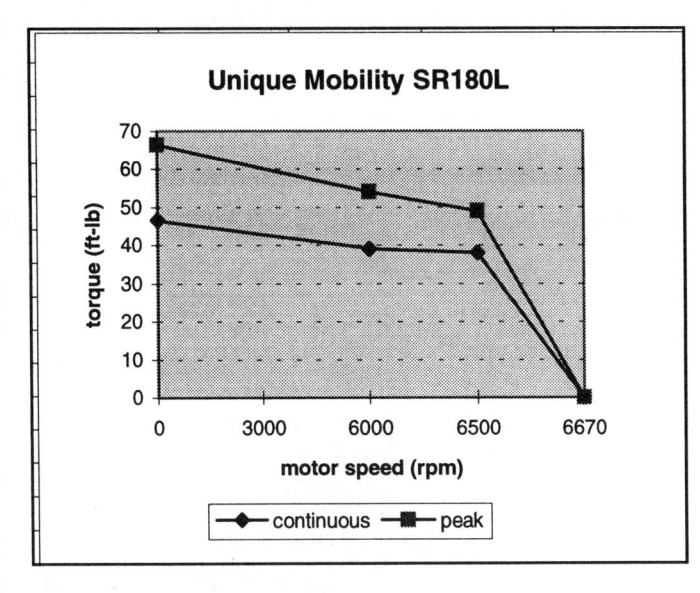

Unique Mobility SR180L

motor speed (rpm)	continuous	peak
0	47	67
3000		
6000	40	55
6500	39	50
6670	0	0

Estimated Curb Weight (lbs):

300	Chassis 200 lbs., body 100 lbs. including paint.
80	Aluminum honeycomb (40 lbs front, 20 lbs each side).
63	Glazing (see p.17 for materials)
80	Suspension, hub carriers, and brakes.
105	Wheels (8.5 lbs. ea.) and tires (12.5 lbs. ea.).
72	Differentials, axles, CV joints, and steering.
204	Electric motors and controllers.
93	IC engine (transmission removed).
825	Battery pack (34 cells @ 24.25 lbs. ea.).
102	Full twelve gallon polymer fuel tank (six lbs. empty).
66	Seats (2 * 27 lbs. front, 2 * 6 lbs. rear).
16	Air bag systems
47	Cooling and heating systems with coolant.
14	Latent heat storage batteries
24	Fuel injection, electronics, and wiring.
15	Foam bolsters, upholstery, and carpet.

2,106 Total Curb Weight

BATTERIES

The 26 kW/hr, 204 volt battery pack consists of 34 Matsushita nickel-metal hydride (Ni-MH) batteries, and weighs a total of 825 lbs. As detailed in the vehicle range section of this paper, the batteries provide a maximum range of 202 miles @ 65 mph and 465 miles @ 30 mph. The 180 W/kg specific power (power density) of the batteries at full charge provides a total of 67,320 watts without a significant drop in discharge efficiency.

Graphic comparison of battery data (see table and graphs on next page) from tests conducted at Argonne National Laboratories and specifications supplied by Matsushita Battery Industrial indicate that the Matsushita Ni-MH batteries have the relatively better overall performance than other current near-term EV batteries. While, the Ovonic Ni-MH "H" cells have slightly better volumetric density and similar specific power relative to the Matsushita cells, the Ovonic cells have poorer energy density and function for less than one-third as many 100% DOD life cycles. Other batteries, such as Na-S, have excellent energy density, but lower levels of performance for the other parameters compared. The Matsushita cells are also designed for recyclability, which is a significant benefit in terms of environmental impact and battery long-run manufacturing costs. If the matsushita batteries were not available or did not perform well in independent testing, the Ovonic "H" cells would be the second choice.

Matsushita Battery Industrial supplied the following information:
Voltage 6V
Capacity .... 130 Ah
Specific Energy .... 70 Wh/kg
Volumetric Energy .... 140 Wh/l
Specific Power @ 20% charge .... 170 W/kg
100% DOD Life Cycles 1,500
Temperature Range .... 20 to 60 deg C
Dimensions (W,D,H) .... 6.3 x 6.3 x 8.25 in

"Energy density of 70 Wh/kg was achieved through increasing the charge ability of the nickel electrode under high temperatures, and improving the composition of the hydrogen absorbing alloy. With this high storage efficiency--twice that of lead storage batteries--the battery component of an electric vehicle can be made smaller and lighter."

"Through a new composition of electrodes, the battery can generate power of up to 180 W/kg, with 170 W/kg even at the end of the discharge."

"Through increased durability of the positive and negative materials and a dual pocket construction with a thin, heat resistant separator for extra protection, the life span of the battery has been extended to 1,500 charges. This eliminates the need for replacing the battery every two to three years, as required by lead storage batteries."

"The application of pressure resistant materials and special seal technology enabled Matsushita to fully close the battery, eliminating the maintenance."

"Not only is the battery itself friendly to the environment, the nickel and hydride substrates and the casing have been designed for recyclability."

The on-board charging system consists of a Hughes inductive charging system, and a regenerative braking system incorporated in the Unique Mobility motor controllers. Approximately 50% of the energy absorbed by the regenerative braking system (including drive train, motors, controllers, and batteries) is actually available for re-use in the form of tractive effort. This, as well as discharge efficiency during acceleration, could be improved with the use of a small pack of ultracapacitors (also called supercapacitors) to load level the battery. Tests conducted at Idaho National Engineering Labs have shown that ultracapacitors can provide round-trip charge/discharge efficiencies of about 95%, even at high current loads. Sufficient data on ultracapacitor performance for detailed design analysis was not, however, available at the time of this writing. The factors influencing regenerative braking are further discussed in the section of this paper devoted to vehicle range.

For service or replacement, the center-tunnel portion of the battery pack is lowered by a light weight system of aircraft cable and pulleys from the underside of the car. The forward portion of the battery pack is accessed by swinging the nose section of the vehicle upward (see component access section for further explanation).

COMPARISON OF BATTERY PERFORMANCE DATA

Battery	Wh/kg	Wh/l	W/kg	Life cycles
Matsushita Ni-MH	70	140	180	1500
Ovonic Ni-MH	55	152	175	380
Asea Brown Boveri Na-S	81	832	152	592
Silent Power Na-S	79	123	90	795
SAFT Li-S	66	133	64	163
SEA Zn-Br	79	56	40	334
Electrochemica Ni-Zn	67	142	105	114
Eagle Picher Ni-Fe	51	118	99	918

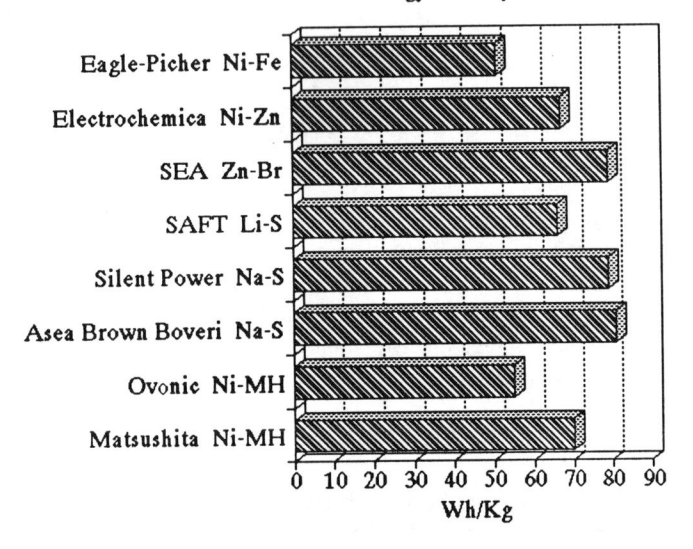

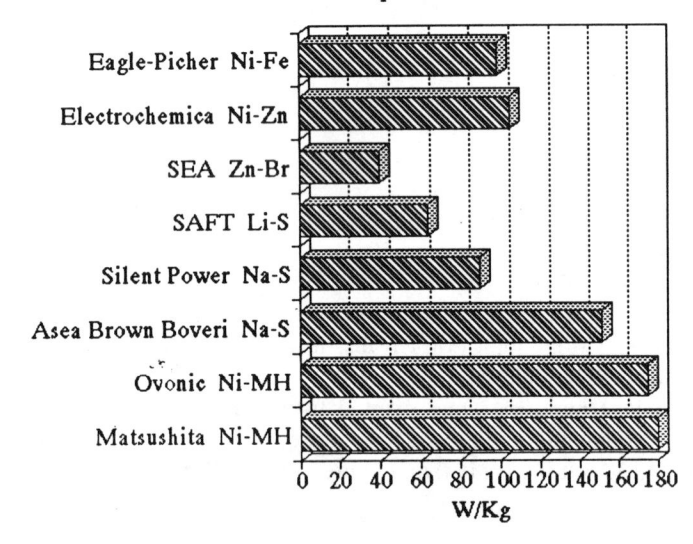

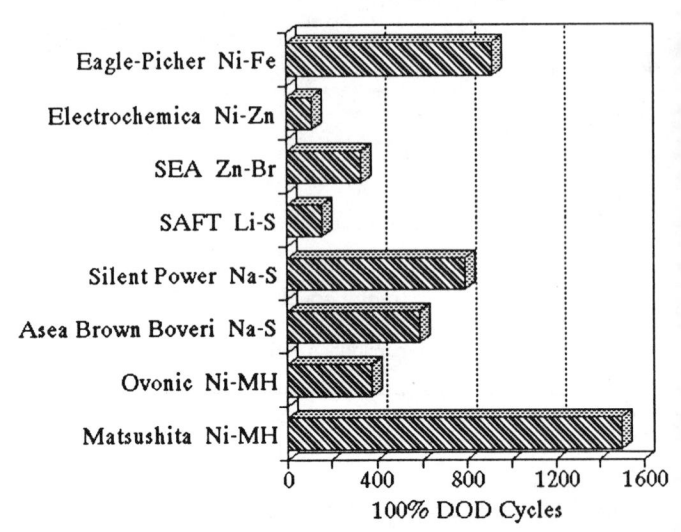

ROLLING RESISTANCE

The coefficient of rolling resistance (r_o) of the Goodyear P165/65R-14 low rolling resistance Aero Radial tires, inflated to 65 psi is 0.006 (Sam Landers, 1991, Goodyear Tire Co., Akron, OH.). With a gross vehicle weight of 2,786 lbs (including four 170 lb passengers) the total rolling resistance is 16.7 lbs force, requiring 2.89 bhp to overcome at 65 mph.

AERODYNAMIC DRAG

The coefficient of aerodynamic drag for the 1:10 scale model was determined using a fixed-ground tunnel at 100 mph. Horizontal forces were measured using a balanced lever-arm system and conductive resistance strain-gauge load-cell. The lever-arms terminated in a triangular configuration of airfoil cross-section posts supporting the model at ride height. An 0.125 in. wide piece of pin-striping tape was placed around the nose at 8.0 scale inches back from the stagnation point to trip the flow, simulating the likely loss of true laminar flow as a result of surface friction.

The horizontal force of aerodynamic drag for the model and standard block is the calculated average of fifteen scans of the strain gauge input (at one second intervals) by an electronic data logger. Vehicle model and standard block force measurements used for analysis were taken from data logged at matched dynamic pressures (+/-0.5 in. H2O).

Projected frontal area was determined by tracing the shadow of the model when placed with its nose against a vertical piece of 0.1 in. square ruled graph paper at 48 feet from a slide projection light source, and then summing the squares (and partial squares) within the outline.

INITIAL WIND TUNNEL MODEL WITH STREAMLINES

The initial 1:10 scale wind tunnel model registered a coefficient of aerodynamic drag (Cd) of 0.19. Streamline tests, using a powdered-chalk slurry, indicated turbulent high-pressure zones at the windshield base and before and after the front wheel opening (shown in the above photo).

Clay was added to move the windshield forward and increase its rake angle from 63 to 67 degrees, smoothing the transition from the hood, and to cover both the front and rear wheel openings (see photo on following page). Subsequent streamline tests detected no significant turbulence forward of the rear bumper except directly to the rear of the tire treads and where the tunnel apparatus connected to the under-side of the model. The clay modifications lowered the measured Cd to 0.176.

The measured Cd of 0.176 was adjusted to 0.18 to account for the estimated interference drag resulting from body panel seams, penta-prism flush-mounted side mirrors, and air intakes and outlets (see text below re: aerodynamic treatment of these items). The projected frontal area of the finished model is 31.66 in.^{2} (22 ft.^{2} full-scale) giving a full-scale CdA of 3.96. Thus, the design requires 7.41 BHP @ 65 mph and 0.73 @ 30 mph to overcome aerodynamic drag.

$$Cd = \frac{Fm / Am}{Fb / Ab} \qquad CdA = Cd * Am$$

$$bhp = \frac{CdA * V^3}{K}$$

where:
Cd = Coefficient of Aero Drag
Fm = Force on vehicle model (g)
Am = Frontal Area of model (ft^2)
Fb = Force on standard block (g)
Ab = Frontal Area of block (ft^2)
V = Velocity (mph)
K = 146,600 (constant)

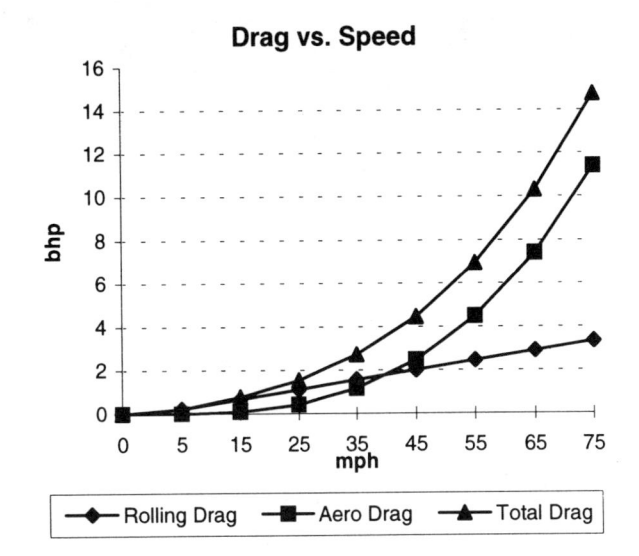

Drag vs. Speed

Speed (mph)	Rolling Drag (bhp)	Aero Drag (bhp)	Total Drag (bhp)
5	0.22	0.00	0.22
15	0.67	0.09	0.76
25	1.11	0.42	1.53
35	1.56	1.16	2.72
45	2.00	2.46	4.46
55	2.45	4.49	6.94
65	2.89	7.41	10.30
75	3.34	11.39	14.73

FINAL VERSION OF WIND TUNNEL MODEL

Consideration of critical factors affecting form drag in fast-back design included nose profile, chord-line angle of attack, and the angle of departure and profile of termination of the rear hatch and underbody. The principle goal was to maintain non-turbulent flow all the way back to the rear bumper in order to minimize the wake area.

The well-radiused nose parts the air without stall or eddies. The smooth transition from the sloping front hood to the 67 degree rake of the windshield almost completely eliminates high-pressure turbulent flow at the windshield base. The surface of the rear hatch slopes at no more than 12 degrees to maintain attached flow all the way back to the trailing edge. With its clean-cut upper edge, the fastback design approaches the aerodynamic efficiency of a fully extended knife edge at the rear. Maximum width is at the front axle, tapering a full 10 in. on each side (5.3 in. to the rear axle and another 4.7 in. to the rear bumper) to reduce turbulence in the wake. The absence of a front air dam allows for higher front-end ground clearance at the nose section. The smooth belly pan slopes upward toward the rear, giving the entire car a negative angle of attack, thereby matching the air flow velocities on the top and bottom of the car to minimize lift and lift-induced drag.

Meticulous attention was paid to each element of the design which might generate interference drag. Composite head lamps have flush mounted lenses. The window glass is flush mounted with drip rails hidden inside of the windshield pillars and door jams. The windshield wipers rest on the lower edge of the windshield glass below the surface level of the hood. A hinged panel keeps the resting wiper arms, and the depression in which they rest, out of the air stream when not in use. Door handles are completely flush with the body sides. Pressing in at the upper edge of the handle exposes the otherwise flush pull grip on the lower edge. All seams around body panels, window glass, lighting components, and door or hatch closures are filled with rubber seals to within 0.08 in. of completely flush (with the exception of the top side portion of the integral door-surround drip rails, which is parallel to the surface air flow).

All four wheel openings are covered by removable body panels. The 3.4 in. clearance between the front wheel cover panels at axle height and the tire sidewalls accommodates a 20 degree steering angle (27.6 ft. turning radius) without the operation of the wheel cover pivot mechanisms. For steering angles from 21 degrees to full-lock the cover panels swing outward at the front or rear seams, pivoting on a pin at their apex, and returning to rest via normally contracted miniature gas charged struts. Elastomeric polyurethane membranes with plastic zip closures prevent turbulence at suspension droop cut-away in the belly pan, while allowing full steering and suspension travel. The suspension droop membranes are held away from the rotating wheel by a tubular aluminum framed guard. The guard extends from the hub carrier around the leading and trailing edges of the front wheels to prevent the membrane from contacting the moving wheel. Extended front wheel hub caps, with thrust bearings in their outer edges, push the pivoting wheel cover panel outward during low speed maneuvering. Fairings fore and aft of the rear wheels part the air flow, and act as vertical underbody fins, moving the lateral center of pressure well to the rear of the center of gravity, to improve lateral stability in cross winds.

A small transverse oval inlet (14 x 2 in. projected frontal area) with generously radiused edges is located just below the nose, to provide fresh air for the electric motors, controllers, battery pack, and the passenger compartment. The heat exchanger (radiator) for cooling the IC engine, combustion air intake, and air conditioning condenser receive fresh air from NACA ducts under the rear seats and fuel tanks. Heated air from the motor/battery heat exchanger normally flows through the upper section of the center tunnel to the rear engine compartment, but can be partially or completely directed to the passenger compartment, where a ceramic positive temperature coefficient heater core and latent heat battery boost the temperature if desired.

A convex, flat-wrap, curved mirror, mounted above the windshield, serves as rear-view and side view mirrors combined. This approach takes advantage of the expansive rear and side window areas beyond the "B" pillar. Penta- prism side mirrors, which require only small streamlined bulges on the vehicle exterior to house a lens, and consequently would add very little aerodynamic drag, could be included to supplement the interior mirror and meet motor vehicle registration requirements. Alternatively, a small, streamlined driver's side mirror with an airfoil-section stalk could be used. Such mirror would be eight inches from the door surface to reduce disturbance of surface air flow.

The design avoids the cost and complexity of active ride height adjustment, which might further reduce aerodynamic drag at freeway speeds if the car weren't designed to match the relative air flow velocities for the upper and lower surfaces.

Exterior Dimensions (in):
Overall length = 173.7   Overall height = 51.0
Max. width at front axle = 76.0   Width at rear axle = 65.4
Ground clearance at low point (front axle) = 6.5

MODE CONTROL STRATEGY

The control strategy is designed to minimize urban vehicle emissions and maximize overall driving range, while providing limited options for mode control by the driver. The limiting factors are the location of the vehicle relative to ZEV zones, as determined by the geographic positioning system, and the state of charge in the battery pack. Up to 40 MPH, only the electric drive system is operational. Reverse is thus also electric only. Above 40 MPH the driver may select the hybrid mode if desired. The accelerator pedal will control the hybrid sub modes via the Electronic control unit (ECU), including low load electric drive cut-out and two-cylinder IC engine operation for the extended range sub-mode. Automatic disabling of hybrid mode in ZEV zones and automatic initiation of hybrid mode or low-battery charging, via low-level regen, at highway speeds will also be controlled by the ECU. The control strategy is intended to maintain enough reserve battery capacity to get the vehicle to a charging facility when entering a ZEV zone with a low battery state of charge.

The shift lever has the following three positions and user labeling:

CITY / HIGHWAY
0 TO 80 MPH
(electric mode)

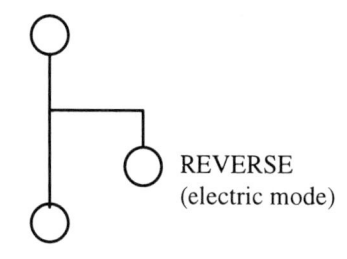

REVERSE
(electric mode)

HIGHWAY / EXTENDED RANGE
40 TO 80 MPH
(hybrid mode)

Overlapping effective accelerator pedal control ranges for the electric mode, hybrid mode, and extended range hybrid sub-mode are as follows (sw1-4 are microswitch locations at indicated percentages of pedal travel and directional orientation):

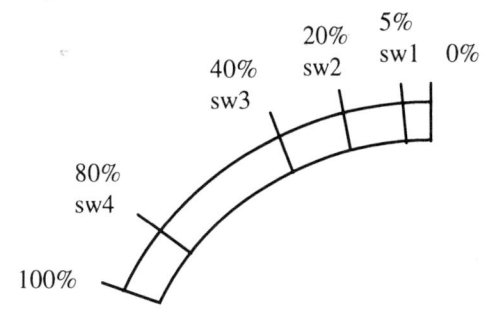

In the electric mode, acceleration is provided by all but the first 5% of pedal travel, which initiates low level regenerative braking or slight forward propulsion depending on vehicle speed and pedal positions. If the vehicle speed is greater than 5 mph, except when using cruise control, releasing the accelerator pedal initiates low-level regenerative braking as a means of simulating conventional engine behavior. If the vehicle speed is less than 5 mph, and the brake pedal is not depressed, slight forward motion or "creep" will take over, simulating a conventional automatic transmission. The latter mode is an important safety feature as it provides a kinesthetic signal that the silent electric motors are engaged and thus the driver's foot should remain on the brake pedal until forward motion is desired.

At 30 seconds after the hybrid mode is selected, the ECU will shift the entire speed range of the electric motors to the last 40% of the accelerator pedal travel. The extended range sub-mode (disabling the fuel injection for two of the four IC engine cylinders) will take over whenever the pedal remains at less than 20% for more than 15 seconds. When the pedal is depressed beyond 80% the extended range mode will be terminated.

The following mode-related LED indicator lights and meters are provided:

- ELECTRIC MODE 0-80 MPH
- HYBRID MODE 40-80 MPH
- EXTENDED RANGE MODE (ICE 2 cyl.)
- ZERO EMISSION ZONE
- Z.E.Z. - HYBRID MODE DISABLED

- LOW BATTERY - USE HYBRID MODE?
- VERY LOW BATTERY - AUTO HYBRID MODE
- BATTERY STATE OF CHARGE (meter)
- LOW FUEL
- FUEL TANK LEVEL (meter)

ECU functions include:

- soft start for electric motor controllers when key is turned to ignition-on position;
- limiting of maximum vehicle speed to 82 mph to prevent over-revving of the electric motors;
- global satellite positioning to prevent use of IC engine in ZEV zones;
- LED indication recommending the use of hybrid mode if battery state of charge is below 15%, vehicle speed is in excess of 55 mph and location is outside of ZEV zone;
- automatic switch to hybrid mode if battery state of charge is below 15%, vehicle speed is in excess of 40 mph, location is outside of ZEV zones and there is fuel remaining in tank (also indicated by an LED);
- LED indication of battery state of charge below 15%, regardless of other variables (flashing indicator below 8%);
- automatic application of low level regenerative braking to charge batteries if battery state of charge is below 8% while in hybrid mode;
- disabling of fuel injection to cylinders two and three of the IC engine when the accelerator is at less than 20% for more than 15 seconds, while in hybrid mode;
- control of electric air pump to aid exhaust catalyst during IC engine warm-up and part-load conditions;
- control of the rear latent-heat battery (LHB) output to aid warm-up of the IC engine;
- control of the front LHB output to aid the initial stage of windshield defrost;
- boosting of LHB temperatures (via positive-temperature-coefficient ceramic cores) when propulsion batteries are being charged (defrost assist LHB only boosted when 24 hour low ambient temperature drops below 50 deg.F);
- passenger compartment climate control.

The underlying assumption is that is that the driver will use the electric mode for all city driving (0 to 50 MPH), and at highway speeds (40 to 80 MPH) when charging between trips is convenient and the pure electric acceleration and load carrying performance is not adequate. When long distance travel without stops for charging or high performance is desired, and the vehicle is not in a ZEV zone, the driver may shift into hybrid mode.

VEHICLE RANGE

Electric Range:
The calculated range in the pure electric mode is 202 miles at 65 mph, and 268 miles at 35 mph.

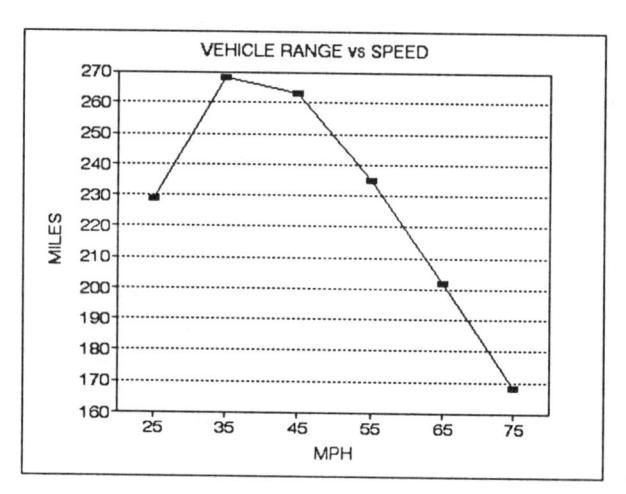

$$\text{Range} = \frac{Bc * Me * De}{(Ad + Rr)746 \text{ watts}} * V$$

where :
Bc = Battery capacity (Wh)
Me = Motor efficiency
De = Driveline efficiency
Ad = Aerodynamic drag (bhp)
Rr = Rolling resistance (bhp)
V = Velocity (mph)

Bc = 26,180 Wh
De = 0.97

Ad = 7.41 bhp at 65 mph
Rr = 2.89 bhp at 65 mph
Me = 0.94 @ 5,160 rpm and 10.3 bhp load

Ad = 0.73 bhp at 35 mph
Rr = 2.89 bhp at 35 mph
Me = 0.91 @ 2,778 rpm and 4.05 bhp load

Regenerative braking would recover roughly 50% of braking energy based on a battery energy efficiency of 80% and combined average motor and controller efficiency of 92%, and driveline efficiency of 97%. The estimated 50% recovery is in terms braking energy which is available again as tractive effort at the wheels. Losses from head winds, heavy loads, or other adverse conditions, as well as the benefits of regenerative braking have not been included in the electric range calculations. With measured "round-trip" regen efficiencies of up to 70% using silver-zinc batteries in a solar race car (Viking XX, Vehicle Research Institute, Western Washington University, 1990) it is reasonable to infer that regenerative braking will substantially offset the losses which are typical of urban driving.

Er = De * Me * Ce * Be * Be * Ce * Me * De

Where:
Er = Energy Recovered
De = Drivetrain efficiency
Me = Motor efficiency
Ce = Controller efficiency
Be = Battery efficiency

IC Range:
Because the design includes extensive modification of the IC engine, estimates of range are approximate (refer to the previous description of IC engine modifications in the Power Plants section of this paper for details). The fuel efficiency while running on M85 has been estimated to be 40 mpg at 65 mph, for a total range of 480 miles on 12 gallons of M85. This is based in part on the 15% gain in fuel efficiency attained by the Vehicle Research Institute (VRI) Viking Six prototype, when running on two out of four cylinders of an 1800cc engine at 50 mph.* The anticipated improvement in fuel economy for the two cylinder extended-range mode in this design is approximately 10% @ 65 mph. The improvement results principally from a reduction in pumping losses, with the throttle further open, and some gains in thermal efficiency. At 65 mph, rather than 50 mph (a significantly higher load), and using a 900cc engine in a vehicle requiring a similar amount of horsepower to maintain highway speeds, there is likely to be less room for improvement by this means, relative to what was achieved with Viking Six. M85 has just over half the BTU's per volume compared to gasoline, however, its high octane rating allows a 14:1 compression ratio. Methanol is also an oxygenated fuel with a high latent heat of vaporization, improving volumetric efficiency. While volumetric efficiency does not directly improve fuel economy, it does mean that a smaller engine can be used, improving thermal efficiency and reducing pumping losses from throttle plate restrictions, without sacrificing power.

---

* The Viking Six was designed and built by the VRI at Western Washington University to demonstrate the feasibility of a road-legal, two-seat commuter vehicle which exceeds federal safety standards, while getting better than 65 miles per gallon on the EPA highway driving cycle and maintaining the ability to accelerate 0-60 in less than 9 seconds. With its most recent powertrain configuration, the Viking Six achieved 110 mpg at 50 mph while meeting 1988 emission standards.

ACCELERATION

0-60 in Electric and Hybrid Modes:

Acceleration times, including correction for acceleration of rotating masses, were calculated using methods described by Colin Campbell in <u>The Sports Car: Its Design and Performance,</u> based on the following assumptions. (calculated acceleration times for each 10 mph increment are totaled in the table that follows)

- Overall gear ratio for electric drive = 5.29:1
- Overall gear ratio for IC drive = 5.62:1
- Net torque = total motor and/or engine torque @ rpm * overall gear ratio * 0.97 driveline efficiency factor (net torque for the IC engine is based on the predicted torque curve after engine modifications)
- Drive tire radius = 0.93 ft.
- Vehicle weight = 2,106 lbs curb weight + 170 lbs for driver = 2,276 lbs.
- Rolling resistance = 13.7 lbs. when gross vehicle weight includes only the driver (no passengers).
- Aerodynamic drag:

$$\text{lbs force} = \frac{C_d * A * V^3}{146{,}600} * \frac{550}{v} \quad \text{see Aero Drag §}$$

- Wheel thrust = Net torque * 0.93 ft tire radius
- Surplus tractive effort (P) = Wheel thrust - Rolling drag - Aero drag.
- Mass correction factor to account for acceleration of rotating masses:

$$f = \frac{\text{motor rpm} * .0132}{V}$$

(for all rotating masses combined except the IC engine)

$$f = \frac{\text{AVG of motor \& engine rpm} * .0133}{V}$$

(for all rotating masses including the IC engine)
where:
f = Mass Correction factor
V = Velocity (mph)

- Effective mass (M') = Vehicle weight * f

- Instantaneous acceleration:

$$a = \frac{P}{M'} * g$$

where:
a = Acceleration (ft/sec^2)
P = Surplus Tractive Effort (lbs)
M'= Effective Mass (Mass-lbs)
g = 32 ft/sec Force of gravity

- Acceleration time:

$$t = \frac{V - V_o}{a}$$

where:
t = Time
a = Instantaneous Acceleration
V - V$_o$ = 14.66 ft/sec for every 10 mph increment

Acceleration vs. Speed

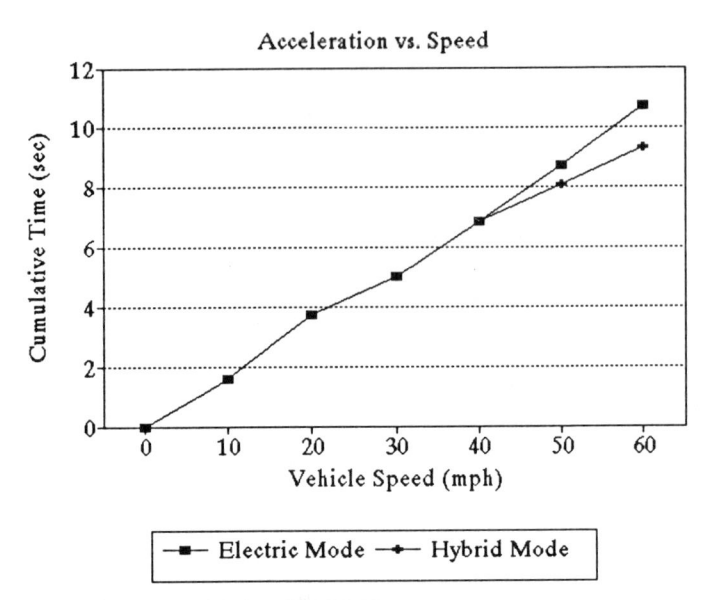

(Graph: Cumulative Time (sec) vs. Vehicle Speed (mph), showing Electric Mode and Hybrid Mode curves)

TABLE OF COMPUTATIONAL INPUTS AND RESULTS FOR 0-60 MPH ACCELERATION

ELECTRIC MODE

Vehicle Speed (mph)	Motor and Engine Speed (rpm)	Net Torque (ft-lbs)	Wheel Thrust (lbs)	Rolling Resistance +Aero. Drag (lbs)	Surplus Tractive Effort (lbs) P	Mass Correction Factor f	Effective Mass (lbs) M'	Instantaneous Acceleration (ft/sec) a	Time for each 10 mph interval (sec) t
10.0	794	657	701	14.7	686	1.05	2,390	9.18	1.60
20.0	1,587	631	673	17.8	655	1.05	2,390	8.77	1.67
30.0	2,381	616	657	22.8	634	1.05	2,390	8.49	1.73
40.0	3,174	595	635	29.9	605	1.05	2,390	8.10	1.81
50.0	3,968	580	619	39.0	580	1.05	2,390	7.77	1.89
60.0	4,761	564	602	50.2	552	1.05	2,390	7.39	1.98
								0-60 Total	10.7 sec

HYBRID MODE

Vehicle Speed (mph)	Motor and Engine Speed (rpm)	Net Torque (ft-lbs)	Wheel Thrust (lbs)	Rolling Resistance +Aero. Drag (lbs)	Surplus Tractive Effort (lbs) P	Mass Correction Factor f	Effective Mass (lbs) M'	Instantaneous Acceleration (ft/sec) a	Time for each 10 mph interval (sec) t
40.0 E	3,174	595							
I.C	3,372	+229	879						
50.0 E	3,968	580							
I.C.	4,215	+294	932	39.0	893	1.07	2,435	11.7	1.25
60.0 E	4,761	564							
I.C.	5,058	+327	950	50.2	900	1.07	2,435	11.8	1.24
								0-60 Total	9.3 sec

CHASSIS AND BODY

The aluminum monocoque chassis/body is designed to exceed federal safety standards while weighing significantly less than current production models. Crashworthiness is principally achieved through rigid passenger compartment design and aluminum crash ridedown and roll protection structures. The most notable achievement of the design is the calculated ability of the chassis to provide a 44 g ridedown for an average weight occupant in a fixed-barrier frontal impact at 56 mph. The estimated weights for chassis and body skin are 200 lbs. and 100 lbs. respectively.

Viking Six Chassis

The weight estimate for the riveted aluminum panel chassis is based on the VRI's Viking Six, which was of similar proportions and construction, and as discussed later in this section, was successfully crash tested at just over 40 mph. The estimated weight of the body skin is based on that of the VRI's Viking Four, which was made from the same alloy, was similar in surface area, and weighed 100 lbs. These estimates are conservative when it is considered that, unlike the Viking Four and Six, the body panels double as the exterior paneling of all the outboard chassis box sections (ie: the body is an integral part of the monocoque chassis).

The passenger compartment is rigid and light weight using 0.06 in. 7075 T6 sheet and 0.08 in. 7075 T6 reinforcement hat sections. The reinforced toe board is 0.10 in. 7075 T6 and serves as the reaction member for the frontal impact crush section. A transverse hat-section reinforced box-beam structure of 0.08 in. 7075 T6 links the door frames, stiffens the front end of the passenger compartment, and supports the instruments, and steering column. A longitudinal central box section houses the batteries, shift and hand brake mechanisms, coolant and break lines, and wiring. To maintain torsional stiffness, this is actually two boxes stacked, with the batteries in the lower compartment and only limited access holes for the remaining components in the upper compartment. Rocker panel box sections stiffen the door frames and the overall passenger compartment structure. All three box sections interlock with common end plates at the toe board and at back side of the rear seats.

The load-bearing structures for the drive systems and suspension at either end are 0.06 in. 6061 T3 with 0.08 7075 T6 reinforcements at the suspension mounting hardpoints. All hard points are located at the intersections of at least two panels, with the most critical brackets for spring mounts and lower control arms located at three plane intersections. Torsional stiffness around component access openings and load transfer to the passenger compartment and roll cage structures are maintained with added hat sections and lips.

The aluminum body panels are Alcoa 5182 alloy with 5% magnesium. This alloy has the unique property of hardening to the same dent resistance as steel during the paint bake process.

All chassis and body panels are fastened together with aluminum rivets and flexible adhesive. The rivets have depressions for drill bit centering when collision repair requires their removal. The use of adhesives along the seam flanges, common in the aircraft industry, improve torsional stiffness, absorbs vibration, and reduces the number of rivets required. A minimum number of rivets are strategically located to prevent peel of the bonded seams. The riveted/bonded sheet and hat section construction is intended for a one-off prototype, but could readily be adapted to a stamp and spot-weld method of high-volume production.

Front and Rear Impact Absorption:

A 23 in. deep crush zone (17.25 in. of actual crush stroke distance, allowing space for compacted materials) of aluminum honeycomb in the nose section with a crush strength of 90 lbs. psi. and an effective frontal area of 1,428 sq. in., maintains a 46 g ridedown for fixed-barrier frontal impacts of up to 44 mph when the gross vehicle weight is 2,786 lbs. (incl. four 170 lb. occupants). Combined nose section structural members, body panels, and honeycomb would be sled tested to assure sufficient support of the 5 mph bumpers while raising the impact forces (over those of the honeycomb alone) to 50 g's at a higher maximum speed of 46 mph. The crushing of the front battery pack, suspension, and chassis forward of the toe board will add 13 in. to the total crush zone. This would add an estimated 8 in. actual crush stroke distance when accounting for density of battery materials, etc... Through sled test analysis, these materials would be engineered to crush at the same rate as the nose section, maintaining the 50 g ridedown of the nose section up to a new maximum speed of 56 mph.

$$\frac{A * S}{W} = g \qquad \frac{v^2}{2a} = d$$

where:
A = Crush Zone Frontal Area (in^2)
S = Crush Strength (lbs/in^2)
W = Gross Vehicle Weight (lbs)
g = Negative Acceleration (g's)
a = g * 32ft/sec
v = Velocity (ft/sec)
d = Crush Stroke Distance (ft)
   (75% of total crush zone depth for aluminum honeycomb)

SYNERGY EHV ELECTRIC HYBRID VEHICLE   SCALE 1:10   3/17/91

15' TURNING RADIUS
96" WHEELBASE
168" OVERALL LENGTH
61" OVERALL WIDTH

BY TIM MOORE

Dynamic Science crash test of Viking Six

The rigid cockpit and crushable nose section with aluminum honeycomb was effectively demonstrated by 40 mph fixed-barrier crash testing of the VRI's Viking Six prototype (see photo.) Test results included head injury criteria and peak chest and femur g loading well below FMVSS 208 injury criteria limits, only 0.16 in. maximum cockpit deformation, and 19 in. static crush of the nose section at an actual frontal crash velocity of 41.2 mph.

Mechanical Fitzpatrick force limiters at the common anchor points of the shoulder and lap belts will further reduce negative acceleration from impact by cold working a steel strap around a small radius. Safety belt force limiters, designed by Fitzpatrick Engineering, successfully reduced peak chest g loading from 56 to 44 g's (stroking 6 of 14 in. total length) in the fixed-barrier crash test of the VRI's Viking Six.

Because the force limiters give at a controlled rate with given force applied, it would be possible to limit g loading for anticipated occupant weights. For example, upper body forces experienced by a 100 lb. occupant in frontal impacts could be limited to 40 g's, with heavier occupants experiencing lower g loading and using a longer portion of the force limiter stroke during the same impact. This trade off between maximum g loading and force limiter stroke would be optimized for average occupants in the 110 to 170 lb. range.

Additional features are provided within the passenger compartment to supplement the crash ridedown characteristics of the chassis. With 4.5 in. aluminum honeycomb covered by 2 in. crash-helmet grade molded urethane or polypropylene foam padding, the box-beam that supports the instruments and steering column functions as a knee restraint to absorb some of the occupants lower-body momentum in a frontal crash. A steering column crush basket reduces the possibility of chest injuries. Air bags for all four occupants (located in the steering wheel, dash board, and seat backs) reduce neck and head injuries. A 1 in. layer of molded urethane foam is used beneath the foam padding on the seats, head liner, and dash panels. Passive restraint three point safety belts are fastened to the force limiters on the center tunnel and to the rigid trailing edge of the doors for forward occupants. The rear seating has manual three point belts which fasten to force limiters and to the intersection of the rear fire wall and roll hoop structures.

Roll, Side, and Rear Impact Protection:

Rollover and side impact protection are provided by a full roll cage, lateral box section through the dash, and generous door sill box sections. Roll hoops are rectangular box sections with internal hat section reinforcements. The full roll cage joins the "A" and "B" pillar roll hoops both at the top of the door opening and through the doors.

Horizontal and diagonal tubular side-guard beams (7075 T6 2 in. OD 0.125 in. wall) are located 1 in. inboard of the door skin to allow for the window mechanism. When the doors are shut the ends of the side-guard beams interlock with the "A" an "B" pillar roll structures. There are also fixed side-guard beams of similar construction forming the triangulation for the "B" pillar roll hoop and protecting the rear seat passengers. Paneling on the inboard side of the side-guard beams is the reaction member for an occupant restraint consisting of 3 in. of 25 lb psi honeycomb and 2 to 5 in. of molded urethane foam, adjacent to the occupants torso and thigh. As the core for the interior door panel, the crushable materials are a bolster, absorbing energy if the occupant impacts the surface. The door sills are 7 in. high 9 in. wide box sections which are hat section reinforced and filled with 200 lb psi honeycomb to improve rigidity and reduce plastic deformation from impacts. The intent is for the side beams and door sills to be stiff enough to force the nose of a striking vehicle to crush, since there is not enough space to absorb the energy from even a 30 mph impact by an average weight automobile.

While the rear quarter panels each contain 2,464 cubic inches of 250 lb psi aluminum honeycomb, it is difficult to predict the weight or angle of impact of a striking vehicle with any degree of accuracy. However, calculations indicate that the honeycomb, rear chassis structure, and components combined would completely absorb an evenly distributed 37 mph impact by a 2700 lb. vehicle at approximately 28 g's in the first 20 in. of crush stroke and rising to approximately 56 g's in the final 20 in. of crush stroke. These figures are most useful for indicating the ability of the vehicle to absorb its own impact if it were to collide with a fixed barrier while skidding backward after a 180 degree spin.

Elastomeric foam bumpers at either end absorb 5 to 6 mph impacts without damage to the chassis. The bumpers are molded two-part polyurethane elastomeric skin backed with kevlar cloth/flex-vinylester resin for support and filled with polypropylene or two-part polyurethane elastomeric foam.

COMPONENT ACCESS

The entire nose section, including the bumper, headlights, and honeycomb crush section, hinges upward from the base of the windshield, providing access to the front electric drive and suspension, steering, electronics, windshield wiper, heating, and ventilation systems. Primary access to the rear drive systems compartment is through a large, interior panel beneath the rear hatch.

Flush mounted access panels in the belly pan are provided for maintenance and repair work. As noted in the discussion of batteries, the main battery pack can be lowered from the underside of the vehicle on a light-weight system of cable, pulleys, and ratchet mechanisms. Front wheel access covers pivot on a pin at their apex and are secured to the gas charged return mechanisms by spring-steel clips for easy removal of the cover panel. Rear wheel access covers are use the same pin and clip system for continuity of operations.

STEERING AND BRAKES

The wheel base of 96 in. is short enough to be responsive and long enough to provide a relatively smooth ride. Front and rear track widths of 61.4 in. and 56.0 in. respectively, along with a low center of gravity and anti-roll bars at either end to control weight transfer, provide stability in hard cornering. Manual rack & pinion steering is employed to give positive response, and to minimize weight and complexity. With a full-lock steering angle of 43 degrees, the minimum turning radius is 15 ft. curb to curb. Wheel opening side cover clearance for various steering angles is discussed in the section of this paper devoted to design elements for reduced aerodynamic drag. Tie rods are equal in length and parallel to the lower control arms to prevent bump steer. Front end king-pin inclination, offset, and caster are 5.5 degrees, 1.0 in., and -5.0 degrees, respectively, for self-straightening, road feel, and minimum chassis lift in tight turns. Toe-in is set at 0.125 in. to prevent toe-out when suspension is stressed during hard braking. Less than full Ackerman angle compensates for increased tire slip angles from lateral weight transfer in hard cornering (virtual steering arm intersection is 1.14 times the wheelbase, measured from the front axle). Rear steer has not been incorporated as production systems are heavy and engineering such a system would introduce unnecessary complexity and cost.

Disc brakes are fitted at all four wheels for efficient heat dissipation per unit weight. A tandem master cylinder actuates dual opposed piston, low-drag, cast aluminum calipers at the front and rear. The rear calipers include a hand brake mechanism which applies pressure directly to the exposed ends of the brake pad backings on either side of the disc. Low drag is achieved by securing the brake pad backings to the caliper pistons for positive retraction from the disc surface. The handbrake mechanisms are also fitted with substantial return springs. G-sensitive pressure relief valves apportion braking effort both laterally and longitudinally.

SUSPENSION

Non-parallel, unequal-length, tubular aluminum upper and lower control arms and coil-over spring/damper units are the essentials of both the front and rear suspension systems. A front track width of 61.4 in. and rear of 56.0 in. are matched to the aerodynamic taper of the body design. The lower "A" arms at the front have a large 16.5 in. base that is rear biased to most efficiently transfer loads during hard braking. At the rear, lower "N" arms have diagonals angled forward toward the chassis to efficiently transfer acceleration forces, since the IC engine output at the rear wheels is in addition to the balanced thrust applied front and rear by the electric motors.

A control arm ratio of .65:1 and virtual swing arm length of roughly twice the track width give moderate and predictable negative camber gain with body roll. Roll-center heights are 1.5 in. and 2.5 in., front and rear respectively. With 4 in. of static deflection at the wheels and a sprung weight of 1926 lbs., the natural frequency is 1.56 oscillations per second. Weight transfer in cornering is controlled by anti-roll bars both front and rear. Anti-roll bar stiffness would be determined by skid-pad cornering and one-wheel-bump tests of the prototype vehicle. Hydraulic links between front and rear dampers on each side reduce dive when braking and provide automatic load leveling to maintain the designed aerodynamic attitude of the car and avoid the necessity of stiffer rear springs to accommodate the occasional heavy load. The hub carriers are machined from cast aluminum to minimize unsprung weight. The 14 x 6 in., 8.5 lb. Alcoa aluminum wheels further reduce unsprung weight. The Goodyear P165/65 R14 low rolling resistance tires weigh 12.5 lbs. each. All cast aluminum parts, such as suspension uprights (hub carriers), brake disc hats, and brake calipers, are 356 A-T6 alloy with strontium modification to improve grain structure.

Front and rear suspension drawings show the geometric analysis on the page that follows.

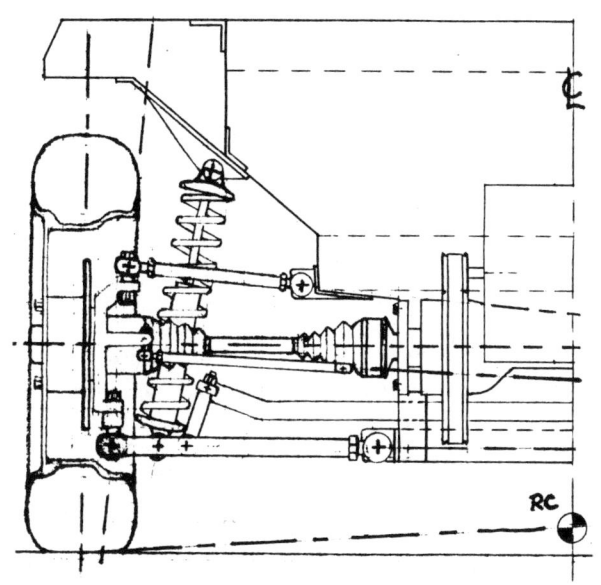

FRONT SUSPENSION

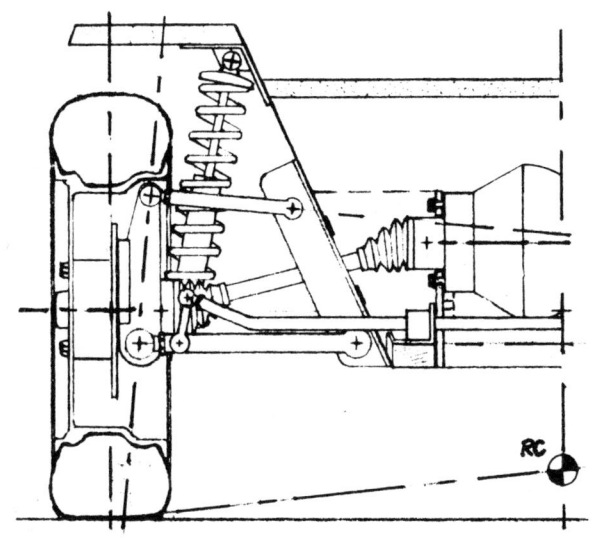

REAR SUSPENSION

PASSENGER COMPARTMENT

The ergonomically designed interior is styled to match the soft organic curves of the body exterior (see illustration on following page). All gauges are analog for easy reading. There are large LED-illuminated indicators for drive system modes, low battery, low fuel, etc... Heat and other controls are minimized and close at hand. Frequently used controls, including turn signals, headlights, windshield wipers, and cruise control, extend from the steering column. A recessed hand brake lever angles away from the driver for good leverage. For driving comfort there is a padded steering wheel, dead pedal on the far left, and fully adjustable front seats. The seats have light weight aluminum frames and a simple hand pump bladders for adjustable lumbar support.

Doors open diagonally, pivoting around the "A" pillar, to allow the opening to extend slightly into the passenger compartment roof area for ease of entry to the relatively low profile vehicle (overall height is 51 in.). Diagonally opening doors have the added advantages of resting fully out of the users path when open, requiring very little side clearance to open, and never scraping on concrete street curbs. The wrap-around rear hatch is lifted by gas charged struts and allows access to the cargo space, spare tire, and rear engine compartment. The liftover-height at the rear is 32 in.

Excellent visibility is afforded by unobstructed wide angles of view in all directions apart from the 1.9 in. "A" pillars and 7.2 in. "B" pillars.

All glazing is PPG Solextria 7010 which blocks 58% of both infra-red and ultra violet radiation, which would otherwise heat the interior space, breakdown interior upholstery, and create glare. Ultimately the glazing would also be angularly selective to screen out high-angle sun, further minimizing heat build-up and glare. Angular selectivity also screens out high-intensity overhead street lighting, which would otherwise illuminate the dashboard top surface, causing reflections on the interior windshield surface in the driver's field of vision. All interior surfaces are light colored to again reduce solar heat gain. A small roof-top solar panel powers a thermostatically controlled ventilation fan which runs whenever the interior temperature exceeds a user selected temperature between 68 and 80 deg.F.

The combination of solar control glass, light interior, and solar vent system should adequately cool the interior in moderate climates. An optional air conditioning system would utilize a Seiko Seiki variable displacement aluminum compressor, weighing 5.4 lbs. and using non-ozone depleting HFC-134a refrigerant, and a light weight, low profile parallel-flow condenser. The compressor would be driven, through a magnetic clutch, by the idler gear between the front electric motor and differential at approximately half the speed of the motor. The variable displacement of the compressor eliminates the usual load increases at higher rpm's and provides continuous operation rather than cycling.

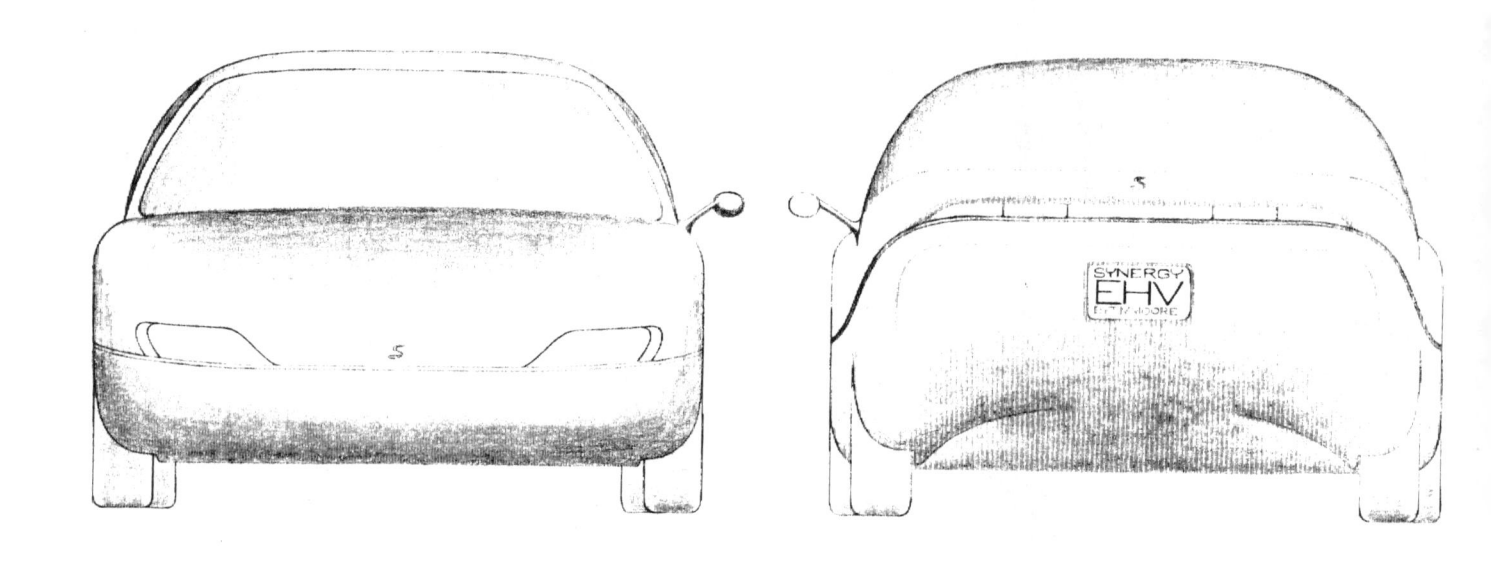

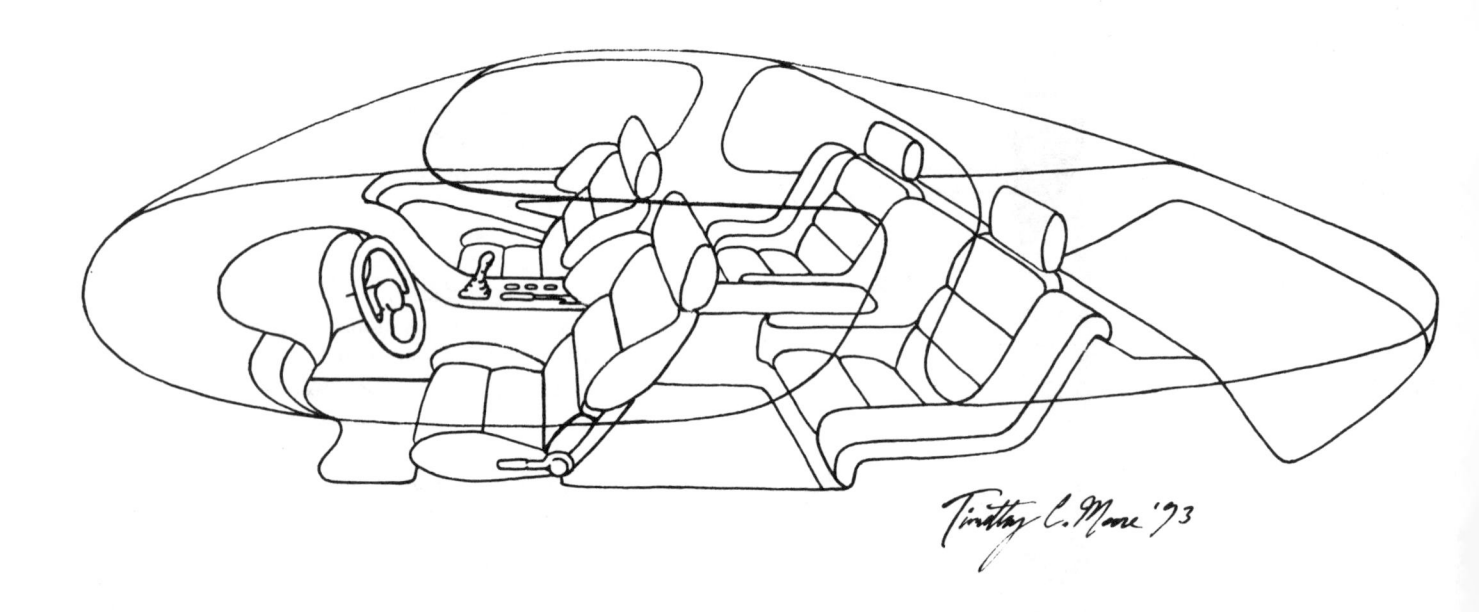

BODY STYLING

The organic styling of the light weight aluminum body skin is the result of careful attention to each detail of the design in seeking an optimum relationship between aerodynamics, function, and overall appearance. For example, only an 0.25 in. wide rubber seal separates the flush-mounted windshield and door glass at the "A" pillar, reducing aerodynamic drag and giving the appearance of a single wrap-around window. The decidedly round wheel-opening side-covers are intended to give the impression of an exposed wheel, in an effort to gain consumer acceptance. The futuristic overall appearance highlights the advanced HEV technology and the functional efficiency of the design.

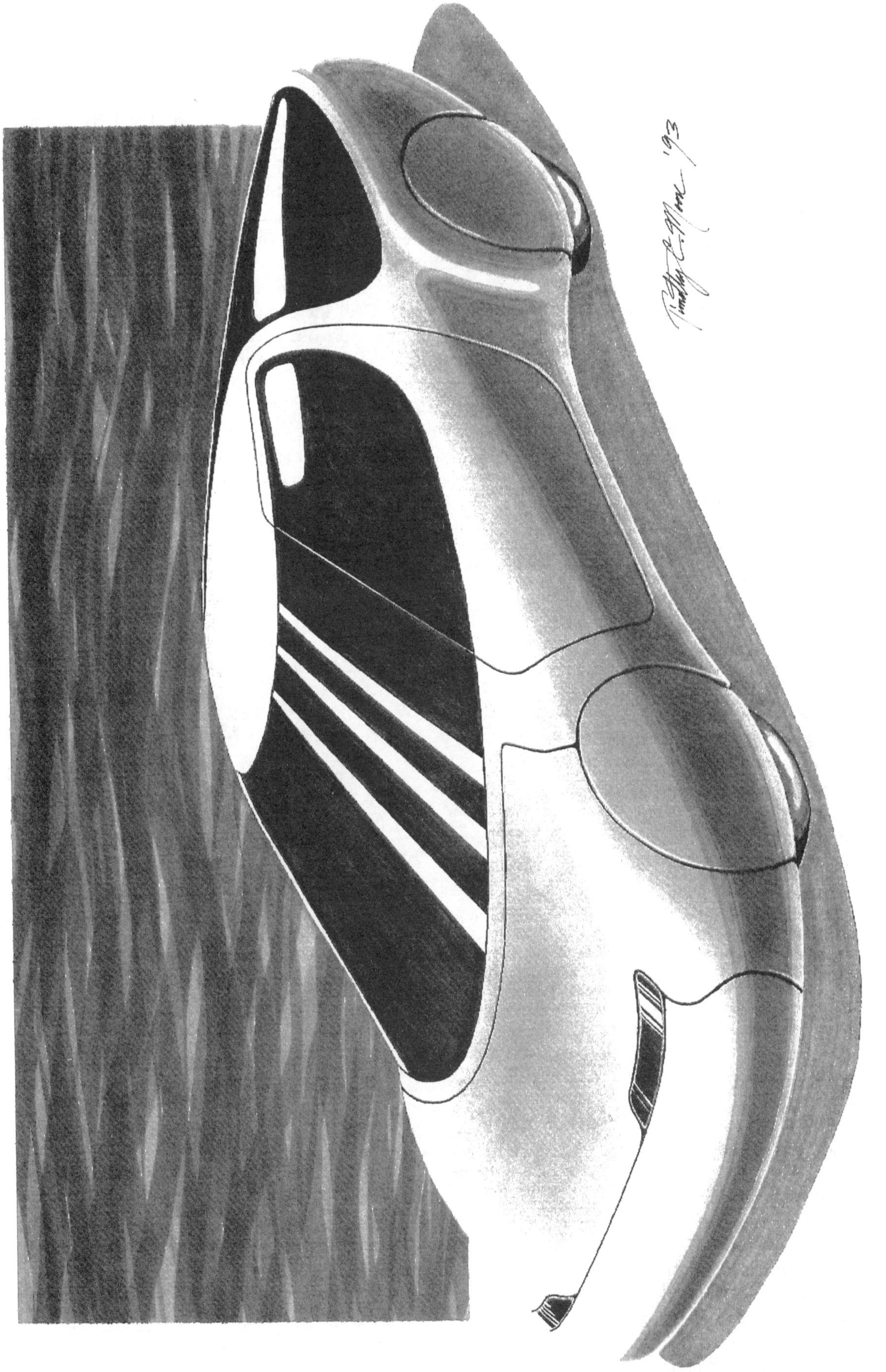

ENVIRONMENTAL IMPACT AND ENERGY CONSUMPTION

Materials:

Composite fiber construction of everything from chassis and body to wheels is feasible and could save a significant amount of weight (lbs/vehicle) and thus reduce energy consumption; however, composites are currently not fully recyclable. Aluminum is readily recyclable (the infrastructure is already in place), and the manufacturing costs associated with the stamping process are less than those of steel, which often requires numerous dies to stamp one part into finished form. While the manufacturing processes for aluminum cars may be more than for composites, the raw material is substantially less expensive and is more likely to be recycled than significantly lighter weight composites, such as carbon fiber, and it's still twice the strength of steel per unit weight. Similarly, although the production of virgin aluminum is highly energy intensive, the recyclability of aluminum has the added advantage of requiring significantly less energy than the production of new metals from ore.

Electric Vehicle Energy Efficiency and Emissions:

Analysis of electric vehicle (EV) emissions impacts, based on efficiencies of 85%, 70%, 90%, and 85% for charger, battery, controller, and motor, respectively, indicates significant improvements in urban air quality, relative to the continued use of gasoline fueled vehicles. Emissions improvements (per unit distance traveled) for EV's relative to gasoline fueled vehicles are estimated to be 99.9% HC, 99.8% CO, and 81.7% NOx in regions where 25% of electricity is generated by fossil fuel burning plants. "Even in regions where all electricity comes from fossil fuel burning plants, the use of EV's would reduce over 90% HC and CO, and over 50% NOx emissions" (Adams et al. Electrochemical Science and Technology Center, Ottawa, Canada). The Electric Power Research Institute, EPRI, arrive at similar figures with 99% HC, 99% CO, 90% NOx, and 70% CO_2 reductions relative to gasoline powered automobiles (based on the mix of U.S. power plants in 1989). Efficiencies of components used in this design are actually better than those assumed in the aforementioned studies, with higher tested averages of 80%, 91%, and 91% or better for batteries, motors, and controller, respectively. Widespread use of EV's would shift automotive emissions from high density urban areas to areas of lower population density. Additionally, EV's would reduce dependence on foreign oil, in turn reducing the risk of spills from ocean-going oil tankers.

IC Engine Emissions and Energy Consumption:

According to Dr. Ing. Ulrich Seiffert, head of research and development at Volkswagen, with the exception of aldehydes, combustion of methanol produces fewer harmful emissions than either gasoline or diesel.

Methanol is most readily made from natural gas which is plentiful relative to other fossil fuels. Ultimately, the IC engine would run on hyper-compressed natural gas (HCNG), which, when stored at 15,000 lbs. psi., provides 2.8 times the BTU's per unit volume of gasoline, as tested by Liberty Natural Gas, Seattle, WA. Methanol, on the other hand, provides only about 0.5 times the BTU's per unit volume of gasoline. The advantages of HCNG would be longer range for a given tank size, more abundant supply since it is the prerequisite source from which methanol is most often obtained, and further reduced exhaust emissions. The primary obstacle to the use of HCNG is the high cost of the infrastructure necessary to compress and deliver it at pressures which provide sufficient vehicle range. The choice of M85, therefore, is based on fuel characteristics relative primarily to gasoline, and on feasibility relative to other alternative fuels. The principal improvements in IC emissions over conventional automobiles will result from the control strategy and latent heat battery. The barium hydroxide latent heat battery speeds the warm-up time of the IC engine (see illustration and EPA graph below). Tests conducted by the US Environmental Protection Agency (EPA) demonstrated 33% HC and 50% CO reductions for M85 powered vehicles with heat batteries to assist cold starting. The control strategy eliminates engine idling (except when coasting at highway speeds), and improves thermal and volumetric efficiency by operating the small IC engine over a limited range of speeds and loads (ie: a small engine map), and shutting down two of four cylinders while cruising.

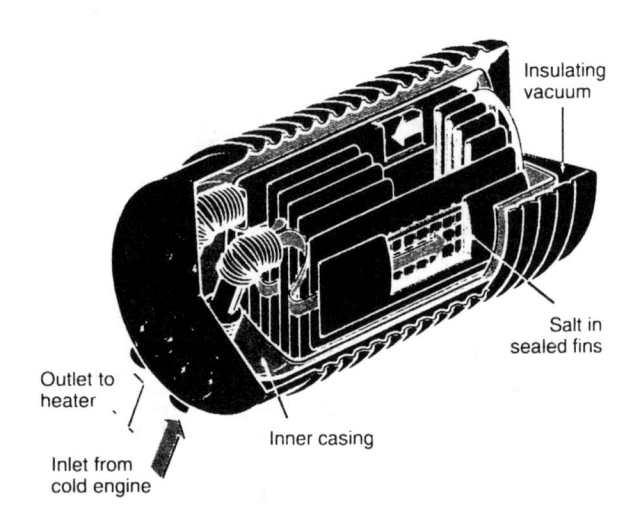

Interior of the heat battery.

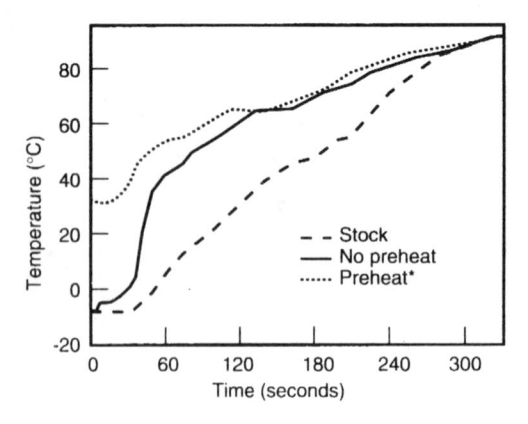

Engine-out coolant temperature (Audi 80 during FTP with M85 fuel).

* Preheat was 60 s

CONCLUSION

The Synergy HEV utilizes current state-of-the-art technology and a systems design approach to achieve high efficiency without sacrificing performance, safety, or practicality. The systems design approach places priority on integrated design features that work together to provide benefits which cannot be derived from the same technologies independently. While reducing rolling resistance through the use of light weight aluminum crash-ridedown and chassis structures, for example, crashworthiness is raised well above that of typical production automobiles. The location of the dual electric motors front and rear with constant-ratio, constant-mesh helical gear drives eliminates the possibility of torque-steer and maintains high efficiency while evenly distributing weight and providing positive full-time four-wheel-drive. Aerodynamic efficiency means a smaller power plant and cooling systems can be used, which require smaller air inlets, further reducing aerodynamic drag, in addition to their lighter weight, both of which further reduce the size of power plant required.

The net result of these and other synergistic relationships, which are the foundation of the Synergy HEV design, is a safe and practical vehicle that generates less pollution and consumes less energy than current production automobiles, without sacrificing performance.